Interpreting Infrared, Raman, and Nuclear Magnetic Resonance Spectra

VOLUME 2 Factors Affecting Molecular Vibrations and Chemical Shifts of Infrared, Raman, and Nuclear Magnetic Resonance Spectra

Interpreting Infrared, Raman, and Nuclear Magnetic Resonance Spectra

VOLUME 2 Factors Affecting Molecular Vibrations and Chemical Shifts of Infrared, Raman, and Nuclear Magnetic Resonance Spectra

RICHARD ALLEN NYQUIST
Nyquist Associates
Midland, MI

ACADEMIC PRESS

A Harcourt Science and Technology Company

San Diego San Francisco New York Boston
London Sydney Tokyo

This book is printed on acid-free paper. ∞

ACADEMIC PRESS
A Harcourt Science and Technology Company
525 B Street, Suite 1900, San Diego, CA 92101–4495, USA
http://www.academicpress.com

Academic Press
Harcourt Place, 32 Jamestown Road, London, NW1 7BY, UK
http://www.academicpress.com

Library of Congress Catalog Number: 00-108478

International Standard Book Number: 0-12-523475-9
Volume 1 International Standard Book Number: 0-12-523355-8
Volume 2 International Standard Book Number: 0-12-523470-8

Printed in the United States of America
01 02 03 04 05 QW 9 8 7 6 5 4 3 2 1

CONTENTS

Acknowledgments		ix
Nyquist's Biography		xi
Preface		xiii
References		xv
Chapter 1	Epoxides and Ethers	1
Chapter 2	Nitriles, Isonitriles, and Dialkyl Cyanamides	27
Chapter 3	Azines, Isocyanates, Isothiocyanates, and Carbodiimides	45
Chapter 4	Thiols, Sulfides and Disulfides, Alkanethiols, and Alkanedithiols (S—H stretching)	65
Chapter 5	Sulfoxides, Sulfones, Sulfates, Monothiosulfates, Sulfonyl Halides, Sulfites, Sulfonamides, Sulfonates, and N-Sulfinyl Anilines	85
Chapter 6	Halogenated Hydrocarbons	119
Chapter 7	Nitroalkanes, Nitrobenzenes, Alkyl Nitrates, Alkyl Nitrites, and Nitrosamines	173
Chapter 8	Phosphorus Compounds	231
Chapter 9	Benzene and Its Derivatives	351

Chapter 10 The Nyquist Vibrational Group Frequency Rule 425

Chapter 11 Infrared Raman and Nuclear Magnetic Resonance (NMR) Spectra-Structure Correlations for Organic Compounds 435

List of Tables 581

List of Figures 589

Index 609

DEDICATION

This book is dedicated to all of the scientists I have been associated with for over 50 years, and to those scientists who find this book useful to them in increasing their skills in the interpretation of infrared (IR), Raman, and Nuclear Magnetic Resonance (NMR) spectra. This dedication especially includes my coauthors and other authors whose manuscripts or books were referenced in the present compilation.

ACKNOWLEDGMENTS

I thank the management of The Dow Chemical Company for providing me with a rewarding career in chemistry for over 41 years. I also thank the management of Sadtler Research Laboratories, a Division of Bio-Rad, for the opportunity to serve as an editorial consultant for several of their spectral collections of IR and Raman spectra.

I thank Marcia Blackson for typing the book manuscript. Her cooperation and editorial comments are appreciated.

Nyquist's Biography

In 1985, Richard A. Nyquist received the Williams-Wright Award from the Coblentz Society for his contributions to industrial IR spectroscopy. He was subsequently named an honorary member of the Coblentz Society for his contributions to vibrational spectroscopy, and in 1989, he was a national tour speaker for the Society of Applied Spectroscopy. The Association of Analytical Chemists honored Dr. Nyquist with the ANACHEM Award in 1993 for his contributions to analytical chemistry. He is listed in Who's Who in Science and Engineering, Who's Who in America, and Who's Who in the World. The Dow Chemical Company, from which Dr. Nyquist retired in 1994, honored him with the V.A. Stenger Award in 1981, and the Walter Graf European Award in 1994 for excellence in analytical chemistry. He has also been a member of ASTM, and received the ASTM Award of Appreciation for his contributions to the Practice of Qualitative Infrared Analysis. In 2000, Dr. Nyquist was awarded honorary membership to the Society of Applied Spectroscopy for his exceptional contributions to spectroscopy and to the Society. Dr. Nyquist received his B.A. in chemistry from Augustana College, Rock Island, Illinois, his M.S. from Oklahoma State University, and his Ph.D. from Utrecht University, The Netherlands. He joined The Dow Chemical Company in 1953. He is currently president of Nyquist Associates, and is the author or coauthor of more than 160 scientific articles including books, book chapters, and patents. Nyquist has served as a consultant for Sadtler Research Laboratories for over 15 years. In 1997 Michigan Molecular Institute, Midland, Michigan selected him as their consultant in vibrational spectroscopy.

PREFACE

My intention in compiling this book is to integrate IR, Raman, and NMR data in order to aid analysts in the interpretation of spectral data into chemical information useful in the solution of problems arising in the real world.

There is an enormous amount of IR and Raman data available in the literature, but in my opinion there has not been enough emphasis on the effects of the physical environment of chemicals upon their molecular vibrations. Manipulation of the physical phase of chemicals by various experiments aids in the interpretation of molecular structure. Physical phase comprises solid, liquid, vapor, and solution phases.

In the solid crystalline phase, observed molecular vibrations of chemicals are affected by the number of molecules in the unit cell, and the space group of the unit cell. In the liquid phase, molecular vibrations of chemicals are affected by temperature, the presence of rotational conformers, and physical interaction between molecules such as hydrogen bonding and/or dipolar interaction. In the vapor phase, especially at elevated temperature, molecules are usually not intermolecularly hydrogen bonded and are free from dipolar interaction between like molecules. However, the rotational levels of the molecules are affected by both temperature and pressure. Induced high pressure (using an inert gas such as nitrogen or argon) will hinder the molecular rotation of molecules in the vapor phase. Thus, the rotational-vibrational band collapses into a band comparable to that observed in a condensed phase. Higher temperature will cause higher rotational levels to be observed in the vibrational-rotational bands observed in the vapor phase.

In solution, the frequencies of molecular vibrations of a chemical are affected by dipolar interaction and/or hydrogen bonding between solute and solvent. In addition, solute-solvent interaction also affects the concentration of rotational conformers of a solute in a solvent.

The number of intermolecular hydrogen-bonded molecules existing in a chain in solution depends upon the solute concentration. In addition, the number of molecules of a solute in solution existing in a cluster in the absence of intermolecular hydrogen bonding also depends upon solute concentration.

INTRODUCTION

Infrared (IR), Raman (R), and Nuclear Magnetic Resonance (NMR) spectroscopy are essential tools for the study and elucidation of the molecular structures of organic and inorganic materials. There are many useful books covering both IR and R spectroscopy (1–14). However, none of these books emphasize the significance of changes in the molecular vibrations caused by changes in the physical state or environment of the chemical substance. One goal of this book is to show how changes in the physical environment of a compound aid in both the elucidation of molecular structure and in the identification of unknown chemical compositions. Studies of a variety of chemicals in various physical states have led to the development of the Nyquist Rule. The Nyquist Rule denotes how the in-phase- and out-of-phase- or symmetric and antisymmetric molecular vibrations (often called characteristic group frequencies) differ with changes to their physical environment. These group frequency shift differences aid the analyst in interpreting the data into useful chemical information. Another goal of this book is to gather information on the nature of solute-solvent interaction, solute concentration, and the effect of temperature. This knowledge also aids the analyst in interpretation of the vibrational data. Another goal of this work was to compile many of the authors' and coauthors' vibrational studies into one compendium.

REFERENCES

1. Herzberg, G. (1945). *Molecular Spectra and Molecular Structure II. Infrared and Raman Spectra of Polyatomic Molecules*, New Jersey: D. Van Nostrand Company, Inc.
2. Wilson, E.B. Jr., Decius, J.C., and Cross, P.C. (1955). *Molecular Vibrations*, New York: McGraw-Hill Book Company, Inc.
3. Colthup, N.B., Daly, L.H., and Wiberley, S.E. (1990). *Introduction to Infrared and Raman Spectroscopy*, 3rd ed., New York: Academic Press.
4. Potts, W.J., Jr. (1963). *Chemical Infrared Spectroscopy*, New York: John Wiley & Sons, Inc.
5. Nyquist, R.A. (1984). *The Interpretation of Vapor-Phase Infrared Spectra: Group Frequency Data*, Phildelphia: Sadtler Research Laboratories, A Division of Bio-Rad.
6. Nyquist, R.A. (1989). *The Infrared Spectra Building Blocks of Polymers*, Philadelphia: Sadtler Research Laboratories, A Division of Bio-Rad.
7. Nyquist, R.A. (1986). *IR and NMR Spectral Data-Structure Correlations for the Carbonyl Group*, Philadelphia: Sadtler Research Laboratories, A Division of Bio-Rad.
8. Griffiths, P.R. and de Haseth, J.A. (1986). Fourier Transform Infrared Spectrometry, *Chemical Analysis*, vol. 83, New York: John Wiley & Sons.
9. Socrates, G. (1994). *Infrared Characteristic Group Frequencies Tables and Charts*, 2nd ed., New York: John Wiley & Sons.
10. Lin-Vien, D., Colthup, N.B., Fateley, W.G., and Grasselli, J.G. (1991). *The Handbook of Infrared and Raman Characteristic Frequencies of Organic Molecules*, San Diego, CA: Academic Press.
11. Nyquist, R.A., Putzig, C.L. and Leugers, M.A. (1997). *Infrared and Raman Spectral Atlas of Inorganic and Organic Salts*, vols. 1–3, San Diego, CA: Academic Press.
12. Nyquist, R.A., and Kagel, R.O. (1997). *Infrared Spectra of Inorganic Compounds*, vol. 4, San Diego, CA: Academic Press.
13. Nakamoto, K. (1997). *Infrared and Raman Spectra of Inorganic and Coordination Compounds, Part A: Theory and Applications in Inorganic Chemistry*, New York: John Wiley & Sons.
14. Nakamoto, K. (1997). *Infrared and Raman Spectra of Inorganic and Coordination Compounds, Part B, Applications in Coordination, Organometallic, and Bioinorganic Chemistry*, New York: John Wiley & Sons.

CHAPTER 1

Epoxides and Ethers

Epoxides	1
Glycidyl Ethers and Glycidylacrylate	4
Cyclic Ethers	5
Open Chain Aliphatic and Aliphatic Aromatic Ethers	5
Vinyl Ethers	6
References	6

Figures		Tables	
Figure 1-1	8 (1)	Table 1-1	15 (1)
Figure 1-2	9 (2)	Table 1-2	16 (2, 3)
Figure 1-3	10 (2)	Table 1-2a	17 (2, 3, 4)
Figure 1-4	10 (3)	Table 1-2b	18 (4)
Figure 1-5	11 (3)	Table 1-3	19 (4)
Figure 1-6	12 (3, 4)	Table 1-4	20 (4)
Figure 1-7	13 (4)	Table 1-5	21 (5)
Figure 1-8	13 (4)	Table 1-6	22 (5)
Figure 1-9	14 (5)	Table 1-7	23 (5)
		Table 1-8	24 (6)
		Table 1-9	25 (6)

*Numbers in parentheses indicate in-text page reference.

EPOXIDES

Ethylene oxide is a cyclic three-membered ring with the molecular formula C_2H_4O, and it has C_{2v} symmetry. The vapor-phase infrared (IR) spectrum of ethylene oxide is shown in Figure 1.1, and Potts has assigned its vibrational spectrum (1). However, assignment of the antisymmetric B_1 ring deformation was not apparent using $1\,cm^{-1}$ resolution. A vapor-phase IR spectrum of ethylene oxide has been recorded using $0.25\,cm^{-1}$ resolution and the antisymmetric ring deformation was observed at $897\,cm^{-1}$ (2). The fundamental vibrations of ethylene oxide are given in Table 1.1.

The three characteristic ring modes in ethylene oxide are ring breathing, symmetric ring deformation, and antisymmetric ring deformation. These three modes are illustrated here:

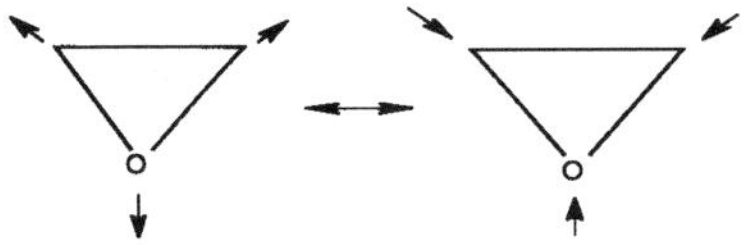

1270.5 cm^{-1}

A_1 ring breathing

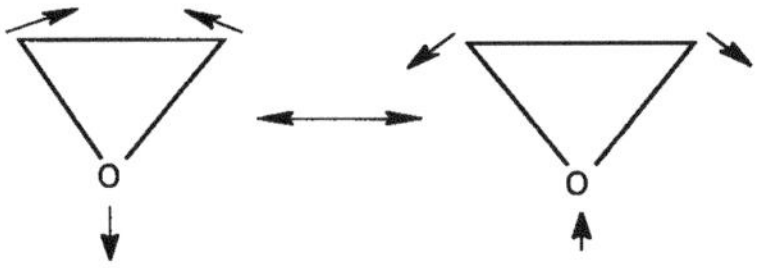

877 cm^{-1}

A_1 symmetric ring deformation

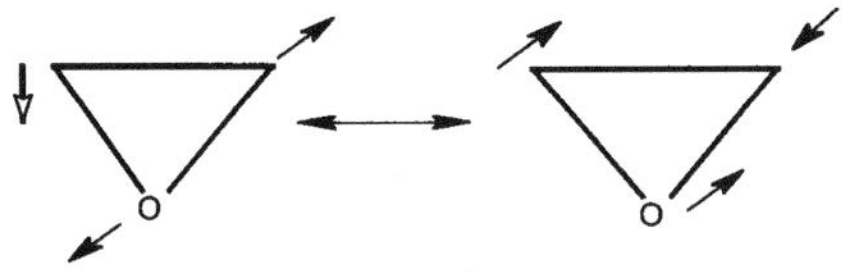

897 cm^{-1}

B_1 antisymmetric ring deformation

Figure 1.2 shows a vapor-phase IR spectrum of 1,2-propylene oxide. This molecule does not even have a plane of symmetry, and its molecular symmetry is C_1. The vapor-phase IR bands near 1269, 839, and 959 cm^{-1} are assigned as ring breathing, symmetric ring deformation, and antisymmetric ring deformation, respectively. For the 1,2-epoxyalkanes (3-alkyl-1,2-ethylene oxide) studied (3), these three fundamentals occur in the ranges 1248–1271, 830–877, and 883–941 cm^{-1}, respectively.

The ring breathing mode decreases in frequency as the 3-carbon atom becomes increasingly branched. Thus, the ring breathing mode occurs at 1265, 1259, 1256, and 1248 cm^{-1} for the 3-methyl, 3-ethyl, 3-isopropyl and 4-tert-butyl analogs respectively, of 3-alkyl-1,2-ethylene oxide in CCl_4 solution. It is suggested that the ring breathing mode frequency decrease is due to the electron release of the 3-alkyl group to the C_2 atom, which increases with increased branching in the C_3 atom and which then weakens the bonds, respectively (see Tables 1.2 and 1.2a).

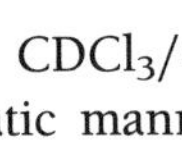

An extensive study of styrene oxide in $CDCl_3/CCl_4$ shows that the epoxy ring breathing mode increases in frequency in a systematic manner from 1252.4 cm^{-1} in CCl_4 solution to 1252.9 cm^{-1} in 52 mol % $CDCl_3/CCl_4$ solution to 1253.5 cm^{-1} in $CDCl_3$ solution. Figure 1.3 shows a plot of mole % $CDCl_3/CCl_4$ vs the epoxy ring breathing mode frequency for styrene oxide, which illustrates the smooth increase in frequency as the mol % $CDCl_3/CCl_4$ is increased (4). Figure 1.4 shows a plot of the symmetric ring deformation frequency for styrene oxide vs mole % $CHCl_3/CCl_4$. This plot shows that symmetric ring deformation decreases in frequency as the mole % $CHCl_3/CCl_4$ is increased (4). Figure 1.5 is a plot of antisymmetric ring deformation frequency for styrene oxide vs $CDCl_3/CCl_4$, and it shows that this mode increases in frequency as the mole % $CDCl_3/CCl_4$ is increased (4). The reaction field becomes larger as the mole % $CDCl_3/CCl_4$ is increased, and, consequently, there is a larger dipolar interaction between styrene oxide as the mole % $CDCl_3/CCl_4$ is increased. The linear frequency shift is attributed to dipolar effects between styrene oxide and the solvent system. Breaks in the plots are attributed to hydrogen bonding to the oxygen atom of the epoxy group and to the π system of the phenyl group of styrene oxide (4).

The compound 2,4-dichlorostyrene oxide exhibits bands at 1249, 989, and 880 cm^{-1}, which are assigned to epoxy ring breathing, asymmetric deformation, and symmetric deformation, respectively. These assignments are comparable to those exhibited by styrene oxide (see Table 1.2a).

The epihalohydrins, or 3-halo-1,2-epoxypropanes, have the empirical structure:

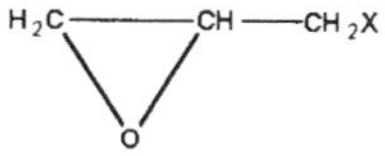

Electron-diffraction studies of epichlorohydrin (5) and epibromohydrin (6) in the vapor phase have shown that both exist in a molecular configuration where the C(3)-X bond eclipses the C(2)—H bond and where the halogen atom is almost trans relative to the midpoint of the C(1)-O bond about the C(2)—C(3) bond.

Table 1.2 lists the characteristic epoxy ring modes for the F, Cl, Br, and I analogs for the epihalohydrins. The RT rotational conformer corresponds to the rotational conformer, which is nearly trans to the midpoint of the C—O bond (7, 8). In solution or neat phases, the IR bands, which decrease in intensity with a decrease in temperature, correspond to the trans rotational conformer (RT). The bands, which increase in intensity with decrease in temperature, are assigned to the rotational conformer (R1). Ring breathing, antisymmetric deformation, and symmetric deformation fall within the same ranges exhibited by the 1,2-epoxyalkanes (see Table 1.2). Complete vibrational assignments for the epihalohydrins have been made using IR and Raman data (see Table 1.2a) (7).

Trans-2,3-epoxybutane (or trans-1,2-dimethyl ethylene oxide) exhibits epoxy ring breathing at 1252 cm^{-1} in CS_2 solution while the corresponding mode for the cis isomer occurs at 1273 cm^{-1}. Both the cis and the trans isomer exhibit epoxy antisymmetric deformation at 885 cm^{-1} in CS_2 solution. In the case of epoxy symmetric deformation, the trans isomer occurs at 810 cm^{-1} and it occurs at 778 cm^{-1} for the cis isomer. Thus, the cis and the trans isomers of 2,3-epoxybutane exhibit symmetric epoxy deformation at lower frequency than those exhibited by the 1,2-epoxyalkanes (883–985 cm^{-1}) (see Table 1.2). The IR vapor-phase spectrum for trans-2,3-epoxybutane is shown in Figure 1.6. The ring breathing, antisymmetric deformation, and symmetric deformation are assigned near 1258, 885, and 817 cm^{-1} in the vapor phase.

Tetrachloroethylene oxide has C_{2v} symmetry and the A_1 ring breathing mode is assigned at 1332 cm^{-1}, the A_1 symmetric ring deformation at 890 cm^{-1}, and the B_1 antisymmetric ring deformation at 978 cm^{-1}. All three of these epoxy in-plane ring modes occur at higher frequency than do the corresponding modes for ethylene oxide (1270, 877, and 897 cm^{-1}, respectively). The rise in frequency for each mode most likely results from the positive inductive effect of the four Cl atoms, which has the effect of strengthening the bonds of the group.

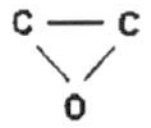

Vibrational assignments for the CH_2X groups for 3-halo-1,2-epoxypropanes are compared to those for 3-halopropynes, 3-halopropenes, and PCH_2X-containing compounds and they are presented in Table 1.3. In several cases, there is spectral evidence for rotational conformers.

Table 1.2b also lists the symmetric CH_3 bending, in-phase $(CH_3)_2$ or $(CH_3)_3$ bending, and out- of-phase $(CH_3)_2$ or $(CH_3)_3$ bending modes for some 1,2-epoxy alkanes. These fundamentals are useful in distinguishing between the 1,2-epoxyalkanes.

Table 1.4 lists both the frequency and absorbance data for oxirane ring breathing, oxirane symmetric in-plane ring deformation, and oxirane asymmetric in-plane ring deformation for 1,2-epoxyalkanes (3). The oxirane CH_2 modes are also listed (3). The oxirane asym. CH_2 stretching mode occurs in the range 3038–3065 cm^{-1}, the oxirane sym. CH_2 stretching mode in the range 2990–3001 cm^{-1}, the oxirane CH_2 bending mode in the range 1479–1501 cm^{-1}, and the oxirane CH_2 wagging mode in the range 1125–1130 cm^{-1}.

Figure 1.6 is a plot of the absorbance data (A) for the oxirane ring breathing mode divided by (A) for the oxirane antisymmetric CH_2 stretching mode vs the number of carbon atoms for the 1,2-epoxy alkanes (the exception is the data point for the ethylene oxide).

Figure 1.7 is a plot of (A) for the ring breathing mode divided by (A) for the oxirane antisymmetric CH_2 stretching mode vs the number of carbon atoms in 1,2-epoxyalkanes. Figure 1.8 is a plot of (A) for oxirane antisymmetric CH_2 stretching divided by (A) for antisymmetric CH_2 stretching of the alkyl group vs the number of carbon atoms in the 1,2-epoxyalkanes (3). The plots for both figures show a relatively smooth correlation and that the absorbance ratios decrease as the length of the alkyl group increases.

GLYCIDYL ETHERS AND GLYCIDYLACRYLATE

The glycidyl group is also named 2,3-epoxypropyl, and the sym. epoxy ring deformation, asym. epoxy ring deformation, and the epoxy ring breathing vibration for four compounds occur in the ranges 840–853 cm^{-1}, 912–920 cm^{-1}, and near 1241–1251 cm^{-1}, respectively, for bis-(2,3-epoxypropyl) ether, 2,3-epoxy propyl phenyl ether, di-glycidyl ether of bis-phenol-A, and 2,3-epoxypropyl acrylate (see Table 1.2a). Strong absorption from a band involving phenyl-oxygen stretching near 1240 cm^{-1} appears to mask the epoxy ring breathing vibrations. Further, the strong absorption from C–C–O stretching in the case of the acrylate also masks the epoxy ring breathing vibration. Monomers of this type are important in the manufacture of epoxy-based paints.

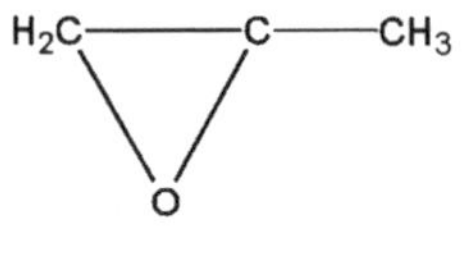

CYCLIC ETHERS

Figure 1.9 is a vapor-phase IR spectrum of tetrahydrofuran. The IR bands near 1071 and 817 cm^{-1} are assigned to antisymmetric and symmetric C—O—C stretching, respectively. In the IR the antisymmetric C—O—C mode is more intense than the symmetric C—O—C mode while the opposite is observed in the Raman spectrum.

Table 1.5 lists correlations for the ring stretching vibrations for cylic ethers. Study of this table shows that there is a trend for the antisymmetric C—O—C stretching vibration to increase in frequency while the symmetric C—O—C stretching vibration decreases in frequency as the number of atoms in the ring increase from 3 to 6.

OPEN CHAIN ALIPHATIC AND ALIPHATIC AROMATIC ETHERS

Dimethyl ether has the basic structure $(CH_3{-})_2O$, and it is the first member of the open chain aliphatic ethers. The antisymmetric $(C{-})_2O$ and symmetric $(C{-})_2O$ vibrations are assigned at 1102 and 929 cm^{-1}, respectively (14). In the vapor phase, asymmetric $(C{-})_2O$ occurs in the range 1111–1140 cm^{-1} in the series $(C_2H_5{-})_2O$ through $(C_{10}\,H_{21}{-})_2O$, and it tends to decrease in frequency as n increases in the $(C_nH_{2n+1+-})_2O$ series (15). Table 1.6 lists IR data for several ethers. The situation becomes more complex when the ether alpha carbon atoms become branched. There is considerable mixing of $(C{-})_nC$ stretching and C—O stretching, and the correlations are not as straightforward as those for $(n\text{-}C_nH_{2n+1})_2O$. In the case of branched ethers such as methyl tert-butyl, ether bands at 1187 and 1070 cm^{-1} involve stretching of the $(CH_3)_3$ C—O and CH_3—O groups.

Both methyl phenyl ether, commonly named anisole (1248 and 1041 cm^{-1}), and ethyl phenyl ether, commonly named phenetole (1245 and 1049 cm^{-1}), exhibit two bands in the IR assigned to the phenyl-O—R group. These may be assigned as antisymmetric and symmetric phenyl-O—R stretching, respectively. However, the 1245–1248 cm^{-1} band involves a complex mode involving phenyl-oxygen stretching. The 1041–1049 cm^{-1} band could be viewed as C—O stretching.

Poly (ethyleneglycol), poly (propyleneglycol), and poly (epichlorohydrin) exhibit a strong IR band near 1100, 1105, and 1105 cm^{-1}, respectively. These bands are assigned to antisymmetric $(C{-})_2O$ stretching in these polymer chains.

In a study of the correspondingly ring-substituted anisoles and phenetoles in the vapor phase, several correlations were developed (16). The phenyl-oxygen-stretching mode for anisole occurs in the range 1243–1280 cm^{-1} and for phenetole in the range 1240–1260 cm^{-1}. Moreover, ring-substituted anisoles occur at higher frequency than those for correspondingly ring-substituted phenetoles. The C—O stretching frequency for the anisoles occurs in the range 1041–1059 cm^{-1} and for phenetoles in the range 1040–1051 cm^{-1}. The C—O stretching frequencies for the phenetoles are not consistently higher in frequency than those for the corresponding ring-substituted anisoles (16).

Table 1.7 lists IR data for 3-X- and 4-X-anisoles in the neat and CS_2 solution phases. The CH_3—O stretching vibration in the neat phase for 3-X- and 4-X-anisoles occurs in the range 1021–1052 cm^{-1}. In the vapor phase, it occurs at higher frequency [e.g., 4-NO_2, 1022 (neat) vs

1041 (vapor)]. Strong IR bands in the region 1198–1270 cm^{-1} are assigned to the complex phenyl-oxygen stretching vibration.

Table 1.8 lists Raman data for vinyl phenyl ether and alkyl vinyl ethers (17). Tentative assignments are presented for these data.

In the case of phenyl vinyl ether, the Raman bands at 1026, 1002 and 615 cm^{-1} whose relative intensities are 4, 9, and 1 are characteristic of mono-substituted benzenes. These bands are very insensitive to substituent groups (18). Infrared spectra and assignments for vinyl ethers are given in Reference 19.

VINYL ETHERS

Infrared bands in the regions 933–972 cm^{-1} and 800–826 cm^{-1} are assigned to vinyl twist and vinyl CH_2 wag, respectively. These vinyl groups exist in cis and gauche forms where C=C stretching vibrations occur in the region 1600–1665 cm^{-1}. The higher frequency band is assigned to the gauche isomer while the lower frequency band is assigned to the cis isomer. Another band present is due to the first overtone of vinyl CH_2 wag being in Fermi resonance with C=C stretching. An IR strong band in the region 1165–1210 cm^{-1} results from C=C—O stretching.

In the case of divinyl ether, gauche and cis C=C stretching are assigned at 1650 and 1626 cm^{-1}, respectively (19).

Aryl vinyl ethers exhibit gauche and cis C=C stretching in the regions 1639–1645 cm^{-1} and 1623–1631 cm^{-1}, respectively (19). In addition to the medium strong IR band in the region 1137–1156 cm^{-1} assigned as C=C—O stretching, the vinyl aryl ethers exhibit a strong IR band at higher frequency (1190–1250 cm^{-1}) assigned to a complex mode involving stretching of the phenyl-oxygen bond. This vibration appears to increase in frequency as the phenyl ring becomes increasingly substituted (19).

Table 1.8 for vinyl ethers shows that the Raman band intensity for the cis C=C stretching conformer is usually more intense than for the gauche C=C stretching conformer while in the case of IR the cis C=C stretching band intensity is also usually more intense than it is for band intensity for the gauche C=C stretching vibration. Temperature studies of 2-ethylhexyl vinyl ether has shown that the cis conformer is the more stable form (20).

Table 1.9 lists vapor-phase IR data for the alkyl group of vinyl alkyl ethers (21). These data are helpful in spectra-structure identification of compounds of this type. The Sadtler collection of vapor-phase IR spectra are essential for spectroscopists utilizing the GC/FT-IR techniques (21).

REFERENCES

1. Potts, W. J. (1965). *Spectrochim. Acta*, **21**: 511.
2. Nyquist, R. A. and Putzig, C. L. (1986). *Appl. Spectrosc.*, **40**: 112.
3. Nyquist, R. A. (1986). *Appl. Spectrosc.*, **40**: 275.
4. Nyquist, R. A. and Fiedler, S. (1994). *Vib. Spectrosc.*, **7**: 149.
5. Igarashi, M. (1955). *Bull. Chem. Soc. Japan*, **28**: 58.

6. Igarashi, M. (1961). *Bull. Chem. Soc. Japan*, **34**: 165.
7. Nyquist, R. A., Putzig, C. L., and Skelly, N. E. (1986). *Appl. Spectrosc.*, **40**: 821.
8. Evans, J. C. and Nyquist, R. A. (1963). *Spectrochim. Acta*, **19**: 1153.
9. Nyquist, R. A., Reder, T. L., Ward, G. R., and Kallos, G. J. (1971). *Spectrochim. Acta*, **27A**: 541.
10. Nyquist, R. A., Reder, T. L., Stec, F. F., and Kallos, G. J. (1971). *Spectrochim. Acta*, **27A**: 897.
11. McLachlan, R. D. and Nyquist, R. A. (1968). *Spectrochim. Acta*, **24A**: 103.
12. Nyquist, R. A. (1968). *Appl. Spectrosc.*, **22**: 452.
13. Nyquist, R. A. (1984). *The Interpretation of Vapor-Phase Infrared Spectra, Group Frequency Data.* Philadelphia: Sadtler Research Laboratories, A Division of Bio-Rad Laboratories, p. 428.
14. Herzberg, G. (1945). *Molecular Spectra and Molecular Structure II. Infrared and Raman Spectra of Polyatomic Molecules.* Princeton, NJ: D. Van Nostrand Company, Inc.
15. Nyquist, R. A. (1984). *The Interpretation of Vapor-Phase Infrared Spectra, Group Frequency Data.* Philadelphia: Sadtler Research Laboratories, A Division of Bio-Rad Laboratories, p. 434.
16. Nyquist, R. A. (1984). *The Interpretation of Vapor-Phase Infrared Spectra, Group Frequency Data.* Sadtler Research Laboratories, A Division of Bio-Rad Laboratories, p. 440.
17. Sadtler Standard Raman Spectral Data.
18. Nyquist, R. A. and Kagel, R. O. (1977). Organic Materials. *Infrared and Raman Spectroscopy*, Part B, *Practical Spectroscopy Series*, vol. 1, E. G. Brame and J. G. Grasselli, eds., Marcel Dekker, Inc., New York, p. 476.
19. Nyquist, R. A. (1989). *The Infrared Spectra Building Blocks of Polymers.* Philadelphia: Sadtler Research Laboratories, Division of Bio-Rad Laboratories Inc., p. 33 and IR spectra 1–158 through 1–171.
20. Owen, N. L. and Sheppard, N. (1964). *Trans. Faraday Soc.*, **60**: 634.
21. (1977). *Sadtler Standard Infrared Vapor Phase Spectra.* Philadelphia: Sadtler Research Laboratories, Division of Bio-Rad Laboratories Inc.

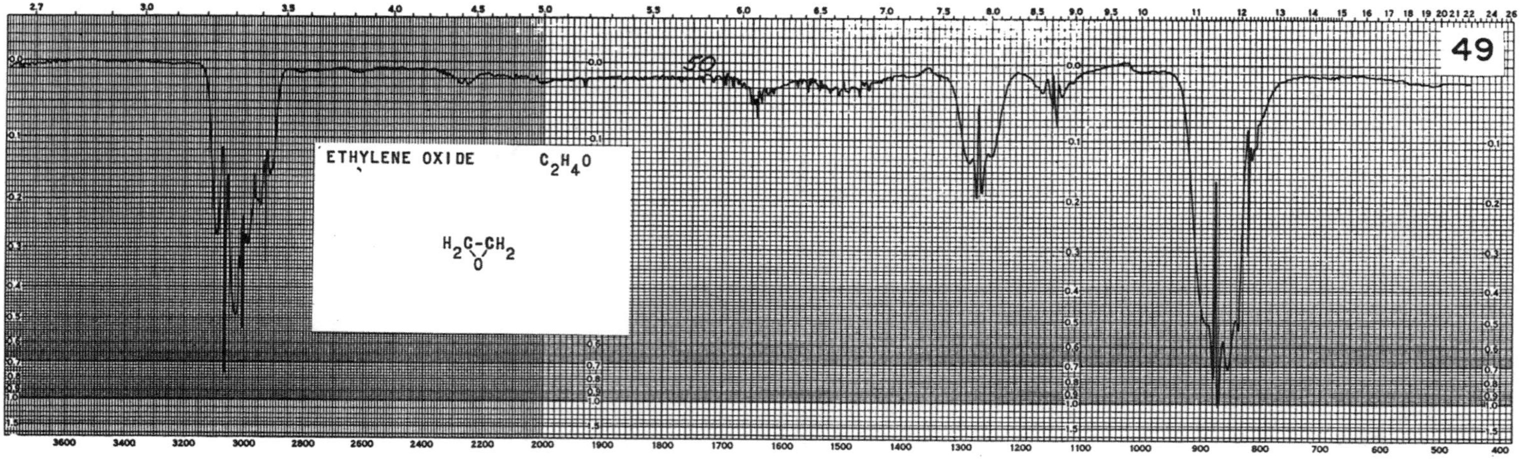

FIGURE 1.1 Vapor-phase IR spectrum of ethylene oxide.

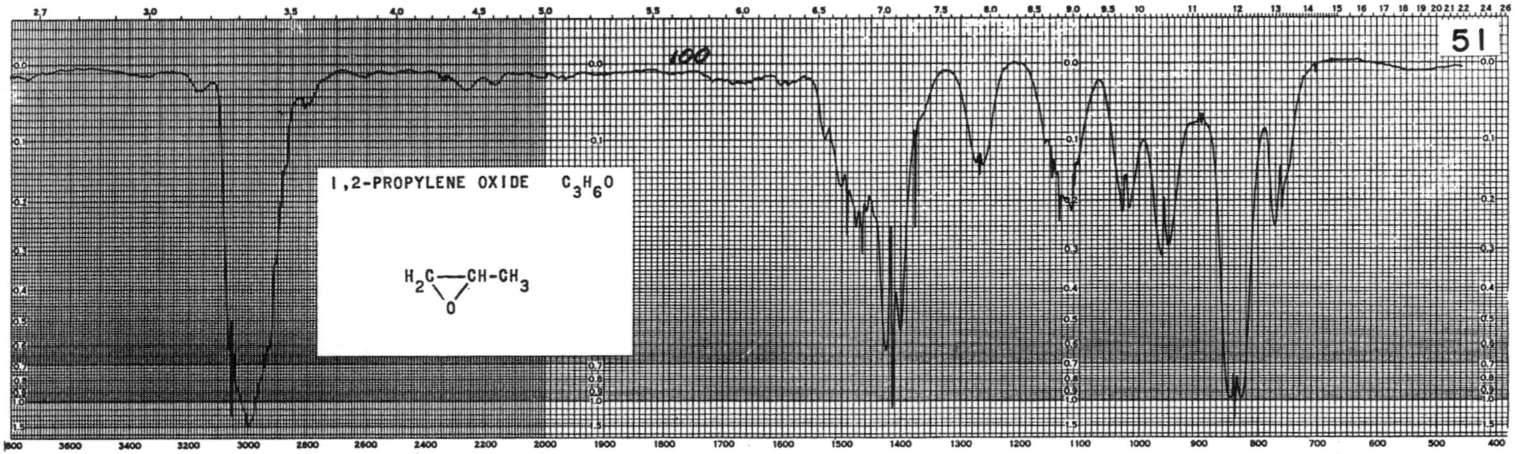

FIGURE 1.2 Vapor-phase IR spectrum of propylene oxide.

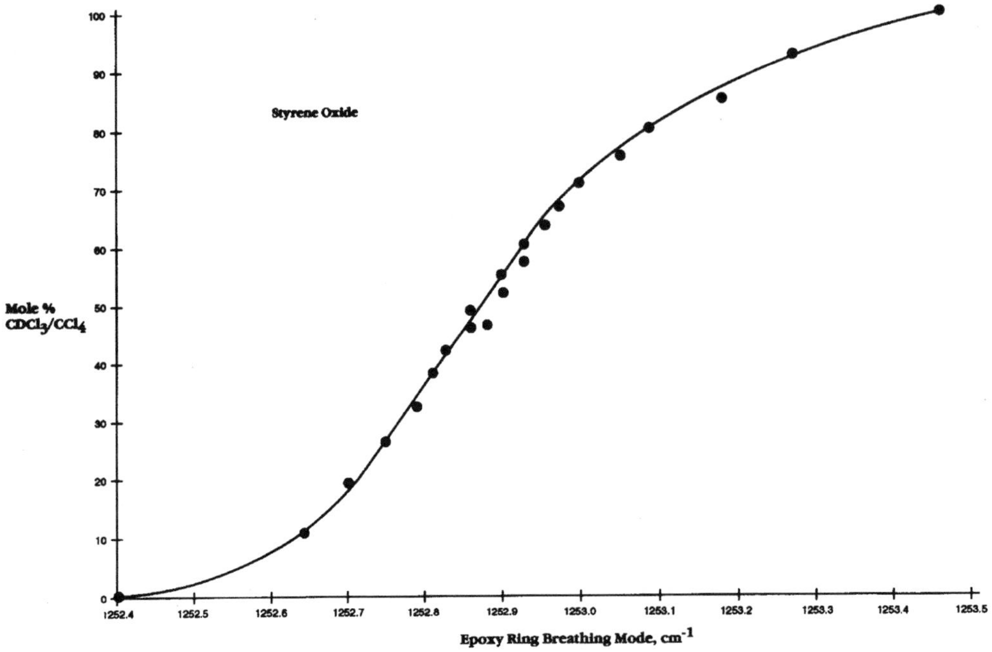

FIGURE 1.3 A plot of the epoxy ring breathing mode frequency for styrene oxide vs the mole % $CDCl_3/CCl_4$ (4).

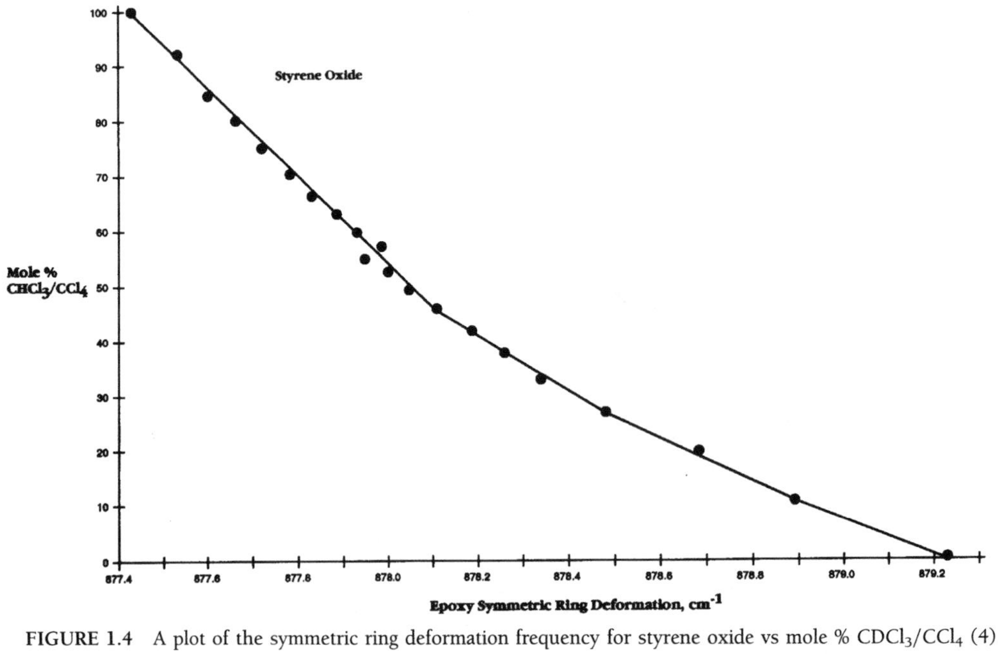

FIGURE 1.4 A plot of the symmetric ring deformation frequency for styrene oxide vs mole % $CDCl_3/CCl_4$ (4).

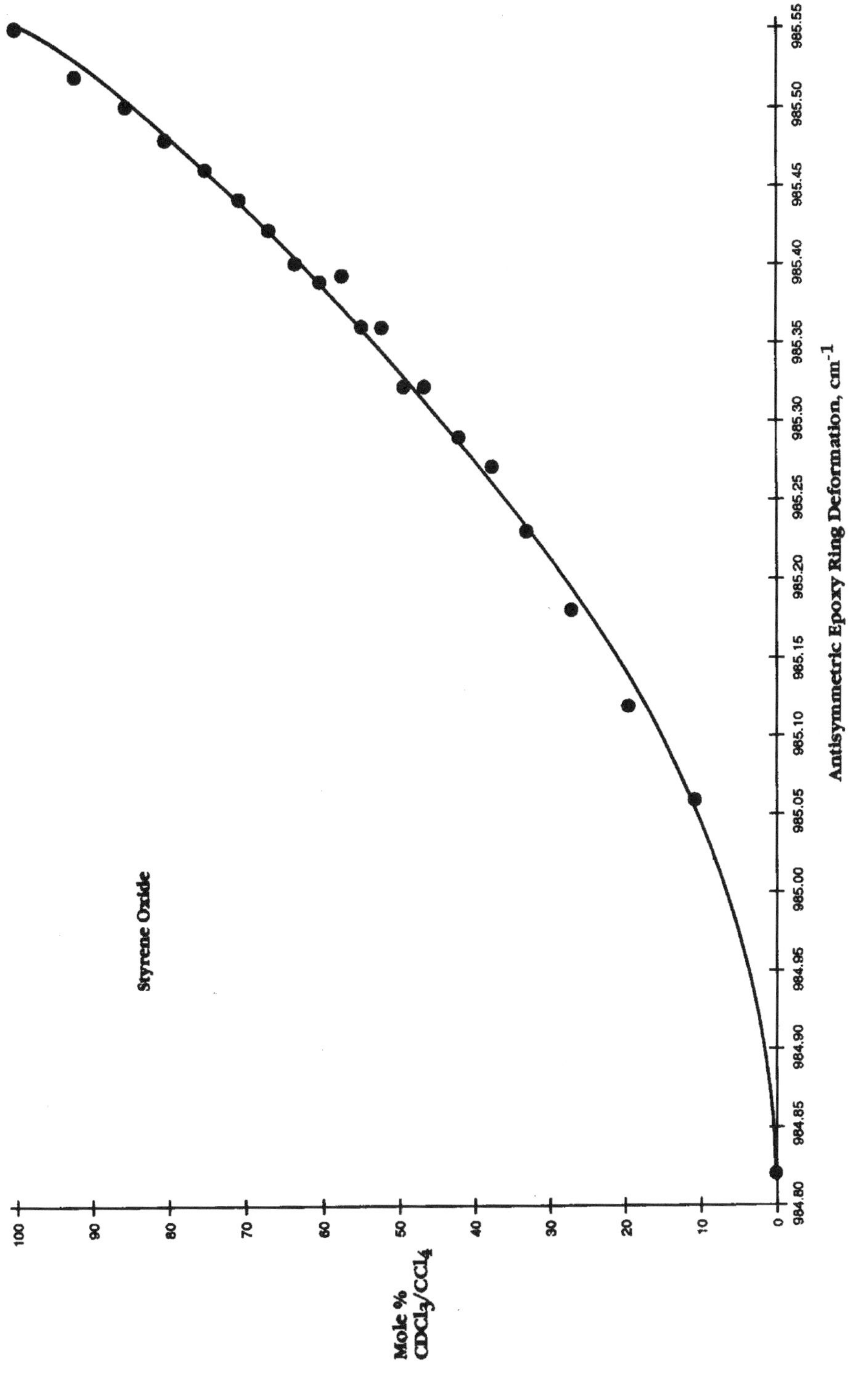

FIGURE 1.5 A plot of the antisymmetric ring deformation frequency for styrene oxide vs mole % $CDCl_3/CCl_4$ (4).

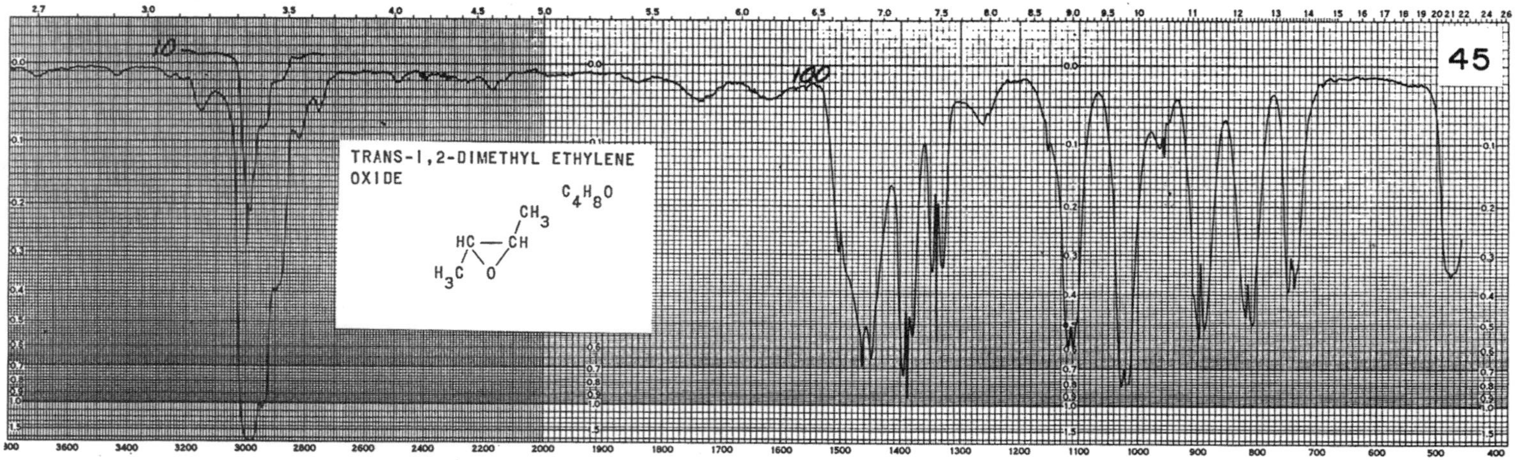

FIGURE 1.6 Vapor-phase IR spectrum of trans-2,3-epoxybutane (or trans-1,2-dimethyl ethylene oxide).

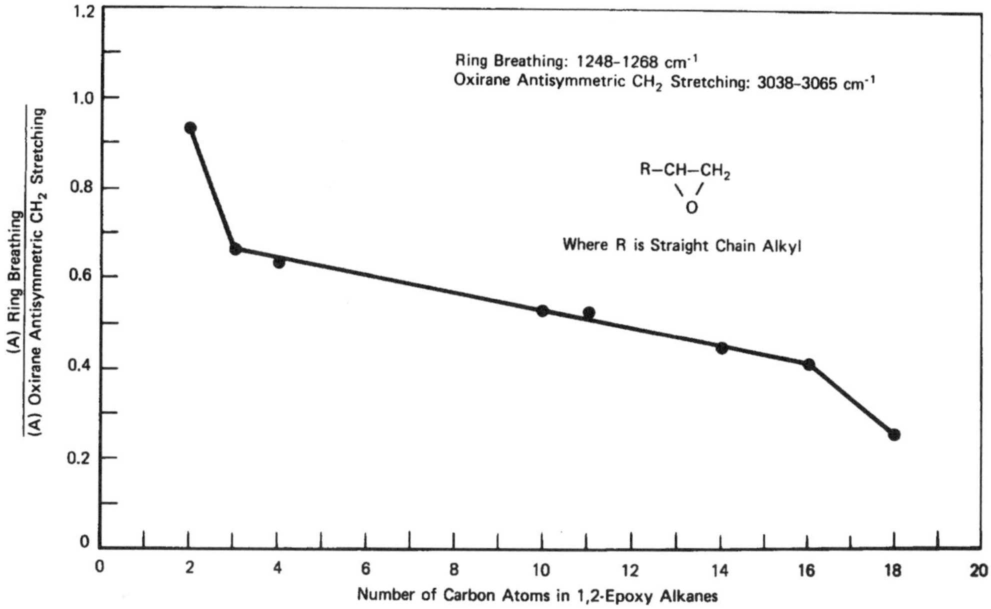

FIGURE 1.7 A plot of the absorbance (A) for the oxirane ring breathing mode divided by (A) for the oxirane antisymmetric CH_2 stretching mode vs the number of carbon atoms for the 1,2-epoxyalkanes (ethylene oxide is the exception).

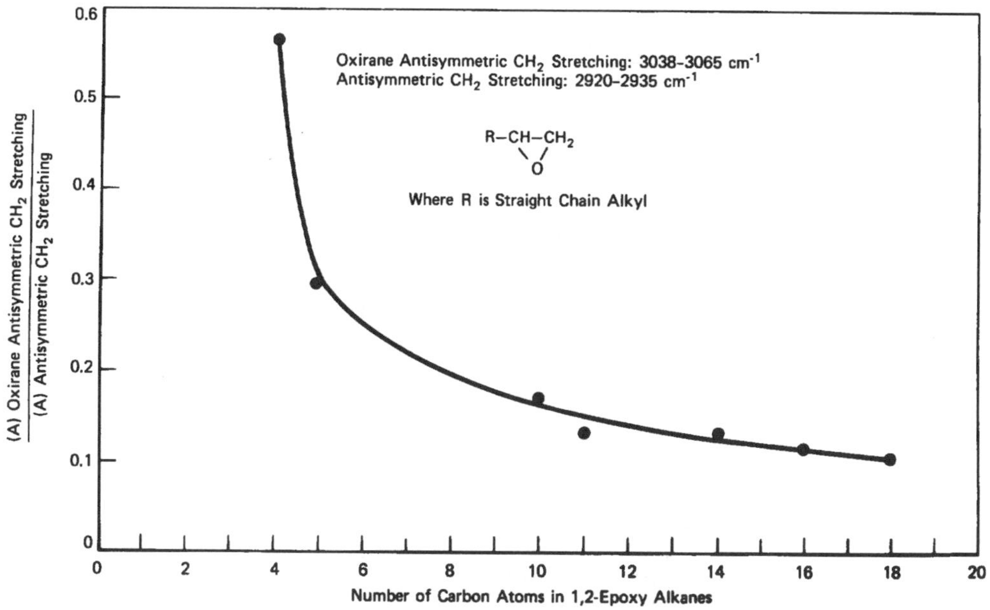

FIGURE 1.8 A plot of (A) for oxirane antisymmetric CH_2 stretching divided by (A) for antisymmetric CH_2 stretching for the alkyl group vs the number of carbon atoms in the 1,2-epoxyalkanes.

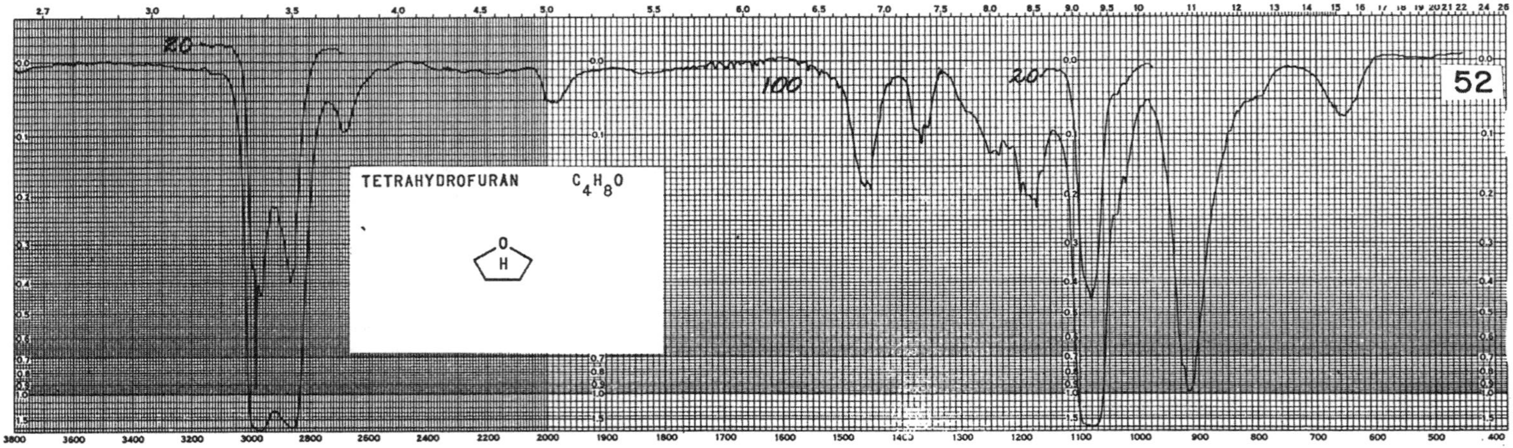

FIGURE 1.9 Vapor-phase IR spectrum of tetrahydrofuran.

TABLE 1.1 Vibrational assignments for ethylene oxide

Species and mode	Ethylene oxide cm^{-1}
A_1	(1,2)
s.CH_2 str.	~3006
CH_2 deformation	1497.5
Ring breathing	1270.5
CH_2 wagging	~1130
s.Ring deformation	897
B_2	
a.CH_2 str.	3065
CH_2 twisting	1412
CH_2 rocking	821.5
B_1	
s.CH_2 str.	3006
CH_3 deformation	1471.5
CH_2 wagging	1151
a.Ring deformation	897
A_2	
a.CH_2 str.	~3065
CH_2 twisting	~1300
CH_2 rocking	~860

TABLE 1.2 Characteristic epoxy ring modes

References	(1,2) Ethylene oxide	(3) 1,2-Epoxy alkanes	(4) Styrene oxide	(5) 1,2-Epoxy-propane 3-fluoro-	(5) 1,2-Epoxy-propane 3-chloro-	(5) 1,2-Epoxy-propane 3-bromo	(5) 1,2-Epoxy-propane 3-iodo-	Rotational isomer	Range
Epoxy ring modes									
Ring breathing	1270.5	1248–1271	1252.4	1257.4	1254.8	1256.1	1260	R1	1248–1271
				[—]	1266.8	1262	1253	RT	
Antisymmetric deformation	897 [5]	883–941	984.8	906.4	905.8	888.2	911	R1	883–985
Symmetric deformation	877	830–877	879.2	860	854	845.8	841	R1	830–877
				838.2	845	833.2	862	RT	
RT = trans isomer									
R1 = gauche isomer?									
Asym. CH_2 str.	(~3065)	3038–3051		3067.7	3063.4	3059.4	3051		3038–3068
Sym. CH_2 str.	(~3006)	2989–3001		3011	3003.9	3000	2991		2989–3011
CH_2 bending	1497.5	1479–1501		1530	1520	1512	1500		1497–1530
CH_2 wagging	(~1130)	1125–1130							1125–1130

	trans-2,3-epoxy-butane [vapor]; [CS_2]	3,4-epoxy-1-butene [vapor]	1,2-epoxy-propane [vapor]	1,2-epoxy-3-isopropoxy-propane [neat]	cis-2,3-epoxy butane [CS_2]	tetrachloro-ethylene oxide [CS_2]	2-methyl-2,3-epoxy-butane [neat]	ethyl 2,3-epoxy-butyrate [CS]	
Ring breathing	[1258]; [1252]	1251	1269	1250	1273	1332	1250	1246?	1246–1332
Antisymmetric deformation	[885]; [885]	915?	959	928	885	978	952	929	885–978
Symmetric deformation	[817]; [810]	825	839	832	778	890	860	864	818–890

? tentative assignment.

TABLE 1.2a Vibrational assignments for 1-halo-1,2-epoxypropanes (or epihalohydrins)

3-Halo,1,2-epoxy-propane or (epihalohydrin)	F	Cl	Br	I	Assignment (7) epoxy group
	3067.7	3063.4	3059.7	3151	a.CH_2 str.
	3011	3003.9	3000	2991	s.CH_2str.
	1484.3	1480.6	1477.4	1475	CH_2 bend*[1]
	1530	1520	1512	1500	2(CH_2 rock)*[1]
	1409.8	1398.2	1395	1392	CH_2 wag
	1164.4	1142	1135.9	1131	CH_2 twist [R1]
	1137.5	1136.6			CH_2 twist [RT]*[2]
	749.1	781.7	779.9	750	CH_2 rock [RT]
	766.4	760.7	756.5	770	CH_2 rock [R1]
	2959.8	2962.8	2967.3	2967	CH str. (see text)
	1252	1208.7	1195.7	1186	CH bend [R1]
		1192.6	1181.7		CH bend [RT]
	1082.4	1081	1081.4	1075	CH wag [R1]
	1113.6				CH wag [RT]
	1257.4	1254.8	1256.1	1260	Ring breath [R1]
		1266.8	1262	1253	Ring breath [RT]
	[906.4]	905.8	888.2	[911]	a.deformation [R1]
	860	854	845.8	841	s.deformation [R1]
	838	845	833.2	862	s.deformation [RT]
					C−CH_2-X group
	3007.1	2999	2991	2991	a.CH_2 str.
	2959.8	2962.8	2967.3	2967	s.CH_2 str.
	1456.2	1432.5	1428.8	1420	CH_2 bend
	1363.3	1276.8	1221.2	1212	CH_2 wag [R1]
	1346.1	1266.8			CH_2 wag [R1]
	1148	1091.8	1061.4	1018	CH_2 twist [R1]
			1036.3		CH_2 twist [RT]
	949.5	961.6	948.6	937	C−C str. [RT]
	[906.4]	926.9	915.7	[911]	C−C str. [R1]
		877	861.2	826	CH_2 rock [RT]
	~983	836		816	CH_2 rock [R1]
	1018.5	695.8	654.9	604	C-X str. [RT]
	992	727.5	643.5		C-X str. [R1]
	601.1	517.9	461.7	417	Skeletal bend [RT]
	496.5	442.3	426.1	395	Skeletal bend [R1]
	365	373	361	359	Skeletal bend [RT]
	397	363	340	318	Skeletal bend [R1]
	260	222	180	177	Skeletal bend
	~150	115	113	105	Torsion

*[1] Solid phase.
*[2] [RT] is the trans rotational conformer.

TABLE 1.2b A comparison of IR data for epoxy ring modes and CH_3, $(CH_3)_2$ and $(CH_3)_3$ bending modes

Epoxy Ring modes	Bis-(2,3-epoxy-) propyl) ether [CS_2] cm^{-1}	2,3-Epoxy-propyl phenyl ether [CS_2] cm^{-1}	Di-glycidyl ether of bis-phenol-A [fluorolube] cm^{-1}	2,3-epoxy-propyl acrylate [CS_2] cm^{-1}	Ethyl 2,3-epoxy-butyrate [CS_2] cm^{-1}	Styrene oxide [CS_2] cm^{-1}	2,4-Dichloro-styrene oxide [neat] cm^{-1}	2,3-Epoxy-2-methyl-4-decyne [neat] cm^{-1}	Range cm^{-1}
Ring breathing	1251	1245 or masked	1241 or masked by C_6H_4-O str.	masked	1246?	1252	1249	1249	1246?–1252
Antisymmetric deformation	912	918	920	913	929	984	989	972	912–989
Symmetric deformation	846	843	840	853	864	879	880	881	840–881
	1,2-epoxy-propane [CS_2]	1,2-epoxy-4-methyl-pentane [CS_2]	1,2-epoxy-butane [CS_2]	1,2-epoxy-3-methyl-butane [CS_2]	1,2-epoxy-3,3-dimethyl-butane [CS_2]	Ethylene oxide [CS_2]	α-Methyl styreneoxide [CS_2]		
Ring breathing	1265	1260	1259	1256	1248	1268	1267		1248–1268
Antisymmetric deformation	893	919	908	932	916	897 [vapor]	997		893–997
Symmetric deformation	829	843	830	878	848	877	861		829–878
Symmetric CH_3 bending	1370 [CCl_4]		1379 [CCl_4]		1381 [CCl_4]				1370–1381
In-phase $(CH_3)_2$ or $(CH_3)_3$ bending		1388 [CCl_4]		1382 [CCl_4]	1366 [CCl_4]				1366–1382
Out-of-phase $(CH_3)_2$ or $(CH_3)_3$ bending		1371 [CCl_4]		1366 [CCl_4]					1366–1371

? tentative assignment.

TABLE 1.3 Vibrational assignments for the CH_2X groups for 3-halo-1,2-epoxypropanes, 3-halopropynes, 3-halopropenes, and PCH_2X-containing compounds

[References] Assignment	(7) 1,2-Epoxy-propane 3-fluoro-	(8) 3-Fluoro-propyne	(8) 1,2-Epoxy-propane 3-chloro-	(8) 3-Chloro-propyne	(9) 3-Chloro-propyne-1-d	(7) 1,2-Epoxy-propane 3-bromo-	(8) 3-Bromo-propyne	(10) 3-Bromo-propyne-1-d	(7) 1,2-Epoxy-propane 3-iodo-	(8) 3-Iode-propyne	Range
Asym. CH_2 str.	3007.1	2955	2999	2968	2969	2991	2976	2978	2991	2958	2958–3007
Sym. CH_2 str.	2959.8	2972	2962.8	3002	2992	2967.3	3006	3008	2967	3008	2959–3008
CH_2 bending	1456.2	1465	1432.5	1441	1442	1428.8	1431	1436	1420	1423	1423–1465
CH_2 wagging [RT]*[1]	1363.3	1381	1276.8	1271	1265	1221.2	1218	1215	1212	1160	1160–1381
CH_2 wagging [R1]*[2]	1346.1		1266.8			[—]			1172		1172–1267
CH_2 twisting [R1]	1148	1242	1091.8	1179	1176	1061.4	1152	1151	1018	1116	1061–1092
CH_2 twisting [RT]	[—]		[—]			1036.3			[—]		1036–[—]
C-C str. [RT]	949.5	940	961.6	960	943	948.6	961	945	937	959	937–962
C-C str. [R1]	906.4		926.9			915.7			911		906–927
CH_2 rocking [RT]	[—]	1018	877	908	908	861.2	866	866	826	810	810–1018
CH_2 rocking [R1]	983		836			[—]			816		816–983
C-X str. [RT]	1018.5	1039	695.8	725	723	654.9	621	634	604	570	570–1019
C-X str. [R1]	992		727.5			643.5			[—]		643–992

References Assignments	(11) 3-Fluoro-propene	(11) 3-Chloro-propene	(11) 3-Bromo-propene	(11) 3-Iodo-propene	(12) $ClCH_2{-}PCl_2$ Cs*[3]; C1*[4]	(12) $ClCH_2{-}P{=}OCl_2$ Cs; C1	(12) $ClCH_2{-}P{=}SCl_2$ Cs; C1	(12) $BrCH_2{-}P{=}OCl_2$ Cs; C1	Range	
Asym. CH_2 str.	2989.6	2990.2	2986.2	2984.4	3000; [—]	2999: 2989	3000; 2993	3011; 3000	2984–2990	2989–3011
Sym. CH_2 str. [g]*[5]	2939	2958.3	2967.7	2967.8	2937; [—]	2933; [—]	2927; [—]	2946; [—]	2939–2968	2927–2946
Sym. CH_2 str. [c]*[6]	2956.9	[—]	[—]	[—]					2957–[—]	
CH_2 bending [g]	1459.2	1445.5	1442.4	1438.8	1391; [—]	1389; [—]	1381; [—]	1373; 1384	1438–1446	1373–1391
CH_2 bending [c]	1467.7	[—]	[—]	[—]					1468–[—]	
CH_2 wagging [g]	1383	1289.5	1245.2	1201	1188; [—]	1212; 1207	1207; [—]	1161; 1169	1201–1333	1161–1212
CH_2 wagging [c]	1239.8	1201	1195	1186.9					1186–1240	
CH_2 twisting [g]	1239.8	1178.2	1154	1088	1098; [—]	1121; 1116	1118; [—]	1072; 1080	1088–1240	1072–1121
CH_2 twisting [c]	989.3	985.2	983.9	980.9					980–989	
C-C str. [g]	935	931.1	925.7	919.9					920–935	
C-C str. [c] or P-C str.	901	[—]	[—]	[—]	812; [—]	818; 811	801; 833	788; 778	901–[—]	778–833
CH_2 rocking [g]	1027	895.5	866.2	825.1	775; [—]	769; [—]	769; [—]	757; 772	778–833	757–775
C-X str. [g]	1005.8	739.4	690.6	669.1	682; [—]	709; [—]	736; [—]	636; 623	669–1006	623–709
C-X str. [c]	989.3	[—]	[—]	[—]				989; [—]		

*[1] [RT] = trans.
*[2] [R1] = gauche.
*[3] Cs = plane of sym.
*[4] C1 = gauche.
*[5] [g] = gauche.
*[6] [c] = cis.

TABLE 1.4 Vapor-phase IR data for the oxirane ring vibrations of 1,2-epoxyalkenes*

1,2-Epoxy-alkane R	Oxirane ring breathing cm^{-1}	A	Oxirane sym. in-plane def. cm^{-1}	A	Oxirane asym. in- plane def. cm^{-1}	A
H	1268	0.362	877	1.677	892	
CH_3	1265	0.401	829	1.755	893	0.156
C_2H_5	1259	0.279	830	1.25	908	1.23
C_3H_7	1259	0.229	843, 831	0.533, 0.644	883	0.48
iso-C_4H_9	1260	0.26	845, 843	0.622, 0.713	919	0.578
iso-C_3H_7	1256	0.285	878	1.357	932	0.779
t-C_4H_9	1248	0.348	848	1.165	916	1.094
C_8H_{17}	1259	0.079	831	0.21	913	0.13
C_9H_{19}	1259	0.06	834	0.153	915	0.099
$C_{12}H_{25}$	1259	0.05	831	0.154	912	0.11
$C_{14}H_{29}$	1259	0.041	830	0.13	911	0.081
$C_{16}H_{33}$	1255	0.023	830	0.065	909	0.047
Range	1255–1268		830–877		883–932	

R	Oxirane a.CH_2 str. cm^{-1}	A	Oxirane s.CH_2 str. cm^{-1}	A	Oxirane CH_2 bending cm^{-1}	A	Oxirane CH_2 wagging cm^{-1}	A
H	3065	0.391	3000	0.429	1491	0.025	1130	
CH_2	3049	0.615	2995	1.305	1501	0.258	1129	
C_2H_5	3044	0.435	masked		1481	0.415	1129	
C_3H_7	3045	0.41	2989	0.736	1481	0.305	1128	
iso-C_4H_9	3049	0.367	2999	0.407	1482	0.232	1126	
iso-C_3H_7	3049	0.359	2996	0.57	1484	0.387	1128	
t-C_4H_9	3051	0.265	3001	0.398	1482	0.886	1129	
C_8H_{17}	3040	0.15	2990	sh	1480	0.1	1129	
C_9H_{19}	3041	0.115	2990	sh	1480	0.084	1129	
$C_{12}H_{25}$	3040	0.111	2990	sh	1480	0.076	1129	
$C_{14}H_{29}$	3040	0.099	2990	sh	1479	0.06	1125	
$C_{16}H_{33}$	3038	0.091	2990	sh	1479	0.06	1125	
Range	3038–3065		2990–3001		1479–1501		1125–1130	

* Reference (12).

TABLE 1.5 Correlations for the ring stretching vibrations for cyclic ethers

Compound or type compound	Ring size atoms in ring	asym. C–O–C stretching [vapor]; (13) cm^{-1}	asym. C–O–C stretching [neat or CS_2] cm^{-1}	sym. C–O–C stretching [vapor]; (13) cm^{-1}	sym. C–O–C stretching [neat or CS_2] cm^{-1}
Ethylene oxides and derivatives	3	883–912	883–912	1246–1332	1246–1322
Tetramethylene oxide	4	[—]	986 [CS_2] IR; stg 980 [neat] R; wk	[—]	1029 [CS_2] IR; wk 1027 [neat] R; stg
Tetrahydrofuran	5	1170	1060 [neat]	919	903 [neat]
Tetrahydrofuran 3-methyl	5	1194	[—]	919	[—]
Tetrahydropyran	6	1095	[—]	820	[—]
Range for four compounds	6	1094–1132	[—]	820–860	[—]

TABLE 1.6 **Infrared data for ethers**

Ether	asym. C−O−C str. neat cm^{-1}	asym. C−O−C str. vapor cm^{-1}	CH_2 rocking neat cm^{-1}	in-phase $(CH_3)_2$ bending cm^{-1}	out-of-phase $(CH_3)_2$ bending cm^{-1}
Methyl	[—]	1102			
Ethyl	1121	~1140			
Propyl	1115	1133			
Butyl	1116	1130	735		
Hexyl	1111	1126	724		
Octyl	1118	1111	725		
Isopropyl	1125	1121		1379	1363
Bis-(2,3-epoxy-propyl)	1100 [CS_2]				
Benzyl	1095	1095			
Bis-(alpha-methyl-benzyl)	1085	[—]			
Phenyl	1235	[—]			
4-Methoxyphenyl	1212; [1030]				
Methyl butyl	1126	1129	734		
Ethyl butyl	1127	1129	742		
Ethyl isobutyl	1120	[—]			
Ethyl octadecyl	1121	[—]	720		
Propyl isopropyl	1125	[—]	751	1378	1361
Methyl tert-butyl	1200; 1085	1209; 1091		1382	1360
Ethyl tert-butyl	1187; 1070	[—]		1388	1356
Isopropyl tert- butyl	1195; 1105			1378; 1368	1355
Methyl benzyl	1100	[—]			
Methyl phenyl	1248; 1041	[—]			
Ethyl phenyl	1245; 1049	[—]			
Propyl phenyl	1229; 1064	[—]			
				CH_2 bending	CH_2 wagging
Poly(ethyleneglycol)	~1100			~1454	~1345
				sym. CH_3 bending	asym. CH_3 stretching
Poly(propyleneglycol)	~1105			1371	2960
				C−Cl stretching	CH_2 wagging
Poly(epichlorohydrin)	~1105			743; 704	~1335

TABLE 1.7 Infrared data for the a and s Aryl-O—R stretching vibrations for 3-X- and 4-X-anisoles in CS_2 solution, vapor, and in the neat phase

Anisole	a.C_6H_4—O—CH_3 str. or C_6H_4—O str. [neat] cm^{-1}	s.C_6H_4—O—CH_3 str. or CH_3—O str. [neat] cm^{-1}	s.C_6H_4—O—CH_3 str. or CH_3—O str. [vapor]; (11b) cm^{-1}	a.C_6H_4—O—CH_3 str. or C_6H_4-—O str. [CS_2 soln.] cm^{-1}
4-X				
NO_2	1262	1022	1041	1264
CH_3SO_2	1260	1021	1032	
CH_3CO	1249	1021		1261
$CO_2C_2H_5$	1252	1028		1258
CN	1240	1024		1259
H	1248	1042		1246
$CH{=}CH_2$	1248	1040	1043	
C_6H_5	1249	1032		
CH_2OH	1245	1030		1248
iso-C_3H_7	1249	1037		
2-ClC_2H_4	1240	1031		
Br	1240	1029	1041	1246
2-BrC_2H_4	1245	1031	1044	
Cl	1242	1032		1247
CH_3	1242	1028		1244
C_2H_5	1248	1037		1248
n-C_3H_7	1244	1038	1041	
t-C_4H_9	1248	1037	1050	1248
CH_3O	1246	1029		1243
$C_6H_5CH_2O$	1230	1033		
NH_2	1238	1034		1241
C_6H_5O	1230	1039		
3-X				
CH_3O	1211	1052		
OH	1200	1043	1046	
NH_2	1198	1029		
Cl	1237	1029		
CH_3	1259	1042	1054	
CH_3CO	1270	1040		
3-BrC_3H_6	1250	1034		

TABLE 1.8 Raman data and assignments for vinyl ethers

Vinyl phenyl ether cm^{-1} (A)	Assignment	Vinyl isobutyl ether cm^{-1} (A)	Vinyl isooctyl ether cm^{-1} (A)	Vinyl decyl ether cm^{-1} (A)	Vinyl docdecyl ether cm^{-1} (A)	Vinyl octadecyl ether cm^{-1} (A)	Vinyl-2-(2-ethoxy ethyl)ethyl ether cm^{-1} (A)	Divinyl ether of butanediol cm^{-1} (A)	Divinyl ether of diethyleneglycol cm^{-1} (A)	Assignment [possible]
		3122 (1)	3121 (0)	3121 (0)	3121 (0)		3120 (1)	3120 (1)	3120 (2)	a.CH_2= str.
3068 (4)	Ring CH str.									
3035 (3)	CH= str.	3046 (3)	3046 (3)	3046 (2)	3046 (1)	3046 (1)	3045 (2)	3044 (4)	3045 (4)	CH= str.
		3023 (2)	3022 (1)	3022 (1)	3022 (1)			3022 (2)		s.CH_2 str.
		2962 (5)	2961 (5)				2975 (4)			a.CH_3 str.
		2943 (4)	2933 (7)				2934 (9)	2924 (4)	2930 (3)	a.CH_2 str.
		2911 (6)	2913 (7)							a.CH_2 str.
1644 (3)	C=C str.			2902 (8)	2895 (8)	2890 (7)				
1593 (2)	Ring 4	2876 (9)	2874 (9)	2874 (9)	2874 (8)		2874 (8)	2876 (4)	2880 (3)	s.CH_3 str.
1311 (2)				2854 (9)	2853 (9)	2852 (9)				s.CH_2 str.
1232 (1)		2724 (1)		2730 (0)	2728 (0)	2725 (0)				s.CH_2 str.
1171 (1)	Ring 6	1638 (2)	1638 (1)	1637 (1)	1638 (0)	1638 (0)	1639 (1)	1639 (3)	1639 (3)	C=C gauche
1157 (1)		1612 (3)	1611 (1)	1611 (2)	1610 (1)	1609 (0)	1620 (2)	1616 (4)	1620 (4)	C=C cis
		1464 (2)						1474 (1)		CH_2 bend
			1450 (3)				1459 (4)		1456 (1)	CH_2 bend
1026 (4)	Ring 8			1439 (4)	1439 (3)	1439 (3)		1436 (1)		CH_2 bend
		1415 (0)						1414 (0)	1415 (0)	CH_2 bend
		1342 (2)								CH_2 wag
1002 (9)	Ring 9	1322 (7)	1320 (4)	1321 (5)	1320 (4)	1320 (2)	1323 (4)	1322 (9)	1323 (9)	CH= rock

						1303 (2)				i.p. $(CH_2)_n$ twist
							1281 (2)			CH_2 wag
		1254 (0)								$C(C)_2$ str.
		1254 (0)						1206 (0)	1207 (0)	a.C–O–C str.
		1181 (1)								$C(C)_2$ str.
		1135 (1)	1142 (0)				1140 (1)			C–C str.
963 (1)				1127 (0)	1128 (0)				1128 (1)	C–C str.
802 (1)				1079 (1)	1079 (1)	1080 (1)				C–C str.
			1041 (0)				1044 (1)		1044 (1)	C–C str.
		1013 (1)								CH_3 rock?
									979 (1)	s.C–O–C?
		956 (1)	957 (1)							s.C–C–C
		941 (1)								C–C str.
								933 (1)		C–C str.
		913 (1)								C–C str.
					892 (1)	891 (1)				s.C–C–C str.
759 (0)	Ring 10			851 (1)	846 (1)		845 (3)	840 (2)		s.C–C–C str.
		833 (5)								s.C–C–C
		808 (2)	815 (1)	813 (1)	812 (1)	811 (1)			827 (2)	CH_2= wag
			768 (1)							CH_2 rock?
615 (1)	Ring 20						490 (1)	498 (0)	499	C–O–C bend
		452 (3)								C–C–C bend
391 (3)										C–O–C bend
		289 (3)								C–C–C def.
269 (1)	C-O-C wag	263 (1)								
			226 (1)	233 (1)	231 (1)		238 (1)	224 (1)	230 (2)	C–O–C wag

TABLE 1.9 Vapor-phase IR data and assignments for the alkyl group of vinyl alkyl ethers*

Vinyl alkyl ether	a.CH_3 str.	s.CH_3 str.	a.CH_3 bend	CH_3 rock	CH_3 rock						
Methyl	2958 (0.310) 2930 (0.240)	2870 (0.100) 2860 (0.095) 2850 (0.110)	1460 (0.152)		1145 (0.260) 1136 (0.190) 1127 (0.210)	1024 (0.186) 1011 (0.175) 997 (0.171)					
	a.CH_3 str.	a.CH_2 str.	s.CH_3 str.	s.CH_2 str.	a.CH_3 bend	s.CH_3 bend	CH_2 wag and or C–C str.	CH_2 twist	CH_3 rock and or C–C str.	CH_2 rock	C–Cl str.	C–Cl str.
Ethyl	2997 (0.434) 2984 (0.370)	2940 (0.270)	2900 (0.260)		1476 (0.087)	1391 (0.225) 1388 (0.200) 1377 (0.210)	1128 (0.481)	1079 (0.226)	1059 (0.271)			
Butyl	2971 (0.790)	2950 (0.690)	2894 (0.410)		1476 (0.135)	1379 (0.190)	1135 (0.249)	1083 (0.270)	1030 (0.177)			
Isobutyl	2965 (0.840)	2925 (0.392)	2897 (0.376)		1475 (0.200)	1388 (0.195)	1145 (0.232)	1080 (0.280)	1019 (0.310)			
2-Ethylhexyl	2970 (1.245)	2938 (1.245)	2882 (0.640)		1470 (0.250)	1380 (0.180)		1079 (0.266)	1015 (0.164)	735 (0.035)		
							a.C–O–C str.					
2-Methoxyethyl	2995 (0.322)	2930 (0.902)	2890 (0.555)	2830 (0.327)	1462 (0.185)	1360 (0.185)	1135 (1.150)	1095 (0.560)	1039 (0.250)			
2-Chloroethyl		2980 (0.105)		2895 (0.065)	1466 (0.060)	1377 (0.100)		1086 (0.240)	1013 (0.145) 1010 (0.138) 1001 (0.142)		764 (0.170)	687 (0.082)? and $CH=CH_2$ wag?
Bis[2-(vinyloxy)ethyl]ether		2938 (0.291)		2880 (0.256)	1460 (0.090)	1359 (0.140)	1140 (0.970)	1092 (0.350)	1010 (0.165) 980 (0.200)			

* Reference (13).

CHAPTER 2

Nitriles, Isonitriles, and Dialkyl Cyanamides

Nitriles	27
Isonitriles	30
Dialkyl Cyanamides	31
Organothiocyanates	31
References	31

Figures		Tables	
Figure 2-1	32 (28)	Table 2-1	39 (27)
Figure 2-2	32 (29)	Table 2-2	40 (29)
Figure 2-3	33 (29)	Table 2-3	40 (29)
Figure 2-4	34 (29)	Table 2-4	41 (29)
Figure 2-5	35 (29)	Table 2-5	42 (30)
Figure 2-6	36 (30)	Table 2-6	42 (30)
Figure 2-7	37 (30)	Table 2-7	43 (31)
Figure 2-8	38 (30)		

* Numbers in parentheses indicate in-text page reference.

NITRILES

Table 2.1 lists IR and/or Raman data for nitriles in different physical phases. In the vapor phase, compounds of form R-CN or NC$-(CH_2)_4-$CN exhibit the CN stretching vibration in the region 2250–2280 cm^{-1}, while in the neat phase the CN stretching vibration for the corresponding compound occurs at a frequency 5 to 22 cm^{-1} lower than in the vapor phase.

Conjugated organonitriles such as benzonitrile, 2-X-, 3-X- or 4-X-substituted benzonitriles, and acrylonitrile exhibit CN stretching in the region 2222–2240 in the vapor phase, and at a frequency 8–19 cm^{-1} lower in the neat phase. Conjugation with the CN group causes the CN stretching mode to vibrate at a lower frequency compared to those for alkylnitriles.

Compounds such as 3-halo-propynonitrile (or 1-cyano-2-haloacetylene) have the following empirical structure: X-C$\equiv$C$-$C$\equiv$N. It is apparent that the CN group is joined to the C$\equiv$C group in a linear manner. Thus, the CN and CC groups are conjugated, and one might expect that CN stretching frequencies would occur at lower frequencies than those for alkanonitriles, benzonitriles, and methyacrylonitrile. However, it is noted that the CN stretching vibration occurs in the region 2263–2298 cm^{-1} (CCl_4 solution) and 2270–2293 cm^{-1} (vapor phase). These CN stretching frequencies are higher than expected. In CCl_4 solution, the C$\equiv$C stretching vibration occurs at 2195, 2122, and 2128 cm^{-1} for the Cl, Br and I analogs of 3-halopropynonitrile, respectively (6). It is likely that there is coupling between the C≡C—C≡N stretching modes, which causes the CN stretching mode to occur at higher frequency and the C≡C stretching vibration to occur at lower frequency than expected.

Figure 2.1 is a plot of the Raman data (CN stretching) for acetonitrile, proprionitrile, isobutyronitrile, pivalonitrile vs the number of protons on the α-carbon atom and this plot shows that the CN stretching vibration decreases in frequency as the number of α-hydrogen atoms decreases from 3 to 0 (7). Tafts σ^* values for CH_3, CH_3CH_2, $(CH_3)_2CH$, and $(CH_3)_3C$ are 0, −0.100, −0.190, and −0.300, respectively (7). Thus, the CN stretching vibration decreases in frequency as the inductive electron release to the nitrile group is increased. This effect would tend to lengthen the CN bond, causing it to vibrate at a lower frequency.

In all cases, in Table 2.1 the CN stretching frequency occurs at a higher frequency in the vapor phase than in the neat or solution phase by 5–47 cm^{-1}. It is of interest to note that the frequency difference between the CN stretching mode in the vapor- and neat phases decreases by 31, 20, 17, and 14 cm^{-1} for acetonitrile, propronitrile, isobutyronitrile, and pivalonitrile, respectively, and this is in decreasing frequency order for both vapor- and neat phases. It is of interest to consider why the frequency difference for CN stretching decreases between the two physical phases progressing in the series from acetonitrile through privalonitrile. It is suggested that this difference is caused by steric factors of the alkyl groups, which alter the amount of dipolar interaction between nitrile groups in the neat phase, while in the vapor phase the dipolar interaction between the nitrile groups is negligible. The steric constant E_s for CH_3, C_2H_5, $(CH_3)_2CH$, and $(CH_3)_3C$ is 0.00, −0.07, −0.47, and −1.54, respectively (9). Thus, as E_s becomes larger the CN groups in the neat phase are spaced farther apart, which weakens the dipolar interaction between nitrile groups. The inductive effect of the alkyl group most likely contributes to some extent to the amount of dipolar interaction, but this is probably a smaller effect because the inductive effect is independent of physical phase.

R-C=N
\+ −
− +
N=C-R

dipolar interaction

The cyanogen halides exhibit the CN stretching vibration at 2290, 2201, 2187, and 2158 cm^{-1} in the neat phase for the F, Cl, Br and I analogs, respectively (4). In the vapor phase the CN stretching for the Cl and Br analogs occurs at frequencies higher by 47 and 13 cm^{-1}, respectively (4, 5).

In this cyanogen halide series the CN stretching vibration decreases as the C≡N bond length increases. For example, in the solid state the CN bond length in the solid state is 1.26, 1.58, 1.77, and 2.03 Å for FCN, ClCN, BrCN, and ICN, respectively (10). In the vapor phase the CN bond is less restricted, and the C≡N bond length is 1.67 and 1.79 Å for ClCN and BrCN, respectively (10). The relative steric factor of F, Cl, Br, CH_3, and I (based on F as zero) is (0.00), −0.31, −0.49, −0.49, and −0.69, respectively. The inductive value σ^* for F, Cl, Br, and I is 1.10, 1.05, 1.00, and 0.85, respectively (11). Thus, the steric factor of the halogen atom increases as the C≡N bond length increases, while the inductive effect of the halogen atom decreases progressing in the series FCN through ICN. Combination of the preceding factors is most likely the cause for the CN stretching frequency for compounds of form R-CN to occur at intermediate frequencies between FCN and ClCN (see Table 2.1). The dipolar interaction between these linear XCN molecules could also be different than for R-CN molecules, which are not linear. For example,

$$\begin{array}{c} \mathrm{X{-}C{\equiv}N} \\ - \;\; + \\ + \;\; - \\ \mathrm{N{\equiv}C{-}C{-}X} \end{array}$$

The larger inductive effect of Cl vs Br, and the large steric effect of Br vs Cl could account for the fact that the frequency difference between the vapor and neat phases for the CN stretching frequency is 47 cm^{-1} for ClCN and 13 cm^{-1} for BrCN.

Table 2.2 lists IR data for acetonitrile 1% wt./vol. in various solvents (12). The solvents are numbered 1–15 in Table 2.2. The CN stretching vibration and the combination tone C—C stretching plus symmetric CH_3 bending are in Fermi resonance (FR). The CN stretching vibration has been corrected for FR in each of the solvents. Figure 2.2 shows a plot for unperturbed $\nu C{\equiv}N$ for acetonitrile vs AN, where AN is the solvent acceptor number (12). This plot shows that the νCN frequency does not correlate well with AN, especially the AN value for dimethyl sulfoxide (DMSO).

Figure 2.3 shows plots of $\nu{\equiv}CN$ (uncorrected and corrected for FR) vs ($\nu{\equiv}CN$ in methyl alcohol) minus ($\nu{\equiv}CN$ in solvent). Both plots are linear, and any set of data plotted in this manner yields a linear mathematical relationship (12). The plots clearly show that FR causes νCN to occur at lower frequency due to resonance with the $\nu C{-}C + \delta$ sym. CH_3 combination tone (CT). It should be noted that the numbering sequence is different in both plots. The extent of FR interaction between νCN and CT ($\nu C{-}C + \delta$ sym. CH_3) is altered in each solvent system because $\nu C{-}C$ and δ sym. CH_3, as well as νCN are affected differently.

Table 2.3 compares IR νCN stretching frequencies for benzonitrile 10 wt./vol.% and 1 wt./vol.% vs those for 1% wt./vol. acetonitrile νCN frequencies corrected for FR in different solvents (12). This table shows that νCN for benzonitrile occurs at lower frequency than νCN for acetonitrile (corrected for FR), from 27 to 34 cm^{-1} in these solvents. The shift to lower frequency in the case of benzonitrile is due to resonance of the CN group with the π system of the phenyl group.

Table 2.3 also shows that νCN for benzonitrile occurs at lower frequency in solution at 10 wt./vol.% than at 1 wt./vol.%. These data are plotted in Figure 2.4. These plots indicate that at higher wt./vol.% solute there is some dipolar interaction between solute molecules, which lowers the νCN frequency.

Figure 2.5 is a plot of unperturbed $\nu C{\equiv}N$, (1% wt./vol. acetonitrile), cm^{-1} vs $\nu C{\equiv}N$ for benzonitrile (1% wt./vol. benzonitrile), cm^{-1} where each compound has been recorded individually in the same solvent. Point 1 for hexane does not fit the essentially linear relationship.

Table 2.4 lists IR data for the νCN frequency for 4-cyanobenzaldehyde in 0 to 100 mol% $CHCl_3/CCl_4$ solutions (1 wt./vol.% solutions). Figure 2.6 shows a plot of $\nu{\equiv}CN$ for 4-cyanobenzaldehyde vs mol% $CHCl_3/CCl_4$ (13). The νCN mode increases in frequency as the mole % $CHCl_3/CCl_4$ increases to ~45%. It then decreases in frequency to ~75%, after which νCN frequency is relatively constant. The νCN frequency for 4-cyanobenzaldehyde is higher in frequency in $CHCl_3$ solution than in CCl_4 solution, and this same observation has been noted for benzonitrile (12, 13). The behavior of νC=O for 4-cyanobenzonitrile has already been discussed; furthermore, its frequency increases as the reaction field is increased (14). Figure 2.6 then shows that νC≡N as well as νC=O are affected, although not in the same manner.

The nitrile group is not always readily detected in the IR. In the case of compounds such as 2,4,4,4-tetrachloro-butyronitrile, the νCN mode is extremely weak, but in the Raman it is readily detected (15). In the vapor phase, benzonitrile and substituted benzonitriles exhibit νCN in the region 2220–2250 cm^{-1}, and the IR band intensity varies from very-weak to weak-medium compared to the most intense IR band in these spectra (2a). In general, the νCN band intensity is significantly higher for atoms or groups with negative Hammett's σ values (OH, OCH_3, NH_2 etc.) than for those with positive values (NO_2, CN, CF_3 etc.). In the vapor phase, the CN group is intramolecularly hydrogen bonded in cases such as 2-hydroxybenzonitrile (νCN, 2225 cm^{-1}) and 2-aminobenzonitrile (νCN, ~2222 cm^{-1}) (2a). The highest vapor-phase frequency is noted at 2250 cm^{-1} for 2,6-difluorobenzonitrile, and this overlaps the region for compounds of form R—CN(νCN, 2250–2260 cm^{-1}) in the vapor phase (2a). The high frequency exhibited by 2,6-difluorobenzonitrile is attributed to a field effect between F and the CN group.

Table 2.5 lists Raman data for the C≡N and C=C groups of organonitriles. In most cases, the νC≡N mode is the strongest band in the Raman spectrum. However, in the case of 1,1-azo bis-(cyclohexane carbonitrile) the symmetric CH_2 stretching vibration is the most intense Raman band. In the case of crotononitrile and 2-methyl crotononitrile the Raman band for νC≡N is approximately twice as strong as the Raman band for νC=C.

ISONITRILES

Methyl isonitrile, ethyl isonitrile, isopropyl isonitrile, and tert-butyl isonitrile exhibit νN≡C at 2183, 2160, 2140, and 2134 cm^{-1}, respectively (16). Figure 2.7 is a plot of the number of protons on the alkyl α-C—N≡ atom vs νN≡C for alkyl isonitriles. This plot shows that νN≡C decreases in frequency with increased branching on the alkyl α-C—N≡ atom (17). Figure 2.8 shows a plot of Taft's σ^* vs νN≡C for alkyl isonitriles (17). Therefore, νN≡C decreases in frequency as the electron release to the isonitrile group is increased. This effect should increase the N≡C bond length. Table 2.6 compares the IR data for organonitriles and organoisonitriles.

In the series of alkanonitriles and alkanoisonitriles presented in Table 2.6, νCN occurs in the range 2236–2249 cm^{-1} and νNC occurs in the range 2134 through 2183 cm^{-1}. The frequency separation between νCN an νNC increases as the electron release of the alkyl group to the CN or NC group is increased (66 to 102 cm^{-1}). In the series 3-x and 4-x substituted benzonitriles and isobenzonitriles presented in Table 2.6, νCN occurs in the range 2226–2240 cm^{-1} and νNC occurs in the range 2116–2125 cm^{-1}. The frequency separation between νCN and νNC (101–122 cm^{-1}) tends to increase somewhat as σ_p or σ_m increase in value. The frequency separation between νCN for 4-chlorobenzonitrile (2233 cm^{-1}) and 2-chlorobenzonitrile (2237 cm^{-1}) is 4 cm^{-1} while for 4-chloroisobenzonitrile (2116 cm^{-1}) and 2-chloroisobenzonitrile (2166 cm^{-1}) is 50 cm^{-1}. The frequency separation between νCN and νNC for 2-chlorobenzonitrile and 2-chloroisobenzonitrile is 71 cm^{-1}. The relatively high νNC frequency exhibited by 2-chloroisobenzonitrile suggests that a field effect of the Cl atom upon the N atom of the NC group causes νNC to shift to higher frequency.

DIALKYL CYANAMIDES

Dialkyl cyanamides have the following empirical structure $(R-)_2N-C\equiv N$. In the vapor phase dimethyl cyanamide and diallyl cyanamide exhibit $\nu C\equiv N$ near $\sim 2228\,cm^{-1}$ and $2221\,cm^{-1}$, respectively (2B). In the neat phase dialkyl cyanamide and dibenzyl cyanamide exhibit $\nu C\equiv N$ at 2200 and $2190\,cm^{-1}$, respectively.

ORGANOTHIOCYANATES

Table 2.7 compares IR data for the vapor and neat phases of organothiocyanates. In the neat phase alkyl thiocyanates (R-S-CN) exhibit νCN in the range $2145–2160\,cm^{-1}$ and in the vapor phase in the range $2161–2175\,cm^{-1}$. In the neat phase arylthiocyanates νCN occurs in the range $2166–2174\,cm^{-1}$, and occur at higher frequency than alkylthiocyanates in the neat phase. The alkyl group releases electrons to the SCN group and the aryl group withdraws electrons from the SCN group. Therefore, νCN for arylthiocyanates occur at higher frequency than νCN for alkylthiocyanates.

REFERENCES

1. Nyquist, R. A. (1984). *The Interpretation of Vapor-Phase Infrared Spectra: Group Frequency Data.* vol. 1. Philadelphia: Sadtler Research Laboratories, Division of Bio-Rad Laboratories, Inc., p. 443.
2. Nyquist, R. A. (1984). Ibid., p. 450.

2a. Nyquist, R. A. (1984). Ibid., p. 443.

2b. Nyquist, R. A. (1984). Ibid., p. 446.

3. *Standard Raman Spectra.* Philadelphia: Sadtler Research Laboratories, Division of Bio-Rad Laboratories, Inc.
4. Colthup, N. B., Daley, L. H., and Wiberley, S. E. (1990). *Introduction to Infrared and Raman Spectroscopy.* 3rd ed., Boston: Academic Press, Inc., p. 183.
5. Nyquist, R. A., Putzig, C. L., and Leugers, M. A. (1997). *Handbook of Infrared and Raman Spectra of Inorganic Compounds and Organic Salts.* Vol. 3, Boston: Academic Press, p. 198.
6. Klaeboe, P. and Kloster-Jensen, E. (1967). *Spectrochim. Acta*, **23A**: 1981.
7. Nyquist, R. A. (1987). *Appl. Spectrosc.*, **41**: 904.
8. Newman, M. S. Editor (1956). *Steric Effects in Organic Chemistry.* New York: John Wiley & Sons, Inc., p. 591; R. W. Taft (1956). *Separation of Polar, Steric, and Resonance Effects in Reactivity.*
9. Newman, M. S. (1956). Ibid., p. 601.
10. Wyckoff, R. (1963). *Crystal Structure.* Vol. 1, New York: John Wiley & Sons, Inc., p. 173.
11. Newman, M. S. (1956). *Steric Effects in Organic Chemistry.* New York: John Wiley & Sons, Inc., p. 595; R. W. Taft (1956). *Separation of Polar, Steric, and Resonance Effects in Reactivity.*
12. Nyquist, R. A. (1990). *Appl. Spectrosc.*, **44**: 1405.
13. Nyquist, R. A., Settineri, S. E., and Luoma, D. A. (1992). *Appl. Spectrosc.*, **46**: 293.
14. Nyquist, R. A., Settineri, S. E., and Luoma, D. A. (1991). *Appl. Spectrosc.*, **45**: 1641.
15. Brame, Jr., E. G. and Grasselli, J. G. (1977). *Infrared and Raman Spectroscopy*, Part B, *Practical Spectroscopy Series.* Vol. 1, New York: Marcel Dekker, Inc., p. 471; R. A. Nyquist and R. O. Kagel (19XX). *Organic Materials.*
16. Bellamy, L. J. (1968). *Advances In Infrared Group Frequencies.* London: Methuen.
17. Nyquist, R. A. (1988). *Appl. Spectrosc.*, **42**: 624.

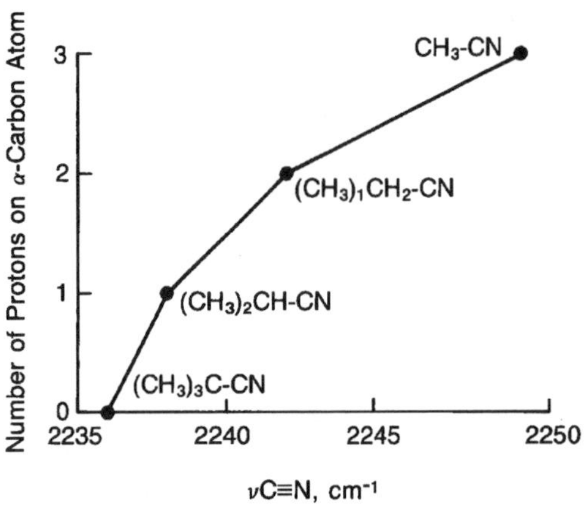

FIGURE 2.1 A plot of νC≡N for alkanonitriles vs the number of protons on the α-carbon atom.

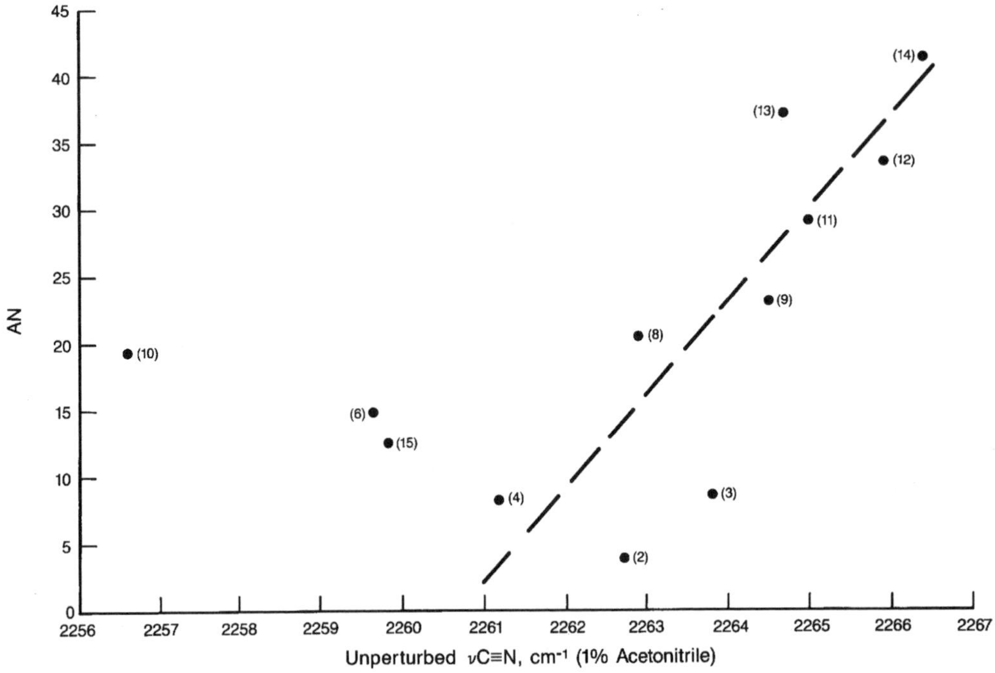

FIGURE 2.2 A plot of unperturbed νC≡N, cm^{-1} (1% wt./vol.) vs AN (the solvent acceptor number).

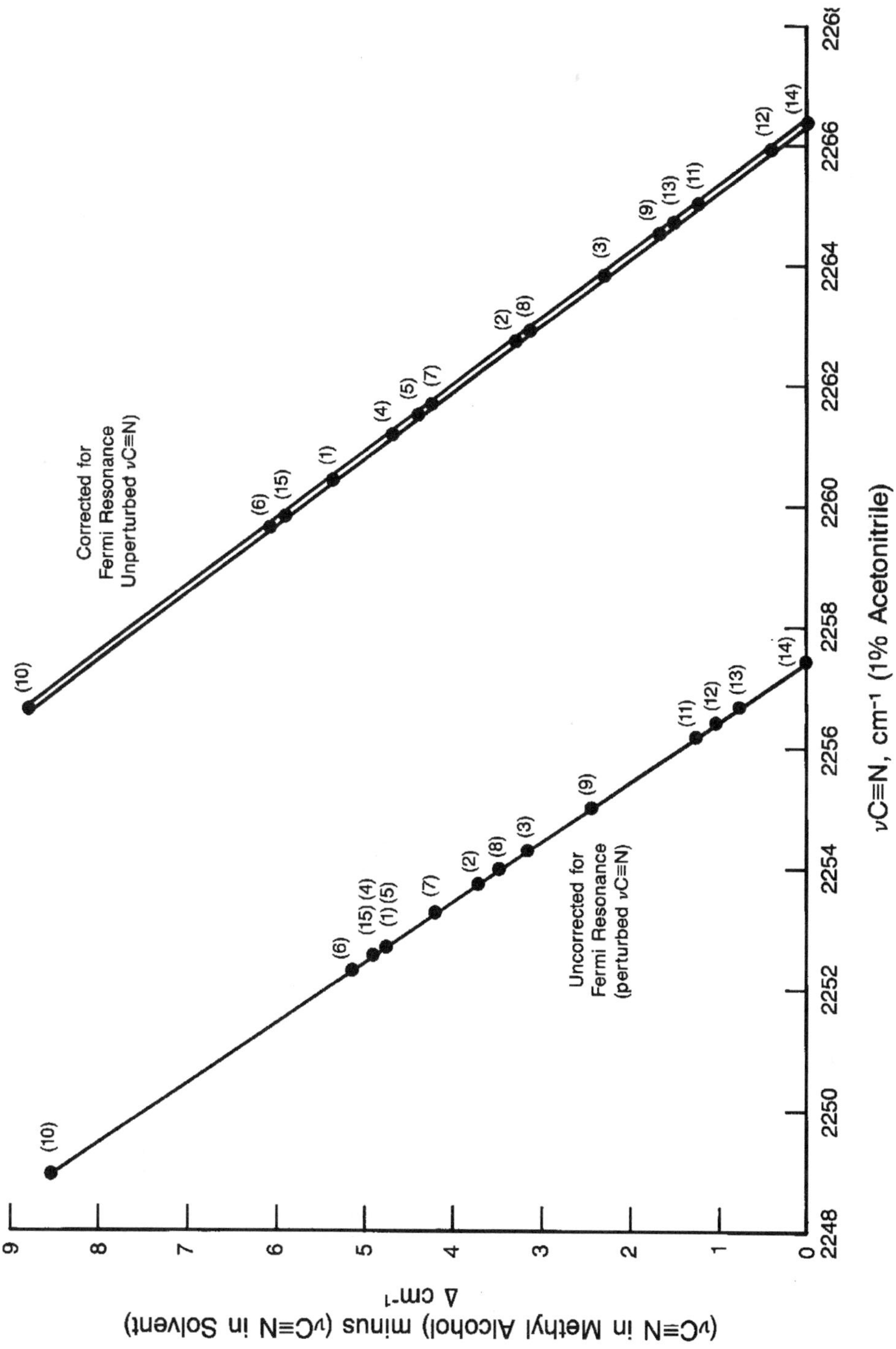

FIGURE 2.3 Plots of νC≡N, cm^{-1} for acetonitrile (1 wt./vol.%) vs (νC≡N in methyl alcohol) minus (νC≡N in another solvent). The two plots represent perturbed and unperturbed νC≡N.

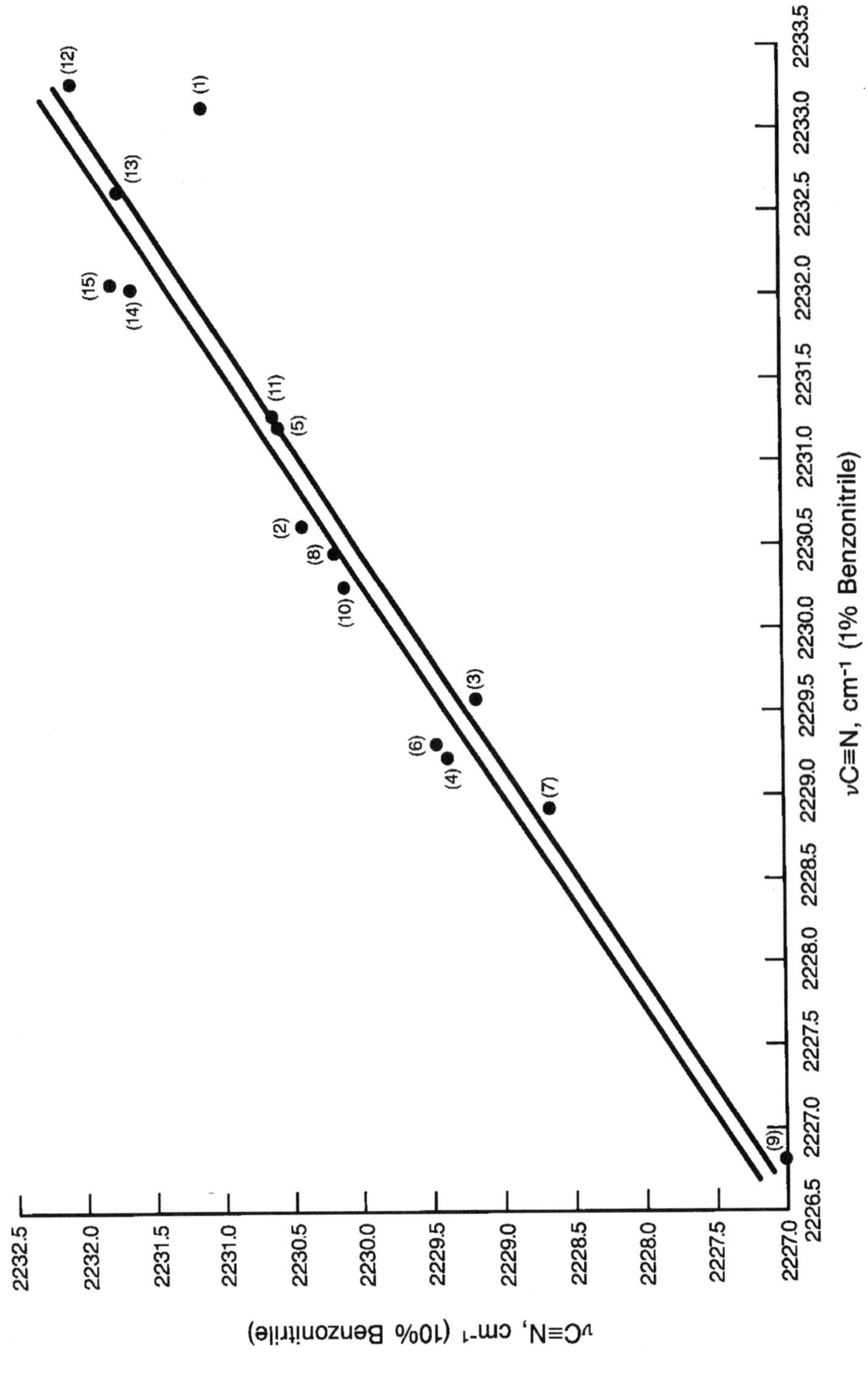

FIGURE 2.4 A plot of νC≡N for 1 wt./vol. % vs νC≡N for 1% wt./vol. in 15 different solvents.

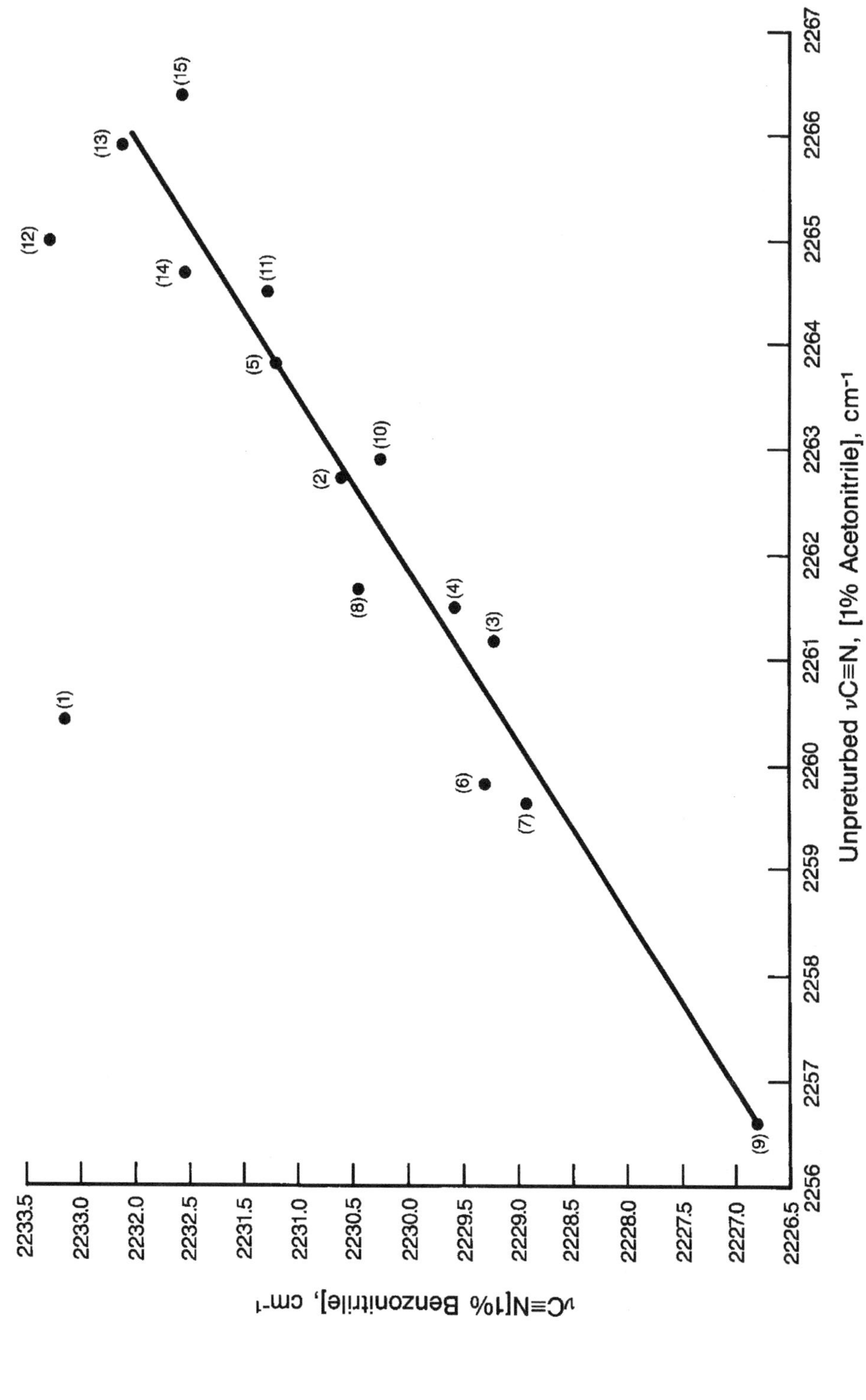

FIGURE 2.5 A plot of unperturbed νC≡N for acetonitrile vs νC≡N for benzonitrile. Both compounds were recorded at 1% wt./vol. separately in each of the 15 solvents.

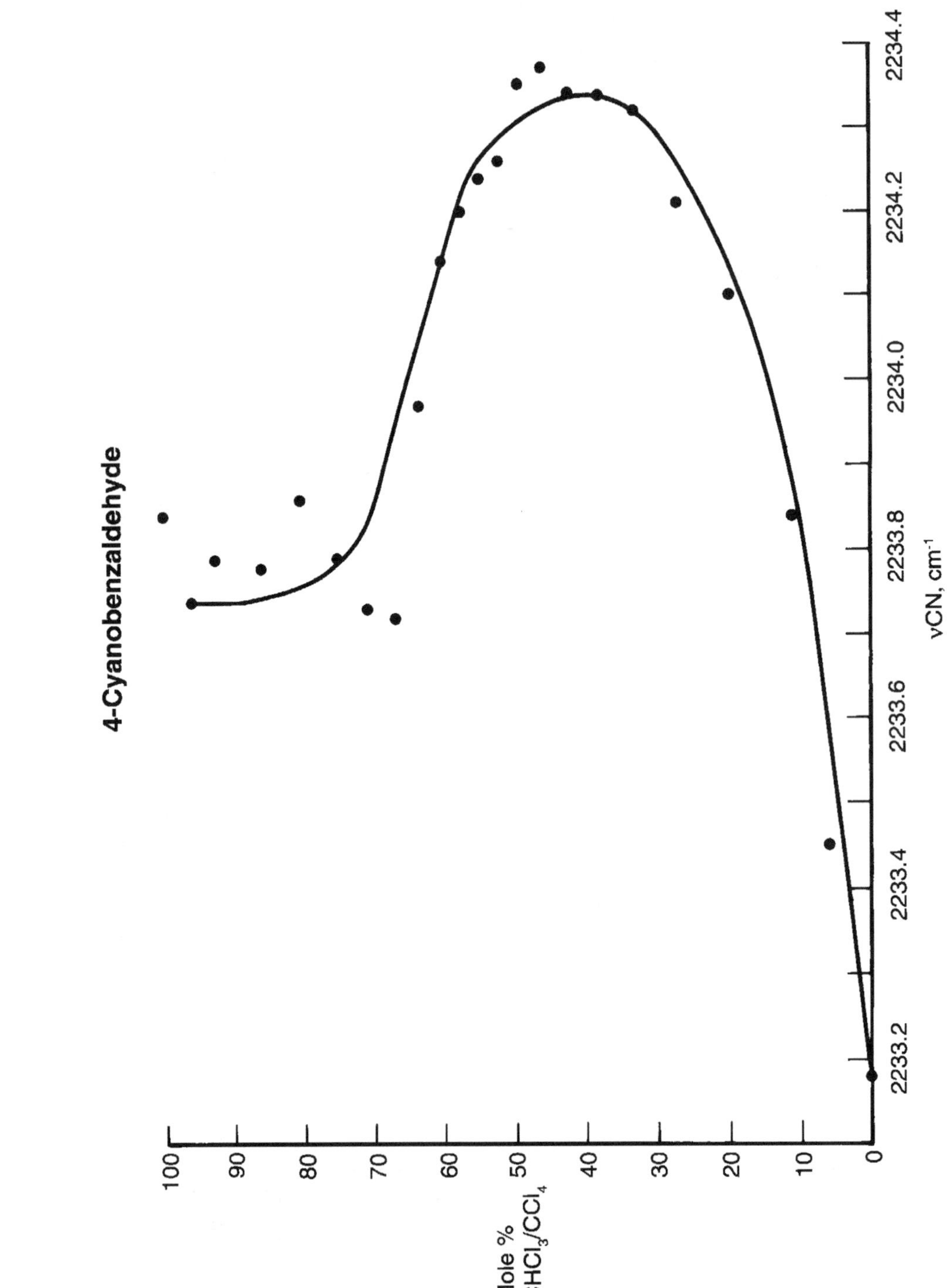

FIGURE 2.6 A plot of $\nu C{\equiv}N$ for 4-cyanobenzaldehyde in cm^{-1} vs mole % $CHCl_3/CCl_4$.

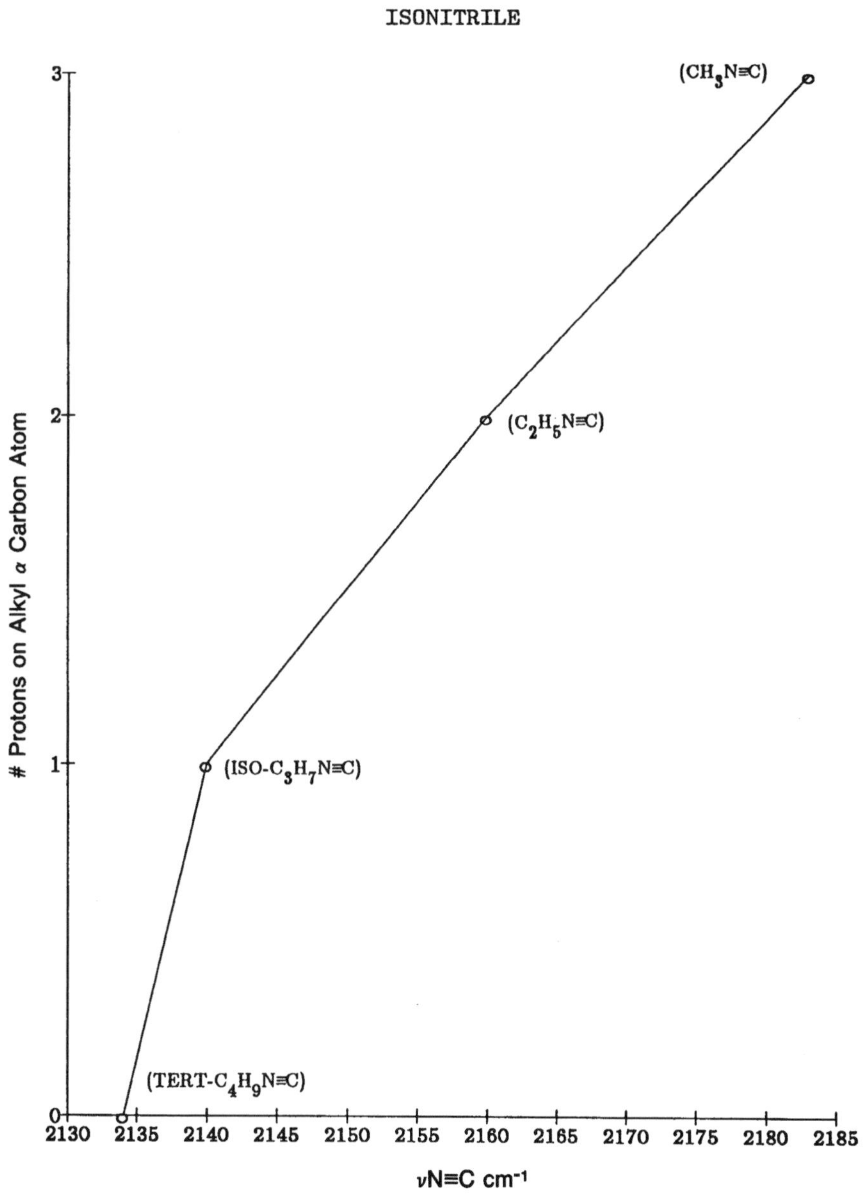

FIGURE 2.7 A plot of the number of protons on the alkyl α-C—N atom vs νNC for alkyl isonitriles.

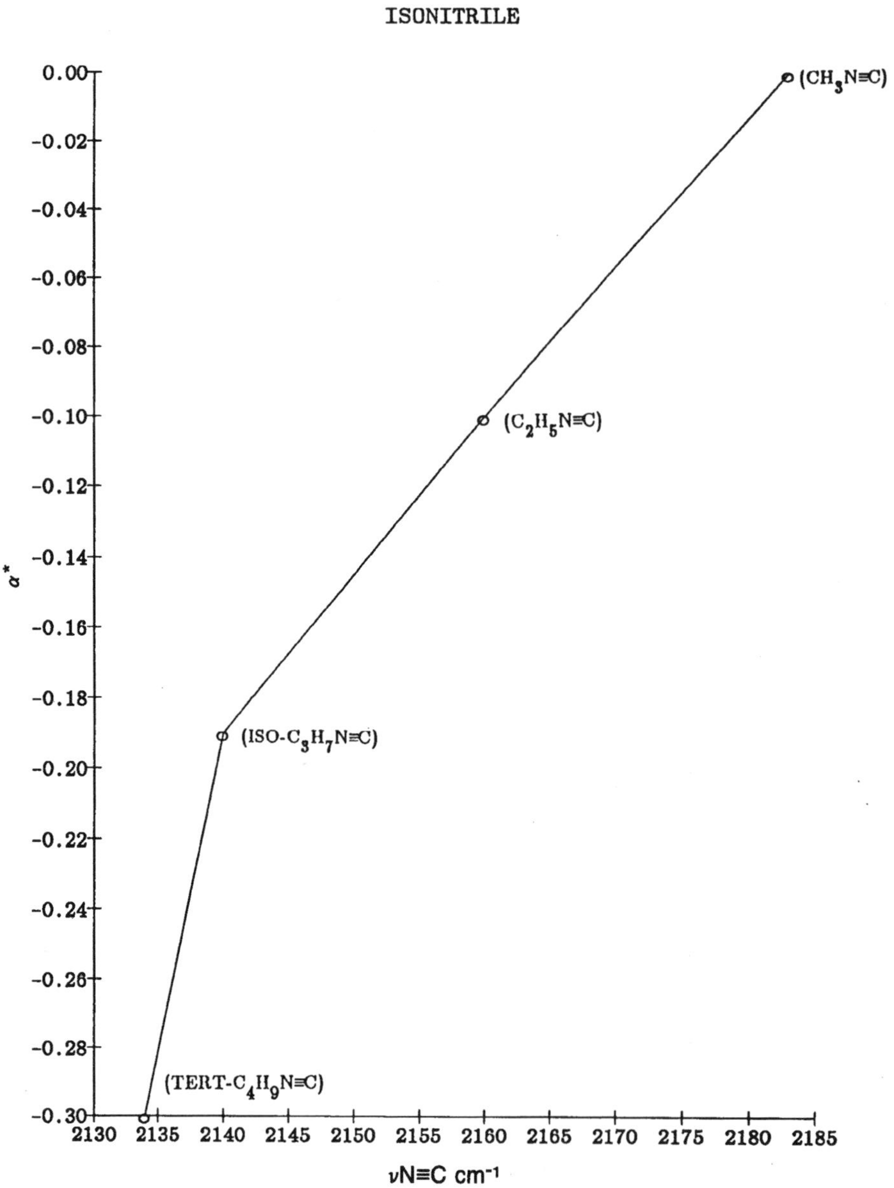

FIGURE 2.8 A plot of Taft's σ^* vs νNC for alkyl isonitriles.

TABLE 2.1 Infrared data for nitriles and cyanogen halides in the vapor and/or neat phases

	CN str. cm^{-1} [vapor]	A	Compound	CN str. cm^{-1} [vapor]	A	Compound	CN str. cm^{-1} [vapor]	A
IR (1)								
	2320	0.078					2244	0.02
acetonitrile	2280	0.11	benzonitrile	2240	0.118	acrytonitrile	2239	
	2252	0.07	benzonitrile				2221	0.012
butyronitrile	2260	0.03	4-hydroxy	2238	0.156	methacrylonitrile	2230	0.025
isobutyronitrile	2255	0.032	4-bromo	2238	0.12	2-chloroacryl-	2241	0.039
valeronitrile	2260	0.029	4-nitro	2240	0.011	nitrile		
isovaleronitrile	2255	0.035	3-hydroxy	2230	0.059			
tetradecanenitrile	2250	0.019	3-chloro	2239	0.115			
adiponitrile	2258	0.051	3-cyano	2241	0.085			
undecanedinitrile	2250	0.049	2-amino	2222	0.21			
			2-hydroxy	2225	0.21			
	[neat] cm^{-1}	[vapor]-[neat] cm^{-1}		[neat] cm^{-1}	[vapor]-[neat] cm^{-1}		[neat] cm^{-1}	[vapor]-[neat] cm^{-1}
butyronitrile	2250	10	benzonitrile	2221	19	acrylonitrile	2224	15
isobutyronitrile	2240	5				methacrylonitrile	2222	8
valeronitrile	2238	22						
isovaleronitrile	2242	13						
	neat cm^{-1}	IR [vapor] cm^{-1}	IR (CCl_4)					
Raman (2, 3)								
acetonitrile	2249	2280	2264*		31			
propionitrile	2242	2262			20			
isobutyronitrile	2238				17			
pivalonitrile	2236	2250			14			
chloroacetonitrile	2258							
trichloroacetonitrile	2250	2255			5			
acrylonitrile	2222	2239						17
methacrylonitrile	2230	2230						0
benzonitrile	2230	2260						10
2-chloroacrylonitrile	2234	2241						
	CCl_4							
1-cyano-2-halo-acetylene								
-2-chloro-[IR; (6)]	2292	2301; 2293			5			
-2-bromo-[IR; (6)]	2278	2295; 2289			14			
-2-iodo-[IR; (6)]	2263	2270			7			
cyanogenhalides [IR]	neat (4)	vapor (5)						
FCN (4)	2290							
ClCN (4)	2201	2248			47			
BrCN (4)	2187	2200			13			
ICN (4)	2158							

* Corrected for Fermi resonance.

TABLE 2.2 Infrared data for acetonitrile in various solvents

Acetonitrile [1 wt./vol. %] Solvent	AN	In FR cm^{-1}	In FR cm^{-1}	C–C str. cm^{-1}	s.CH_3 bend cm^{-1}	CN str. cor. for FR cm^{-1}	[C–C str.] + [s.CH_3 bend] cor. for FR cm^{-1}
1. Hexane	0	2292.1	2252.7			2260.4	2284.4
2. Diethyl ether	3.9	2292.1	2253.7			2262.7	2283.1
3. Carbon tetrachloride	8.6	2292.5	2254.3	918.6	1374.7	2263.8	2283
4. Benzene	8.2	2290.8	2252.5	917.9	1373.8	2261.2	2282.2
5. Toluene		2291.2	2252.5			2261.5	2282.3
6. Nitrobenzene	14.8	2291.4	2252.3			2259.6	2284.1
7. Nitromethane		2291.2	2253.2			2261.7	2282.8
8. Methylene chloride	20.4	2292.1	2254	918.1	1373.2	2262.9	2283.2
9. Chloroform	23.1	2292.4	2255	919.8	1373	2264.5	2282.9
10. Dimethyl sulfoxide	19.3	2293.4	2248.9			2256.6	2285.7
11. tert-Butyl alcohol	29.1	2294.9	2256.2			2265	2286.1
12. Isopropyl alcohol	33.5	2294.9	2256.4			2265.9	2285.4
13. Ethyl alcohol	37.1	2295.2	2256.7			2264.7	2287.1
14. Methyl alcohol	41.3	2295.5	2257.4			2266.4	2286.5
15. Acetone	12.5	2293.3	2252.7			2259.8	2286.1

TABLE 2.3 A comparison of the IR νCN stretching frequencies for acetonitrile [corrected for Fermi resonance] with those for benzonitrile

Solvent	Benzonitrile 10 wt./vol. % cm^{-1}	Benzonitrile 1 wt./vol. % cm^{-1}	Acetonitrile 1 wt./vol. % cor. for FR cm^{-1}	[Acetonitrile]-[Benzonitrile] 1 wt./vol. % cm^{-1}
Hexane	2231.12	2233.14	2260.42	27.28
Diethyl ether	2230.41	2230.61	2262.72	32.11
Benzene	2229.38	2229.22	2261.17	31.95
Toluene	2229.18	2229.57	2261.49	31.92
Carbon tetrachloride	2230.58	2231.21	2263.8	32.59
Acetone	2229.46	2229.32	2259.81	30.5
Nitrobenzene	2228.67	2228.92	2259.63	30.71
Nitromethane	2230.18	2230.45	2261.66	31.21
Dimethyl sulfoxide	2227.01	2226.81	2256.6	29.79
Methylene chloride	2230.11	2230.25	2262.89	32.64
Chloroform	2230.62	2231.28	2264.49	33.21
tert-Butyl alcohol	2232.06	2233.28	2264.98	31.7
Isopropyl alcohol	2231.73	2232.62	2265.9	33.28
Ethyl alcohol	2231.64	2232.04	2264.67	32.63
Methyl alcohol	2231.78	2232.07	2266.37	34.3

TABLE 2.4 The CN stretching frequency for 4-cyanobenzaldehyde in 0 to 100 mol% $CHCl_3/CCl_4$ solutions [1 wt./vol. % solutions]

4-Cyanobenzaldehyde 1% (wt./vol.)	CN str. cm^{-1}	A
Mole % $CHCl_3/CCl_4$		
0	2233.2	0.065
5.68	2233.5	0.071
10.74	2233.8	0.076
19.4	2234.1	0.083
26.53	2234.2	0.085
32.5	2234.3	0.089
37.57	2234.3	0.092
41.93	2234.3	0.094
45.73	2234.4	0.094
49.06	2234.4	0.097
52	2234.3	0.097
54.62	2234.2	0.098
57.22	2234.2	0.1
60.07	2234.1	0.101
63.28	2234	0.109
66.74	2233.7	0.111
70.65	2233.7	0.112
75.06	2233.8	0.112
80.05	2233.9	0.114
85.75	2233.8	0.118
92.33	2233.8	0.125
96.01	2233.7	0.126
100	2233.8	0.123
Δ CN str.	1.2	
	−0.6	
Δ A		0.061

TABLE 2.5 Raman data for the C≡N group and C=C group of organonitriles*

Compound	CN str. cm^{-1}	C=C str. cm^{-1}
Isobutyl 2-cyanoacrylate	2239 (9)	
2,2-Azobis(4-methoxy-2,4-dimethylvaleronitrile)	2239 (8)	
1,1-Azobis(cyclohexane carbonitrile)	2236 (2)	
2,2-Azobis(2- methylbutyronitrile)	2241 (9)	
4,4′-Azobis(4-cyanovaleric acid)	2246 (9)	
Cyanoethylated cellulose	2251 (9)	
Crotononitrile	2223 (9)	1639 (4)
		1629 (5)
2-Methyl crotononitrile	2218 (9)	1646 (4)

* Reference (3).

TABLE 2.6 A comparison of infrared data for organonitriles vs organoisonitriles

Compound	CN str. cm^{-1}	NC str. cm^{-1}	Compound	σ_p or σ_m	σ_p- σ'	σ^*	CN str.- NC str. cm^{-1}
Benzonitrile			Benzoisonitrile				
4-CH_3O	2226	2125	4-CH_3O	[−0.27]	[−0.50]	[0.52]	101
4-CH_3	2229	2125	4-CH_3	[−0.17]	[−0.13]	[−0.10]	104
3-CH_3	2229	2125	3-CH_3	[−0.07]	[−0.02]	[−0.10]	104
4-H	2229	2123	4-H	[0.00]	[0.00]	[0.00]	106
4-Cl	2233	2116	4-Cl	[0.23]	[−0.24]	[1.05]	117
4-NO_2	2238	2116	4-NO_2	[0.78]	[0.15]	[1.40]	122
3-NO_2	2240	2120	3-NO_2	[0.71]	[0.08]	[1.40]	120
2-CH_3	2226	2122	2-CH_3				104
2-Cl	2237	2166	2-Cl				71
Alkanonitrile			Alkanoisonitrile				
Acetonitrile	2249	2183	Acetoisonitrile			[0.000]	66
Propionitrile	2242	2160	Propionoisonitrile			[−0.100]	82
Isobutyronitrile	2238	2140	Isobutyroisonitrile			[−0.190]	98
Pivalonitrile	2236	2134	Pivaloisonitrile			[−0.300]	102

TABLE 2.7 A comparison of the infrared data for organothiocyanates in the vapor and neat phases

Thiocyanate	Vapor cm^{-1}	Neat cm^{-1}	Vapor-heat cm^{-1}	Neat cm^{-1}
Methyl	2175	2160	15	
Ethyl	2165	2160	5	
Pentyl	2165	[—]	[—]	
Isopentyl	2164	[—]	[—]	
Octyl	2161	2145	16	
Decyl	2164	2150	14	
Chloromethyl	2170	2150	20	Raman*
Benzyl	2164	2150	14	2150 (40)
Range	2161–2175	2145–2160		
2,6-Dichlorobenzyl Thiocyanate	[—]	2115	[—]	
Phenyl	[—]	2170	[—]	
4-Nitrophenyl	[—]	2174	[—]	
4-Aminophenyl	[—]	2166	[—]	
Range	[—]	2166–2174	[—]	
	Phenyl ring cm^{-1}	Phenyl ring cm^{-1}	Phenyl ring cm^{-1}	
	1029 (10)	1003 (80)	631 (5)	

* Relative intensity.

CHAPTER 3

Azines, Isocyanates, Isothiocyanates, and Carbodiimides

Azines	45
Isocyanates	46
Isothiocyanates	49
Carbodiimides	50
References	50

Figures	
Figure 3-1	51 (47)
Figure 3-2	51 (47)
Figure 3-3	52 (47)
Figure 3-4	52 (47)
Figure 3-5	53 (47)
Figure 3-6	53 (47)
Figure 3-7	54 (47)
Figure 3-8	54 (49)
Figure 3-9	55 (49)
Figure 3-10	55 (49)
Figure 3-11	56 (49)

Tables	
Table 3-1	57 (46)
Table 3-2	58 (46)
Table 3-3	59 (46)
Table 3-4	60 (48)
Table 3-5	60 (49)
Table 3-6	61 (49)
Table 3-6a	62 (49)
Table 3-7	63 (50)

* Numbers in parentheses indicate in-text page reference.

AZINES

Infrared and Raman studies of aldehyde and ketone azines have been summarized by Dollish *et al.* (1). Azines have the following empirical structures:

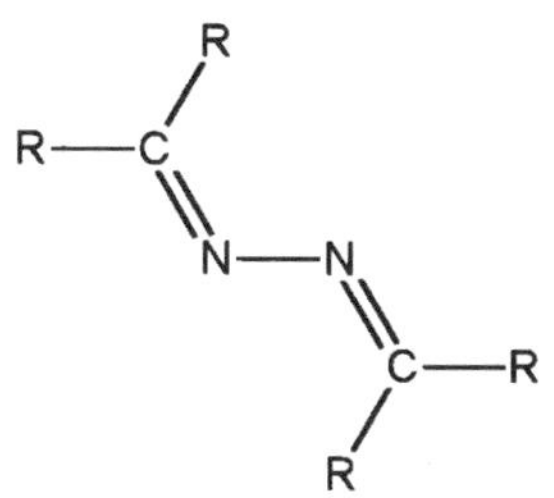

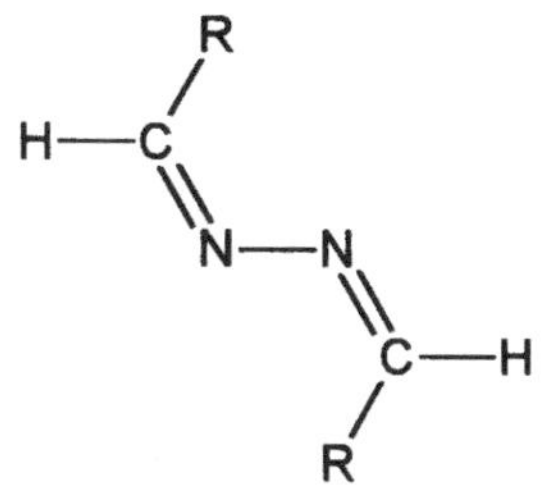

(Where R can be hydrogen, alkyl, and/or aryl in the case of aldehyde azines, and in the case of ketone azines R can be alkyl and/or aryl.)

These molecules exist in an *s*-transoid configuration and have a center of symmetry (2–7). In these cases, the antisymmetric $(C{=}N-)_2$ stretching vibration is only IR active while the symmetric $(C{=}N-)_2$ stretching vibration is only Raman active. The compound, $(F_2C{=}N-)_2$, is reported to have a trans planar structure (6). Infrared and Raman studies of benzaldehyde azines are also reported to have a center of symmetry (7).

Table 3.1 lists IR and Raman data for the benzaldehyde azines (7) and also summarizes the ν asym. $(C{=}N-)_2$ and ν sym. $(C{=}N-)_2$ vibrations for other azines (2–7). The IR bands assigned to ν asym. $(C{=}N-)_2$ were not apparent in the Raman spectrum and Raman bands assigned to ν sym. $(C{=}N-)_2$ were not apparent in the IR spectrum. The ν sym. $(C{=}N-)_2$ mode yields the most intense Raman band in each azine spectrum. For example, ν sym. NO_2 in nitrobenzenes has high Raman band intensity. In the Raman spectrum of 2-nitrobenzaldehyde azine, ν sym. $(C{=}N-)_2$ at 1560 cm^{-1} has twice the intensity of the ν sym. NO_2 vibration at 1348 cm^{-1} (7). With the exception of 2,6-dichlorobenzaldehyde azine, ν sym. $(C{=}N-)_2$ occurs in the region 1539–1563 cm^{-1} while the 2,6-dichloro analog exhibits ν sym. $(C{=}N-)_2$ at 1587 cm^{-1}. This higher frequency is intermediate between arylaldehyde azines and alkylaldehyde azines (7). This intermediate ν sym. $(C{=}N-)_2$ frequency is due to the fact that the 2,6-Cl_2 atoms prevent the two 2,6-dichlorophenyl groups from being coplanar with the $(C{=}N-)_2$ group. Thus, the resonance effect of the phenyl group, which lowers ν sym. $(C{=}N-)_2$ frequency, is absent in the case of the 2,6-dichloro analog. Further, only the inductive effect operates in this case, causing ν sym. $(C{=}N-)_2$ to occur at the intermediate 1587 cm^{-1} frequency.

In the IR benzaldehyde azines exhibit ν asym. $(C{=}N-)_2$ at lower frequency than those for alkylaldehyde azines (1606–1632 vs 1636–1663 cm^{-1}, respectively) due to resonance of the phenyl groups with the $(C{=}N-)_2$ group (7).

It is interesting to note that in the case of $(F_2C{=}N-)_2$ the ν sym. $(C{=}N-)_2$ frequency (1758 cm^{-1}) occurs at a higher frequency than that for ν asym. $(C{=}N-)_2$ at 1747 cm^{-1} (6). In all other cases ν asym. $(C{=}N-_2)$ occurs at a higher frequency than that for ν sym. $(C{=}N-)_2$.

ISOCYANATES

Table 3.2 lists IR data for alkyl and aryl isocyanates in various physical phases. In the liquid phase, methyl, ethyl and isopropyl isocyanate exhibit ν asym. N=C=O at 2288, 2280, and 2270 cm^{-1}, respectively, and ν sym. N=C=O at 1437, 1432, and 1421 cm^{-1}, respectively (8). Phenyl isocyanate and Cl_3Si isocyanate in the liquid phase exhibit ν asym. N=C=O at 2285 and 2311 cm^{-1}, respectively (9, 10).

In CCl_4 solution, *n*-butyl, *n*-pentyl, isobutyl and sec-butyl isocyanate exhibit ν asym. N=C=O at 2273, 2274, 2263, and 2261, respectively (11). Aryl isocyanates in CCl_4 solution exhibit ν asym. N=C=O in the region 2242–2278 cm^{-1} (11) (see Table 3.2). In the case of phenyl isocyanate, two IR bands are observed in the region expected for ν asym. N=C=O and these two bands are reported to arise from ν asym. N=C=O in Fermi resonance (FR) with a combination tone of the same symmetry species (12). The ν sym. N=C=O mode for phenyl isocyanate has been assigned at 1448 cm^{-1} (10).

Table 3.3 lists IR data for alkyl isocyanates in 0 to 100 mol % $CHCl_3/CCl_4$ in 0.5 wt./vol.% solutions (13, 14). In the case of methyl isocyanate, IR bands are noted at 2285.6 cm^{-1} (A is

1.176), 2256.3 cm^{-1} (A is 0.716), and 2318.3 cm^{-1} (A is 0.314) in CCl_4 solution, and 2286.0 cm^{-1} (A is 0.478), 2253.3 cm^{-1} (A is 0.478), and 2315.9 cm^{-1} (A is 0.519) in $CHCl_3$ solution. The data listed in Table 3.3 have been corrected for FR by the method developed in Reference (14), and these frequencies corrected for FR have been plotted vs mole % $CHCl_3/CCl_4$ as shown in Figure 3.1. These plots show that unperturbed ν asym. N=C=O and an unperturbed combination tone $\nu(C_\alpha{-}N) + \delta$ sym. CH_3 both increase in frequency while the combination $\nu(C_\alpha{-}N) + \nu$ sym. N=C=O decreases in frequency as the mole % $CHCl_3$ is increased (14). Figure 3.2 shows a plot of these same bands in FR (14). In the case of the uncorrected IR data where all three bands are in FR, the band with the most intensity has the most contribution from ν asym. N=C=O. This band decreases only 0.5 cm^{-1} in going from solution in CCl_4 to solution in $CHCl_3$ while after correction for FR ν asym. N=C=O decreases 3 cm^{-1}, as might be expected due to intermolecular hydrogen bonding between the $CHCl_3$ proton and the N=C=O group and the increase in the field effect of the solvent system. Ethyl isocyanate and propyl isocyanate also show that ν asym. N=C=O is in FR with a combination tone (CT), while isopropyl isocyanate and tert-butyl isocyanate exhibit only unperturbed ν asym. N=C=O (14).

Figure 3.3 shows plots of ν asym. N=C=O frequencies for *n*-butyl, isopropyl and tert-butyl isocyanate, and of the frequencies for the most intense IR band for ν asym. N=C=O (uncorrected for FR) for methyl, ethyl and *n*-propyl isocyanate vs the mole % $CHCl_3/CCl_4$ (13). Figure 3.4 shows plots of unperturbed ν asym. N=C=O frequencies for the alkyl isocyanates vs mole % $CHCl_3/CCl_4$ solutions (in this case the ν asym. N=C=O mode for the methyl, ethyl and *n*-propyl isocyanates have been corrected for FR). In this case, the six plots decrease in frequency in the order methyl, *n*-butyl, ethyl, *n*-propyl, isopropyl and tert-butyl isocyanate. The *n*-butyl analog occurs at a higher frequency than would be predicted by the reasons given here. Figure 3.5 show plots of ν asym. N=C=O for alkyl isocyanates in CCl_4 solution and in $CHCl_3$ solution vs σ^* (the inductive value of the alkyl group). Figure 3.6 show plots of ν asym. N=C=O for alkyl isocyanates in CCl_4 solution and in $CHCl_3$ solution vs E_s (the steric parameter of the alkyl group). Figure 3.7 show plots of ν asym. N=C=O frequencies for alkyl isocyanates vs σ^* times E_s. Figures 3.5–3.7 show that ν asym. N=C=O for these six alkyl isocyanates occur at higher frequency in $CHCl_3$ solution than in CCl_4 solution (13). In addition, the data points for *n*-butyl isocyanate do not correlate with the other five alkyl isocyanates. In CCl_4 solution, the five alkyl isocyanates apparently exist as a complex such as *A* or *B*. A complex such as *A* or *B*:

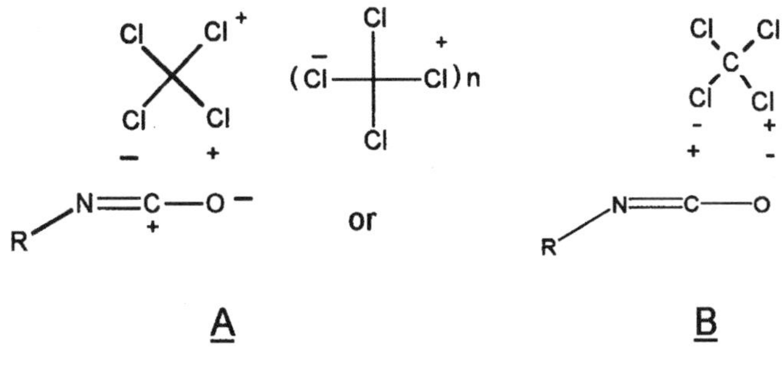

would weaken the N=C=O bond, and it would vibrate at lower frequency (13).

In the case of N-butyl isocyanate, *v* asym. N=C=O decreases in frequency in the order $CHCl_3$ ($2278.7\,cm^{-1}$), CCl_4 ($2274.5\,cm^{-1}$), and C_6H_{14} ($2270.5\,cm^{-1}$), and this is the reverse solvent effect order exhibited upon carbonyl stretching frequencies (*v* C=O). Therefore, a complex between $CHCl_3$ or CCl_4 must cause the *v* asym. N=C=O to vibrate at a higher frequency than occurs in solution with hexane. It is suggested that the complex between $CHCl_3$ and *n*-butyl isocyanate is either a steric or intramolecular hydrogen-bonded complex such as *C*, which stabilizes the N=C=O group and thus prevents a hydrogen-bonding complex from being formed between the Cl_3CH proton and the isocyanate nitrogen atom.

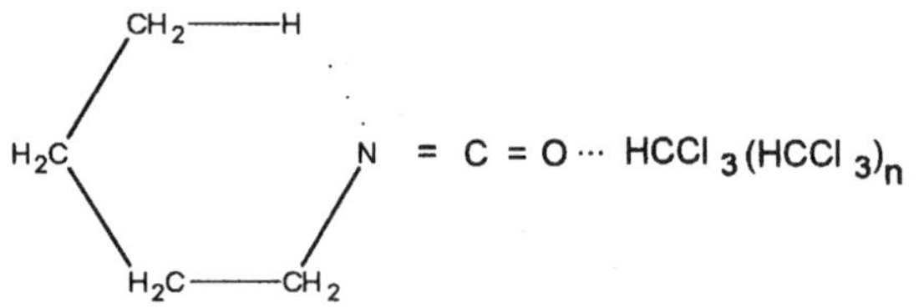

C

In the case of tert-butyl isocyanate, the tert-butyl group would have more of an electron release toward the isocyanate group, and it would have the most contribution from the resonance form *D*. The steric factor

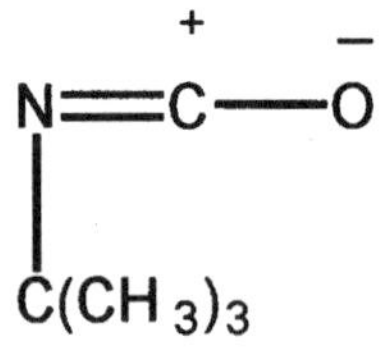

D

of the tert-butyl group is also higher than that of the other alkyl groups. Thus, structures *E* and *A* would help contribute the lower *v* asym. N=C=O frequencies exhibited by the tert-butyl analog than when in hexane solution

Table 3.4 lists IR data for *v* asym. N=C=O for the alkyl isocyanates in both CCl_4 and $CHCl_3$ solutions. Examination of Table 3.4 clearly shows that *v* asym. N=C=O in $CHCl_3$ solution always occurs at higher frequency than when in CCl_4 solution (13).

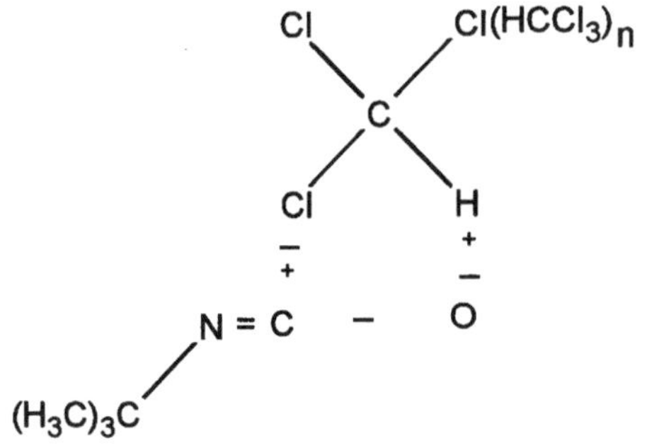

E

Table 3.5 lists IR and Raman data for alkyl isocyanates. The last column lists the Raman data for these compounds (15). These data show that the frequency of ν sym. N=C=O decreases, progressing in the order of methyl isocyanate through tert-butyl isocyanate; this is the order of electron release of the alkyl group to the isocyanate group.

ISOTHIOCYANATES

Table 3.6 lists IR data for 1% wt./vol. alkyl isothiocyanates in 0 to 100 mol % $CHCl_3/CCl_4$ or $CDCl_3/CCl_4$ solution. The unperturbed ν asym. N=C=S frequencies for the alkyl isothiocyanates occur at higher frequency in $CHCl_3$ or $CDCl_3$ solution than in CCl_4 (16). Figure 3.8 and Figure 3.9 show plots of ν asym. N=C=S in FR resonance and ν asym. N=C=S corrected for FR, respectively. The propyl analog was not corrected for FR and is not included in Figure 3.9. Both plots show that ν asym. N=C=S generally increases in frequency as the mole % $CHCl_3/CCl_4$ or $CDCl_3/CCl_4$ is increased (16). Examination of Figure 3.9 shows that ν asym. N=C=S frequency for the methyl, ethyl and tert-butyl analogs decrease in frequency as the σ* values decrease (increasing electron release to the C=N=S group). The exception is the *n*-butyl analog. It is suggested that the explanation used here to explain the behavior of ν asym. N=C=O for *n*-butyl isocyanate can also be used to explain the behavior of ν asym. N=C=S for *n*-butyl isothiocyanate (16).

Figure 3.10 shows a plot of ν asym. N=C=S and 2νC–N in FR and the same two modes corrected for FR for methyl isothiocyanate vs mole % $CHCl_3CCl_4$ (16). Figure 3.11 shows a plot of νC–N for methyl isothiocyanate vs mole % $CHCl_3/CCl_4$. Figure 3.11 shows that νC–N decreases in frequency as the mole % $CHCl_3/CCl_4$ is increased. Figure 3.10 shows that 2νC–N decreases in frequency while the ν asym. N=C=S increases in frequency as the mole % $CHCl_3/CCl_4$ is increased. In CCl_4 solution, the amount of FR between ν asym. N=C=S and 2νC-N is the least; it is the most in $CHCl_3$ solution because the IR band intensity ratio of perturbed ν N=C=O/perturbed 2νC–N is 0.808/0.200 = 4.04 in CCl_4 solution and 0.690/0.371 = 1.86 in $CHCl_3$ solution. In addition, that perturbed ν asym. N=C=S increases 18.0 cm^{-1} while perturbed 2νC–N decreases 10.7 cm^{-1} in going from solution in CCl_4 to solution in $CHCl_3$. Another way to look at these data is that the frequency separation between perturbed ν asym. N=C=S and perturbed 2νC–N is 116.5 cm^{-1} in CCl_4 solution and 87.8 cm^{-1} in $CHCl_3$ solution, while the frequency separation between ν asym. N=C=S and 2νC–N corrected for FR is 70.2 cm^{-1} in CCl_4 solution and 26.4 cm^{-1} in $CHCl_3$. In other words, the closer in frequency that unperturbed ν asym. N=C=S and 2νC–N occur, the larger the amount of Fermi interaction between the fundamental and the first overtone. The first overtone gains intensity from the fundamental during the FR interaction.

The correction for FR is readily performed by application of the equation presented here (17, 18):

$$W^o = \frac{W_a + W_b}{2} \pm \frac{W_a - W_b}{2} \cdot \frac{I_a - I_b}{I_a + I_b}$$

where W_a and W_b are the observed band frequencies, I_a and I_b are their intensities, and the two values of W^o calculated by this equation will be approximately the unperturbed frequencies.

Table 3.6a lists vibrational data for alkyl and aryl isothiocyanates in different physical phases. In the case of heptyl isothiocyanate, ν asym. N=C=S gives a depolarized Raman band at

2090 cm^{-1} and ν sym. N=C=S gives a polarized Raman band at 1070 cm^{-1}. Raman data for the three aryl isothiocyanates are assigned as ν asym. N=C=S in the range 2070–2150 cm^{-1} and ν sym. N=C=S in the range 1245–1250 cm^{-1}.

CARBODIIMIDES

Table 3.7 lists vapor- and neat-phase infrared data for dialkyl and diaryl carbodiimides (19). In the vapor phase the compounds of form R-N=C=N-R exhibit ν asym. N=C=N in the region 2118–2128 cm^{-1} and compounds of form ϕ-N=C=N-ϕ exhibit a doublet. One band of the doublet occurs in the regions 2100–2120 cm^{-1} and the other band of the doublet occurs in the region 2150–2170 cm^{-1}. In the neat phase, the compounds of form R-N=C=N-R exhibit a doublet. One band of the doublet occurs in the region 2040–2098 cm^{-1} and the other band of the doublet occurs in the region 2120–2158 cm^{-1} (19). The higher frequency band in the doublet always has more intensity than the lower frequency band. These two bands result from ν asym. N=C=N in FR with a combination tone or an overtone (18). The unperturbed ν asym. N=C=N frequencies for the diaryl carbodiimides occur at higher frequency (2124–2145 cm^{-1}) than do those for the dialkyl carbodiimides (2105–2125 cm^{-1}). The higher frequency for the diaryl analog is attributed to the positive inductive effect of the two aryl groups vs the negative inductive effect of the two alkyl groups. In addition, the ν asym. N=C=N mode occurs at higher frequency in the vapor phase than in the neat phase.

REFERENCES

1. Dollish, F. R., Fateley, W. G., and Bentley, F. K. (1974). *Characteristic Raman Frequencies of Organic Compounds.* New York: Wiley.
2. Kirrman, A. (1943). *Comp Rend.*, **217**: 148.
3. West, W. and Killingsworth, R. B. (1938). *J. Chem. Phys.*, **6**: 1.
4. Kitaev, Yu. P., Nivorozhkin, L. E., Plegontov, S. A., Raevskii, O. A., and Titova, S. Z. (1968). *Dokl. Acad. Sci. USSR*, **178**: 1328.
5. Ogilivie, J. F. and Cole, K. C. (1971). *Spectrochim. Acta*, **27A**: 877.
6. King, S. T., Overend, J., Mitsch, R. A., and Ogden, P. H. (1970). *Spectrochim. Acta*, **26A**: 2253.
7. Nyquist, R. A. and Peters, T. L. (1978). *Spectrochim. Acta*, **34A**: 503.
8. Hirschmann, R. P., Kniseley, R. N., and Fassel, V. A. (1965). *Spectrochim. Acta*, **21**: 2125.
9. Koster, D. F. (1968). *Spectrochim Acta*, **24A**: 395.
10. Stephenson, C. V., Coburn, Jr. W. C., and Wilcox, W. S. (1961). *Spectrochim Acta*, **17**: 933.
11. Hoyer, H. (1956). *Chem. Ber.*, **89**: 2677.
12. Ham, N. S. and Willis, J. B. (1960). *Spectrochim. Acta*, **16**: 279.
13. Nyquist, R. A., Luoma, D. A., and Putzig, C. L. (1992). *Appl. Spectrosc.*, **46**: 972.
14. Nyquist, R. A., Putzig, C. L., and Hasha, D. L. (1989). *Appl. Spectrosc.*, **43**: 1049.
15. Herzberg, G. and Reid, C. (1950). *Discuss. Faraday Soc.*, **9**: 92.
16. Nyquist, R. A. and Puehl, C. W. (1993). *Appl. Spectrosc.*, **47**: 677.
17. Nyquist, R. A., Fouchea, H. A., Hoffman, G. A., and Hasha, D. L. (1991). *Appl. Spectrosc.*, **45**: 860.
18. Nyquist, R. A. (1984). *The Interpretation of Vapor-phase Infrared Spectra: Group Frequency Data.* Vol. 1, Philadelphia: Sadtler Research Laboratories, Division of Bio-Rad Laboratories, Inc., p. 43.
19. Nyquist, R. A. (1984). *Ibid.*, p. 461.

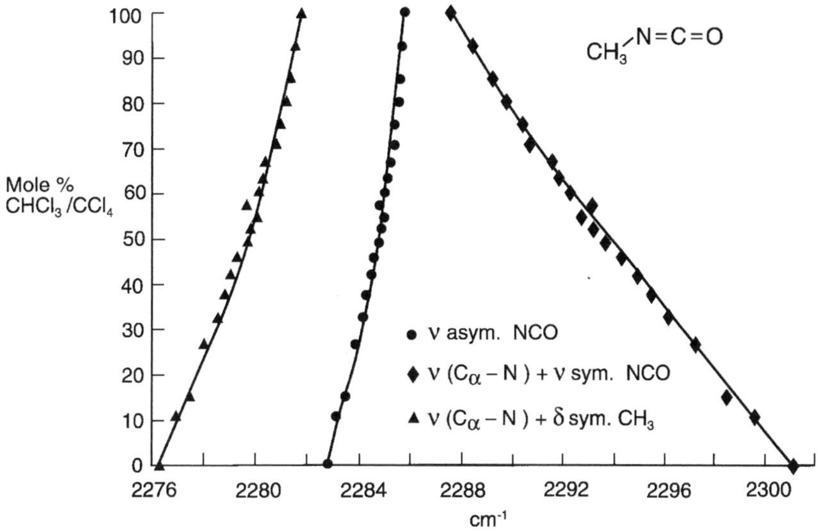

FIGURE 3.1 Plots of ν asym. N=C=O and the combination tone ν (C_α—N) + ν sym. N=C=O and ν (C_α—N) + sym. CH_3 for methyl isocyanate all corrected for Fermi resonance.

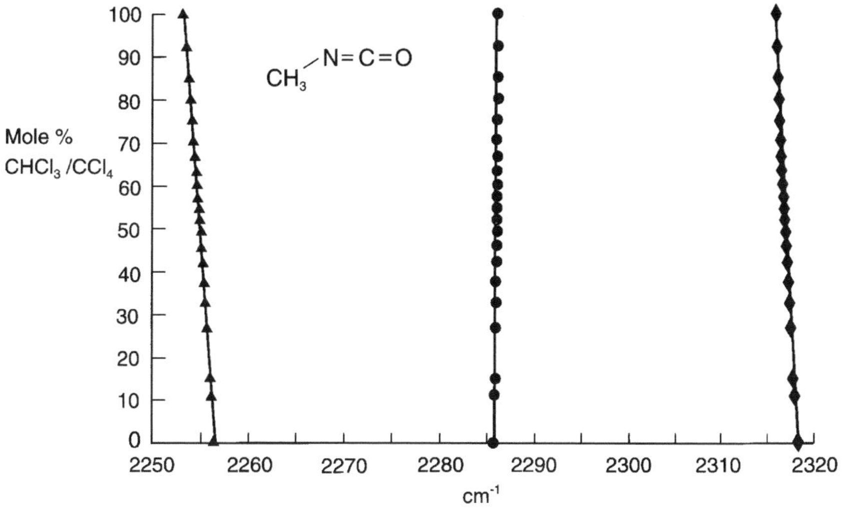

FIGURE 3.2 Plots of the three observed IR bands for methyl isocyanate occurring in the region 2250–2320 vs mole % $CHCl_3/CCl_4$.

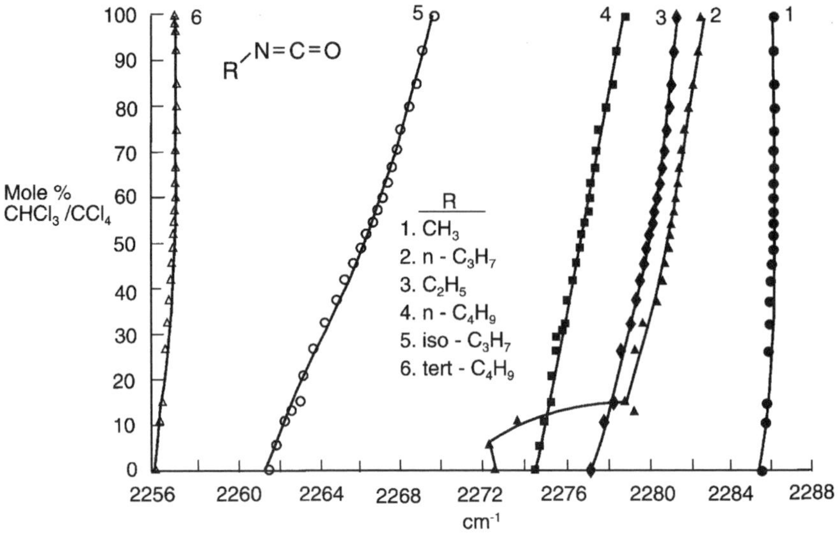

FIGURE 3.3 Plots of the ν asym. NCO frequencies for *n*-butyl, isopropyl and tert-butyl isocyanate and of the frequencies of the most intense IR band for ν asym. N=C=O in FR (uncorrected for FR) for methyl, ethyl and *n*-propyl isocyanate vs mole % $CHCl_3/CCl_4$.

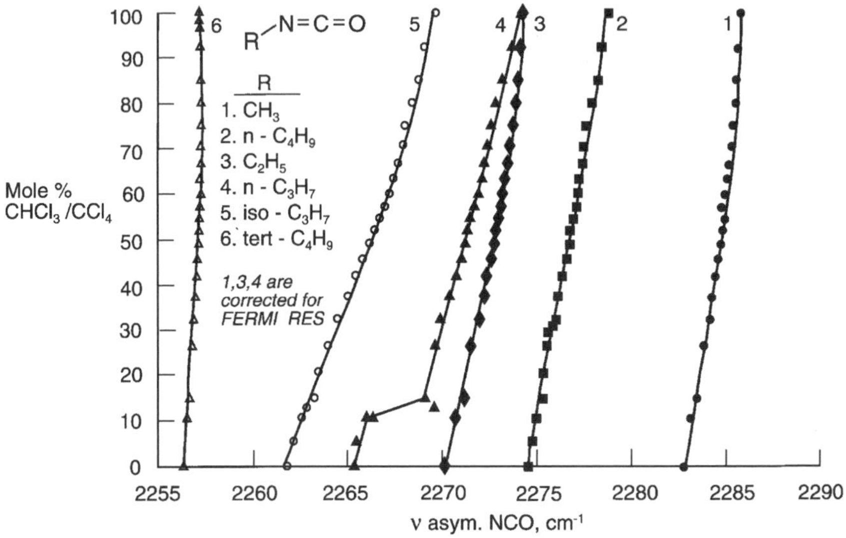

FIGURE 3.4 Plots of unperturbed ν asym. N=C=O for the alkyl isocyanates vs mole % $CHCl_3/CCl_4$.

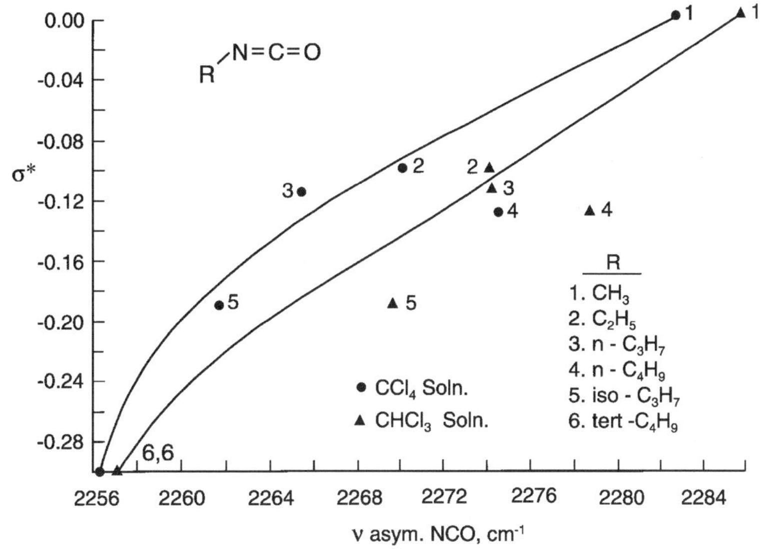

FIGURE 3.5 Plots of ν asym N=C=O frequencies for alkyl isocyanates in CCl_4 solution and in $CHCl_3$ solution vs σ^*. (The inductive release value of the alkyl group.)

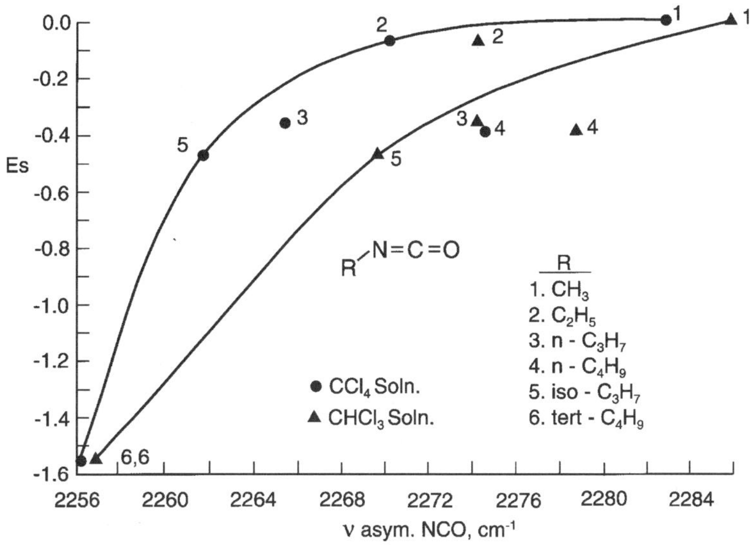

FIGURE 3.6 Plots of ν asym. N=C=O frequencies for alkyl isocyanates in CCl_4 solution and in $CHCl_3$ solution vs E_s. (The stearic parameter of the alkyl group.)

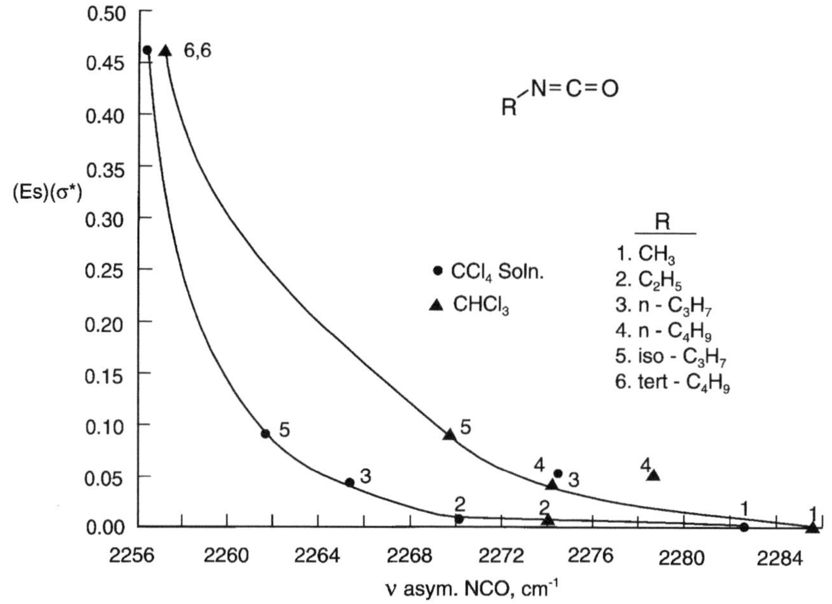

FIGURE 3.7 Plots of *ν* asym. N=C=O frequencies for alkyl isocyanates in CCl_4 solution and in $CHCl_3$ solution vs (E_s) (σ*).

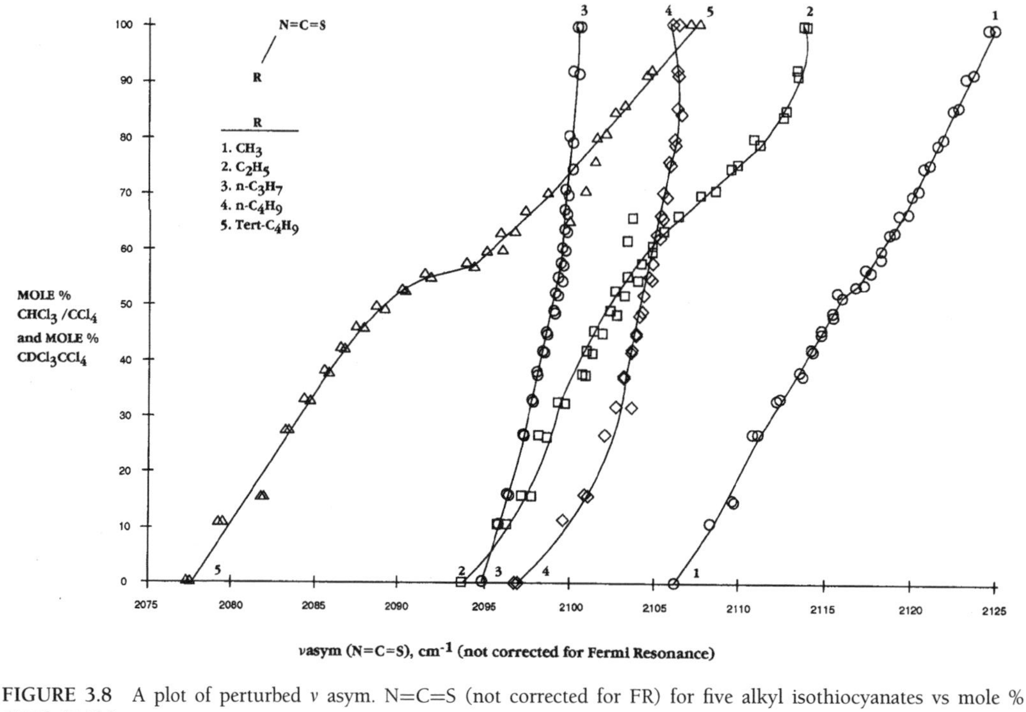

FIGURE 3.8 A plot of perturbed *ν* asym. N=C=S (not corrected for FR) for five alkyl isothiocyanates vs mole % $CHCl_3/CCl_4$).

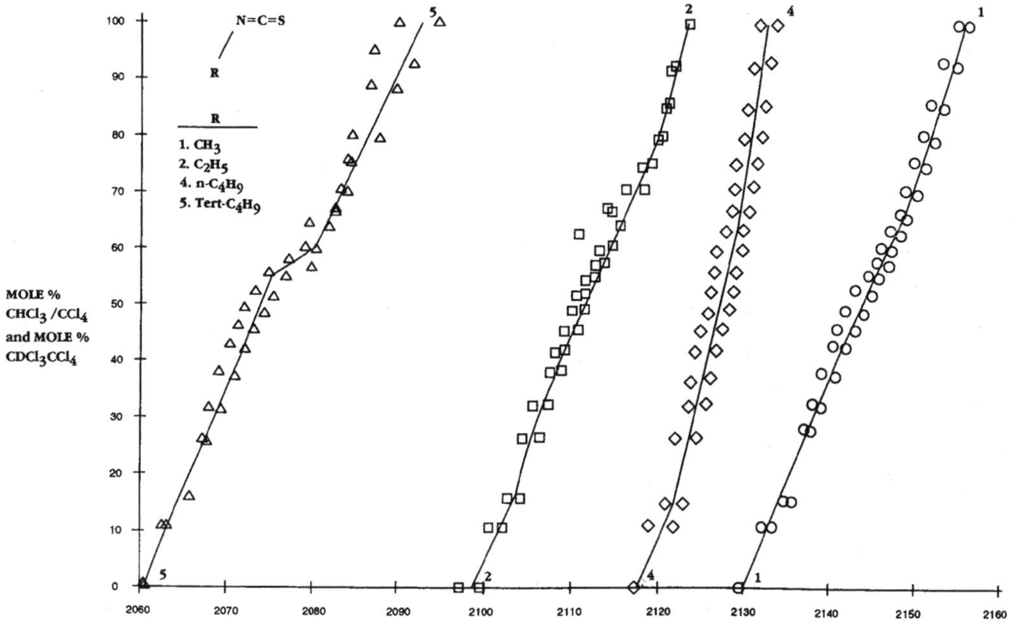

FIGURE 3.9 A plot of unperturbed ν asym. N=C=S (corrected for FR) for four alkyl isothiocyanates vs mole % $CHCl_3/CCl_4$ and $CDCl_3/CCl_4$.

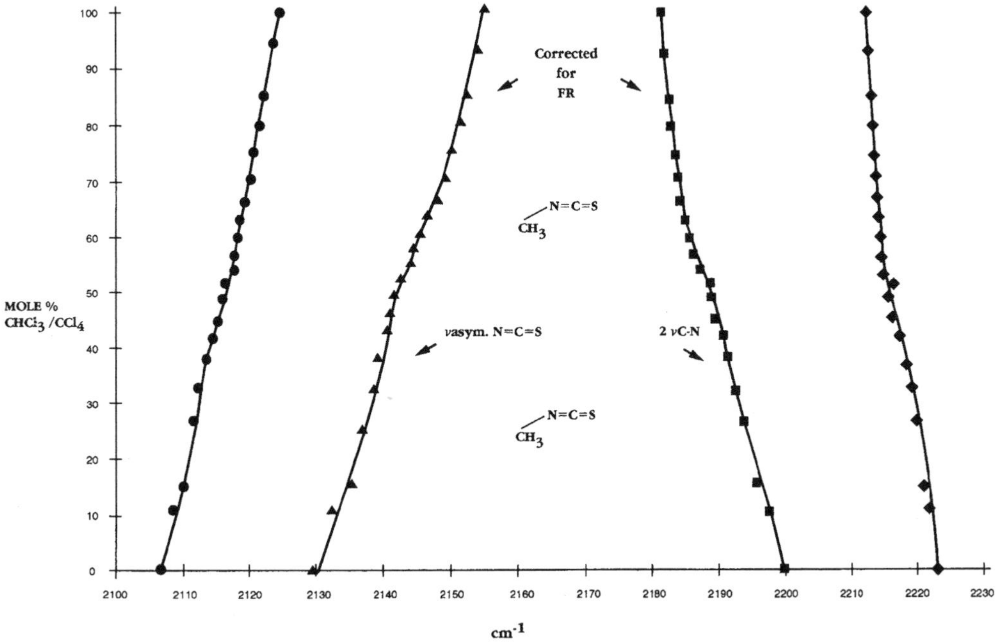

FIGURE 3.10 A plot of perturbed 2ν C—N and perturbed ν asym. N=C=S and unperturbed 2ν C—N and unperturbed ν asym, N=C=S vs mole % $CHCl_3/CCl_4$.

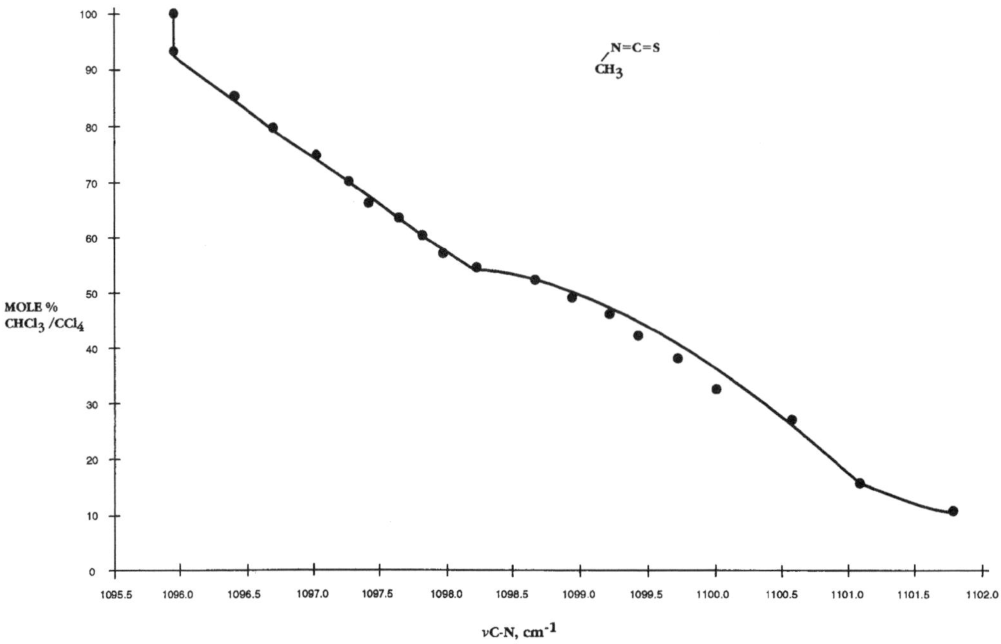

FIGURE 3.11 A plot of ν C—N for methyl isothiocyanate vs mole % $CHCl_3/CCl_4$.

TABLE 3.1 Infrared and Raman data and assignments for the $(C{=}N-)_2$ antisymmetric and symmetric stretching vibrations for azines

Aldehyde Azines	asym. $(C{=}N-)_2$ str. IR active cm^{-1}	sym. $(C{=}N)_2$ str. Raman active cm^{-1}	asym. $(C{=}N)_2$ str.- sym. $(C{=}N)_2$ str. cm^{-1}
4-X-benzaldehyde			
X			
dimethylamino	1608	1539	69
methoxy	1605	1553	52
methyl	1623	1553	70
hydrogen	1628	1556	72
fluoro	1633	1561	72
chloro	1627	1547	80
acetoxy	1631	1562	69
trifluoromethyl	1631	1561	70
cyano	1624	1541	83
2-X-benzaldehyde			
X			
methoxy	1619	1552	67
chloro	1618	not recorded	[–]
hydroxy	1630	1555	75
nitro	1627	1560	67
3-X-benzaldehyde			
X			
nitro	1629	1551	78
benzaldehyde			
2,4-dimethoxy	1617	1546	71
3,4-dimethoxy	1626	1557	69
2-hydroxy-4-methoxy	1630	1554	76
2,4-dihydroxy	1632	1563	69
3-methoxy-4-hydroxy	1629	1558	71
2,6-dichloro	1629	1587	42
Summary			
Azine Type			
benzaldehyde	1606–1632	1539–1563	52–83
2,6-Cl_2-benzaldehyde	1629	1587	42
alkylaldehyde and alkyketone	1636–1663	1608–1625	28–38
$CF_2{=}$	1747	1758	[−11]
Range	1606–1747	1539–1758	

TABLE 3.2 The symmetric and/or antisymmetric stretching frequencies for alkyl and aryl isocyanates in various physical phases

Alkyl and phenyl isocyanate R-N=C=O and $C_6H_5-N=C=O$	a.N=C=O cm^{-1}	s.N=C=O cm^{-1}	Physical phase	Reference
Methyl	2288	1437	L*1	8
Ethyl	2280	1432	L	8
Isopropyl	2270	1421	L	8
t-Butyl	2270		L	9
Phenyl	2285	1448	L	10
Cl_2Si	2311	1467	L	8
H	2274	1318	V*2	13
n-Butyl	2273		CCl_4*3	11
n-Pentyl	2274		CCl_4	11
Isobutyl	2263		CCl_4	11
sec-Butyl	2261		CCl_4	11
	2262			11
Cyclohexyl	2255		CCl_4	11
Benzyl	2266		CCl_4	11
	2265			
Acetyl	2246		CCl_4	11
Phenyl	2260		CCl_4	11
	2278			
4-X-phenyl				
X				
Ethoxy	2274		CCl_4	11
Chloro	2266		CCl_4	11
Methyl	2263		CCl_4	11
Nitro	2261		CCl_4	11
Cyano	2258		CCl_4	11
3-X-phenyl				
Methoxyl	2267		CCl_4	11
Methyl	2266		CCl_4	11
Chloro	2265		CCl_4	11
2-X-phenyl				
Methyl	2273		CCl_4	11

*1 L is liquid.
*2 V is vapor.
*3 CCl_4 is in CCl_4 solution.

TABLE 3.3 Infrared data for alkyl isocyanates in 0 to 100 mol % $CHCl_3/CCl_4$ in 0.5 wt./vol % solutions

Methyl isocyanate $CH_3-N=C=O$ [0.5% solutions] Mole % $CHCl_3/CCl_4$	cm^{-1}	cm^{-1}	cm^{-1}	Corrected for FR a.N=C=O str. cm^{-1}	Corrected for FR C–N str. + s.CH_3 bend cm^{-1}	Corrected for FR C–N str. + s.N=C=O str. cm^{-1}
0	2285.56	2256.26	2318.28	2282.8	2276.3	2301.1
10.74	2285.7	2256.08	2317.87	2283.1	2276.9	2299.6
15.07	2285.8	2255.93	2317.7	2283.5	2277.5	2298.5
26.53	2285.87	2255.74	2317.51	2283.8	2278	2297.2
32.5	2285.93	2255.59	217.31	2284.1	2278.6	2296.1
37.5	2285.92	2255.4	2317.21	2284.3	2278.8	2295.5
41.93	2285.99	2255.3	2317.09	2284.4	2279.1	2294.9
45.73	2285.99	2255.1	2316.99	2284.5	2279.3	2294.3
49.06	2286.04	2255.08	2316.96	2284.7	2279.7	2293.7
52	2286.05	2254.93	2316.84	2284.8	2279.8	2293.2
54.62	2286.06	2254.86	2316.76	2284.9	2280.1	2292.7
57.22	2286.08	2254.78	2316.7	2284.8	2279.7	2293.1
60.07	2286.08	2254.67	2316.62	2285	2280.1	2292.3
63.28	2286.09	2254.56	2316.54	2285.1	2280.3	2291.8
66.73	2286.1	2254.51	2316.49	2285.2	2280.4	2291.5
70.65	2286.1	2254.2	2316.36	2285.3	2280.8	2290.7
75.06	2286.1	2254.23	2316.28	2285.4	2280.9	2290.4
80.06	2286.12	2254.1	2316.21	2285.5	2281.2	2289.7
85.05	2286.12	2253.93	2316.21	2285.6	2281.4	2289.2
92.33	2286.07	2253.6	2316	2285.6	2281.6	2288.4
100	2286.03	2253.27	2315.87	2285.8	2281.8	2287.6
Δ [CCl_4]−[$CHCl_3$]	0.47	−2.99	−2.41	3.02	5.56	−13.51
$C_2H_5N=C=O$						
0	2277.25	2220.81		2270.1	2227.9	
100	2281.25	2217.65		2274.1	222.8	
Δ [CCl_4]−[$CHCl_3$]	4	−3.13		4	−3.2	
$C_3H_7-N=C=O$						
Δ [CCl_4]−[$CHCl_3$]						
0	2272.63	2259.31		2265.4	2266.6	
100	2282.47	2264.8		2274.3	2273	
Δ [CCl_4]−[$CHCl_3$]	9.84	5.49		8.9	6.5	
$C_4H_9-N=C=O$						
0	2274.57					
100	2278.8					
Δ [CCl_4]−[$CHCl_3$]	4.23					

TABLE 3.4 Infrared data for the antisymmetrical N=C=O stretching frequency and two combination tones for alkyl isocyanates in $CHCl_3$ and CCl_4 solutions

Alkyl isocyanate	cm^{-1}	cm^{-1}	cm^{-1}	Corrected for FR a.N=C=O cm^{-1}	Corrected for FR C–N str. + s.CH_3 bend or CT cm^{-1}	Corrected for FR C–N str. + s.N=C=O str. cm^{-1}	Solvent
Methyl	2285.6	2256.3	2318.3	2282.8	2276.3	2301.1	CCl_4 soln.
	2286	2253.3	2315.9	2285.8	2281.8	2287.6	$CHCl_3$ soln.
Ethyl	2277.3	2220.8		2270.1	2227.9		CCl_4 soln.
	2281.3	2217.7		2274.1	2224.8		$CHCl_3$ soln.
Propyl	2272.6	2259.3		2265.4	2266.6		CCl_4 soln.
	2282.5	2264.8		2274.3	2273		$CHCl_3$ soln.
	ν asym. N=C=O						
n-Butyl	2274.6						CCl_4 soln.
	2278.8						$CHCl_3$ soln.
Isopropyl	2261.7						CCl_4 soln.
	2269.7						$CHCl_3$ soln.
tert-Butyl	2256.3						CCl_4 soln.
	2257.1						$CHCl_3$ soln.

[CT is a combination tone].

TABLE 3.5 Infrared and Raman data for alkyl isocyanates

Raman data for Alkyl isocyanate	IR (13) a.N=C=O str. C_6H_{14} soln. cm^{-1}	IR (13) a.N=C=O str. CCl_4 soln. cm^{-1}	IR (13) a.N=C=O str. $CHCl_3$ soln. cm^{-1}	Raman (15) s.N=C=O str. liquid cm^{-1}	E_N*
Methyl	2283.8	2282.8	2285.8	1437	0.000–0.00
Ethyl	2270.8	2270.1	2274.1	1431.7	−0.100–0.07
n-Propyl	2266.9	2265.4	2274.3	1431	−0.115–0.36
n-Butyl	2270.5	2274.5	2278.7	1431	−0.130–0.39
Isopropyl	2262.9	2261.7	2269.7	1422.9	−0.0190–0.47
tert-Butyl	2258.8	2256.3	2257.1	1396.9	−0.300–1.54

* Raman NC=O.

TABLE 3.6 Infrared data for 1 wt./vol. % alkyl isothiocyanates in 0 to 100 mol % $CHCl_3/CCl_4$ solutions [the ν asym. N=C=S and the first overtone of C–N stretching frequencies in Fermi resonance]

CH_3–N=C=S [1 wt. % solutions] Mole % $CHCl_3/CCl_4$	cm^{-1}	cm^{-1}	Corrected for Fermi res. N=C=S str. cm^{-1}	Corrected for Fermi res. 2(C–N str.) cm^{-1}
0	2106.35	2222.81	2129.5	2199.7
10.74	2108.14	2221.46	2131.9	2197.7
15.07	2109.64	2220.65	2134.4	2195.9
26.53	2111.07	2219.47	2136.4	2194.2
32.5	2112.17	2218.74	2137.9	2193
37.57	2113.01	2217.54	2138.9	2191.7
41.93	2113.95	2216.91	2140.2	2190.7
45.73	2114.6	2215.93	2141.1	2189.4
49.06	2115.27	2215.42	2141.9	2188.8
52	2115.8	2215.8	2142.9	2188.7
54.62	2117.04	2214.33	2144.6	2186.8
57.22	2117.31	2214.1	2145	2186.4
60.07	2117.96	2213.76	2145.8	2186
63.28	2118.48	2213.57	2146.5	2185.6
66.73	2119.41	2213.34	2148.1	2184.7
70.65	2119.95	2213.16	2148.6	2184.5
75.06	2120.67	2213.01	2141.6	2184.1
80.06	2121.49	2212.73	2150.9	2183.3
85.05	2122.37	2212.55	2152.1	2182.8
92.33	2123.33	2212.35	2153.5	2182.2
100	2124.38	2212.14	2155.1	2181.5
Mole % $CDCl_3/CCl_4$				
0	2222.81	2106.3	2129.4	2199.8
10.65	2221.58	2108.1	2129.4	2196.7
14.94	2220.78	2109.75	2135.3	2195.2
26.31	2219.62	2110.87	2137.1	2193.4
32.23	2219.04	2112.03	2138.8	2192.3
37.26	2218.06	2113.27	2140.6	2190.7
41.58	2217.4	2114	2141.7	2189.7
45.35	2216.82	2114.66	2142.8	2188.7
48.65	2216.34	2115.31	2143.6	2188
51.57	2215.73	2115.96	2144.4	2187.3
54.17	2215.31	2116.68	2145.4	2186.6
56.74	2214.96	2117.48	2146.5	2185.9
59.57	2214.64	2117.87	2146.9	2185.6
62.75	2214.51	2118.34	2147.8	2185
66.18	2214.23	2118.82	2148.6	2184.4
70.06	2214	2119.69	2149.8	2183.9
74.44	2213.78	2120.41	2150.8	2183.4
79.39	2213.52	2121.15	2152	2182.7
84.34	2213.35	2122.07	2153.3	2182.1
91.56	2213.12	2122.95	2154.7	2183.3
100	2213.01	2124.66	2157	2180.7

(continues)

TABLE 3.6 (*continued*)

$CH_3-N=C=S$ [1 wt. % solutions] Mole % $CHCl_3/CCl_4$	cm^{-1}	cm^{-1}	Corrected for Fermi res. N=C=S str. cm^{-1}	Corrected for Fermi res. 2(C−N str.) cm^{-1}
Mole % $CHCl_3/CCl_4$				
$C_2H_5-N=C=S$				
0	2093.85		2099.6	
100	2113.9		2124	
$C_3H_7-N=C=S$				
0	2094.8		(not corrected	
100	2100.09		for FR)	
$C_4H_9-N=C=S$				
0	2096.65		2117.6	
100	2106.1		2133.9	
tert $C_4H_9-N=C=S$				
0	2077.1		2060.4	
100	2106.7		2090.2	

TABLE 3.6a **Vibrational data for organoisothiocyanates**

Compound	Phase	a.N=C=S str. cm^{-1}	2(C−N str.) cm^{-1}	s.N=C=S str. or 2(s.N=C=S str.) cm^{-1}	N=C=S bend cm^{-1}
$CH_3-N=C=S$	vapor [IR]	2085 (1.240)	2228 (0.233)	~1090 (0.020)	685 (0.020)
$(CH_3)_3C-N=C=S$	liquid [Raman]	2082 (dep.)		1005 (pol.)	625 (pol.)
$C_7H_{15}-N=C=S$	liquid [Raman]	2090 (dep.)	2170	1070 (pol.)	655 (pol.)
$C_7H_{15}-N=C=S$	vapor [IR]	2061 (1.250)		1110 (0.020)	
	Δ [v-l]	[29]		[40]	
$C_6H_5-N=C=S$	liquid [Raman]	2150 (dep.)		1245 (pol.)	
4-Br−$C_6H_4-N=C=S$	solid [Raman]	2070	2170sh	1250	
2,4,6−$(CH_3)_3-C_6H_2-N=C=S$	solid [Raman]	2150	2120sh	1245	

TABLE 3.7 Vapor- and neat-phase infrared data for dialkyl and diaryl carbodiimides

Carbodiimide	a.N=C=N str. vapor cm^{-1}	A	Not corrected for FR a.N=C=N str. OT or CT neat cm^{-1}	A	Corrected for FR a.N=C=N str. CT or OT cm^{-1}
dicyclohexyl	2128	1.245	2110	1.179	2090.4 (neat)
			2045	0.51	2064.6 (neat)
tert-butyl	2118	1.225	2110	1.895	2105.8 (neat)
triphenylmethyl			2040	0.122	2044.2 (neat)
	Not corrected for FR a.N=C=N OT or CT				
bis-(2-methylphenyl)	2150	1.246	2139	1.18	2124.5 (neat)
	2120	0.498	2105	0.88	2119.5 (neat)
					2141.4 (vapor)
					2122.6 (vapor)
bis-(2,6-diethylphenyl)	2170	1.231	2158	1.168	2145 (neat)
	2100	0.212	2098	0.323	2111 (neat)
					2147.4 (vapor)
					2122.6 (vapor)

CHAPTER 4

Thiols, Sulfides and Disulfides, Alkanethiols, and Alkanedithiols (S–H stretching)

Alkanethiols (S–H stretching)	65
Benzenethiols (S–H stretching)	66
Alkanethiols and Alkanedithiols (C–S stretching)	66
Alkanethiol, Alkane Sulfides, and Alkane Disulfides	66
Carbon Hydrogen Modes	67
Dialkyl Sulfides (rotational conformers)	67
Dialkyl Disulfides, Aryl Alkyl Disulfides, and Diaryldisulfide	67
4-Chlorobenzenthiol	67
Phosphorodithioates (S–H stretching)	68
Alkyl Groups Joined to Sulfur, Oxygen, or Halogen	68
Methyl (Methylthio) Mercury	69
References	69

Figures	
Figure 4-1	71 (68)
Figure 4-2	73 (69)
Figure 4-3	73 (69)

Tables	
Table 4-1	74 (65, 66)
Table 4-1a	75 (66)
Table 4-2	76 (66, 67)
Table 4-3	78 (67)
Table 4-4	79 (67)
Table 4-5	80 (68)
Table 4-6	81 (68)

*Numbers in parentheses indicate in-text page reference.

ALKANETHIOLS (S–H STRETCHING)

Table 4.1 lists IR and Raman S–H stretching frequency data for alkanethiols and benzenethiols in different physical phases.

In the vapor phase ν S–H for alkanethiols and alkanedithiols exhibit a weak IR which occurs in the region 2584–2598 cm^{-1}. In CCl_4 solution ν S–H occurs at lower frequency in the region 2565–2583 cm^{-1} (1), and in the liquid phase 2554–2585 cm^{-1} (2, 3). There is a decrease in the ν S–H frequency in going from vapor to CCl_4 solution, with this most likely resulting from intermolecular hydrogen bonding between S–H and CCl_4 to form S–H$\cdots$ClCCl$_3$.

In going from CCl_4 solution to the liquid phase, the ν S—H frequency decreases 13–21 cm^{-1} for alkanethiols, with the exception of tert-butylthiol, which increases in frequency by 6 cm^{-1}. This decrease in frequency is attributed to intermolecular hydrogen bonding between S—H groups $(S—H)_n$. In the case of tert-butylthiol the steric factor of the tert-butyl group apparently prevents such a strong intermolecular hydrogen bond from forming between S—H groups.

The Raman band assigned to ν S—H (2571–2583 cm^{-1}, liquid) has weak-medium to strong intensity and it is polarized (2).

BENZENETHIOLS (S—H STRETCHING)

Benzenethiols exhibit ν S—H in the region 2560–2608 cm^{-1}1 in the vapor phase, with most absorbing in the region 2582–2608 cm^{-1} (1). The compounds 2-aminobenzethiol and toluene-3,4-dithiol exhibit ν S—H···NH_2 at 2560 cm^{-1} and ν S—H···SH at 2570 cm^{-1}. These relatively low frequencies are the result of intramolecular hydrogen bonding between S—H···NH_2 and S—H···SH groups, respectively (1).

ALKANETHIOLS AND ALKANEDITHIOLS (C—S STRETCHING)

The C—S stretching (ν C—S) vibration decreases in frequency as the branching on the sulfur α-carbon atom is increased, and the ν C—S mode is often observed as a doublet in both the IR and Raman due to the existence of rotational conformers (5). For example, Table 4.1 shows that ν C—S for methanethiol occurs at 703 cm^{-1}, butanethiol at 651 cm^{-1}, isopropylthiol at 629 and 616 cm^{-1}, and tert-butyl thiol at 587 cm^{-1}. This decrease in frequency is attributed to the increased electron release to the C—S bond progressing in the order CH_3—S to $(CH_3)_3CSH$, which weakens the C—S bond.

The major rotational conformer has strong Raman band intensity and the lesser rotational conformer has weak- to medium Raman band intensity depending upon the concentrations of the two rotational conformers. In both cases the ν S—H band is polarized (2). Table 4.1a lists Raman liquid-phase data for some compounds containing SH groups (6).

ALKANETHIOL, ALKANE SULFIDES, AND ALKANE DISULFIDES

Table 4.2 lists IR vapor phase data for alkanethiols, alkane sulfides, and alkane disulfides. The compounds are compared (methanethiol vs dimethyl sulfide vs dimethyl disulfide, etc.) in Table 4.2. In the series for the methyl analog, ν C—S decreases in frequency progressing in the order CH_3SH, $(CH_3)_2S$, and $(CH_3S)_2$. In the case of the dialkyl sulfides vs dialkyl disulfides, weak IR bands are noted in the regions 760–786 cm^{-1} and 732–751 cm^{-1}. The higher frequency IR band has more intensity than the lower frequency band, and these IR bands are assigned to ν asym.$(C)_2S$ and ν sym.$(C)_2S$, respectively. In the IR ν asym.$(C)_2S$ and ν sym.$(C)_2S$ are weak, and these bands are not readily apparent in the higher molecular weight dialkyl sulfides and dialkyl disulfides.

CARBON HYDROGEN MODES

Table 4.2 also lists vibrational assignments for the alkyl groups for the alkanethiols, alkane sulfides, and alkane disulfides as a convenience to the reader.

DIALKYL SULFIDES (ROTATIONAL CONFORMERS)

Table 4.3 lists IR and Raman data for organic sulfides and disulfides (2). Examination of these data show that ν asym.$(C)_2$ and ν sym.$(C)_2$ occur at approximately the same frequency in either CS_2 solution or the liquid phase. In the case of methyl ethyl sulfide an IR band at 729 cm^{-1} and the depolarized medium Raman band at 732 cm^{-1} are assigned to ν asym.$(C)_2S$. The IR band at 652 cm^{-1} and the polarized Raman band at 661 cm^{-1} are assigned to ν sym.$(C)_2S$ for the low-temperature conformer. The 672 cm^{-1} IR band and the 684 cm^{-1} polarized medium Raman band are assigned to ν sym.$(C)_2S$ for the other rotational conformer. In the case of diethyl sulfide, there is both IR and Raman evidence to support the presence of three rotation conformers. One has C_1 symmetry, one has C_2 symmetry, and one has C_{2v} symmetry.

DIALKYL DISULFIDES, ARYL ALKYL DISULFIDES, AND DIARYLDISULFIDE

Table 4.3 lists IR and Raman data for dialkyl disulfides, aryl alkyl disulfide, and diaryl disulfide (2). A weak IR band in the region 502–528 cm^{-1} and a polarized weak-medium to strong Raman band in the region 501–529 cm^{-1} are assigned to ν S—S (2). In some cases ν S—S is observed as a doublet in both the IR and Raman spectra due to the existence of rotational conformers.

4-CHLOROBENZENTHIOL

Table 4.4 compares the vibrational assignments of 4-chlorobenzenethiol (7) and 1,4-dichlorobenzene (8). Sulfur and chlorine masses are 32 and ~35, respectively, the 30 ring modes occur at very similar frequencies, and they have comparable IR and Raman band intensities. The ν S—H mode is observed at 2579 cm^{-1} in the solid-phase IR spectrum and at 2569 cm^{-1} in the liquid Raman spectrum (7). Thus, it is helpful in spectra-structure identification of chlorinated benzenethiol isomers to have IR and Raman spectra of comparable chlorinated benzene isomers for reference. Vibrational data and assignments are available for the chlorinated benzenes (8, 9, 10), and normal coordinates data for these chlorinated benzenes have also been determined (11). In-plane vibrations have also been assigned for a large number of differently substituted benzenes whose IR spectra have been recorded in the vapor phase (12), and for the out-of-plane deformations and their combination and overtones (13).

PHOSPHORODITHIOATES (S–H STRETCHING)

Table 4.5 lists the ν S–H frequencies for O,O-(dialkyl) phosphorodithioate and O,O-bis-(2,4,5-trichlorophenyl) phosphorodithioate (14). The ν S–H rotational conformer 1 is assigned to a band in the region 2575–2588 cm^{-1} and ν S–H rotational conformer 2 is assigned in the region 2549–2550 cm^{-1} (14). Figure 4.1 shows IR spectra for O,O-bis-(2,4,5-trichlorophenyl) phosphorodithioate in CS_2 solution in the region 2700–2400 cm^{-1} for temperatures ranging from 29 °C to −100 °C. The higher frequency band is more intense than the lower frequency band at 29 °C but the lower frequency band steadily increases in intensity as the temperature is decreased and at −100 °C it is more intense than the higher frequency band (14). These data support the postulation that ν S–H occurs as a doublet due to rotational conformers (15). Rotational conformers 1 and 2 are assigned to the following empirical structures:

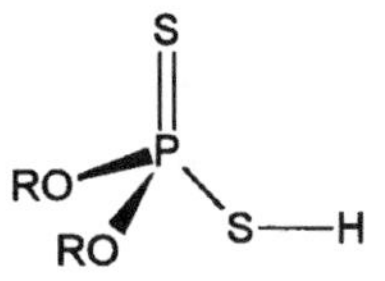

rotational conformer 1
2575-2588 cm^{-1}

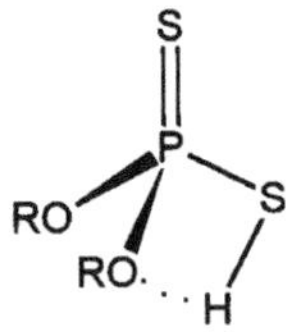

rotational conformer 2
2549-2550 cm^{-1}

The ν S–H···OR mode for conformer 2 is lower in frequency than ν S–H for conformer 1 due to intramolecular hydrogen bonding between the S–H proton and the OR oxygen atom S–H···OR (14).

In summary, the ν S–H or ν S–H···X vibrations are compared here:

	ν S–H cm^{-1}	ν S–H···X cm^{-1}
Phosphorodithioates	2575–2588	2549–2550
Benzenethiols	2582–2608	2560–2570
Alkane thiols	2565–2583	
Range:	2565–2608	2549–2570

ALKYL GROUPS JOINED TO SULFUR, OXYGEN, OR HALOGEN

Table 4.6 compares IR vapor-phase data for compounds where the same alkyl group is joined to S, O, or halogen (1). These data show that it is an aid in spectra structure identification of unknown chemicals to be able to compare the unknown spectrum with spectra of other chemical compounds containing the same alkyl group.

METHYL (METHYLTHIO) MERCURY

Methyl (methylthio) mercury has the empirical structure $CH_3{-}Hg{-}S{-}CH_3$ (16). Figure 4.2 (top) is a liquid-phase IR spectrum of methyl (methylthio) mercury between KBr plates, and Figure 4.2 (bottom) is a liquid-phase IR spectrum of methyl (methythio) mercury between polyethylene plates. The band at $\sim 72\,cm^{-1}$ is a lattice mode for polyethylene. Figure 4.3 (top) is a Raman liquid-phase spectrum of methyl (methylthio) mercury in a small glass capillary tube. The sample was positioned perpendicularly to both the laser beam and the optical axis of the spectrometer. The bottom Raman spectrum is the same as the top one except that the plane of polarization of the incident beam was rotated through 90° (16).

The $692\,cm^{-1}$ IR band and the $700\,cm^{-1}$ Raman band are assigned to ν C–S. The $533\,cm^{-1}$ IR band and $537\,cm^{-1}$ Raman band are assigned to ν C–Hg (16). The ν S–Hg vibration is assigned to the $333\,cm^{-1}$ IR band and the $329\,cm^{-1}$ Raman band. Characteristic CH_3 vibrations for the CH_3–S and CH_3–Hg group are:

CH_3–S cm^{-1}	Assignment	CH_3–Hg cm^{-1}
2984	ν asym. CH_3	2984
2919	ν sym. CH_3	2919
1432	δasym. CH_3	1408
1309	δsym. CH_3	1177
956	ρ CH_3	765

REFERENCES

1. Nyquist, R. A. (1984). *The Interpretation of Vapor-Phase Infrared Spectra: Group Frequency Data.* Philadelphia: Bio-Rad Laboratories Sadtler Div., p. 468.
2. Nyquist, R. A. and Kagel, R. O. (1977). *Infrared and Raman Spectroscopy.* Part B, Chap. 6, E. G. Brame, Jr. and J. G. Grasselli, eds., New York: Marcel Dekker, Inc., p. 497.
3. Simons, W. W. (ed.) (1978). *The Sadtler Handbook of Infrared Spectra.* Philadelphia: Bio-Rad Laboratories, Sadtler Div., p. 360.
4. Nyquist, R. A. and Evans, J. C. (1961). *Spectrochim. Acta,* **17**: 795.
5. Sheppard, N. (1950). *Trans. Faraday Soc.,* **46**: 429.
6. (19XX). *Sadtler Raman Spectra.* Philadelphia: Bio-Rad Laboratories Sadtler Div.
7. Nyquist, R. A. and Evans, J. C. (1961). *Spectrochim. Acta,* **17**: 795.
8. Stojilykovic, A. and Whiffen, D. H. (1958). *Spectrochim. Acta,* **12**: 47.
9. Scherer, J. R. and Evans, J. C. (1963). *Spectrochim. Acta,* **19**: 1793.
10. Scherer, J. R., Evans, J. C., Muelder, W. W., and Overend, J. (1962). *Spectrochim. Acta,* **18**: 57.
11. Scherer, J. R. (1963). *Planar Vibrations of Chlorinated Benzenes.* Midland, MI: The Dow Chemical Company.

12. Nyquist, R. A. (1984). *The Interpretation of Vapor-Phase Infrared Spectra: Group Frequency Data*. Philadelphia: Bio-Rad Laboratories Sadtler Div., p. 708.
13. Nyquist, R. A. (1984). *The Interpretation of Vapor-Phase Infrared Spectra: Group Frequency Data*. Philadelphia: Bio-Rad Laboratories Sadtler Div., p. 635.
14. Nyquist, R. A. (1969). *Spectrochim Acta*, **25A**: 47.
15. Popov, E. M., Kabachnik, M. I., and Mayants, L. S. (1961). *Russian Chem. Rev.*, **30**: 362.
16. Nyquist, R. A. and Mann, J. R. (1972). *Spectrochim. Acta*, **28A**: 511.

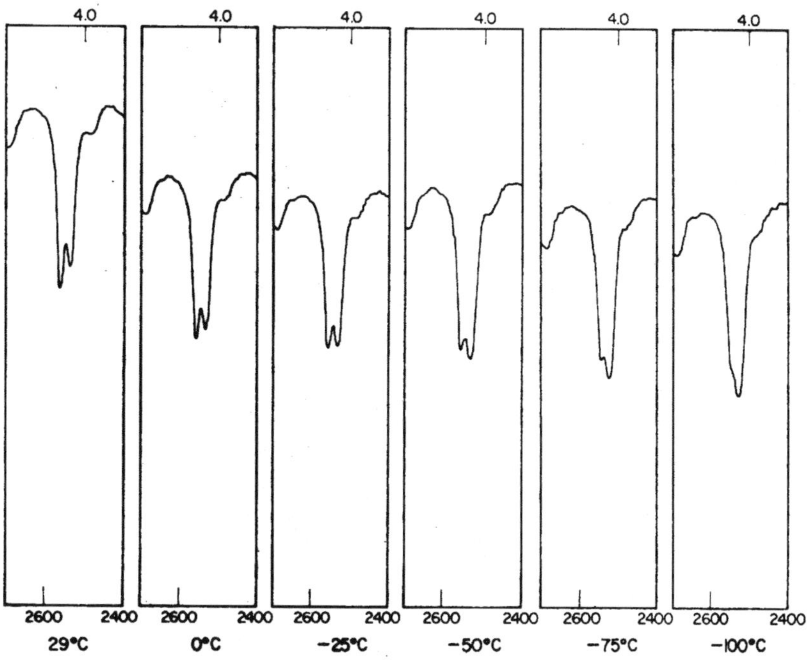

FIGURE 4.1 Infrared spectra of O,O-bis-(2,4,5-trichlorophenyl) phosphorodithioate in 5 wt/vol in CS_2 solution (2700–2400 cm^{-1}) at temperatures ranging from 29 to −100 °C.

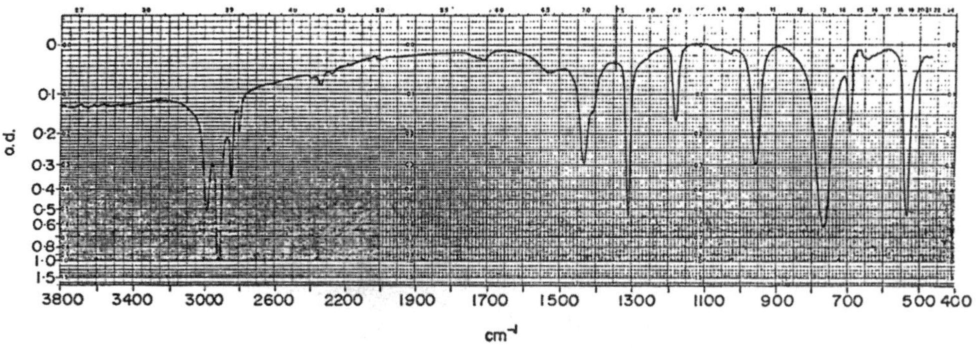

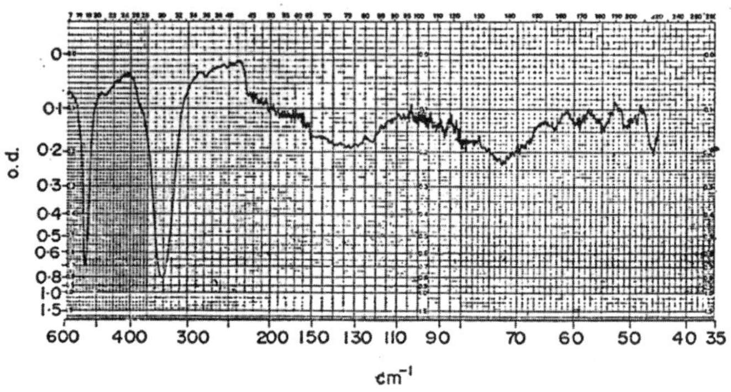

FIGURE 4.2 top: Liquid-phase IR spectrum of methyl (methylthio) mercury between KBr plates in the region 3800–450 cm^{-1}. bottom: Liquid-phase IR spectrum of methyl (methylthio) mercury between polyethylene plates in the region 600–45 cm^{-1}. The IR band near 72 cm^{-1} is due to absorbance from poly (ethylene).

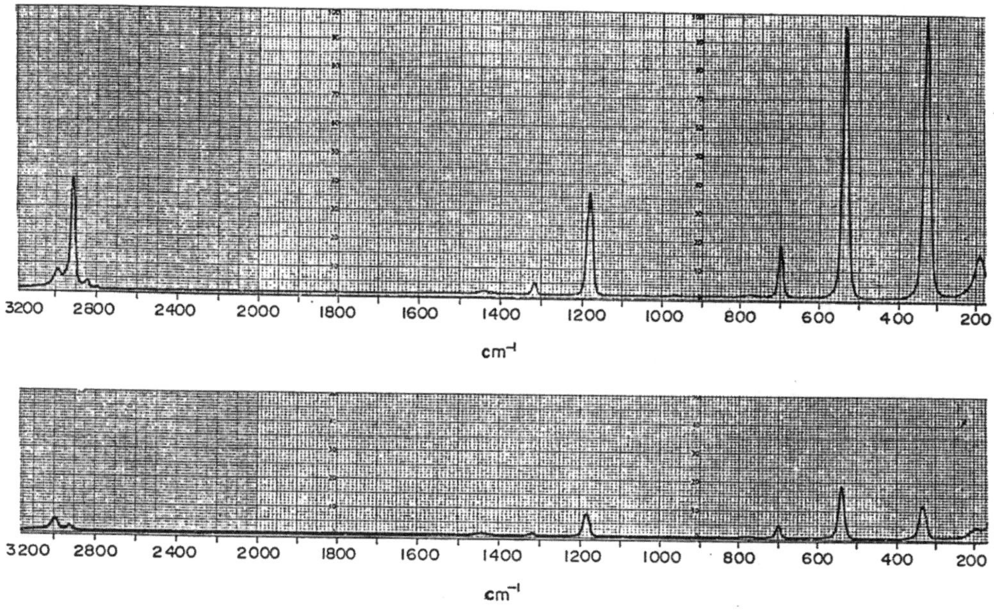

FIGURE 4.3 top: Raman liquid-phase spectrum of methyl (methylthio) mercury in a glass capillary tube. The sample was positioned perpendicularly to both the laser beam and the optical axis of he spectrometer. bottom: Same as top except that the plane of polarization of the incident beam was rotated 90°.

TABLE 4.1 Infrared and Raman data for alkanethiols and benzenethiols

Alkanethiol	[1] (A) IR [vapor] SH str. cm^{-1}	[2] (B) IR [CCl_4] SH str. cm^{-1}	(A)–(B) and [(A)–(C)] cm^{-1}	[3] (C) IR [liqud] SH str. cm^{-1}	(B)–(C) cm^{-1}	(B)–(D) cm^{-1}	[2] (D) R [liquid] SH str. cm^{-1}	(2) (E) RI	[2] (F) Dep. Ratio	[2] (G) IR [CS_2] CS str. cm^{-1}	[2] (H) IR [liquid] CS str. cm^{-1}	[2] (I) R [liquid] CS str. cm^{-1}	(2) (J) RI	[2] (K) Dep. Ratio
R														
CH_3	2601	2584	17							703				
C_2H_5	~2600													
n-C_4H_9		2580		2560	20	0	2580	68.5	0.1	651	652	657	100	0.11
n-C_5H_{11}		2576		2555	21	−4	2580	73	0.1	653	654	657	100	0.14
iso-C_3H_7		2579		2562	17	1	2578	31.4	0.09	629; 616sh	631; 620sh	635; 622sh	100; 61.0	0.09; 0.10
s-C_4H_9	~2590	2572	[18] 31	2559	13	−10	2582	54.1	0.13	608	610; 619sh	625; 618sh	100; 85.0	0.13; 0.14
tert-C_4H_9		2579		2585	−6	−4	2583	100	0.13			587	100	p
$HSCH_2CH_2$		2569		2550	19	3	2566	100	0.24	635; 667	635; 667	641; 675	89.5; 20.0	0.15; 0.68
Alkanedithiol														
R														
2,6-$(CH_3)_2$		~2550												
2,6-octane														
Benzenethiol														
2-CH_3				~2560										
2-NH_2	2560		[50]	~2510										
3-CH_3	2595		[33]	~2562										
3-NH_2				~2550										
4-CH_3	2590		[20]	~2570										
4-tert-C_4H_9	2595		[45]	~2550										
4-Cl [4]		2589		[2579]	10	20	2569	55	p					
4-Br				~2562										
2,5-Cl_2	2600		[25]	~2575 [solid]										

TABLE 4.1A Raman data for organic thiols

Compound	SH str. cm^{-1} (RI)	C=O str. cm^{-1} (RI)
Ethylene bis-(thioglycolate)	2574 (9)	1734 (3)
Ethylene bis-(thiopropionate)	2572 (7)	1734 (3)
Thiophenol	2569 (1)	
Toluenethiol isomers	2570 (4)	
Thioxylenol isomers	2570 (6)	

TABLE 4.2 Vapor-phase infrared data for alkanethiols, alkane sulfides, and alkane disulfides

Compound	a.CH_3 str. cm^{-1}(A)	s.CH_2 str. cm^{-1}(A)	s.CH_3 str. cm^{-1}(A)	s.CH_2 str. cm^{-1}(A)	OT in FR or CT cm^{-1}(A)	S–H str. cm^{-1}(A)	a.CH_3 bend and or CH_2 bend cm^{-1}(A)	s.CH_3 bend cm^{-1}(A)	cm^{-1}(A)	CH_2 wag cm^{-1}(A)	cm^{-1}(A)	CH_3 rock cm^{-1}(A)	C–C str. or a.CCC str. cm^{-1}(A)	s.CCC str. cm^{-1}(A)	CH_2 twist cm^{-1}(A)	CH_2 rock cm^{-1}(A)	C–S str. cm^{-1}(A)	a.$(C)_2S$ str. cm^{-1}(A)	s.$(C)_2$ S. str. cm^{-1}(A)
	3028		2882		2822	2820	1470	1348				1089					728		
	(0.240)		(1.240)		(0.140)	(0.022)	(0.140)	(0.150)				(0.135)					(0.030)		
Methanethiol*	3018		2864		2864	2601	1452	1331				1071					710		
	(0.250)		(1.040)		(0.100)	(0.015)	(0.090)	(0.060)				(0.095)					(0.037)		
			2858		2828	2592	1445	1319				1060							
			(0.740)		(0.130)	(0.020)	(0.151)	(0.132)				(0.155)							
	3000																		
	(0.370)																		
Methyl sulfide	2968		2935		2860		1439	1311				1011					695		
	(0.890)		(1.250)		(0.290)		(0.432)	(0.112)				(0.121)					(0.015)		
			2910		2838														
			(1.150)		(0.220)														
Methyldisulfide	3000		2962		2920		1430										688		
	(0.305)		(1.240)		(0.130)		(0.305)										(0.030)		
Ethanethiol	2991	2960	2900	2880		2598	1455	1390		1280		1100	982			866	659		
	(1.240)	(1.035)	(0.270)	(0.244)		(0.041)	(0.190)	(0.052)		(0.301)		(0.056)	(0.102)			(0.030)	(0.050)		
Ethyl sulfide	2975	2940		2890			1458	1389		1260		1075	975					776	651
	(1.250)	(1.050)		(0.380)			(0.170)	(0.090)		(0.380)		(0.020)	(0.095)					(0.039)	(0.005)
Ethyl disulfide	2987	2940		2884	2838		1454	1383		1251		1045	970					760	645
	(1.250)	(0.930)		(0.315)	(0.060)		(0.190)	(0.110)		(0.470)		(0.055)	(0.081)					(0.081)	(0.010)
Propyl sulfide	2970	2942		2890			1460	1390	1340	1295	1237	1092		896	845			784	742
	(1.235)	(1.035)		(0.585)			(0.185)	(0.104)	(0.030)	(0.149)	(0.202)	(0.040)		(0.095)	(0.019)			(0.050)	(0.039)
Propyl disulfide	2970	2942		2882		2882	1460	1388	1339	1290	1230	1088	1054	894	825			785	732
	(1.240)	(0.610)		(0.340)		(0.040)	(0.140)	(0.078)	(0.045)	(0.139)	(0.170)	(0.020)	(0.019)	(0.020)	(0.015)			(0.040)	(0.019)
1-Butanethiol	2965	2942	2895	2880		2600	1460	1396		1288	1235	1104		960				779	740
	(1.250)	(1.050)	(0.410)	(0.418)		(0.021)	(0.139)	(0.050)		(0.121)	(0.060)	(0.040)		(0.025)				(0.020)	(0.030)
Butyl sulfide	2975	2950		2885			1465	1389	1347	1270	1227	1098	1050	918		880		786	751
	(1.150)	(1.150)		(0.750)			(0.241)	(0.058)	(0.041)	(0.100)	(0.080)	(0.010)	(0.005)	(0.019)		(0.011)		(0.020)	(0.030)
Butyl disulfide	2970	2940		2880			1466	1385	1347	1272	1218	1097	1065	914		876		782	749
	(1.250)	(1.050)		(0.550)			(0.200)	(0.100)	(0.070)	(0.140)	(0.140)	(0.031)	(0.015)	(0.042)		(0.042)		(0.029)	(0.062)

1-Pentanethiol	2972 (1.140)	2940 (1.240)	2880 (0.370)		2600 (0.020)	1462 (0.101)	1385 (0.032)	1348 (0.030)	1277 (0.072)	1226 (0.032)		1102 (0.020)				730 (0.024)
Pentyl disulfide	2965 (1.122)	2938 (1.230)	2872 (0.385)			1465 (0.240)	1385 (0.060)	1348 (0.090)	1271 (0.080)	1259 (0.080)	1210 (0.050)	1105 (0.010)	1075 (0.010)			730 (0.035)
Hexanethiol	2962 (0.950)	2938 (1.240)	2868 (1.000)		2596 (0.020)	1460 (0.150)	1384 (0.058)	1350 (0.052)	1295 (0.090)	1254 (0.058)	1214 (0.053)					740 (0.043)
Hexyl disulfide	2972 (1.010)	2940 (1.220)	2870 (0.491)			1460 (0.151)	1289 (0.061)	1340 (0.070)	1290 (0.081)	1260 (0.075)	1208 (0.030)					730 (0.035)
Heptyl sulfide	2970 (0.580)	2938 (1.220)	2865 (0.420)			1465 (0.120)	1385 (0.031)	1351 (0.049)	1300 (0.040)	1248 (0.048)		1105 (0.005)		900 (0.005)		728 (0.015)
Heptyl disulfide	2970 (0.560)	2935 (1.250)	2865 (0.410)			1460 (0.131)	1380 (0.040)	1350 (0.040)	1300 (0.042)	1240 (0.050)						720 (0.030)
1-Octanethiol	2970 (0.690)	2935 (1.235)	2864 (0.552)	2680 (0.030)	2595 (0.021)	1463 (0.140)	1385 (0.040)	1355 (0.040)	1289 (0.061)	1249 (0.040)	1175 (0.011)	1110 (0.011)				727 (0.037)
Octyl sulfide	2970 (0.364)	2935 (1.204)	2862 (0.244)			1465 (0.080)	1382 (0.029)	1352 (0.029)	1280 (0.028)	1230 (0.025)						
Octyl disulfide	2972 (0.660)	2935 (1.250)	2862 (0.550)			1461 (0.140)	1385 (0.059)	1350 (0.050)	1294 (0.069)	1238 (0.049)		1109 (0.010)				720 (0.030)
Nonyl sulfide	2970 (0.290)	2936 (1.240)	2862 (0.368)			1461 (0.090)	1380 (0.020)	1350 (0.020)	1301 (0.040)	1255 (0.250)	1191 (0.040)					719 (0.020)
1-Decanethiol	2970 (0.500)	2935 (1.250)	2860 (0.500)		2595 (0.100)	1464 (0.100)	1380 (0.025)	1352 (0.030)	1308 (0.038)			1110 (0.011)		900 (0.010)		716 (0.024)
Decyl sulfide	2970 (0.400)	2935 (1.250)	2864 (0.366)			1460 (0.091)	1381 (0.030)									722 (0.018)
Decy disulfide	2970 (0.271)	2935 (1.245)	2862 (0.290)			1460 (0.060)		1350 (0.027)	1299 (0.020)							
Dodecyl sulfide	2970 (0.320)	2935 (1.220)	2862 (0.400)			1460 (0.080)	1380 (0.005)	1351 (0.026)	1300 (0.020)							720 (0.010)
1-Tetradecanethiol	2970 (0.120)	2938 (1.250)	2862 (0.330)		2595 (0.005)	1464 (0.053)		1355 (0.022)	1305 (0.020)							
1-Octadecanethiol	2970 (0.250)	2935 (1.250)	2861 (0.550)		2595 (0.010)	1460 (0.100)			1300 (0.040)							715 (0.025)
Cyclohexanethiol			2860 (0.040)		2596 (0.020)	1453 (0.120)		1345 (0.039)		1266 (0.020)	1210 (0.064)	1129 (0.010)	1000 (0.058)	884 (0.031)	818 (0.029)	735 (0.20)

* [SH bend, 958 (0.058)?].

TABLE 4.3 Infrared and Raman data for organic sulfides and disulfides

R-S-R′ R; R′	IR [CS_2] a.CSC str. cm^{-1}	IR [neat] a.CSC str. cm^{-1}	IR [CS_2] s.CSC str. cm^{-1}	IR [neat] s.CSC str. cm^{-1}	R [neat] a.CSC str. cm^{-1}	RI	Dep. ratio	R [neat] s.CSC str. cm^{-1}	RI	Dep. ratio
CH_3; CH_3	743	746	691	693	745	15.3	0.81	696	100	0.11
CH_3; C_2H_5		729		652; 672	732	43.8	0.75	661; 684	100; 40.8	0.19; 0.25*
C_2H_5; C_2H_5	761	762	651	652	766	5.9	0.29	651	0.25; C_1	
	782	781	637	638	782	5.7	0.29	641	0.23; C_2	
	694	694	694	694	696	55.8	0.66	696	0.66; C_{2v}	
C_2H_5; n-C_4H_9	743	743	651	661	750	28.4	0.72	660	100	0.16
								643	38.6	0.19
								694	41.5	0.54
n-C_4H_9; n-C_4H_9	745	746	650	660	750	78.6	0.47	659	100	0.25
								670	81.3	0.29
								688	48.6	0.5

R-S—S-R′ R; R′	IR S—S str. cm^{-1}	R S—S str. cm^{-1}	RI	Dep. ratio
CH_3; CH_3	511	511	100	0.06
C_2H_5; C_2H_5	511; 526	511; 528	98.0; 49.2	0.06; 0.12
iso-C_4H_9; iso-C_4H_9	512; 528	516; 528	58,8; 37.5	0.06; 0.14
tert-C_4H_9; tert-C_4H_9	502	515	7.2	0.09
CH_3; C_6H_5	507; 525	512; 529 [solid]	24.8; 23.0	0.22; 0.21
C_6H_5; C_6H_5	not observed	546	26.8	

* Low-temperature conformer.

TABLE 4.4 Vibrational assignments for 4-chlorobenzenethiol and 1,4-dichlorobenzene

4-Chlorobenzenethiol cm^{-1}	1,4-Dichlorobenzene cm^{-1}	Interpretation
a1	a1	
3060	3072	C—H stretch
3050	3050	C—H stretch
1575	1573	C—C stretch
1481	1475	Ring stretch
1178	1169	C—H deformation
1098	1106	Ring stretch
1067	1087	Ring stretch
1017	1013	C—H deformation
745	747	Ring deformation
545	546	C—X stretch
329	405	C—X stretch
a2	a2	
951	951	C—H deformation
817	811	C—H deformation
403	405	Skeleton deformation
b1	b1	
935	934	C—H deformation
812	816	C—H deformation
692	689	Skeleton deformation
485	484	Skeleton deformation
293	299	C—X deformation
122	125	C—X deformation
b2	b2	
3078	3095	C—H stretch
3050	3072	C—H stretch
1575	1573	C—C stretch
1397	1393	C—C stretch
1298	1291	C—H deformation
1260	1260	C—C stretch
1104	1104	C—H deformation
629	628	Ring deformation
342	351	C—X deformation
221	226	C—X deformation
2589		S—H stretch
915		S—H deformation
?		S—H free rotation

TABLE 4.5 The P=S, P—S, and S—H stretching frequencies for O,O-dialkyl phoshorodithioate and O,O-bis-(aryl)phosphorodithoate

$(CH_3{-}O{-})_2P(=S)SH$ cm^{-1}	$(C_2H_5{-}O{-})_2P(=S)SH$ cm^{-1}	$(2,4,5\text{-}Cl_3{-}C_6H_4{-}O{-})_2$ $P(=S)SH$ cm^{-1}	Assignment
2588	2582	2575	S—H str., rotational conformer 1
2550	2550	2549	S—H str., rotational conformer 2
670sh	670sh		P=S str., rotational conformer 1
659	659		P=S str., rotational conformer 2
524	535		P—S str. rotational conformer 1
490	499		P—S str., rotational conformer 2

TABLE 4.6 A comparison of the alkyl groups joined to sulfur, oxygen, or halogen

Compound	a.CH_3 str.	a.CH_2 str.	s.CH_3 str.	s.CH_2 str.	a.CH_3 bend	CH_2 bend	o.p.[CH_3]2 bend	i.p.[CH_3]2 bend												
Isobutyl disulfide	2970 (1.250)	2962 (0.581)	2905 (0.390)	2880 (0.430)	1469 (0.189		1385 (0.180)	1375 (0.160)	1323 (0.080)	1240 (0.150)	1215 (0.080)	1170 (0.101)	1109 (0.021)	1075 (0.030)	946 (0.030)	926 (0.030)	856 (0.030)	800 (0.030)		
																			C-X str.	C-X str.
Isobutyl chloride	2980 (1.040)	2962 (1.240)		2880 (0.310)	1471 (0.200)	1444 (0.170)	1389 (0.160)	1381 (0.170)	1334 (0.095) 1329 (0.090) 1322 (0.100)	1272 (0.085) 1267 (0.080) 1260 (0.051)	1220 (0.020)	1168 (0.030)	1100 (0.020)		951 (0.070)	938 (0.090)	889 (0.030) 880 (0.025) 87190.030)	812 (0.090) 808 (0.080) 802 (0.090)	744 (0.310)	698 (0.090)
Isopentyl sulfide	2962 (1.250)	2925 (0.500)		2882 (0.330)	1470 (0.120)	1444 (0.069)	1390 (0.97)	1373 (0.086)	1350 (0.050)	1280 (0.075)	1226 (0.060)	1170 (0.060)	1110 (0.010)			920 (0.010)	885 (0.007)	745 (0.015)		
Isopentyl disulfide																				
Isopentyl chloride	2965 (1.250)	2930 (0.058)		2885 (0.460)	1471 (0.172) 1469 (0.165) 1462 (0.172)		1382 (0.141)	1367 (0.100)	1351 (0.094)	1296 (0.161)	1250 (0.060)	1170 (0.059)				930 (0.025)	872 (0.060)		741 (0.180)	670 (0.135)
Isopentyl bromide	2965 (1.250)	2940 (0.550)		2885 (0.421)	1470 (0.180)		1385 (0.149)	1370 (0.110)	1351 (0.080)	1267 (0.251)	1220 (0.180)	1170 (0.080)		1025 (0.029)	950 (0.015)	922 (0.029)	860 (0.055)	750 (0.020)	652 (0.121)	570 (0.081)
	a.CH_3 str.	s.CH_3 str.	2 (a.CH_3 bend) in F.R. (s.CH_3)	a.CH_3 bend	i.p.[CH_3]2 bend		o.p[CH_3]2 bend						s.C[C]2 str.	C-X str.						
Isopropanethiol*[1]	2979 (1.240)	2930 (0.490)	2898 (0.290)	1470 (0.180) 1459 (0.240) 1444 (0.140)	1390 (0.190) 1382 (0.170)		1370 (0.120)		1262 (0.240) 1251 (0.240) 1242 (0.190)	1165 (0.101)	1090 (0.090)	940 (0.020)	861 (0.040)	625 (0.040)						
Isopropyl disulfide	2970 (1.250)	2935 (0.700)	2880 (0.450)	1454 (0.220)			1375 (0.220)	1311 (0.050)	1238 (0.470)	1155 (0.230)	1047 (0.190)	928 (0.021)	876 (0.019)	620 (0.010)						
	2995 (1.240)			1470 (0.180)					1276 (0.330)	1166 (0.160)	1075 (0.190)		897 (0.122)	639 (0.270)						

(continues)

TABLE 4.6 (continued)

Compound	a.CH_3 str.	a.CH_2 str.	s.CH_3 str.	s.CH_2 str.	a.CH_3 bend	CH_2 bend	o.p.[CH_3]2 bend	i.p.[CH_3]2 bend							
Isopropyl chloride	2980 (1.250)	2941 (0.361) 2925 (0.0140)	2885 (0.160) 2870 (0.120)	1450 (0.240) 1440 (0.120)	1390 (0.270) 1384 (0.250)		1372 (0.170)	1315 (0.035)	1265 (0.430) 1253 (0.350)	1161 (0.190) 1150 (0.172)	1061 (0.240) 1053 (0.160)	945 (0.020)		885 (0.140) 87590.120)	630 (0.358) 620 (0.340)
Isopropyl bromide	2998 (1.240) 2982 (1.240)	2932 (0.630)	2885 (0.291)	1470 (0.270)	1385 (0.380)		1372 (0.260)	1330 (0.037)	1235 (0.882) 1229 (0.932) 1220 (0.912)	1160 (0.520)	1051 (0.172) 1042 (0.210) 1035 (0.182)	945 (0.025)		889 (0.142) 880 (0.160)) 875 (0.150)	540 (0.345)
Isopropyl iodide	2998 (1.040) 2980 (1.240)	2922 (0.600)	2890 (0.420)	1465 (0.220) 1459 (0.260) 1440 (0.170)	1382 (0.330)		1372 (0.220)	1330 (0.090)	1210 (0.690) 1203 (0.840) 1197 (0.670)	1150 (1.030)	1021 (0.159	950 (0.060)		874 (0.090)	495 (0.150)
Isopropyl alcohol*[2]	2980 (1.250)	2890 (0.460)		1471 (0.140) 1460 (0.151)	1380 (0.510)				1250 (0.275)	1147 (0.540)	1090 (0.320) 1080 (0.300) 1075 (0.320)	968 (0.380) 951 (0.400) 941 (0.380)		817 (0.070)	
Isopropyl ether															
	a.CH_3 str.	s.CH_3 str.	O.T.	a.CH_3 bend	i.p.[CH_3]3 bend		o.p.[CH_3]3 bend	CH_3 rock	a.C[C]3 str.			s.C[C]3 str.		C-X str.	
tert-Butyl sulfide	2975 (1.250)	2930 (0.650) 2910 (1.250) 2891 (0.660)	2721 (0.029)	1470 (0.230)	1399 (0.100)		1370 (0.360)	1208 (0.100)	1156 (0.560)	1024 (0.010)	926 (0.010)	810 (0.010)		604 (0.021)	
tert-Butyl disulfide	2970 (1.250)	2925 (0.610) 2902 (0.550) 2890 (0.440)	2720 (0.021)	1461 (0.241)	1392 (0.090)		1368 (0.410)	1214 (0.080)	1160 (0.620)	1017 (0.011)	931 (0.011)	800 (0.005)		[or S–S str.] 564 (0.030)	
tert-Butyl chloride	2995 (1.224) 2980 (1.139)	2938 (0.500)	2720 (0.020)	1470 (0.170) 1462 (0.220) 1450 (0.150)	1388 (0.329)		1378 (0.631) 1365 (0.290)	1238 (0.180) 1228 (0.212)	1154 (0.550)	1018 (0.015)	930 (0.041)	819 (0.039) 810 (0.045)		581 (0.180) 570 (0.340)	500 (0.070)
tert-Butyl bromide	2990 (0.930)	2924 (0.459)	2738 (0.020)	1459 (0.190)	1391 (0.240)		1378 (0.440)	1238 (0.1300 1228 (0.120)	1151 (1.240)		918 (0.020)	810 (0.070) 802 (0.090)	790 (0.070)	520 (0.172)	

tert-Butyl alcohol*[3]	2980 (1.250)	2900 (0.320)		1490 (0.075) 1472 (0.125) 1460 (0.099)	1389 (0.440)		1371 (0.550) 1360 (0.350)		1214 (0.450)	1141 (0.450)		1018 (0.060)	931 (0.290) 919 (0.400)		760 (0.010) 750 (0.034) 735 (0.020)					
tert-Butyl hydroperoxide*[4]	2990 (1.250)	2945 (0.298)		1478 (0.086) 1469 (0.099) 1452 (0.070)	1379 (0.340) 1369 (0.591)		1369 (0.591)	1330 (0.440) 1320 (0.465)	1248 (0.381)	1210 (0.401) 1202 (0.451)	1140 (0.053)			855 (0.050) 845 (0.150) 835 (0.060)	755 (0.025) 746 (0.040) 732 (0.025)	520 (0.050)				
Di-tert-butyl peroxide	2990 (1.245)	2941 (0.320)		1477 (0.081) 1469 (0.070) 1460 (0.068)	1388 (0.170)		1369 (0.590)		1245 (0.281)	1200 (1.041)		1025 (0.030)	920 (0.025)	880 (0.380)	750 (0.060)	531 (0.020)				
tert-Butyl amine*[5]	2965 (1.250)	2900 (0.480)	[a.NH2 str.] 3402 (0.010)	1470 (0.190)	1380 (0.370)		1368 (0.540)		1235 (0.440)		1125 (0.030) 1110 (0.030)	1032 (0.040)	941 (0.060)	[NH2 wag] 810 (0.898)	740 (0.160)					
tert-Butyl nitrite*[6]	2985 (0.520)	2945 (0.150)	[2(trans N=O str.)] 3280 (0.025)	1472 (0.040)	1388 (0.160)		1376 (0.220)		1250 (0.109)	1200 (0.211)	1142 (0.063)	1035 (0.032)		810 (0.501)	[trans N-O str.] 759 (1.240)					
2,2-Dimethylpropio-nitrile	2995 (1.245)	2940(0240)	[CN str.] 2250 (0.020)	1470 (0.190)	1390 (0.190)		1380 (0.230)		1246 (0.150)	1215 (0.170)	1140 (0.020)	1039 (0.011)			759 (0.009)					
2-Methyl-2-nitropropane*[7]	2999 (0.600)	2948 (0.260)	[a.NO2 str.] 1555 (1.250)	1495 (0.240) 1482 (0.280) 1472 (0.260)	1417 (0.180) 1409 (0.250)		1376 (0.500) 1360 (0.450)		1248 (0.230)	1199 (0.200)		1036 (0.010)	935 (0.011)	800 (0.050)	[C-N str.] ? 860 (0.180)	[NO2 rock] 568 (0.061)				
2-Butanethiol*[8]	2979 (1.240)	2938 (0.830)		s.CH_2 str. 2890 (0.261)	CH_3 bend 1459 (0.205)		s. CH_3 bend 1390 (0.141)		1290 (0.096)	1234 (0.112)		1150 (0.055)	1078 (0.043)	1005 (0.060)		965 (0.040)	835 (0.040)	792 (0.040)	670 (0.020)	625 (0.020)
2-Chlorobutane	2975 (1.250)	2942 (0.760)		2890 (0.368)	1462 (0.240)	1450 (0.265)	1389 (0.220)		1293 (0.210)	1250 (0.220)		1150 (0.110)	1068 (0.090)	1000 (0.100)		962 (0.128)	838 (0.130)	800 (0.110)	670 (0.130)	625 (0.200)
2-Bromobutane																			611 (0.070)	529 (0.080)
2-Iodobutane																			590 (0.041)	490 (0.041)

"1[SH str.,2580 (0.046)]"	*2 [OH str.]	*3 [OH str.]	*4 [OH str.]	*5 [s.NH2 str.]	*5 [NH2 bend]	*6 [trans N=O str.]	*7 [s.NO2 str.]	*8 [SH str.]
	3660 (0.130)	3645 (0.114)	3600 (0.230)	3325 (0.040)	1625 (0.201)	1662 (0.850) 1652 (1.150) 1645 (0.750)	1350 (0.445)	2598 (0.022)

CHAPTER 5

Sulfoxides, Sulfones, Sulfates, Monothiosulfates, Sulfonyl Halides, Sulfites, Sulfonamides, Sulfonates, and N-Sulfinyl Anilines

S=O Stretching Frequencies	86
SO_2 Stretching Frequencies	87
Diakyl Sulfones and Dimethyl Sulfoxide	87
Diaryl Sulfones	88
Sulfate and Thiosulfate Phenoxarsine Derivatives	88
Dialkyl Sulfites	89
Primary Sulfonamides	89
Secondary and Tertiary Sulfonamides	89
Organic Sulfonates	90
Organosulfonyl Chlorides	91
Organosulfonyl Fluorides	91
Summary of S=O and SO_2 Stretching Frequencies in Different Physical Phases	92
ν asym. SO_2 and ν sym. SO_2 Correlations	92
SO_2 Stretching Vibrations in $CHCl_3/CCl_4$ Solutions	92
Calculated ν SO_2 Frequencies	93
Sulfinyl Anilines	93
References	93

Figures

Figure 5-1	96 (86)	Figure 5-5	100 (92)
Figure 5-2	97 (92)	Figure 5-6	101 (92)
Figure 5-3	98 (92)	Figure 5-7	102 (93)
Figure 5-4	99 (92)	Figure 5-8	103 (93)

Tables

Table 5-1A	104 (86)	Table 5-3	109 (88)
Table 5-1B	105 (87)	Table 5-4	110 (89)
Table 5-1C	107 (87)	Table 5-5	111 (89)
Table 5-1D	108 (88)	Table 5-6	111 (89)
Table 5-2	109 (88)	Table 5-6A	112 (89)

Tables

Table 5-7	112 (90)	Table 5-10	116 (92)
Table 5-8	113 (91)	Table 5-11	117 (92)
Table 5-9	114 (91)	Table 5-12	117 (92)
Table 5-9A	115 (91)		

*Numbers in parentheses indicate in text page reference.

S=O STRETCHING FREQUENCIES

Table 5.1a lists infrared and Raman data for compounds containing an S=O group. In the vapor phase ν S=O occurs in the range 1092–1331 cm^{-1}, and in the neat phase ν S=O usually occurs at lower frequency in the range 1035–1308 cm^{-1}. The decrease in the ν S=O frequency in going from the vapor to the neat phase is attributed to dipolar interaction between the S=O groups of these molecules in the neat phase. In solution, this type of intermolecular interaction is called a reaction field (RF). The equation for determining RF has been presented previously. This equation includes both the dielectric constants of all molecules present and their refractive indices. Because the contribution of the refractive indices is minor compared to the contribution of the dielectric constants in chemical mixtures, the refractive indices are usually ignored in the RF calculations. The ν S=O vibration is highly dependent upon the dielectric surroundings of the S=O groups and the steric factors, which determine the spatial distance between S=O groups or the spatial distance between S=O groups and solvent molecules. Examples are presented here:

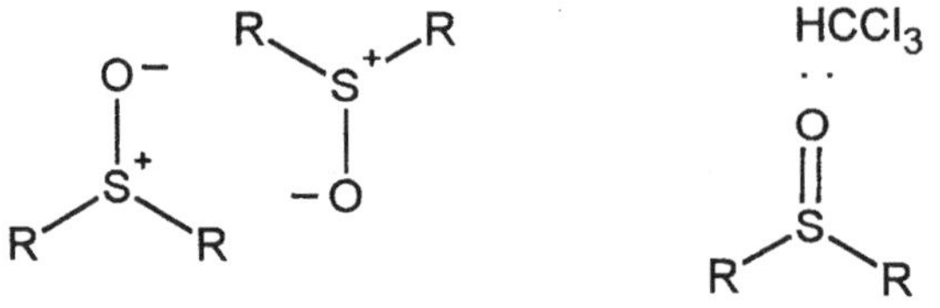

Perhaps a better descriptive name for RF is a name such as intermolecular force field association (IFFA).

Figure 5.1 shows plots of the sum of the σ^* constants for the atoms or groups joined to the S=O or SO_2 groups vs the ν S=O frequencies and the mean average of the ν sym. SO_2 and ν sym. SO_2 frequencies (1). The plot for the ν S=O frequencies for $(CH_3)_2S{=}O$, $(CH_3O)_2S{=}O$, and $F_2S{=}O$ vs $\sigma^{*\prime}$ is linear, and the plot for ν S=O for $(CH_3O)_2S{=}O$, $(CH_3O)(Cl)S{=}O$, and $Cl_2S{=}O$ vs $\Sigma\sigma'$ is essentially linear. It is also noted that there is a good correlation between the mean average of ν asym. SO_2 and ν sym. SO_2 for the $(CH_3)_2SO_2$, $(CH_3O)_2SO_2$, and $F_2S{=}O$ vs $\Sigma\sigma'$. Both ν S=O for $Cl_2S{=}O$ and the mean average for the ν SO_2 modes occur at lower frequency than do the linear or pseudolinear plots for the three other S=O analogs. This suggests that mass is also a factor in affecting the ν S=O frequencies. Chlorine is heavier than C, O, and F by a factor of 2 to 3. During a cycle of νC=O, the X-S-X bond angle opens and closes as depicted here,

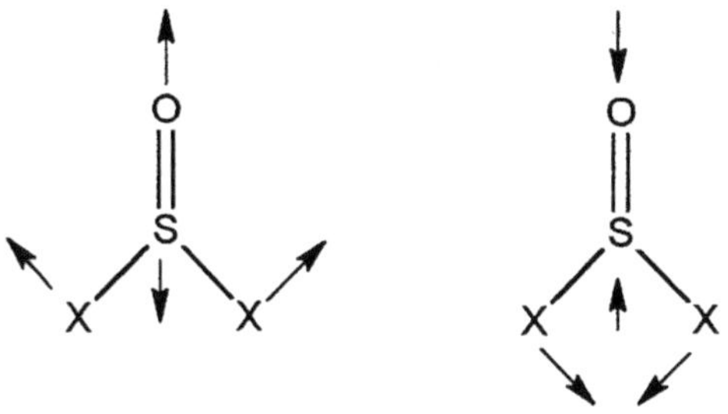

This increased mass of Cl would cause ν S=O to vibrate at lower frequency, as the Cl atoms impedes the motion of the S atom during a cycle of S=O stretching.

SO_2 STRETCHING FREQUENCIES

Table 5.1b compares the ν asym. SO_2 and ν sym. SO_2 frequencies for a variety of chemical compounds in the vapor and neat phases. In the vapor phase the ν asym. SO_2 vibration occurs in the range 1335–1535 cm^{-1} and in the neat phase in the range 1295–1503 cm^{-1}. In the vapor phase the ν sym. SO_2 vibration occurs in the range 1141–1300 cm^{-1} and in the neat phase in the range 1125–1270 cm^{-1} (1–42).

Study of the last two columns in Table 5.1b shows that in all but two cases the ν asym. SO_2 shifts more to lower frequency than ν sym. SO_2 does in going from the vapor phase to the neat phase. One exception is Cl_2SO_2 where the shift is less for ν asym. SO_2 (24 cm^{-1}) than for ν sym. SO_2 (28 cm^{-1}). The effect of the mass of Cl upon the ν SO_2 mean average was already discussed here, and this may be the reason for the deviation from this group frequency spectra-structure correlation. It should be noted that the shift to lower frequency for ν asym. SO_2 is 18 cm^{-1} and for ν sym. SO_2 is 19 cm^{-1} for Cl_2SO_2 in going from the vapor phase to a CCl_4 solution.

DIALKYL SULFONES AND DIMETHYL SULFOXIDE

Table 5.1c lists vapor-phase IR data for methyl sulfoxide and dialkyl sulfones. The ν asym. $(C)_2S$ and ν sym. $(C)_2S$ vibrations are assigned at 755 cm^{-1} and 669 cm^{-1} for methyl sulfoxide, respectively. Corresponding vibrations for dimethyl sulfone are assigned at 742 cm^{-1} and 680 cm^{-1}, respectively. Thus, the ν asym. $(C)_2S$ vibration decreases 13 cm^{-1} while the ν sym. $(C)_2S$ vibration increases 11 cm^{-1} when S=O is substituted $SO_2[(CH_3)_2S{=}O$ to $(CH_3)_2SO_2]$. The ν asym. $(C)_2S$ (742–812 cm^{-1}) and ν sym. $(C)_2S$ (680–702 cm^{-1}) vibrations increase in frequency, progressing in the series $(CH_3)_2SO_2$ through $(n\text{-}C_6H_{13})SO_2$. However, ν sym. SO_2 decreases in frequency (1160–1140 cm^{-1}), progressing in the series $(CH_3)_2SO_2$ through $(n\text{-}C_6H_{13})SO_2$. The ν asym. SO_2 mode also decreases in frequency (1351–1335 cm^{-1}), progressing in the series $(CH_3)_2SO_2$ through $(n\text{-}C_6H_{15})SO_2$, but not in a decreasing order. It is possible that the ν sym. $(C)_2S$ (680–702 cm^{-1}) and ν sym. SO_2 vibrations (1160–1140 cm^{-1}) are coupled, because both modes belong to the A_1 symmetry species [assuming $(CH_3)_2SO_2$ has C_{2v} symmetry] together with the fact that as ν sym. $(C)_2$ increases in frequency, ν sym. CO_2 decreases in frequency. The ν asym. $(C)_2S$ and ν asym. SO_2 vibration can not couple, because ν

asym. $(C)_2S$ and ν asym. SO_2 belong to the B_1 and B_2 symmetry species, respectively. The only mode that could couple with ν asym. SO_2 would be a ρ $(C)_2$ B_2 vibration.

Table 5.1d lists vapor- and solid-phase IR data for dimethyl sulfoxide and dialkyl sulfones. When dimethyl sulfoxide goes from the vapor to the liquid phase the ν asym. $(C)_2S$ and ν S=O vibrations decrease by $60\,cm^{-1}$ and $50\,cm^{-1}$ while the ν sym. $(C)_2S$ vibration decreases by $5\,cm^{-1}$. In going from the vapor to the solid phase the ν asym. $(C)_2S$ and ν sym. $(C)_2S$ vibrations decrease in frequency by 8 and $5\,cm^{-1}$, respectively. In addition, in going from the vapor to solid phase the ν asym. SO_2 and ν sym. SO_2 vibrations decrease in frequency by $65\,cm^{-1}$ and $36\,cm^{-1}$, respectively. These data are consistent with the spectra-structure correlation that the antisymmetric vibrations of the same group shift more in frequency than the symmetric vibration in going from the vapor phase to the solid or liquid phase.

DIARYL SULFONES

Table 5.2 lists IR vapor- and solid-phase data for diaryl sulfones. In the vapor phase ν asym. SO_2 and ν sym. SO_2 occur in the range 1348–1354 cm^{-1} and 1161–1170 cm^{-1}, respectively. In the solid phase, ν asym. SO_2 and ν sym. SO_2 occur in the ranges 1318–1325 cm^{-1}, and 1145–1170 cm^{-1}, respectively. In all cases, ν asym. SO_2 shifts more to lower frequency than ν sym. SO_2 in going from the vapor to the solid phase.

SULFATE AND THIOSULFATE PHENOXARSINE DERIVATIVES

Phenoxarsine has the empirical structure presented here:
Table 5.3 lists IR data for these phenoxarsine derivatives in CS_2 solution.

Both of these types of phenoxarsine derivatives exhibit ν asym. SO_2 in the region 1320–

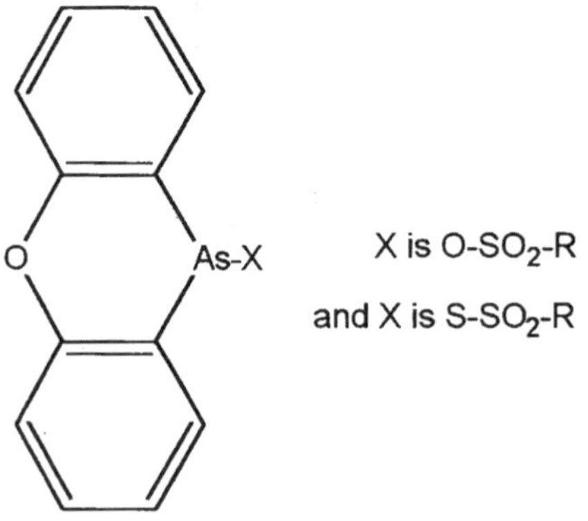

1328 cm^{-1} (35). On the other hand, the ν sym. SO_2 mode for the O–SO_2-R analog occurs in the range 1161–1171 cm^{-1} while the S–SO_2-R analog exhibits ν sym. SO_2 in the range 1125–1141 cm^{-1}.

Table 5.4 lists IR data for the SO_2 stretching vibrations for compounds in $CHCl_3$ and CCl_4 solutions (37). Table 5.4 lists data for sulfate, a benzenesulfonate, benzenesulfonyl chloride, and two sulfones. In all cases the ν asym. SO_2 vibration shifts more in frequency than the ν sym. SO_2 vibration in going from solution in CCl_4 to solution in $CHCl_3$. In $CHCl_3$ solution part of the decrease in both ν SO_2 frequencies is the result of intermolecular hydrogen bonding of the form $SO_2 \cdots HCCl_3$ and/or $SO_2(\cdots HCCl_3)_2$ (see discussion on SO_2 containing compounds in mole % $CHCl_3/CCl_4$ solutions). This set of data is consistent with previous discussions about the shifts in ν SO_2 frequencies with change in physical phase.

DIALKYL SULFITES

Table 5.5 lists IR data for dialkyl sulfites (see Reference 5). The ν S=O frequency decreases in frequency by 6 to 13 cm^{-1} in going from the vapor phase to the liquid phase. The IR bands in the ranges 955–1010 cm^{-1} and 685–748 cm^{-1} are attributed to C—O—S stretching vibrations. These IR bands also shift with change of phase.

PRIMARY SULFONAMIDES

Tables 5.6 and 5.6a list IR data for primary sulfonamides in the vapor and solid phases. In the vapor phase, ν asym. SO_2 occurs in the range 1370–1382 cm^{-1}, and in the solid phase in the range 1308–1355 cm^{-1}. In the vapor phase ν sym. SO_2 occurs in the range 1170–1178 cm^{-1} and in the solid phase it occurs in the range 1140–1158 cm^{-1}. In all cases the ν asym. SO_2 vibration shifts more to lower frequency than does the ν sym. SO_2 vibration in going from the vapor phase to the solid phase. Part of the reason both modes shift to lower frequency in the solid phase is that the NH_2 group is intermolecularly hydrogen-bonded to the SO_2 group of another molecule ($NH_2 \cdots O_2S$).

In the vapor phase, the ν asym. NH_2 mode for these primary sulfonamides occurs in the range 3456–3479 cm^{-1} and in the solid phase it occurs in the range 3323–3397 cm^{-1}. The ν sym. NH_2 mode occurs in the range 3365–3378 cm^{-1} in the vapor phase and in the range 3228–3275 cm^{-1} in the solid phase. The compounds methanesulfonamide, benzenesulfonamide, and 4-toluenesulfonamide all show that ν asym. NH_2 decreases more in frequency than ν sym. NH_2 in going from the vapor phase to the solid phase. The decrease in frequency varies between 137–155 cm^{-1} for ν asym. NH_2 and between 120–137 cm^{-1} for ν sym. NH_2 in going from the vapor phase to the solid phase. The situation is reversed in the case of 2-toluenesulfonamide. The decrease in the ν sym. NH_2 (90 cm^{-1}) vibration is more than the ν asym. NH_2 (68 cm^{-1}) in going from the vapor phase to the solid phase. Moreover, these frequency decreases are much less than those for the other three sulfonamides. These data indicate that the strength of the intermolecular hydrogen bond ($NH_2 \cdots O_2S$) is less in the case of 2-toluenesulfonamide due to the steric factor of the 2-methyl group, which increases the spatial distance between the $NH_2 \cdots O_2S$ groups in the solid phase as compared to this case for the other sulfonamides. The data for 2,4,6-trimethylbenezenesulfonamide presented in Table 5.6a shows the same correlation as for 2-toluenesulfonamide.

The ν S—N mode is assigned in the range 846–918 cm^{-1} and NH_2 bending is assigned in the range 1545–1569 cm^{-1}. The ν S—N frequency occurs at lower frequency in the vapor phase than in the solid phase. In the case of NH_2 bending there is no consistent spectra-structure correlation with change in physical phase.

SECONDARY AND TERTIARY SULFONAMIDES

The νN–H frequency for N-butyl 4-toluenesulfonamide and 4-toluenesulfonanilide in the vapor phase are assigned at 3419 cm^{-1} and 3402 cm^{-1}, respectively, as listed in Table 5.6a.

Table 5.7 lists IR data for secondary and tertiary sulfonamides. In the vapor phase, compounds of form R-SO_2–NHR′ (where R and R′ can be both alkyl, both aryl, or one aryl and one alkyl group) exhibit ν asym. SO_2 in the range 1344–1360 cm^{-1}. The compound whose empirical structure is presented here:
appears to be intramolecularly hydrogen-bonded in the vapor phase, because its ν asym. SO_2 frequency occurs at 1325 cm^{-1}. Its ν sym. SO_2 vibration is assigned at 1150 cm^{-1} in the vapor phase, and ν sym. SO_2 for the other compound occurs at higher frequency in the region 1160–1175 cm^{-1}. In the case of the one secondary sulfonamide studied in both the vapor and solid

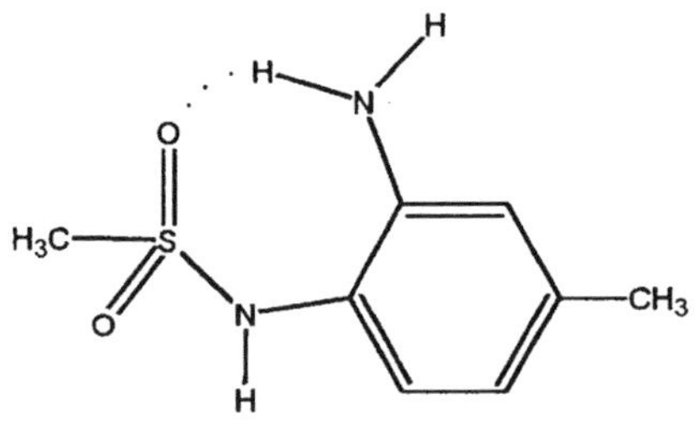

states, ν asym. SO_2 decreases 66 cm^{-1} while ν sym. SO_2 decreases 10 cm^{-1}.

Tertiary sulfonamides of form R-$SO_2N(CH_3)_2$, where R is either CH_3 or aryl, exhibit ν asym. SO_2 in the range 1362–1365 cm^{-1} in the vapor phase and in the range 1324–1334 cm^{-1} in the solid phase. The ν asym. SO_2 vibration occurs in the range 1165–1169 cm^{-1} in the vapor phase and in the range 1142–1164 cm^{-1} in the solid phase. Again ν asym. SO_2 vibration shifts more the ν sym. SO_2 vibration in going from the vapor phase to the liquid phase.

The secondary sulfonamides exhibit ν S–N in the range 825–920 cm^{-1}, and the tertiary sulfonamides in the range 770–815 cm^{-1} from the limited spectra included in the study.

The νNH frequencies for the secondary sulfonamides in the vapor phase (excluding the intramolecularly hydrogen-bonded compound just discussed) occur in the range 3400–3422 cm^{-1}. In the series, 4-$CH_3C_6H_4SO_2NHR$ (where R is CH_3, n-CH_4H_9 and tert-C_4H_9) νN–H decreases in the order 3422 cm^{-1}1, 3418 cm^{-1}, and 3400 cm^{-1}. This is the order of increased branching on the nitrogen α-carbon atom which is also on the order of increasing electron release of the alkyl group to the nitrogen atom. The increasing electron release steadily weakens N–H bond, causing νN–H to vibrate at increasingly lower frequency.

ORGANIC SULFONATES

Table 5.8 lists IR data for organic sulfonates. Sulfonates of form CH_3SO_2OR and $C_6H_5SO_2OR$ in the vapor-phase exhibit ν asym. SO_2 in the ranges 1368–1380 cm^{-1} and 1389–1391 cm^{-1}, respectively. Moreover, these same two classes of sulfonates in the vapor phase exhibit ν sym. SO_2 in the ranges 1185–1190 cm^{-1} and 1188–1195 cm^{-1}, respectively.

The two examples in which vapor-phase and liquid phase data are compared show that the shift in frequency in going from the vapor phase to the liquid phase for these sulfonates is larger for the ν asym. SO_2 vibration than for the ν sym. SO_2 vibration.

Vibrations involving stretching of the C—O—S bonds are assigned in the ranges 934–1019 cm^{-1} and 751–810 cm^{-1}.

The frequency separation between ν asym. SO_2 and ν sym. SO_2 in the vapor phase increases in the order: CH_3SO_2OR, 188–192 cm^{-1}; $C_6H_5SO_2OR$, 195–197 cm^{-1}; and 4-$CH_3C_6H_4SO_2OR$, 200–202 cm^{-1}.

ORGANOSULFONYL CHLORIDES

Table 5.9 lists IR data for organosulfonyl chlorides. Compounds of forms RSO_2Cl and $ArSO_2Cl$ in the vapor phase exhibit ν asym. SO_2 in the range 1390–1408 cm^{-1} and ν sym. SO_2 in the range 1179–1192. In the liquid or solid phase, these same sulfonyl chlorides exhibit ν asym. SO_2 and ν sym. SO_2 in the regions 1362–1382 cm^{-1} and 1170–1174 cm^{-1}. With change in physical phase from vapor to neat phases, the ν asym. SO_2 vibration shifts more in frequency than the ν sym. SO_2 vibration.

An IR band in the region 529–599 cm^{-1} is assigned to ν SCl, and the frequency separation between ν asym. SO_2 and ν sym. SO_2 varies between 208–236 cm^{-1} in the vapor phase and 191–212 cm^{-1} in the neat phase.

ORGANOSULFONYL FLUORIDES

Table 5.9a lists IR data for organosulfonyl fluorides of form $ArSO_2F$. In the vapor phase, ν asym. SO_2 and ν sym. SO_2 occur in the ranges 1428–1441 cm^{-1} and 1211–1235 cm^{-1}, respectively. An exception here is the sulfonyl fluoride substituted in the 2-position with an NH_2 group, because ν asym. SO_2 and ν sym. SO_2 occur at 1411 and 1204 cm^{-1}, respectively. These normal modes occur at lower frequency due to intramolecular hydrogen bonding as presented here:

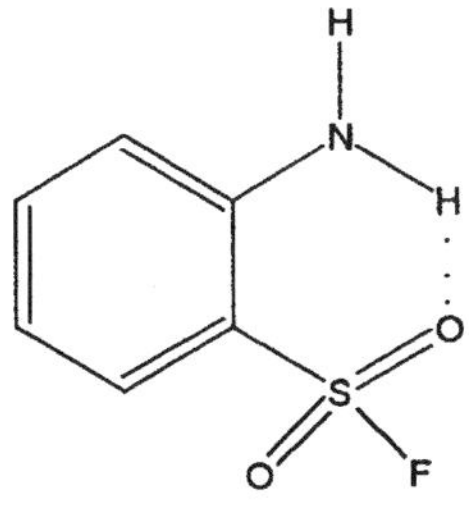

A

In the liquid phase, ν asym. SO_2 and ν sym. SO_2 occur in the ranges 1399–1420 cm^{-1} and 1186–1214 cm^{-1}, respectively. Compound *A* in the liquid phase exhibits ν asym. SO_2 and ν sym. SO_2 at 1384 cm^{-1} and 1186 cm^{-1}, respectively. In all cases the ν asym. SO_2 vibration shifts more in frequency than the ν sym. SO_2 vibration in going from the vapor phase to the liquid phase.

The band in the region 750–811 cm^{-1} is assigned to ν S—F in either the liquid or vapor phase. However, ν S—F occurs at higher frequency in the vapor phase compared to the liquid phase. The frequency separation between ν asym. SO_2 and ν sym. SO_2 varies between 207 and 219 cm^{-1} in the vapor phase and between 190 and 206 cm^{-1} in the liquid phase.

SUMMARY OF S=O AND SO_2 STRETCHING FREQUENCIES IN DIFFERENT PHYSICAL PHASES

Tables 5.10 and 5.11 list IR data for S=O and SO_2 containing compounds in different physical phases. In all cases the ν S=O or ν SO_2 vibrations occur at lower frequency in the neat phase than in the vapor phase.

ν asym. SO_2 AND ν sym. SO_2 CORRELATIONS

Figure 5.2 is a plot of ν asym. SO_2 vs ν sym. SO_2 vapor-phase frequencies for 13 compounds. This plot shows that there is an essentially linear relationship between these two ν SO_2 vibrations (24).

Figures 5.3 and 5.4 are plots of ν asym. SO_2 and ν sym. SO_2, respectively, vs the summation of the inductive σ' values of the atoms or groups joined to the SO_2 group. These plots show that in general both ν SO_2 vibrations decrease in frequency as the electron release of the two groups joined to the SO_2 group decreases in value (24).

SO_2 STRETCHING VIBRATIONS IN $CHCl_3/CCl_4$ SOLUTIONS

Figure 5.5 shows plots of the ν asym. and ν sym. SO_2 frequencies for 1 wt./vol. % in mole % $CHCl_3/CCl_4$ solutions (36). Both ν SO_2 vibrations decrease in frequency as the mole % $CHCl_3/CCl_4$ is increased. The ν asym. SO_2 vibration occurs in the range 1317.7–1325.7 cm^{-1} and the ν sym. SO_2 vibration occurs in the range 1153.2–1157.4 cm^{-1}. Over the 0 to 100 mol % $CHCl_3/CCl_4$ the ν asym. SO_2 frequency shifts to lower frequency by ~8 cm^{-1} while the ν sym. SO_2 frequency shifts to lower frequency by ~4 cm^{-1}. With the first introduction of $CHCl_3$ to CCl_4 the ν SO_2 vibrations show a marked decrease in frequency, and after ~35 mol % $CHCl_3/CCl_4$ both plots are essentially linear. This results from the formation of intermolecular hydrogen bonding between $SO_2 \cdots HCCl_3$, which lowers the ν SO_2 frequencies. The continued decrease in both ν SO_2 frequencies is the result of the increased RF surrounding methyl phenyl sulfone molecules. Figure 5.6 shows a plot of ν asym. SO_2 vs ν sym. SO_2 for these same solutions. This plot is not linear over the 0–100 mol % $CHCl_3/CCl_4$. The break in linearity reflects the formation of different complexes with the methyl phenyl sulfone molecules. This could be where $CHCl_3$ also intermolecularly hydrogen bonds with the π system of the phenyl group as the mole % $CHCl_3/CCl_4$ is increased.

Figure 5.7 shows plots of ν asym. SO_2 and ν sym. SO_2 frequencies for 1 wt./vol. % in mole % $CDCl_3/CCl_4$ solutions (36). Both ν SO_2 frequencies decrease as the mole % $CDCl_3/CCl_4$ is increased. With the first addition of $CDCl_3$, intermolecular deutero bonds are formed between SO_2 groups of dimethyl sulfate and $CDCl_3$ (viz. $SO_2 \cdots DCCl_3$). After formation of the intermolecular $SO_2 \cdots DCCl_3$ complexes, the ν SO_2 vibrations decrease essentially in a linear manner due to the increased RF surrounding dimethyl sulfate molecules.

Figure 5.8 shows a plot of ν asym. SO_2 vs ν sym. SO_2 for 1% solutions of dimethyl sulfate in 0 to 100 mol % $CDCl_3/CCl_4$. This plot shows that both ν asym. SO_2 and ν sym. SO_2 decrease essentially in a linear manner as the mole % $CDCl_3/CCl_4$ is increased.

CALCULATED ν SO_2 FREQUENCIES

Using dimethyl sulfone ν SO_2 vapor-phase frequencies as the standard, two equations were developed for calculating ν asym. SO_2 and ν sym. SO_2 frequencies within $\pm 7\,cm^{-1}$ (43). These equations are:

$$\nu\text{asym. } SO_2 = 1351\ cm^{-1} + \Sigma(X + Y \text{ shift constants in } cm^{-1})$$

$$\nu\text{sym. } SO_2 = 1160 + \Sigma(X + Y \text{ shift constants in } cm^{-1})$$

For those interested in applying these equations to predict ν SO_2 frequencies, the series of shift constants are available in Reference 43.

SULFINYL ANILINES

These sulfinyl anilines have the empirical structure presented here:
These molecules have ν asym. N=S=O and ν sym. N=S=O vibrations, where the ν asym. N=C=O has strong IR band intensity and the ν sym. N=C=O IR band has medium intensity. The ν asym. N=S=O and ν sym. N=S=O vibrations occur in the ranges 1254–1290 cm^{-1} and

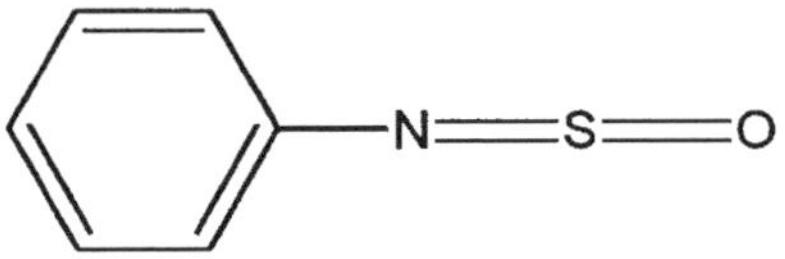

B

1157–1173 cm^{-1}, respectively (see Table 5.12 for details). In the cases where both modes of a sulfinyl aniline derivative were recorded in two physical phases, ν asym. N=S=O shifts more in frequency (24 cm^{-1}) than ν sym. N=S=O (14 cm^{-1}) in going from the vapor phase to CS_2 solution.

REFERENCES

1. Nyquist, R. A. (1984). *The Interpretation of Vapor-Phase Infrared Spectra: Group Frequency Data*. Philadelphia: Bio-Rad Lab. Inc., Sadtler Div., p. 509.
2. Dollish, F. R., Fateley, W. G., and Bentley, F. F. (1974). *Characteristic Raman Frequencies of Organic Compounds*. New York: John Wiley & Sons, pp. 46–53.
3. Bellamy, L. J. (1975). *The Infrared Spectra of Complex Molecules*. New York: John Wiley & Sons, pp. 394–410.
4. Lin-Vien, D., Colthup, N. B., Fateley, W. G., and Grasselli, J. G. (1991). *The Handbook of Infrared and Raman Characteristic Frequencies of Organic Molecules*. Boston: Academic Press, Inc., p. 241.
5. Nyquist, R. A. (1984). *The Interpretation of Vapor-Phase Infrared Spectra: Group Frequency Data*. Philadelphia: Bio-Rad Lab. Inc. Sadtler Div., p. 495.
6. Nyquist, R. A. (1984). pp. 479, 517.
7. Klaeboe, P. (1968). *Acta Chem. Scand.*, **22**: 2817.
8. Paetzold, R. and Ronsch, E. (1970). *Spectrochim. Acta*, **26A**: 569.
9. Nyquist, R. A., Putzig, C. H., and Leugers, M. A. (1997). *Handbook of Infrared and Raman Spectra of Inorganic Compounds and Organic Salts*. Vol. 3, Boston: Academic Press, p. 285.
10. Nyquist, R. A. *et al.* (1997). *Ibid.*, p. 235.
11. Nyquist, R. A. (1984). *The Interpretation of Vapor-Phase Infrared Spectra: Group Frequency Data*. Philadelphia: Bio-Rad Lab. Inc., Sadtler Div., p. 480.
12. Bender, P. and Wood, Jr. J. M. (1955). *J. Chem. Phys.*, **22**: 1316.
13. Martz, D. E. and Lagemann, R. T. (1956). *J. Chem. Phys.*, **25**: 1277.
14. Stammreich, H., Forneris, R., and Tavares, Y. (1956). *J. Chem. Phys.*, **25**: 1277.
15. Geisler, G. and Hanschmann, G. (1972). *J. Mol. Struct.*, **11**: 283.
16. Detoni, S. and Hadzi, D. (1957). *Spectrochim. Acta*, **11**: 601.
17. Ubo, T., Machida, K., and Hanai, K. (1971). *Spectrochim. Acta*, **27A**: 107.
18. Fawcett, A. H., Fee, S., Stuckey, M., and Walkden, P. (1987). *Spectrochim. Acta*, **43A**: 797.
19. Nyquist, R. A., Putzig, C. H., and Leugers, M. A. (1997). *Handbook of Infrared and Raman Spectra of Inorganic Compounds and Organic Salts*, Vol. 13, Boston: Academic Press, p. 225.
20. Nyquist, R. A. (1984). *The Interpretation of Vapor-Phase Infrared Spectra: Group Frequency Dates*, Philadelphia: Bio-Rad Lab. Inc., Sadtler Div., p. 495.
21. Nyquist, R. A. (1984). *Ibid.*, p. 496.
22. Nyquist, R. A. (1984). p. 489.
23. Nyquist, R. A. (1984). p. 490.
24. Nyquist, R. A. (1984). p. 497.
25. Nyquist, R. A. (1984). p. 474.
26. Nyquist, R. A. (1984). p. 485.
27. Nyquist, R. A. (1984). p. 484.
28. Bouquet, M., Chassaing, G., Corset, J., Favort, J., and Limougi, J. (1981). *Spectrochim. Acta*, **37A**: 727.
29. Joshi, U. C., Joshi, M., and Singh, R. N. (1981). *Ind. J. Pure & Appl. Phys.*, **19**: 1226.
30. Hanai, K., Okuda, T., Uno, T., and Machida, K. (1975). *Spectrochim. Acta*, **31A**: 1217.
31. Uno, T., Machida, K., and Hanai, K. (1966). *Spectrochim. Acta*, **22**: 2065.
32. Tanka, Y., Tanka, Y., and Saito, Y. (1983). *Spectrochim. Acta*, **39A**: 159.
33. Tanka, Y., Yanka, Y., Saito, Y., and Machida, K. (1978). *Bull. Chem. Soc. Jpn.*, **51**: 1324.
34. Hanai, K. and Okuda, T. (1975). *Spectrochim. Acta*, **31A**: 1227.
35. Kanesaki, I. and Kawai, K. (1970). *Bull. Chem. Soc. Jpn.*, **43**: 3298.
36. Nyquist, R. A., Sloane, H. J., Dunbar, J. E., and Strycker, S. J. (1966). *Appl. Spectrosc.*, **20**: 90.

37. Nyquist, R. A. (1990). *Appl. Spectrosc.*, **44**: 594.

38. (1977) *Sadtler Standard Infrared Vapor Phase Spectra*. Philadelphia: Bio-Rad Inc., Sadtler Div.

39. Nyquist, R. A., Putzig, C. H., and Leugers, M. A. (1997). *Handbook of Infrared and Raman Spectra of Inorganic Compounds and Organic Salts*. Vol. 3, Boston: Academic Press, p. 231.

40. Nyquist, R. A. *et al.* (1997). *Ibid.*, p. 232.

41. CS_2 Solution Data. The Dow Chemical Company, Analytical Laboratories.

42. Sportouch, S., Clark, R. J. H., and Gaufres, R. (1974). *J. Raman Spectrosc.*, **2**: 153.

43. Nyquist, R. A. (1984). *The Interpretation of Vapor-Phase Infrared Spectra: Group Frequency Data*. Philadelphia: Bio-Rad Lab. Inc., Sadtler Div., p. 506.

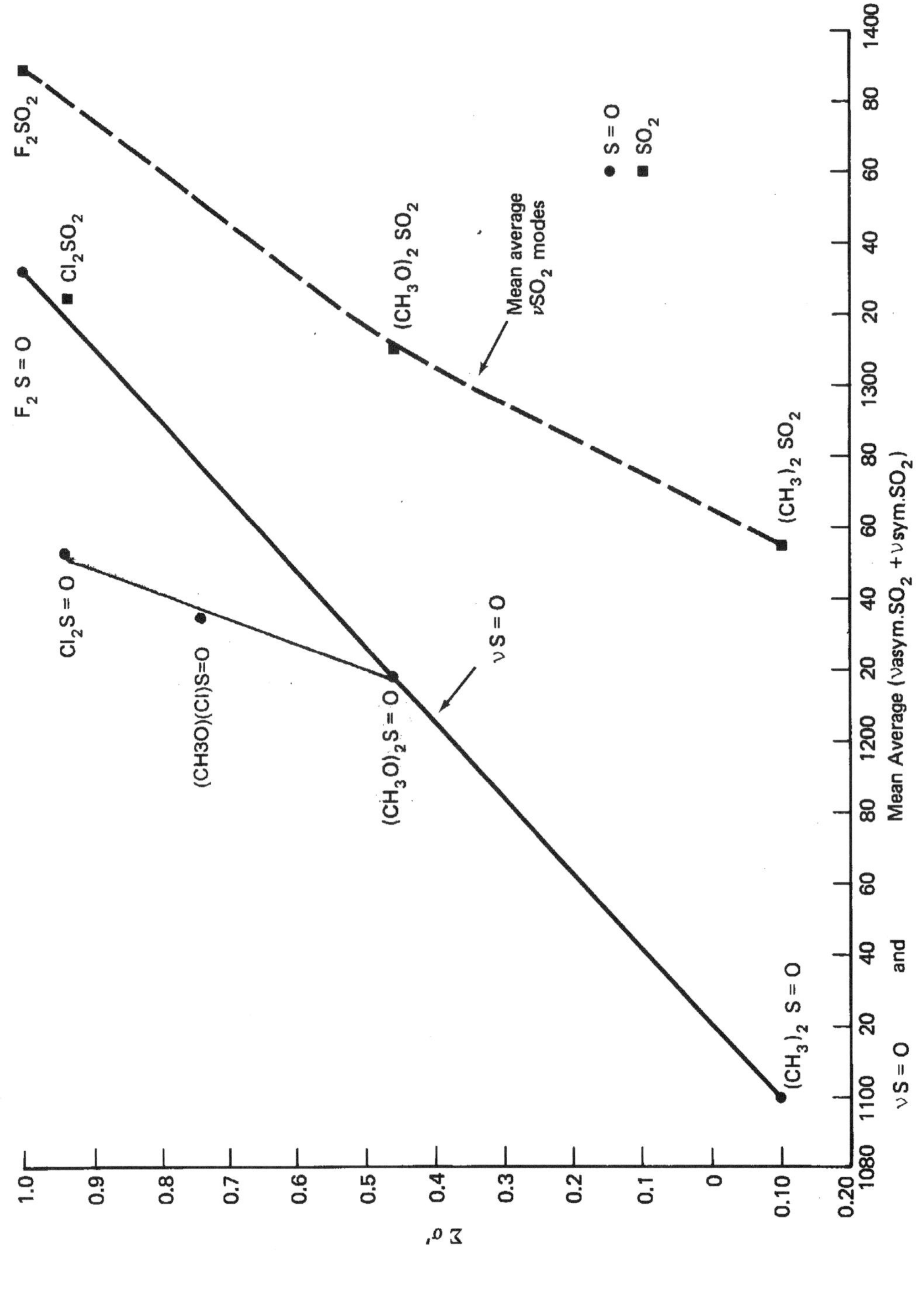

FIGURE 5.1 Plots of ν S=O and the mean average of (ν asym. SO_2 + ν sym. SO_2) vapor-phase frequencies for compounds containing the S=O or SO_2 group vs $\Sigma\sigma'$.

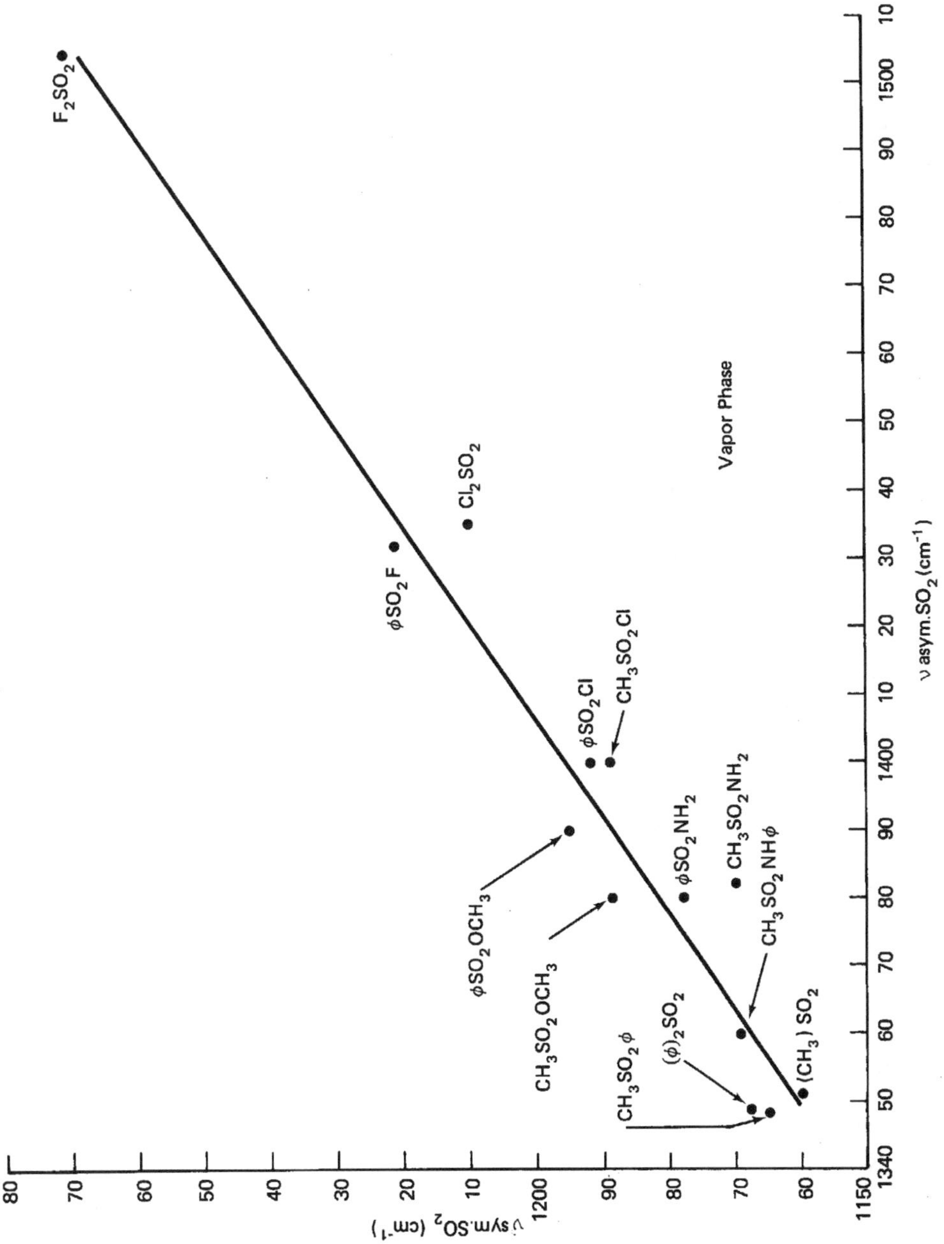

FIGURE 5.2 A plot of ν asym; SO_2 vs ν sym. SO_2 vapor-phase frequencies for a variety of compounds containing the SO_2 group.

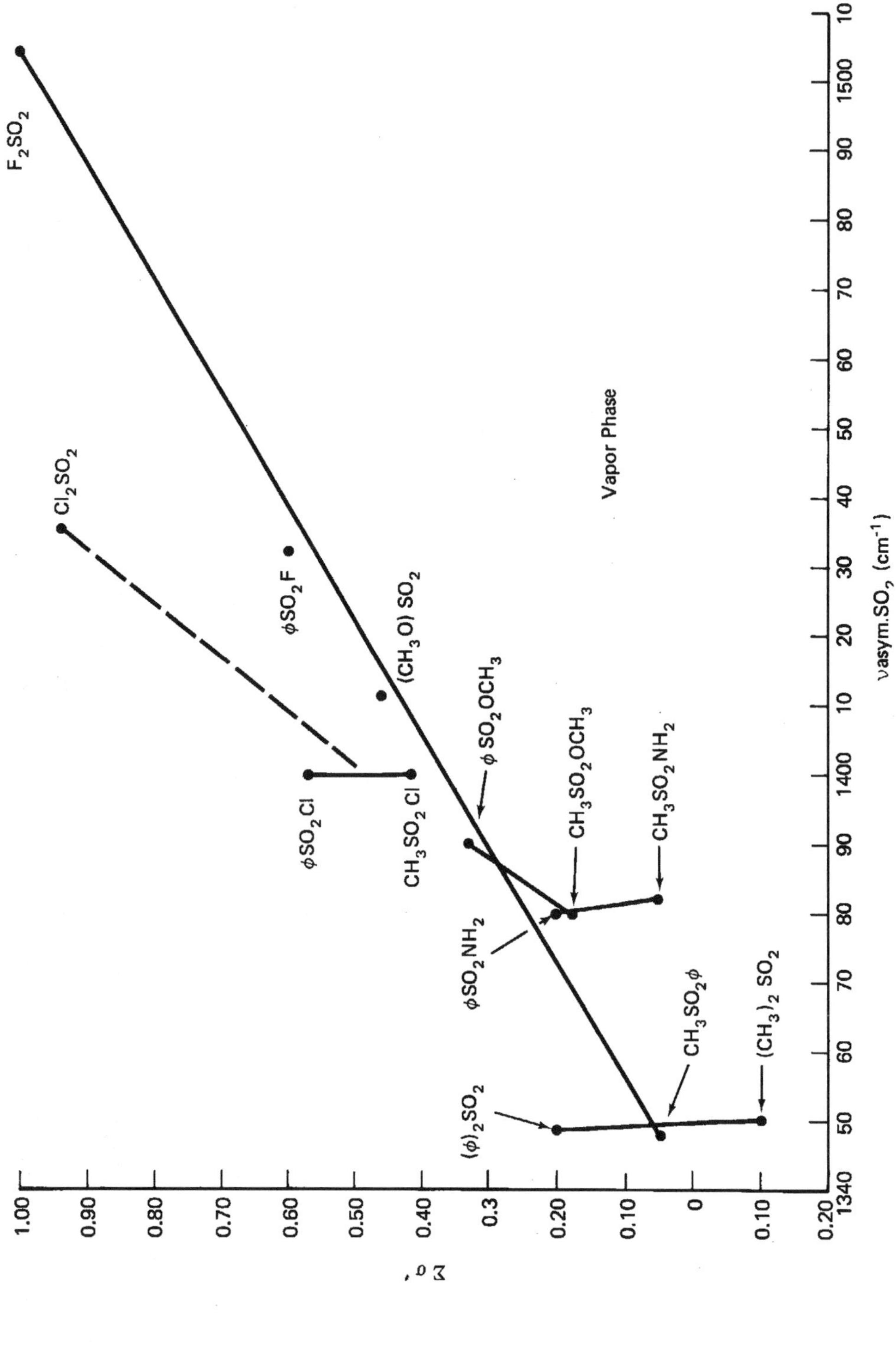

FIGURE 5.3 Plots of ν asym. SO_2 vapor-phase frequencies for a variety of compounds containing a SO_2 group vs $\Sigma\sigma'$.

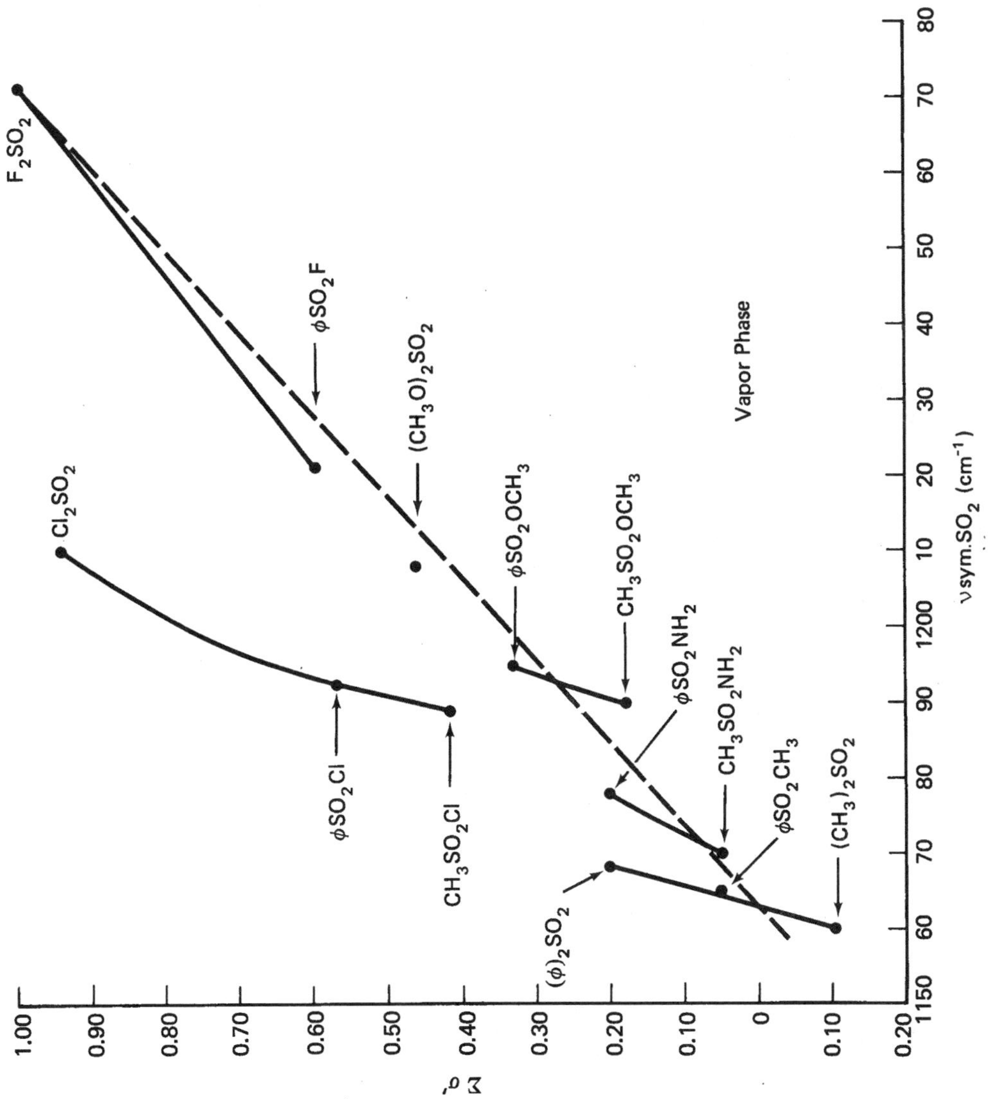

FIGURE 5.4 Plots of ν sym. SO_2 vapor-phase frequencies for a variety of compounds containing the SO_2 group vs $\Sigma\sigma'$.

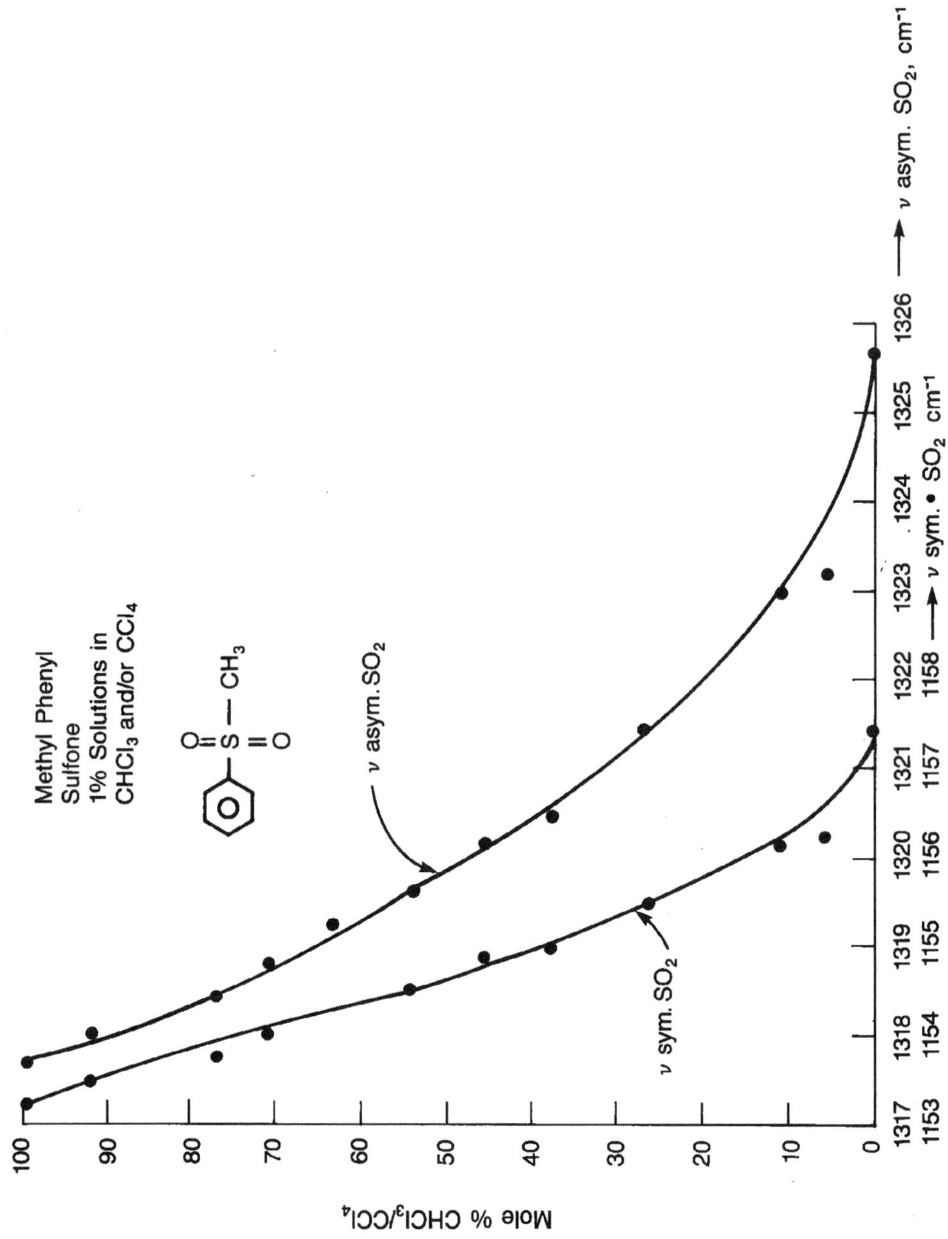

FIGURE 5.5 Plots of ν asym. and ν sym. SO_2 for methyl phenyl sulfone in 10 wt./vol. solvent vs mole % $CHCl_3/CCl_4$.

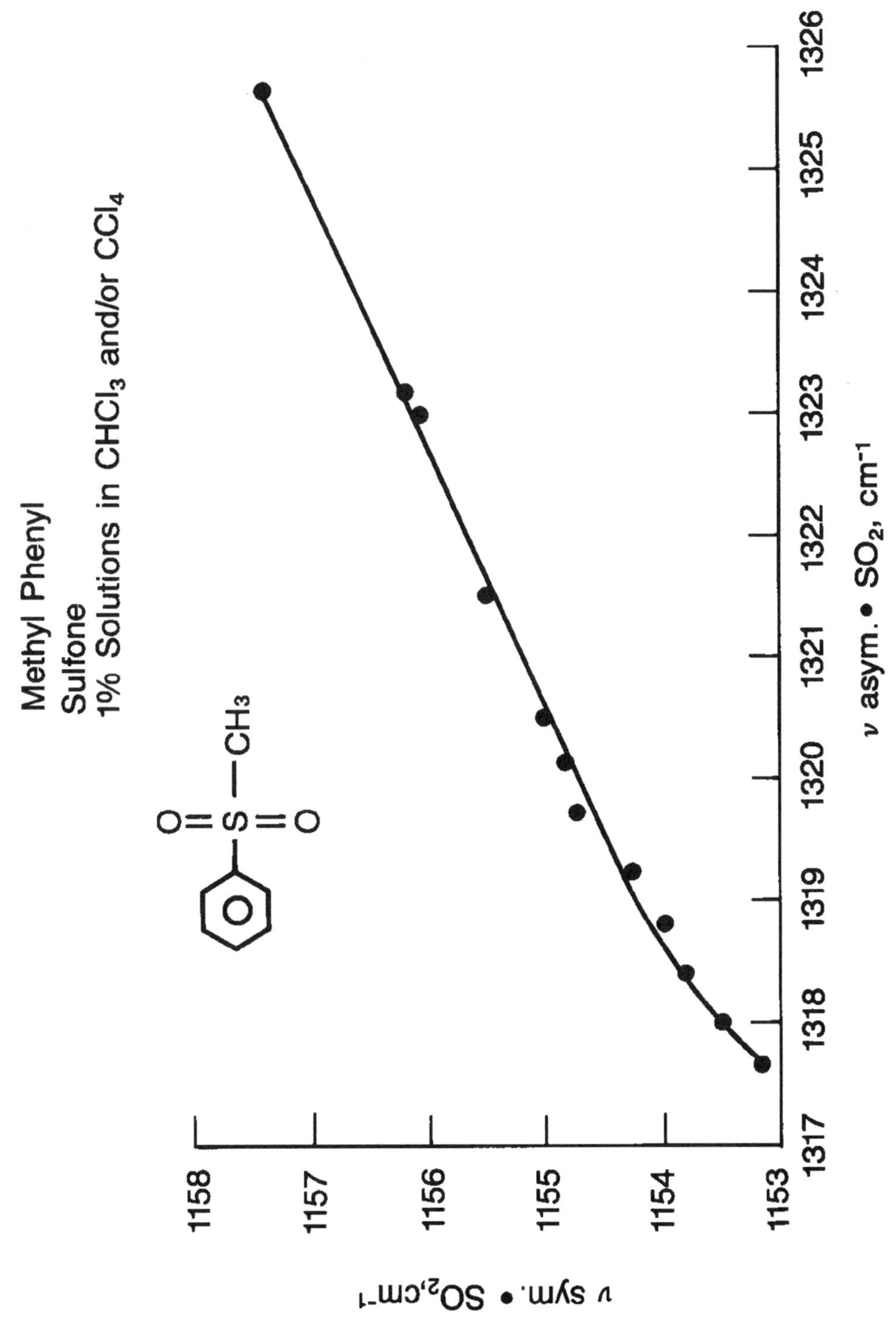

FIGURE 5.6 A plot of ν asym. SO_2 vs ν sym. SO_2 for methyl phenyl sulfone in 1% wt./vol. solvent vs mole % $CHCl_3/CCl_4$.

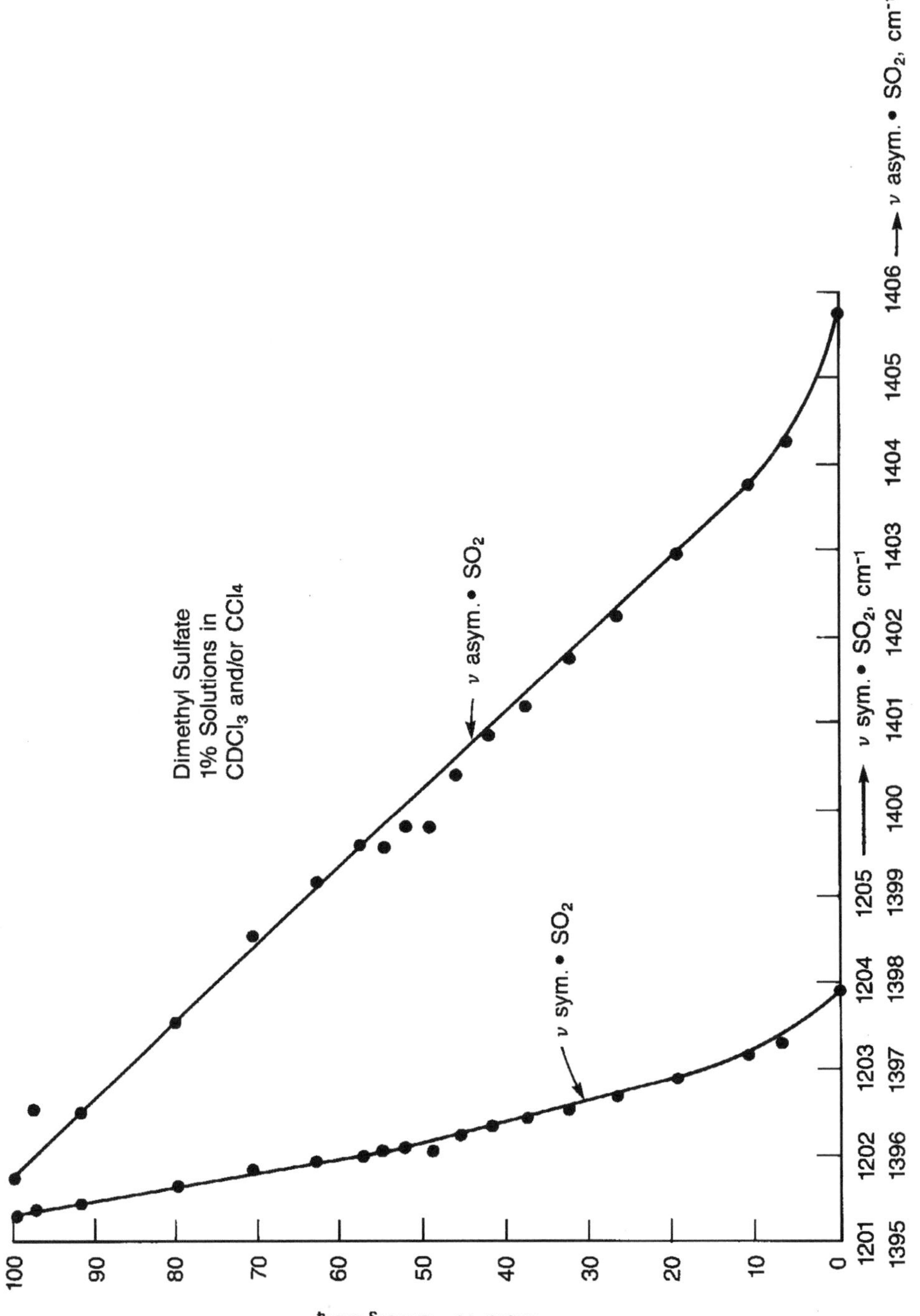

FIGURE 5.7 Plots of ν asym. SO_2 and sym. SO_2 for dimethyl sulfate in 1% wt./vol. solvent vs mole % $CDCl_3/CCl_4$.

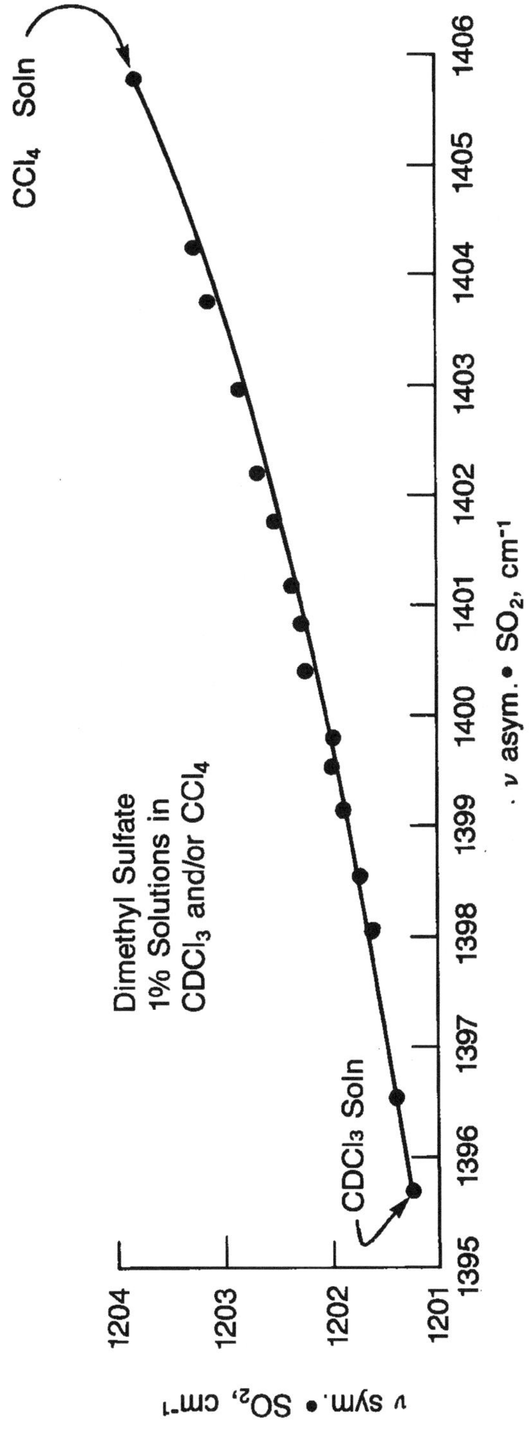

FIGURE 5.8 A plot of ν asym. SO_2 vs ν sym. SO_2 for dimethyl sulfate 1% wt./vol. solvent vs mole % $CDCl_3/CCl_4$.

TABLE 5.1a A comparison of the S=O stretching frequencies for S=O containing compounds

Compound or compound type	νS=O vapor cm^{-1}	References	νS=O Ref. 3 cm^{-1}	IR	R	References
$(R\text{-})_2S{=}O$	[—]		1035–1070	vs	w-m	2, 3, 4
$(CH_3\text{—})_2S{=}O$	1100	5				
$(CD_3\text{—})_2S{=}O$	1092	5				
$(R\text{-})S({=}O)(\text{-}Ar)$	[—]		1040–1050	vs	w-m	4
$(Ar\text{—})_2S{=}O$	[—]		1035–1042	vs	w-m	2, 3, 4
$(CH_3\text{—}O\text{—})_2S{=}O$	1218	6	1207	vs	vs, p	7
$(R\text{-}O\text{—})_2S{=}O$	1209–1218	6	1198–1209	s		2, 3, 4
$(CH_2\text{—}O)_2S{=}O$	1240	6				
	1230					
$[(CH_3)_2N]_2S{=}O$	[—]		1108	vs	m	8
$(CH_3)_2NS({=}O)Cl$	[—]		1185	vs	m	8
$(R\text{-}O\text{—})S({=}O)Cl$	[—]		1214–1221			2, 4
$(CH_3\text{—}O\text{—})S({=}O)Cl$	1232	9				
$F_2S{=}O$	1331	10	1308			12
$Cl_2S{=}O$	1251	10	1253			13
$Br_2S{=}O$	[—]		1121			14

TABLE 5.1b A comparison of asym. SO_2 and sym. SO_2 stretching frequencies in different physical phases, and asym. N=S=O and sym. N=S=O frequencies in CS_2 solution

Compound or compound Type	ν asym. SO_2 vapor cm^{-1}	ν sym. SO_2 vapor cm^{-1}	Ref.	ν asym. SO_2 Ref. 3 cm^{-1}	IR	R	ν sym. SO_2 Ref. 3 cm^{-1}	IR SO_2	R SO_2	Ref.	[ν asym. SO_2 vapor]-[ν asym. neat or CCl_4] cm^{-1}	[ν sym. SO_2 vapor]-[ν sym. neat or CCl_4] cm^{-1}
SO_2	1360	1150	19									
$(R-)_2SO_2$	1335–1371	1141–1160	20	1295–1330	vs	w-m	1125–1152	vs	vs, p	2, 3, 15–18	[40]–[41]	[9]–[8]
$(Ar-)SO_2(-R)$	1346–1349	1159–1172	21	1325–1334	s	w	1150–1160	s	w	3, 28	[21]–[15]	[9]–[12]
$(Ar)_2SO_2$	1351–1357	1158–1169	21	1328		vs	1162		vs, p	29	[23]–[29]	[−4]–[−7]
$(R-)SO_2(-OH)$	[—]	[—]		1342–1352	vs		1150–1165			16		
$(R-)SO_2(-OR)$	[—]	[—]		1352–1358		m	1165–1172		vs, p	2, 4		
$(CH_3-)SO_2(-OR)$	1368–1380	1185–1190	22									
$(Ar-)SO_2(-OR)$	1390–1400	1188–1195	22	1338–1363		w	1185–1192		vs	2, 4	[52]–[37]	[3]–[3]
$(Ar-)SO_2(-OAr)$	1402–1409	1178–1190	23									
$Cl-SO_2(-OR)$				1401–1406		m-s	1184–1191	vs	2, 4			
$(R-)SO_2(-SR)$				1305–1344	m-s	1126–1128		s	2, 4			
$(RO-)_2SO_2$	1408–1420	1200–1219	24	1372–1388		s	1188–1196		vs	2, 4	[36]–[32]	[12]–[26]
$(Ar-As-O-)SO_2(-OR$ or $-OAr)$				1325	s		1161–1171	s		34	[69]	[25]
$(CH_3-)SO_2(-NH_2)$	1382	1170	25	1315	vs	m	1145	vs	vs	28		
$(C_2H_5-)SO_2(NH_2)$	1375	1161	25									
$(C_2H_5-)SO_2(NH_2)$	1380	1178	25				1157		s	2, 4	[?]	[21]
$(Ar-)SO_2(NH_2)$	1370–1380	1172–1178	25									
$(CH_3-)SO_2(-NHC_6H_5)$	1360	1160	25									
$(CH_3-)SO_2$ $(-2NH_2-4-C_6H_3)$	1325	1150	25									
$(Ar-)SO_2(-NHR)$	1344–1366	1160–1175	26									
$(CH_3-)SO_2(-N(CH_3)_2)$	1362	1165	26									
$(Ar-)SO_2(-N(CH_3)_2)$	1365	1169	26									
$(NH_2)_2SO_2$				1358	vs		1156	vs		31		
$((CH_3)_2N-)SO_2(-NH_2)$				1335	vs	m	1140	s	s	32		
$((CH_{32}N-)SO_2(-Cl)$	1411	1190	36	1385	vs	m, dp	1177	vs	s, p	33	[26]	[13]
$(CH_3-)SO_2(-F)$				1401		m	1186		vs	2, 4		
$(Ar-)SO_2(-F)$	1428–1442	1211–1229	26	1402–1412		w-m	1167–1197		vs	2, 4	[18]–[32]	[11]–[27]

(*continues*)

TABLE 5.1b (*continued*)

Compound or compound Type	ν asym. SO_2 vapor cm^{-1}	ν sym. SO_2 vapor cm^{-1}	Ref.	ν asym. SO_2 Ref. 3 cm^{-1}	IR	R	ν sym. SO_2 Ref. 3 cm^{-1}	IR SO_2	R SO_2	Ref.	[ν asym. SO_2 vapor]-[ν asym. neat or CCl_4] cm^{-1}	[ν sym. SO_2 vapor]-[ν sym. neat or CCl_4] cm^{-1}
(R-)SO_2(−Cl)	1400–1402	1177–1189	27	1366	vs	w	1171	vs	s, p	2, 4	[34]–[36]	[6]–[5]
(Ar−)SO_2(−Cl)	1390–1410	1172–1192	27	1361–1384		w-m	1169–1184		vs	2, 4	[29]–[25]	[3]–[8]
(Cl−)SO_2(−N=C=O)				1442	vs	m	1182	vs	m	35		
(F−)$_2SO_2$				1497; [1503]		w	1263; [1270]		vs	2, 4	[42]	
(Cl−)$_2SO_2$	1438	1210	39	1414		s	1182		s	2, 3	[24]*	[28]*
(Cl−)$_2SO_2$	[1420, CCl_4 soln.]	[1191, CCl_4 soln.]	40								[18]*	[19]*
				ν asym. N=S=O cm^{-1}			ν sym. N=S=O cm^{-1}					
(Ar−)N=S=O [in CS_2 soln.]				1254–1290	s		1158–1175	m		41		

* An exception (see text).

TABLE 5.1c Vapor-phase infrared data for dimethyl sulfoxide and dialkyl sulfones

Sulfone Di	$a.CH_3$ str. cm^{-1} (A)	$a.CH_2$ str. cm^{-1} (A)	$s.CH_3$ str. cm^{-1} (A)	$s.CH_2$ cm^{-1} (A)	$a.CH_3$ bend cm^{-1} (A)	CH_2 bend cm^{-1} (A)	$s.CH_3$ bend cm^{-1} (A)	CH_3 rock cm^{-1} (A)	$a.(C)_2S$ str. cm^{-1} (A)	$s.(C)_2S$ str. cm^{-1} (A)	$a.SO_2$ str. cm^{-1} (A)	$s.SO_2$ str. cm^{-1} (A)	Ratio (A) $s.SO_2$ str./ $s.CH_2$ str.	Ratio (A) $s.SO_2$ str./ $a.CH_3$ str.	Ratio (A) $s.SO_2$ str./ $s.(C)_2$ str.	Ratio (A) $s.SO_2$ str./ $a.(C)_2$ str.	Ratio (A) $s.(C)_2$ str./ $a.(C)_2$ str.
Methyl	3022 (0.026)		2942 (0.039)		1425 (0.044)		1311 (wk)	932 (0.460)	742 (0.270)	680 (0.093)	1351 (1.250)	1160 (1.040)		40.01	11.18	3.85	0.34
Propyl	2979 (0.930)	2943 (1.230)		2890 (0.480)	1465 (0.185)	1415 (0.125)			779 (0.195)	699 (0.205)	1335 (1.250)	1145 (1.250)	2.61	1.34	6.09	6.41	1.05
Butyl	2975 (1.240)	2942 (0.770)		2885 (0.430)	1464 (0.180)	1414 (0.120)			810 (0.130)	700 (0.130)	1338 (1.240)	1142 (1.040)	2.42	0.84	8.01	8.01	1.01
Hexyl	2970 (0.835)	2935 (1.240)		2875 (0.395)	1461 (0.150)	1412 (0.070)			812 (0.080)	702 (0.080)	1338 (0.680)	1140 (0.580)	1.47	0.69	7.25	7.25	1.01
											[SO str.]		Ratio (A) SO str./ $s.CH_3$ str.	Ratio (A) SO str./ $a.CH_3$ str.			
Sulfoxide																	0.45
Methyl	3000 (0.152)		2921 (0.115)		1441 (0.170)	1420 (0.135)	1310 (0.086)	930 (0.135)	755 (0.020)	669 (0.169)	[1100 (1.220)]		10.61	8.03			

TABLE 5.1d Vapor- and solid-phase infrared data for dimethyl sulfoxide and dialkyl sulfones

Sulfone Di	Phase	a.$(C)_2$ str. cm^{-1} (A)	s.$(C)_2S$ str. cm^{-1} (A)	a.SO_2 str. cm^{-1} (A)	s.SO_2 str. cm^{-1} (A)	[a.SO_2 str.]-[s.SO_2 str.] cm^{-1}	Ratio (A) A[s.SO_2 str.]/A[s.$(C)_2S$ str.]	Ratio (A) A[s.SO_2 str.]/A[a.$(C)_2$ str.]	Ratio (A) A[s.$(C)_2$ str.]/A[a.$(C)_2$ str.]
Methyl	vapor	742 (0.270)	680 (0.093)	1351 (1.250)	1160 (1.040)	191	11.18	3.85	0.34
Methyl	solid	750 (0.765)	685 (0.245)	1286 (1.145)	1124 (1.031)	162	4.21	1.35	0.33
	Δ [v-s]	[8]	[5]	[−65]	[−36]	[−29]			
Propyl	vapor	779 (0.195)	699 (0.205)	1335 (1.250)	1145 (1.250)	190	6.09	6.41	1.05
Propyl	solid	781 (0.300)	699 (0.170)	1309 (stg)	1127 (stg)	182			0.57
				1280 (stg)		153			
	Δ [v-s]	[2]	[0]	[−26]; [−55]	[−18]	[−8]; [−37]			
Butyl	vapor	810 (0.130)	700 (0.130)	1338 (1.240)	1142 (1.040)	196	8.01	8.01	1.01
Butyl	solid	810 (0.350)	699 (0.150)	1311 (stg)	1117 (stg)	194			0.43
				1290 (stg)		173			
	Δ [v-s]	[0]	[−1]	[−28]; [−48]	[−25]	[−2]; [−23]			
Hexyl	vapor	812 (0.080)	702 (0.080)	1338 (0.680)	1140 (0.580)		7.25	7.25	1.01
Sulfoxide				[SO str.]					
Methyl	vapor	755 (0.020)	669 (0.169)	[1100 (1.220)]					0.45
Methyl	liquid	695 (0.240)	664 (0.094)	1050 (1.050)					0.39
	Δ [v-l]	[−60]	[−5]	[−50]					

TABLE 5.2 Vapor- and solid-phase infrared data for diaryl sulfones

Compound Diaryl sulfone	Phase	a.SO_2 str. cm^{-1} (A)	s.SO_2 str. cm^{-1} (A)	SO_2 bend cm^{-1} (A)	SO_2 wag cm^{-1} (A)	[a.SO_2 str.]-[s.SO_2 str.] cm^{-1}	A[s.SO_2 str.]/A[a.SO_2 str.]
(4-CH_3–C_6H_4–)SO_2	vapor	1348 (0.850)	1165 (1.240)	678 (1.024)	558 (0.950)	183	1.46
(4-CH_3–C_6H_4–)SO_2	solid	1318 (0.840)	1152 (1.150)	677 (0.930)	554 (0.830)	166	1.37
		1301 (0.611)			546 (0.835)	149	1.88
	Δ [v-s]	[−30]; [−47]	[−13]	[−1]	[−4]; [−12]	[−17]; [−34]	
(4-F–C_6H_4–)$_2SO_2$	vapor	1351 (0.470)	1161 (1.240)	680 (0.321)	550 (1.250)	190	2.64
(4-F–C_6H_4–)$_2SO_2$	solid	1325 (0.860)	1155 (1.168)	671 (0.392)	550 (1.032)	170	1.36
	Δ [v-s]	[−26]	[−6]	[−9]	[0]	[−20]	
(4-Cl–C_6H_4–)$_2SO_2$	vapor	1352 (0.534)	1169 (0.773)	631 (1.010)	581 (0.180)	183	1.45
(4-Cl–C_6H_4–)$_2SO_2$	solid	1319 (0.840)	1145 (1.240)	628 (0.855)	574 (0.575)	174	1.48
	Δ [v-s]	[−33]	[−24]	[−3]	[−7]	[−9]	
(4-NO_2–C_6H_4–)SO_2–	vapor	1354 (0.790)	1170 (0.750)	607 (1.240)	564 (0.255)	184	0.95
C_6H_5*	solid	1318 (0.650)	1164 (1.150)	608 (0.950)	564 (0.766)	154	1.77
		1301 (0.839)				137	
	Δ [v-s]	[−36]; [−53]	[−6]	[1]	[0]	[−30]; [−47]	

		a.NO_2 str.	s.NO_2 str.	A[s.NO_2 str.]/A[a.NO_2 str.]	[a.NO_2 str.]-[s.NO_2 str.]
(4-NO_2–C_6H_4)SO_2–C_6H_5	vapor	1548 (0.500)	1354 (0.970)	1.94	194
(4-NO_2–C_6H_4)SO_2–C_6H_5	solid	1529 (1.150)	1351 (0.950)	0.83	178
	Δ [v-s]	[−19]	[−3]		[−16]

TABLE 5.3 Infrared data for phenoxarsine derivatives containing the S(SO_2)R and O(SO_2)R groups in CS_2 solution

Phenoxarsine [CS_2]	a.SO_2 str.	s.SO_2 str.	C–S str.	SO_2 bend	SO_2 rock
X=S–SO_2-R					
R					
Methyl	1328	1137	740		
Ethyl	1322	1125	751	601	548
n-butyl	1320	1126	753	601	537
Phenyl	1322	1141	715	591	538
X=O–SO_2-R					
R					
Methyl	1325	1161			
Phenyl	1325	1171			

TABLE 5.4 Infrared data for the SO_2 stretching vibrations for compounds in $CHCl_3$ and CCl_4 solutions

Compound	a.SO_2 str. CCl_4 soln. cm^{-1}	a.SO_2 str. $CHCl_3$ soln. cm^{-1}	[CCl_4 soln.]–[$CHCl_3$ soln.] cm^{-1}	s.SO_2 str. CCl_4 soln. cm^{-1}	s.SO_2 str. $CHCl_3$ soln. cm^{-1}	[CCl_4 soln.]-[$CHCl_3$ soln.] cm^{-1}	[a.SO_2 str.]-[s.SO_2 str.] CCl_4 soln. cm^{-1}	[a.SO_2 str.]-[s.SO_2 str.] $CHCl_3$ soln. cm^{-1}	[CCl_4 soln.]-[$CHCl_3$ soln.] cm^{-1}
Dimethyl sulfate	1405.7	1395.7	−10	1203.8	1201.8	−2	201.9	193.9	−8
4′-Chlorophenyl 4-chloro-benzenesulfonate	1392.4	1382.9	−9.5	1178.9	1176.8	−2.1	213.5	206.1	−7.4
Benzenesulfonyl chloride	1387.4	1380.3	−7.1	1177.8	1177.1	−0.7	209.6	203.2	−6.4
Diphenyl sulfone	1326.8	1319.1	−7.7	1161.4	1157.6	−3.8	165.4	161.5	−3.9
Methyl phenyl sulfone	1325.7	1317.7	−8	1157.4	1153.2	−4.2	168.3	164.5	−3.8

TABLE 5.5 Infrared data for dialkyl sulfites

Compound R-O–SO–O-R R	Phase	S=O str. cm^{-1}	A	a.C–O–S str. cm^{-1}	A	s.C–O–S str. cm^{-1}	A
CH_3	vapor	1218	0.588	979	1.25	685	1.142
CH_3	liquid	1205	0.85	990	0.721	730	0.48
				955	0.81	685	0.58
	Δ [v-l]	[−13]		[11]; [−24]		[45]; [0]	
C_2H_5	vapor	1210	0.499	1010	0.84	695	0.735
C_2H_5	liquid	1204		1019		748	
				1001		717	
	Δ [v-l]	[−6]		[9]; [−9]		[55]; [22]	

TABLE 5.6 Infrared data for primary sulfonamides in the vapor and solid phases

Compound $R\text{-}SO_2NH_2$	Phase	a.NH_2 str. cm^{-1}	A	s.NH_2 str. cm^{-1}	A	NH_2 bending cm^{-1}	A	A(NH_2 bending)/ A(a.NH_2 str.)	A(NH_2 bending)/ A(s.NH_2 str.)
Methane	vapor	3472	0.19	3378	0.173	1553	0.237	1.25	1.37
Methane	solid	3335	0.98	3258	0.62	1569	0.09	0.09	0.15
	Δ [v-s]	[−137]		[−120]		[16]			
Benzene	vapor	3479	0.094	3378	0.135	1552	0.136	1.45	1.01
Benzene	solid	3339	0.659	3241	0.721	1547	0.225	0.34	0.31
	Δ [v-s]	[−140]		[−137]		[−5]			
4-Toluene	vapor	3478	0.048	3370	0.074	1552	0.08	1.67	1.08
4-Toluene	solid	3323	1.02	3238	0.94	1545	0.154	0.15	0.16
	Δ [v-s]	[−155]		[−132]		[−7]			
2-Toluene	vapor	3465	0.073	3365	0.103	1550	0.14	1.92	1.36
2-Toluene	solid	3397	0.56	3275	0.635	1558	0.175	0.31	0.28
	Δ [v-s]	[−68]		[−90]		[8]			

Compound	Phase	[a.NH_2 str.]-(s.NH_2 str.] cm^{-1}	a.SO_2 str. cm^{-1}	A	s.SO_2 str. cm^{-1}	A	SN str. cm^{-1}	A	A[s.SO_2 str.]/ A[a.SO_2 str.]	[a.SO_2 str.]-[s.SO_2 str.] cm^{-1}
Methane	vapor	94	1382	1.262	1170	1.035	856	0.402	0.82	212
Methane	solid	77	1308	1.101	1140	0.82	871	0.135	0.75	168
			1355	0.841	1158	0.65			0.31	197
	Δ [v-s]		[−74]		[−30]		[15]			
	Δ [v-s]		[−27]		[−12]					
Benzene	vapor	101	1380	0.848	1178	1.265	851	0.359	1.49	202
Benzene	solid	98	1325	0.97	1152	0.88	900	0.21	0.91	198
	Δ [v-s]		[−55]		[−26]		[45]			
4-Toluene	vapor	108	1380	0.523	1172	1.248	851	0.231	2.39	208
4-Toluene	solid	85	1325	0.96	1150	0.875	909	0.29	0.91	175
	Δ [v-s]		[−55]		[−22]		[58]			
2-Toluene	vapor	100	1370	1.1	1172	1.245	846	0.312	1.13	198
2-Toluene	solid	122	1311	1.03	1148	1.1	918	0.155	1.03	163
	Δ [v-s]		[−59]		[−24]		[72]			

TABLE 5.6a Infrared data for some NH and NH_2 vibrations for compounds containing SO_2NH_2 or SO_2NH groups in different physical phases

Amides	a.NH_2 str. cm^{-1}	s.NH_2 str. cm^{-1}	NH_2 bend cm^{-1}	Temperature	Physical phase
Methanesulfonamide	3470	3375	1552	280°C	vapor
	3335	~3255	~1568		KBr
	135	120	[−16]		vapor-KBr
Benzenesulfonamide	3480	3370	1551	280°C	vapor
	~3335	~3242	~1548		KBr
	145	128	[−17]		vapor-KBr
Benzenesulfonamide	3460	3360	1552		vapor
2,4,6-trimethyl	3370	3258	1551		KBr
	90	102	1		vapor-KBr
	NH str.				
p-Toluenesulfonamide	3419			280°C	vapor
N-butyl	~3370				KBr
	49				vapor-KBr
p-Toluenesulfonanilide	3402				vapor
					KBr
					vapor-KBr

In KBr it is $NH_2:O_2S$; str. or $NH:O_2S$ str.

TABLE 5.7 Infrared data for secondary and tertiary sulfonamides

Compound R-SO_2−NH-R′ R;R′	Phase	NH str. cm^{-1}	A	a.SO_2 str. cm^{-1}	A	s.SO_2 str. cm^{-1}	A	SN str. cm^{-1}	A	A[s.SO_2 str.]/ A[a.SO_2 str.]	[a.SO_2 str.]-[s.SO_2 str.] cm^{-1}
CH_3; C_6H_5	vapor	3420	0.07	1360	0.17	1169	1.241	890	0.18	7.3	191
CH_3; 2-NH_2-4-CH_3−C_6H_3	vapor	3298	0.189	1325	0.67	1150	1.202	881	0.135	1.79	175
C_6H_5; *n*-C_4H_9	vapor	3420	0.062	1355	0.363	1170	1.21	840	0.09	3.33	185
4-CH_3−C_6H_4; CH_3	vapor	3422	0.09	1355	0.49	1175	1.248	825	0.35	2.55	180
4-CH_3−C_6H_4; *n*-C_4H_9	vapor	3418	0.083	1356	0.42	1169	1.246	839	0.138	2.97	187
4-CH_3−C_6H_4; tert-C_4H_9	vapor	3400	0.09	1344	0.58	1160	1.22	852	0.2	2.1	184
4-CH_3−C_6H_4; C_6H_5	vapor	3402	0.07	1403	0.54	1170	1.22	888	0.18	2.56	233
4-CH_3−C_6H_4; C_6H_5	solid	3255	0.33	1337	0.32	1160	1.07	920	0.25	3.34	177
	Δ [v-s]	[−1.47]		[−66]		[−10]		[32]			
R-SO_2−N(−CH_3)$_2$											
R											
CH_3	vapor			1362	1.25	1165	1.049	770	0.7	0.84	197
CH_3	liquid			1324	stg.	1142	stg.	770	mstg.		182
	Δ [v-l]			[−38]		[−23]		[0]			
4-CH_3−C_6H_4	vapor			1365	0.7	1169	1.25	770	0.87	1.79	196
4-CH_3−C_6H_4	solid			1334	stg.	1164	mstg.	815	m		
	Δ [v-s]			[−31]		[−5]		[45]			

TABLE 5.8 Infrared data for organic sulfonates

Compound CH_3-SO_2-O-R R	Phase	a.SO_2 str. cm^{-1}	A	s.SO_2 str. cm^{-1}	A	A[s.SO_2 str.]/ A[a.SO_2 str.]	[a.SO_2 str.]-[s.SO_2 str.] cm^{-1}	a.COS str. cm^{-1}	A	s.COS str. cm^{-1}	A
CH_3	vapor	1380	1.25	1190	1.15	0.92	190	1011	1.2	795	0.95
C_2H_5	vapor	1378	1.25	1190	1.15	0.92	188	1020	0.76	790	0.47
n-C_4H_9	vapor	1377	1.16	1185	1.25	1.08	192	950	1.25	800	0.47
n-C_4H_9	liquid	1350	stg.	1173	mstg.		177	934	mstg.	807	m
	Δ [v-l]	[−27]		[−12]			[−15]	[−16]		[7]	
$nC_{18}H_{37}$	vapor	1368	0.089	1190	0.17	1.91	178	960	0.087	810	0.04
$C_6H_5-SO_2-O-R$											
R											
CH_3	vapor	1390	0.53	1195	1.25	2.36	195	1015	0.69	775	0.7
C_2H_5	vapor	1391	0.619	1194	1.24	2	197	1019	0.65	770	0.41
n-C_4H_9	vapor	1390	0.49	1194	1.24	2.53	196	950	0.57	789	0.28
4-$CH_3-C_6H_4-SO_2-$ O-R											
R											
CH_3	vapor	1389	0.531	1189	1.251	2.36	200	1010	0.68	752	0.781
C_2H_5	vapor	1390	0.5	1189	1.24	2.48	201	1012	0.51	751	0.385
n-C_4H_9	vapor	1390	0.275	1188	1.24	4.51	202	949	0.261	772	0.15
n-C_4H_9	liquid	1355	1.14	1173	1.15	1.01	182	940	0.95	785	0.5
	Δ [v-l]	[−35]		[−15]				[−9]		[13]	

TABLE 5.9 Infrared data for organosulfonyl chlorides

Compound $R-SO_2-Cl$	Phase	$a.SO_2$ str. cm^{-1} (A)	$s.SO_2$ str. cm^{-1} (A)	SCl str. cm^{-1} (A)	CS str. cm^{-1} (A)	[$a.SO_2$ str.]-[$s.SO_2$ str.] cm^{-1}	A[$s.SO_2$ str.]/A[$a.SO_2$ str.]
CH_3	vapor	1400 (1.240)	1189 (0.690)	540 (1.040)	746 (0.200)	211	0.56
CH_3	liquid	1362 (1.059)	1170 (1.041)	530 (0.656)	742 (0.370)	192	0.98
	Δ [v-l]	[−38]	[−19]	[−10]	[−4]	[−19]	
C_2H_5	vapor	1402 (1.250)	1177 (0.870)	541 (0.480)	704 (0.400)	225	0.7
Aryl-SO_2-Cl							
C_6H_5	vapor	1400 (0.581)	1192 (0.679)	549 (0.459)		208	1.17
C_6H_5	liquid	1372 (0.724)	1181 (0.770)	548 (0.810)		191	1.06
	Δ [v-l]	[−28]	[−11]	[1]		[−17]	
4-F$-C_6H_4$	vapor	1401 (0.579)	1191 (0.910)	525 (0.485)		210	1.57
4-F$-C_6H_4$	liquid	1372 (0.910)	1174 (0.815)	529 (0.775)		198	0.9
	Δ [v-l]	[−29]	[−17]	[14]		[−12]	
4-$CH_3-C_6H_4$	vapor	1399 (0.836)	1189 (1.234)	533 (0.401)		210	1.48
4-$CH_3-C_6H_4$	solid	1379 (0.940)	1189 (0.620)	530 (0.915)		190	0.66
		1364 (0.920)	1172 (0.670)			192	0.73
	Δ [v-s]	[−20]; [−35]	[0]; [−17]	[−33]		[−20]; [−18]	
2,5-$(CH_3)_2-C_6H_5$	vapor	1390 (0.650)	1179 (1.060)	562 (0.921)		211	1.63
2,5-$(CH_3)_2-C_6H_5$	liquid	1369 (0.965)	1171 (0.980)	569 (0.958)		198	0.81
	Δ [v-l]	[−21]	[−8]	[7]		[−13]	
4-Cl,3-$NO_2-C_6H_3$*	vapor	1408 (1.149)	1190 (1.250)	559 (0.411)		218	1.09
4-Cl,3-$NO_2-C_6H_3$*	solid	1382 (1.310)	1170 (1.040)	551 (0.810)		212	0.79
	Δ [v-s]	[−26]	[−20]	[−8]		[−6]	
2,3,4-$(Cl)_3-C_6H_2$	vapor	1408 (1.250)	1172 (0.730)	599 (0.481)		236	0.58
2,4,5-$(Cl)_3-C_6H_2$	vapor	1408 (0.850)	1189 (0.939)	560 (1.240)		219	1.1
2,4,5-$(Cl)_3-C_6H_2$	liquid	1371 (1.010)	1170 (0.992)	561 (0.859)		201	0.98
	Δ [v-l]	[−37]	[−19]	[1]		[−18]	

		$a.NO_2$ str.	$s.NO_2$ str.			[$a.NO_2$ str.]-[$s.NO_2$ str.]	A[$s.NO_2$ str.] A[$a.NO_2$ str.]
4-Cl,3-$NO_2-C_6H_3$*	vapor	1560 (0.620)	1350 (0.465)			210	0.75
4-Cl,3-$NO_2-C_6H_3$*	solid	1542 (1.310)	1360 (0.910)			182	0.69
	Δ [v-s]	[−18]	[10]			[−28]	

* reference 38.

* Sadtler Standard Infrared Spectra, Philadelphia, PA.

TABLE 5.9a Infrared data for organosulfonyl fluorides

Compound Aryl-SO_2F	Phase	asym.SO_2 cm^{-1} (A)	sym.SO_2 cm^{-1} (A)	SF str. cm^{-1} (A)	[asym.SO_2]-[sym.SO_2] cm^{-1}	A[sym.SO_2]/A[asym.SO_2]
C_6H_5	vapor	1429 (0.718)	1221 (1.240)	782 (1.232)	208	1.73
C_6H_5	liquid	1404 (1.100)	1204 (1.250)	780 (0.735)	200	1.14
	[v-l]	[25]	[17]	[2]	[8]	
4-CH_3–C_6H_4	vapor	1430 (1.085)	1220 (1.240)	775 (1.095)	210	1.14
4-CH_3–C_6H_4	solid	1399 (1.011)	1209 (0.920)	750 (0.905)	190	0.91
			1200 (9.30)			0.92
	[v-s]	[31]	[11]	[25]	[20]; [11]	
4-NH_2–C_6H_4	vapor	1428 (0.608)	1218 (1.265)	777 (1.027)	210	2.08
3-NH_2–C_6H_4	vapor	1430 (0.812)	1211 (1.245)	780 (0.976)	219	1.53
3-NH_2,4–ClC_6H_3	vapor	1431 (1.050)	1219 (1.150)	785 (1.250)	212	1.1
3-NH_2,4–ClC_6H_3	liquid	1399 (0.790)	1208 (0.639)	771 (1.020)	191	0.81
	[v-l]	[32]	[11]	[14]	[21]	
2-NH_2–C_6H_4	vapor	1411 (0.598)	1204 (0.950)	785 (1.229)	207	1.59
2-NH_2–C_6H_4	liquid	1384 (0.919)	1186 (0.919)	775 (0.690)	198	1
	[v-l]	[27]	[18]	[10]	[9]	
2-NO_2–C_6H_4*[1]	vapor	1438 (0.821)	1228 (1.240)	811 (0.808)	210	1.51
2-NO_2–C_6H_4	liquid	1411 (1.124)	1210 (1.113)	800	201	0.99
	[v-l]	[18]	[11]	[11]	[9]	
3-NO_2–C_6H_4*[2]	vapor	1440 (0.931)	1235 (1.131)	800 (0.932)	205	1.21
3-NO_2–C_6H_4	liquid	1411 (0.770)	1208 (0.870)	790 (0.621)	203	1.13
	[v-l]	[29]	[27]	[10]	[2]	
3-NO_2-4-Cl–C_6H_3*[3]	vapor	1441 (0.640)	1222 (1.240)	801 (0.850)	219	1.94
3-NO_2-4-Cl–C_6H_3	liquid	1420 (0.680)	1214 (0.715)	796 (0.610)	206	1.05
	[v-l]	[21]	[8]	[5]	[13]	

		asym.NO_2 str.	sym.NO_2 str.	[a.NO_2 str.]–[s.NO_2 str.]	A[s.NO_2 str.]/A[a.NO_2 str.]
2-NO_2–C_6H_4*[1]	vapor	1568 (0.939)	1358 (0.349)	210	0.37
2-NO_2–C_6H_4	liquid	1543 (1.270)	1354 (0.974)	189	
	[v-l]	[25]	[4]	[21]	
3-NO_2–C_6H_4*[2]	vapor	1551 (0.745)	1352 (0.750)	199	1.01
3-NO_2–C_6H_4	liquid	1530 (0.869)	1347 (0.830)	183	0.96
	[v-l]	[21]	[5]	[16]	
3-NO_2-4-Cl–C_6H_3*[3]	vapor	1562 (0.360)	1351 (0.264)	211	0.73
3-NO_2-4-Cl–C_6H_3	liquid	1545 (0.669)	1350 (0.619)	195	0.93
	[v-l]	[17]	[1]	[16]	

*[1] reference 38
*[2] reference 38
*[3] reference 38

TABLE 5.10 Infrared data for organosulfur compounds containing SO_4, SO_3, SO_2N, and SO_2X groups in different physical phases

Compound	a.SO_2 str. cm^{-1} Vapor	s.SO_2 str. cm^{-1} Vapor	Temperature	a.SO_2 str. cm^{-1} Neat	a.SO_2 str. cm^{-1} Neat	s.SO_2 str. cm^{-1} Vapor-Neat	IR or cm^{-1} Vapor-Neat	Raman
Dimethyl sulfate	1401	1208	200 °C	1391	1200	10	17	IR
Methanesulfonate								
methyl	1380	1189	200 °C	1340	1169	40	20	IR
ethyl	1378	1188	240 °C	1344	1168	34	20	IR
Propanesultone	1395	1190	280 °C	1344	1165	51	25	IR
p-toluenethiolsulfonic acid, S-methyl ester	1360	1155	175 °C	~1319	~1134	41	21	IR
Methanesulfonamide	1382	1170	280 °C	1308 KBr	1140	42	30	IR
Benzenesulfonamide	1380	1175	280 °C	1328 KBr	1151	52	24	IR
Benzenesulfonamide 2,4,6-trimethyl	1368	1170	280 °C	1337 KBr	1152	31	18	IR
p-Toluenesulfonamide N-butyl	1410	1170	280 °C	1315	1150	95	20	IR
p-Toluenesulfonanilide	1404	1170	275 °C	1336	1160	68	10	IR
Methanesulfonyl chloride	1393	1189		1375	1172	18	17	IR; R
Benzenesulfonyl chloride	1400	1192		1381	1177	18	15	IR; R
p-Toluenesulfonyl chloride	1399	1188	240 °C	1378	1188 or 1175	21	0 or 13	IR
p-Bromobenzenesulfonyl	1401	1180	200 °C	1370	1162	31	18	IR
chloride		1200			1183		17	IR
m-fluorosulfonylbenzene	1410	1192	200 °C	1390	1184	20	8	IR
sulfonyl chloride	1442	1228		1430	1221	12	7	IR
Benzenesulfonyl fluoride	1429	1221		1411	1215	18	6	IR; R
p-Toluenesulfonyl fluoride	1430	1220	280 °C	1400	1200	30	20	IR

TABLE 5.11 Infrared data for sulfones and sulfoxides in different physical phases

Compound	a.SO_2 str. cm^{-1} Vapor	s.SO_2 str. cm^{-1} Vapor	Temperature	a.SO_2 str. cm^{-1} Neat	s.SO_2 str. cm^{-1} Neat	a.SO_2 str. cm^{-1} Vapor-Neat	s.SO_2 str. cm^{-1} Vapor-Neat
Sulfone							
Di-butyl	1336	1142	280 °C	1321CS_2	1132CS_2	15*[1]	10*[1]
Methyl *p*-tolyl	1350	1164	220 °C	1300	1147	50	17
Di-phenyl	1348	1165	280 °C	1324CS_2	1160CS_2	24*[1]	5*[1]
Di-phenyl	1348	1165	280 °C	1311	1159	37	6
Phenyl *p*-tolyl	1347	1167	240 °C	1309	1155	38	12
Di-(*P*-tolyl)	1348	1165	280 °C	1327CS_2	1158CS_2	21*[1]	7*[1]

Sulfoxide	SO str. cm^{-1} Vapor	Temperature	SO str. cm^{-1} Neat	SO str. cm^{-1} Vapor-Neat	SO str. cm^{-1} CCl_4 soln.	SO str. cm^{-1} Vapor-CCl_4 soln.
Dimethyl	1100	200 °C	1050	50	1070	30
Dimethyl-d_3	1095	200 °C	1060; 1031	35; 64	1068	27
Phenyl methyl	1102	160 °C	1050	52	1057	45
p-Bromophenyl methyl	1101	280 °C	1050	51	[—]	[—]
Di-phenyl	1100	280 °C	1046	~54	1052	48
Di-(*p*-tolyl)	1098	280 °C	1040	~58		
Di-(*o*-tolyl)	1088	280 °C	1028 KBr	~60*[2]	[—]	[—]
Di-(*p*-chlorophenyl)	1090	280 °C	1042 Nujol	~48*[3]	[—]	[—]

*[1] CS_2 soln.
*[2] KBr.
*[3] Nujol.

TABLE 5.12 Infrared data for N-sulfinyl-4-X-anilines and N,N′-disulfinyl-p-phenylenediamine

N-Sulfinyl 4-X-aniline X	asym. N=S=O CS_2 soln. cm^{-1}	sym. N=S=O CS_2 soln. cm^{-1}	[asym. N=S=O]-[sym. N=S=O] CS_2 soln. cm^{-1}	asym. N=S=O vapor-CS_2 soln. cm^{-1}	sym. N=S=O vapor-CS_2 soln. cm^{-1}
CH_3–O	1254	1158	96		
Cl	1272	1162 or 1173	110 or 99		
CH_3	1283	1157	126		
H	1274; [1298 vapor]	1162; [1176 vapor]	114; [112 vapor]	24	14
NO_2	1290	1175	115		
	Nujol mull cm^{-1}	Nujol mull cm^{-1}	Nujol mull cm^{-1}		
N,N′-disulfinyl *p*-phenylenediamine	1271	1166	105		

CHAPTER 6

Halogenated Hydrocarbons

Halogenated Methanes	120
1-Haloalkanes	121
2-Haloalkanes	122
Tertiary Butyl Halides	122
Ethylene Propyne, 1,2-Epoxypropane, and Propadiene Halogenated Analogs	125
Halopropadienes	127
Halogenated Methanes with T_d and C_{3v} Symmetry	127
References	128

Figures

Figure 6-1	129 (120)	Figure 6-21	151 (125)
Figure 6-2	130 (120)	Figure 6-22	152 (125)
Figure 6-3	131 (121)	Figure 6-23	153 (125)
Figure 6-4	132 (122)	Figure 6-24	154 (125, 127)
Figure 6-5	133 (123)	Figure 6-25	154
Figure 6-6	134	Figure 6-26	155
Figure 6-7	135	Figure 6-27	155
Figure 6-8	136	Figure 6-28	156
Figure 6-9	137	Figure 6-29	156
Figure 6-10	138	Figure 6-30	157
Figure 6-11	139	Figure 6-31	157
Figure 6-12	140	Figure 6-32	157
Figure 6-13	141	Figure 6-33	158
Figure 6-14	142 (125)	Figure 6-34	158
Figure 6-15	143 (125)	Figure 6-35	159
Figure 6-16	144, 145 (125)	Figure 6-36	160
Figure 6-17	146, 147 (125)	Figure 6-37	161
Figure 6-18	148 (125)	Figure 6-38	162
Figure 6-19	149 (125)	Figure 6-39	163
Figure 6-20	150 (125)		

Tables

Table 6-1	164 (120, 127)	Table 6-5	168 (123, 127)
Table 6-2	164 (121, 127)	Table 6-6	169 (123, 127)
Table 6-2a	165 (121, 127)	Table 6-7	170 (123, 127)
Table 6-3	166 (121, 127)	Table 6-8	171 (125)
Table 6-4	167 (122, 127)		

*Numbers in parentheses indicate in-text page reference.

HALOGENATED METHANES

The IR and Raman data for halogenated hydrocarbons have been summarized (1, 2), and some examples will be discussed in this chapter.

Table 6.1 lists vapor-phase IR data for the methyl halides (also study Figs. 6.1–6.3). The ν C—X stretching frequencies for the F, Cl, Br, and I analogs occur at 1044, 732, 608, and 530 cm^{-1}, respectively. These ν C—X vibrations decrease in frequency as the mass of the halogen atom increases, as the C—X bond length increases, as the value of the inductive parameter σ' for X decreases, and as the value of the C—X force constants decreases in value in the order F through I. In general, other carbon-halogen stretching frequencies behave similarly. However, in many cases the situation becomes more complex due to the presence of rotational conformers, and also due to the presence of ν asym. CX_n and ν sym. CX_n vibrations.

The methylene halides CH_2X_2 exhibit ν asym. CX_2 and ν sym. CX_2 vibrations. The ν asym. CX_2 (F-I) decrease in frequency in the order 1180, 742, 640, and 570 cm^{-1}, respectively (also study Figs. 6.4 and 6.5). The ν sym. CX_2 (F-I) decrease in frequency in the order 1110, 704, 577, and 480 cm^{-1}, respectively. It is noted that in all cases the ν asym. CX_2 mode always occurs at a higher frequency than does the ν sym. CX_2 mode for each CX_2 analog. Moreover, the Raman band intensity for ν sym. CX_2 is higher than the Raman band intensity for ν asym. CX_2 and in both cases of ν CX_2 the Raman band intensities increase in the order F, Cl, Br, and I (3).

Comparison of the ν asym. CX_2 and ν sym. CX_2 frequency shifts in going from the vapor phase to the liquid phase shows that the ν sym. CX_2 vibration shifts more in frequency (9 to -30 cm^{-1}) compared to ν asym. CX_2 (-1 to -5 cm^{-1}). In other cases discussed in previous chapters, the opposite spectra-structure correlation has been observed. The asymmetric vibration also shifted more in frequency than did the symmetric vibration with change in physical phase. In the case of the CH_2X_2 compounds, the ν asym. CX_2 vibrations occur at lower frequency in the liquid phase than in the vapor phase. However, the ν sym. CX_2 vibration for CH_2Cl_2 occurs at a higher frequency in the liquid phase than in the vapor phase, while for CH_2Br_2 and CH_2I_2 the ν sym. CX_2 vibrations occur at increasingly lower frequencies in the liquid phase than in the vapor phase, progressing in the order of Br_2 to I_2. In other words, it is apparently easier for ν sym. X_2 for the Br_2 and I_2 analogs to vibrate in the liquid phase than in the vapor phase. Mass and the decreasing C—X force constants progressing in the series F_2 through I_2 apparently account for this frequency behavior during expansion and contraction of surrounding molecules during the ν sym. CX_2 modes. In the case of ν asym. CX_2, there is an equal tradeoff in energy as the one C—X bond expands toward neighboring molecules and the other C—X bond contracts from neighboring molecules. Hence, there is not as much of a shift in frequency with a change in physical phase for ν asym. CX_2 as there is for ν sym. CX_2. With a lesser force constant for C—Br and C—I, it is relatively easier for the C—Br or C—I bonds to vibrate without exerting as much pressure against neighboring molecules in the liquid state.

Table 6.2 lists IR and Raman data for trihalomethane and tetrahalomethane (also study Figs. 6.6 through 6.12). The ν asym. F_3 vibration for CHF_3 and the ν asym. CF_4 vibration for CF_4 occur at 1376 cm^{-1} and 1283 cm^{-1}, respectively. The ν sym. F_3 vibration for CHF_3 and the ν sym. F_4 vibration for CF_4 occur at 1165 cm^{-1} and 908 cm^{-1}, respectively. The ν asym. CX_4 mode for the CF_4, CCl_4, and CBr_4 occurs at 1283, 797, and 671 cm^{-1}, respectively. In the liquid phase, the ν sym. CX_4 mode for CF_4 through Cl_4 occurs at 908, 790, 662, and 560 cm^{-1}, respectively.

Table 6.2a lists a comparison of CX, CX_2, CX_3, and CX_4 stretching frequencies. The C—F_n, C—Cl_n, C—Br_n, and C—I_n stretching frequencies in this methane series occur in the regions 908–1376 cm^{-1}, 667–797 cm^{-1}, 541–662 cm^{-1}, and 480–572 cm^{-1}, respectively. Moreover, the ν CX_2 vibrations occur at higher and lower frequencies than is the case for the ν C—X vibrations. The ν asym. CX_3 vibrations occur at higher frequencies than those of the ν asym. CX_2 vibrations. The other ν CX_n modes do not correlate as well as the ones just discussed.

1-HALOALKANES

Haloalkanes exist as rotational conformers, and Shipman *et al.* have specified the notation for describing these rotational conformers (4). The notation P, S, and T denote primary, secondary and tertiary carbon atoms to which the halogen atom is joined. Atom X is joined to the β-carbon atom in a trans position to the α-halogen atom in a zig-zag plane together with subscript *C* or *H* for the trans X atom denotes the specific rotational conformer. Applying this nomenclature for the planar skeleton trans rotational conformer for *n*-propyl chloride,

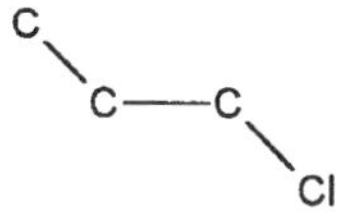

, the notation is P_C, and

for the gauche skeleton rotational conformer,

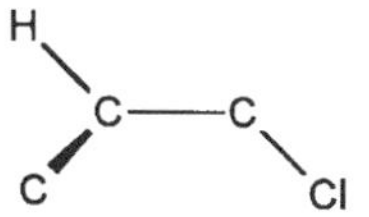

, the notation is P_H

Table 6.3 lists vapor- and liquid-phase IR data for the carbon-halogen stretching frequencies for 1-haloalkanes. The ν CCl (P_c rotational conformers) for the 1-chloroalkanes (C_3–C_{18}) occur in the region 738–750 cm^{-1} in the vapor phase and in the region 732–758 cm^{-1} in the liquid phase. The ν CCl P_H gauche (skeletal rotational conformer) occurs in the region 660–664 cm^{-1} in the vapor phase and in the region 650–659 cm^{-1} in the liquid phase. Study of the absorbance data for A(ν CCl P_c)/A(ν CCl P_H) for these 1-chloroalkanes shows that the ratio is less in the liquid phase than in the vapor phase. (Another band in this region in the case of the 1-chloroalkanes results from $(CH_2)_n$ rocking.) These data show that the concentration of P_c and P_H rotational conformers change with change in phase. Assuming that the extinction coefficient for the absorbance values are equal for both the P_c and P_H conformers, the concentrations for 1-chloropropane are 55.6% P_c and 44.4% P_H in the vapor phase and 49.1% P_c and 50.9% P_H in liquid phase.

In the case of 1-bromalkanes, C_4 through C_{19}, ν CBr(P_c) occurs in the region 645–653 cm^{-1}, and ν CBr(P_H) occurs in the region 562–572 cm^{-1} in the vapor phase. In all cases, these ν CBr modes occur at lower frequency in the liquid phase.

In the case of 1-iodoalkanes, C_3–C_{16}, the ν Cl(P_c) rotational conformer (P_c) occurs in the region 591–598 cm^{-1} and the ν Cl(P_H) rotational conformer occurs in the region 500–505 cm^{-1} in the vapor phase. Both ν Cl(P_c) and ν Cl(P_H) occur at lower frequency in the liquid phase and the concentration of the rotational conformer P_c decreases in going from the vapor to the liquid phase. 1-Fluorodecene exhibits ν CF(P_c) at 1050 cm^{-1} and ν CF(P_H) at 1032 cm^{-1} in the vapor phase and ν CF(P_c) at 1042 and ν CF(P_H) at 1005 cm^{-1} in the liquid phase.

2-HALOALKANES

The 2-haloalkanes exist as three rotational conformers, and these are S_{CH}, $S_{HH'}$, and S_{HH}. Vapor- and liquid-phase IR data are listed in Table 6.4 for 2-halobutane and tert-butyl halide. These data for the 2-halobutenes are also shown here under the rotational conformer for which the ν CX vibration is assigned.

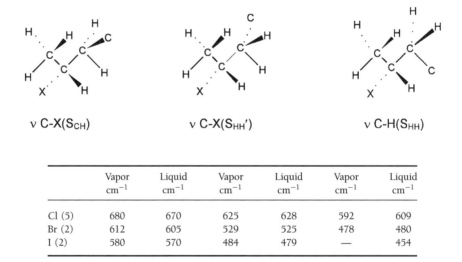

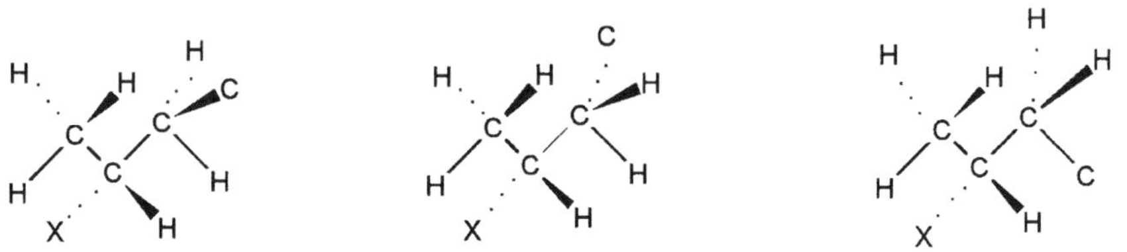

	Vapor cm^{-1}	Liquid cm^{-1}	Vapor cm^{-1}	Liquid cm^{-1}	Vapor cm^{-1}	Liquid cm^{-1}
Cl (5)	680	670	625	628	592	609
Br (2)	612	605	529	525	478	480
I (2)	580	570	484	479	—	454

The H′ notation is used to specify a rotation of the carbon skeleton away from the zigzag carbon plane at the 3-position where the trans hydrogen is located. In the S_{HH} conformer, the first four carbon atoms are in a planar zigzag configuration while only the first three carbon atoms are in a planar zigzag configuration in the case of conformer $S_{HH'}$.

TERTIARY BUTYL HALIDES

In the liquid phase the tert-butyl halides exhibit ν C–X(T_{HHH}) at Cl (570 cm^{-1}), Br(520 cm^{-1}), and I (492 cm^{-1}). In the case of 2-halo-2-methylbutanes, ν CX(T_{HHH}) occurs at Cl (566 cm^{-1}), Br (510 cm^{-1}), and ν CX(T_{CHH}) occurs at Cl (621 cm^{-1}), Br (577 cm^{-1}) (2).

Table 6.5 lists vapor- and liquid-phase IR data for 1-halocycloalkanes. 1-Halocycloalkanes containing four or more ring carbon atoms can exist in an equatorial or axial configuration as presented here for 1-halocyclohexanes. In general, the ν CX(S_{cc}) vibrations

occur at higher frequencies than ν CH($S_{H'H'}$) vibrations. In the case of 1-halocyclohexane in the vapor phase ν CX(S_{cc}) occurs at 740 cm^{-1} (Cl), 692 cm^{-1} (Br), and 664 cm^{-1} (I), and for ν CX($S_{H'H'}$) it occurs at 690 cm^{-1} (Cl), 661 cm^{-1} (Br), and 635 cm^{-1} (I). The frequency separation between ν CX(S_{cc}) and ν CX($S_{H'H'}$) decrease, progressing in the series Cl through I in both the vapor and liquid phases, where

vapor is Cl (50 cm^{-1}), Br (31 cm^{-1}), and I (29 cm^{-1}), and
liquid Cl (48 cm^{-1}), Br (27 cm^{-1}), and I (18 cm^{-1}).

In addition, study of the absorbance ratios for (A)ν CX($S_{H'H'}$) shows that their concentration changes in going from the vapor to the liquid state. The concentration of conformer S_{cc} is higher than the concentration of conformer $S_{H'H'}$ in the vapor phase than in the liquid phase. If we assume for purposes of comparison that the extinction coefficients for (A) for each conformer are identical in both the liquid and vapor states, the following calculations can be performed for 1-halocyclohexane:

	Vapor % ν CX(S_{cc})	Vapor % ν CS($S_{H'H'}$)	Liquid % ν CX(S_{cc})	Liquid % ν CX($S_{H'H'}$)
Cl	72.3	27.7	70.3	29.7
Br	78.1	21.9	65.3	34.7
I	84.2	15.8	72.9	27.1

In the vapor phase, there is only IR evidence for the presence of ν CX(S_{cc}) for 1-halocyclopentane and their ν CX(S_{cc}) frequencies occur at 620 cm^{-1} (Cl), 520 cm^{-1} (Br), 476 cm^{-1} (I). In the liquid phase ν CCl($S_{H'H'}$) is assigned to a very weak Raman band at 588 cm^{-1} (8). No spectral evidence is reported for the presence of ν CBr($S_{H'H'}$) or ν CH($S_{H'H'}$) conformers.

Table 6.6 lists vapor- and liquid-phase IR data for primary dihaloethane through dihalohexane. These P,P-dihaloalkanes also exist as rotational conformers. It is interesting to compare the highest ν CX vapor-phase frequency in each series:

	ν CCl cm^{-1}	ν CBr cm^{-1}	ν CI cm^{-1}	
1,2-dihaloethane	730	594	485*	(also study Fig. 6.13)
1,3-dihalopropane	740	655	602	
1,4-dihalobutane	778	659	599	
1,5-dihalopentane	748	654	584*	
1,6-dihalopentane	740	654	600	
Range:	730–778	594–659	485*–602	

* Liquid.

These ν CX frequencies are affected by the number of $(CH_2)_n$ groups between the two CH_2X groups. The ν CX frequencies increase in the order $n = 0, 1, 2$, and then decrease or stay relatively constant in the order $n = 3, 4$.

It is also interesting to compare the lowest ν CX vapor-phase frequency in each series:

	ν CCl cm^{-1}	ν C Br cm^{-1}	ν CI cm^{-1}
1,2-dihaloethene	660	545*	—
1,3-dihalopropane	660	559	526
1,4-dihalobutane	660	572	505
1,5-dihalopentane	662	571	492*
1,6-dihalohexane	660	570	509
Range:	660–662	545–572	492*–526

* Liquid.

The ν CX rotational conformer that occurs between the high and low ν CX rotational conformer is not noted in all of the spectra studied. In this case only the liquid-phase data are compared for the intermediate ν CX rotational conformer.

	ν CCl* cm^{-1}	ν C Br* cm^{-1}	ν CI* cm^{-1}
1,2-dihaloethane	662	—	—
1,3-dihalopropane	—	585	—
1,4-dihalobutane	735	—	—
1,5-dihalopentane	720	—	604

* Liquid.

A study of the absorbance ratios in Table 6.6 shows that the concentrations of these conformers change with their physical phase. Take for example, 1,3-dibromopropane (assume the extinction coefficient for (A) is equal). Then the % concentrations of the 3-rotational conformers in the vapor and liquid phases are listed here for 1,3-dibromopropane.

Rotational Conformer, cm^{-1}	Vapor Phase %	Liquid Phase %	Rotational Conformer, cm^{-1}
655	30.5	24.8	645
599	26.5	31.4	585
559	43.0	43.8	543

Table 6.7 lists Raman data for the methyl halides and IR and Raman data for tetrabromoalkanes. Vapor-phase IR data for the ν CX frequencies are presented in Table 6.1. Tentative assignments for the ν CBr_2 and ν CBr vibrations are compared here.

	ν_a CBr_2 cm^{-1}	ν_a CBr cm^{-1}	νs CBr_2 cm^{-1}
1,1,2,2-Br_4 ethane	714	664	537
	ν CBr cm^{-1}	ν CBr cm^{-1}	ν CBr cm^{-1}
1,2,3,4-Br_4 butane	702	567	532

ETHYLENE, PROPYNE, 1,2-EPOXYPROPANE, AND PROPADIENE HALOGENATED ANALOGS

Table 6.8 lists carbon halogen stretching frequencies for ethylene, propyne, 1,2-epoxypropane, and propadiene analogs. The first six examples, 3-halopropene through 1,3-dihalopropyne all contain an isolated CH_2—X group. (Study Figs. 6.14 through 6.24.) The ranges for ν CF, ν CCl, ν CBr, and ν Cl for the CH_2X group are listed here.

ν CF cm^{-1}	ν CCl cm^{-1}	ν CBrν cm^{-1}	Cl cm^{-1}
989–1006	695–740	613–691	570–670

The ν CX vibrations for 1-fluoroethylene (vinyl fluoride) and 1-chloroethylene (vinyl chloride) are assigned at 1157 cm^{-1} and 719 cm^{-1}, respectively. In the case of the 1,1-dihaloethylenes,

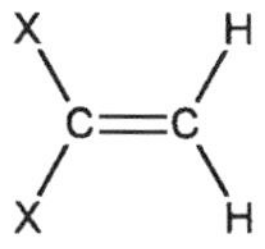

the two vibrations are ν asym. CX_2 and ν sym. CX_2. These molecules have C_{2v} symmetry, and ν asym. CX_2 vibration belongs to the B_2 symmetry species and the ν sym. CX_2 vibration belongs to the A_1 symmetry species. Both of these vibrations decrease in frequency, progressing in the series CF_2CH_2 through CBr_2CH_2; ν asym. CX_2 occurs in the region 1301–698 cm^{-1}, and ν sym. CX_2 occurs in the region 922–474 cm^{-1}. The compounds, 1-bromo-1-chloroethylene, has C_s symmetry, and both modes belong to the A′ symmetry species (study Figs. 6.25 and 6.28) (19).

The tetrahaloethylenes have ν_h symmetry, and the four C_2X_4 carbon halogen stretching vibrations are depicted here (see Figs. 6.36 through 6.39).

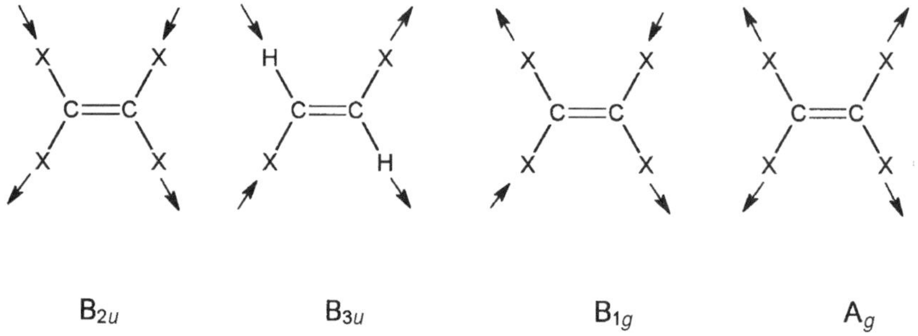

The B_{2u} and B_{3u} C_2X_4 stretching vibrations are allowed in the IR, and the B_{1g} and A_g C_2X_4 stretching vibrations are allowed in the Raman. In the series C_2F_4 through C_2Br_4, the B_{2u} ν C_2X_4 mode occurs in the region 909–1340 cm^{-1} and the B_{3u} ν C_2X_4 mode occurs in the region 632–875 cm^{-1}.

In the case of 1,1-dichloro-2,2-difluoroethylene, the molecular symmetry is C_{2v}, and the two A_1 and two B_2

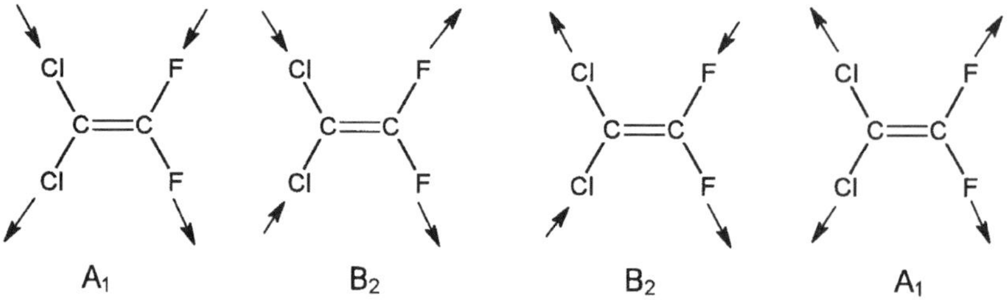

$C_2Cl_2F_2$ stretching vibrations are IR active. These four modes are assigned in the region 890–1219 cm^{-1}.

The ethylenes of form C_2X_3Y have C_s symmetry. The four C_2X_3Y stretching frequencies belong to the A′ symmetry species. Trichloroethylene has three C_2Cl_3 stretching frequencies and one ν CH mode. Three of the C_2X_3Y stretching vibrations occur in the ranges 1329–1330 cm^{-1}, 1202–1211 cm^{-1}, and 1025–1052 cm^{-1} for both the C_2F_3Cl and C_2F_3Br analogs, and they occur in the ranges 931–1181 cm^{-1}, 852–987 cm^{-1}, and 639–869 cm^{-1} for both C_2Cl_3F and C_2HCl_3.

HALOPROPADIENES

The 1-halopropadienes have C_s symmetry and the ν CX vibration decrease in frequency, progressing in the series Cl, 767 cm^{-1}, Br, 681 cm^{-1}, and I, 609 cm^{-1} (study Figs. 6.29–6.35). Evidently these ν CX vibrations are complex, because ν CBr for 1-bromopropadiene-1-d occurs at 636 cm^{-1} while ν CBr for 1-bromopropadiene occurs at 681 cm^{-1}, a shift of 45 cm^{-1} by substitution of D for H on the same C—Br carbon atom. The ν CF vibration for 1-fluoropropadiene is estimated to occur at 1050 cm^{-1} (23).

Halogen atoms joined to a carbon–carbon triple bond occur at relatively low frequency. In the case of 1-halopropyne and 1,3-dihalopropyne the $\nu \equiv C-X$ vibrations occur in the range Cl, 574–617 cm^{-1}, Br, 464–512 cm^{-1}, and I, 403 cm^{-1} (study Figs. 6.30 and 6.39) (16, 24).

Figure 6.24 shows the approximate skeletal bending modes of propyne, 3-halopropynes, and 1,3-dihalopropynes. Vibrations involving bending of the carbon–halogen bonds occur at low frequencies (25).

HALOGENATED METHANES WITH T_d AND C_{3v} SYMMETRY

Compounds for form CX_4 have T_d symmetry. The ν asym. CX_4 mode is triply degenerate and decreases in frequency, progressing in the series CF_4, 1265 cm^{-1}, CCl_4, 776 cm^{-1}, and CB_4, 672 cm^{-1}. The ν sym. CX_4 mode decreases in frequency, progressing in the series CF_4, 904 cm^{-1}, CCl_4, 458 cm^{-1}, and CBr, 267 cm^{-1} (26).

Compounds of form CX_3Y have C_{3v} symmetry. The ν asym. CX_3 mode is doubly degenerate. These molecules also have a ν sym. CX_3 vibration and a ν CY vibration. These three stretching frequencies also decrease in frequency, progressive in the order F through Br.

Compound: (22)	ν asym. CX_3 cm^{-1}	ν sym. X_3 cm^{-1}	ν CY cm^{-1}
CF_3Cl	1210	1101	780
CF_3Br	1201	1080	759
CCl_3F	840	930	1080

Study of Tables 6.1 through 6.8 shows that different ν CX frequencies overlap in several cases in the Cl through I series. More extensive coverage of the vibrational spectra and frequencies of the halogenated alkanes can be found in References 1–10.

Standard vapor and neat IR spectra and standard Raman spectra of these halogenated materials as well as other organic compounds are readily available from Bio-Rad Sadtler Division, and these spectra are valuable in identifying unknown chemical compositions.

Figures 6.1 through 6.39 are included as a convenience to the reader, since some of the data discussed in this chapter were obtained from these IR and Raman spectra.

REFERENCES

1. Lin-Vien, D., Colthup, N. B., Fateley, W. G., and Grasselli, J. G. (1991). *The Infrared and Raman Characteristic Frequencies of Organic Molecules*. Boston: Academic Press, Inc., p. 29.
2. Nyquist, R. A. (1984). *The Interpretation of Vapor-Phase Infrared Spectra: Group Frequency Data*. Philadelphia: Bio-Rad Laboratories, Sadtler Div.
3. Schrader, B. (1989). Raman/Infrared Atlas of Organic Compounds. 2nd Edition, New York, VCH-Verl.-Ges. Weinkeim.
4. Shipman, J. J., Folt, V. L., and Krimm, S. (1962). *Spectrochim. Acta*, **18**: 1603.
5. George, W. O., Goodfield, J. E., and Maddams, W. F. (1985). *Spectrochim. Acta*, **41A**: 1243.
6. Rothschild, W. G. (1966). *J. Chem. Phys.*, **45**: 1214.
7. Durig, J. R., Karriker, J. M., and Wertz, D. M. (1969). *J. Mol. Spectrosc.*, **31**: 237.
8. Ekejiuba, I. O. C. and Hallam, H. E. (1969). *Spectrochim. Acta*, **26A**: 59.
9. Rey-Lafon, M., Rouffi, C., Camiade, M., and Forel, M. (1970). *J. Chim. Phys.*, **67**: 2030.
10. Wurrey, C. J., Berry, R. J., Yeh, Y. Y., Little, T. S., and Kalasinsky, V. J. (1983). *J. Raman Spectrosc.*, **14**: 87.
11. McLachlan, R. D. and Nyquist, R. A. (1968). *Spectrochim. Acta*, **24A**: 103.
12. Nyquist, R. A., Putzig, C. L., and Skelly, N. E. (1986). *Appl. Spectrosc.*, **40**: 821.
13. Evans, J. C. and Nyquist, R. A. (1963). *Spectrochim. Acta*, **19**: 1153.
14. Nyquist, R. A., Reder, T. L., Ward, G. F., and Kallos, G. J. (1971). *Spectrochim. Acta*, **27A**: 541.
15. Nyquist, R. A., Stec, F. F., and Kallos, G. J. (1971). *Spectrochim. Acta*, **27A**: 897.
16. Nyquist, R. A., Johnson, A. L., and Lo, Y.-S. (1965). *Spectrochim. Acta*, **21**: 77.
17. Nyquist, R. A. (1984). *The Interpretation of Vapor-Phase Infrared Spectra: Group Frequency Data*. Philadelphia: Bio-Rad Laboratories, Sadtler Div.
18. Nyquist, R. A. and Thompson, J. W. (1977). *Spectrochim. Acta*, **33A**: 63.
19. Joyner, P. and Glockler, G. (1952). *J. Chem. Phys.*, **20**: 302.
20. Winter, F. (1970). *Z. Naturforsch.*, **25a**: 1912.
21. Scherer, J. R. and Overend, J. (1960). *J. Chem. Phys.*, **32**: 1720.
22. Nyquist, R. A. (1989). *The Infrared Spectra Building Blocks of Polymers*. Philadelphia: Bio-Rad, Sadtler Div.
23. Nyquist, R. A., Lo, Y.-S., and Evans, J. C. (1964). *Spectochim. Acta*, **20**: 619.
24. Nyquist, R. A. (1965). *Spectrochim. Acta*, **21**: 1245.
25. Herzberg, G. (1945). *Molecular Spectra and Molecular Structure II. Infrared and Raman Spectra of Polyatomic Molecules*. New York: Van Nostrand Co., Inc., p. 167.
26. Erley, D. S. and Blake, B. H. (1965). Infrared Spectra of Gases and Vapors, Vol. II. *Grating Spectra*. The Dow Chemical Co.

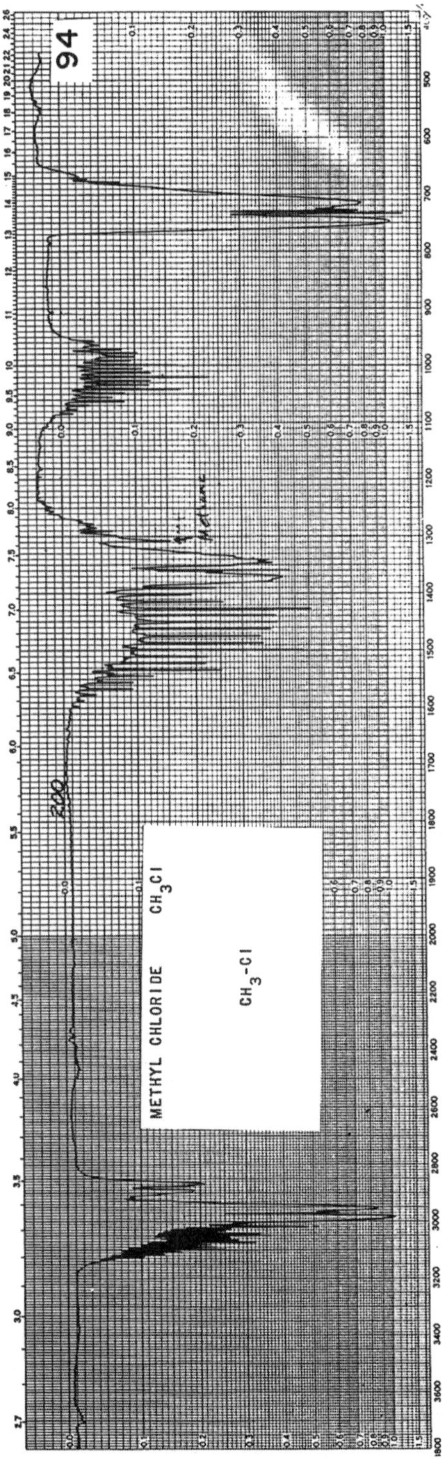

FIGURE 6.1* Methyl chloride (200-mm Hg sample) (26).

*Those vapor-phrase infrared spectra figures for Chapter 6 with an asterisk following the figure number have a total vapor pressure of 600-mm Hg with nitrogen (N_2), in a 5-cm KBr cell. The mm Hg sample is indicated in each figure.

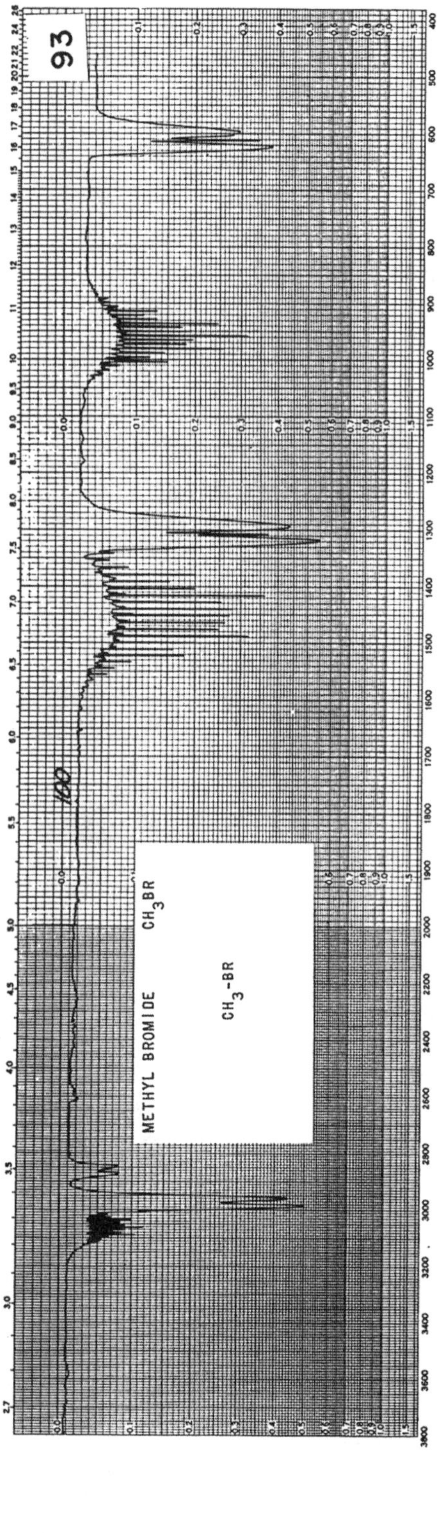

FIGURE 6.2* Methyl bromide (100-mm Hg sample) (26).

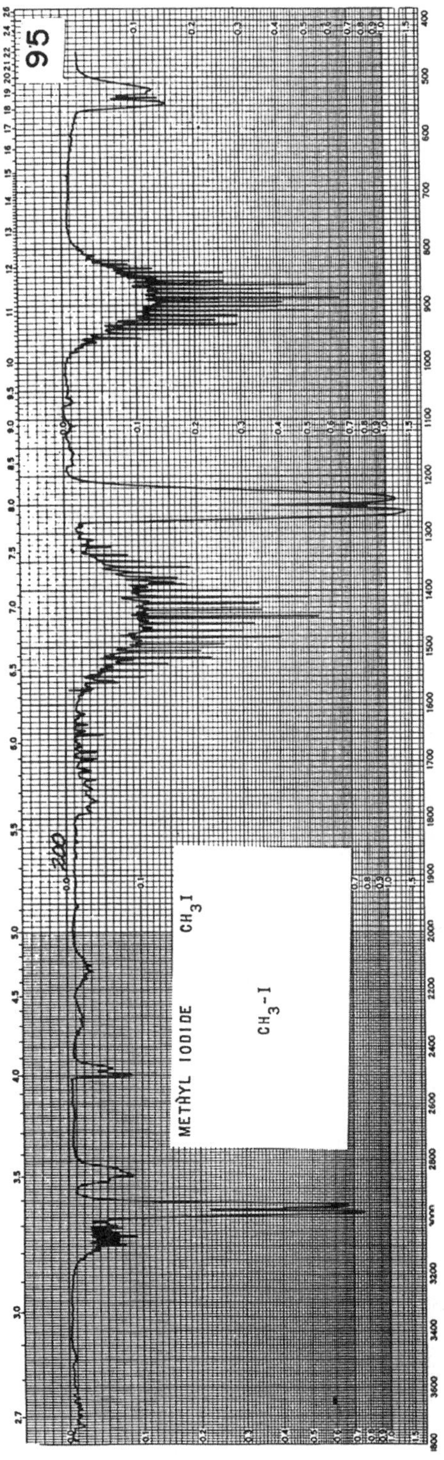

FIGURE 6.3* Methyl iodide (200-mm Hg sample) (27).

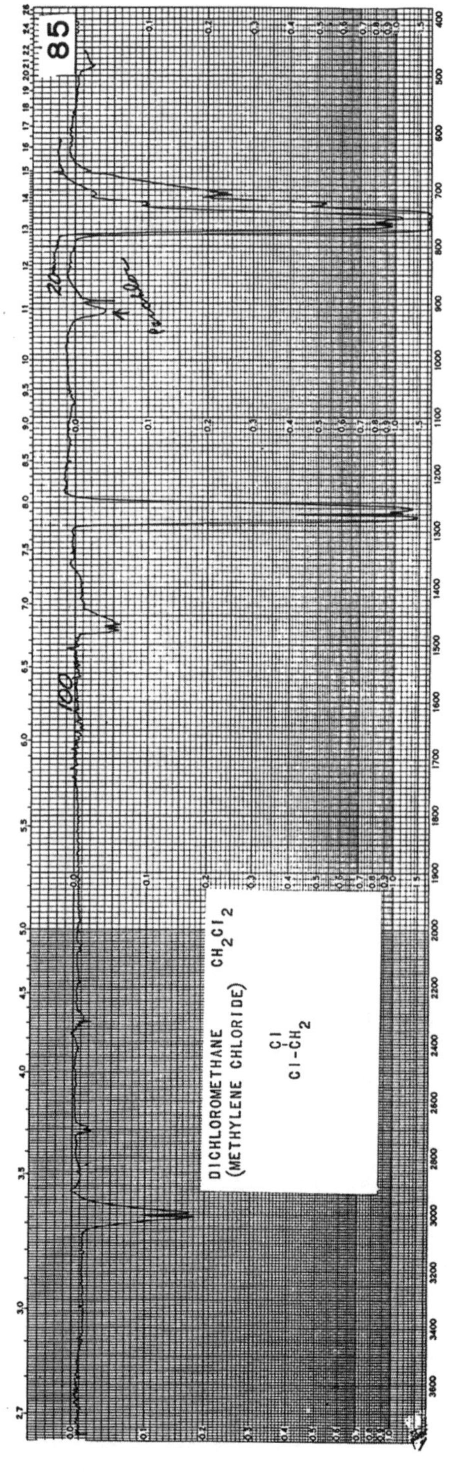

FIGURE 6.4* Methylene chloride (20- and 100-mm Hg sample) (26).

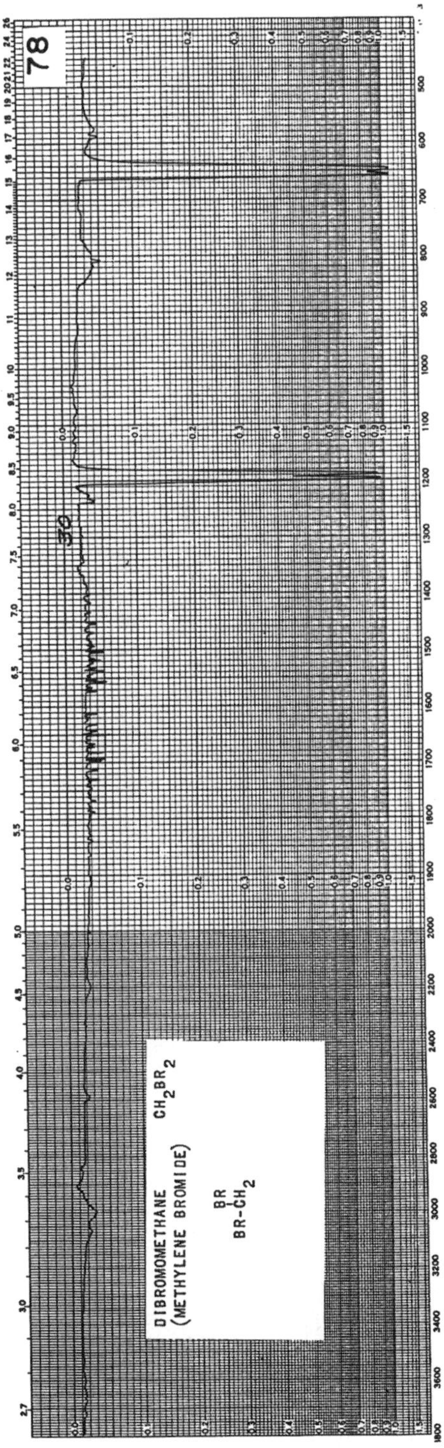

FIGURE 6.5* Methylene bromide (30-mm Hg sample) (26).

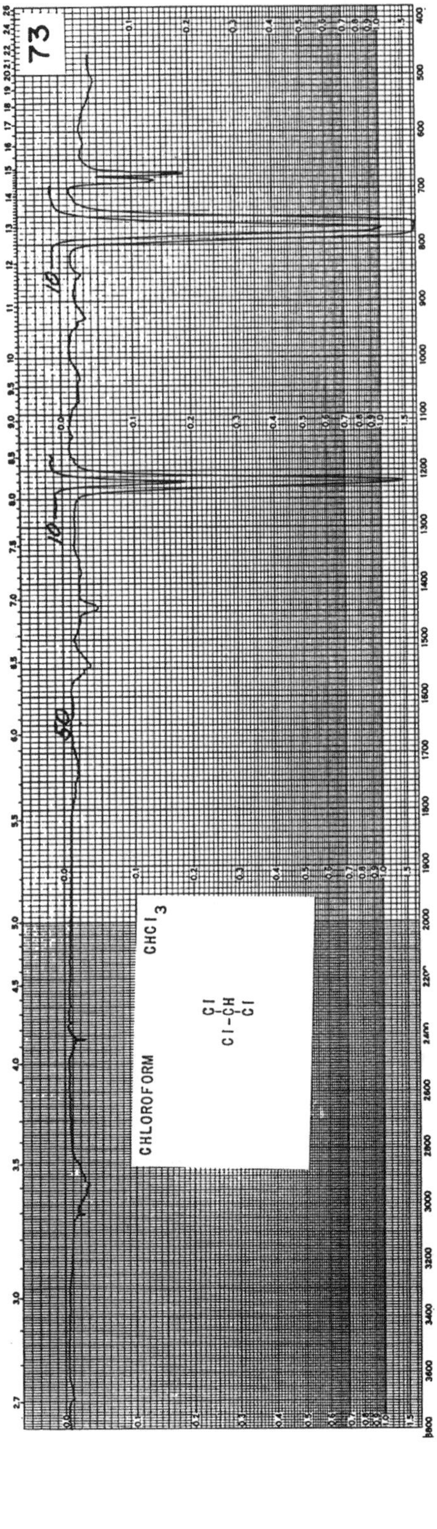

FIGURE 6.6* Trichloromethane (chloroform) (10- and 50-mm Hg sample) (26).

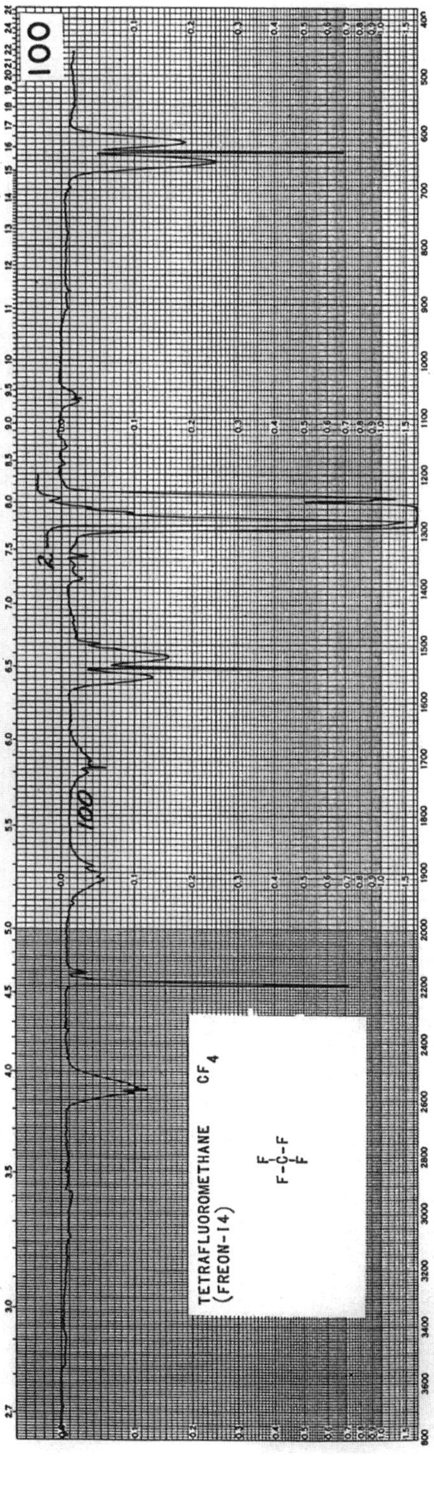

FIGURE 6.7 Tetrafluoromethane (Freon 14) (2 and 100 Hg sample) (26)

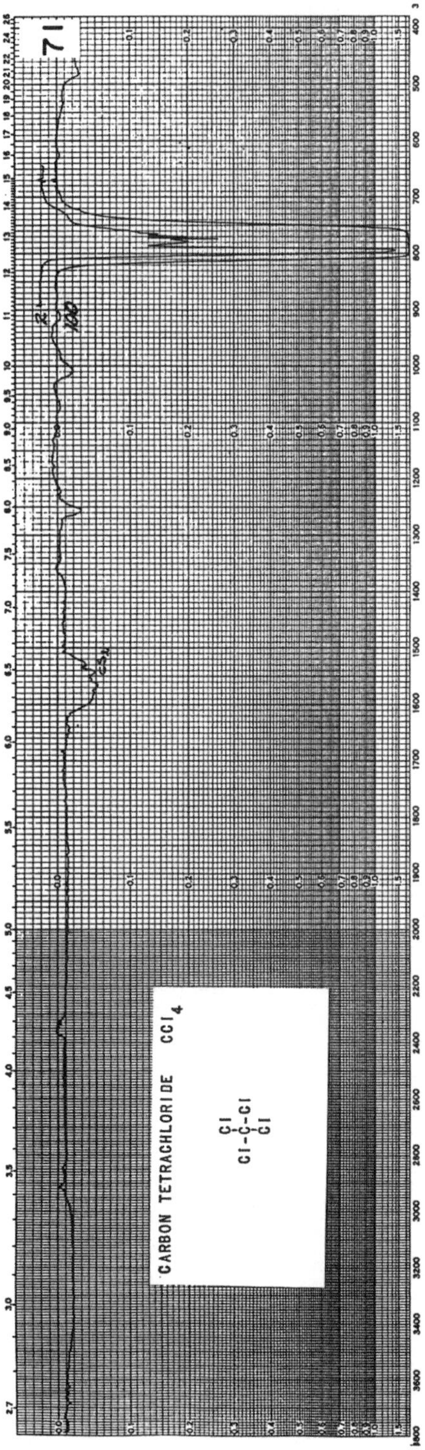

FIGURE 6.8* Tetrachloromethane (carbon tetrachloride) (2- and 100-mm Hg sample) (26).

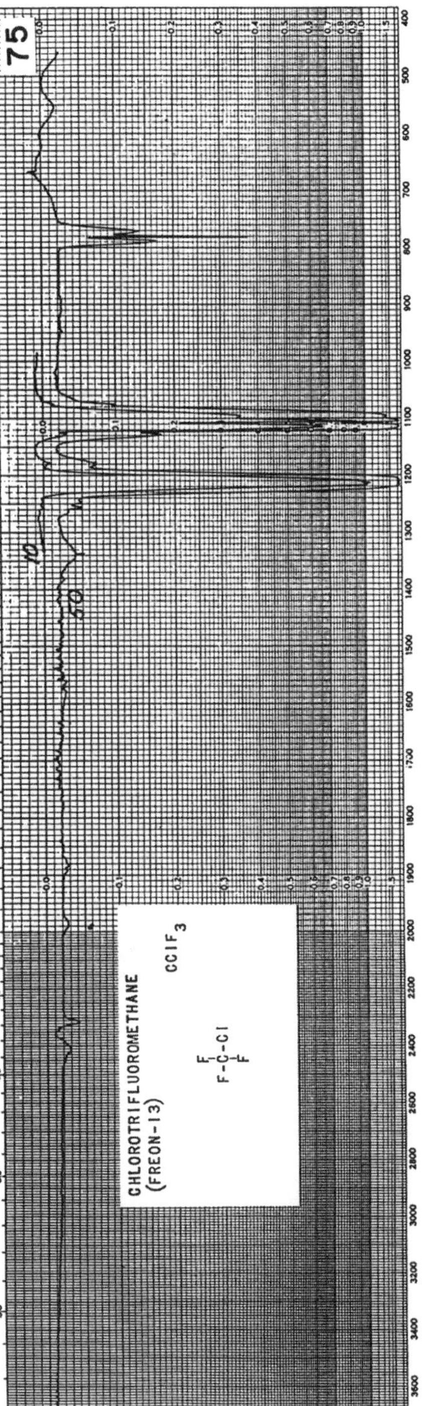

FIGURE 6.9* Chlorotrifluoromethane (10- and 50-mm Hg sample) (26).

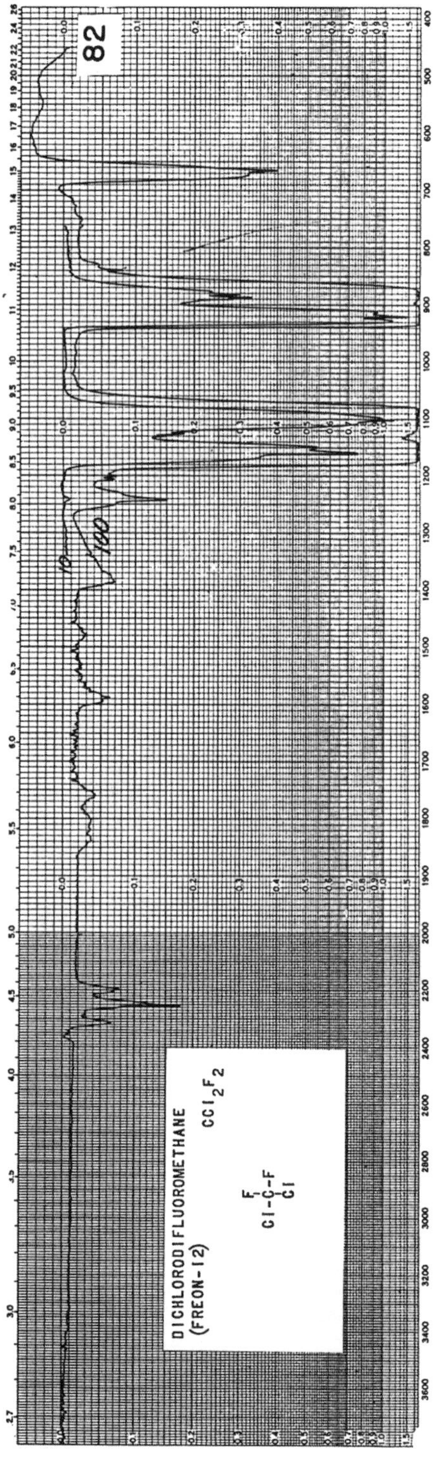

FIGURE 6.10* Dichlorodifluoro methane (10- and 100-mm Hg sample) (26).

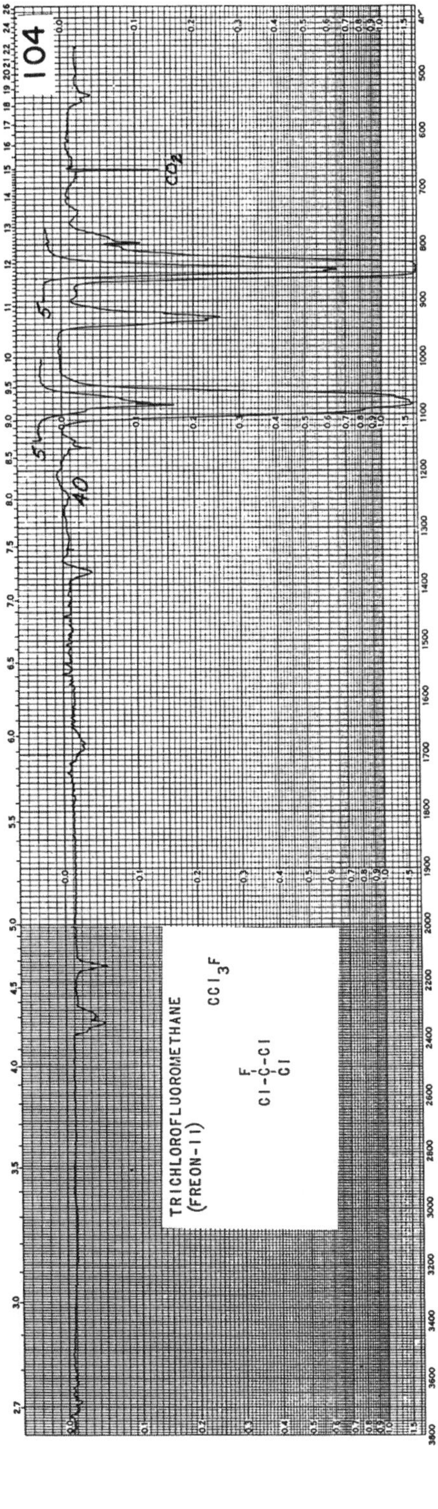

FIGURE 6.11* Trichlorofluoromethane (f and 40-mm Hg sample) (26).

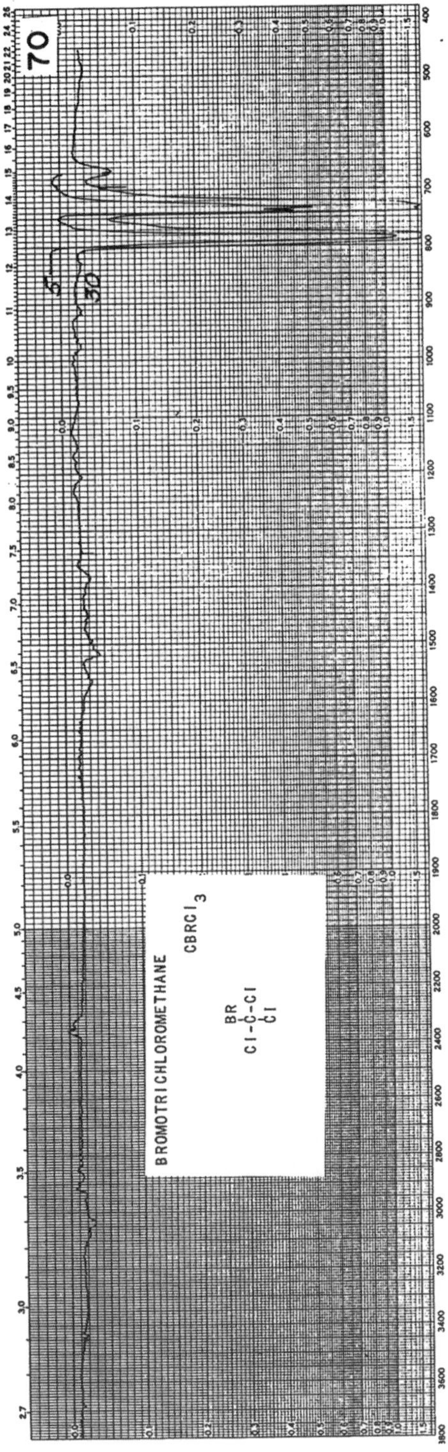

FIGURE 6.12* Bromotrichloromethane (5- and 30-mm Hg sample) (27).

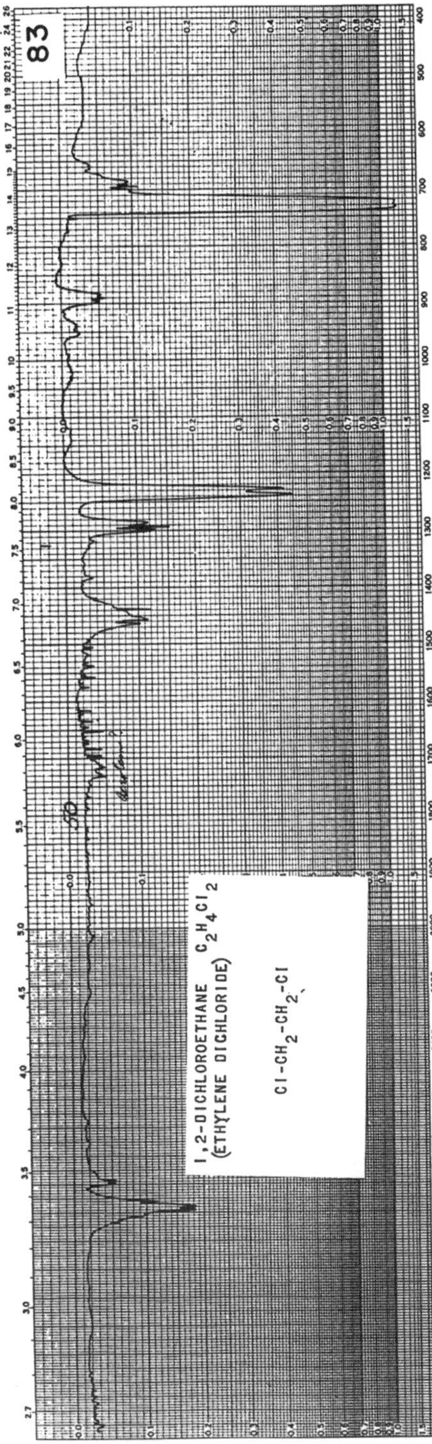

FIGURE 6.13* 1,2-Dichloroethane (ethylene dichloride) (50-mm Hg sample) (27).

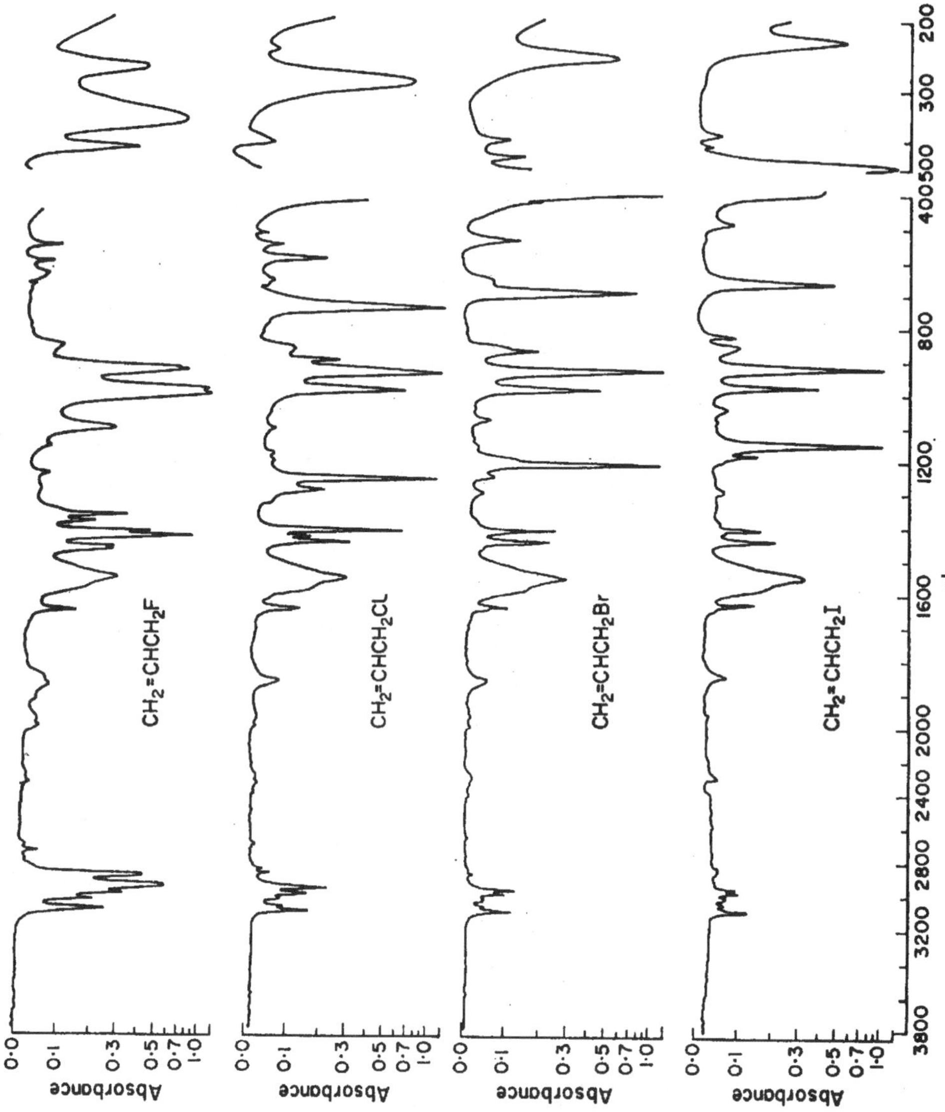

FIGURE 6.14* Infrared spectra of 3-halopropenes (allyl halides) in CCl_4 solution (3800–1333 cm^{-1}) (133–400 cm^{-1}) (12).

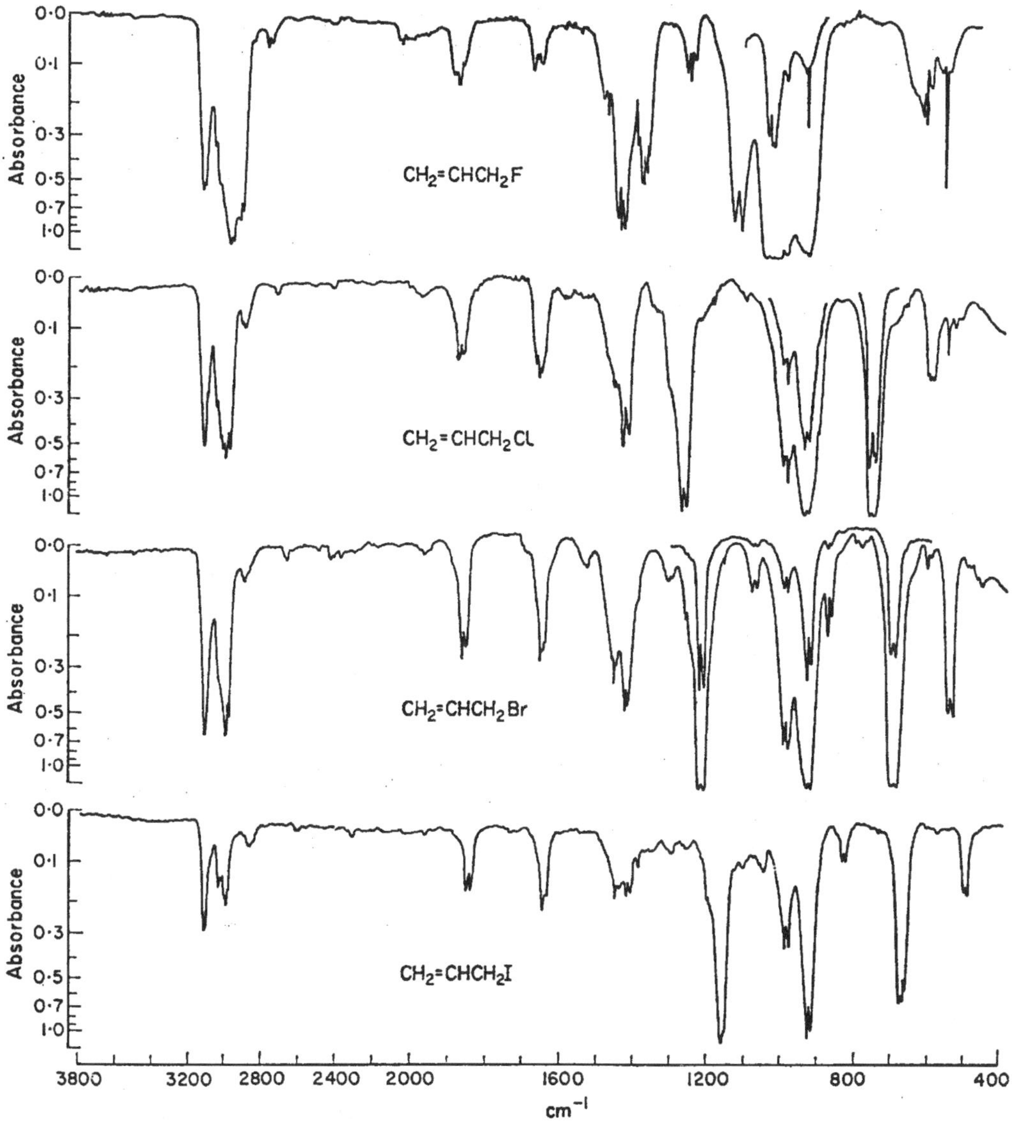

FIGURE 6.15 Vapor-phase infrared spectra of 3-halopropenes (allyl halides) (12).

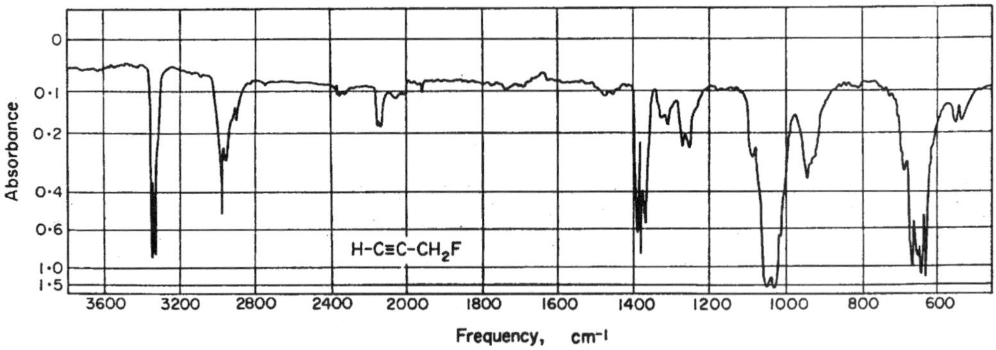

FIGURE 6.16a Vapor-phase IR spectrum of 3-fluoropropyne in a 5-cm KBr cell (50-mm Hg sample).

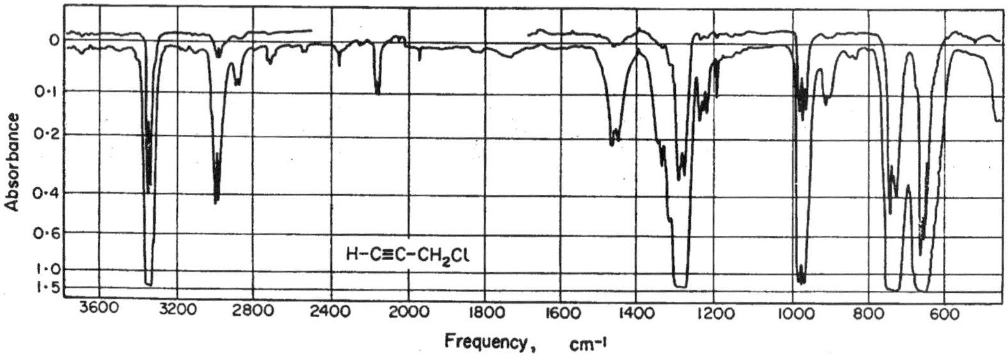

FIGURE 6.16b Vapor-phase IR spectrum of 3-chloropropyne in a 5-cm KBr cell (vapor pressure at −10 and 25 °C samples).

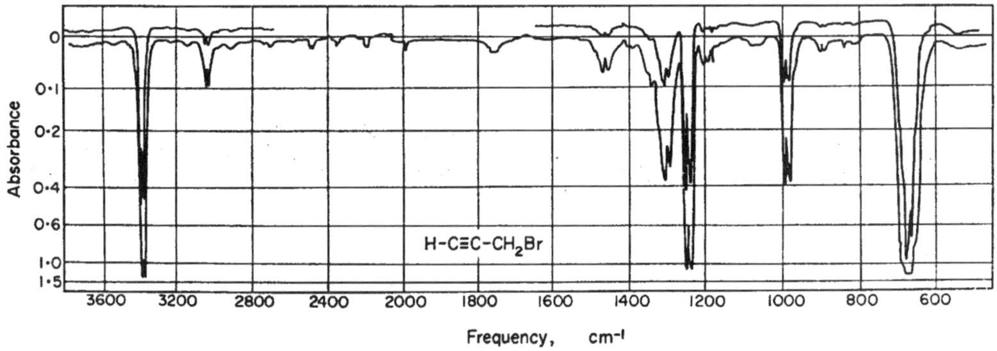

FIGURE 6.16c Vapor-phase IR spectrum of 3-bromopropyne in a 5-cm KBr cell (vapor pressure at 0 and 25 °C samples).

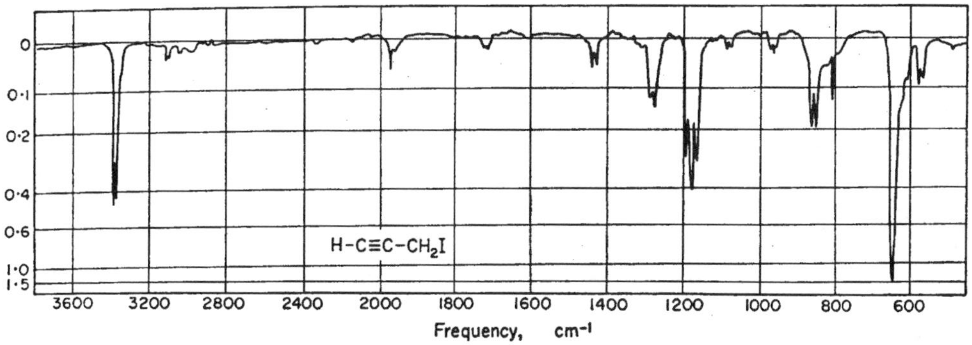

FIGURE 6.16d Vapor-phase IR spectrum of 3-iodopropyne in a 15-cm KBr cell (~8-mm Hg sample).

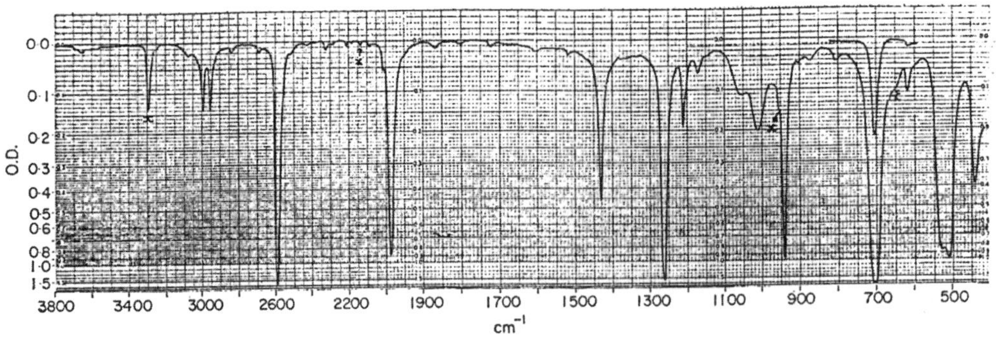

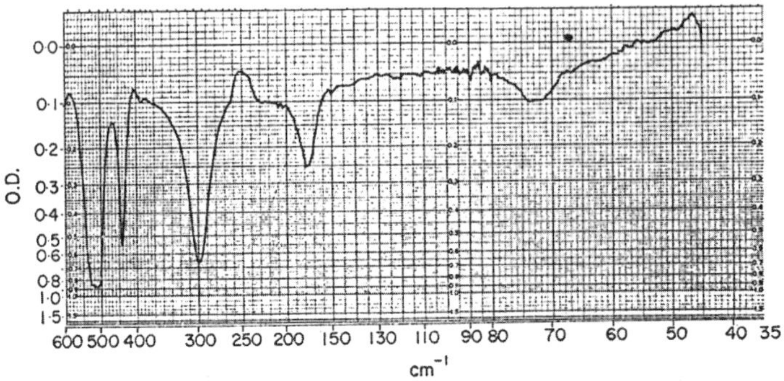

FIGURE 6.17a Top: Liquid-phase IR spectrum of 3-chloropropyne-1-d in a 0.023-mm KBr cell. Bottom: Liquid-phase IR spectrum of 3-chloropropyne-1-d in a 0.1-mm polyethylene cell.

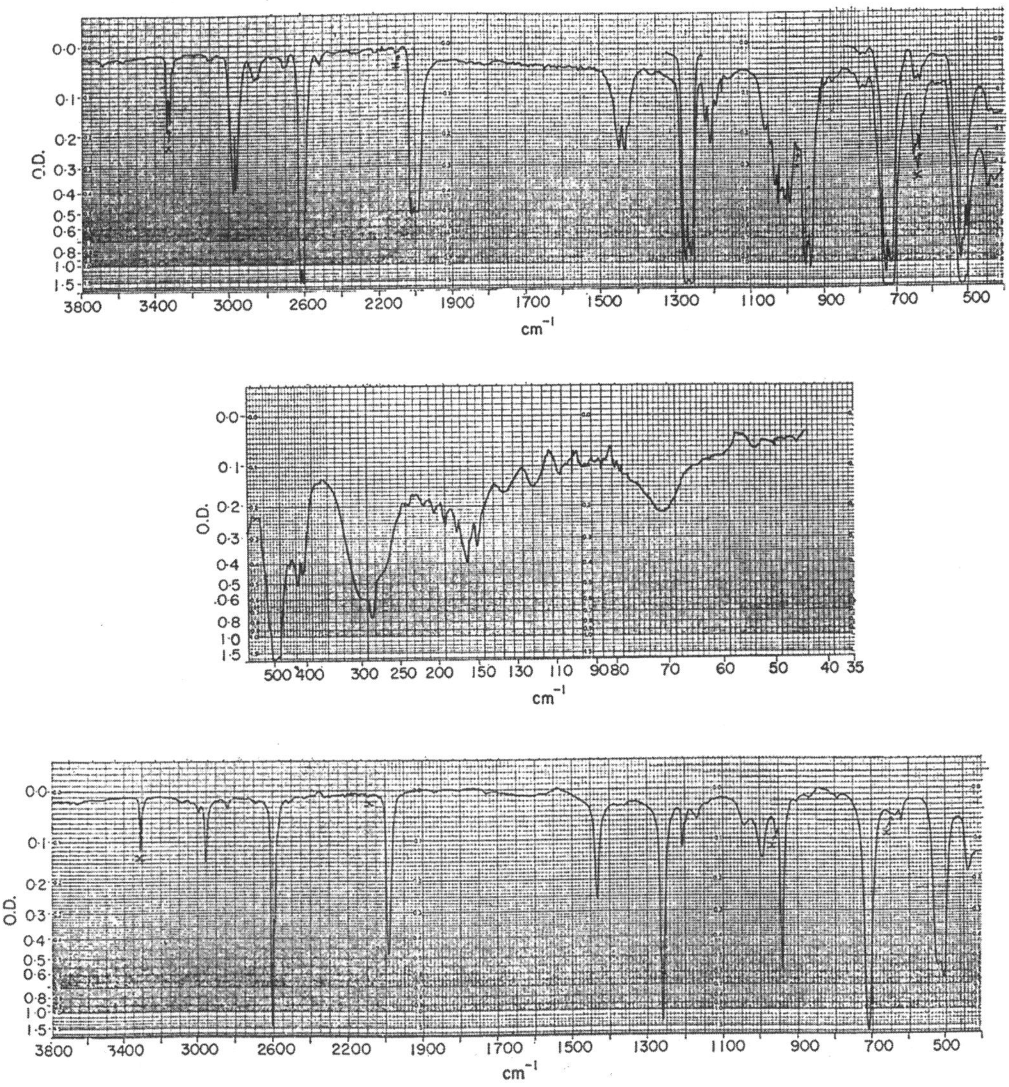

FIGURE 6.17b Top: Vapor-phase IR spectrum of 3-chloropropyne-1-d in a 10-cm KBr cell (33- and 100-mm Hg sample); Middle: 3-chloropropyne-1-d in a 10-cm polyethylene cell. Bottom: Solution-phase IR spectrum of 3-chloropropyne-1-d in 10% wt./vol. CCl_4 (3800–1333 cm^{-1}) and 10% wt./vol. in CS_2 (1333–400 cm^{-1}) using 0.1-mm KBr cells. Bands marked with X are due to 3-chloropropyne.

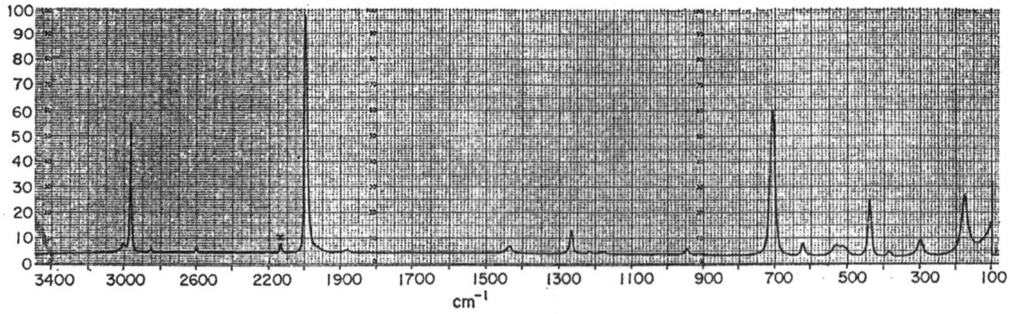

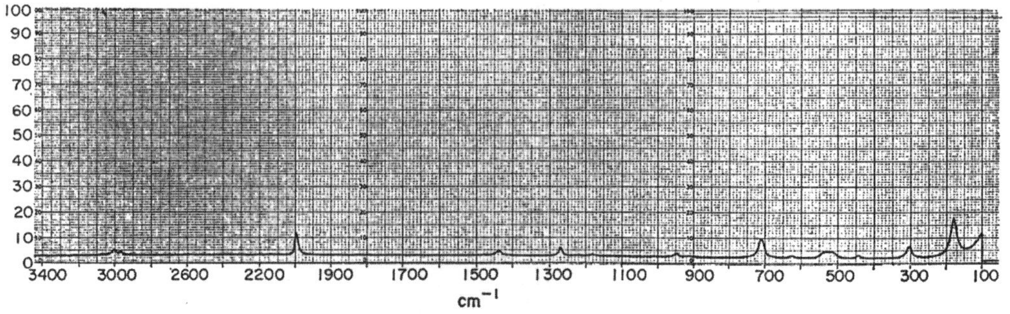

FIGURE 6.18 Top: A Raman liquid-phase spectrum of 3-chloropropyne-1-d. Bottom: A Raman polarized liquid-phase spectrum of 3-chloropropyne-1-d. Some 3-chloropropyne is present (15).

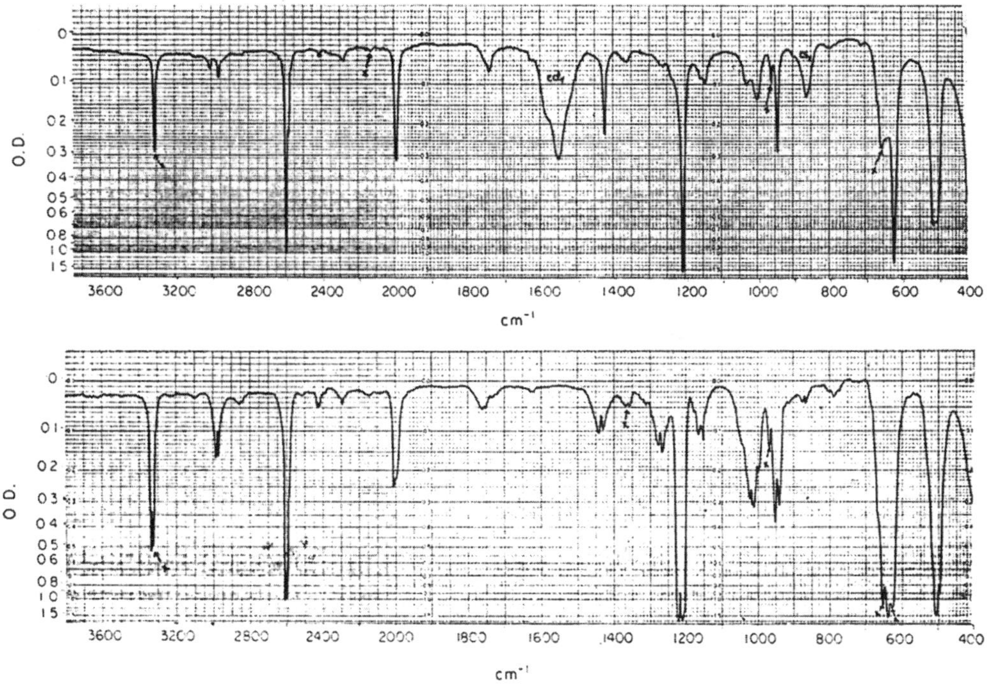

FIGURE 6.19 Top: Solution-phase IR spectrum of 3-bromopropyne-1-d in 10% wt./vol. in CCl_4 (3800–1333 cm^{-1}) and 10% wt./vol. in CS_2 (1333–450 cm^{-1}) using 0.1-mm KBr cells (16). Bottom: A vapor-phase IR spectrum of 3-bromopropyne-1-d in a 10-cm KBr cell (40-mm Hg sample). Infrared bands marked with X are due to the presence of 3-bromopropyne (16).

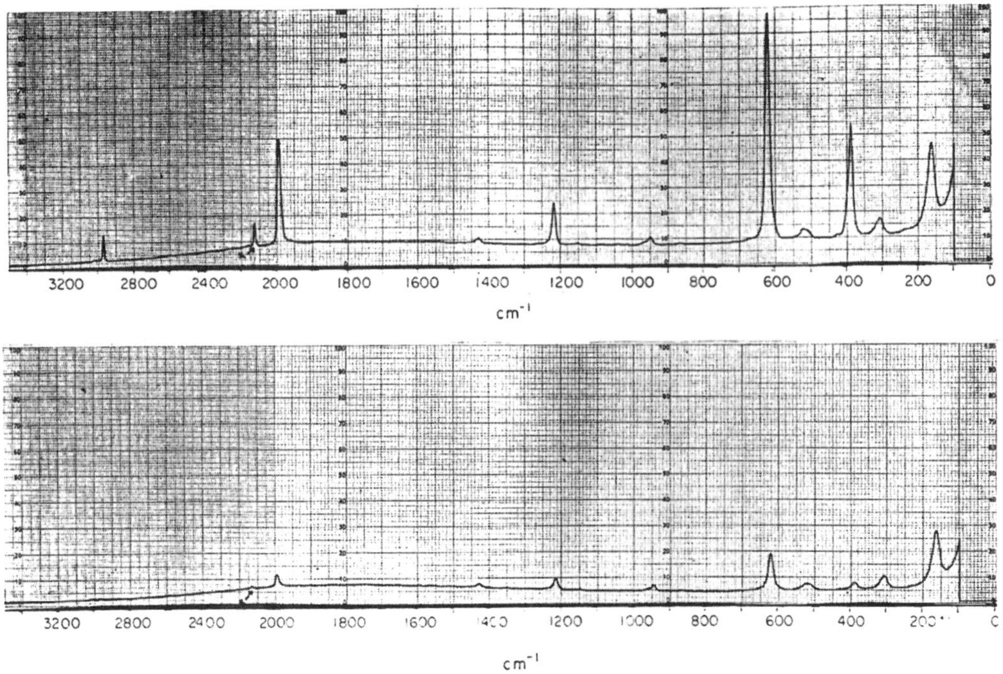

FIGURE 6.20 Top: Raman spectrum of 3-bromopropyne-1-d. Bottom: Polarized Raman spectrum of 3-bromopropyne-1-d. Infrared bands marked with X are due to the presence of 3-bromopropyne (16).

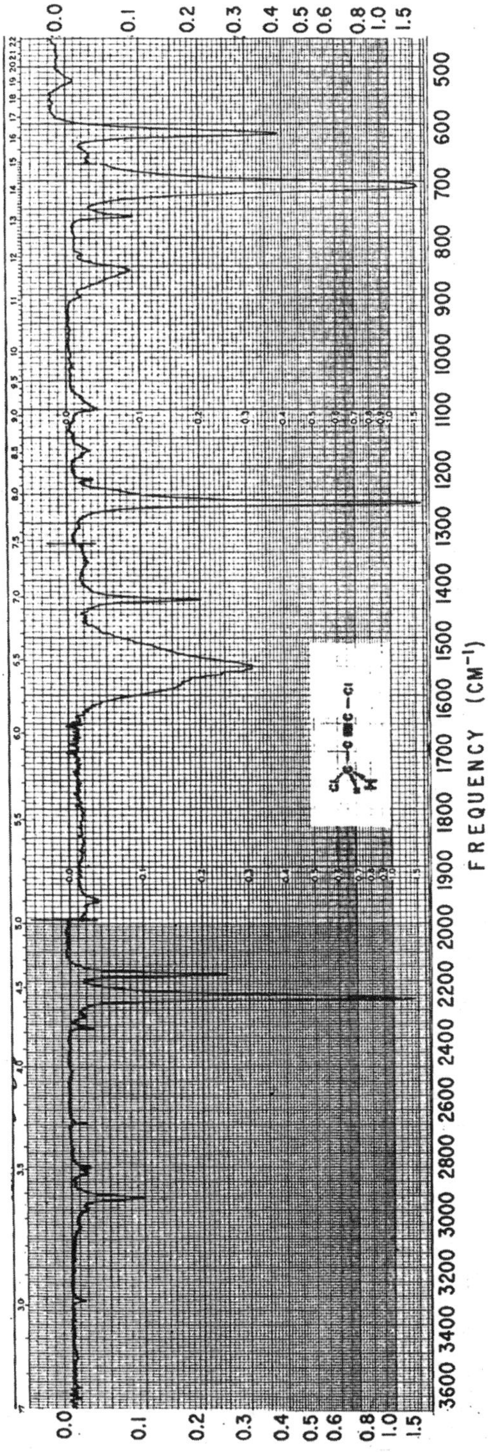

FIGURE 6.21 An IR spectrum of 1,3-dichloropropyne in 10% wt./vol. CCl_4 solution (3800–1333 cm^{-1}) and in CS_2 solution (1333–450 cm^{-1}) using 0.1-mm NaCl and KBr cells, respectively. Infrared bands at 1551 and 1580 cm^{-1} are due to CCl_4 and the IR band at 858 cm^{-1} is due to CS_2 (17).

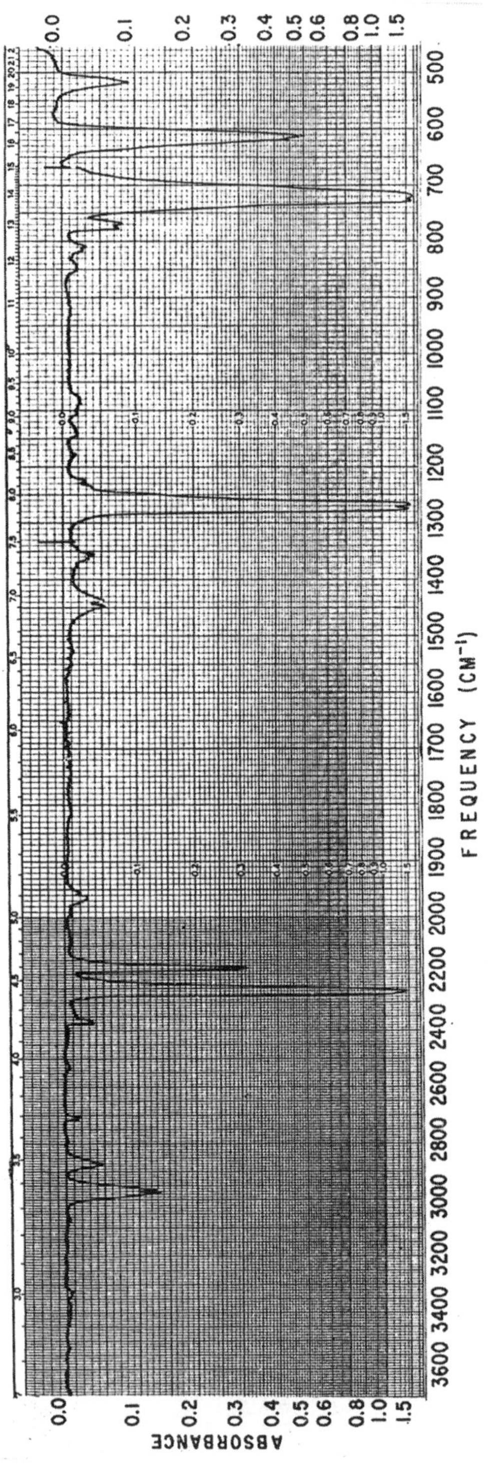

FIGURE 6.22 Vapor-phase IR spectrum of 1,3-dichloropropyne (ambient mm Hg sample at 25 °C in a 12.5-cm KBr cell) (17).

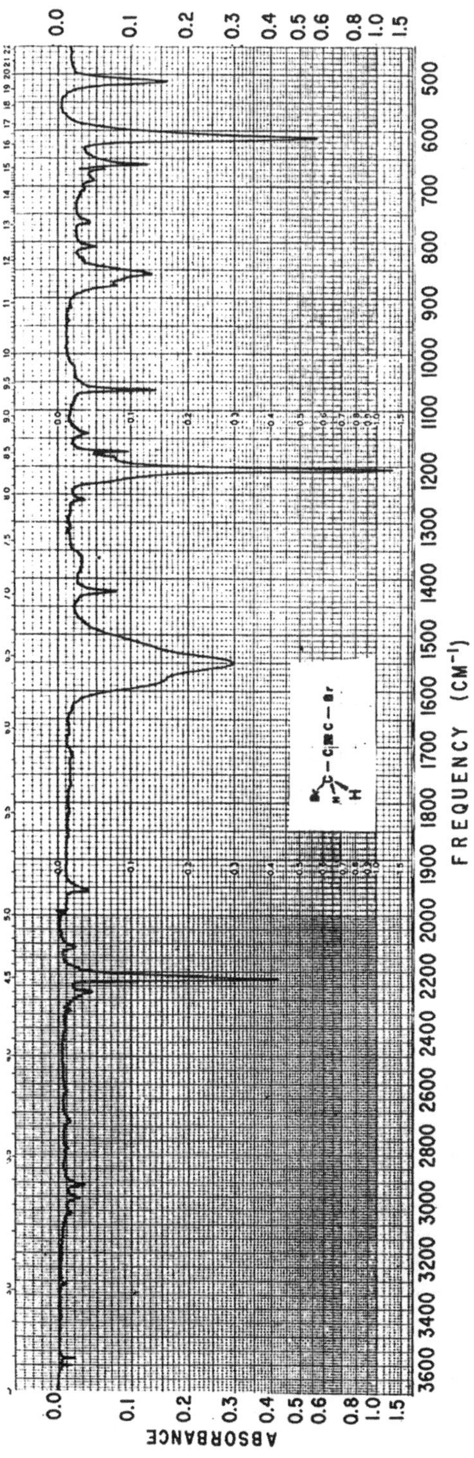

FIGURE 6.23 Solution-phase IR spectrum of 1,3-dibromopropyne in 10% wt./vol. in CCl_4 (3800–1333 cm^{-1}) and in CS_2 solution (1333–450 cm^{-1}) using 0.1-mm NaCl and KBr cells, respectively. Infrared bands at 1551 and 1580 cm^{-1} are due to CCl_4, and the IR band at 858 cm^{-1} to CS_2 (17).

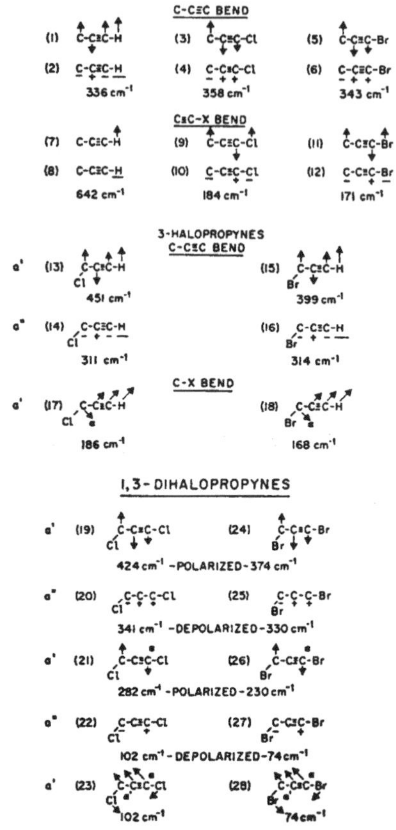

FIGURE 6.24 Approximate normal modes for propyne, 3-halopropynes, and 1,3-dihalopropynes.

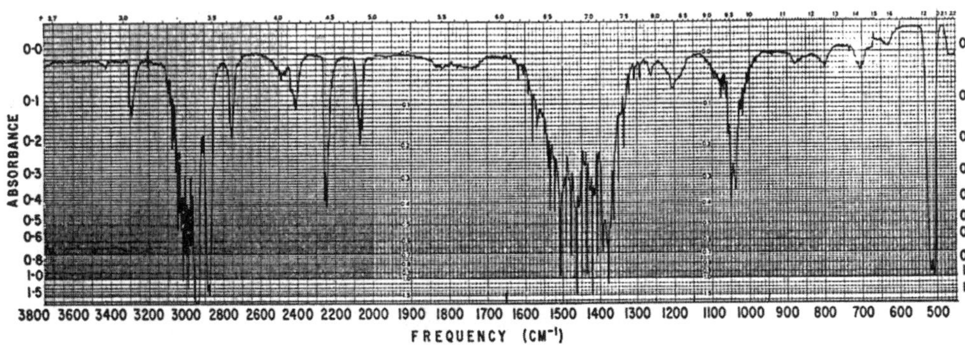

FIGURE 6.25 Vapor-phase IR spectrum for 1-bromopropyne in a 12.5-cm KBr cell. The weak IR band at 734 cm^{-1} is due to an impurity. The 1-bromopropyne decomposes rapidly in the atmosphere (25).

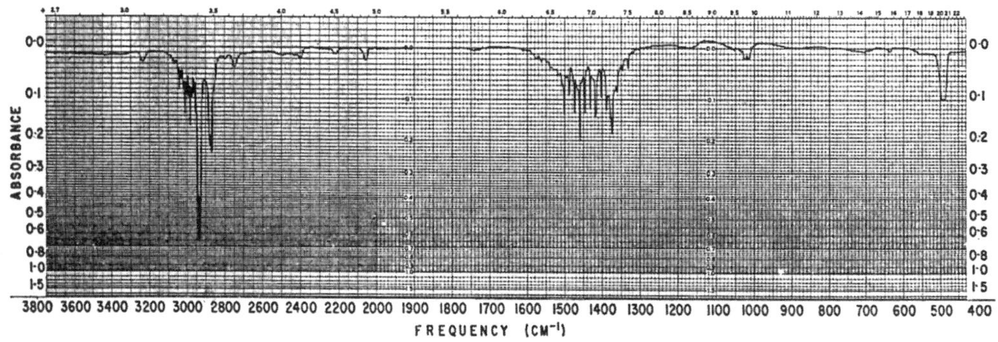

FIGURE 6.26 Infrared vapor spectrum for 1-iodopropyne in a 12.5-cm KBr cell (25).

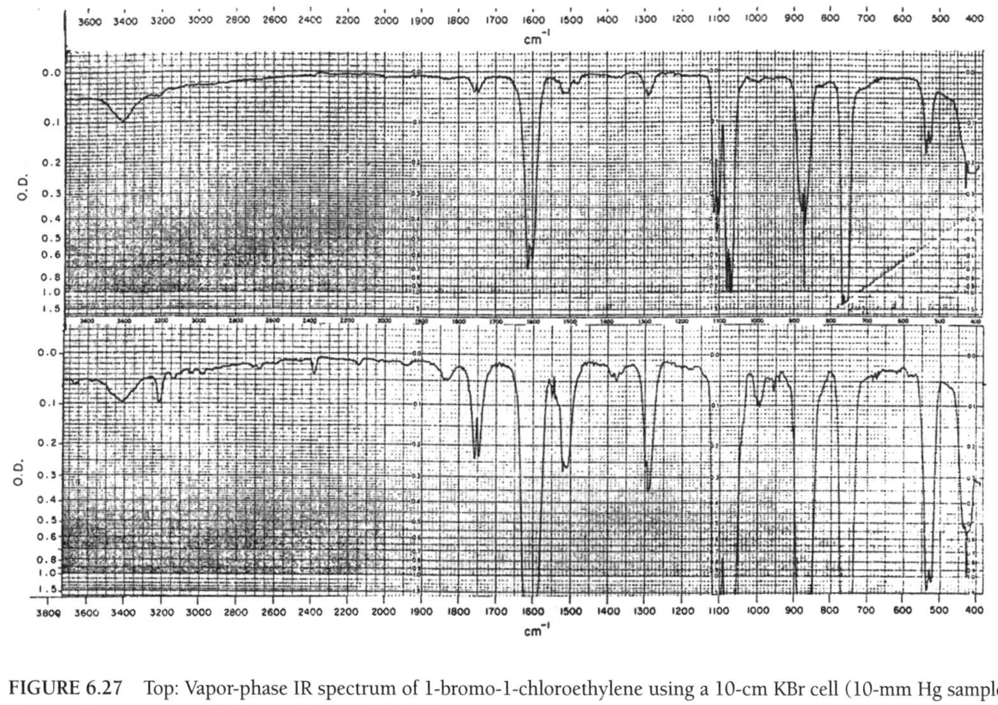

FIGURE 6.27 Top: Vapor-phase IR spectrum of 1-bromo-1-chloroethylene using a 10-cm KBr cell (10-mm Hg sample). Bottom: Same as upper (100-mm Hg sample) (19).

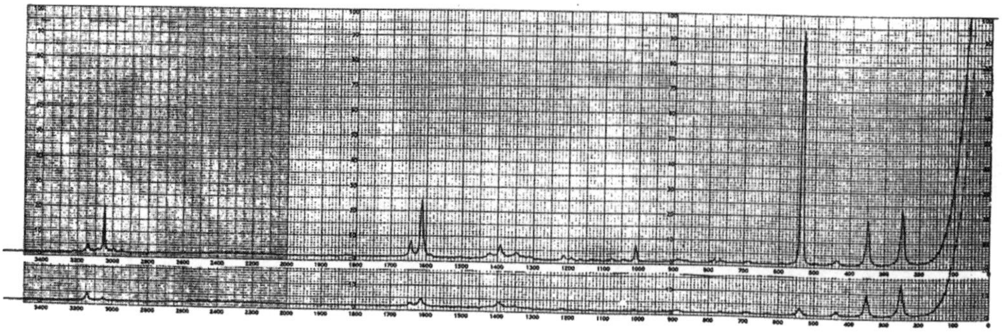

FIGURE 6.28 Raman liquid-phase spectrum of 1-homo-1-chloroethylene. top: Parallel polarization. bottom: Perpendicular polarization.

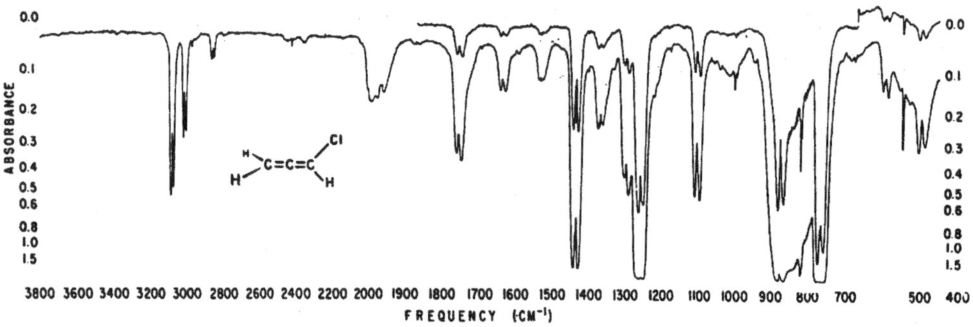

FIGURE 6.29 Vapor-phase IR spectrum of 1-chloropropadiene in a 12.5-cm KBr cell (50- and 100-mm Hg sample) (23).

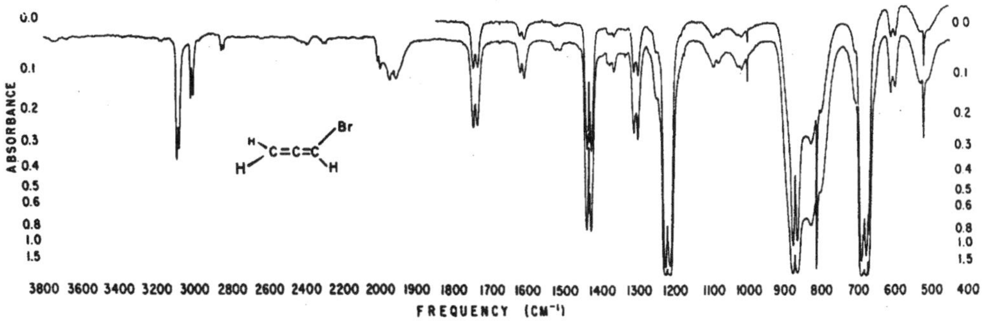

FIGURE 6.30 Vapor-phase IR spectrum of 1-bromopropadiene in a 12.5-cm KBr cell (50- and 100-mm Hg sample).

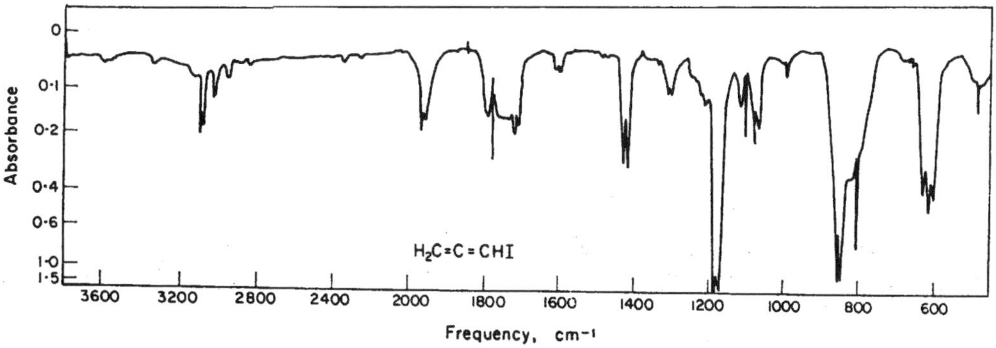

FIGURE 6.31 Vapor-phase IR spectrum of 1-iodopropadiene in a 5-cm KBr cell (vapor pressure at 25 °C). Bands at 1105 and 1775 cm^{-1} are due to the presence of an impurity.

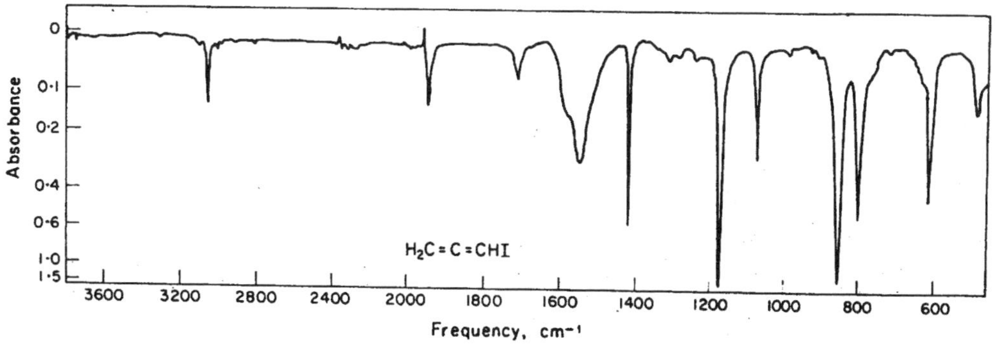

FIGURE 6.32 Infrared spectrum of 1-iodopropadiene in 10% wt./vol. CCl_4 solution (3800–1333 cm^{-1}) and 10% wt./vol. CS_2 solution (1333–450 cm^{-1}) using NaCl and KBr cells, respectively.

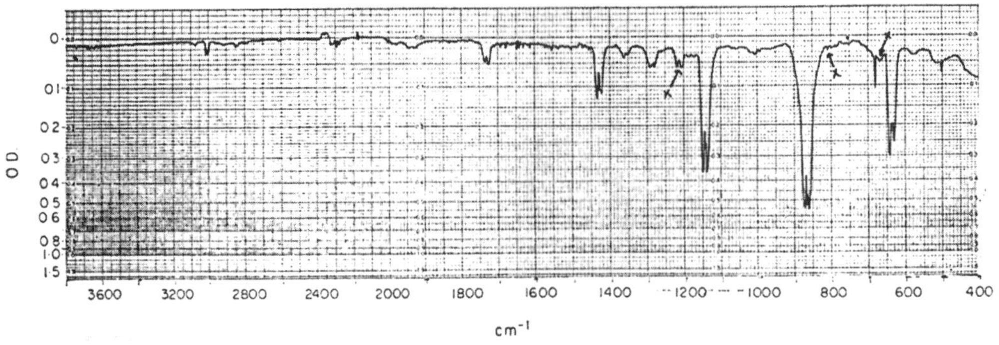

FIGURE 6.33 Vapor-phase IR spectrum of 1-bromopropadiene-1-d in a 12.5-cm KBr cell (50- and 100-mm Hg sample) (16).

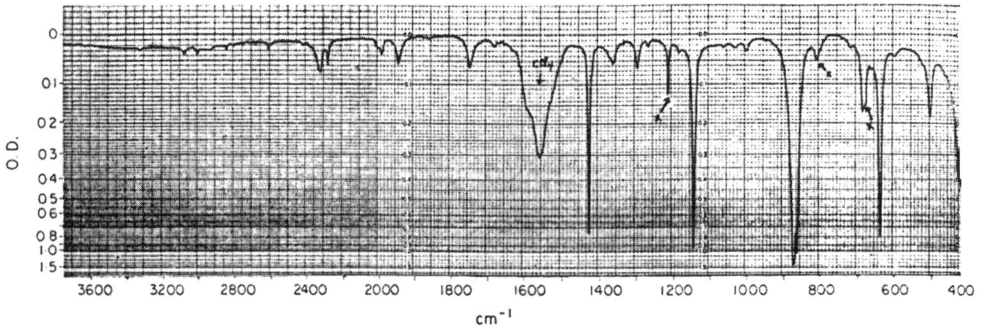

FIGURE 6.34 Solution-phase IR spectrum of 1-bromopropadiene-1-d 10% wt./vol. in CCl_4 (3800–1333 cm^{-1}) and in CS_2 solution using 0.1-mm KBr cells. Infrared bands marked with X are due to the presence of 1-bromopropadiene (16).

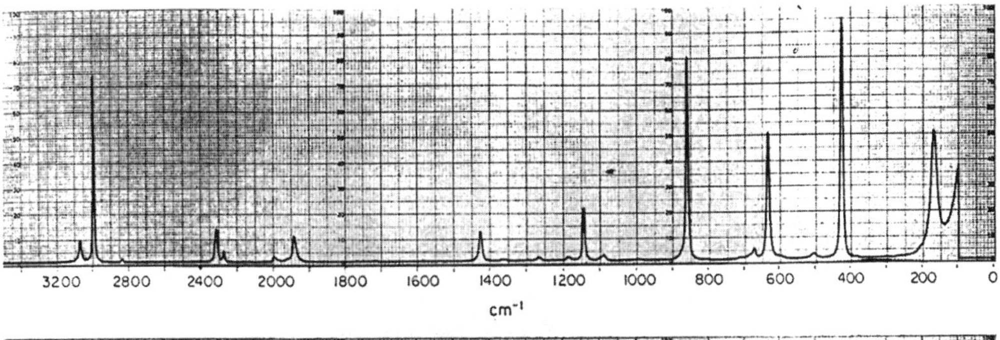

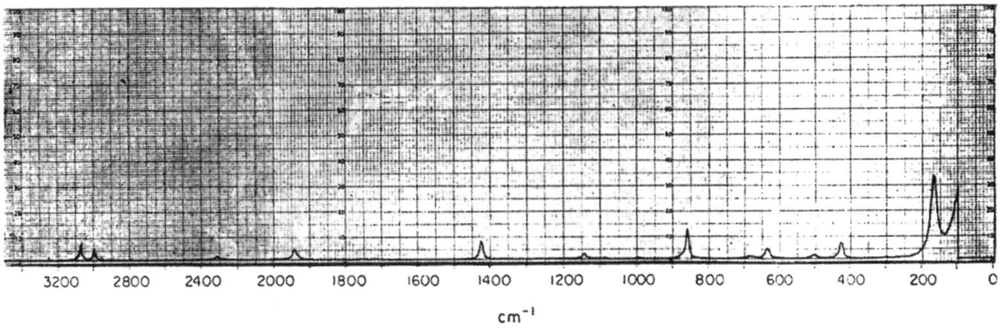

FIGURE 6.35 Top: Raman spectrum of 1-bromopropadiene-1-d using a capillary tube. Bottom: Polarized Raman spectrum of 1-bromopropadiene-1-d (16).

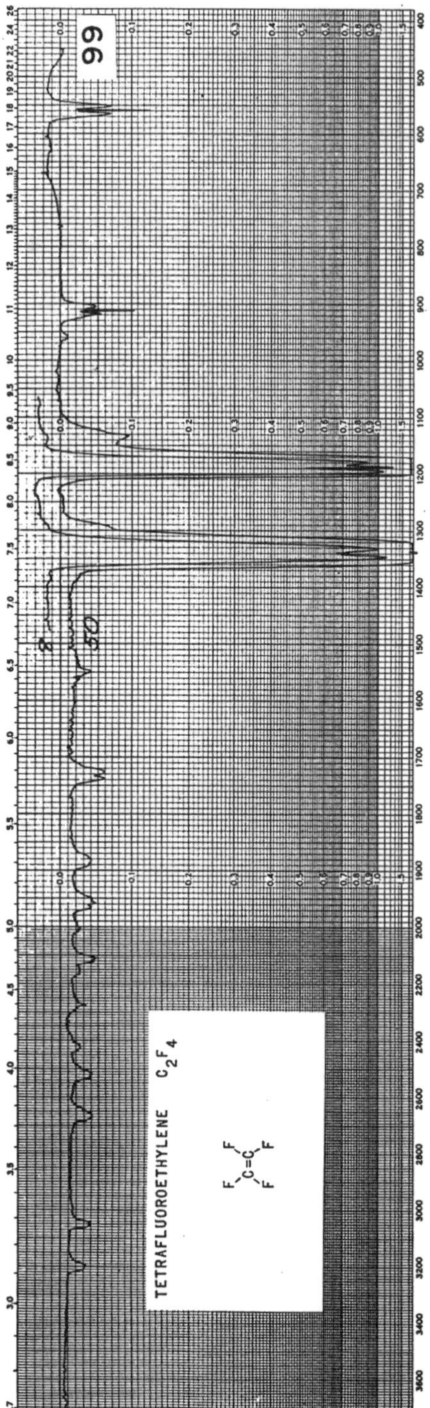

FIGURE 6.36* Vapor-phase IR spectrum of tetrafluoroethylene (8- and 50-mm Hg samples) (26).

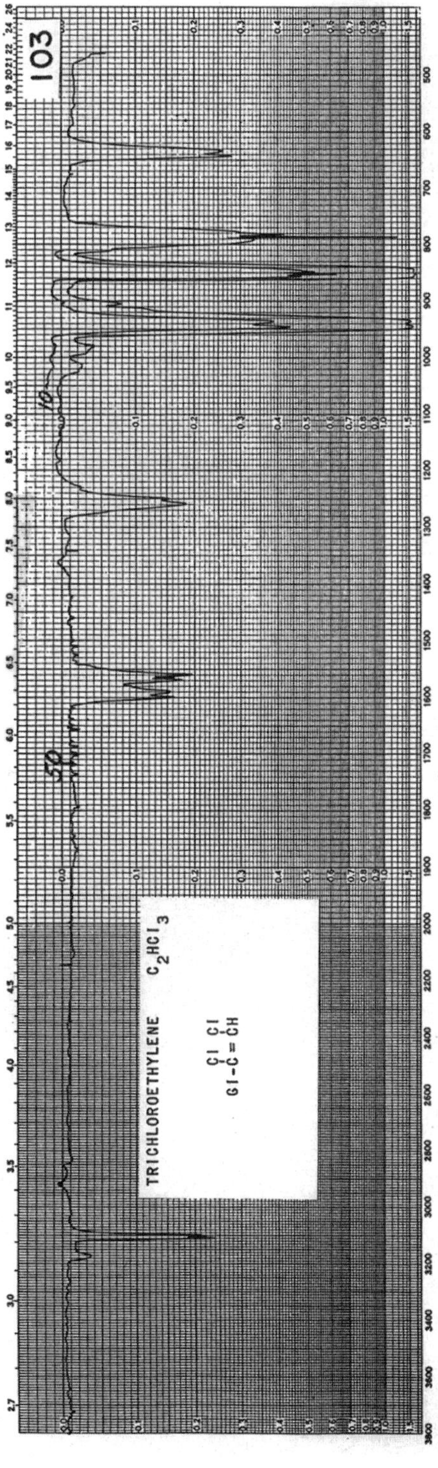

FIGURE 6.37* Vapor-phase IR spectrum of tetrachloroethylene (13-mm Hg sample) (26).

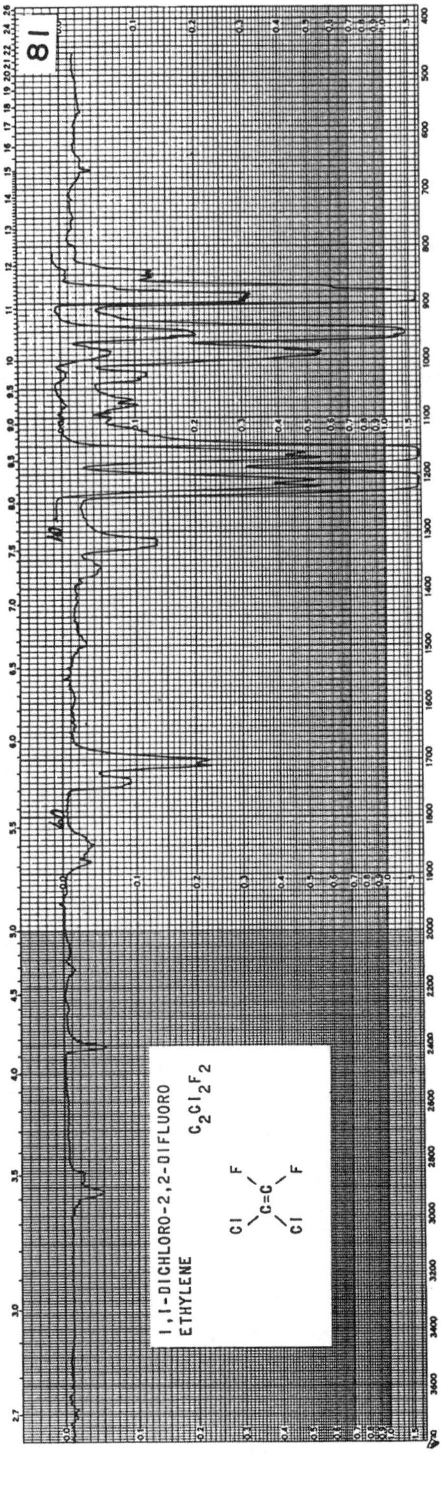

FIGURE 6.38* Vapor-phase IR spectrum of 1,1-dichloro-2,2-difluoroethylene (10- and 60-mm Hg samples) (26).

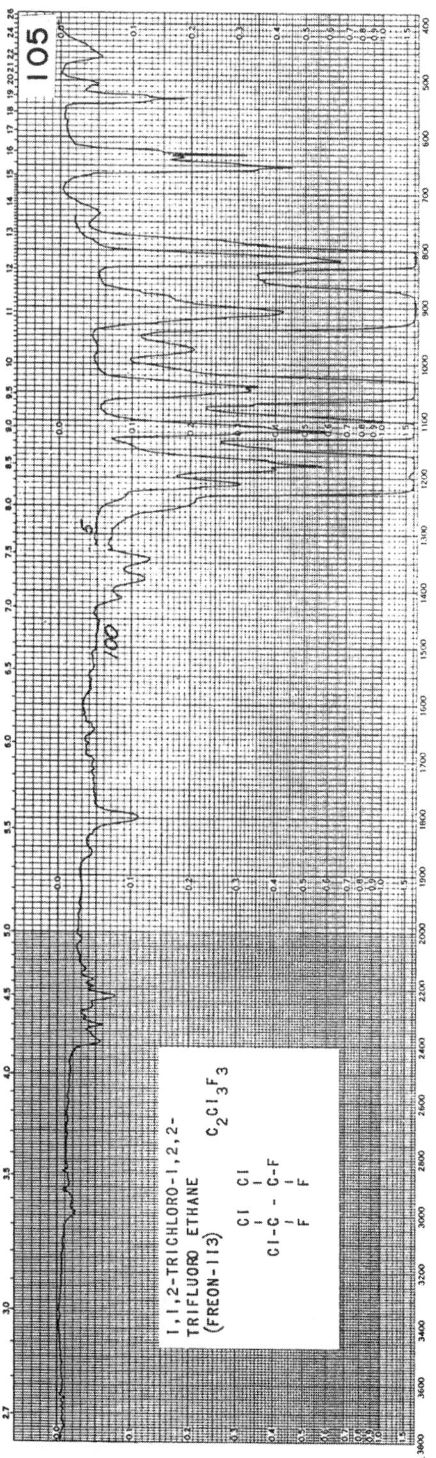

FIGURE 6.39* Vapor-phase IR spectrum of 1,1,2-trichloro-1,2,2-trifluoroethane (5 and 100 mm Hg samples) (26).

TABLE 6.1 Infrared and Raman data for the methylene halides

Compound	a.CH_2 str. cm^{-1} (RI)	s.CH_2 str. cm^{-1} (RI)	a.C–X_2 str. cm^{-1} (RI)	s.C–X_2 str. cm^{-1} (RI)	[a.C–X_2 str.]-[s.C–X_2 str.] cm^{-1}	[a.CH_2 str.]-[s.CH_2 str.] cm^{-1}
Methylene fluoride (3)	3027 (23,p)	2950 (41,p)	1180	1110 (18,p)	70	77
Methylene chloride (3)	3058 (3,p)	2989 (41,p)	742 (4,p)	704 (40,p)	38	69
Methylene bromide (3)	3065 (4,p)	2987 (22,p)	640 (8,p)	577 (88,p)	63	78
Methylene iodide (3)	3048 (1,p)	2967 (14,p)	570 (11,p)	480 (80,p)	90	81
	a.C–X_2 str. [vapor] cm^{-1}	a.C–X_2 str. [liquid] cm^{-1}	a.C–X_2 str. [v-p] cm^{-1}	s.C–X_2 str. [v- p]	s.C–X_2 str. [vapor] cm^{-1}	s.C–X_2 str. [liquid] cm^{-1}
Methylene chloride	744	742	−2	9	695	704
Methylene bromide	641	640	−1	−3	580	577
Methylene iodide	575	570	−5	−30	510	480
	C–X str. cm^{-1}	[C–F str.]-[C–X str.] cm^{-1}	σ p	[C–I str.]-[C–X str.] cm^{-1}	Mass X atomic mass	C–X bond length angstroms
Methyl fluoride [IR vap.] (2)	1044	0	0.52	−514	19	0.64
Methyl chloride [IR vap.] (2)	732	312	0.47	−436	35.457	0.99
Methyl bromide [IR vap.] (2)	608	436	0.45	−312	79.916	1.14
Methyl iodide [IR vap.] (2)	530	514	0.39	0	126.92	1.33

TABLE 6.2 Raman and infrared data for trihalomethane and tetrahalomethane

Compound	C–H str. cm^{-1} (RI)	a.CX_3 str. cm^{-1} (RI)	s.CX_3 str. cm^{-1} (RI)	CX_3 bend cm^{-1} (RI)
Trifluoromethane	3036 (8,p)	1376 (0.05)	1165 (1,p)	697 (4,p)
Triiodomethane	2975 (9,p)		572 (18,p)	
	a.CX_4 str. cm^{-1} (RI)	s.CX_4 str. cm^{-1} (RI)		
Tetrafluoromethane (3)	1283 (2,p)	908 (46,p)		
Tetrachloromethane (3)	797	790 (8,p)		
Tetrabromomethane (3)	671 (6,p)	662 (7,p)		
Tetraiodomethane (3)		560 (4,p)		
[IR vapor]	cm^{-1}			
Tetrafluoromethane (2)	1282			
Tetrachloromethane (2)	790			

TABLE 6.2a A comparison of C–X, CX_2, CX_3, and CX_4 stretching frequencies

X	C–X str. cm^{-1}	a.CX_2 str. cm^{-1}	s.CX_2 str. cm^{-1}	a.CX_3 str. cm^{-1}	s.CX_3 str. cm^{-1}	a.CX_4 str. cm^{-1}	s.CX_4 str. cm^{-1}	Range cm^{-1}
F	1044	1180	1110	1376	1165	1283	908	908–1376
Cl	732	742	704	[755]	[667]	797	790	667–797
Br	608	640	577	[649]	541	671	662	541–662
I	530	570	480	571	572	[—]	560	480–572

TABLE 6.3 Vapor- and liquid-phase infrared data for 1-haloalkanes

1-Haloalkane F	PC [planar trans rotational conformer] vapor cm^{-1} (A)	PH [gauche skeletal rotational conformer] vapor cm^{-1} (A)	PC liquid cm^{-1} (A)	PH liquid cm^{-1} (A)	[PC-PH] vapor cm^{-1}	[PC-PH] liquid cm^{-1}	A[PC]/A[PH] vapor	A[PC]/A[PH] liquid	PC cm^{-1}	PH cm^{-1}
$C_{10}H_{21}$	1050 (0.212)	1032 (0.239)	1042 (0.289)	1005 (0.300)	18	37	0.89	0.96	−8	−27
Cl										
C_3H_7	741 (0.150)	661 (0.120)	732 (0.262)	654 (0.320)	80	78	1.25	0.82	−9	−7
C_4H_9	750 (0.210)	661 (0.132)	735 (0.341)	659 (0.460)	79	76	1.59	0.74	−15	−2
C_5H_{11}	740 (0.210)	661 (0.132)	743 (0.250)	665 (0.160)	79	79	1.59	1.56	3	4
C_7H_{15}	740 (0.107)	661 (0.080)	758 (0.028)	652 (0.102)	79	106	1.34	0.27	18	−9
			725 (0.140)			73		1.37		
$C_{10}H_{21}$	740 (0.080)	662 (0.050)	749 (0.010)	650 (0.080)	78	99	1.6	0.13	9	−12
			721 (0.089)			71		1.11		
$C_{11}H_{23}$	740 (0.100)	660 (0.071)	751 (0.070)	645 (0.159)	80	106	1.41	0.44		
			711 (0.169)			66		1.06		
$C_{16}H_{33}$	738 (0.061)	660 (0.040)	758 (0.030)	659 (0.098)	78	99	1.53	0.31		
			725 (0.135)			66		1.38		
$C_{18}H_{37}$	~740 (0.029)	~664 (0.020)	750 (0.034)	650 (0.080)	76	100	1.45	0.43		
			~720 (0.129)			~70		1.61		
Br										
C_4H_9	650 (0.090)	564 (0.074)	638 (0.200)	554 (0.155)	86	84	1.22	1.29	−12	−10
C_5H_{11}	648 (0.118)	568 (0.090)	641 (0.280)	564 (0.241)	80	77	1.31	1.16	−7	−4
C_6H_{13}	652 (0.139)	569 (0.129)	642 (0.305)	565 (0.280)	83	77	1.08	1.09	−10	−4
C_7H_{15}	652 (0.071)	569 (0.090)	645 (0.070)	562 (0.070)	83	83	0.79	1	−7	−7
C_8H_{17}	652 (0.072)	569 (0.060)	640 (0.090)	559 (0.090)	83	81	0.9		−12	−10
$C_{11}H_{23}$	653 (0.060)	562 (0.050)	642 (0.060)	560 (0.043)	91	82	1.2	1.4	−11	−2
$C_{12}H_{25}$	650 (0.039)	561 (0.050)	640 (0.082)	560 (0.072)	89	80	0.78	1.14	−10	−1
$C_{14}H_{29}$	649 (0.050)	568 (0.036)	640 (0.079)	559 (0.066)	81	81	1.39	1.2	−9	−9
$C_{16}H_{33}$	651 (0.026)	572 (0.019)	646 (0.060)	565 (0.050)	79	81	1.37	1.2	−5	−7
$C_{19}H_{39}$	~645 (0.019)	~568 (0.020)	650 (0.059)	574 (0.039)	~77	76	0.95	1.51	5	6
I										
C_3H_7	591 (0.080)	500 (0.052)	590 (0.131)	498 (0.110)	91	92	1.53	1.19	−1	−2
C_4H_9	593 (0.090)	505 (0.040)	581 (0.121)	495 (0.061)	88	86	2.3	1.98	−12	−10
C_9H_{19}	598 (0.039)	504 (0.010)	593 (0.071)	493 (0.040)	94	100	3.9	1.78	−5	−11
$C_{16}H_{33}$	593 (0.030)	500 (0.010)	600 (0.046)	499 (0.031)	93	101	3	1.48	7	−1

TABLE 6.4 Vapor- and liquid-phase infrared data for 2-halobutane and tert-butyl halide

Rotational conformer 2-Haloalkane X	S(CH) C–X str.[1] vapor cm^{-1} (A)	S(CH) C–X str.[1] liquid cm^{-1} (A)	S(HH′) C–X str.[2] vapor cm^{-1} (A)	S(HH′) C–X str.[2] liquid cm^{-1} (A)	S(HH) C–X str.[3] vapor cm^{-1} (A)	S(HH) C–X str.[3] liquid cm^{-1} (A)	A[1]/A[2] vapor	A[1]/A[2] liquid	A[1]/A[3] vapor	A[1]/A[3] liquid
2-Halobutane										
X										
Cl	680 (0.130)	670 (0.392)	625 (0.200)	628 (0.341)	592 (0.045)	609 (0.460)	0.65	1.1	2.9	0.85
Br	612 (0.069)	605 (0.055)	529 (0.070)	525 (0.090)	478 (0.020)	480 (0.030)	0.99	0.61	3.5	2.3
I	580 (0.041)	570 (0.050)	489 (0.041)	479 (0.072)		454 (0.040)	1	0.69		1.02

Rotational conformer tert-Butyl halide X	T(HHH)	T(HHH)	C–X str. [v-l] cm^{-1}
Cl	580	570	−10
Br	521	520	−1
I		492	

2-Halobutane X	C–X str.[1] [v-l]	C–X str.[2] [v-l]	C–X str.[3] [v-l]
Cl	−10	3	17
Br	−7	−4	2
I	−10	−10	

	[1]–[2] [C–Cl str.]- [C–X str.] vapor cm^{-1}	[1]–[2] [C–Cl str.]- [C–X str.] liquid cm^{-1}	[1]–[3] [C–Cl str.]- [C–X str.] vapor cm^{-1}	[1]–[3] [C–Cl str.]- [C–X str.] liquid cm^{-1}
Cl	55	42	88	61
Br	83	80	134	125
I	91	91		116

TABLE 6.5 Vapor- and liquid-phase infrared and Raman liquid-phase data for 1-halocycloalkanes

Compound 1-X-cyclopentane	C–X str. vapor cm^{-1} (A)	C–X str. liquid cm^{-1} (A)	C–X str. [v- l] cm^{-1}	A[equatorial]/ A[axial] vapor	A[equatorial]/ A[axial] liquid	[C–Cl str.]- [C–X str.] vapor cm^{-1}	[C–Cl str.]- [C–X str.] liquid cm^{-1}	C–X str. Raman liquid cm^{-1}	RI	Ref.
X										
Cl [equatorial]	620 (0.046)	595 (0.115)	−25					624	vw	2, 6, 7
Cl [axial]								588	vw	8
Br [equatorial]	520 (0.060)	519 (0.110)	−1					516	s	2, 6
Br [axial]										
I [equatorial]	476 (0.041)	481 (0.140)	5					487	s	2, 8
I [axial]										
1-X-cyclohexane										
X										
Cl [equatorial]	740 (0.125)	729 (1.110)	−11	5.3	2.4	0	0	733	s	2, 9, 10
Cl [axial]	690 (0.048)	681 (0.470)	−9					688	m	2, 9
Δ [e-a]	[−50]	[−48]								
Br [equatorial]	692 (0.100)	686 (0.471)	−6	3.6	2	−48	−43	689	s	2, 9
Br [axial]	661 (0.028)	659 (0.250)	−2			−29	−22	660	m	2, 9
Δ [e-a]	[−31]	[−27]								
I [equatorial]	664 (0.160)	659 (0.510)	−5	5.3	2.7	−76	−70	654		2, 6
I [axial]	635 (0.030)	641 (0.190)	6			−55	−40			2
Δ [e-a]	[−29]	[−18]								

TABLE 6.6 Vapor- and liquid-phase infrared data for primary, primary dihaloalkanes

Compound 1,2-Dihaloethane X,X	C–X str.[1] vapor cm^{-1} (A)	C–X str.[1] liquid cm^{-1} (A)	C–X str.[2] vapor cm^{-1} (A)	C–X str.[2] liquid cm^{-1} (A)	C–X str.[3] vapor cm^{-1} (A)	C–X str.[3] liquid cm^{-1} (A)	C–X str.[1] [v-l] cm^{-1}	C–X str.[2] [v-l] cm^{-1}	C–X str.[3] [v-l] cm^{-1}	A[1]/ A[3] vapor	A[1]/ A[3] liquid	A[1]/ A[2] vapor	A[1]/ A[2] liquid
Cl,Cl	720 (1.230)	710 (0.940)		662 (0.320)	660 (0.070)	655 (0.380)	−10		−5	17.6	2.5		
Br,Br	594 (1.210)	590 (0.440)				545 (0.100)	−4						
I,I		485 (0.785)											
1,3-Dihalopropane													
X,X													
Cl,Cl	740 (0.240)				660 (0.540)								
Br,Br	655 (0.163)	645 (0.210)	599 (0.142)	585 (0.265)	559 (0.230)	543 (0.370)	−10	−14	−16	0.71	0.57	1.1	0.81
I,I	602 (0.070)	590 (0.105)			526 (0.050)	512 (0.105)	−12		−14	1.4	1		
						483 (0.110)							
1,4-Dihalobutane													
X,X													
Cl,Cl	778 (0.524)	780 (0.320)	742 (0.434)	735 (0.300)	660 (0.511)	655 (0.415)	2	−7	−5	1.03	0.77	1.21	1.07
Br,Br	659 (0.178)	649 (0.160)			572 (0.311)	560 (0.260)	−10		−12	0.31	0.62		
I,I	599 (0.080)	591 (0.052)			505 (0.078)	502 (0.120)	−8		−3	1.02	0.43		
1,5-Dihalopentane													
X,X													
Cl,Cl	748 (0.389)	740 (0.285)		720 (0.255)	662 (0.298)	651 (0.379)	−8		−11	1.3	0.75		
Br,Br	654 (0.299)	647 (0.525)			571 (0.309)	570 (0.580)	−7		−1	0.97	0.91		
I,I		584 (0.140)		604 (0.109)		492 (0.140)							
1,6-Dihalohexane													
X,X													
Cl,Cl	740 (0.350)				660 (0.260)								
Br,Br	654 (0.169)				570 (0.157)								
I,I	600 (0.090)	590 (0.190)			509 (0.047)	498 (0.070)	−10		−11	1.9	2.7	2.7	

TABLE 6.7 Raman data for methyl halides and infrared and Raman data for tetrabromoalkanes

Compound	Phase	a.CH_3 str. cm^{-1} (RI)	s.CH_3 str. cm^{-1} (RI)	2(CH_3 bend) cm^{-1} (RI)	C–X str. cm^{-1} (RI)					
Methyl fluoride (3)	vapor		2967 (41,p)	2865 (26,p)	1043 (5,p)					
Methyl chloride (3)	vapor	3052 (4)	2968 (95,p)	2879 (7,p)	729 (36,p)					
Methyl bromide (3)	vapor		2972 (43,p)	2862 (4,p)	609 (42,p)					
Methyl iodide (3)	liquid		3042 (2,p)	2946 (44,p)	523 (96,p)					
	Raman		CH str. cm^{-1} (RI)	s.CH_2 str. cm^{-1} (RI)	a.CBr_2 str. cm^{-1} (RI)	a.CBr_2 str. cm^{-1} (RI)	s.CBr_2 str. cm^{-1} (RI)	CBr_2 bend cm^{-1} (RI)	CBr_2 wag cm^{-1} (RI)	CBr_2 twist cm^{-1} (RI)
1,1,2,2-tetrabromoethane	liquid				714 (2)	664 (1)	537 (2)	451 (0)	219 (9)	176 (2)
	Raman				C–Br str.	C–Br str.	C–Br str.	CCBr bend	CCBr torsion	
1,2,3,4-tetrabromobutane	liquid		2967 (1)	2935 (0)	702 (3)	567 (8)	532 (3)	305 (1)	191 (0)	
	IR				cm^{-1} (A)	cm^{-1} (A)	cm^{-1} (A)			
1,1,2,2-tetrabromoethane (2)	vapor				710 (1.240)	642 (0.560) 589 (0.250)	538 (0.135) 619 (1.030)			

TABLE 6.8 Carbon halogen stretching frequencies for ethylene propyne, 1,2-epoxpropane, and propadiene derivatives

Compound	ν C–F cm^{-1}	ν C–Cl cm^{-1}	ν C–Br cm^{-1}	ν C–I cm^{-1}	Rotational conformer				Ref.
3-Halopropene	1005.8	739.4	690.6	669.1	gauche				11
	989.3				cis				11
3-Halo,1,2-epoxypropane	1018.5	695.8	654.9	604	RT				12
	992	727.5	643.5		R1				12
3-Halopropyne	1039	725	621	570					13
3-Chloropropyne-1d	[—]	723	[—]	[—]					14
3-Bromopropyne-1d	[—]	[—]	634	[—]					15
1,3-Dihalopropyne	[—]	709	613	[—]					16
1-Haloethylene	1157	719	[—]	[—]					17
	F,F ν$a.CF_2$ cm^{-1}	Cl,Cl ν$a.CCl_2$ cm^{-1}	Cl,Br ν$a.CClBr$ cm^{-1}	Br,Br ν$a.CBr_2$ cm^{-1}	F,F ν$s.CF_2$ cm^{-1}	Cl,Cl ν$s.CCl_2$ cm^{-1}	Cl,Br ν$s.CClBr$ cm^{-1}	Br,Br ν$s.CBr_2$ cm^{-1}	
1,1-Dihaloethylene	1301	788	765	698	922	601	531	474	18–22
	F_4	F_4	Cl_4	Cl_4	Br_4	Br_4			
Tetrahaloethylene	1340	1189	909	875	768	632	[see text]		17
Difluorodichloroethylene	1219	1179	956	890			[see text]		21
Trifluorochloroethylene	1330	1211	1032				[see text]		22
Trifluorobromoethylene	1329	1202	1025				[see text]		22
Trichlorofluoroethylene	1181	987	869				[see text]		22
Trichloroethylene	931	852	639				[see text]		22
	ν C–F cm^{-1}	ν C–Cl cm^{-1}	ν C–Br cm^{-1}	ν C-I cm^{-1}					
1-Halopropadiene	est. [1050]	767	681	609					23
1-Bromopropadiene-1d	[—]	[—]	636	[—]					15
1-Halopropyne	[—]	574	464	403					24
1,3-Dihalopropyne	[—]	617	512	[—]					16

CHAPTER 7

Nitroalkanes, Nitrobenzenes, Alkyl Nitrates, Alkyl Nitrites, and Nitrosamines

Nitroalkanes	174
Nitroalkanes: Vapor vs Liquid-Phase Data	176
Tetranitromethane	176
Nitrobenzenes	177
Nitrobenzene in Different Physical Phases	177
4-Nitrobenzaldehyde	178
3-X-Nitrobenzenes	179
2-Nitrobenzenes	180
4-X-Nitrobenzenes in CCl_4 and $CHCl_3$ Solutions	181
Alkyl Nitrates	182
Ethyl Nitrate vs Nitroalkanes and Nitrobenzene	183
Alkyl Nitrites	183
Raman Data for Organonitro Compounds	184
A Summation of ν asym. NO_2 and ν sym. NO_2 in Different Physical Phases	184
Nitrosamines	185
References	185

Figures

Figure 7-1	186 (175)	Figure 7-12	196 (182)
Figure 7-2	187 (175)	Figure 7-13	197 (182)
Figure 7-3	188 (175)	Figure 7-14	198 (182)
Figure 7-4	189 (175)	Figure 7-15	199 (182)
Figure 7-5	190 (178)	Figure 7-16	200 (183)
Figure 7-6	190 (178)	Figure 7-17	201 (183)
Figure 7-7	191 (181)	Figure 7-18	202 (184)
Figure 7-8	192 (182)	Figure 7-19	203 (184)
Figure 7-9	193 (182)	Figure 7-20	204 (184)
Figure 7-10	194 (182)	Figure 7-21	205 (184)
Figure 7-11	195 (182)		

Tables

Table 7-1	206 (175)	Table 7-2	208 (175)
Table 7-1a	207 (175)	Table 7-3	209 (176)

Table 7-4	209 (177)	Table 7-19	221 (181)
Table 7-5	210 (177)	Table 7-20	222 (181)
Table 7-6	211 (177)	Table 7-21	223 (181)
Table 7-7	212 (177)	Table 7-22	224 (182)
Table 7-8	213 (177)	Table 7-23	224 (182)
Table 7-9	214 (178)	Table 7-24	225 (183)
Table 7-10	215 (179)	Table 7-25	226 (183)
Table 7-11	216 (179)	Table 7-26	226 (183)
Table 7-12	216 (180)	Table 7-27	227 (184)
Table 7-13	217 (180)	Table 7-28	227 (184)
Table 7-14	218 (180)	Table 7-29	228 (184)
Table 7-15	218 (180)	Table 7-30	229 (184)
Table 7-16	219 (180)	Table 7-31	229 (184)
Table 7-17	220 (180)	Table 7-32	230 (184)
Table 7-18	221 (180)	Table 7-33	230 (185)

*Numbers in parentheses indicate in-text page reference.

NITROALKANES

The variations in frequencies of antisymmetric NO_2 stretching, ν asym. NO_2, and symmetric NO_2 stretching, ν sym. NO_2 have been attributed to inductive and/or resonance effects (1, 2). Empirical correlations have been developed for the calculation of ν asym. NO_2 and ν sym. NO_2 frequencies in unknown compounds containing this functional group. These correlations relate to the substituent groups joined to the α-carbon atom of nitroalkanes and the wavenumber values of ν asym. NO_2 and ν sym. NO_2 for nitromethane (1).

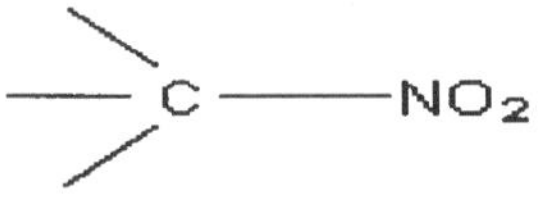

α

Vapor phase
$$\left[\begin{array}{l} \nu \text{ asym. } NO_2 = 1582 + \Sigma(Y_1 + X_1) \quad (1) \\ \nu \text{ sym. } NO_2 = 1397 + \Sigma(Y_2 + X_2) \quad (2) \end{array}\right.$$

ν asym. NO_2
$$\left[\begin{array}{l} Y_1 = -7\ cm^{-1} \text{ for each } CH_3 \text{ or } C_2H_5 \text{ group joined to the } \alpha\text{-carbon atom} \\ X_1 = 10\ cm^{-1} \text{ for each Cl atom joined to the } \alpha\text{-carbon atom} \end{array}\right.$$

ν asym. NO_2
$$\left[\begin{array}{l} Y_2 = -17\ cm^{-1} \text{ for each } CH_3 \text{ or } C_2H_5 \text{ group joined to the } \alpha\text{-carbon atom} \\ X_2 = -29\ cm^{-1} \text{ for each Cl atom joined to the } \alpha\text{-carbon atom} \end{array}\right.$$

Application of Eqs. (1) and (2) allow one to estimate the observed frequencies to within $16\,cm^{-1}$ in the vapor phase.

Equations 3 and 4 were developed to estimate ν asym. NO_2 and ν sym. NO_2 frequencies in the liquid phase (1). The 1558 cm^{-1} and 1375 cm^{-1} are for nitromethane in

$$\nu \text{ asym. } NO_2 = 1558 \text{ cm}^{-1} + \Sigma\Delta R \quad (3)$$

$$\nu \text{ sym. } NO_2 = 1375 + \Sigma\Delta R \quad (4)$$

the liquid phase (1). The ΔR values are for atoms or groups joined to the α- and β- carbon atoms, and their $\Delta\, cm^{-1}$ values are presented here in Table 7.1.

Figures 7.1 and 7.2 show plots of ν asym. NO_2 (observed) vs ν asym. NO_2 (calculated) and ν sym. NO_2 (observed) vs ν sym. NO_2 (calculated), respectively, and the agreement is essentially linear. These correlations do not distinguish between primary, secondary or tertiary nitroalkanes.

Applying Eqs. 3 and 4 to determine the $\nu\, NO_2$ vibrations for CF_3NO_2 is shown here:

$$\nu \text{ asym. } NO_2 = 1558 + 3(17) = 1609 \text{ cm}^{-1} \text{ vs } 1607 \text{ cm}^{-1} \text{ observed}$$

$$\nu \text{ sym. } NO_2 = 1375 + 3(-23) = 1306 \text{ cm}^{-1} \text{ vs } 1311 \text{ cm}^{-1} \text{ observed}$$

Figure 7.3 is a vapor-phase IR spectrum for nitromethane. The IR band at 919 cm^{-1} is assigned to C—N stretching, ν C—N, and the IR band at 659 cm^{-1} is assigned to NO_2 bending, $\delta\, NO_2$. The weak band at 605 cm^{-1} is assigned to NO_2 wagging, W NO_2, and an IR band at 472 cm^{-1} (liquid phase) is assigned to NO_2 rocking, $\rho\, NO_2$ (3). Figure 7.4 is a vapor-phase IR spectrum for 2-nitropropane. Presumably the IR band at ~625 cm^{-1} results from $\delta\, NO_2$, and the W NO_2 mode at 535 cm^{-1}.

Table 7.1a lists the vapor-phase IR data for nitroalkanes. Most primary nitroalkanes exist as a mixture of trans and gauche conformers due to rotation of the C—CNO_2 moiety. The trans ν C—N vibration has been assigned in the range 895–914 cm^{-1} and the gauche ν C—N vibration has been assigned in the range 868–894 cm^{-1} (4). Figure 7.4 shows that the absorbance for the 901 cm^{-1} trans ν C—N vibration is much less than the absorbance for the 851 cm^{-1} gauche ν C—N vibration. The gauche ν C—N vibration for 1-nitrobutane, 1-nitropentane, and 1-nitrohexane are assigned in the range 850–859 cm^{-1}.

The $\delta\, NO_2$ vibration for these *n*-alkanes listed in Table 7.1a is assigned in the range 603–659 cm^{-1}.

The frequency separation between ν asym. NO_2 and ν sym. NO_2 for these nitroalkanes in the vapor phase varies between 185 and 205 cm^{-1}, and the absorbance ratio (A) ν asym. NO_2/(A) ν sym. NO_2 varies between 1.94 and 3.47.

Table 7.2 lists the vapor-phase IR data for nitroalkanes. The ν asym. CH_3 vibrations occur in the range 2975–2995 cm^{-1}, the ν asym. CH_2 vibrations occur in the range 2940–2950 cm^{-1}, and the ν sym. CH_2 vibration in the range 2880–2910 cm^{-1}. In the case of 2-nitropropane it appears as though ν sym. CH_3 is in Fermi resonance (FR) with 2δ asym. CH_3. The $\delta\, CH_2$ and $\rho\, CH_2$ vibrations occur in the ranges 1446–1458 cm^{-1} and 1113–1129 cm^{-1}, respectively.

If one calculates the absorbance ratios: (A) ν sym. NO_2/(A) ν sym. CH_2; (A) ν sym. NO_2/(A) ν asym. CH_2; and (A) ν sym. NO_2/(A) ν asym. CH_3, the calculated values decrease as the number of carbon atoms in the nitroalkanes increase. Absorbance ratios such as these aid in identifying specific nitroalkanes.

NITROALKANES: VAPOR VS LIQUID-PHASE DATA

Table 7.3 compares vapor- and liquid-phase IR data for the ν NO_2 vibrations for 12 nitroalkanes. In both vapor and liquid phases, ν asym. NO_2 occur in the ranges 1555–621 cm^{-1} and 1535–1601 cm^{-1}, respectively. Thus, the ν asym. NO_2 vibration for these nitroalkanes decreases in frequency by 17 to 32 cm^{-1} in going from the vapor to the liquid phase. In the case of ν sym. NO_2, this vibration occurs in the range 1310–1397 cm^{-1} in the vapor and in the range 1310–1381 cm^{-1} in the liquid phase. Thus, the ν sym. NO_2 vibration remains constant in the case of trichloronitromethane or decreases in frequency by 1–22 cm^{-1} in going from the vapor to the liquid phase. Thus, the ν asym. NO_2 vibration decreases much more in frequency than does the ν sym. NO_2 vibration in going from the vapor to the liquid phase.

TETRANITROMETHANE

Tetranitromethane has a maximum symmetry of Td. In the vapor phase, IR bands are noted at 1651 cm^{-1} (A = 0.180), 1623 cm^{-1} (A = 1.240), 1278 cm^{-1} (A = 0.220), 801 cm^{-1} (A = 0.330), and very weak bands at ~665 cm^{-1}, and 604 cm^{-1} (1). In the liquid phase, IR bands are noted at 1642 cm^{-1} (A = 0.490), 1610 cm^{-1} (A = 0.840), 1268 cm^{-1} (A = 0.490), 799 cm^{-1} (A = 0.611), 662 cm^{-1} (A = 0.070), and 602 cm^{-1} (A = 0.113).

Nitroalkanes have been reported to exhibit ν asym. NO_2 in the range 1555–1621 cm^{-1} (1). The absorbance ratio of the 1623 cm^{-1}/1651 cm^{-1} bands in the vapor phase is 6.89, and the absorbance ratio of the 1642 cm^{-1}/1610 cm^{-1} bands in the liquid phase is 1.71. These two IR bands must result from ν asym. $(NO_2)_4$ vibrations. It is possible that the 1623 cm^{-1} (VP) and 1610 cm^{-1} (LP) bands result from out-of-phase ν asym. $(NO_2)_4$ and the 1651 cm^{-1} (VP) and 1642 cm^{-1} (LP) bands result from in-phase ν asym. $(NO_2)_4$. The ratio of the out-of-phase ν asym. $(NO_2)_4$ vibrations to the in-phase ν asym. $(NO_2)_4$ vibrations is higher in the vapor phase at elevated temperature. There are no other spectral features to suggest that these two bands are the result of ν asym. $(NO_2)_4$ being in FR with a combination tone.

Nitroalkanes in the vapor phase exhibit ν sym. NO_2 in the range 1310–1397 cm^{-1} (1). It is suggested that the 1278 cm^{-1} (VP) and the 1268 cm^{-1} (LP) bands result from out-of-phase ν sym. $(NO_2)_4$. With Td symmetry, the in-phase ν sym. $(NO_2)_4$ vibration would be IR inactive (Raman active). A weak IR bond occurs at 2885 cm^{-1} in the vapor phase and this can be assigned to the combination tone (1623 + 1278 = 2901 cm^{-1}). In the liquid phase the combination tone is (1610 + 1268 = 2878 cm^{-1} vs 2870 cm^{-1} observed). This assignment is for the out-of-phase ν asym. $(NO_2)_4$ + out-of-phase ν sym. $(NO_2)_4$ combination tone. A weak IR band is observed at 2970 cm^{-1} in the vapor phase. If we assume that the in-phase ν asym. $(NO_2)_4$ and in-phase ν sym. $(NO_2)_4$ also exhibit a combination tone (the 2970 cm^{-1} IR band), the in-phase ν sym. $(NO_2)_4$ vibration is calculated to occur at 1319 cm^{-1}. Raman data is required to determine if this assumption is correct.

Weak IR bands are observed at 2550 cm^{-1} (VP) and 2532 cm^{-1} (LP), and these can be assigned to the overtone of out-of-phase ν asym. $(NO_2)_4$.

Infrared bands at 801 cm^{-1} (VP), 799 cm^{-1} (A = 0.611) (LP), ~665 cm^{-1} (VP), 666 cm^{-1} (A = 0.070) (LP) and 604 (VP), and 602 cm^{-1} (A = 0.113) (LP) are tentatively assigned to the

$\delta(NO_2)_4$, $W(NO_2)_4$, and $\rho(NO_2)_4$ vibrations. The decrease in frequency for out-of-phase ν asym. $(NO_2)_4$ and out-of-phase ν sym. $(NO_2)_4$ in going from the vapor phase to the liquid phase is 131 cm^{-1} and 10 cm^{-1}, respectively. The data for tetranitromethane was read from Sadtler IR vapor spectra and standard condensed phase IR spectra (1).

It is suggested that the range for ν asym. NO_2 for nitroalkanes extends over the range 1555–1651 cm^{-1} and that ν sym. NO_2 extends over the range 1278–1397 cm^{-1}.

NITROBENZENES

Study of 131 vapor-phase IR spectra of nitrobenzenes shows that ν asym. NO_2 occurs in the range 1530–1580 cm^{-1} and ν sym. NO_2 in the range 1325–1371 cm^{-1}. Comparisons of these nitrobenzene data with those data for nitroalkanes show that both ν asym. NO_2 and ν sym. NO_2 vibrations overlap. Comparison of the ν asym. NO_2 frequencies for nitrobenzene (1540 cm^{-1}), nitromethane (1582 cm^{-1}), and trimethylnitromethane (1555 cm^{-1}) show that this mode occurs at lower frequency in the case of nitrobenzene. This decrease in frequency for ν asym. NO_2 is attributed to resonance effects of the phenyl group with the NO_2 group (5). Table 7.4 lists IR data for 4-X-nitrobenzenes in the vapor phase. The ν asym. NO_2 mode occurs at 1530 cm^{-1} for 4-nitroanilines and at 1567 cm^{-1} for 1,4-dinitrobenzene and the σ_p values are -0.66 and $+0.78$, respectively. However, there is no apparent smooth correlation for ν asym. NO_2 vs σ_p. Neither is there a smooth correlation for ν sym. NO_2 vs σ_p.

Comparison of the ν sym. NO_2 frequencies for nitrobenzene (1351 cm^{-1}), nitromethane (1397 cm^{-1}), and trimethylnitromethane (1349 cm^{-1}) show that these vibrations occur at similar frequencies. Thus, it would be helpful to have another parameter to help distinguish between nitrobenzenes and nitroalkanes. A distinguishing feature is the absorbance ratio (A) ν asym. NO_2/(A) ν sym. NO_2. The band intensity ratio for nitrobenzene is 0.9 compared to 1.9–5.9 for the nitroalkanes. The absorbance ratio for (A) ν asym. NO_2/(A)ν sym. NO_2 varies between 0.9 and 1.9 for these 4-X-nitrobenzenes.

NITROBENZENE IN DIFFERENT PHYSICAL PHASES

Tables 7.5 through 7.8 list IR frequency data for nitrobenzenes in different physical phases. In all cases ν asym. NO_2 occurs at lower frequency in $CHCl_3$ solution or in the neat liquid or solid phase than it occurs in the vapor phase. However, ν sym. NO_2 usually occurs at lower frequency in $CHCl_3$ than in the vapor phase. In certain cases ν sym. NO_2 occurs at higher frequency in the neat phase than in the vapor phase (4-chloronitrobenzene for example, 1360 cm^{-1} vs 1349 cm^{-1}; see Table 7.5). In all cases, the frequency separation between ν asym. NO_2 and ν sym. NO_2 is larger in the vapor phase than it is in $CHCl_3$ solution or the neat phases.

Table 7.7 lists IR ν NO_2 data for 4-X-nitrobenzenes and 3-X-nitrobenzenes in CCl_4 and $CHCl_3$ solution. In all cases the frequency separation between ν asym. NO_2 and ν sym. NO_2 is larger in CCl_4 solution than in $CHCl_3$ solution. In addition, the ν asym. NO_2 frequency always decreases in going from CCl_4 solution to $CHCl_3$ solution, while ν sym. NO_2 generally increases in frequency. The exceptions are for 4-nitroaniline, 4-nitroanisole, and 4-nitrodiphenyl oxide. The decrease in frequency for ν asym. NO_2 in going from solution in CCl_4 to solution in $CHCl_3$

is attributed to intermolecular hydrogen bonding between the $CHCl_3$ proton and the oxygen atoms of the NO_2 group as shown here:

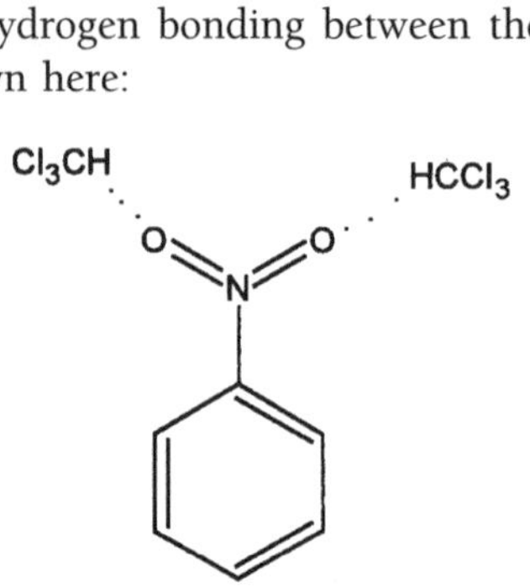

In the case of ν sym. NO_2 it is more difficult for the oxygen atoms to vibrate in phase, compared to the ν asym. NO_2 vibration which causes an increase in frequency due to intermolecular hydrogen bonding. In contrast there is a trade-off in energy during ν asym. NO_2 because one NO_2 oxygen atom is pulling away from the $CHCl_3$ proton while the other oxygen atom is expanding toward the $CHCl_3$ proton. The overall effect is a decrease in the ν asym. NO_2 frequency.

Table 7.8 shows that the frequency separation between ν asym. NO_2 in the vapor and $CHCl_3$ solution phase is larger than it is between ν asym. NO_2 in the vapor and the CCl_4 solution phase, and again this is the result of intermolecular hydrogen bonding in $CHCl_3$ solution. The reaction field of the solvent also affects ν NO_2 frequencies, and part of the decrease in frequency in $CHCl_3$ solution is attributed to the $CHCl_3$ reaction field. In the case of CCl_4, the decrease in frequency is attributed to the reaction field of CCl_4.

4-NITROBENZALDEHYDE

Table 7.9 lists IR data for ν asym. NO_2 and ν sym. NO_2 frequencies of 4-nitrobenzaldehyde 1% wt./vol. in 0 to 100 mol% $CHCl_3/CCl_4$ solutions (6). The ν asym. NO_2 decreases 4.8 cm^{-1} while ν sym. NO_2 increases 1.5 cm^{-1} in going from solution in CCl_4 to solution in $CHCl_3$. Figure 7.5 show plots of ν asym. NO_2 and ν C=O frequencies for 4-nitrobenzaldehyde vs mole% $CHCl_3/CCl_4$ (6). Both ν asym. NO_2 and ν C=O decrease in frequency in a linear manner as the mole% $CHCl_3/CCl_4$ is increased from 0 to 100 mol%. Moreover, both modes decrease in frequency at approximately the same amount (viz. ν asym. NO_2, 4.8 cm^{-1} vs ν C=O, 4.5 cm^{-1}). In contrast the ν C=O decrease in frequency for 4-dimethylaminobenzaldehyde is 9.8 cm^{-1} as the mole% is increased from 0 to 100 mol% $CHCl_3/CCl_4$. These data indicate that the strength of the intermolecular hydrogen formed between the CCl_3H proton and the $C{=}O\cdots HCCl_3$ or $NO_2\cdots(HCCl_3)_{1\text{ or }2}$ depends upon the basicity of the oxygen atoms. As these compounds are both 1,4-disubstituted benzenes, this is not a steric factor altering the distance between the site of the oxygen atom and the CCl_3H proton.

Figure 7.6 shows a plot of ν sym. NO_2 for 4-nitrobenzaldehyde vs mole% $CHCl_3/CCl_4$ (6). The plot shows that ν sym. NO_2 increases in frequency from 0 to 10.74 mol% $CHCl_3/CCl_4$, then decreases in frequency from 19.4 to 26.5 mol% $CHCl_3/CCl_4$, and then steadily increases in frequency to 100 mol% $CHCl_3/CCl_4$. This suggests that at mole% $CHCl_3/CCl_4$ < 10.74 the intermolecular hydrogen bonding in the case of 4-nitrobenzaldehyde is as shown here. At higher concentrations of $CHCl_3$, the intermolecular hydrogen bonding is as shown here:

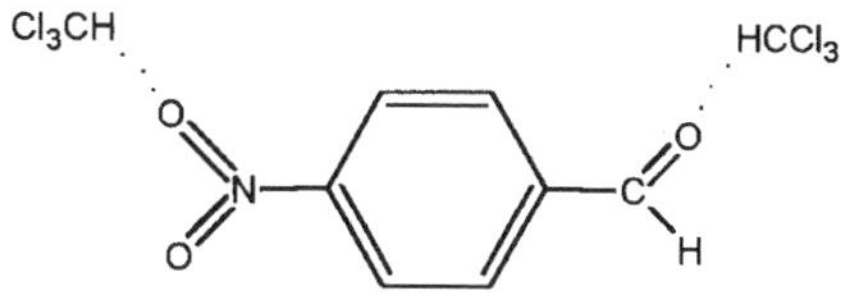

The steady increase in frequency is due to the increased energy required to expand and contract the NO_2 oxygen atoms away from the CCl_3H protons. This effect is larger than it appears in this plot, because the reaction field increases as the mole % $CHCl_3/CCl_4$ is increased, which has the effect of decreasing ν asym. NO_2 and ν C=O frequencies.

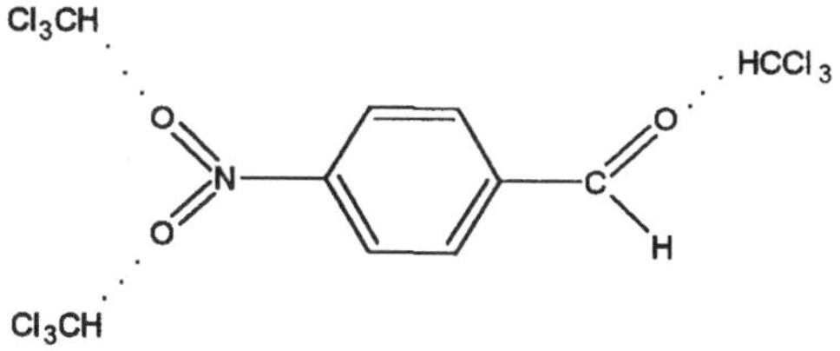

Table 7.10 is a comparison of IR ν asym. NO_2 and ν sym. NO_2 frequency data for nitromethane and nitrobenzene in a series of 13 different solvents (7). In both cases, the ν asym. NO_2 frequency is highest when in solution with hexane (CH_3NO_2, 1569 cm^{-1}; $C_6H_5NO_2$, 1535.8 cm^{-1}), and the frequency is lowest in dimethyl sulfoxide (CH_3NO_2, 1552.8 cm^{-1}; $C_6H_5NO_2$, 1524.7 cm^{-1}). In this series of solvents, the frequency difference between ν asym. NO_2 for nitromethane and nitrobenzene varies between 28.1 cm^{-1} and 34.3 cm^{-1}.

In several cases the ν sym. NO_2 vibration for nitromethane is masked by the solvent, and in two cases for nitrobenzene. The ν sym. NO_2 mode for nitrobenzene occurs 26.2 to 28 cm^{-1} lower in frequency than ν sym. NO_2 for nitromethane. Both ν asym. NO_2 and ν sym. NO_2 for nitrobenzene occur at lower frequency than the corresponding vibrations for nitromethane, and this is attributed to resonance of the phenyl group with the nitro group.

The frequency difference for ν asym. NO_2 for nitromethane in hexane and each of the other solvents is more than the frequency difference for ν asym. NO_2 for nitrobenzene in hexane and each of the other solvents, except for the solvents nethylene chloride and chloroform. These exceptions may be attributed to intermolecular hydrogen bonding between the NO_2 oxygen atoms and $CHCl_3$ or CH_2Cl_2 protons, which is apparently stronger in the case of nitrobenzene.

3-X-NITROBENZENES

The vapor-phase IR ν asym. NO_2 and ν sym. NO_2 frequencies for 3-X-nitrobenzenes occur in the ranges 1540–1553 cm^{-1} and 1349–1360 cm^{-1}, respectively (see Table 7.11). In addition the frequency difference between ν asym. NO_2 and ν sym. NO_2 ranges between 187 and 204 cm^{-1}, and the absorbance ratio (A)ν asym. NO_2/(A)ν sym. NO_2 varies between 1.07 and 2.16. The frequency separation between ν asym. NO_2 and ν sym. NO_2 appears to decrease in the order vapor, $CHCl_3$, $CHCl_3$, and neat or solid phase (see Tables 7.12 and 7.13).

2-NITROBENZENES

Table 7.14 lists vapor-phase IR data for 2-nitrobenenes (1). Their ν asym. NO_2 and ν sym. NO_2 frequencies occur in the ranges 1540–1560 cm^{-1} and 1350–1360 cm^{-1}, respectively. The compound 2-nitrophenol is an exception because ν sym. NO_2 occurs at 1335 cm^{-1}. This relatively low ν sym. NO_2 for 2-nitrophenol is attributed to intramolecular hydrogen bonding between the OH proton and the NO_2 oxygen atom as illustrated here:

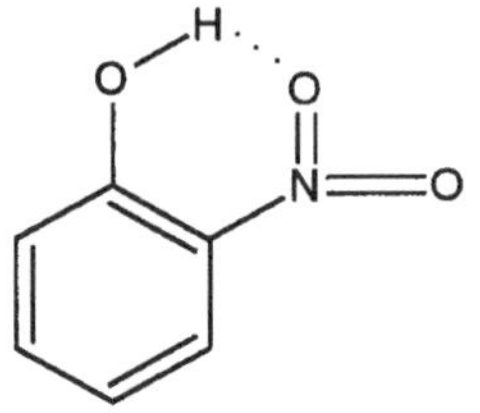

Intramolecular hydrogen bonding alters the band intensity ratio (A)ν asym. NO_2/(A) sym. NO_2 because this ratio is 0.52 for 2-nitrophenol and varies between 1.09 to 2.59 for the 13 other compounds included in this study. In addition the frequency separation between ν asym. NO_2 and ν sym. NO_2 is 210 cm^{-1} for 2-nitrophenol and varies between 186 and 200 cm^{-1} for the other 13 compounds included in this study.

Tables 7.15 and 7.16 compare the ν asym. NO_2 and ν sym. NO_2 frequency data for 2-X-nitrobenzenes in different physical phases. The frequency separation between ν asym. NO_2 and ν sym. NO_2 is larger in the vapor phase than in either $CHCl_3$ solution or the neat or solid phases. Moreover, the ν asym. NO_2 vibration decreases more in frequency in going from the vapor phase to the $CHCl_3$ solution phase or to the neat or solid phase than does the ν sym. NO_2 vibration.

Table 7.17 lists IR data for 1-X-, 3-X- and 4-X-nitrobenzenes in the solid phase, and there does not appear to be a consistent trend in ν asym. NO_2 and ν sym. NO_2 frequencies in the three sets of these substituted nitrobenzenes in the solid phase.

Table 7.18 lists vapor-phase IR data for the ν asym. NO_2 and ν sym. NO_2 frequencies of 2,5- and 2,6-X,Y-nitrobenzenes. The ν asym. NO_2 and ν sym. NO_2 frequencies occur in the ranges 1541–1560 cm^{-1} and 1349–1361 cm^{-1}, respectively. The exceptions are for those 2,5-X,Y-nitrobenzenes with an OH or an NHCO CH_3 group in the 2-position. These ν sym. NO_2 vibrations occur at 1325 cm^{-1}, 1330 cm^{-1}, and 1340 cm^{-1}, and occur at lower frequency as a result of intramolecular hydrogen bonding. It is also interesting to note that the absorbance ratio (A)ν asym. NO_2/(A) ν sym. NO_2 is 1.07, 0.74, and 0.37 for these same compounds.

In the case of the 2,6-X-Y-nitrobenzenes, the compound 2-nitro-6-methylphenol exhibits ν sym. NO_2 at 1349 cm^{-1} and the absorbance ratio (A)ν asym. NO_2/(A)ν sym. NO_2 is 0.88. These data indicate that the NO_2 group is intramolecularly hydrogen bonded with the phenolic OH group.

In the case of 2,6-dichloronitrobenzene the ν asym. NO_2 frequency occurs at 1568 cm^{-1} and the absorbance ratio (A) ν asym. NO_2/(A)ν sym. NO_2 is 3.60. These data are comparable to those exhibited by the nitroalkanes. The reason for this is that the NO_2 group and the 2,6-dichlorophenyl group are not coplanar in the case of 2,6-dichloro-1-nitrobenzene. Therefore, the resonance effect upon the NO_2 is no longer possible.

Table 7.19 is a comparison of the frequency difference between ν asym. NO_2 and ν sym. NO_2 in the vapor, neat or solid phases for 2,5-X-Y-nitrobenzenes. These data show that the ν asym. NO_2 decrease more in frequency than ν sym. NO_2 in going from the vapor to the neat or solid phase.

Table 7.20 lists IR data for tri-, tetra- and pentasubstituted nitrobenzenes. The highest ν asym. NO_2 frequencies are exhibited by 1,2-dinitro-3,6-dichlorobenzene, 1580 cm^{-1}, and 1,2-dinitro-3,6-dibromobenzene, 1578 cm^{-1} (1). These two compounds also have the highest absorbance ratio (A)ν asym. NO_2/(A)ν sym. NO_2, 4.17 and 3.88 for the 3,6-Cl_2 and 3,6-Br_2 analogs, respectively. These data indicate that the NO_2 groups are not coplanar with the 3,6-dihalophenyl group.

In the case of nitrobenzenes where there is at least 2,6-dichloro atoms, the NO_2 group is not coplanar with the 2,6-dichloro phenyl group. Thus, compounds such as 2,4,6-trichloronitrobenzene, 2,3,5,6-tetrachloronitrobenzene, and pentachloronitrobenzene exhibit ν asym. NO_2 at 1562, 1569, and 1568 cm^{-1}, respectively. In the same compound order, ν sym. NO_2 is assigned at 1361, 1333, and 1332 cm^{-1}, respectively, and the absorbance ratio for (A)ν asym. NO_2/(A)ν sym. NO_2 is 2.67, 1.25, and 1.48, respectively.

4-X-NITROBENZENES IN CCl_4 AND $CHCl_3$ SOLUTIONS

Table 7.21 lists the ν asym. and ν sym. NO_2 frequencies for 21 4-X-nitrobenzenes in 1% or less wt./vol. in CCl_4 and $CHCl_3$ solutions (8). Figure 7.7 is a plot of ν asym. NO_2 frequencies in CCl_4 solution vs ν asym. NO_2 frequencies in $CHCl_3$ solution. This linear plot increases from a low of 1513.9 cm^{-1} vs 1505.3 cm^{-1} for 4-nitroaniline to a high of 1556.1 cm^{-1} vs 1554.4 cm^{-1} for 1,4-dinitrobenzene. At all frequency points, the ν asym. NO_2 vibration occurs at lower frequency in $CHCl_3$ solution than in CCl_4 solution. As stated previously, the shift to lower frequency in $CHCl_3$ solution is due to intermolecular hydrogen bonding [$NO_2(\cdots HCCl_3)_2$] and an increased value for the reaction field.

It has been suggested that ν sym. NO_2 couples with an in-plane ring mode as approximated here:

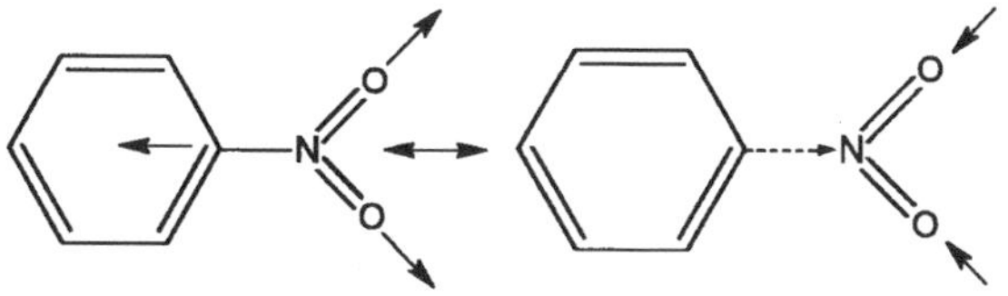

This is because ν sym. NO_2 is not affected in the same manner as ν asym. NO_2.

In general, the frequency separation between ν asym. NO_2 and ν sym. NO_2 increases progressing in the series 1–21 in both solvents. This is attributed to the nature of the NO_2 bonds. Bellamy has pointed out that the ν asym. NO_2 for 4-X-nitrobenzenes is directly related to the electron donor or acceptor property of the 4-X substituent (9). For example, substituent groups with negative σ_p values would contribute to a structure such as presented here:

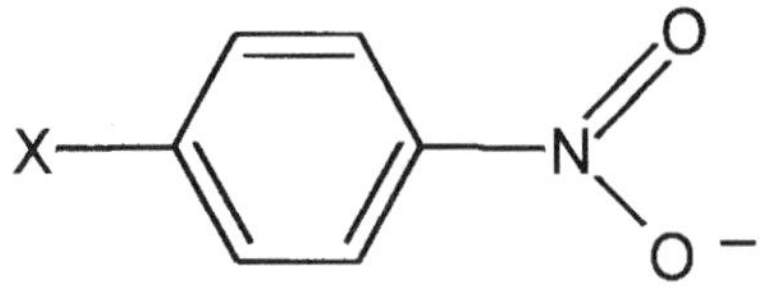

Substituent groups with positive σ_p values would contribute to a structure such as presented here:

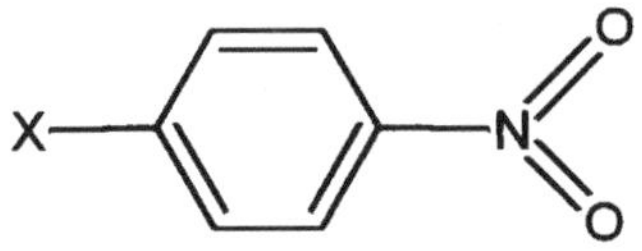

Figures 7.8 and 7.9 are plots of ν asym. NO_2 for 4-X-nitrobenzenes in CCl_4 solution and in $CHCl_3$ solution vs Hammett σ_p values for the 4-X substituent group, respectively (8). The Hammett σ_p values include both inductive and resonance contributions of the 4-substituent groups (10). Both of these plots show pseudolinear relationships. The most deviant point is 6 for 4-nitrophenol. In this case the OH group is intermolecularly hydrogen bonded to the Cl atom of each solvent (8).

Resonance parameters have been derived for substituted benzenes. When the ring is unperturbed, it is σ_R. When the ring is electron poor, it is σ_R+. When the ring is electron rich, it is σ_R-. When the ring is in resonance with a carboxylic acid group, it is σ_R (11).

Figures 7.10 and 7.11 are plots of ν asym. NO_2 frequencies vs σ_R+ for 4-X-nitrobenzenes in CCl_4 and $CHCl_3$ solutions, respectively (8). These pseudo-linear relationships show that as σ_R+ becomes more positive, the ν asym. NO_2 vibration increases in frequency.

Figures 7.12 and 7.13 are plots of ν asym. NO_2 vs σ_R for 4-X-nitrobenzenes in CCl_4 and $CHCl_3$ solutions, respectively (8). These plots show the same trend as for Figs. 7.10 and 7.11.

Figures 7.14 and 7.15 are plots of ν asym. NO_2 for 4-X-nitrobenzenes vs σ_I, a measure of the inductive power of the 4-X group (10, 11) in CCl_4 and $CHCl_3$ solutions, respectively (8). These figures show that ν asym. NO_2 does not correlate well with the inductive parameter of the 4-X group. In conclusion, the ν asym. NO_2 frequencies for 4-X-nitrobenzenes correlate best with σ_p values, which include both resonance and inductive parameters.

ALKYL NITRATES

Tables 7.22 and 7.23 list IR data and vibrational assignments for alkyl nitrates. Depending upon the physical phase, ν asym. NO_2 and ν sym. NO_2 for alkyl nitrates occur in the ranges 1617–1662 cm^{-1} and 1256–1289 cm^{-1}, respectively. The absorbance for ν asym. NO_2 is higher than the absorbance for ν sym. NO_2. The ν asym. NO_2 vibration decrease more in frequency then the ν sym. NO_2 vibration in going from the vapor phase to CCl_4 solution or neat phase. Both ν asym. NO_2 and ν sym. NO_2 exhibit a first overtone, and 2νasym. NO_2 decreases more in frequency than 2νsym. NO_2 in going from the vapor phase to the neat or solution phase.

The ν N—O vibration occurs in the range 854–865 cm^{-1}, and $\gamma\, NO_2$ and $\delta\, NO_2$ are assigned near 760 cm^{-1} and 700 cm^{-1}, respectively. A vapor-phase IR spectrum for ethyl nitrate is shown in Fig. 7.16.

ETHYL NITRATE VS NITROALKANES AND NITROBENZENE

Table 7.24 lists IR data for ethyl nitrate, nitroalkanes, and nitrobenzene in CCl_4 and $CHCl_3$ solutions (7). In all cases, the ν asym. NO_2 vibration decreases in frequency and the ν sym. NO_2 vibration increases in frequency in going from CCl_4 solution to $CHCl_3$ solution. The decrease in the ν asym. NO_2 frequency in going from solution in CCl_4 to solution in $CHCl_3$ is attributed to intermolecular hydrogen bonding $[NO_2(\cdots HCCl_3)_2]$ and an increased reaction field. In the case of ν sym. NO_2, the situation is reversed. The increased reaction is expected to lower the ν sym. NO_2 frequency. This reversal is due to the fact that it requires more energy to expand both NO_2 oxygen atoms against the $HCCl_3$ protons $[NO_2(\cdots HCCl_3)_2]$ during a cycle of ν sym. NO_2. In the case of ν asym. NO_2, the energy required to expand one NO_2 oxygen atom toward the $HCCl_3$ proton is canceled by contraction of the other NO_2 oxygen atom away from the $HCCl_3$ proton.

Another correlation for these compounds is that as the ν asym. NO_2 vibration decreases in frequency from 1637 through 1531 cm^{-1}, the ν sym. NO_2 vibration tends to increase in frequency. The frequency separation between ν asym. NO_2 and ν sym. NO_2 decreases in the order ethyl nitrate through nitrobenzene in both CCl_4 and $HCCl_3$ solutions (355.8–183.3 cm^{-1} CCl_4 and 348.8–177.6 cm^{-1} in $CHCCl_3$) (7).

ALKYL NITRITES

Alkyl nitrites have the empirical structure R—O—N=O. However, these compounds exist in a cis and trans structure as depicted here:

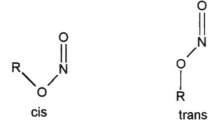

cis trans

Therefore, alkyl nitrites exhibit IR bands for cis ν N=O and trans ν N=O. Tables 7.25 and 7.26 list characteristic IR group frequency data for alkyl nitrites. The data in Table 7.26 are from Tarte (12). The ν N=O frequency data reported by Tarte are higher than those reported by Nyquist (1). An IR spectrum for *n*-butyl nitrite is shown in Fig. 7.17. The trans ν N=O vibration (1653–1681 cm^{-1}) occurs at higher frequency than the cis ν N=O vibration (1610–1625 cm^{-1}), and the absorbance ratio (A) trans ν N=O/(A) cis ν N=O varies from 0.95 for methyl nitrite to ~50 for tert-butyl nitrite. As the steric factor of the R group becomes larger, the preferred structure is apparently the trans configuration.

The trans ν N—O vibration occurs in the range 751–814 cm^{-1}, and the cis δ O—N=O and trans δO—N=O vibrations are assigned in the ranges 617–689 cm^{-1} and 564–625 cm^{-1}, respectively (12).

Figure 7.18 shows plots of cis ν N=O and trans ν N=O for alkyl nitrites vs σ^* (the inductive parameter of the alkyl group), and both plots show a pseudo-linear relationship (1).

Figure 7.19 shows a plot of trans ν N—O vs σ^* and this plot also shows a pseudolinear relationship (1).

Figure 7.20 shows a plot of trans ν N=O vs trans ν N—O, and this plot also exhibits a pseudolinear relationship (1).

Figure 7.21 shows plots of the absorbance ratio (A) trans ν N=O/(A) cis ν N=O for 10 alkyl nitrites vs σ^*. These data show that the concentration of the trans conformer increases as the electron release of the alkyl group to the O—N=O group is increased (1).

RAMAN DATA FOR ORGANONITRO COMPOUNDS

Table 7.27 lists Raman data for some organonitro compounds (13). The eight compounds listed show that ν sym. NO_2 is a medium to strong Raman band in the range 1312–1350 cm^{-1}, and a weak to strong Raman band assigned to δ NO_2 in the range 831–882 cm^{-1}. The frequency separation between these two Raman bands varies between 454 and 492 cm^{-1}. The relative Raman band intensity for the absorbance ratio (RI) ν sym. NO_2/(RI) δ NO_2 varies between 0.56 and 9.

Table 7.28 compares IR and Raman data for nitroalkanes in different physical phases. These data also show that ν asym. NO_2 occurs at lower frequency in the neat phase than in the vapor phase, and it also decreases more in frequency than ν sym. NO_2, which either decreases or in some cases increases in frequency.

Table 7.29 lists IR and Raman data for nitrobenzenes in different physical phases. Not listed in this Table but worth mentioning is that in the Raman spectrum the ν sym. NO_2 mode is always more intense than in the IR spectrum.

A SUMMATION OF ν asym. NO_2 AND ν sym. NO_2 IN DIFFERENT PHYSICAL PHASES

Tables 7.30–7.32 list IR and/or Raman data in various physical phases. In all cases ν asym. NO_2 changes more in going from the vapor phase to the solution or neat phases than does the ν sym. NO_2, which sometimes even increases in frequency. Intermolecular hydrogen bonding and the increased reaction field lowers ν NO_2 frequencies. The increased energy required during expansions of the NO_2 oxygen atoms against the $CHCl_3$ protons offset the increase in the reaction field and hydrogen bonding effects.

NITROSAMINES

Nitrosamines have the empirical structure presented here:

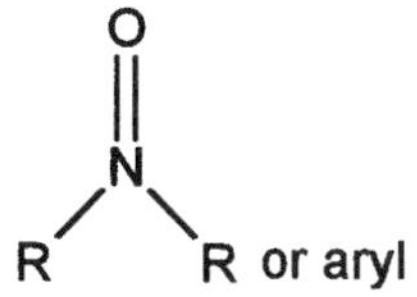

Table 7.33 lists IR data for nitrosamines in different physical phases. These compounds exhibit ν N=O in the region 1482–1492 cm^{-1} in the vapor phase, and in the region 1438–1450 cm^{-1} in the liquid phase. Thus, there is a decrease in frequency of between 32–54 cm^{-1} in going from the vapor to the liquid phase. This frequency decrease is attributed to dipolar interaction between the N=O groups in the condensed phase. Another way to look at this ν N=O frequency decrease in going from the vapor to the neat phase is that there is an increase in the reaction field.

The compound illustrated here exhibits ν N=O at 1534 cm^{-1} in the vapor phase and at 1509 cm^{-1} in the neat phase:

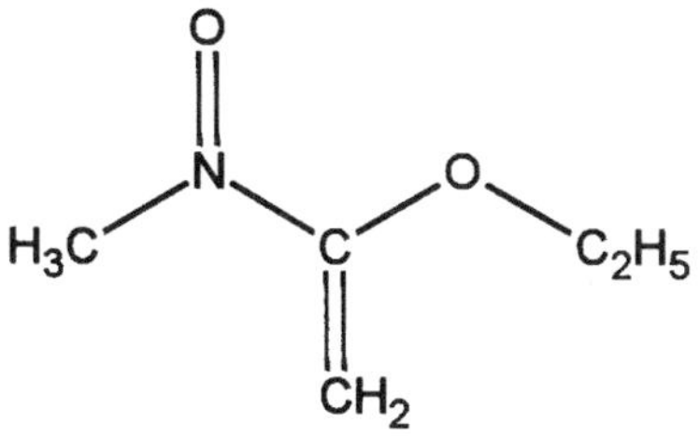

REFERENCES

1. Nyquist, R. A. (1984). *The Interpretation of Vapor-Phase Infrared Spectra: Group Frequency Data.* Philadelphia: Sadtler Research Laboratories, pp. 550–608; see also Sadtler Standard vapor- and neat-phase IR collections.
2. Lunn, W. H. (1960). *Spectrochim. Acta*, **16**: 1088.
3. Smith, D. C., Pan, C. Y., and Nielsen, J. R. (1950). *J. Chem. Phys.* **18**: 706.
4. Geiseler, G. and Kesler, G. (1964). *Ber. Bunsenges. Phys. Chem.* **68**: 571.
5. Bellamy, L. J. (1968). *Advances In Infrared Group Frequencies*. London: Methuen & Co. Ltd., p. 231.
6. Nyquist, R. A., Settineri, S. E., and Luoma, D. A. (1991). *Appl. Spectrosc.* **45**: 1641.
7. Nyquist, R. A. (1990). *Appl. Spectrosc.* **44**: 594.
8. Nyquist, R. A. and Settineri, S. E. (1990). *Appl. Spectrosc.* **44**: 1552.
9. Bellamy, L. J. (1968). *Advances in Infrared Group Frequencies*. London: Methuen & Co., p. 228.
10. Taft, R. W. (1956). *Steric Effects in Organic Chemistry.* M. S. Newman, ed., New York: John Wiley.
11. Brownlee, R. T. C. and Topsom, R. D. (1975). *Spectrochim. Acta* **31A**: 1677.
12. Tarte, P. (1952). *J. Chem. Phys.* **20**: 1570.
13. *Sadtler Laboratories Standard Collection of Raman Data for Organic Compounds*. Philadelphia, PA.
14. Bellamy, L. J. (1968). *Advances in Infrared Group Frequencies*. London: Methuen & Co.

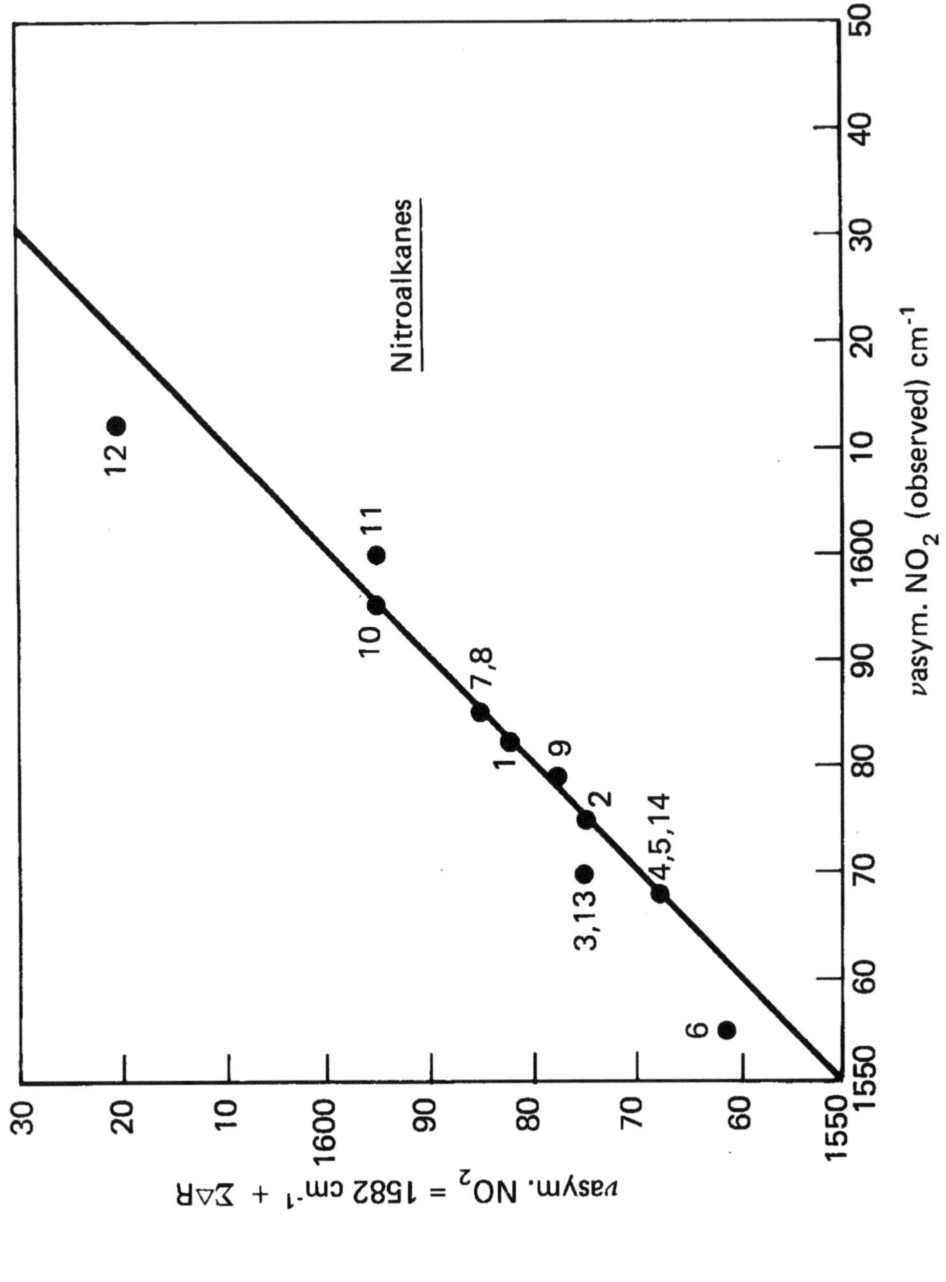

FIGURE 7.1 A plot of the observed ν asym. NO_2 frequencies vs the calculated ν asym. NO_2 frequencies for nitroalkanes using the equation ν asym. $NO_2 = 1582\,cm^{-1} + \Sigma\Delta R$.

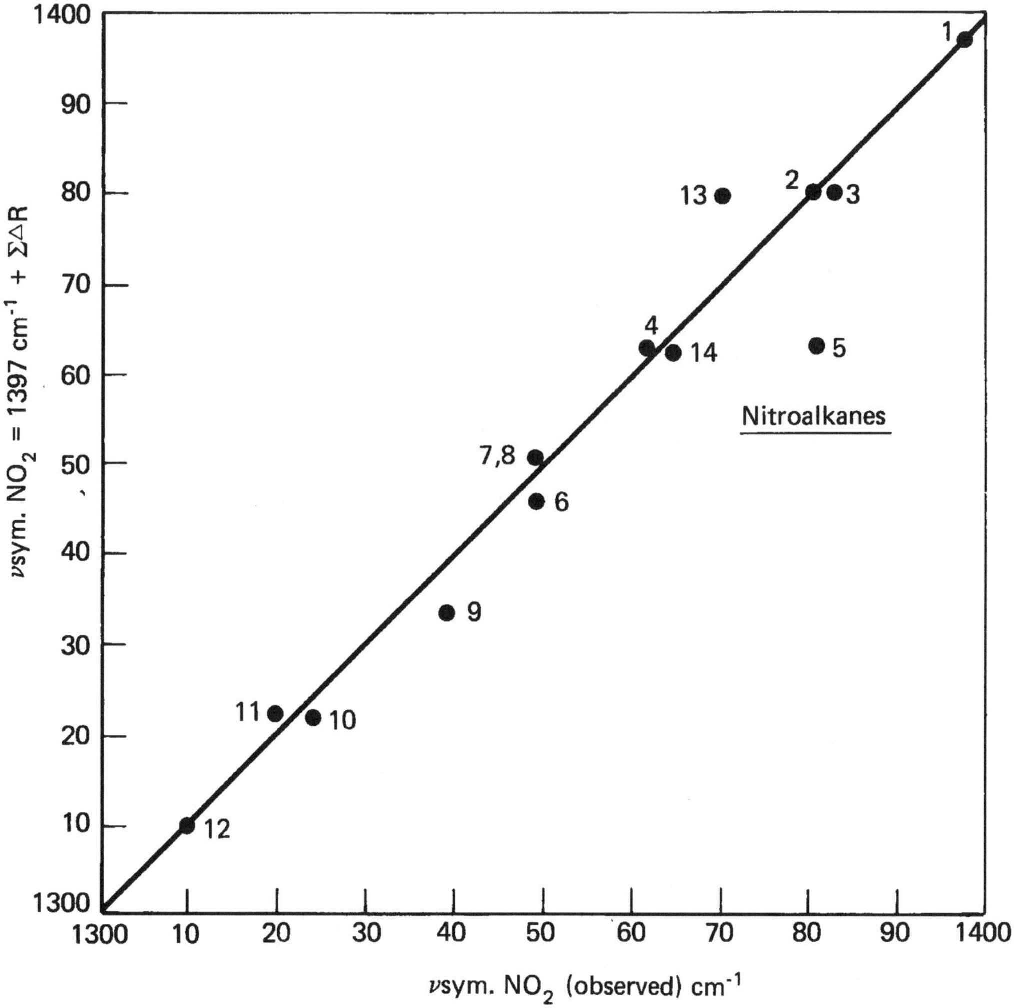

FIGURE 7.2 A plot of the observed ν sym. NO_2 frequencies vs the calculated ν sym. NO_2 frequencies using the equation ν sym. $NO_2 = 1397\,cm^{-1} + \Sigma\Delta R$.

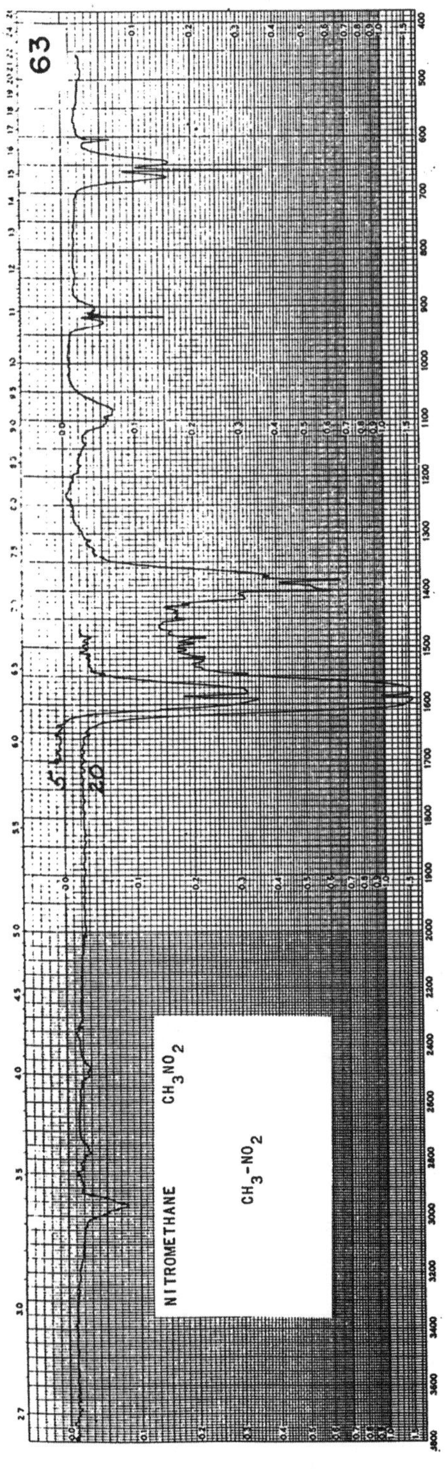

FIGURE 7.3 Vapor-phase IR spectrum for nitromethane in a 5-cm KBr cell (5 and 20 mm Hg sample to 600 mm Hg with N_2).

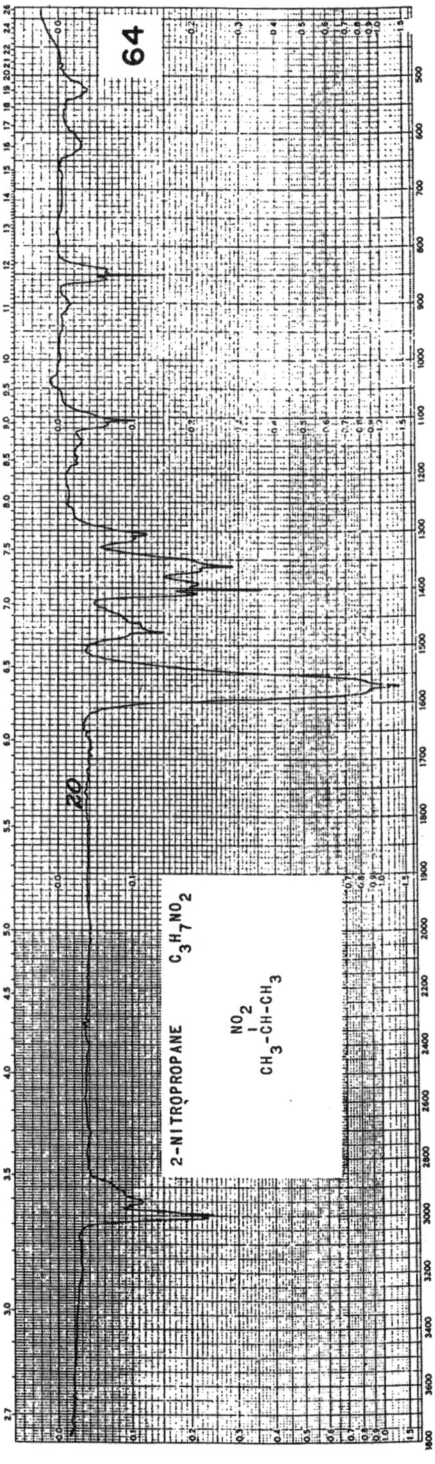

FIGURE 7.4 Vapor-phase IR spectrum for 2-nitropropane in a 5-cm KBr cell (20 mm Hg sample to 600 mm Hg with N_2).

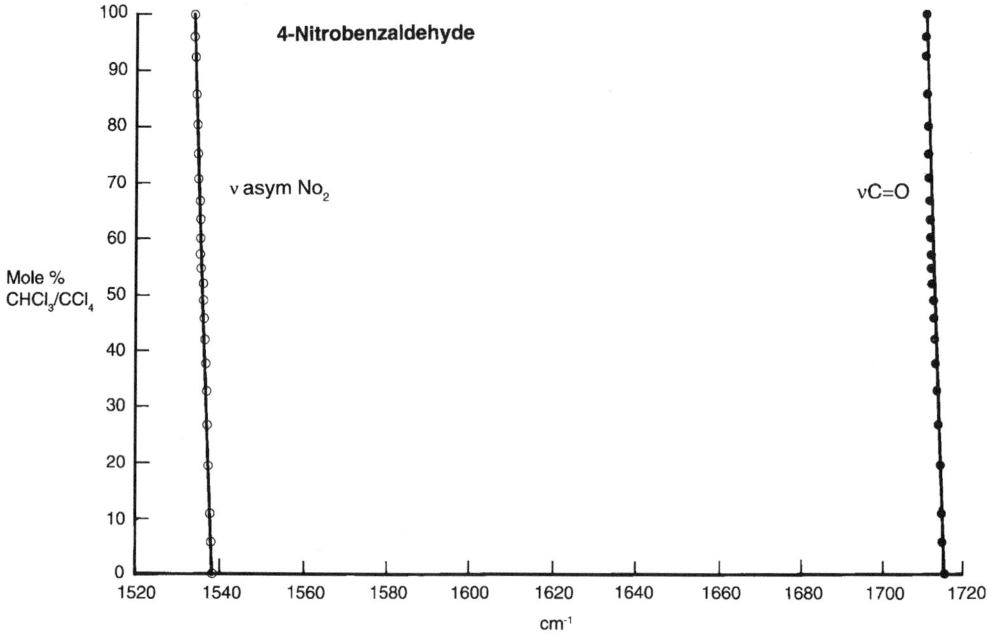

FIGURE 7.5 Plots of ν asym. NO_2 and ν C=O for 4-nitrobenzaldehyde vs mole % $CHCl_3/CCl_4$.

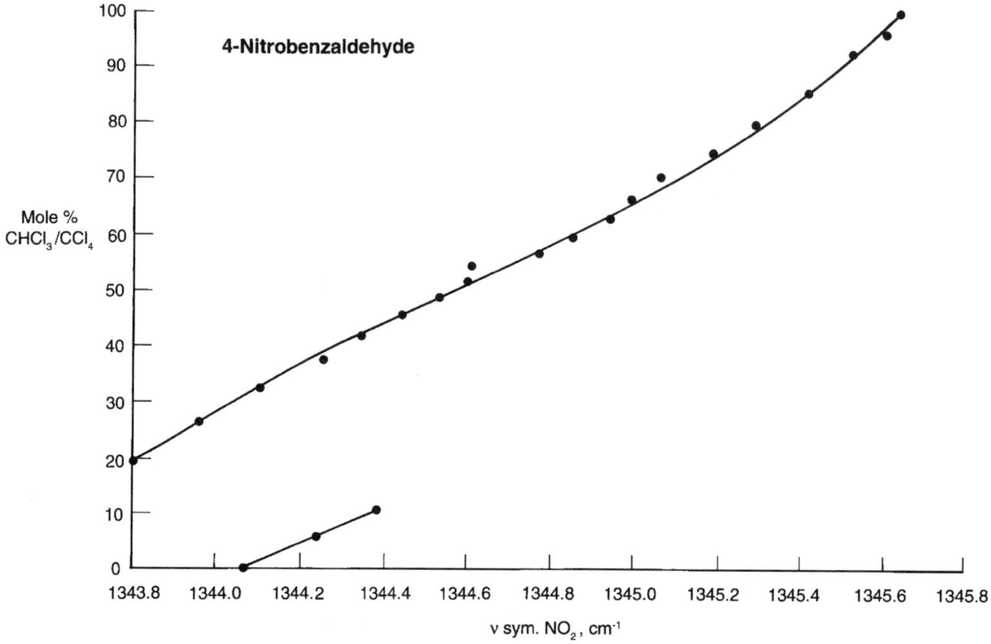

FIGURE 7.6 Plots of ν sym. NO_2 for 4-nitrobenzaldehyde vs mole % $CHCl_3/CCl_4$.

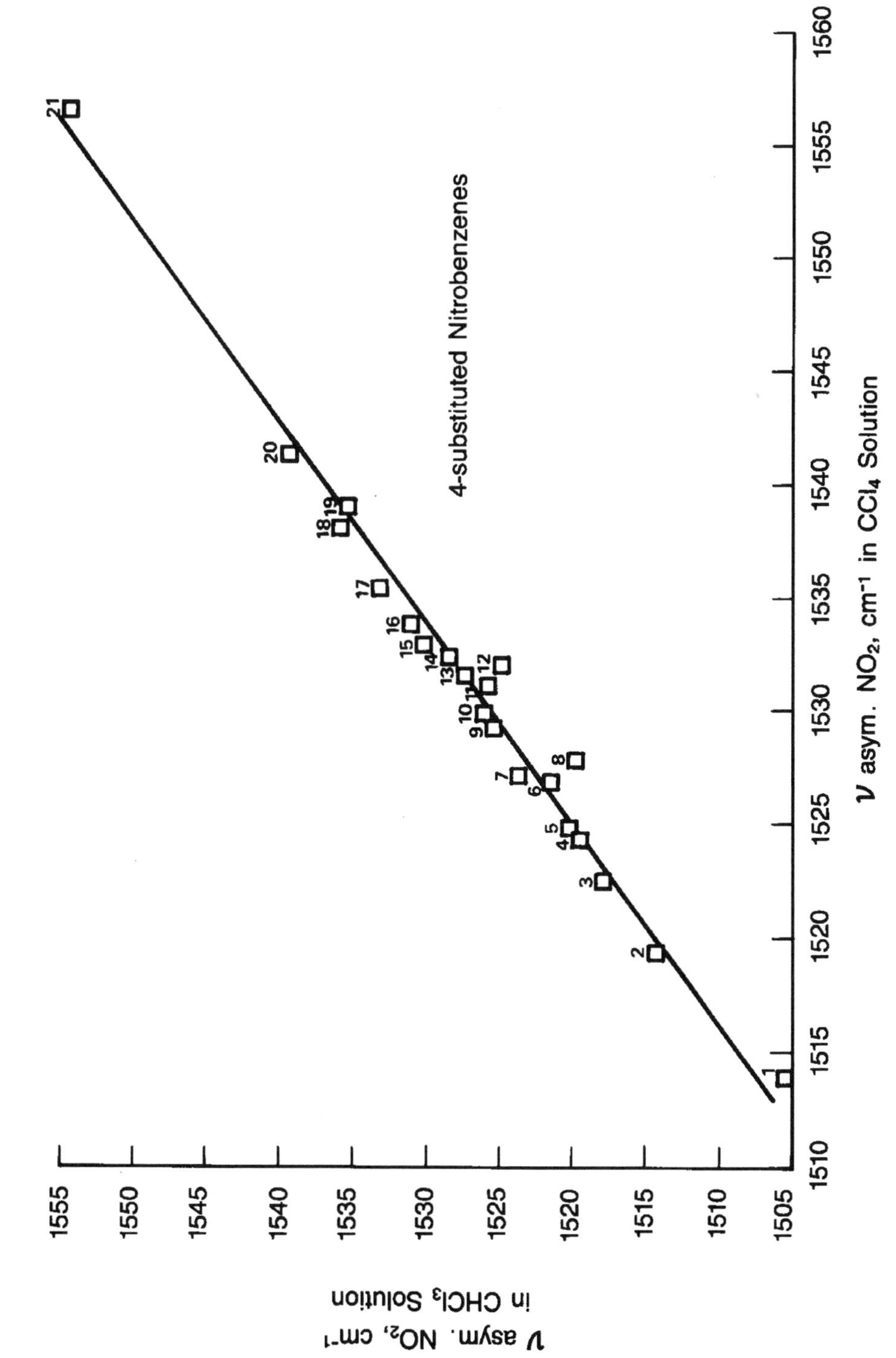

FIGURE 7.7 A plot of ν asym. NO_2 for 4-X-nitrobenzenes in CCl_4 solution vs ν asym. NO_2 in $CHCl_3$ solution.

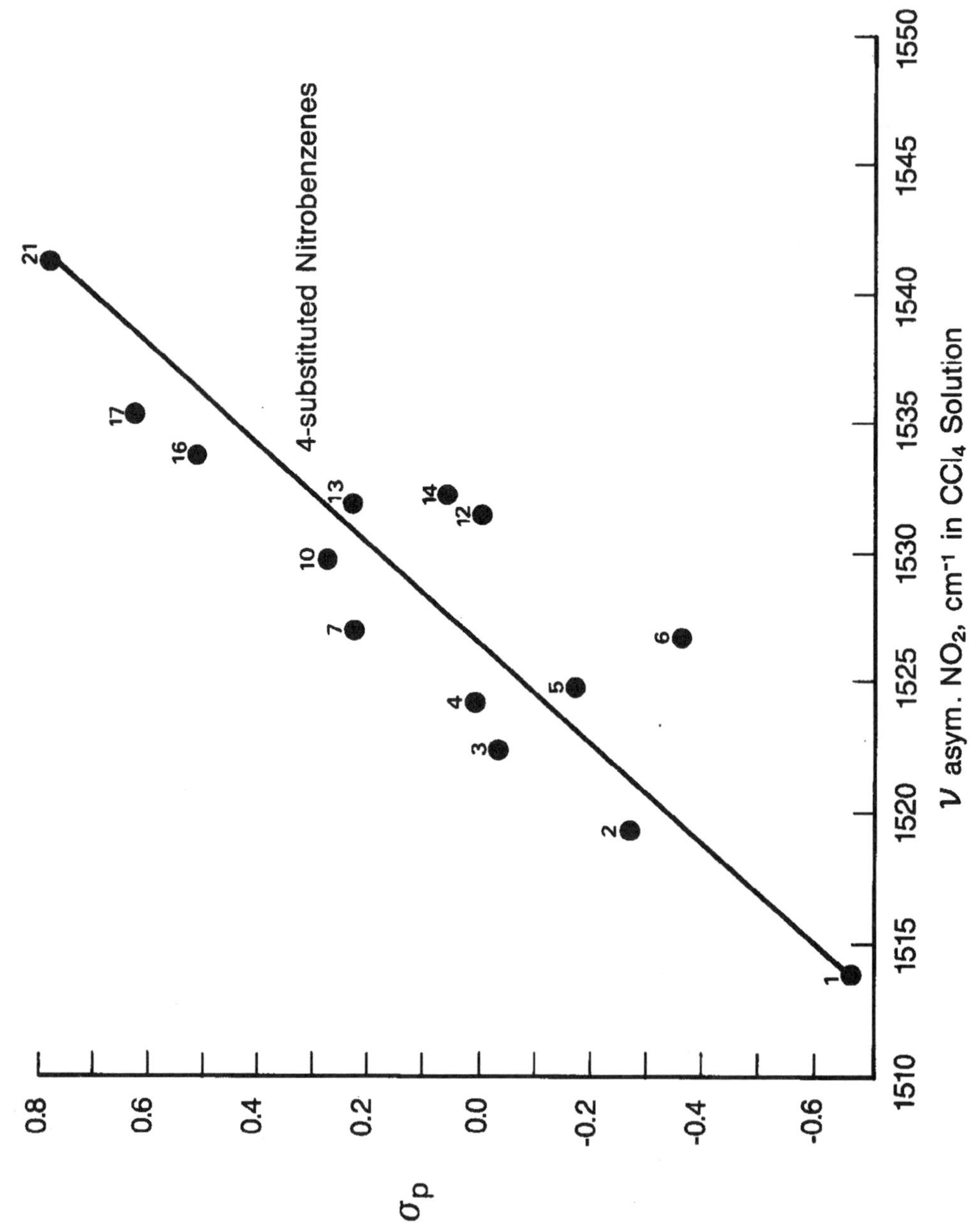

FIGURE 7.8 A plot of ν asym. NO_2 for 4-X-nitrobenzenes in CCl_4 solution vs σ_p.

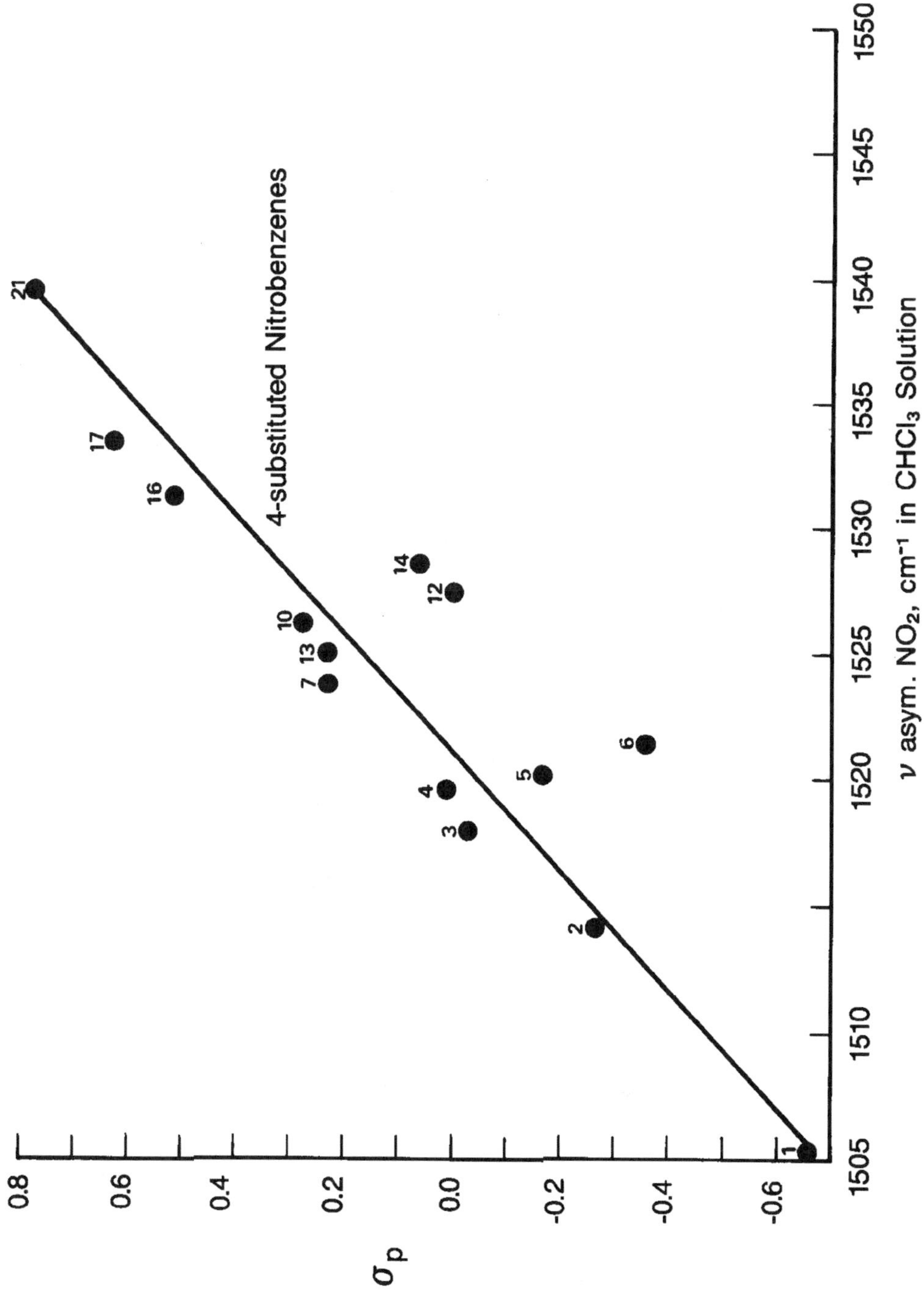

FIGURE 7.9 A plot of ν asym. NO_2 for 4-X-nitrobenzenes in $CHCl_3$ solution vs σ_p.

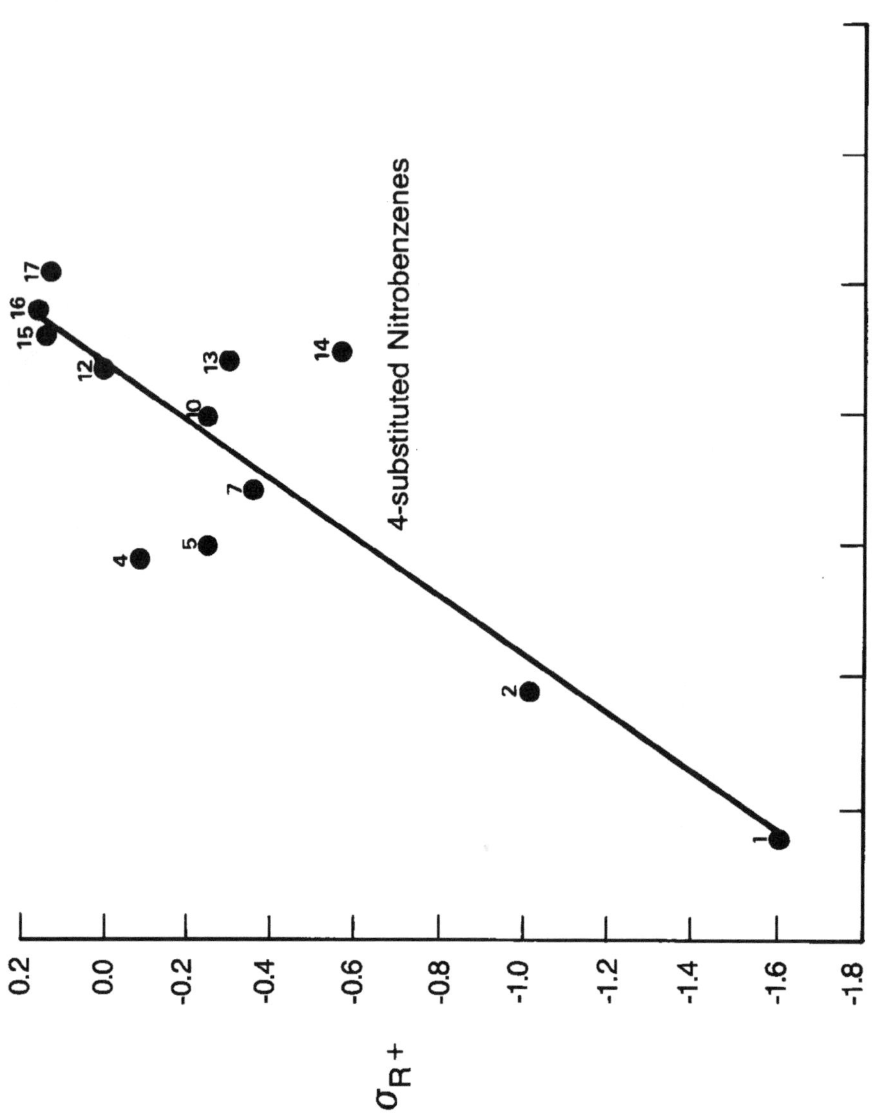

FIGURE 7.10 A plot of ν asym. NO_2 for 4-X-nitrobenzenes in CCl_4 solution vs σ_{R^+}.

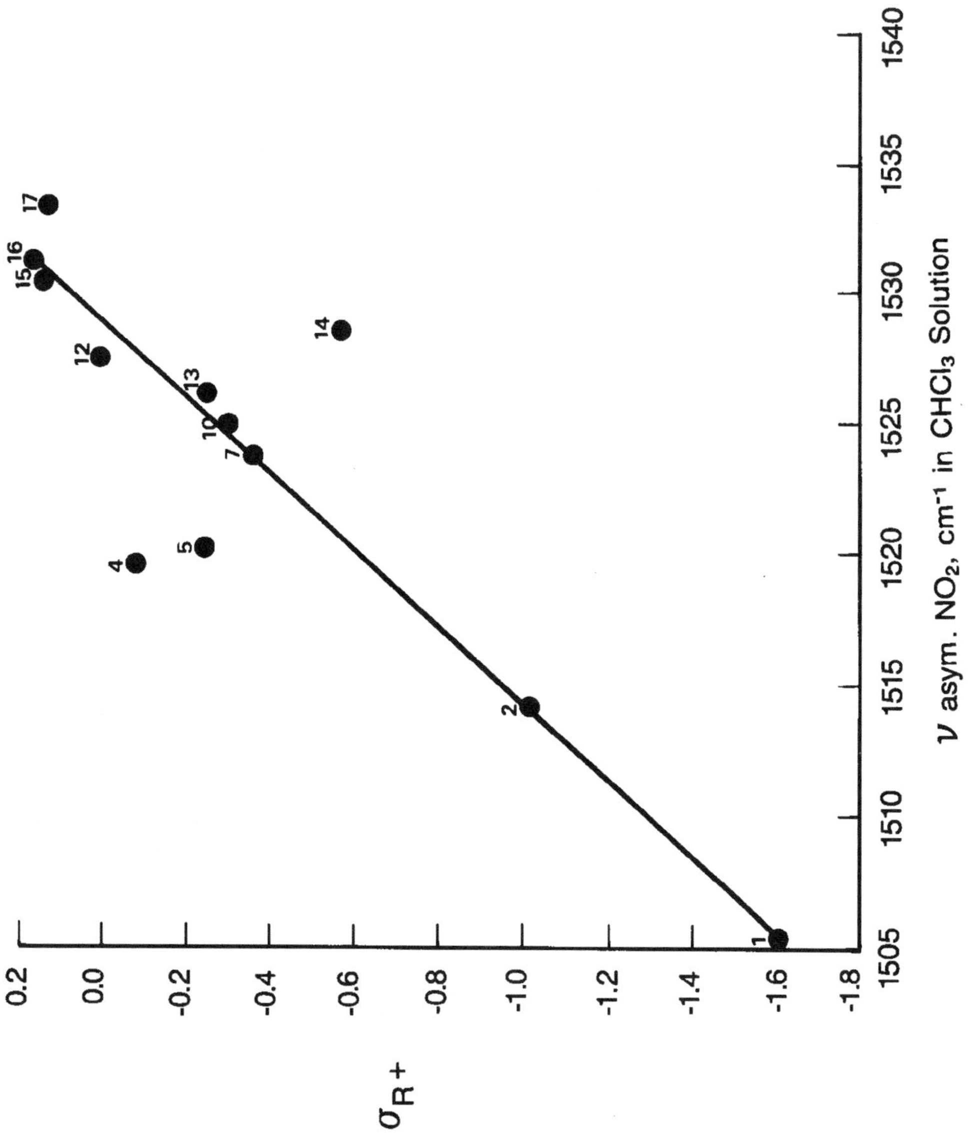

FIGURE 7.11 A plot of ν asym. NO_2 for 4-X-nitrobenzenes in $CHCl_3$ solution vs σ_R+.

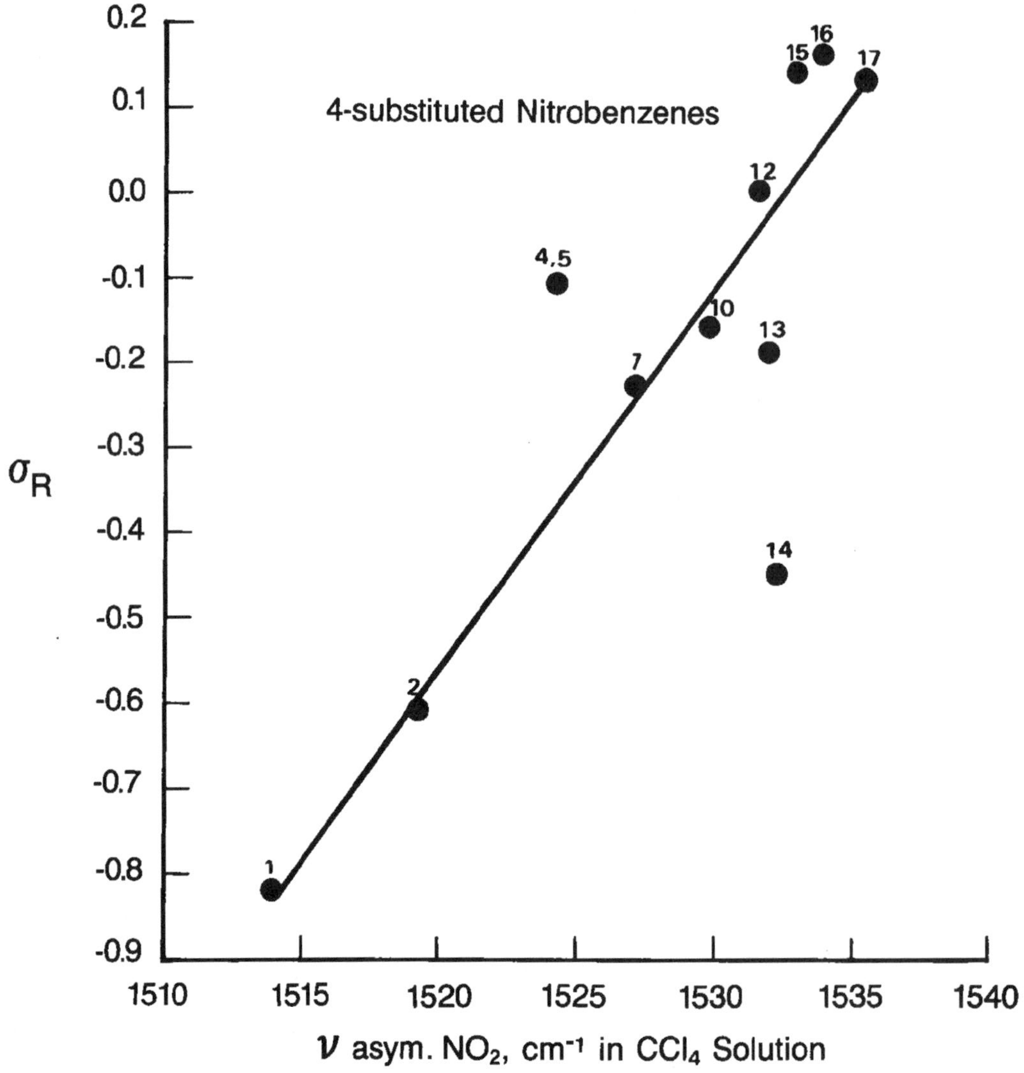

FIGURE 7.12 A plot of ν asym. NO_2 for 4-X-nitrobenzenes in CCl_4 solution vs σ_R.

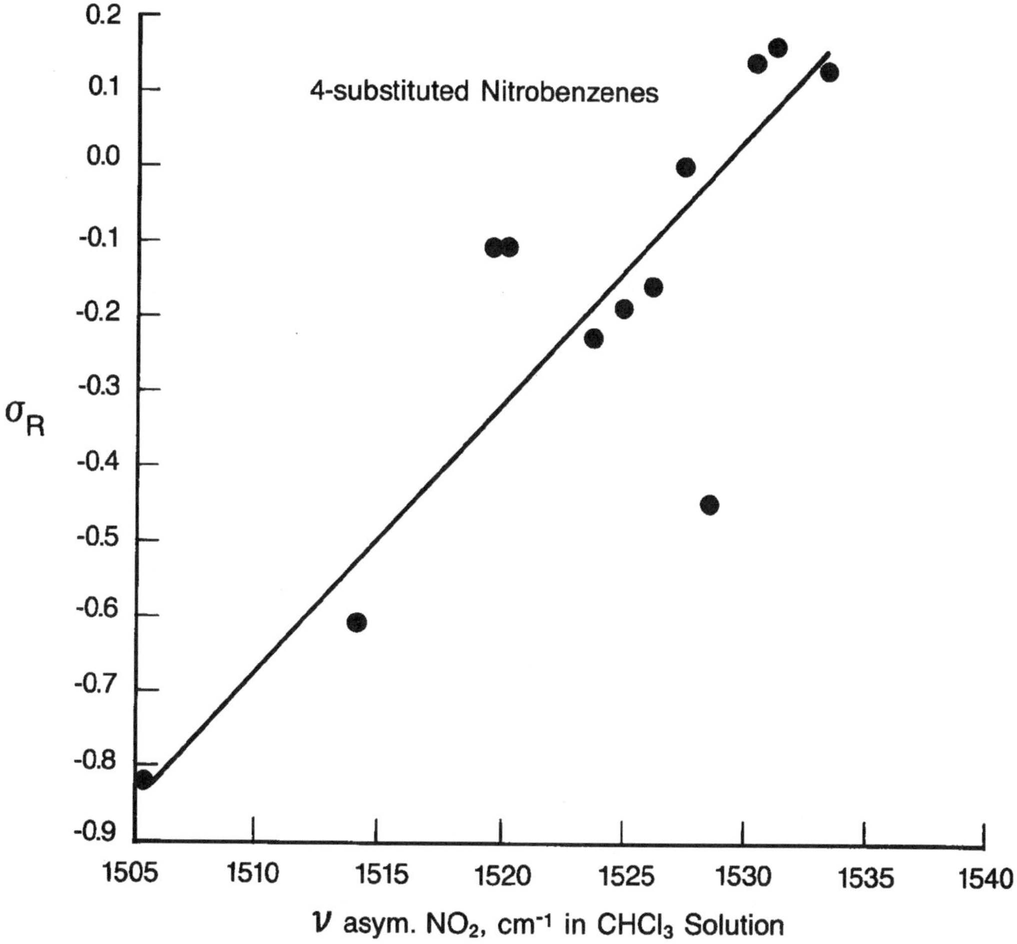

FIGURE 7.13 A plot of ν asym. NO_2 for 4-X-nitrobenzenes in $CHCl_3$ solution vs σ_R.

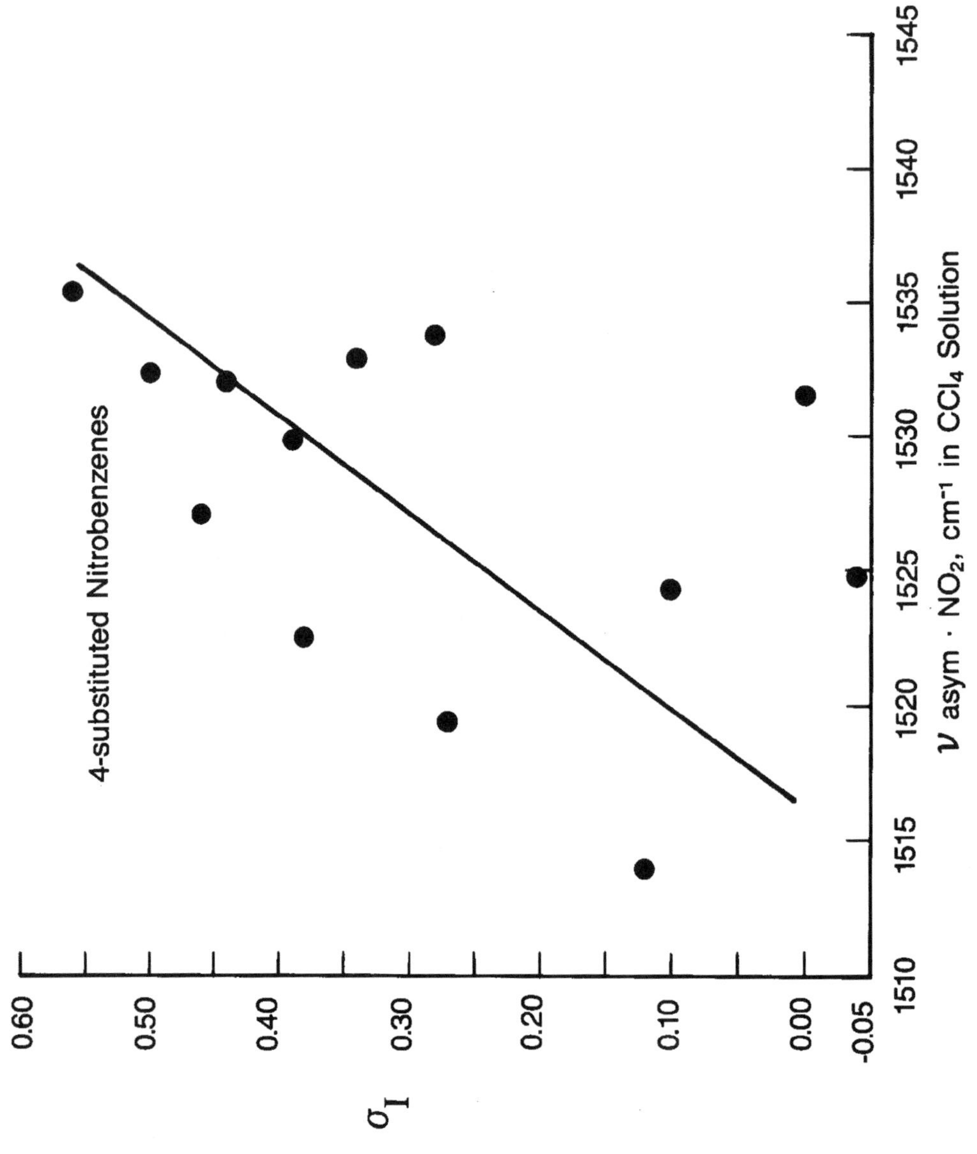

FIGURE 7.14 A plot of ν asym. NO_2 for 4-X-nitrobenzenes in CCl_4 solution vs σ_I.

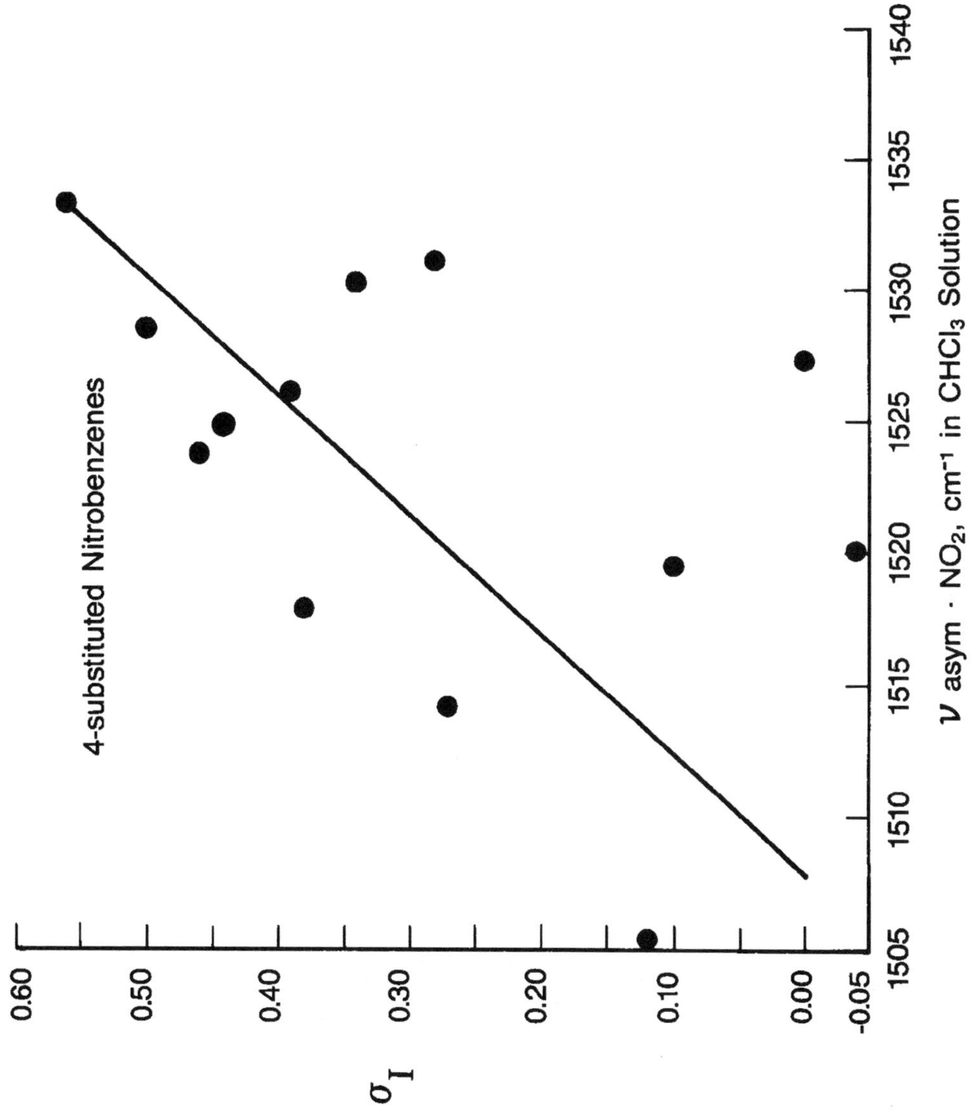

FIGURE 7.15 A plot of ν asym. NO_2 for 4-X-nitrobenzenes in $CHCl_3$ solutions vs σ_I.

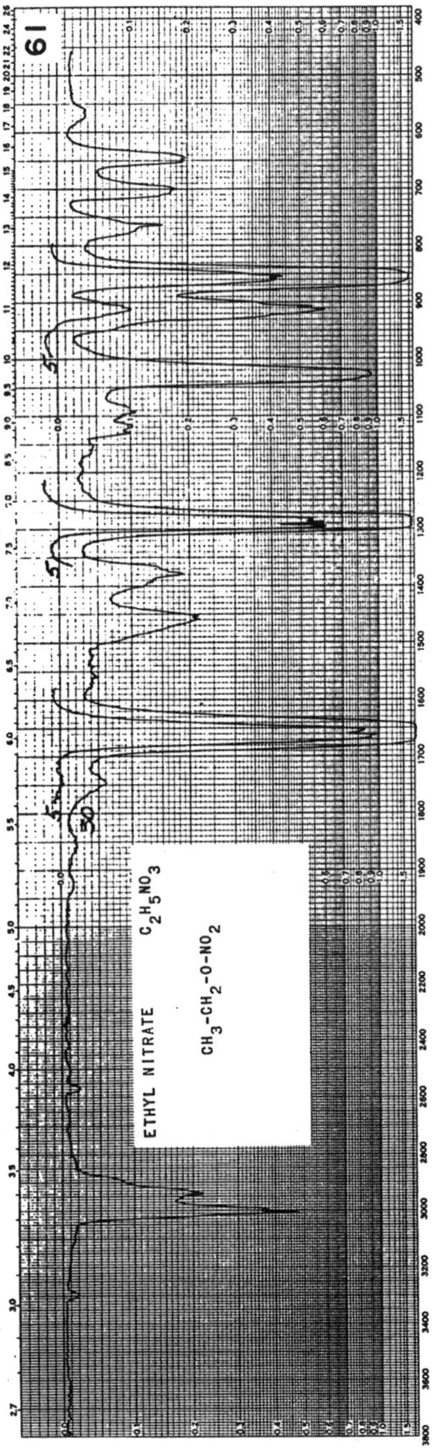

FIGURE 7.16 Vapor-phase IR spectrum for ethyl nitrate in a 5-cm KBr cell (5 and 30 mm Hg sample to 600 mm Hg with N_2).

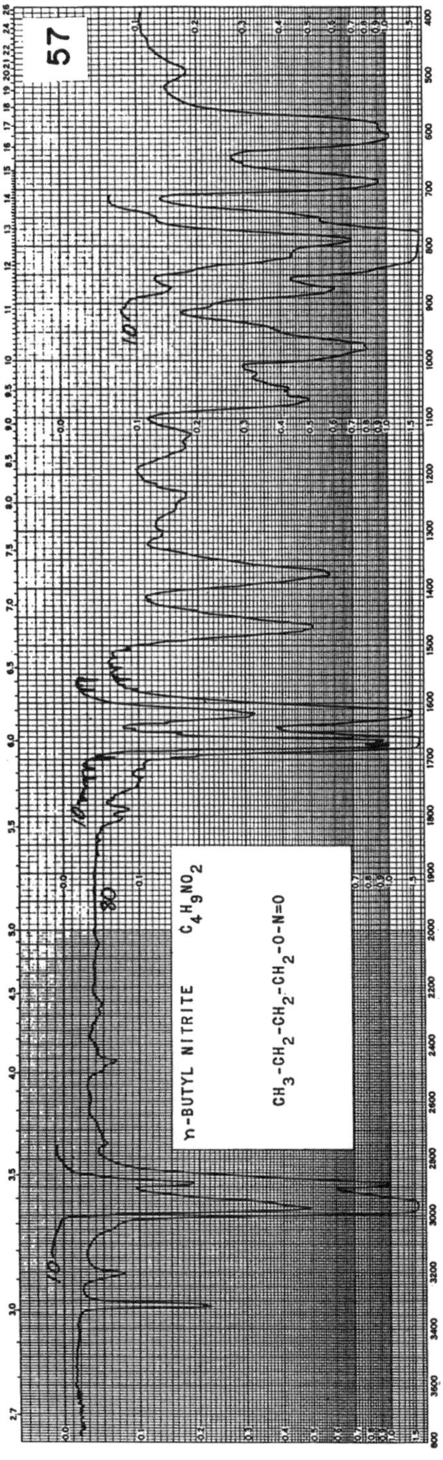

FIGURE 7.17 Vapor-phase IR spectrum for *n*-butyl nitrite in a 5-cm KBr cell (10 and 80 mm Hg sample to 600 mm Hg with N_2).

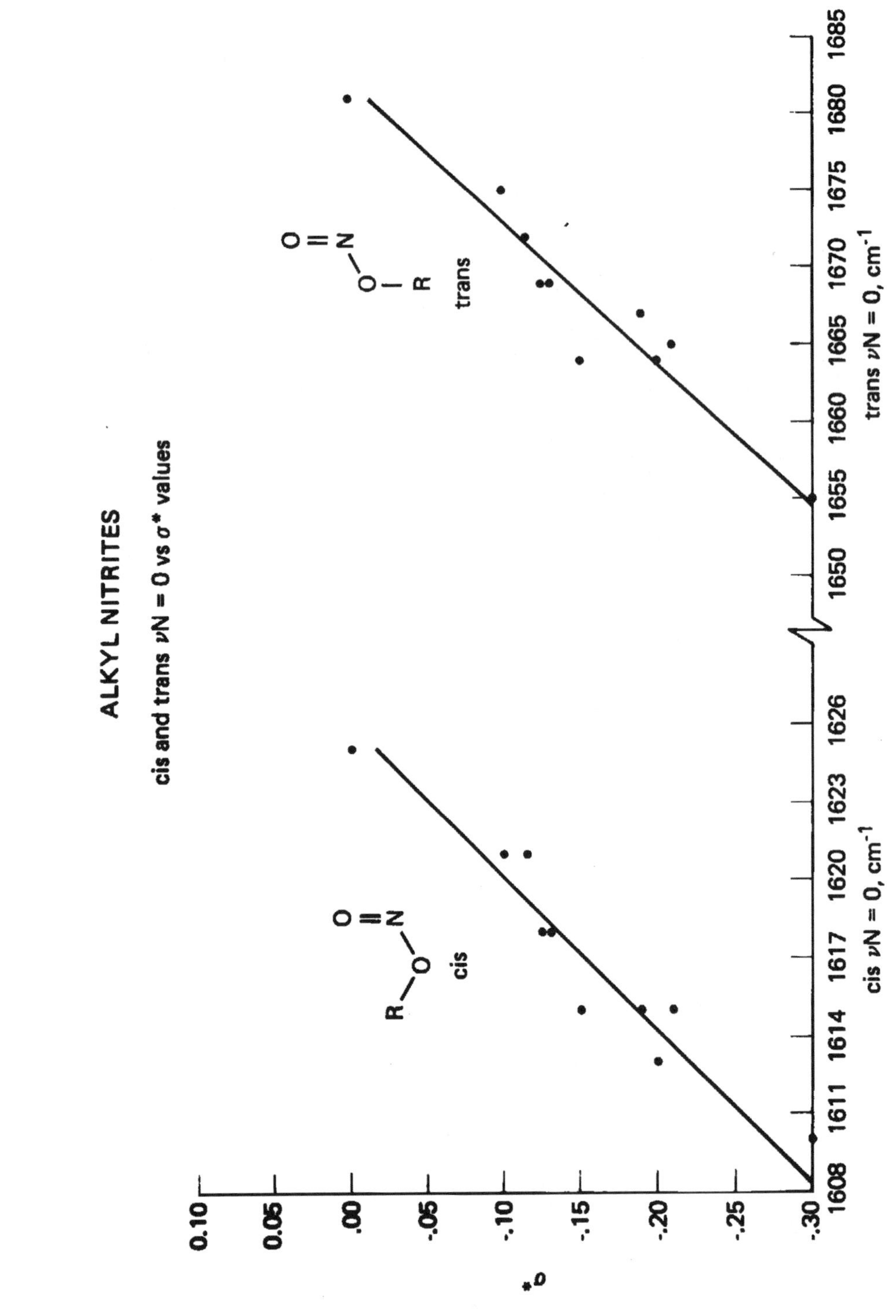

FIGURE 7.18 Plots of cis ν N=O and trans ν N=O for alkyl nitrites vs σ^*.

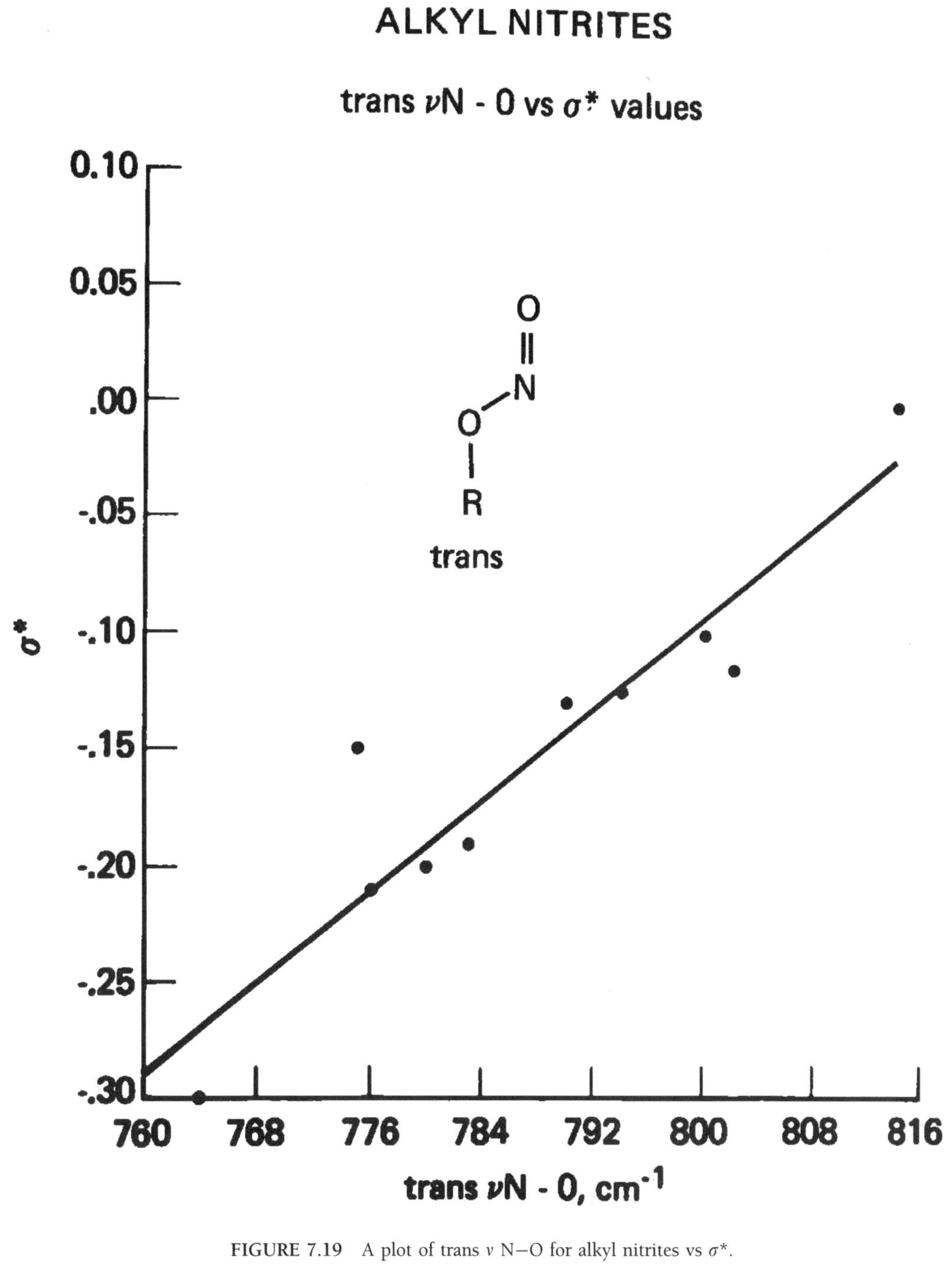

FIGURE 7.19 A plot of trans ν N–O for alkyl nitrites vs σ^*.

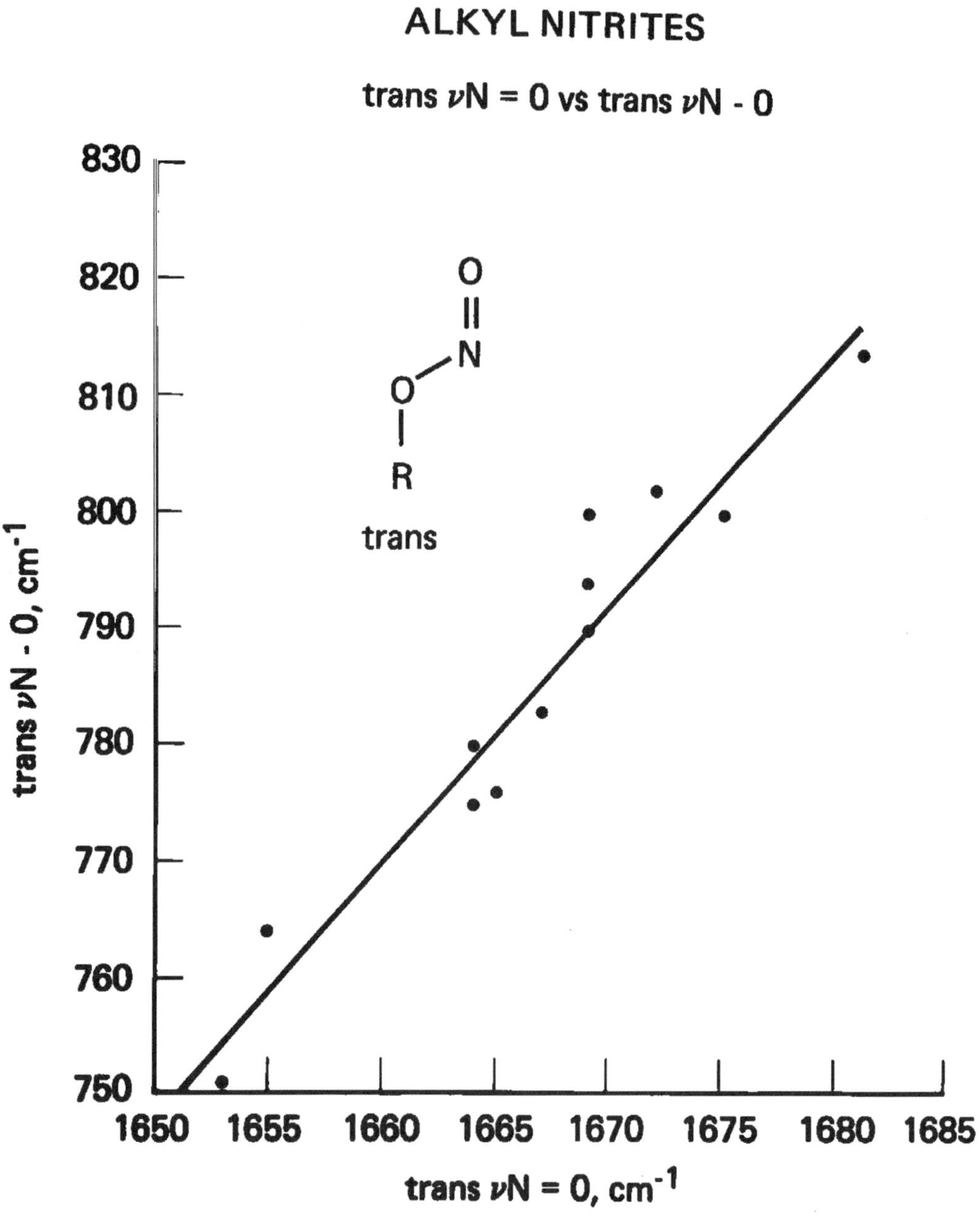

FIGURE 7.20 A plot of trans ν N=O vs trans ν N—O for alkyl nitrites.

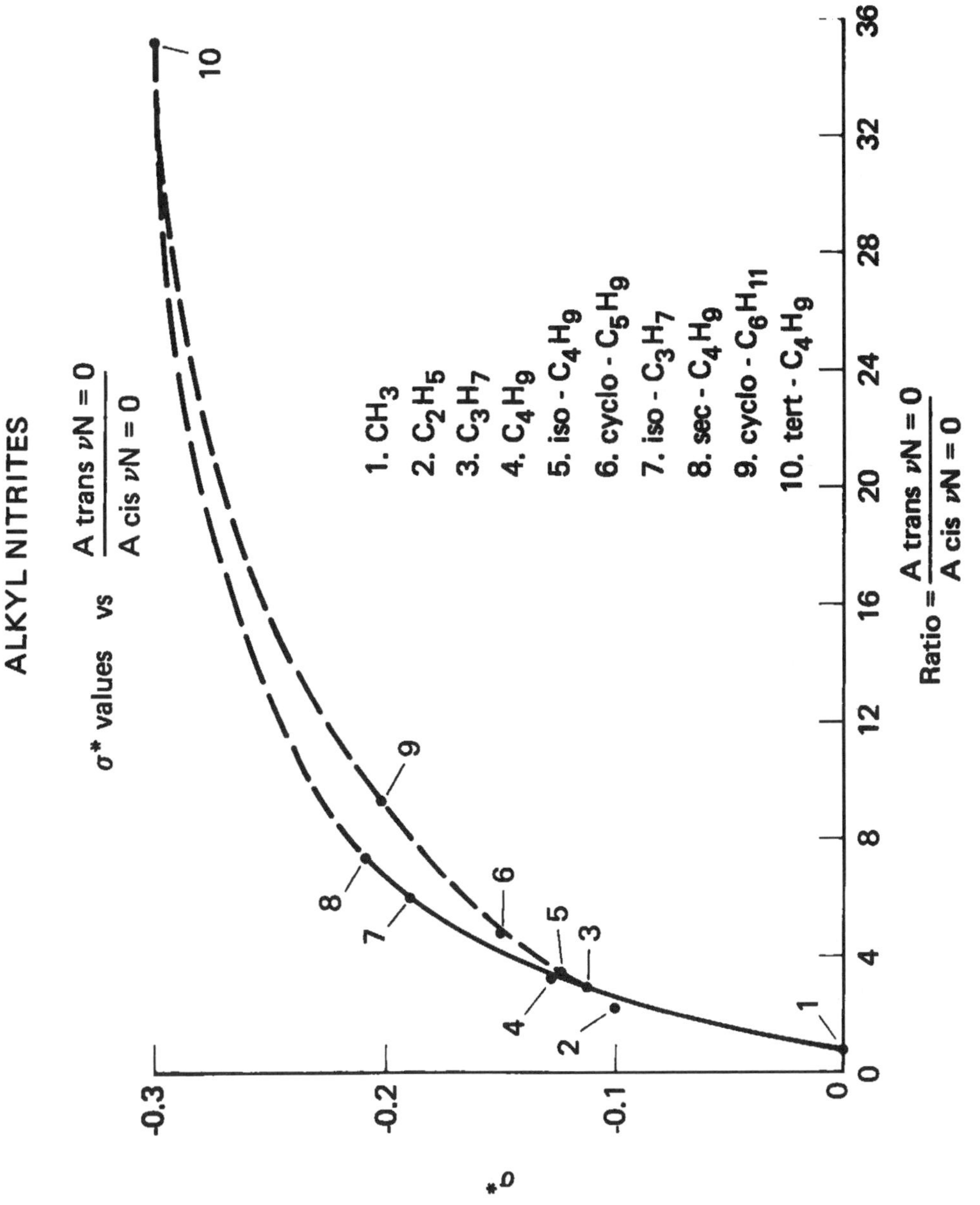

FIGURE 7.21 Plots of the absorbance ratio (A) trans ν N=O/(A) cis ν N=O vs σ^*.

TABLE 7.1 The ν asym. NO_2 and ν sym. NO_2 frequency shifts of substituted nitro compounds from those for nitromethane in the liquid phase

CH_3NO_2	ν asym. NO_2: 1558 cm^{-1} ν asym. NO_2	ν sym. NO_2: 1375 cm^{-1} ν sym. NO_2
α-substituent	ΔR	ΔR
H	0	0
CH_3	−5	−8
C_2H_5	−5	−8
C_6H_5	—	−8
C=O (ester)	10	−8
F	17	−23
Cl	17	−23
Br	17	−23
NO_2	29	−29
β-substituent		
H	0	0
OH	−4	−6
Cl	2	−4
Br	2	−4
NO_2	2	−15

TABLE 7.1a Vapor-phase IR data for nitroalkanes

Compound $R-NO_2$ R	C—N st. cm^{-1} (A)	NO_2 wag cm^{-1} (A)	a.NO_2 str. cm^{-1} (A)	s.NO_2 str. cm^{-1} (A)	[a.NO_2 str.]-[s.NO_2 str.] cm^{-1}	Ratio (A) a.NO_2 str./s.NO_2 str.	Ratio (A) s.NO_2 str./s.CH_2 str.	Ratio (A) s.NO_2 str./a.CH_2 str.	Ratio (A) s.NO_2 str./a.CH_3 str.
Methane	918 (0.060)	659 (0.230)	1582 (1.049)	1397 (0.541)	185	1.94			13.53
Ethane	872 (0.152)	619 (0.080)	1575 (1.152)	1380 (0.442)	195	2.61	6.31	3.13	2.75
1-()Propane	891 (0.040)	603 (0.050)	1570 (1.250)	1382 (0.405)	188	3.09	3.38	1.69	1.23
1-()Butane	859 (0.049)	610 (0.050)	1576 (1.250)	1382 (0.360)	194	3.47	2.01	1.09	0.73
1-()Pentane	850 (0.060)	610 (0.040)	1575 (1.250)	1371 (0.390)	204	3.21	1.32	0.67	0.57
1-()Hexane	850 (0.049)	605 (0.040)	1571 (1.250)	1382 (0.380)	189	3.29	1.06	0.45	0.51
2-()Propane	851 (0.100)*1	618 (0.060)	1565 (1.250)	1360 (0.450)	205	2.78	4.09	2.25	1.09
	901 (0.030)*2								
Cyclohexane	1152 (0.032)		1567 (1.250)	1379 (0.429)	188	2.91	1.03	0.41	

*1 gauche.
*2 trans.

TABLE 7.2 Vapor-phase IR data for nitroalkanes

Nitro	a.CH_3 str.	a.CH_2 str.	s.CH_3 str.	s.CH_2 str.	a.CH_3 bend	CH_2 bend	CH_3 rock	C–N st.	NO_2 wag	a.NO_2 str.	s.NO_2 str.	Ratio (A) a.NO_2 str./ s.NO_2 str.	Ratio (A) s.NO_2 str./ s.CH_2 str.	Ratio (A) s.NO_2 str./ a.CH_2 str.	Ratio (A) s.NO_2 str./ a.CH_3 str.
Methane	2990 (0.040)		2960 (0.070)					918 (0.060)	655 (0.230)	1582 (1.049)	1397 (0.541)	1.94			13.53
Ethane	3000 (0.161)	2960 (0.141)		2910 (0.070)		1458 (0.170)	1113 (0.100)	872 (0.152)	619 (0.080)	1575 (1.152)	1380 (0.442)	2.61	6.31	3.13	2.75
1-()Propane	2982 (0.330)	2950 (0.240)		2900 (0.120)		1450 (0.120)	1125 (0.040)	891 (0.040)	603 (0.050)	1570 (1.250)	1382 (0.405)	3.09	3.38	1.69	1.23
1-()Butane	2980 (0.490)	2950 (0.330)		2895 (0.180)	1469 (0.100)	1450 (0.170)	1129 (0.049)	859 (0.049)	610 (0.050)	1576 (1.250)	1382 (0.360)	3.47	2.01	1.09	0.73
1-()Pentane	2975 (0.685)	2945 (0.585)		2885 (0.295)	1465 (0.150)	1447 (0.210)	1125 (0.040)	850 (0.060)	610 (0.040)	1575 (1.250)	1371 (0.390)	3.21	1.32	0.67	0.57
1-()Hexane	2975 (0.740)	2940 (0.840)		2880 (0.360)	1465 (0.160)	1446 (0.200)	1126 (0.040)	850 (0.040)	605 (0.040)	1571 (1.250)	1382 (0.380)	3.29	1.06	0.45	0.51
			2(a.CH_3 bend)												
2-()Propane	2995 (0.411)	2952 (0.200)*	2880 (0.110)*		1460 (0.209)			851 (0.100) 901 (0.030)	618 (0.060)	1565 (1.250)	1360 (0.450)	2.78	4.09	2.25	1.09
Cyclohexane		2950 (1.070)		2875 (0.418)		1460 (0.232)		1152 (0.032)		1567 (1.250)	1379 (0.429)	2.91	1.03	0.41	

* In Fermi Resonance.

TABLE 7.3 Vapor- and liquid-phase IR data for nitroalkanes

Compound $R\text{-}NO_2$ R	a.NO_2 str. vapor cm^{-1}	a.NO_2 str. liquid cm^{-1}	a.NO_2 str. Δ [v-l] cm^{-1}	s.NO_2 str. Δ [v-l] cm^{-1}	s.NO_2 str. vapor cm^{-1}	s.NO_2 str. liquid cm^{-1}	Taft σ^*
CH_3	1582	1558	−24	−22	1397	1375	[0]
C_2H_5	1575	1555	−20	−17	1380	1363	[−0.100]
C_3H_7	1570	1549	−21	−1	1382	1380	[−0.115]
$(CH_3)_2CH$	1568	1550	−18	−1	1360	1359	[−0.190]
C_6H_{11}	1567	1535	−32	−9	1379	1370	[−0.150]
$(CH_3)_3C$	1555	1535	−20	−3	1349	1346	[−0.300]
CH_3ClCH	1589	1565	−24	−2	1349	1347	[0.95]
C_2H_5ClCH	1585	1564	−21	10	1348	1358	[0.95]
$(CH_3)_2ClC$	1579	1554	−25	−3	1339	1336	[0.86]
CH_3Cl_2C	1600	1581	−19	−2	1324	1322	[1.84]
$C_2H_5Cl_2C$	1595	1578	−17	−8	1320	1312	[1.82]
Cl_3C	1621	~1601	−20	0	1310	1310	[2.65]
Range:	[1555–1621]	[1535–1601]			[1310–1397]	[1310–1381]	
Δ Range:	66	66			87	71	

TABLE 7.4 Vapor-phase IR data for 4-X-nitrobenzenes

Compound 4-X-Nitrobenzene X	a.NO_2 str. cm^{-1} (A)	s.NO_2 str. cm^{-1} (A)	[a.NO_2 str.]-[s.NO_2 str.] cm^{-1}	A[s.NO_2 str.]/A[a.NO_2 str.]	σ_p
NH_2	1530 (0.756)	1350 (1.273)	180	1.68	[−0.66]
OH	1539 (0.800)	1350 (1.222)	189	1.53	[−0.36]
OCH_3	1535 (0.530)	1348 (0.749)	187	1.41	[−0.27]
OC_6H_5	1538 (0.669)	1350 (0.521)	188	0.78	[−0.03]
Cl	1540 (1.230)	1349 (1.225)	191	1	[0.23]
CH_3	1538 (1.152)	1355 (1.239)	183	1.08	[−0.17]
C_6H_5	1538 (0.669)	1352 (1.240)	186	1.85	[0.01]
H	1540 (1.248)	1351 (1.141)	189	0.91	[0]
CN	1542 (1.145)	1349 (1.242)	193	1.08	[0.86]
NO_2	1567 (1.275)	1340 (0.974)	227	0.76	[0.78]
CH_3CO	1541 (1.250)	1349 (1.250)	192	1	[0.52]
F	1542 (1.030)	1352 (1.240)	190	1.2	[0.06]
Range:	[1530–1567]	[1340–1355]			
Δ Range:	[37]	[15]			

TABLE 7.5 Infrared data for 4-X-nitrobenzenes in different phases

Compound 4-X-nitrobenzenes X	a.NO_2 str. vapor cm^{-1}	a.NO_2 str. $CHCl_3$ cm^{-1}	a.NO_2 str. neat or solid cm^{-1}	[a.NO_2 str.]-[s.NO_2 str.] vapor cm^{-1}	[a.NO_2 str.]-[s.NO_2 str.] $CHCl_3$ cm^{-1}	[a.NO_2 str.]-[s.NO_2 str.] neat or solid cm^{-1}	s.NO_2 str. vapor cm^{-1}	s.NO_2 str. $CHCl_3$ cm^{-1}	s.NO_2 str. neat or solid cm^{-1}
$N(CH_3)_2$		1487			169			1318	
OCH_3	1535	1510		187	171		1348	1339	
OH	1539	1517		189	179		1350	1338	
F	1542		1524	190		177	1352		1347
Cl	1540	1522	1522	191	179	162	1349	1343	1360
					180			1342	
Br		1527		172				1355	
				181				1346	
I			1510			160			1350
						175			1335
CH_3	1538	1520	1510	183	174	159	1355	1346	1351
C_6H_5	1538		1520	186		169	1352		1351
CH_2Cl		1527			179		1348		
CN	1542	1536		193	188		1349	1348	
CO_2CH_3		1528			180			1348	
CHO		1535			192			1343	
CH_3CO	1541	1530		192	188		1349	1342	
NO_2	1567	1555		229	217		1340	1338	

TABLE 7.6 A comparison of the frequency differences between asym. NO_2 str. and sym. NO_2 str. in the vapor and $CHCl_3$ solution and in the vapor and neat or solid phases

Compound 4-X-nitrobenzene X	a.NO_2 str. vapor cm^{-1}	a.NO_2 str. $CHCl_3$ cm^{-1}	a.NO_2 str. neat or solid cm^{-1}	[a.NO_2 str.] [v-$CHCl_3$] cm^{-1}	[s.NO_2 str.] [v-$CHCl_3$] cm^{-1}	[a.NO_2 str.] [v-n or -s] cm^{-1}	[s.NO_2 str.] [v-n or -s] cm^{-1}	s.NO_2 str. vapor cm^{-1}	s.NO_2 str. $CHCl_3$ cm^{-1}	s.NO_2 str. neat or solid cm^{-1}
OCH_3	1535	1510		−25	−9			1348	1339	
OH	1539	1517		−27	−12			1350	1338	
F	1542		1524			−18	−5	1352		1347
Cl	1540	1522	1522	−18	−6	−18	11	1349	1343	1360
						−7			1342	
CH_3	1538	1520	1510	−18	−28	−9	−4	1355	1346	1351
C_6H_5	1538		1520		−18	−1		1352		1351
CN	1542	1536		−6	−1			1349	1348	
CH_3CO	1541	1530		−11	−7			1349	1342	
NO_2	1567	1555		−12	−2			1340	1338	

TABLE 7.7 Infrared data for the asym. NO_2 and sym. NO_2 stretching frequencies for 3-X and 4-X-nitrobenzene in CCl_4 and $CHCl_3$ solutions

4-X-Nitrobenzene [1 wt. % solutions or saturated at 1 wt. %] X	a.NO_2 str. [CCl_4]	s.NO_2 str. [CCl_4]	[a.NO_2 str.]-[s.NO_2 str.] [CCl_4]	[a.NO_2 str.]-[s.NO_2 str.] [$CHCl_3$]	a.NO_2 str. [$CHCl_3$]	s.NO_2 str. [$CHCl_3$]	[a.NO_2 str. (CCl_4)]-[a.NO_2 str. ($CHCl_3$)]	[s.NO_2 str. (CCl_4)]-[s.NO_2 str. ($CHCl_3$)]
NH_2	1513.94	1338.08	175.86	170.19	1505.34	1335.15	8.6	2.93
CH_3O	1519.32	1343.07	176.25	171.36	1514.12	1342.76	5.2	0.31
C_6H_5O	1522.43	1344.98	177.45	172.97	1517.85	1344.88	4.58	0.1
C_6H_5	1524.26	1346.67	177.59	171.73	1519.47	1347.74	4.79	−1.07
CH_3	1524.8	1346.18	178.62	173.17	1520.1	1346.93	4.7	−0.75
HO	1526.74	1344.45	182.29	178.95	1521.41	1342.46	5.33	1.99
Cl	1527.01	1344.3	182.71	178.25	1523.64	1345.39	3.37	−1.09
iso-C_3H_7	1527.68	1347.11	180.57	171.72	1519.8	1348.08	7.8	−0.97
CH_2Cl	1529.17	1348.01	181.16	175.35	1525.44	1350.09	3.73	−2.08
	1529.78	1348.7	181.08	175.78	1526.08	1350.3	3.7	−1.6
C_6H_5-C=O	1531.06	1353.19	177.87	170.54	1525.77	1355.23	5.29	−2.04
H	1531.53	1348.24	183.29	177.61	1527.37	1349.76	4.16	−1.52
Br	1531.95	1350.1	179.68	173.42	1524.88	1351.46	7.07	−1.36
F	1532.31	1346.41	185.9	180.22	1528.48	1349.26	3.83	−1.85
CH_3CO_2	1532.87	1348	184.87	180.23	1530.26	1350.03	2.62	−2.03
CH_3C=O	1533.76	1345.71	188.05	183.6	1531.06	1347.46	2.7	−1.75
CN	1535.33	1345.37	189.96	185.65	1533.21	1347.56	2.12	−2.19
CH_3SO_2	1537.97	1348.2	189.77	185.3	1535.9	1350.6	2.07	−2.4
CF_3	1538.91	1354.35	184.56	179.1	1535.44	1356.34	3.47	−1.99
SO_2Cl	1541.25	1346.06	195.19	191.4	1539.26	1347.86	1.99	−1.8
NO_2	1556.41	1338.68	217.73	213.77	1554.37	1340.6	2.04	−1.92
Δ cm^{-1}	42.47	16.27	41.87	43.58	49.03	20.08		
3-X-Nitrobenzene X								
$N(CH_3)_2$	1535.29	1349.47	185.82	181.83	1531.64	1349.81	3.65	−0.34
CH_3	1532.78	1350.88	181.9	177.12	1529.16	1352.04	3.62	−1.16
Br	1536.44	1348.28	188.16	183.12	1533.09	1349.97	3.35	−1.69
I	1533.69	1346.97	186.72	181.8	1530.42	1348.62	3.27	−1.65
NO_2	1544.05	1345.29	198.76	194.19	1541.63	1347.44	2.42	−2.15
Δ cm^{-1}	11.27	5.59	16.86	17.07	12.47	4.6		

TABLE 7.8 Infrared data for the asym. NO_2 and sym. NO_2 stretching frequencies for 4-X-nitrobenzenes in the vapor, CCl_4 and $CHCl_3$ solution phases

4-X-Nitrobenzene X	σ_p	a.NO_2 str. [vapor] cm^{-1}	a.NO_2 str. [CCl_4 soln.] cm^{-1}	a.NO_2 str. [$CHCl_3$ soln.] cm^{-1}	a.NO_2 str. [solid] cm^{-1}	[vapor]-[CCl_4 soln.] [-cm^{-1}]	[vapor]-[$CHCl_3$ soln.] [-cm^{-1}]	[vapor]-[solid] [-cm^{-1}]
NH_2	−0.66	1530	1513.94	1505.34		16	25	
HO	−0.36	1539	1526.74	1521.41		12	18	
CH_3O	−0.27	1535	1519.32	1514.12		15	21	
CH_3	−0.17	1538	1524.8	1520.1	1510	13	18	28
C_6H_5O	−0.03	1538	1522.43	1517.85		16	20	
H	0	1540	1531.53	1527.37		9	13	
C_6H_5	0.01	1538	1524.26	1519.47	1520	14	19	18
F	0.06	1542	1532.31	1528.48	1524	10	14	18
Cl	0.23	1540	1527.01	1523.64	1522	13	16	18
$CH_3C{=}O$	0.52	1541	1532.87	1530.26		8	11	
NO_2	0.78	1567	1556.41	1554.37		11	13	
CN	0.86	1542	1535.33	1533.21		7	9	
X		s.NO_2 str. [vapor]	s.NO_2 str. [CCl_4 soln.]	s.NO_2 str. [$CHCl_3$ soln.]	s.NO_2 str. [solid]	[vapor]-[CCl_4 soln.]	[vapor]-[$CHCl_3$ soln.]	[vapor]-[solid]
NH_2	−0.66	1350	1338.08	1335.15		11.9	14.9	
OH	−0.36	1350	1344.45	1342.46		5.6	7.5	
CH_3O	−0.27	1348	1343.07	1342.76		4.9	5.2	
CH_3	−0.17	1355	1346.18	1346.93	1351	8.8	8.1	4
C_6H_5O	−0.03	1350	1344.98	1344.88		5	5.1	
H	0	1351	1348.2	1349.76		2.8	1.2	
C_6H_5	0.01	1352	1346.67	1347.74	1350	5.3	4.3	2
F	0.06	1352	1346.41	1349.26	1347	5.6	2.7	5
Cl	0.23	1349	1344.3	1345.39	1342	4.7	1.5	7
$CH_3C{=}O$	0.52	1349	1345.71	1347.41		3.3	1.6	
NO_2	0.78	1340	1338.68	1340.6		1.3	−0.6	
CN	0.86	1349	1348.37	1347.56		3.6	1.5	

TABLE 7.9 Infrared data for the asym. NO_2 and sym. NO_2 stretching frequencies of 4-nitrobenzaldehyde 1 wt./vol % in 0 to 100 mol % $CHCl_3/CCl_4$ solutions

4-Nitrobenzaldehyde 1% (wt./vol.) mole % $CHCl_3/CCl_4$	a.NO_2 str. cm^{-1}	s.NO_2 str. cm^{-1}
0	1538.4	1344.1
5.68	1538.1	1344.2
10.74	1537.8	1344.4
19.4	1537.4	1343.8
26.53	1537.1	1344.1
32.5	1537.1	1344.1
37.57	1536.9	1344.3
41.93	1536.6	1344.3
45.73	1536.4	1344.4
49.06	1536.2	1344.5
52	1536.1	1344.6
54.62	1535.4	1344.6
57.22	1535.4	1344.8
60.07	1535.3	1344.9
63.28	1535.1	1344.9
66.74	1535.2	1345.1
70.65	1535.1	1345.1
75.06	1534.9	1345.2
80.05	1534.6	1345.3
85.75	1534.3	1345.4
92.33	1534.1	1345.5
96.01	1533.8	1345.6
100	1533.6	1345.6
Δ C=O	−4.8	1.5

TABLE 7.10 A comparison of infrared data for nitromethane vs nitrobenzene in various solvents

Solvent	Nitromethane a.NO_2 str. cm^{-1}	Nitrobenzene a.NO_2 str. cm^{-1}	[CH_3NO_2]-[$C_6H_5NO_2$] cm^{-1}	Nitromethane s.NO_2 str. cm^{-1}	Nitrobenzene s.NO_2 str. cm^{-1}	[CH_3NO_2]-[$C_6H_5NO_2$] cm^{-1}	Nitromethane [Hexane]-[Solvent] a.NO_2 str. cm^{-1}	Nitrobenzene [Hexane]-[Solvent] a.NO_2 str. cm^{-1}	Nitrobenzene [Hexane]-[Solvent] s.NO_2 str. cm^{-1}
Hexane	1569	1535.8	33.2	masked	1348.2	[—]	0	0	0
Diethyl ether	1563.6	1532.6	31	masked	1349	[—]	5.4	3.2	0.8
Carbon tetrachloride	1564.5	1531.5	33	1374.4	1348.2	26.2	4.5	4.3	0
Benzene	1560.8	1528.5	32.3	1374.6	1348.3	26.3	8.2	7.3	−0.1
Acetonitrile	1561	1529.4	31.6	1374.2	1350.9	23.3	8	6.4	−2.7
Benzonitrile	1558.7	1526	32.7	1376.2	1348.5	27.7	10.3	9.8	−0.3
Methylene chloride	1562.2	1527.9	34.3	1376.7	1349.7	27	6.8	7.9*	−1.5
t-Butyl alcohol	1562.3	1531.9	30.4	1376.1	1349.5	26.6	6.7	3.9	−1.5
Chloroform	1562.4	1527.4	35	1376.2	1349.8	26.4	6.6	8.4*	−1.6
Isopropyl alcohol	1562.5	1531.6	30.9	1376.4	masked	[—]	6.5	4.2	[—]
Ethyl alcohol	1561.7	1530.7	31	masked	masked	[—]	7.3	5.1	[—]
Methyl alcohol	1561	1529.9	31.1	masked	1350.6	[—]	8	5.9	−2.4
Dimethyl sulfoxide	1552.8	1524.7	28.1	1376.1	1348.1	28	16.2	11.1	−0.1

* (see text).

TABLE 7.11 Vapor-phase infrared data for 3-X-nitrobenzenes

Compound 3-X-Nitrobenzenes X	a.NO_2 str. cm^{-1} (A)	s.NO_2 str. cm^{-1} (A)	[a.NO_2 str.]-[s.NO_2 str.] cm^{-1}	A[s.NO_2 str.]/A[a.NO_2 str.]	A[a.NO_2 str.]/A[s.NO_2 str.]
OH	1547 (1.230)	1360 (0.870)	187	0.71	1.41
OCH_3	1550 (1.235)	1353 (0.672)	197	0.54	1.84
OC_6H_5	1550 (1.220)	1353 (0.566)	197	0.46	2.16
Cl	1551 (1.242)	1351 (0.758)	200	0.61	1.64
CH_3	1545 (1.234)	1358 (1.139)	187	0.92	1.08
C_6H_5	1545 (1.245)	1357 (0.830)	188	0.67	1.5
H	1540 (1.248)	1351 (1.141)	189	0.91	1.09
CH_2Br	1550 (1.231)	1355 (0.923)	195	0.75	1.33
NO_2	1553 (1.271)	1349 (0.904)	204	0.71	1.41
$CO_2C_2H_5$	1550 (0.672)	1351 (0.562)	199	0.84	1.2
CH_3CO	1549 (1.229)	1358 (1.148)	191	0.93	1.07
Range:	[1540–1553]	[1349–1360]			
Δ Range:	[13]	[11]			

TABLE 7.12 Infrared data for 3-X-nitrobenzenes in different phases

Compound 3-X-nitro-benzenes X	a.NO_2 str. vapor cm^{-1}	a.NO_2 str. $CHCl_3$ cm^{-1}	a.NO_2 str. neat or solid cm^{-1}	[a.NO_2 str.]-[s.NO_2 str.] vapor cm^{-1}	[a.NO_2 str.]-[s.NO_2 str.] $CHCl_3$ cm^{-1}	[a.NO_2 str.]-[s.NO_2 str.] neat or solid cm^{-1}	s.NO_2 str. vapor cm^{-1}	s.NO_2 str. $CHCl_3$ cm^{-1}	s.NO_2 str. neat or solid cm^{-1}
OCH_3	1550	1526	[—]	197	178	[—]	1353	1348	[—]
OH	1547	1529	[—]	187	177	[—]	1360	1352	[—]
Cl	1551	1527	1520	200	177	176	1351	1350	1344
			1540			196			
Br	[—]	1532	[—]	[—]	152	[—]	[—]	1380	[—]
I	[—]		1521	[—]	[—]	181	[—]	1340	[—]
CH_3	1545	1531	1521	187	181	174	1358	1350	1347
C_6H_5	1545		1529	188	[—]	178	1357		1351
CH_2Cl	[—]	1532	[—]	[—]	179	[—]	[—]	1353	[—]
CN	[—]	1538	[—]	[—]	186	[—]	[—]	1352	[—]
NO_2	1553	1539	1522	204	193	174	1349	1346	1348

TABLE 7.13 A comparison of the frequency difference between asym. NO_2 str. and sym. NO_2 for 3-X-nitrobenzenes in the vapor and $CHCl_3$ solution and in the vapor and solid or neat phases

Compound 3-X-nitro-benzene X	a.NO_2 str. vapor cm^{-1}	a.NO_2 str. $CHCl_3$ cm^{-1}	a.NO_2 str. neat or solid cm^{-1}	[a.NO_2 str.] [v-$CHCl_3$] cm^{-1}	[s.NO_2 str.] [v-$CHCl_3$] cm^{-1}	[a.NO_2 str.] [v-n or -s] cm^{-1}	[s.NO_2 str.] [v-n or -s] cm^{-1}	s.NO_2 str. vapor cm^{-1}	s.NO_2 str. $CHCl_3$ cm^{-1}	s.NO_2 str. neat or solid cm^{-1}
OCH_3	1550	1526		−24	−5			1353	1348	
OH	1547	1529		−18	−8			1360	1352	
Cl	1551	1527	1540	−24	−11	−11	−7	1351	1350	1344
			1520			−31				
Br		1532							1380	
I			1521							1340
CH_3	1545	1531	1521	−14	−8	−24	−11	1358	1350	1347
C_6H_5	1545		1529			−16	−6	1357		1351
CH_2Cl		1532							1353	
CN		1538							1352	
NO_2	1553	1539	1522	−14	−3	−31	−1	1349	1346	1348

TABLE 7.14 Vapor-phase infrared data for 2-X-nitrobenzenes

Compound 2-X-Nitrobenzenes X	a.NO_2 str. cm^{-1} (A)	s.NO_2 str. cm^{-1} (A)	[a.NO_2 str.]-[s.NO_2 str.] cm^{-1}	A[s.NO_2 str.]/A[a.NO_2 str.]	A[a.NO_2 str.]/A[s.NO_2 str.]
OH	1545 (0.640)	1335 (1.239)	210	1.94	0.52
OCH_3	1550 (1.190)	1360 (0.920)	190	0.77	1.29
OC_2H_5	1547 (1.230)	1360 (0.920)	187	0.75	1.34
OC_4H_9	1545 (1.240)	1359 (0.479)	186	0.39	2.59
Cl	1551 (1.220)	1359 (0.700)	192	0.57	1.74
Br	1553 (1.240)	1359 (0.770)	194	0.62	1.61
CH_3	1542 (1.242)	1350 (0.960)	192	0.77	1.29
I	1550 (1.241)	1353 (0.640)	197	0.52	1.94
C_6H_5	1545 (1.240)	1358 (0.638)	187	0.51	1.94
H	1540 (1.248)	1351 (1.141)	189	0.91	1.09
CF_3	1560 (1.141)	1360 (0.662)	200	0.58	1.72
CO_2CH_3	1551 (1.240)	1357 (0.818)	194	0.66	1.52
$CO_2C_2H_5$	1551 (1.150)	1352 (0.720)	199	0.63	1.6
F	1549 (1.250)	1354 (1.131)	195	0.9	1.11
Range:	[1540–1560]	[1335–1360]			
Δ Range:	[20]	[25]			

TABLE 7.15 Infrared data for 2-X-nitrobenzenes in different phases

Compound 2-X-nitrobenzene X	a.NO_2 str. vapor cm^{-1}	a.NO_2 $CHCl_3$ cm^{-1}	a.NO_2 neat or solid cm^{-1}	[a.NO_2 str.]-[s.NO_2 str.] vapor cm^{-1}	[a.NO_2 str.]-[s.NO_2 str.] $CHCl_3$ cm^{-1}	[a.NO_2 str.]-[s.NO_2 str.] neat or solid cm^{-1}	s.NO_2 str. vapor cm^{-1}	s.NO_2 str. $CHCl_3$ cm^{-1}	s.NO_2 str neat or solid cm^{-1}
OCH_3	1550	1530		190	173		1360	1357	
OH	1545	1537		210	202		1335	1335	
F	1549		1520	195		180	1354		1340
Cl	1551	1537	1528	192	180	176	1359	1357	1352
Br	1553	1536	1520	194	179	175	1359	1356	1345
CH_3	1542	1527	1520	192	173	175	1350	1354	1345
C_6H_5	1545		1520	187		170	1358		1350
CO_2CH_3	1551	1537		194	184		1357	1353	
CHO		1532			185			1347	
NO_2			1546			176			1370
			1529			175			1354

TABLE 7.16 A comparison of the frequency differences between asym. NO_2 and sym. NO_2 str. for 2-X-nitrobenzenes in the vapor, $CHCl_3$ solution, and in the vapor, neat or solid phases

Compound 2-X-nitro-benzene X	a.NO_2 str. vapor cm^{-1}	a.NO_2 str. $CHCl_3$ cm^{-1}	a.NO_2 str. neat or solid cm^{-1}	[a.NO_2 str.] [v-$CHCl_3$] cm^{-1}	[s.NO_2 str.] [v-$CHCl_3$] cm^{-1}	[a.NO_2 str.] [v-n or -s] cm^{-1}	[s.NO_2 str.] [v-n or -s] cm^{-1}	s.NO_2 str. vapor cm^{-1}	s.NO_2 str. $CHCl_3$ cm^{-1}	s.NO_2 str. neat or solid cm^{-1}
OCH_3	1550	1530		−20	−3			1360	1357	
OH	1545	1537		−8	0			1335	1335	
F	1549		1520			−29	−14	1354		1340
Cl	1551	1537	1528	−14	−2	−23	−7	1359	1357	1352
Br	1553	1536	1520	−17	−3	−33	−14	1359	1356	1345
CH_3	1542	1527	1520	−15	4	−22	−5	1350	1354	1345
C_6H_5	1545		1520			−25	−8	1358		1350
CO_2CH_3	1551	1537		−14	−4			1357	1353	
CHO		1532							1347	
NO_2			1546							1370
			1529							1354

TABLE 7.17 Infrared data for nitrobenzenes in the solid and CCl_4 solution phases

1-X-2-Nitro-benzene a.NO_2 str. [solid] cm^{-1}	1-X-3-Nitro-benzene a.NO_2 str. [solid] cm^{-1}	1-X-4-Nitro-benzene a.NO_2 str. [solid] cm^{-1}	X	1-X-2-Nitro-benzene s.NO_2 str. [solid] cm^{-1}	1-X-3-Nitro-benzene s.NO_2 str. [solid] cm^{-1}	1-X-4-Nitrobene-benzene s.NO_2 str. [solid] cm^{-1}	[a.NO_2 str.]-[s.NO_2 str.] 1,2- [solid] cm^{-1}	[a.NO_2 str.]-[s.NO_2 str.] 1,3- [solid]; [CCl_4-solid] cm^{-1}	[a.NO_2 str.]-[s.NO_2 str.] 1,4- [solid]; [CCl_4-solid] cm^{-1}
1534*1	1512*2	1500*3	OH	1330	1351	1324	204	170	176
1500	1522		NHC_2H_5	1337	1348		163		
1520	1521	1510	CH_3	1344	1347	1351	176	174; 181.9	159; 178.6
1524		1509	C_2H_5	1349		1334	175		175
1520	1529	1520	C_6H_5	1350	1351	1350	170		170; 177.6
1520		1524	F	1340		1347	180		177; 185.9
1529	1521	1522	Cl	1351	1345	1342	178	180; 182	
1520			Br	1343			177		
	1520	1510	I	1366; 1355	1339	1328			182; 181*1
1548; 1530	1522		NO_2		1348		182; 165		

1-X-2,4-di-Nitrobenzene a.NO_2 str. [solid] cm^{-1}	4-X-2,4-di-Nitrobenzene a.NO_2 str. [solid] cm^{-1}	1-X-3,5-di-Nitrobenzene a.NO_2 str. [solid] cm^{-1}	X	1-X-2,4-di-Nitrobenzene s.NO_2 str. [solid] cm^{-1}	4-X-di-2,4-Nitrobenzene s.NO_2 str. [solid] cm^{-1}	1-X,3,5-di-Nitrobenzene s.NO_2 str. [solid] cm^{-1}	[a.NO_2 str.]-[s.NO_2 str.] cm^{-1}	2-OH,3-Cl-Nitrobenzene*4 a.NO_2 str.; s.NO_2 str. [solid] cm^{-1}	2-OH,3,5-di-Nitrobenzene*5
1538			F	1345			193		
1540			Br	1341			199		
	1534		$N(CH_3)_2$		1310		224		
		1544	NO_2			1348	196		
			OH; Cl				180	1515; 1335	
			OH; NO_2; NO_2	188				1528; 1340	
				240				1550; 1310	

*1 OH; O_2N str. 3260.
*2 OH str. 3380.
*3 OH str. 3325.
*4 OH; O_2N str. 3230.
*5 OH; NO_2 str. ~3230 (broad).

TABLE 7.18 Vapor-phase infrared data for 2,5- and 2,6-X,Y-nitrobenzenes

Compound 2,5- X,Y-nitrobenzenes X,Y	a.NO_2 str. cm^{-1} (A)	s.NO_2 str. cm^{-1} (A)	[a.NO_2 str.]-[s.NO_2 str.] cm^{-1}	A[s.NO_2 str.]/A[a.NO_2 str.]	A[a.NO_2 str.]/A[s.NO_2 str.]
CH_3, NH_2	1541 (1.250)	1360 (0.508)	181	0.4	2.46
OH, CH_3	1545 (1.240)	1325 (1.155)	220	0.93	1.07
CH_3, CH_3	1541 (1.241)	1359 (0.610)	182	0.49	2.03
CH_3, CH_2OH	1542 (1.240)	1355 (0.660)	187	0.53	1.88
CH_3, NO_2	1550 (1.240)	1351 (1.225)	199	0.99	1.01
OH, NO_2	1552 (0.671)	1345 (1.210)	207	1.8	0.55
		1330 (0.910)	222	1.36	0.74
Cl, CH_3	1558 (1.230)	1361 (0.661)	197	0.54	1.86
Cl, Cl	1555 (1.250)	1352 (0.849)	203	0.7	1.47
Cl, NO_2	1553 (1.348)	1349 (1.140)	204	0.85	1.18
Cl, CF_3	1560 (0.341)	1357 (0.175)	203	0.51	1.9
F, CF_3	1560 (0.417)	1352 (0.219)	208	0.53	1.9
F, NO_2	1559 (1.230)	1349 (1.225)	210	1	1
$NHCOCH_3$, NO_2	1541 (0.460)	1340 (1.230)	201	2.67	0.37
CH_3O, CH_3CO	1552 (0.668)	1360 (0.542)	192	0.81	1.23
F, CH_3	1555 (1.240)	1351 (0.590)	204	0.48	2.1
2,6-X,Y-nitrobenzenes X,Y					
OH, CH_3	1547 (0.590)	1349 (0.672)	198	1.14	0.88
CH_3, CH_3	1545 (1.341)	1371 (0.590)	174	0.44	2.27
Cl, Cl	1568 (1.230)	1360 (0.342)	208	0.28	3.6

TABLE 7.19 A comparison of the frequency difference between asym. NO_2 str. and sym. NO_2 str. in the vapor, neat or solid phase

Compound 2,5-X,Y-nitrobenzene X,Y	a.NO_2 str. vapor cm^{-1}	a.NO_2 str. neat or solid cm^{-1}	[a.NO_2 str.] [v-n or -s] cm^{-1}	[s.NO_2 str.] [v-n or -s] cm^{-1}	s.NO_2 str. vapor cm^{-1}	s.NO_2 str. neat or solid cm^{-1}
CH_3, CH_3	1541	1514	−27	−19	1359	1340
F, NO_2	1559	1537	−22	−3	1349	1346

TABLE 7.20 Vapor-phase infrared data for tri-X,YZ-, W,X,Y,Z-tetra-, and V,W,X,Y,Z-penta-nitrobenzenes

Compound 2,3,4-X,Y,Z-nitrobenzenes X,Y,Z	a.NO_2 str.	s.NO_2 str.	[a.NO_2 str.]-[s.NO_2 str.]	A[s.NO_2 str.]/A[a.NO_2 str.]	A[a.NO_2 str.]/A[s.NO_2 str.]
Cl, CH_3, Cl	1552 (1.241)	1359 (0.735)	193	0.59	1.69
Cl, Cl, Cl	1560 (1.050)	1351 (1.232)	209	1.17	0.85
2,3,5-X,Y,Z-nitrobenzenes X,Y,Z					
OH, Br, Br	1545 (1.230)	1338 (0.678)	207	0.55	1.81
OH, Cl, Cl	1550 (1.240)	1335 (0.740)	215	0.6	1.68
OH - s- C_4H_9, NO_2	1561 (0.680)	1348 (1.240)	213	1.82	0.55
OH, CH_3, NO_2	1564 (0.750)	1347 (1.240)	217	1.65	0.6
Br, NO_2, Br	1561 (1.241)	1340 (0.541)	221	0.44	2.29
Cl, NO_2, Cl	1567 (1.220)	1350 (0.570)	217	0.47	2.14
2,3,6-X,Y,Z-nitrobenzenes X,Y,Z					
NH_2, Cl, Cl	1541 (1.250)	1359 (0.442)	182	0.35	2.83
NO_2, Br, Br	1578 (1.242)	1348 (0.320)	230	0.26	3.88
NO_2, Cl, Cl	1580 (1.250)	1350 (0.300)	230	0.24	4.17
2,4,5-X,Y,Z-nitrobenzenes X,Y,Z					
Br, NO_2, Br	1552 (1.250)	1339 (0.405)	213	0.32	3.09
NH_2, Cl, Cl	1561 (0.531)	1340 (0.585)	221	1.1	0.91
Cl, NO_2, Cl	1565 (1.230)	1340 (0.530)	225	0.43	2.32
Cl, Cl, Cl	1565 (1.240)	1335 (0.512)	230	0.41	2.42
2,4,6-X,Y,Z-nitrobenzenes X,Y,Z					
CH_3, OH, CH_3	1538 (1.229)	1362 (0.404)	176	0.33	3.04
Cl, Cl, Cl	1562 (1.248)	1361 (0.488)	201	0.38	2.67
3,4,5-X,Y,Z-nitrobenzenes X,Y,Z					
CH_3, OH, CH_3	1543 (0.660)	1357 (1.201)	186	1.82	0.55
CH_3, CH_3O, CH_3	1540 (0.780)	1359 (1.210)	181	1.55	0.64
2,3,4,5-W,X,Y,Z-nitrobenzenes W,X,Y,Z					
Cl, Cl, Cl, Cl	1560 (0.922)	1332 (1.240)	228	1.34	0.74
F, F, F, F	1570 (1.330)	1360 (0.625)	210	0.47	2.13

(*continues*)

TABLE 7.20 (*continued*)

Compound 2,3,4-X,Y,Z-nitrobenzenes X,Y,Z	a.NO_2 str.	s.NO_2 str.	[a.NO_2 str.]-[s.NO_2 str.]	A[s.NO_2 str.]/A[a.NO_2 str.]	A[a.NO_2 str.]/A[s.NO_2 str.]
2,3,5,6-W,X,Y,Z- nitrobenzenes					
Cl, Cl, Cl, Cl	1569 (1.199)	1333 (0.961)	236	0.8	1.25
2,3,4,5,6-V,W,X,Y,Z-penta-nitrobenzenes V,W,X,Y,Z					
Cl, Cl, Cl, Cl, Cl	1568 (1.239)	1332 (0.838)	236	0.68	1.48

TABLE 7.21 **Infrared data for 4-X-nitrobenzenes in CCl_4 and $CHCl_3$ solutions (1% wt./vol. or less)**

4-X-Nitrobenzene	ν asym.NO_2 CCl_4 soln. cm^{-1}	ν sym.NO_2 CCl_4 soln. cm^{-1}	ν asym.NO_2 $CHCl_3$ soln. cm^{-1}	ν sym.NO_2 $CHCl_3$ soln. cm^{-1}
X				
1 NH_2	1513.9	1338.1	1505.3	1335.2
2 OCH_3	1519.3	1343.1	1514.1	1342.8
3 OC_6H_5	1522.4	1345	1517.9	1344.9
4 C_6H_5	1524.3	1346.7	1519.5	1347.7
5 CH_3	1524.8	1346.2	1520.1	1346.9
6 OH	1526.7	1344.5	1521.4	1342.5
7 Cl	1527	1344.3	1523.6	1345.4
8 iso-C_3H_7	1527.7	1347.1	1519.8	1348.1
9 CH_2Cl	1529.2	1348	1525.4	1350.1
10 I	1529.8	1348.7	1526.1	1350.3
11 (C=O)C_6H_5	1531.1	1353.2	1525.8	1355.2
12 H	1531.5	1348.2	1527.4	1349.8
13 Br	1532	1350.1	1524.9	1351.5
14 F	1532.3	1346.4	1528.5	1349.3
15 CO_2CH_3	1532.9	1348	1530.3	1350
16 (C=O)CH_3	1533.8	1345.7	1531.1	1347.5
17 CN	1535.3	1345.4	1533.2	1347.6
18 SO_3CH_3	1538	1348.2	1535.9	1350.6
19 CF_3	1538.9	1354.4	1535.4	1356.3
20 SO_2Cl	1541.3	1346.1	1539.3	1347.9
21 NO_2	1556.4	1338.7	1554.4	1340.6
Δ cm^{-1}	42.5	16.3	49	20.1

TABLE 7.22 **Infrared data and assignments for alkyl nitrates**

Compound $R-O-NO_2$ R	$a.NO_2$ str. cm^{-1}	$s.NO_2$ str. cm^{-1}	A[$s.NO_2$ str.]/ A[$a.NO_2$ str.]	A[$a.NO_2$ str.]/ A[$s.NO_2$ str.]	NO str. cm^{-1}	A[NO str.]/ A[$a.NO_2$ str.]	γ NO_2 cm^{-1}	Δ NO_2 cm^{-1}
C_2H_5 [vapor]	1662 (0.932)	1289 (0.625)	0.67	1.49	854 (0.455)	0.49	764	700
iso-C_3H_7 [vapor]	1651 (1.265)	1282 (0.619)	0.49	2.04	860 (0.525)	0.42	760	700
C_3H_7 [liquid]	1622 (stg)	1268 (stg)			861 (stg)		760 (wm)	699 (wm)
C_5H_{11} [liquid]	1617 (stg)	1256 (stg)			865 (stg)		760 (wm)	700 (wm)

TABLE 7.23 **Infrared data for alkyl nitrates in various phases**

Compound $R-O-NO_2$ R	$a.NO_2$ str. vapor cm^{-1}	$a.NO_2$ str. CCl_4 or neat cm^{-1}	[$a.NO_2$ str.]- [v-CCl_4 or −neat]	[$s.NO_2$ str.]- [v-CS_2 or −neat]	$s.NO_2$ str. vapor cm^{-1}	$s.NO_2$ str. CS_2 or neat
C_2H_5 [CCl_4 and CS_2]	1662	1637	−25	−9	1289	1280
Iso-C_3H_7 [neat]	1651	1620	−31	−8	1282	1274
		[$a.NO_2$ str.]- [$s.NO_2$ str.] vapor cm^{-1}	[$a.NO_2$ str.]- [$s.NO_2$ str.] CCl_4; CS_2 or neat cm^{-1}	[2($a.NO_2$ str.)]- [2($s.NO_2$ str.)] vapor cm^{-1}	[2($a.NO_2$ str.)]- [2($s.NO_2$ str.)] CCl_4; CS_2 or neat cm^{-1}	
C_2H_5 [CCl_4 and CS_2]		373	357	731	691	
Iso-C_3H_7 [neat]		369	346	730	672	
	2($a.NO_2$ str.) vapor cm^{-1}	2($a.NO_2$ str.) CCl_4 or neat cm^{-1}	[2($a.NO_2$ str.)]- [v-CCl_4 or neat] cm^{-1}	[2($s.NO_2$ str.)- [v-CS_2 or neat] cm^{-1}	2($s.NO_2$ str.) vapor cm^{-1}	2($s.NO_2$ str.) CS_2 or neat cm^{-1}
C_2H_5 [CCl_4 and neat]	3299	3250	−49	−9	2568	2559
[calculated]	[3334]	[3274]	−60	−12	[2578]	[2560]
[Δ obs. −calc.]	[−35]	[−24]			[−10]	[−1]
Iso-C_3H_7 [neat]	3290	~3210	−80	−22	~2560	2538
[calculated]	[3302]	[3240]	−62	−16	[2564]	[2548]
[Δ obs. −calc.]	[−12]	[−30]			[−4]	[−10]

TABLE 7.24 Infrared data for ethyl nitrate, nitroalkanes, and nitrobenzene in CCl_4 and $CHCl_3$ solutions

Compound	a.NO_2 str. CCl_4 soln. cm^{-1} cm^{-1}	a.NO_2 str. $CHCl_3$ soln. cm^{-1} cm^{-1}	[CCl_4 soln.]-[$CHCl_3$ soln.] cm^{-1}	s.NO_2 str. CCl_4 soln. cm^{-1}	s.NO_2 str. $CHCl_3$ soln. cm^{-1}	[CCl_4 soln.]-[$CHCl_3$ soln.] cm^{-1}	[a.NO_2 str.]-[s.NO_2 str.] CCl_4 soln. cm^{-1}	[a.NO_2 str.]-[s.NO_2 str.] $CHCl_3$ soln. cm^{-1}	[CCl_4 soln.]-[$CHCl_3$ soln.] cm^{-1}
Ethyl nitrate	1637.3	1631.7	−5.6	1281.5	1282.9	1.4	355.8	348.8	−7
Trichloronitromethane	1608.2	1606.5	−1.7	1307.5	1309.6	2.1	300.7	296.8	−3.8
Tribromonitromethane	1595.1	1593.3	−1.8	1305.9	1308.7	2.8	289.2	284.6	−4.6
Nitromethane	1564.5	1562.4	−2.1	1374.4	1376.2	1.8	190.1	182.2	−3.9
Nitrobenzene	1531.5	1527.4	−4.1	1348.2	1349.8	1.6	183.3	177.6	−5.7

TABLE 7.25 **Vapor-phase infrared data for alkyl nitrites**

Compound R-O–N=O R	cis N=O str. cm^{-1} (A)	trans N=O str. cm^{-1} (A)	[trans N=O str.]-[cis N=O str.] cm^{-1}	cis N–O str. cm^{-1} (A)	trans N–O str. cm^{-1} (A)	cis C–O str. cm^{-1} (A)	trans C–O str. cm^{-1} (A)
CH_3	1619 (0.726)	1676 (0.426)	57	835 (0.230)	795 (0.821)	1020 (w)	1055 (w)
C_3H_7	1614 (0.300)	1669 (0.850)	55		794 (0.710)		
iso-C_4H_9	1609 (0.170)	1668 (0.690)	59	849 (0.280)	781 (0.560)	982 (w)	1039 (w)
sec-C_4H_9	1611 (0.211)	1660 (0.721)	49	850 (0.220)	766 (0.870)	?	1120 (w)
tert-C_4H_9	~1610 (0.020)	1653 (1.148)	43		759 (1.240)		
tert-C_4H_9 [liquid]	1554	1619	65		760 (stg)		1190 (m)
Δ [vap.−liq.]	[−56]	[−34]	[22]		[1]		

TABLE 7.26 **Vapor-phase infrared data of the characteristic vibrations of alkyl nitrites***

Compound R-O–N=O R	N=O str. cis cm^{-1}	N=O str. trans cm^{-1}	N–O str. trans cm^{-1}	Δ O–N=O cis cm^{-1}	Δ O–N=O trans cm^{-1}	σ†	A[trans N=O str.]/A[cis N=O str. ~values
CH_3	~1625	1681	814	617	565	[0.000]	0.95
C_2H_5	1621	1675	~800	691	581	[−0.100]	2.3
C_3H_7	1621	1672	802	687	602	[−0.115]	3
C_4H_9	1618	1669	790	689	610	[−0.130]	3.3
iso-C_4H_9	1618	1669	794	680	625	[−0.125]	3.5
iso-C_5H_{11}	1618	1669	800	687	617		305
iso-C_3H_7	1615	1667	783	688	605	[−0.190]	6
sec-C_4H_9	1615	1665	776	678	600	[−0.210]	7.2
sec-C_5H_{11}	1618	1664	775	678	594		10
cyclo-C_5H_9	1613	1664	780	682	604	[−0.150]	4.7
cyclo-C_6H_{11}	1615	1664	775			[−0.200]	9.3
tert-C_4H_9	1610	1655	764		621?	[−0.300]	~30
tert-C_5H_{11}	~1613	1653	751		613?		~50

* See Tarte (12).
† σ* is a polar value for the alkyl group.

TABLE 7.27 Raman data for organonitro compounds

Compound	s.NO_2 str. cm^{-1} (RI)	NO_2 bend cm^{-1} (RI)	[s.NO_2 str.]-[NO_2 bend] cm^{-1}	RI[s.NO_2 str.]/RI[NO_2 bend]
2-Nitro-2-methyl propyl methacrylate	1350 (5)	858 (8)	492	0.63
4-Nitrostyrene	1343 (9)	859 (2)	484	4.5
Poly(4-nitrostyrene)	1347 (9)	859 (1)	488	9
3-Nitrostyrene	1349 (9)			
4- Nitrophenyl acrylate	1349 (9)	866 (3)	483	3
4-Nitrophenyl methacrylate	1351 (9)	866 (3)	485	3
4,4′-Diamino-3,3-dinitrophenyl ether	1336 (5)	882 (9)	454	0.56
1,3-Diamino-4,6-dinitrobenzene	1312 (9)	831 (4)	481	2.3

TABLE 7.28 Infrared and Raman data for nitroalkanes in different physical phases

Compound	asym.NO_2 str. [vapor] cm^{-1}	asym.NO_2 str. [neat] cm^{-1}	asym.NO_2 str. [vapor]-[neat] cm^{-1}	sym.NO_2 str. [vapor]-[neat] cm^{-1}	sym.NO_2 str. [vapor] cm^{-1}	sym.NO_2 str. [neat] cm^{-1}	IR or Raman
CH_3NO_2	1582	1558	24	22	1397	1375	IR
		1562	20	13		1384	R
$C_2H_5NO_2$	1575	1555	20	17	1380	1363	IR
		1559	16	5		1375	R
$(CH_3)_2CHNO_2$	1568	1550	18	1	1360	1359	IR
		1552	16	[−2]		1362	R
$(CH_3)_3CNO_2$	1555	1535	20	3	1349	1346	IR
$(CH_3)_2ClCNO_2$	[—]	1564	[—]	[—]	[—]	1346	R
$CH_3ClHCNO_2$	1589	1565	24	2	1349	1347	IR
		1572	17	[−7]		1356	R
$CH_3Cl_2CNO_2$	1600	1581	19	2	1324	1322	IR
		1587	13	[−6]		1330	R
$C_2H_5ClHCNO_2$	1585	1564	21	[−10]	1348	1358	IR
		1570	15	[−14]		1362	R
$C_2H_5Cl_2CNO_2$	1595	1578	17	8	1320	1312	IR
		1585	10	[−3]		1323	R
Cl_3NO_2	1621	~1601	20	0	1310	1310	IR

TABLE 7.29 Infrared and Raman data for nitrobenzenes in different physical phases

Group	4-Nitro-benzene [vapor] asym.NO_2 str. cm^{-1}	4-Nitro-benzene [$CHCl_3$] asym.NO_2 str. cm^{-1}	4-Nitro-benzene [neat] asym.NO_2 str. cm^{-1}	3-Nitro-benzene [vapor] asym.NO_2 str. cm^{-1}	3-Nitro-benzene [$CHCl_3$] asym.NO_2 str. cm^{-1}	3-Nitro-benzene [neat] asym.NO_2 str. cm^{-1}	2-Nitro-benzene [vapor] asym.NO_2 str. cm^{-1}	2-Nitro-benzene [$CHCl_3$] asym.NO_2 str. cm^{-1}	2-Nitro-benzene [neat] asym.NO_2 str. cm^{-1}	IR or Raman IR
CH_3O	1535	1510	[—]	1550	1526	[—]	1550	1530	[—]	IR
HO	1539	1517	[—]	1547	1529	[—]	1545	1537	[—]	IR
Cl	1540	1522	1522	1551	1527	1530	1551	1537	1528	IR
		1525		1530						R [neat]
F	1542									IR
		1528								R
CH_3	1538	1520	1510	1545	1531	1522	1542	1527	1521	IR
		1520					1525	R		R [neat]
NO_2	1567	1555	[—]	1553	1539	1522	[—]	[—]	1530 [KBr]	IR
	sym.NO_2 str.	sym.NO_2 str.	sym.NO_2 str.	sym.NO_2 str.	sym.NO_2 str.	sym.NO_2 str.	sym.NO_2 str.	sym.NO_2 str.	sym.NO_2 str.	
CH_3O	1348	1339	[—]	1353	1348	[—]	1360	1357	[—]	IR
HO	1350	1338	[—]	1360	1352	[—]	1335	1335	[—]	IR
					1350					
Cl	1350	1343	1342	1351	1350	1345	1359	1357	1352	IR
F	1352									IR
		1347								R [neat]
CH_3	1355	1346	1351	1358	1350	1350	1350	1354	1344	IR
		1350					1350			R [neat]
NO_2	1340	1338	[—]	1349	1346	1344	[—]	[—]	[—]	IR

TABLE 7.30 A comparison of the frequency separation between asym. NO_2 stretching and sym. NO_2 stretching [vapor-phase data minus $CHCl_3$ solution data] and [vapor-phase data minus neat-phase data] for nitrobenzenes

Group	4-Nitro-benzene asym.NO_2 str. [vapor]-[$CHCl_3$] cm^{-1}	4-Nitro-benzene sym.NO_2 str. [vapor]-[$CHCl_3$] cm^{-1}	3-Nitro-benzene asym.NO_2 str. [vapor]-[$CHCl_3$] cm^{-1}	3-Nitro-benzene asym.NO_2 str. [vapor]-[$CHCl_3$] cm^{-1}	2-Nitro-benzene asym.NO_2 str. [vapor]-[$CHCl_3$] cm^{-1}	2-Nitro benzene asym.NO_2 str. [vapor]-[$CHCl_3$] cm^{-1}	IR or Raman
CH_3O	25	9	24	5	20	3	IR
HO	22	12	18	8	8	0	IR
Cl	18	6	24	1	14	2	IR
CH_3	18	9	14	8	14	[−4]	IR
NO_2	12	2	14	3	[—]	[—]	IR
	[vapor]-[neat] cm^{-1}	[vapor]-[neat] cm^{-1}	[vapor]-[neat] cm^{-1}	[vapor]-[neat] cm^{-1}	[vapor]-[neat] cm^{-1}	[vapor]-[neat] cm^{-1}	
Cl	18	7	21	6	23	7	IR
	15	0					IR; R
F	14	5					IR; R
CH_3	28	4	23	12	21	6	IR
	18	5					IR; R
NO_2	[—]	[—]	14	0	[—]	[—]	IR

TABLE 7.31 Infrared data for organonitro compounds in different physical phases

Nitro-Compounds	a.NO_2 str. cm^{-1} vapor	s.NO_2 str. cm^{-1} vapor	Tempera-ture	a.NO_2 str. cm^{-1} neat	s.NO_2 str. cm^{-1} neat	a.NO_2 str. cm^{-1} vapor-neat	s.NO_2 str. cm^{-1} vapor-neat
Benzene							
Nitro	1540	1352	20 °C	1520	1345	20	7
1,2-Dinitro	1563	1353	280 °C	1529	1353	34	0
1,3-Dinitro	1555	1349	280 °C	1532	1349	23	0
1,4-Dinitro	1565	1340	240 °C	1562	1339	3	1
Phenol 2-Nitro	1545	1333	280 °C	1530 Melt	1322	15	11
				CCl_4	CCl_4	Vapor-CCl_4	Vapor-CCl_4
Benzene							
1,2,4,5-Cl_4-3-NO_2	1569	1333		1556	1337	13	[−4]
				neat			
				1559	1339	10	[−6]
Cl_5—NO_2	1569	1335		1556	1338	13	[−3]
1,2,3,4-Cl_4—5-NO_2	1562	1338		1532	1338	30	0
1-Cl—2-NO_2	1551	1359		~1535	1352	16	1
1-Cl—3-NO_2	1551	1351		1542	1351	9	0
1-Cl—4-NO_2	1540	1349		1527	1344	13	5

TABLE 7.32 Infrared and Raman data for organonitrates, organonitrites, and 1,4-dinitropiperazine in different physical phases

Nitrate	asym.NO_2 str. [vapor] cm^{-1}	asym.NO_2 str. [CCl_4] cm^{-1}	asym.NO_2 [vapor]-[CCl_4] cm^{-1}	sym.NO_2 [vapor]-[CCl_4] cm^{-1}	sym.NO_2 str. [vapor] cm^{-1}	sym.NO_2 str. [CCl_4] cm^{-1}	IR or Raman
C_2H_5	1660	1637	23	9	1289	1280	IR
		[neat] cm^{-1}	[vapor]-[neat] cm^{-1}	[vapor]-[neat] cm^{-1}		[neat] cm^{-1}	IR
$(CH_3)_2CH$	1651	1620	31	8	1282	1274	IR
C_3H_7	[—]	1625	[—]	[—]	1277	[—]	IR
		1635 wk			1282 stg		R
C_5H_{11}	[—]	1624	[—]	[—]	1275	[—]	IR
	trans N=O str. [vapor] cm^{-1}	trans N=O str. [CCl_4] cm^{-1}	trans N=O str. [vapor]-[CCl_4] cm^{-1}	cis N=O str. [vapor]-[CCl_4] cm^{-1}	cis N=O str. [vapor] cm^{-1}	cis N=O str. [CCl_4] cm^{-1}	
CH_3	1681	1685	[−4]	[−5]	1625	1630	
C_4H_9	1673	[—]	[—]	[—]	1618	[—]	IR
iso-C_5H_{11}	1669	1672	[−3]	[−2]	1618	1620	IR
tert-$(CH_3)_3C$	1655	1661	[−6]	0	1610	~1610	IR
N–NO_2		asym.NO_2 str. [neat] cm^{-1}			sym.NO_2 str. [neat] cm^{-1}		
1,4-Dinitropiperazine		1543			1234		IR

TABLE 7.33 The N=O stretching frquencies for nitrosamines in different physical phases

X,Y	XYNN=O vapor N=O str. cm^{-1}	Temp. °C	XYNN=O N=O str. cm^{-1}	State	XYNN=O* CCl_4 soln. N=O str. cm^{-1}	Vapor-state Δ N=O str. cm^{-1}	Vapor-CCl_4 soln. Δ N=O str. cm^{-1}
CH_3, CH_3	1485	200 °C	1438	neat	1460	47	25
C_2H_5, C_2H_5	1482	200 °C	1450	neat	1454	32	28
C_3H_7, C_3H_7	1482	200 °C	1449	neat	[—]	33	[—]
C_4H_9, C_4H_9	1482	200 °C	1448	neat	[—]	34	[—]
CH_3, C_6H_5	1492	200 °C	1438	neat	[—]	54	[—]
CH_3, O=C–OC_2H_5	1534	200 °C	1509	neat	[—]	25	[—]
iso-C_3H_7, iso-C_3H_7	[—]	[—]	[—]	[—]	1438	[—]	[—]
sec-C_4H_9, sec-C_4H_9	[—]	[—]	[—]	[—]	1437	[—]	[—]

* See Reference 14.

CHAPTER 8

Phosphorus Compounds

Introduction 233
Phosphorus-Halogen 234
Phosphorus Halogen Stretching F or Organophosphorus Halides 235
O-Methyl Phosphorodichloridothioate 236
O-Ethyl Phosphorodichloridothioate, O-Ethyl-1,1-d_2 Phosphorodichloridothioate, and O-Ethyl-2,2,2-d_3 Phosphorodichloridothioate 236
O,O-Dimethyl Phosphorochloridothioate 236
P—Cl Stretching 237
CH_3, CD_3, C_2H_5, CH_3CD_2 and CD_3CH_2 Vibrational Assignments for (R-O)P=OCl_2 and (RO)P(=S)Cl_2 Analogs 237
P=O Stretching, ν P=O 237
O,O-Dimethyl O-(2-Chloro 4-X-Phenyl) Phosphate 238
Phenoxarsine Derivatives 238
ν P=O vs ν P=S 239
P=S Stretching, ν P=S 239
S-Methyl Phosphorodichloridiothioate 239
Skeletal Modes of the $(C-O-)_3P$ Group of Trimethyl Phosphite and Trimethyl Phosphate 240
ν P=S, ν P—S, and ν S—H 240
Dialkyl Hydrogenphosphonate and Diphenyl Hydrogenphosphonate 240
The C—O—P Stretching Vibrations 241
C—P Stretching, ν C—P 241
Compounds Containing P—NH—R Groups 242
O-Alkyl O-Aryl N-Methylphosphoramidate vs O-Alkyl O-Aryl N-Methylphosphoramidothioate 243
Primary Phosphoramidothioates, P(=S)NH_2 244
P(=S)NH_2, P(=S)NH, P(=S)ND_2, P(=S)NHCH_3, and R(=S)NDCH_3 244
O-Methyl O-(2,4,5-trichlorophenyl) N-alkylphosphoramidates 245
Summary of PNHR and PNH_2 Vibrations 245
Summary of Vibrational Assignments for N-Alkyl Phosphoramidodichloridothioate and Deuterated Analogs 245
O,O-Dimethyl O-(2,4,5-Trichlorophenyl) Phosphorothioate, and Its P=O and $(CD_3O)2$ Analogs 245
A Comparison of IR Data for O,O-Dialkyl Phosphorochlorothioate and O,O,O-Trialkyl Phosphorothioate in Different Physical Phases 246
A Comparison of IR Data for Organophosphates and Organohydrogenphosphonates in Different Physical Phases 246
A Comparison of O-alkyl Phosphorodichloridothioate and S-alkyl Phosphorodichloridothioate in Different Physical Phases 247

Infrared Data for O,O-diethyl *N*-alkylphosphoramidates in Different Physical Phases 247
Vibrational Assignments for $CH_3-PO_3^{2-}$, $CD_3-PO_3^{2-}$, $H-PO_3^{2-}$, and PO_4^{3-} 247
Vibrational Data for Sodium Dimethylphosphinate, Potassium Dimethylphosphinate, and Sodium Dialkylphosphinate 248
Solvent Effects P=O Stretching, ν P=O 248
PCl_3 and PCl_2 Vibrations 250
Absorbance Ratios for $ROP(=O)Cl_2$ Molecules 250
Alkyl Group Vibrations 250
The ν C–C Mode for the C_2H_5OP Group 251
The ν COP Group 251
ν C–D for $CDCl_3$ in CCl_4 Solutions 252
Reference Spectra 252
References 253

Figures

Figure 8.1 255 (236)
Figure 8.2 256 (236)
Figure 8.3 257 (236)
Figure 8.4 258 (236)
Figure 8.5 259 (236)
Figure 8.6 260 (236)
Figure 8.7 261 (236)
Figure 8.8 262 (236)
Figure 8.9 263 (236)
Figure 8.10 264 (236)
Figure 8.11 265 (236)
Figure 8.12 266 (236)
Figure 8.13 267 (237)
Figure 8.14 268 (239)
Figure 8.15 269 (239)
Figure 8.16 270 (239)
Figure 8.17 271 (240)
Figure 8.18 272 (240)
Figure 8.19 272 (240)
Figure 8.20 273 (240)
Figure 8.21 273 (240)
Figure 8.22 274 (241)
Figure 8.23 274 (241)
Figure 8.24 275 (241)
Figure 8.25 275 (241)
Figure 8.26 276 (241)
Figure 8.27 277 (242)
Figure 8.28 278 (242)
Figure 8.29 279 (242)
Figure 8.30 279 (242)
Figure 8.31 280 (243, 244)
Figure 8.32 281 (243)
Figure 8.33 282 (243)
Figure 8.34 283 (243)
Figure 8.35 283 (243)
Figure 8.36 284 (243)
Figure 8.37 284 (243)
Figure 8.38 285 (243)
Figure 8.39 285 (244)
Figure 8.40 286 (244)
Figure 8.41 287 (244)
Figure 8.42 288 (244)
Figure 8.43 289 (244)
Figure 8.44 290 (244)
Figure 8.45 291 (246)
Figure 8.46 292 (246)
Figure 8.47 293 (248)
Figure 8.48 294 (247)
Figure 8.49 295 (247)
Figure 8.50 296 (247)
Figure 8.51 297 (248)
Figure 8.52 298 (248)
Figure 8.53 299 (248)
Figure 8.53a 300 (248)
Figure 8.53b 300 (249)
Figure 8.54 301 (249)
Figure 8.55 301 (249)
Figure 8.56 302 (249)
Figure 8.57 302 (249)
Figure 8.58 303 (249)
Figure 8.59 303 (249)
Figure 8.60 304 (250)

Figure 8.61	304 (250)	Figure 8.74	311 (251)
Figure 8.62	305 (250)	Figure 8.75	311 (251)
Figure 8.63	305 (250)	Figure 8.76	312 (251)
Figure 8.64	306 (250)	Figure 8.77	312 (251)
Figure 8.65	306 (250)	Figure 8.78	313 (251)
Figure 8.66	307 (250)	Figure 8.79	313 (251)
Figure 8.67	307 (250)	Figure 8.80	314 (251)
Figure 8.68	308 (250)	Figure 8.81	314 (252)
Figure 8.69	308 (250)	Figure 8.82	315 (252)
Figure 8.70	309 (251)	Figure 8.83	315 (252)
Figure 8.71	309 (251)	Figure 8.84	316 (252)
Figure 8.72	310 (251)	Figure 8.85	316 (252)
Figure 8.73	310 (251)	Figure 8.86	317 (252)

Tables

Table 8.1	318 (234)	Table 8.18	333 (240)
Table 8.2	319 (234)	Table 8.19	334 (240)
Table 8.3	320 (234)	Table 8.20	336 (241)
Table 8.4	321 (235)	Table 8.21	338 (241)
Table 8.5	322 (237)	Table 8.22	338 (241)
Table 8.5a	323 (237)	Table 8.23	339 (243)
Table 8.6	323 (237)	Table 8.24	340 (244)
Table 8.7	324 (237)	Table 8.25	341 (245)
Table 8.8	325 (237)	Table 8.26	342 (245)
Table 8.8a	326 (237)	Table 8.27	343 (245)
Table 8.9	326 (238)	Table 8.28	344 (245)
Table 8.10	327 (238)	Table 8.29	345 (246)
Table 8.11	327 (239)	Table 8.30	346 (246)
Table 8.12	328 (239)	Table 8.31	347 (246)
Table 8.13	329 (239)	Table 8.32	348 (247)
Table 8.14	331 (239)	Table 8.33	349 (247)
Table 8.15	331 (239)	Table 8.34	349 (247)
Table 8.16	332 (240)	Table 8.35	350 (248)
Table 8.17	332 (240)		

*Numbers in parentheses indicate in-text page reference.

INTRODUCTION

There has been interest in phosphorus compounds for many years, because these compounds have been found to be useful as chlorinating agents, flame retardants, antioxidants, fertilizers, and pesticides. On the other hand certain phosphorus derivatives have been manufactured for poisonous war gases, and this application is not positive for mankind. Therefore, many articles covering IR and Raman spectral data and assignments for these materials have been published and reviewed (1–55).

PHOSPHORUS-HALOGEN

Tables 8.1 through 8.3 list vibrational data and assignments for compounds of forms PX_3, $P(=S)X_3$, and $P(=O)X_3$. The ν PX_3, ν PX_2 and ν PX vibrations for these inorganic phosphorus compounds decrease in frequency in the X order: F, Cl, Br, and I. The overall frequency ranges for the ν PX_n modes for PX_3, $P(=S)X_3$ and $P(=O)X_3$ analogs are:

PF_{1-3}, 817–981 cm^{-1}

PCl_{1-3}, 431–618 cm^{-1}

PBr_{1-3}, 299–466 cm^{-1}

PI_3, 303–325 cm^{-1}

The ν asym. PF_3 modes for $P(=O)F_3$, $P(=S)F_3$, and PF_3 decrease in the order 982 cm^{-1}, 981 cm^{-1}, and 840 cm^{-1}, and the ν sym. PF_3 modes for $P(=S)F_3$, PF_3, and $P(=O)F_3$ decrease in the order 981 cm^{-1}, 890 cm^{-1}, and 875 cm^{-1}. Thus, the compound order for the ν asym. PF_3 frequency decrease $P(=O)F_3$, $P(=O)F_3$, and PF_3 is not the same as the compound order P(=S)F3, PF3, and P(=O)F3 for ν sym. PF3 (Table 8.1). In the case of the P=O analog, the ν P=O vibration occurs at 1405 cm−1 (see the Table 8.8 discussion on page 237), and ν P=O and ν sym. PF_3 belong to the A_1 species. Therefore, it is possible for ν P=O and ν sym. PF_3 to couple, causing ν P=O to occur at higher frequency and ν sym. PF_3 to occur at a lower frequency than ν sym. PF_3 for both the $P(=S)F_3$ and PF_3 analogs. On the other hand, the ν P=S vibration for $P(=S)F_3$ occurs at 695 cm^{-1}, and ν P=S and ν sym. PF_3 both belong to the A_1 species. Therefore, it also appears that these two modes are coupled, causing ν sym. PF_3 to occur at a higher frequency and ν P=S to occur at a lower frequency than otherwise expected. A similar argument has been given for the frequency behavior of ν PCl_2 for the compound $P(=S)Cl_2F$ (9).

The ν asym. PCl_3 and ν asym. PBr_3 vibrations and the ν sym. PCl_3 and ν sym. PBr_3 vibrations for the PX_3, $P(=O)X_3$, and $P(=S)X_3$ are compared here:

ν asym. PX_3 cm^{-1}	Compound	ν sym. PX_3, cm^{-1}
484	PCl_3	511
547	$P(=S)Cl_3$	431
581	$P(=O)Cl_3$	486
400	PBr_3	380
438	$P(=S)Br_3$	299
488	$P(=O)Br_3$	340
325	PI_3	303

The ν asym. PX_3 vibration increases in frequency in the order PX_3, $P(=S)X_3$, and $P(=O)X_3$ for both the PCl_3 and PBr_3 analogs. The ν asym. PX_3 belongs to the *e* species and the ν P=O and ν P=S vibration belong to the a_1 species. Therefore, it is not possible for these two modes to couple. On the other hand, it is possible for ν sym. PX_3 and ν P=S to couple, because both modes belong to the a_1 species. Coupling between ν P=S and ν sym. PX_3 causes ν sym. PCl_3 and ν sym. Br_3 to occur at lower frequency in the case of the $P(=S)X_3$ analogs than in the case of the $P(=O)X_3$ analogs.

Frequencies in parentheses in Tables 8.2 and 8.3 are estimated based upon spectra-structure correlations presented in Nyquist *et al.* (51).

PHOSPHORUS HALOGEN STRETCHING FOR ORGANOPHOSPHORUS HALIDES

Table 8.4 lists the ν asym. PCl_2 and ν sym. PCl_2 frequencies for compounds of form $XPCl_2$ $XP(=O)Cl_2$, and $XP(=S)Cl_2$. It is of interest to compare these vibrations for a series of analogs:

ν asym. PCl_2, cm^{-1}	Compound	ν sym. PCl_2, cm^{-1}
506	CH_3OPCl_2	453
579/607	$CH_3OP(=O)Cl_2$	515/548
531/560	$CH_3OP(=S)Cl_2$	456/478
505	CD_3OPCl_2	456
531/552	$CD_3OP(=S)Cl_2$	452/471
576/600	$C_2H_5OP(=O)Cl_2$	516/550
528/558	$C_2H_5OP(=S)Cl_2$	472–490
525/550	$CH_3CD_2OP(=S)Cl_2$	465/484
531/552	$CD_3CH_2OP(=S)Cl_2$	462–483
560	$CH_3NHP(=O)Cl_2$	517
512	$CH_3NHP(=S)Cl_2$	450/468
560	$(CH_3)_2NP(=O)Cl_2$	517
512	$(CH_3)_2NP(=S)Cl_2$	450

In all cases, the ν asym. PCl_2 and ν sym. PCl_2 vibrations occur at higher frequency for the P=O analog than for the P=S analog. It is suggested that ν P=S and ν sym. PCl_2 are coupled to some degree. It is interesting to note that both ν asym. PCl_2 and ν sym. PCl_2 are affected by substitution of CD_3O for CH_3O, and CH_3CD_2O and CH_3CH_2O for C_2H_5O in compounds of form $ROP(=S)Cl_2$. These data show that these vibrations involve more than just stretching of the PCl_2 bonds. In addition, two frequencies are listed for the ν asym. PCl_2 and ν sym. PCl_2 vibrations for many compounds, and these doublets are due to the existence of rotational conformers (22). The low-temperature rotational conformer is always listed first in each set.

O-METHYL PHOSPHORODICHLORIDOTHIOATE

Figure 8.1 illustrates how a model compound such as dichlorofluorophosphorothioate can be used to help assign the vibrational assignments for the nine $OP(=S)Cl_2$ skeletal vibrations for $CH_3OP(=S)Cl_2$. In addition, substitution of CD_3O for CH_3O aids in the vibrational assignments for the CH_3 and CD_3 groups. Variable temperature experiments aided in showing which bands resulted from each rotational conformer (21).

Figures 8.1 and 8.2 show the IR spectrum for $P(=S)Cl_2F$ in the vapor and solution phases and a complete vibrational assignment is presented in Reference 9.

Figure 8.3 shows the IR spectrum for phosphoryl chloride, $(P{=}O)Cl_3$. The vibrational assignment for $P(=S)Cl_2F$ was aided by comparison with the vibrational assignment for $P(=O)Cl_3$ (see Fig. 8.4). The a_1 modes for $P(=O)Cl_3$ correspond to a' modes for $P(=S)Cl_2F$, and the doubly degenerate e modes for $P(=O)Cl_3$ correspond to a' and a'' modes for $P(=S)Cl_2F$(9).

Figure 8.5 shows IR spectra for $CH_3OP(=S)Cl_2$, $CD_3OP(=S)Cl_2$, and $CH_3OP(=O)Cl_2$ (21). Figure 8.6 shows an IR spectrum for $CH_3OP(=S)Cl_2$, and Figure 8.7 shows an IR spectrum for $CD_3OP(=S)Cl_2$ (25). Vibrational assignments for these compounds have been reported (21, 25).

Because these molecules exist as rotational conformers, several of the vibrational modes for $CH_3OP(=S)Cl_2$ and $CD_3OP(=S)Cl_2$ appear as doublets. Rotational conformer 1 is assigned to the sets of IR bands that increase in intensity (A) with decrease in temperature (T), and rotational conformer 2 is assigned to IR bands that decrease in intensity (A) with a decrease in temperature (T).

Figure 8.8 compares the vibrational assignments for $CH_3OP(=S)Cl_2$, $CD_3OP(=S)Cl_2$, $P(=S)Cl_2F$, $CH_3OP(=O)Cl_2$, and $P(=O)Cl_3$ (25).

O-ETHYL PHOSPHORODICHLORIDOTHIOATE, O-ETHYL-1,1-d$_2$ PHOSPHORODICHLORIDOTHIOATE, AND O-ETHYL-2,2,2-d$_3$ PHOSPHORODICHLORIDOTHIOATE

Complete vibrational assignments have been reported for $C_2H_5OP(=S)Cl_2$, $CH_3CD_2OP(=S)Cl_2$, and $CD_3CH_2OP(=S)Cl_2$. These compounds exist as rotational conformers and conformers 1 and 2 were determined experimentally in the sample manner as those reported for $CH_3OP(=S)Cl_2$ and $CD_3OP(=S)Cl_2$ (22, 25).

Figures 8.9, 8.10, and 8.11 show IR spectra for $C_2H_5OP(=S)Cl_2$, $CH_3CD_2OP(=S)Cl_2$, and $CD_3CH_2OP(=S)Cl_2$, respectively (22). Comparison of Figs. 8.9–8.11 with Figs. 8.5–8.7 should help the reader identify the $ROP(=S)Cl_2$ skeletal vibrations.

O,O-DIMETHYL PHOSPHOROCHLORIDOTHIOATE

The compound $P(=S)ClF_2$ was used as a model compound for the vibrational assignments for $(CH_3)_2P(=S)Cl$. Vibrational assignments for $P(=S)ClF_2$ used are presented by Durig and Clark (11). Figure 8.12 shows IR spectra for O,O-dimethyl phosphorochloridothioate in the solution and liquid phases. Figure 8.13 shows IR spectra for O,O-dimethyl-d_6 phosphorochloridothioate in the solution and liquid phases. Figure 8.14 shows IR spectra of $(CH_3O)_2P(=S)Cl$ and $(CD_3O)_2P(=S)Cl$ in solution (27). Tables 8.5 and 8.5a list the vibrational assignments for these three compounds. Several of the vibrations for the $(CO)_2P(=S)Cl$ analogs appear as doublets due to the existence of rotational conformers (27).

P—Cl STRETCHING

Table 8.6 lists the ν P—Cl frequencies for P=O and P=S derivatives. All but the $[(CH_3)2N]_2$ and $[(C_2H_5)_2N]_2$ derivatives exhibit ν P—Cl as a doublet due to the existence of rotational conformers. The P=O analogs exhibit ν P—Cl at higher frequency(ies) than corresponding P=S analogies [e.g., $(CH_3O)_2P(=O)Cl$, 553/598 cm^{-1} vs $(CH_3O)_2P(=S)Cl$, 486/525 cm^{-1}] (27). For the compounds studied, ν P—Cl for the P=O series occur in the range 532–557 cm^{-1} and for the P=S series in the range 470–543 cm^{-1}.

CH_3, CD_3, C_2H_5, CH_3CD_2 AND CD_3CH_2 VIBRATIONAL ASSIGNMENTS FOR (R-O)P=OCl_2 AND (RO)P(=S)Cl_2 ANALOGS

Table 8.7 lists vibrational assignments of the alkyl groups of O-methyl phosphorodichloridothioate, O-methyl-d_3 phosphorodichloridothioate, methyl phosphorodichloridate, O-ethyl phosphorodichloridothioate, O-ethyl-1,1-d_2 phosphorodichloridothioate, and O-ethyl-2,2,2-d_3 phosphorodichloridothioate (21, 22, 25, 27). The ratios $\nu\, CH_2/\nu\, CD_2$ and $\nu\, CH_3/\nu\, CD_3$ vary from 1.14 to 1.52 vs a theoretical value of 1.414.

P=O STRETCHING, ν P=O

Table 8.8 lists the P=O stretching, ν P=O, frequencies for a variety of compounds. For the compounds studied, ν P=O occurs in the range 1243–1405 cm^{-1}. Many of the phosphate esters exhibit ν P=O as a doublet due to the existence of rotational conformers, and the frequency separation for these doublets varies between 7 and 29 cm^{-1}.

Table 8.8a lists ν P=O frequencies for organophosphorus compounds in different physical phases. In the liquid phase many of the organophosphorus esters exhibit only a single ν P=O frequency; however, in CS_2 solution ν P=O is observed as a doublet due to the existence of rotational conformers. The ν P=O frequency difference between these rotational conformers varies between 7 and 22 cm^{-1}.

In the liquid phase the structural configuration may be stabilized by the intermolecular association shown here:

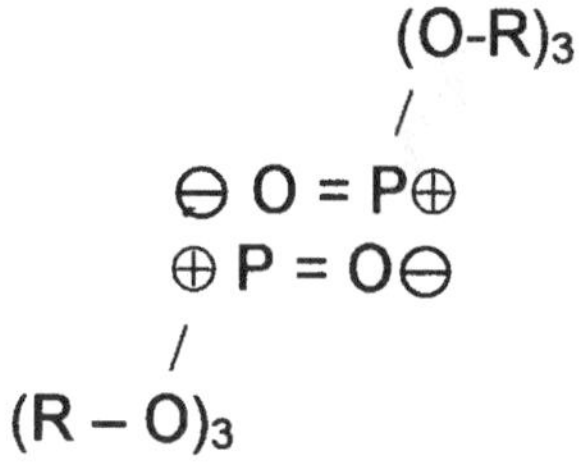

O,O-DIMETHYL O-(2-CHLORO 4-X-PHENYL) PHOSPHATE

Table 8.9 lists the ν P—O frequencies for the rotational conformers for the X-analogs of O,O-dimethyl O-(2-chloro 4-X-phenyl) phosphate (20). The Hammett σ_p values for the 4-X group are also listed.

The frequency separation between the ν P=O rotational conformers varies between 15.6 and 18.4 cm^{-1}. The high frequency ν P=O conformer occurs in the range 1303.2–1308.5 cm^{-1}, and the low frequency P=O conformer occurs in the range 1285.8–1291.1 cm^{-1}. There is a trend that the ν P=O vibration decreases in frequency as σ_p changes from positive to negative.

Thomas and Chittenden have developed an equation to predict ν P=O frequencies for a variety of organophosphorus compounds. Their equation is presented here (30):

$$\nu \text{ P=O}(cm^{-1}) = 930 + 40\ \Sigma\ \pi$$

Table 8.10 lists π constants for group G for compounds of form G-P(=O)Cl_2. The calculated and observed frequencies vary between 2 and 28 cm^{-1} of the observed frequencies. In the case of organophosphorus compounds in the vapor phase, the 930 cm^{-1} constant should most likely be raised 10 to 20 cm^{-1} to account for the higher ν P=O frequencies observed in the vapor phase. In general, the ν P=O frequencies are lower in the neat phase, and this is attributed to the higher field effect in the neat phase as compared to the field effect in solution or the relative absence of the field effect in the vapor phase.

PHENOXARSINE DERIVATIVES

These compounds have the empirical structure presented here:

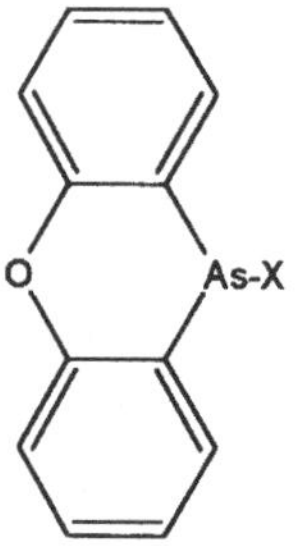

where X is S-P(=S)$(OR)_2$

and X is O-P(=O)$(OR)_2$

The ν P=S vibration occurs in the range 645–648 cm^{-1}, and the ν P=O vibration occurs in the range 1228–1235 cm^{-1} (31). See Table 8.11 for more specific data.

ν P=O VS ν P=S

A normal coordinate analysis of phosphorus oxyhalides and their derivatives has shown that the ν P=O frequencies exhibit an accidental relationship between mass and electronegativity for the halogen atoms. The good agreement obtained between the calculated and observed ν P=O frequencies for all the phosphorus oxyhalides indicates that kinetic energy does have a significant effect on the ν P=O frequency. The calculated potential energy distribution indicates that the P=O fundamental in each molecule results from over 85% ν P=O, and mixes only slightly with other molecular vibrations of the same symmetry species (47). These calculations show that the force constant for ν P=O for $CH_3P(=O)Cl_2$ is closer to that for $P(=O)Cl_3$ than the force constant for $(CH_3)_3P(=O)$.

Table 8.12 lists IR data for ν P=O and ν P=S for $P(=O)X_3$ ν and $P(=S)X_3$-type compounds. This table shows that ν P=O decreases in frequency progressing in the series $P(=O)F_3$, 1415 cm^{-1} to P(=O)Cl Br_2, 1275 cm^{-1}. The ν P=S vibration does not show the same trend and, as stated before, ν P=S couples with other fundamental vibrations.

P=S STRETCHING, ν P=S

The ν P=S vibration couples with other fundamentals as discussed previously. The characteristic ν P=S vibration occurs in the range 565–742 cm^{-1}. These limits are set by $[(CH_3)_2N]_3$ P=S and $(HCCCH_2O)P(=S)Cl_2$ as listed in Table 8.13. Many of these P=S containing compounds exist as rotational conformers as shown in Table 8.13.

The frequency separation between these rotational conformers varies between 11 and 47 cm^{-1}.

S-METHYL PHOSPHORODICHLORIDIOTHIOATE

This compound has the empirical structure $CH_3SP(=O)Cl_2$ (23). This compound also exists as rotational conformers. The ν P=O frequencies are 1275/1266 cm^{-1} and the vibrational assignments are presented in Table 8.14.

Figure 8.15 (top) is an IR spectrum of S-methyl phosphorodichloridiothioate in the range 3800–400 cm^{-1}. Figure 8.15 (bottom) is a spectrum of S-methyl phosphorodichloridiothioate in the range 600–45 cm^{-1}. Figure 8.16 (top) is a Raman spectrum of S-methyl phosphorodichloridiothioate, and Fig. 8.16 (bottom) is a polarized spectrum of S-methyl phosphorodichloridiothioate in the range 3000–100 cm^{-1}. The skeletal vibrations for $S\text{-}P(=O)Cl_2$ are similar to those for $P(=O)Cl_3$ (23). Consequentially, the ν sym. PCl_3, a_1 mode for $P(=O)Cl_3$ at 483 cm^{-1} corresponds to the rotational conformers at 450 and 471 cm^{-1} for the $SP(=O)Cl_2$ skeletal vibration. The doubly degenerate ν asym. PCl_3, e mode for $P(=O)Cl_3$ corresponds to the 593/579 cm^{-1} a'' rotational conformers (23).

SKELETAL MODES OF THE $(C-O-)_3P$ GROUP OF TRIMETHYL PHOSPHITE AND TRIMETHYL PHOSPHATE

Table 8.15 compares the $P(-O-C)_3$ vibrations for $(CH_3O)_3P$ and $(CH_3O)_3P{=}O$. These skeletal vibrations occur at similar frequencies. The vibrational assignments for the CH_3 vibrations for trimethyl phosphite are listed in Table 8.16. Infrared spectra of trimethyl phosphite in different physical phases are presented in Figs. 8.17–8.19.

ν P=S, ν P–S, AND ν S–H

The ν P=S, ν P–S and ν S–H vibrations in compounds of form $(RO)_2P(=S)SH$ and $(ArO)_2P(=S)SH$ exist as a doublet due to the existence of rotational conformers (see Table 8.17). The ν SH rotational conformers were discussed in Chapter 4 in the section on thiols. The lower frequency ν S–H rotational conformer was assigned to the rotational conformer where the S–H group was intramolecularly hydrogen bonded to the free pair of electrons of the POR group (viz. SH···OR). The ν P=S rotational conformers 1 and 2 are assigned 670/659 cm^{-1} and ν P–S rotational conformers are assigned 524–538 cm^{-1}/490–499 cm^{-1}, respectively (32).

Figure 8.20 compares the IR spectrum for O,O-dimethyl phosphorodithioic acid with the IR spectrum for O,O-dimethyl phosphorochloridothioate. Figure 8.21 compares the IR spectrum for O,O-diethyl phosphorodithioic acid with the IR spectrum for O,O-diethyl phosphorochloridothioate. The IR group frequencies just discussed are readily apparent in these spectra.

DIALKYL HYDROGENPHOSPHONATE AND DIPHENYL HYDROGENPHOSPHONATE

These compounds are often named phosphites, but the phosphorus atom is actually pentavalent in these cases and has the empirical structure $(RO)_2P(=O)H$.

In the neat phase ν P–H occurs in the range 2410–2442 cm^{-1}, δ P–H in the range 955–1028 cm^{-1}, and ν P=O in the range 1241–1260 cm^{-1}. In the vapor phase ν PH occurs 19–25 cm^{-1} higher in frequency, δ P–H 8–20 cm^{-1} higher in frequency, and ν P=O 27–41 cm^{-1} higher in frequency than they occur in the neat phase. These frequency changes going from the neat to vapor phase suggest that in the neat phase these compounds are intermolecularly hydrogen bonded, such as in the one illustrated here:

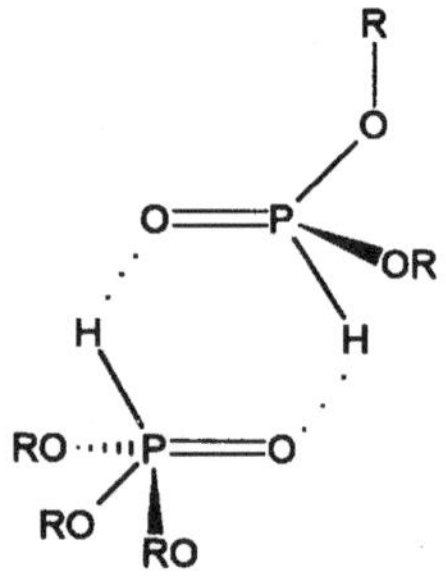

In other compounds ν P=O does not shift as much to lower frequency in going from the vapor to the liquid phase as do these hydrogenphosphonates (see Table 8.18).

In CS_2 solution, compounds of form $(RO)_2P(=O)H$ exhibit ν P=O in the range 1273–1283 cm^{-1} and ν P=O in the range 1257–1266 cm^{-1}. The higher frequency ν P=O is often seen as a shoulder. It is possible that the higher frequency ν P=O is due to unassociated $(RO)_2P(=O)H$ molecules.

Figures 8.22 and 8.23 show the IR spectra of dimethyl hydrogenphosphonate and dimethyl deuterophosphonate in solution. Figures 8.24 and 8.25 show the IR spectra of diethyl hydrogenphosphonate and diethyl deuterophosphonate in solution (32). Figure 8.26 shows Raman spectra of dimethyl hydrogenphosphonate in the liquid phase (32). These spectra aid the reader in recognizing the group frequencies discussed in this chapter.

THE C—O—P STRETCHING VIBRATIONS

The C—O—P group frequencies are complex, and are often described as "C—O stretching", ν C—O, and "P—O stretching", ν P—O. Tables 8.19 and 8.20 list IR group frequency assignments for ν C—O, ν P—O for C—O—P and ν aryl-O and ν P—O for aryl-O—P. The ν C—O frequencies occur in the range 990–1065 cm^{-1}, and the ν P—O frequencies occur in the range 739–861 cm^{-1} for the P—O—C group. The ν aryl-O frequencies occur in the range 1160–1261 cm^{-1} and ν P—O in the range 920–962 cm^{-1} for the aryl-O—P group.

Table 8.21 lists the aryl-O stretching frequencies for the O-methyl O-(X-phenyl N-methyl-phosphoramidates, and ν aryl-O occurs in the range 1207–1259 cm^{-1} (39).

C—P STRETCHING, ν C—P

Table 8.22 lists the ν C—P frequencies for a variety of compounds containing this group, with ν C—P occurring in the range 699–833 cm^{-1}. In compounds such as $ClCH_2P(=O)Cl_2$ and $ClCH_2P(=S)Cl_2$, ν C—P exhibits a doublet (811/818 cm^{-1} at 25 °C) for ν C—P due to the presence of rotational conformers:

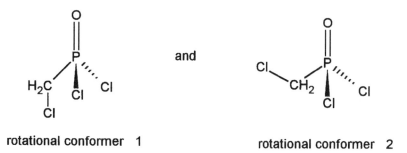

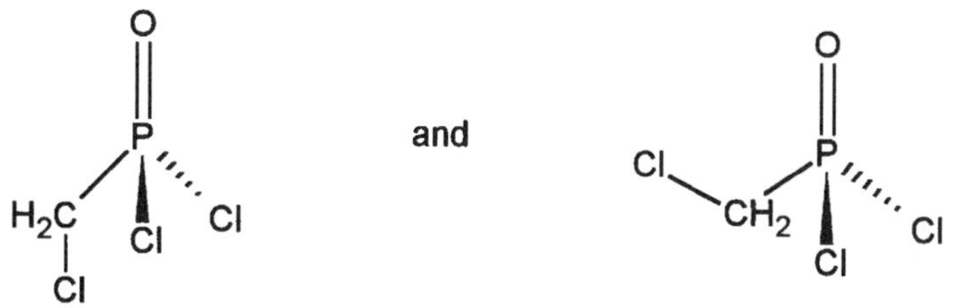

rotational conformer 1 rotational conformer 2

In the case of a compound such as $(CH_3)_2PO_2Na$ the IR bands at 737 cm^{-1} and 700 cm^{-1} are assigned to ν asym. PC_2 and ν sym. PC_2, respectively (41).

Figure 8.27 shows the IR spectrum of chloromethyl phosphonic dichloride in CS_2 solution at O °C and at −75 °C. Note that several bands occur as doublets due to the presence of rotational conformers, and that one band in each set of doublets increases in intensity (A) with a decrease in temperature (T). The higher frequency band of the doublet in the range 800–820 cm^{-1} increases in (A) with a decrease in (T). The lower frequency band in the range 1270–1300 cm^{-1} increases in (A) with a decrease in (T). Thus, the lower frequency ν P=O band near 1280 cm^{-1} and the band near 818 cm^{-1} for ν C−P are assigned to rotational conformer 2. The IR bands near 1282 cm^{-1} and 810 cm^{-1} are assigned to ν P=O and ν C−P for rotational conformer 1. On this same basis, ν asym. PCl_2 rotational conformers 1 and 2 are assigned near 559 cm^{-1} and 570 cm^{-1}, respectively and the ν sym. PCl_2 rotational conformers 1 and 2 are assigned near 490 cm^{-1} and 510 cm^{-1}, respectively.

Figures 8.28 and 8.29 are IR spectra for $ClCH_2P(=O)Cl_2$ and $ClCH_2P(=S)Cl_2$, respectively (17). Vibrational assignments have been reported for $ClCH_2PCl_2$, $CH_3P(=O)Cl_2$, $ClCH_2P(=S)Cl_2$, $CH_3P(=S)Cl_2$, and $BrCH_2P(=O)Br2$ (17). The ν P=S frequency for $CH_3P(=S)Cl_2$ occurs at 696 cm^{-1}, and the ν P=S rotational conformers 1 and 2 for $ClCH_2P(=S)Cl_2$ occur at 683 and 657 cm^{-1} (17). The ν P=O rotational conformers 1 and 2 occur at 1264 and 1275 cm^{-1} for $BrCH_2P(=O)Br_2$, and ν asym. PBr2 rotational conformers 1 and 2 occur at 463 and 481 cm^{-1} (17) (see Fig. 8.30).

COMPOUNDS CONTAINING P−NH-R GROUPS

The compounds containing the R-NH−P(=O) group exist in a cis configuration in dilute solution as shown here:

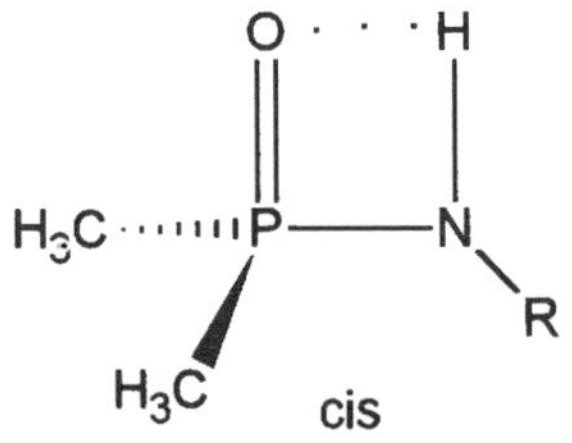

This cis configuration is stabilized by the intramolecular bonding between the N−H proton and the free pair of electrons on the P=O group (viz. N−H···O=P). In the case of compounds containing the R-NH−P(=S) group the N−H group in dilute solution exists in both a cis and trans configuration as shown here:

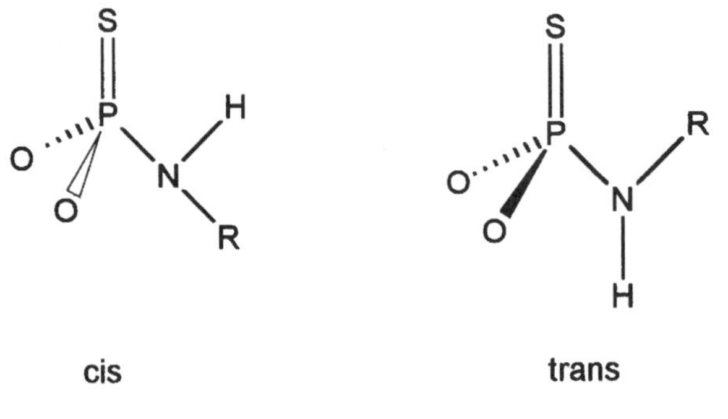

The cis ν N—H frequencies occur in the range 3388–3461 cm^{-1} and trans ν N—H frequencies occur in the range 3380–3434 cm^{-1} (see Table 8.23). The frequency separation between cis ν N—H and trans ν N—H varies between 18 and 44 cm^{-1}.

Figure 8.31 shows the solution-phase IR spectrum of N-methyl phosphoramidodichloridothioate. The cis νN—H frequency is assigned at 3408 cm^{-1} and that shoulder 3300 cm^{-1} is assigned to intermolecularly bonded ν N—H. The bands at 1094 and 838 cm^{-1} are assigned to "ν C—N" and "P—N" for the P—N—C group. Infrared bands at 725 and 690 cm^{-1} are assigned to ν P=S rotational conformers 1 and 2, respectively (24). Figure 8.32 show an IR spectrum of N-methyl-d_3 phosphoramidodichloridothioate (24). The ν N—H frequency is assigned at 3409 cm^{-1} and intermolecularly hydrogen bonded ν N—H at 3301 cm^{-1}. The "ν C—N" and "ν P—N" for the P—N—C group are assigned at 1117 cm^{-1} and 785 cm^{-1}, respectively. The ν P=S rotational conformers 1 and 2 are assigned at 711 cm^{-1} and 683 cm^{-1}, respectively.

Figure 8.33 is an IR spectrum for the N—D analog of N-methyl phosphoramidodichloridothioate (a small amount of the N—H analog is present as an impurity). The ν N—D mode is assigned at 2522 cm^{-1} and the intermolecular, ν N—D mode is assigned at 2442 cm^{-1}. The ν NH/ND ratio 3408 cm^{-1}/2522 cm^{-1} is 1.35, and the intermolecularly bonded ν NHν ND ratio 3300 cm^{-1}/2442 cm^{-1} is 1.35. In this case, "ν C—N" and "ν P—N" for the group C—N—P occur at 1107 cm^{-1} and 846 cm^{-1}/825 cm^{-1}, respectively (24). The ν P=S conformers 1 and 2 are 728 cm^{-1}/681 cm^{-1}. Figure 8.34 is an IR spectrum for the N—D analog of N-methyl-d_3 phosphoramidodichloridothioate (a small amount of the N—H analog is also present). The "ν P—N" mode is assigned at 780 cm^{-1} and "ν C—N" at 1107 cm^{-1}. The ν P=S rotational conformers 1 and 2 are assigned at 701 cm^{-1}/673 cm^{-1} (24).

O-ALKYL O-ARYL N-METHYLPHOSPHORAMIDATE VS O-ALKYL O-ARYL N-METHYLPHOSPHORAMIDOTHIOATE

In Fig. 8.35 IR spectrum 1 is for a 10% CCl_4 solution spectrum of O-alkyl O-aryl N-methylphosphoramidate in a 0.1-mm NaCl cell and IR spectrum 2 is for a 10% CCl4 solution of O-alkyl O-aryl N-methyl-phosphoramidothioate. In Fig. 8.31, spectrum 1, the strong IR band at 3235 cm^{-1} is assigned to intermolecularly hydrogen-bonded ν N—H, and the IR band 3442 cm^{-1} from cis ν N—H (42). In Figure 8.35, spectrum 2, the IR bands at 3440 cm^{-1}, 3402 cm^{-1}, and 3320 cm^{-1} are assigned to cis ν N—H, trans ν N—H, and intermolecularly hydrogen-bonded ν N—H, respectively. Study of Fig. 8.35 shows that it is a rather simple task to distinguish between CH_3NH—P=O and CH_3NH—P=S groups (42).

Figure 8.36 is a plot of the ν N—H frequencies of O-alkyl O-aryl N-alkylphosphoramidates vs an arbitrary assignment of one for each proton joined to the N-α-carbon atom. Figure 8.37 is a comparable plot for the O-alkyl O-aryl N-alkylphosphoramidothioates. However, in this case (with the exception of the N-tert-butyl analog) both cis ν N—H and trans ν N—H frequencies are observed (42). In the case of the N-tert-butyl analog, only the cis configuration is possible, because the N-tert-butyl group is sterically prevented from being in a cis position with the P=S group (42).

Figure 8.38 shows IR spectra in the range 3300–3500 cm^{-1} for O-alkyl O-aryl N-alkylphosphoramidothioates as obtained in ~0.01 molar or less CCl_4 solutions using a 14-mm NaCl cell. In spectrum A the N—R group is N—CH_3, B is N—C_2H_5, C is N—C_3H_7, D is N-iso-C_4H_9, E is

N-$_4H_9$, F is N-iso-C_3H_7, G is N-sec-C_4H_9, and H is N-tert-C_4H_9. Spectra I and J are O, O-dialkyl N-methyphosphoramidothioate and O,O-dialkyl N-isopropylphosphoramidothioate, respectively. These data show that the ν NH vibrations decrease in frequency with increased branching on the N-α-carbon atom. The electron release of the N-alkyl group to the nitrogen atom increases in the order methyl, ethyl, isopropyl, and tert-butyl. Thus, as the electron release to the nitrogen atom increases, the N—H bond weakens, causing it to vibrate at a lower frequency.

Figure 8.39 shows IR spectra of O-alkyl N, N′-dialkylphosphorodiamidothioate in the range 3300–3500 cm^{-1}. Spectrum A is where N,N′ is dimethyl, B is N,N′-diethyl, E is N,N′-dipropyl, D is N,N′-dibutyl, E is N,N′-diisopropyl, F is N,N′-di-sec-butyl, and G is the N,N′-dibenzyl analog. The H and I are for O-aryl N,N′-di-isopropylphosphorodiamidothioate and O-aryl N-isopropyl, N′-methyl phosphorodiamidothioate, respectively. These diamido compounds differ from the mono-amido compounds in that the intensity ratio is higher for the lower-frequency ν N—H band to the higher frequency ν N—H band. This indicates that there is a higher fraction of trans ν N—H in the diamides, but the cis ν N—H isomic is still prodominant. Spectra A through J are 0.01 molar or less solutions in 3-mm NaCl cells. Spectra D and J are for the N,N′-dibutyl analog; however, spectrum J is for a 10% CCl4 solution using a 0.1-mm NaCl cell. The broad band in J is due to intermolecularly hydrogen-bonded ν N—H (42).

PRIMARY PHOSPHORAMIDOTHIOATES, P(=S)NH_2

Figure 8.40 shows an IR spectrum (coded R on the figure) of O-alkyl O-aryl phosphoramidothioate and an IR spectrum of O,O-dialkyl phosphoramidothioate (coded S) in the range 3300–3500 cm^{-1}. Samples were prepared as 0.01 molar CCl_4 solutions, and the spectra were recorded utilizing a 3-mm NaCl cell. The IR band near 3490 cm^{-1} is assigned as ν asym. NH_2 and the IR band near 3390 cm^{-1} is assigned as ν sym. NH_2 (42).

P(=S)NH_2, P(=S)NH, P(=S)ND_2, P(=S)$NHCH_3$, AND R(=S)$NDCH_3$

Figures 8.41, 8.42, and 8.43 are IR spectra of O,O-dimethyl phosphoramidothioate and its ND_2 analog, O,O-diethyl phosphoramidothioate and its D_2 analog, and O,O-diethyl N-methylphosphoramidothioate, and its N—D analog. The vibrational assignments for the NH_2, ND_2, NHD vibrations are presented in Table 8.24. The values for the $\nu NH_2/\nu ND_2$ and $\nu NH/\nu ND$ are also presented in Table 8.24. The ν asym. NH_2 and asym. ND_2 modes occur near 3480 cm^{-1} and 2600–2611 cm^{-1}, respectively, and ν sym. NH_2 and ν sym. ND_2 occur near 3390 cm^{-1} and 2490 cm^{-1}, respectively. Both cis and trans ν NH and ν ND modes are also assigned.

The δNH_2 and δND_2 modes are assigned near 1542 cm^{-1} and 1171 cm^{-1}, respectively (37).

The IR spectrum for O,O-dimethyl N-methylphosphoramidothioate is given in Fig. 8.44. Compare Fig. 8.44 with Fig. 8.43. upper spectrum for O,O-dimethyl N-methylphosphoramidothioate. The IR band at 954 cm^{-1} results from ν C—C for the P—O—C_2H_5 group in Fig. 8.43 (not from a ν N—C mode.

O-METHYL O-(2,4,5-TRICHOROPHENYL) N-ALKYLPHOSPHORAMIDATES

Table 8.25 lists the N–H and N–D frequencies for O-methyl O-(2,4,5-trichlorophenyl) N-alkyl phosphoramidates in CCl_4 solution. The frequency separation between νN–H and νN–D is $\sim$200 cm^{-1} while for the νN–H and νN–D intermolecularly bonded species it varies between 144 and 155 cm^{-1}. In the case of the O-methyl O-(2,4,5-trichlorophenyl) N-alkylphosphoramidothioates the frequency separation between the cis νN–H and trans νN–H is in the range of 31–34 cm^{-1}. The cis νN–H frequencies decrease in the order of N–CH_3, N-ethyl, N-isopropyl, and N-tert-butyl (3442, 3429, 3419, and 3402 cm^{-1}, respectively). The trans νN–H frequencies decrease in the same order, N–CH_3, N-ethyl, and N-isopropyl (3409, 3398, and 3385 cm^{-1}, respectively). The trans νN–H mode is not observed for the N-tert-butyl analog.

SUMMARY OF PNHR AND PNH_2 VIBRATIONS

Table 8.26 lists the cis and trans νNH frequencies for compound types containing the P(=O)NHR, P(=S)NHR groups, and P(=S)NH_2 groups (37). The N–H bending mode, νN–H, is assigned in the range 1372–1416 cm^{-1}. In all cases cis νN–H occurs at higher frequency than trans νNH by 16 through 36 cm^{-1}.

SUMMARY OF VIBRATIONAL ASSIGNMENTS FOR N-ALKYL PHOSPHORAMIDODICHLORIDOTHIOATE AND DEUTERATED ANALOGS

Table 8.27 lists a summary of the vibrational assignments for N-alkyl phosphoramidodichloridothioates (24). For a more detailed discussion of these vibrational frequency assignments, the reader is referred to Reference 24.

O,O-DIMETHYL O-(2,4,5-TRICHLOROPHENYL) PHOSPHOROTHIOATE, AND ITS P=O AND $(CD_3O)_2$ ANALOGS

Table 8.28 lists vibrational assignments for O,O-dimethyl O-(2,4,5-trichlorophenyl) phosphorothioate, and its P=O and $(CD_3O)_2$ analogs (43). O,O-dimethyl O-(2,4,5-trichlorophenyl) phosphorothioate has the empirical structure presented here:

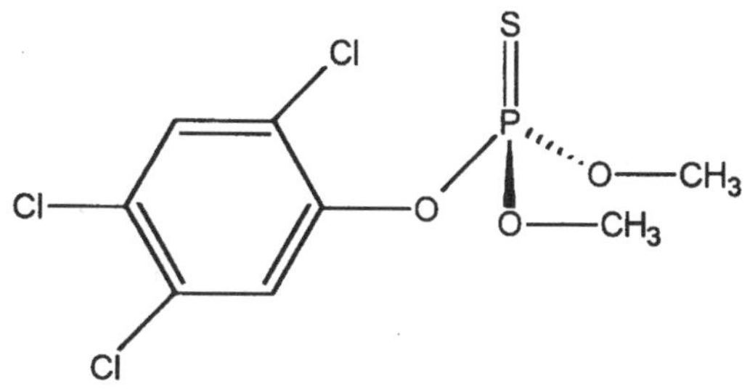

This phosphorus compound has 66 fundamental vibrations. Eighteen fundamentals result from vibrations with the two methyl groups. Thirty fundamentals result from vibration within the 2,4,5-trichlorophenoxy group, and 18 fundamentals result from vibrations within the $(C{-}O{-})_2$(aryl-O−)P=S group.

Vibrational assignments for the 2,4,5-trichlorophenoxy group were aided by comparison with the vibrational assignments for 1-fluoro-2,4,5-trichlorobenzene (52). The assignments for these ring modes are compared in Table 8.29.

Figure 8.45 (top) is an IR spectrum for O,O-dimethyl O-(2,4,5-trichlorophenyl) phosphorothioate, and Fig. 8.45 (bottom) is an IR spectrum for O,O-dimethyl O-(2,4,5-trichlorophenyl) phosphate. Figure 8.46 (top) is an IR spectrum for O,O-dimethyl-d_6 O-(2,4,5-trichlorophenyl) phosphorothioate, and Fig. 8.46 (bottom) is an IR spectrum for O,O-dimethyl-d_6 O-(2,4,5-trichlorophenyl)phosphate. It is easy to distinguish between the P=S and P=O analogs in this case. The ν P=O rotational conformers 1 and 2 are readily apparent by the doublet in the range 1275–1310 cm^{-1}. Note that these IR bands are not present in the P=S analogs. It should be noted that the ν P=O rotational conformer vibrations occur at lower frequency (1300/1285 cm^{-1}) in the case of the $(CH_3{-}O{-})_2$ analog than for the $(CD_3{-}O{-})_2$ analog (1306 cm^{-1}/1292 cm^{-1}). Thus, this frequency difference shows that ν P=O involves motion other than simple stretching of the P=O bond.

The strongest Raman band in the spectrum of O,O-dimethyl O-(2,4,5-trichlorophenyl) phosphorothioate and its O,O-dimethyl-d_6 analog occur at 616 cm^{-1} and 617 cm^{-1} in the solid phase, respectively. This Raman band is assigned to ν P=S. The out-of-plane ring mode (25 for the 2,4,5-trichlorophenoxy group occurs as a weak IR band in the range 627–629 cm^{-1} (43). However, another weak IR band is noted at 615 cm^{-1} for O,O-dimethyl O-(2,4,5-trichlorophenyl) phosphorothioate and at 607 cm^{-1} for the O,O-dimethyl-d_6 analog in CS_2 solution. These weak IR bands are assigned to ν P=S.

A COMPARISON OF IR DATA FOR O,O-DIALKYL PHOSPHOROCHLOROTHIOATE AND O,O,O-TRIALKYL PHOSPHOROTHIOATE IN DIFFERENT PHYSICAL PHASES

Table 8.30 lists IR data for O,O-dialkyl phosphorochlorothioate and O,O,O-trialkyl phosphorothioate in different physical phases. It is interesting to note that the combination tone ν C−O + ν P−O occurs at higher frequency in the vapor phase than in the neat phase. In the vapor phase at 200 °C, O,O-dimethyl phosphorochloridothioate and O,O,O-trimethyl phosphorothioate each exist as one rotational conformer.

A COMPARISON OF IR DATA FOR ORGANOPHOSPHATES AND ORGANOHYDROGENPHOSPHONATES IN DIFFERENT PHYSICAL PHASES

The ν P=O rotational conformers generally occur at higher frequency in the vapor phase than in CS_2 solution (see Table 8.31).

A COMPARISON OF O-alkyl PHOSPHORODICHLORIDOTHIOATE AND S-alkyl PHOSPHORODICHLORIDOTHIOATE IN DIFFERENT PHYSICAL PHASES

Table 8.32 compares IR data for O-alkyl phosphorodichloridothioates and S-alkyl phosphorodichloridothioates in different physical phases. These data show that some molecular vibrations increase in frequency while other molecular vibrations decrease in frequency with change of physical phase.

INFRARED DATA FOR O,O-diethyl *N*-ALKYLPHOSPHORAMIDATES IN DIFFERENT PHYSICAL PHASES

Table 8.33 lists IR data for O,O-diethyl N-alkylphosphoramidates in different physical phases. In the vapor phase ν P=O occurs in the region 1274–1280 cm^{-1} and in the neat phase ν P=O : H occurs in the region 1210–1240 cm^{-1}. The decrease in the ν P=O frequency in going from the vapor phase is attributed mainly to intermolecular hydrogen bonding between the free pair of electrons on the P=O oxygen atom and the N–H proton (viz. P=O···H – N). The ν N–H mode occurs in the region 3420–3460 cm^{-1} in the vapor phase, and occurs in the region 3198–3240 cm^{-1} in the neat phase. Thus, intermolecular hydrogen bonding causes ν N–H···O=P to occur 205–225 cm^{-1} lower in frequency than it occurs in the vapor phase.

VIBRATIONAL ASSIGNMENTS FOR $CH_3{-}PO_3^{2-}$, $CD_3{-}PO_3^{2-}$, $H{-}PO_3^{2-}$, AND PO_4^{3-}

Table 8.34 lists vibrational assignments for $CH_3{-}PO_3Na_2$, $CD_3{-}PO_3Na_2$, HPO_3^{2-}, and PO_4^{3-}. Fig. 8.47 (top) is an IR spectrum of disodium methanephosphonate in the solid phase and Fig. 8.47 (bottom) is an IR spectrum of disodium methane-d_3-phosphonate in the solid phase. Fig. 8.48 (top) is an IR spectrum of disodium methanephosphonate in water solution, and Fig. 8.48 (bottom) is an IR spectrum of disodium methane-d_3-phosphonate in water solution (40). Figure 8.49 (top) is an IR spectrum of disodium methanephosphonate in the solid phase, Fig. 8.49 (middle) is an IR spectrum of disodium methane-d_3-phosphonate in the solid phase, and Figure 8.49 (bottom) is an IR spectrum of dipotassium methanephosphonate in the solid phase (40). Figure 8.50 is a solid-phase IR spectrum for disodium n-octadecanephosphonate (40).

The model compounds containing the $H{-}PO_3^{2-}$ and PO_4^{3-} anions (44,45) aid in the vibrational assignments for the CPO_3^{2-} skeletal vibrations. Vibrational assignments for the CH_3 and CD_3 groups are presented in Table 8.34. It is apparent that the *e* fundamentals for the PO_3^{2-} groups are split in the case of the CPO_3^{2-} skeletal modes. Thus, ν asym. PO_3^{2-} for $CH_3PO_3Na_2$ occurs near 1110 and 1080 cm^{-1} and near 1100 and 1070 cm^{-1} in the case of

$CD_3PO_3Na_2$. The ν sym. PO_3 vibration occurs at 993 cm^{-1} for the CH_3 analog and at 972 cm^{-1} for the CD_3 analog.

In H_2O solution ν asym. PO_3^{2-} occurs at 1060 cm^{-1} and ν sym. PO_3^{2-} occurs at 975 cm^{-1} for the CH_3 analog, and at 1060 cm^{-1} and 969 cm^{-1} for the CD_3 analog. Comparison of the IR band near 2330 cm^{-1} for $CH_3{-}PO_3Na_2$ with the IR band near 185 cm^{-1} for $CH_3{-}PO_3K_2$ indicates that these bands arise from a lattice vibration (see Figure 8.48 (top and bottom). In the case of n-$C_{18}H_{37}{-}PO_3Na_2$, the IR bands in the region 1050–1130 are assigned to ν sym. PO_3^{2-} vibrations, the IR band near 987 cm^{-1} is assigned to (sym. PO_3^{2-}, and the IR band near 771 cm^{-1} is assigned as ν C–P.

VIBRATIONAL DATA FOR SODIUM DIMETHYLPHOSPHINATE, POTASSIUM DIMETHYLPHOSPHINATE, AND SODIUM DIALKYLPHOSPHINATE

Table 8.35 lists IR data and assignments for the $(CH_{3\_})_2PO_2^{1-}$ anion (41). Figure 8.51 upper is an IR spectrum of sodium dimethylphosphinate, $(CH_{3\_})_2PO_2Na$ in the solid phase and Fig. 8.51 (bottom) is an IR spectrum of the same compound in water solution. Figure 8.52 (top) is an IR spectrum for $(CH_{3\_})_2$ PO_2Na and Fig. 8.52 (bottom) is an IR spectrum for $(CH_3)_2$ PO_2K. Both spectra were recorded in the solid phase. Figure 8.53 gives Raman spectra of $(CH_{3\_})_2$ PO_2Na saturated in water solution.

Vibrational assignments for these dialkyl phosphinate salts were aided by vibrational assignments for H_2PO_2K (44), $(CH_{3\_})_3P$ (53), and $(CH_{3\_})_3P{=}O$ (54).

The IR bands for $(CH_{3\_})_2PO_2Na$ at 1169 cm^{-1} and 1065 cm^{-1} in the solid phase and at 1128 cm^{-1} and 1040 cm^{-1} in water solution are assigned to ν asym. PO_2 and ν sym. PO_2, respectively (41). The IR band for $(CH_{3\_})_2PO_2Na$ in the solid phase at 725 cm^{-1} and 695 cm^{-1}, and at 738 cm^{-1} (depolarized Raman band) and 700 cm^{-1} (polarized Raman band) in water solution are assigned as ν asym. PC_2 and ν sym. PC_2, respectively. The ν asym. PO_2 and ν sym. PO_2 near 1150 cm^{-1} and 1050 cm^{-1} for sodium diheptyl phosphinate and sodium dioctylphosphinate are readily apparent in Fig. 8.53a.

The IR bands near 1300 cm^{-1}, and in the region 840–920 cm^{-1} are assigned to δ sym. CH_3 modes and ρCH_3 modes, respectively. For more detailed vibrational assignments see Reference 41.

SOLVENT EFFECTS P=O STRETCHING, ν P=O

Solvent effects aid the spectroscopist in interpreting the vibrational spectra of organophosphorus compounds. These studies also allow one to gain information upon solute solvent interactions. Intermolecular hydrogen bonding between a solvent and a basic site affects the vibrational spectrum. Field effects of a solvent system also alter the vibrational spectrum. Often both of these effects operate simultaneously on the vibrational frequencies of chemical molecular vibrations (26, 55).

Figure 8.54 shows separate plots of ν P=O frequencies for P(=O)Cl_3 vs the mole % CCl_4/C_6H_{14}, mole % $C_6H_{14}/CHCl_3$, and mole % $CHCl_3/CCl_4$ solvent systems. These are 1% wt./vol. solutions of P(=O)Cl_3, and at this concentration the field effect of P(=O)Cl_3 is minimal and constant in these solvent systems. In all three solvent systems ν P=O decreases in frequency as the mole % solvent system increases. The CCl_4/C6H_{14} solvent system has the least effect upon ν P=O, and ν P=O decreases as the mole % CCl_4/C_6H_{14} increases. Thus, the ν P=O frequency decreases as the field effect of the solvent system increases. The same trend is noted for the $C_6H_{14}/CHCl_3$ and $CHCl_3/CCl_4$ solvent systems and the ν P=O frequency decreases in the order of the increased field effect of the solvent system. However, there is a distinct difference in the solvent systems containing CCl_3H. The ν P=O frequency is rapid with the first addition of $CHCl_3$ into the solvent system, and this is attributed to the formation of intermolecular hydrogen bonding between the CCl_3H proton and the free pair of electrons of the P=O oxygen atom (viz. $CCl_3H\cdots$O=P) (26).

Figure 8.55 show a plot of ν P=O for methanephosphonic dichloride $CH_3P(=O)Cl_2$ vs mole % $CDCl_3$ and plots of ν P=O for 1,1-dimethylethane-phosphonic dichloride vs mole % $CHCl_3/CCl_4$ and vs mole % $CDCl_3/CCl_4$. The ν P=O vibration for $(CH_3)_3CP(=O)Cl_2$ occurs at lower frequency than ν P=O for $CH_3P(=O)Cl_2$ at all mole % concentrations of $CDCl_3/CCl_4$. The P=O group is more basic in the case of the $(CH_3)_3C$ analog than for the CH_3 analog, because the electron release of the $(CH_3)_3C$ group to the P=O group is larger than that of the CH_3 group to the P=O group. With the first addition of $CHCl_3$ or $CDCl_3$ to the solvent system ν P=O decreases in frequency due to intermolecular hydrogen bonding (viz. $CCl_3D\cdots$O=P) (26). In the case of $(CH_3)3CP(=O)Cl_2$ there appears to be a second break in the plot after ~60 mol % $CDCl_3/CCl_4$. Perhaps this is the result of a second intermolecular hydrogen bond between the $CDCl_3$ or $CHCl_3$ and the P=O group (viz. $(CCl_3H)_2\cdots$(O=P). The second intermolecular hydrogen bond may occur because the P=O group is more basic than the P=O group for $CH_3P(=O)Cl_2$. This is apparently unique because the steric factor of the $(CH_3)_3C$ group is significantly larger than the steric factor for the CH_3 group.

Figure 8.56 shows plots of the ν P=O rotational conformers 1 and 2 for O-methyl phosphorodichloridate and O-ethyl phosphorodichloridate vs mole % $CHCl_3$/CCl4 (26). Both ν P=O rotational conformers 1 and 2 decrease in frequency with the first addition of $CHCl_3$, and this is attributed to intermolecular hydrogen bonding (viz. $CCl_3H\cdots O=P)_2$. The ν P=O frequency separation between rotational conformers 1 and 2 is larger in CCl_4 solution than in $CHCl_3$ solution, and this reflects the effect of hydrogen bonding. There appears to be a second intermolecular hydrogen bond formed in all but rotational conformer 2 for $C_2H_5OP(=O)Cl_2$. In all of these plots the general decrease in the ν P=O frequencies is due to the increased field effect of the solvent system.

Figure 8.57 shows plots of ν P=O rotational conformers 1 and 2 for trimethyl phosphate vs mole % $CHCl_3/CCl_4$. The IR band in the range 1273.4 cm^{-1} through 1289.9 cm^{-1} is assigned to ν P=O rotational conformer 1, and the IR band in the range 1258 cm^{-1} through 1270.9 cm^{-1} is assigned to ν P=O rotational conformer 2. Complexes such as $CCl_3H\cdots$O=P and $(CCl_3H\cdots)_2$O=P are apparently formed between solute and solvent.

Figures 8.58 and 8.59 show plots of ν P=O conformers 1 and 2 for triethyl phosphate and tributyl phosphate, respectively. The apparent decrease in the ν P=O frequencies for rotational conformers 1 and 2 reflects the initial intermolecular hydrogen-bond formation. As the mole % $CHCl_3/CCl_4$ is increased ν P=O for the rotational conformer increases in frequency for both $(C_2H_5{-}O)_3P{=}O$ and $(C_4H_9O)_3P{=}O$ while rotational conformer 1 for $(C_4H_9O)_3P{=}O$ remains relatively constant. These data suggest that as the alkyl group becomes larger it sterically shields the P=O group from the solvent field effect. Thus, the ν P=O frequencies are not continually lowered in frequency due to the field effect of the solvent system.

PCl_3 AND PCl_2 VIBRATIONS

Figures 8.60 through 8.62 show plots of ν asym. PCl_3 or ν asym. PCl_2 vs mole % solvent system for $P({=}O)Cl_3$, $CH_3P({=}O)Cl_2$, $(CH_3)_3CP({=}O)Cl_2$, and $(RO)P({=}O)Cl_2$. In all cases, the ν asym. PCl_3 or ν sym. PCl_2 vibration increases in frequency as the mole % CCl_4/C_6H_{14}, mole % $CHCl_3/C_6H_{14}$, or $CHCl_3/CCl_4$ is increased. In the case of $(CH_3O)P({=}O)Cl_2$ or $(C_2H_5O)P({=}O)Cl_2$, the ν asym. PCl_2 mode is a doublet due to the presence of rotational conformers 1 and 2.

Figures 8.63–8.66 show plots of ν sym. PCl_3 or ν sym. PCl_2 for $P({=}O)Cl_3$, $CH_3P({=}O)Cl_2$, $(CH_3)_3CP({=}O)Cl_2$, and $(RO)P({=}O)Cl_2$. With the exception of the solvent system CCl_4/C_6H_{14}, the ν sym. PCl_3 or ν sym. PCl_2 vibration increases in frequency as the mole % solvent system is increased. Therefore, both the ν sym. $PCl_{3\,or\,2}$ and ν asym. $PCl_{3\,or\,2}$ vibrations in mole % $CHCl_3/CCl_4$ solutions increase in frequency as the ν P=O vibrations decrease in frequency. Thus, as the field effect of the solvent increases, it requires more energy to vibrate these PCl_3 or PCl_2 bonds.

Figure 8.66 shows plots of ν asym. PCl_2 vs ν sym. PCl_2 for $CH_3OP({=}O)Cl_2$, $C_2H_5OP({=}O)Cl_2$, $(CH_3)_3CP({=}O)Cl_2$, and $CH_3P({=}O)Cl_2$. These plots show that ν asym. PCl_2 occurs within the range 545–615 cm^{-1} and that ν sym. PCl_2 occurs within the range 495–555 cm^{-1}.

ABSORBANCE RATIOS FOR $ROP({=}O)Cl_2$ MOLECULES

Figures 8.67 and 8.68 show plots of the absorbance ratios A(conformer 1)/A(conformer 2) for the ν P=O, ν COP, ν asym. PCl_2 and ν sym. PCl_2 vibrations vs mole % $CHCl_3/CCl_4$ for $(CH_3O)P({=}O)Cl_2$ and $(C_2H_5O)P({=}O)Cl_2$, respectively. In all cases the ratio A(conformer 1)/A(conformer 2) increases as the mole % $CHCl_3/CCl_4$ increases. These data indicate that as the field effect of the solvent system increases, the concentration of rotational conformer 1 increases while the concentration of conformer 2 decreases.

ALKYL GROUP VIBRATIONS

Figure 8.69 shows a plot of ν sym. CH_3 for $CH_3P({=}O)Cl_2$ vs mole % $CHCl_3$/CCl4. This symmetric CH_3 bending vibration increases in frequency in an essentially linear manner as the mole % $CHCl_3/CCl_4$ increases. The effect is small, as the frequency difference is only $\approx 0.5\ cm^{-1}$ (26).

Figure 8.70 is a plot of $\rho\,CH_3$ for $CH_3P(=O)Cl_2$ vs mole % $CHCl_3/CCl_4$. The CH_3 rocking vibration increases in frequency in a nonlinear manner as the mole % $CHCl_3/CCl_4$ increases. The effect is small, as the frequency difference is $< 0.5\ cm^{-1}$.

Figure 8.71 shows plots of the in-phase δ sym. $(CH_3)_3$ and out-of-phase δ sym. $(CH_3)_3$ frequencies for $(CH_3)_3CP(=O)Cl_2$ vs mole % $CHCl_3/CCl_4$. The in-phase symmetric $(CH_3)_3$ deformation decreases in frequency while the out-of-phase symmetric $(CH_3)_3$ deformation increases in frequency as the mole % $CHCl_3/CCl_4$ increases. These solvent effects are small because the frequency differences are $\approx 1\ cm^{-1}$.

Figure 8.72 shows plots of the ρCH_3 modes for $(C_2H_5O)P(=O)Cl_2$ vs mole % $CHCl_3/CCl_4$. The a′ CH_3 rocking vibration ($\sim 1103\ cm^{-1}$) decreases in frequency while the *a″* CH_3 rocking vibration (~1165 in) increases in frequency as the mole % $CHCl_3/CCl_4$ increases. Figure 8.73 shows plots of ρCH_3 modes for $(C_2H_5O)_3P{=}O$ vs mole % $CHCl_3/CCl_4$. The *a″* $\rho(CH_3$ mode decreases in frequency while the *a″* ρCH_3 mode decreases in frequency until ~16 mol % $CHCl_3/CCl_4$, and then the frequency remains relatively constant.

THE ν C–C MODE FOR THE C_2H_5OP GROUP

Figures 8.74 and 8.75 show a plot of ν C–C for $(C_2H_5O)P(=O)Cl_2$ and for $(C_2H_5O)3P{=}O$ vs mole % $CHCl_3/CCl_4$, respectively. This plot shows that the C–C stretching vibration for the C_2H_5OP group increases in frequency as the mole % $CHCl_3/CCl_4$ is increased (26, 55).

THE ν COP GROUP (26, 55)

Figure 8.76 shows plots of ν C–O for the COP group of $CH_3OP(=O)Cl_2$ and $C_2H_5OP(=O)Cl_2$ vs mole % $CHCl_3/CCl_4$. The ν CO frequency for rotational conformer 1 of the COP group decreases in frequency as the mole % $CHCl_3/CCl_4$ is increased. In case of rotational conformer 2, ν C–O for $CH_3OP(=O)Cl_2$ increases in frequency while ν C–O for $C_2H_5OP(=O)Cl_2$ decreases in frequency (26, 55).

Figure 8.77 shows plots of ν P=O for rotational conformers 1 and 2 for $(CH_3O)3$ P=O vs mole % $CHCl_3/CCl_4$. The erratic behavior of these two plots shows the effect of intermolecular hydrogen bonding upon the $\nu\,(C{-}O)_3$ modes of the $(CH_3O)_3$ groups. These data suggest that the molecular configuration of the rotational conformers is changing. Figure 8.78 shows a plot of ν P–O for the $(C{-}O)_3P$ groups for $(CH_3O)_3$ P=O vs mole % $CHCl_3/CCl_4$. The $\nu\,P(-O)_3$ vibrations first decrease in frequency and then increase in frequency as the mole % $CHCl_3/CCl_4$ is increased. These data indicate that the rotational configuration of the rotational conformers is changing as the mole % $CHCl_3/CCl_4$ is increased.

Figures 8.79 and 8.80 show plots of $\nu\,\phi$-O and ν O–P for the ϕ-O–P groups of $(C_6H_5O)_3P{=}O$ vs mole % $CHCl_3/CCl_4$, respectively. These plots show that as $\nu\,\Phi$-O decreases in frequency, ν P–O increases in frequency to a point in the range near 50 mol % $CHCl_3/CCl_4$. At higher mole % $CHCl_3/CCl_4$ concentrations, $\nu\,\phi$-O increases in frequency while ν P–O decreases in frequency. These plots indicate that the rotational conformers are changing as the mole % $CHCl_3/CCl_4$ is increased.

Figure 8.81 shows plots of ν P=O rotational conformers 1 and 2 for triphenyl phosphate. Both ν P=O rotational conformers decrease in frequency as the mole % $CHCl_3/CCl_4$ is increased. Figure 8.82 show a plot of the absorbance ratio A (ν P=O, conformer 1)/A(ν P=O, conformer 2) for triphenyl phosphate vs mole % $CHCl_3/CCl_4$. A break is noted in the plot near 10 mol % $CHCl_3/CCl_4$ in Figure 8.8, which corresponds to the break in the plot of ν P=O, conformer 1. This break is attributed to the formation of the intermolecular hydrogen bond for rotational conformer 1. The absorbance ratio decreases as the mole % $CHCl_3/CCl_4$ is increased, which shows that the concentration of conformer 2 increases while the concentration of conformer 1 decreases (see Figure 8.82).

Figure 8.83 shows a plot of the in-plane hydrogen deformation for the phenyl groups of triphenyl phosphate vs mole % $CHCl_3/CCl_4$. This vibration increases in frequency at a more rapid rate below 10 mol % $CHCl_3/CCl_4$ with the formation of the intermolecular hydrogen bonds, and then increases in a linear manner as the mol % $CHCl_3/CCl_4$ is increased.

Figure 8.84 shows a plot of the out-of-plane ring deformations for the phenyl groups of triphenyl phosphate vs mole % $CHCl_3/CCl_4$. This out-of-plane ring deformation increases in frequency as the field effect of the solvent system increases.

ν C—D FOR $CDCl_3$ IN CCl_4 SOLUTIONS

Figure 8.85 shows a plot of ν C—D for $CDCl_3$ vs mole % $CDCl_3/CCl_4$. This plot shows that ν C—D increases in frequency as the mole % $CDCl_3/CCl_4$ is increased in essentially a linear manner (55). The ν C—D frequency decreases as the concentration of CCl_4 is increased. The Cl atoms of CCl_4 are more basic than the Cl atoms of $CDCl_3$. Thus, ν C—D of $CDCl_3$ decreases as the surrounding field of CCl_4 molecules is increased in the CCl_3-D: CCl_4 equilibrium.

Figure 8.86 shows a plot of ν C—D for $CDCl_3$ containing 1 % wt./vol. triethyl phosphate in $CDCl_3/CCl_4$ solutions. In this case ν C—D does not decrease in a linear manner as the mole % $CDCl_3/CCl_4$ is increased. The more rapid decrease in the ν C—D frequency at lower mole % $CDCl_3/CCl_4$ concentrations is due to intermolecular hydrogen bonding with $(CH_3O)_3P{=}O$ molecules in conjunction with competing intermolecular hydrogen bonding with CCl_4 molecules. Thus, in all cases of $CHCl_3/CCl_4$ there is an equilibrium of $(CCl_3H)_n \cdots (ClCCl_3)_m$ with intermolecular hydrogen bonding with solute molecules.

For further discussion of solvent effects on phosphorus compounds the reader is referred to References 26 and 55.

REFERENCE SPECTRA

Reference IR and Raman spectra are available for identification of unknown samples containing phosphorus. These spectra are available in references such as 48–51, or from the Sadtler Research Laboratories. The book *Analytical Chemistry of Phosphorus Compounds* is an excellent source for other techniques required for the solution of chemical problems involving phosphorus derivatives (50).

REFERENCES

1. Wilson, M. K. and Polo, S. R. (1952) *J. Chem. Phys.* **20**: 16.
2. Delwaulle, M. L. and Francois, F. (1949) *J. Chem. Phys.* **46**: 87.
3. Gutowwsky, H. S. and Liehr. A. D. (1952). *J. Chem. Phys.* **20**: 1652.
4. Wilson, M. K. and Polo, S. R. (1953) *J. Chem. Phys.* **21**: 1426.
5. Eucken, A., Hellwege, K. H. and Landolt-Bornstein, I. (1951). *Band, Atomic Und Molekulorphsk. 2 Terl, Molekeln I.* Berlin: Springer-Verlag, p. 247.
6. Stammereich, H., Foreris, R. and Tavares, Y. (1956). *J. Chem. Phys.* **25**: 580.
7. Durig, J. R. and Clark, J. W. (1967) *J. Chem. Phys.* **46**: 3057.
8. Delwaulle, M. L. and Francois, F. (1946) *Comp. Rend.* **22**: 550.
9. Nyquist, R. A., (1967) *Spectrochim. Acta.* **23A**: 845.
10. Gerding, H. and Westrick, R. (1942). *Rec. Trav. Chem.* **61**: 842.
11. Durig, J. R. and Clark, J. W. (1967) *J. Chem. Phys.* **46**: 3057.
12. Delwaulle, M. L. and Francois, F. (1948). *Comp. Rend.* **226**: 894.
13. Delwaulle, M. L. and Francois, F. (1946). *Comp. Rend.* **22**: 550.
14. Eucken, A., Hellewege, K. H. and Landolt-Bornstein, I. (1951). *Band, Atomic Und Moleknlorphsk. 32 Terl. Molcheln I.* Berlin: Springer-Verlage, p. 278.
15. Delwaulle, M. L. and Francois, F. (1945) *Comp. Rend.* **220**: 817.
16. Nyquist, R. A. (1987). *Appl. Spectrosc.* **41**: 272.
17. Nyquist, R. A. (1968). *Appl. Spectrosc.* **22**: 452.
18. Durig, J. R., Black, F. and Levin, J. W. (1965). *Spectrochim. Acta* **21**: 1105.
19. Durig, J. and Di Yorio, J. (1960) *Chem. Phys.* **48**: 4154.
20. The Dow Chemical Company Data in CS_2 solution.
21. Nyquist, R. A. and Muelder, W. W. (1966) *Spectrochem. Acta.* **22**: 1563.
22. Nyquist, R. A., Muelder, W. W. and Wass, M. N. (1970). *Spectrochim. Acta* **26A**: 769.
23. Nyquist, R. A. (1971). *Spectrochim. Acta* **27A**: 697.
24. Nyquist, R. A., Wass, M. N. and Muelder, W. W. (1970). *Spectrochim. Acta* **26A**: 611.
25. Nyquist, R. A. (1967). *Spectrochim. Acta* **23A**: 1499.
26. Nyquisit, R. A. and Puehl, C. W. (1999). *Appl. Spectrosc.* **46**: 1552; Mortimer, F. S. (1957). *Spectrochim. Acta.* **9**: 270.
27. Nyquist, R. A. and Muelder, W. W. (1968) *Spectrochim. Acta* **24A**: 187.
28. Dow Chemical Company data in CCl_4 solution.
29. Taft, R. W. (ed.). (1976). *Progress in Physical Organic Chemistry*, Vol. 12, New York: John Wiley & Sons, p. 74.
30. Thomas, L. C. and Chittenden, R. A. (1964). *Spectrochim. Acta* **20**: 489.
31. Nyquist, R. A., Sloane, H. J., Dunbar, J. E. and Stycker, S. J. (1966). *Appl. Spectrosc.* **20**: 90.
32. Nyquist, R. A. (1969). *Spectrochim. Acta* **25A**: 47.
33. Durig, J. R. and Clark, J. W. (1967) *J. Chem. Phys.* **46**: 3057.
34. Delwaulle, M. L. and Francois, F. (1946). *Comp. Rend.* **22**: 550.
35. Nyquist, R. A. and Muelder, W. W. (1968). *J. Mol. Structure* **2**: 465.
36. Nyquist, R. A. (1966) *Spectrochim. Acta* **22**: 1315.
37. Nyquist, R. A., Blair, E. H. and Osborne, D. W. (1967) *Spectrochim. Acta* **23A**: 2505.
38. Herrail, F. (1965). *Comp. Rend.* **261**: 3375.
39. Neely, W. B., Unger, I., Blair, E. H. and Nyquist, R. A. (1964). *Biochemistry* **3**: 1477.
40. Nyquist, R. A. (1968) *J. Mol. Structure* **2**: 123.
41. Nyquist, R. A., (1968) *J. Mol. Structure* **2**: 111.

42. Nyquist, R. A. (1963) *Spectrochim. Acta* **19**: 713.
43. Nyquist, R. A. and Muelder, W. W. (1971). *Appl. Spectrosc.* **25**: 449.
44. Tsubio, M. (1957). *J. Amer. Chem. Soc.* **79**: 1351.
45. Herzberg, G. (1945). *Infrared and Raman Spectra of Polyatomic Molecules.* New York: Van Nostrand.
46. Corbridge, D. E.C. (1969). *Topics in Phosphorus Chemistry*, Vol. 6, M. Grayson and E. J. Griffiths, eds, *The Infrared Spectra of Phosphorus Compounds.* New York: Interscience, p. 235.
47. King, S. T. and Nyquist, R. A. (1970). *Spectrochim. Acta* **26A**: 1481.
48. Nyquist, R. A. and Craver, C. D. (1977). In *The Coblentz Society Desk Book of Infrared Spectra.* (C. D. Craver, ed.), Kirkwood, MO: The Coblentz Society.
49. Lin-Vien, D., Cotthup, N. B., Fateley, W. C. and Grasselli, J. G. (1991). *The Handbook of Infrared and Raman Characteristic Frequencies of Organic Molecules.* San Diego, CA: Academic Press, Inc.
50. Nyquist, R. A. and Potts, Jr., W. J. (1972). Vibrational Spectroscopy of Phosphorus Compounds, Ch. 5, in *Analytical Chemistry of Phosphorus Compounds.* Vol. 37, M. Halmann, ed., New York: Wiley-Interscience.
51. Nyquist, R. A., Putzig, C. L., Leugers, M. A. and Kagel, R. O. (1997). *Handbook of Infrared and Raman Spectra of Inorganic Compounds and Organic Salts.* Vols. 1–4, San Diego: Academic Press.
52. Nyquist, R. A. (1970) *Spectrochim. Acta* **26A**: 849.
53. Bouquet, S. and Bigorne, M. (1967). *Spectrochim. Acta* **23A**: 1231.
54. Daasch, L. W. and Smith, D. C. (1951). *J. Chem. Phys.* **19**: 22.
55. Nyquist, R. A. and Puehl, C. W. (1992). *Appl. Spectrosc.* **46**: 1564.

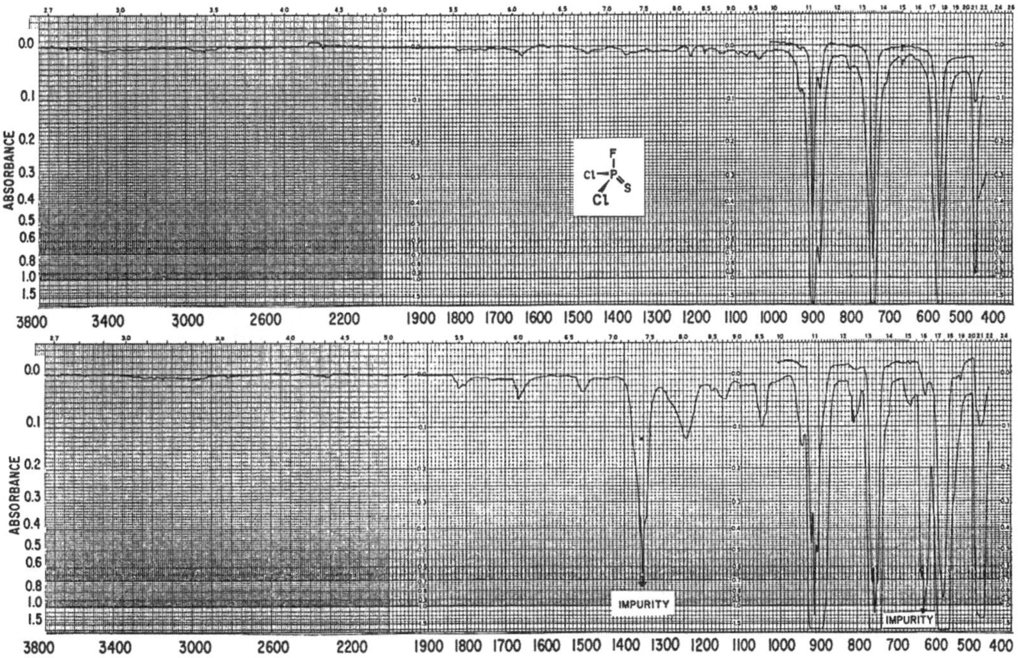

FIGURE 8.1 Top: Infrared spectrum of thiophosphoryl dichloride fluoride 10% wt./vol. in CCl_4 solution (3800–1333 cm^{-1}) and 10% wt./vol. in CS_2 solution. Bottom: Vapor-phase spectrum of thiophosphoryl dichloride fluoride in the region 3800–450 cm^{-1} (9).

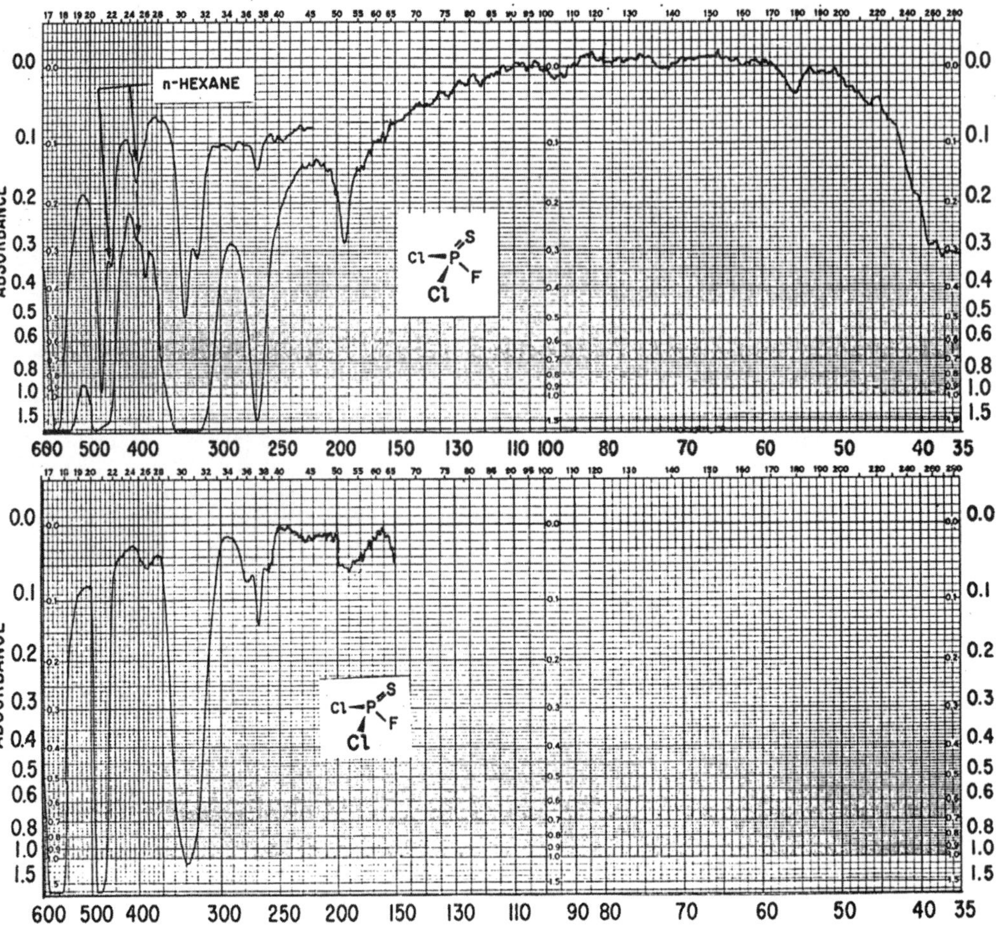

FIGURE 8.2 Top: Infrared spectrum of thiophosphoryl dichloride fluoride in hexane solution using polyethylene windows in the region 600–40 cm^{-1}. Bottom: Vapor-phase IR spectrum of thiophosphoryl dichloride fluoride using polyethylene windows in the region 600–150 cm^{-1} (9).

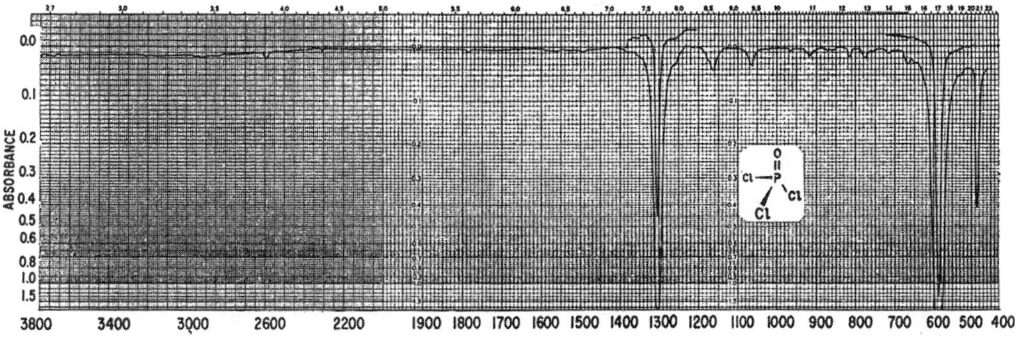

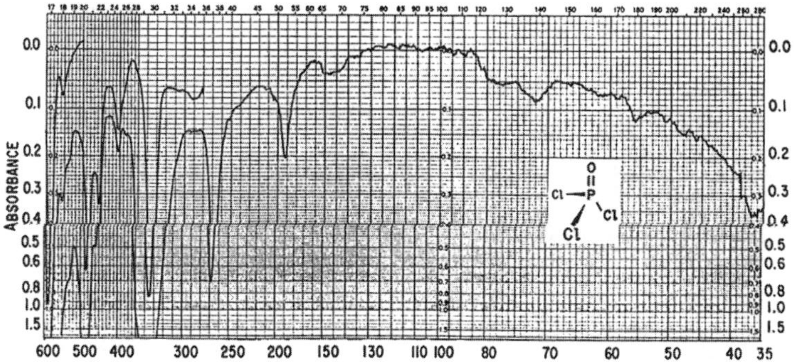

FIGURE 8.3 Top: Infrared spectrum of phosphoryl chloride in CCl_4 solution (3800–1333 cm^{-1}) and in CS_2 solution (1333–45 cm^{-1}). Bottom: Infrared spectrum of phosphoryl chloride in hexane solution (9).

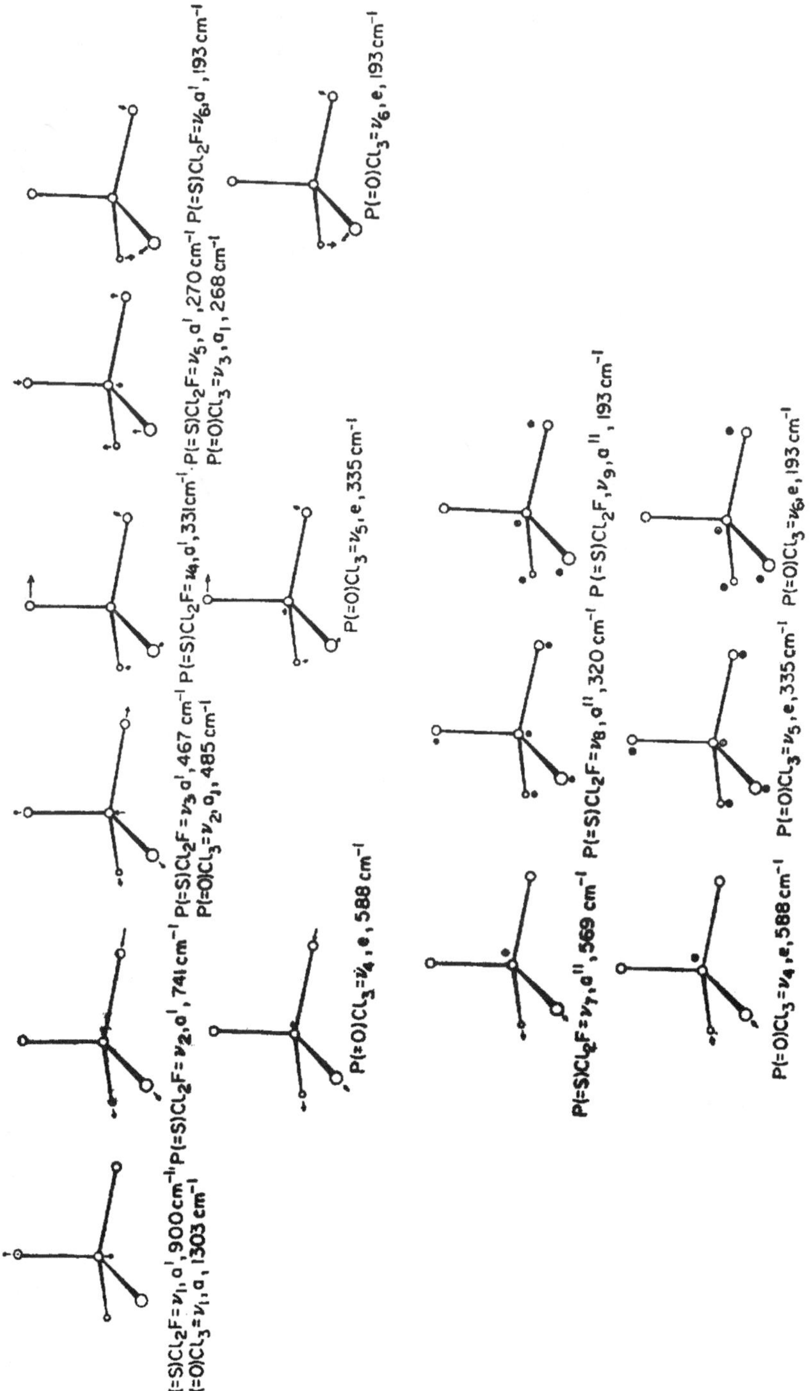

FIGURE 8.4 A comparison of the approximate normal vibrations of thiophosphoryl dichloride fluoride vs those for phosphoryl chloride (9).

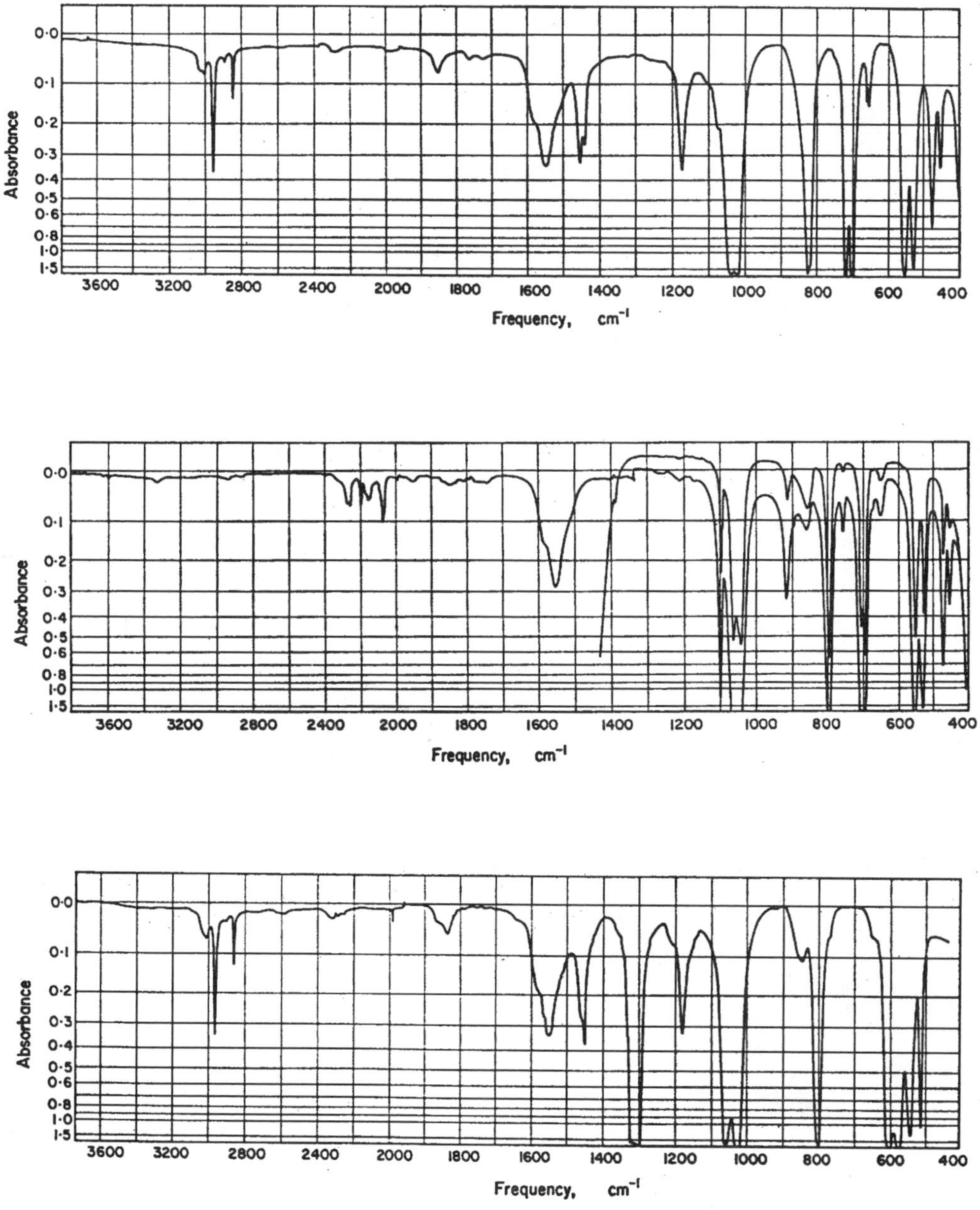

FIGURE 8.5 Top: Infrared spectrum of O-methyl phosphorodichloridothioate in 10 wt./vol. % in CCl_4 solution (3800–1333 cm^{-1}) and in 10 wt./vol. % in CS_2 solution (1333–400 cm^{-1}) using 0.1-mm KBr cells. The weak band at ~752 cm^{-1} is due to the presence of a trace amount of $P(=S)Cl_3$. The band at 659 cm^{-1} is due to the presence of ~4% O,O-dimethyl phosphorochloridothioate. Middle: Infrared spectrum of O-methyl-d_3 phosphorodichloride-thioate in 10 wt./vol. % in CCl_4 solution (3800–1333 cm^{-1}) and 10 + 2 wt./vol. % in CS_2 solution (1333–400 cm^{-1}) using 0.1-mm KBr cells. Bottom: Infared spectrum of O-methyl phosphorodichloridate in 10 wt./vol. % in CCl_4 solution (3800–1333 cm^{-1}) and 10 wt./vol. % in CS_2 solution (1333–450 cm^{-1}) using 0.1-mm KBr cells (21).

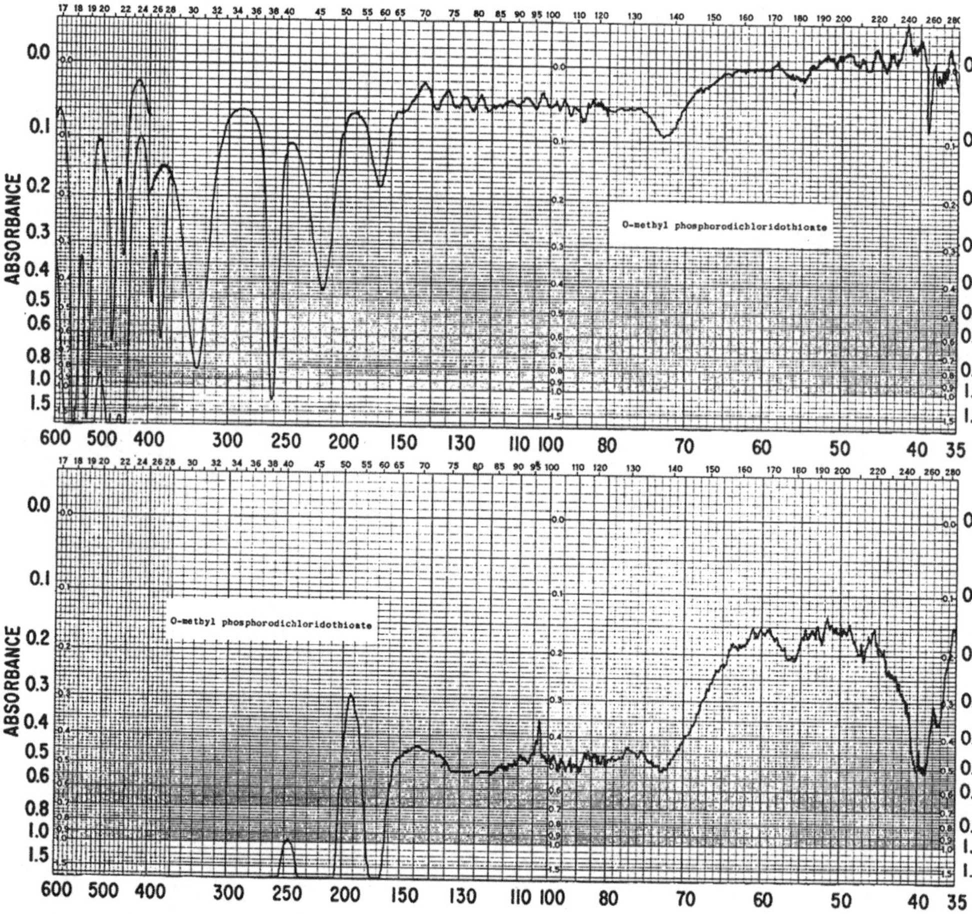

FIGURE 8.6 Top: Infrared spectrum of O-methyl phosphorodichloridothiate in 10 wt./vol. % in hexane solution in a 1-mm polyethylene cell. Bottom: Infrared spectrum of O-methyl phosphorodichloridothioate in 25 wt./vol. % in hexane solution in a 2-mm polyethylene cell (25).

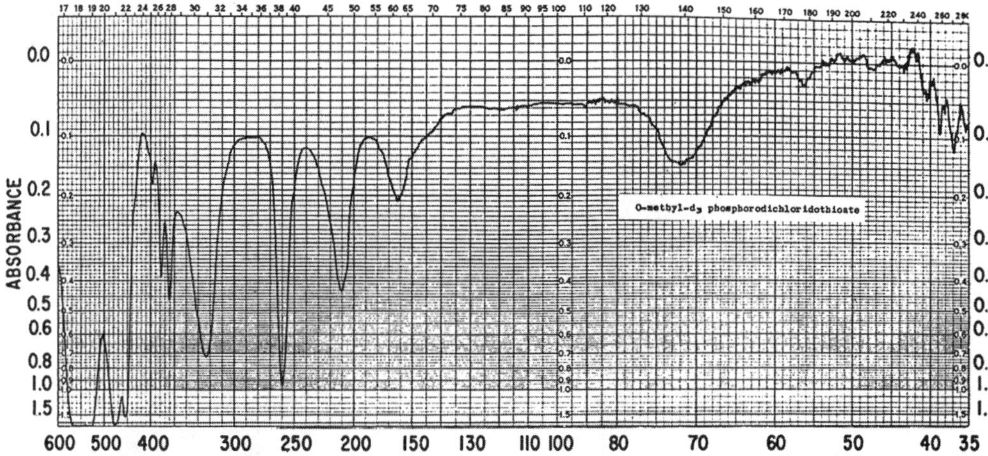

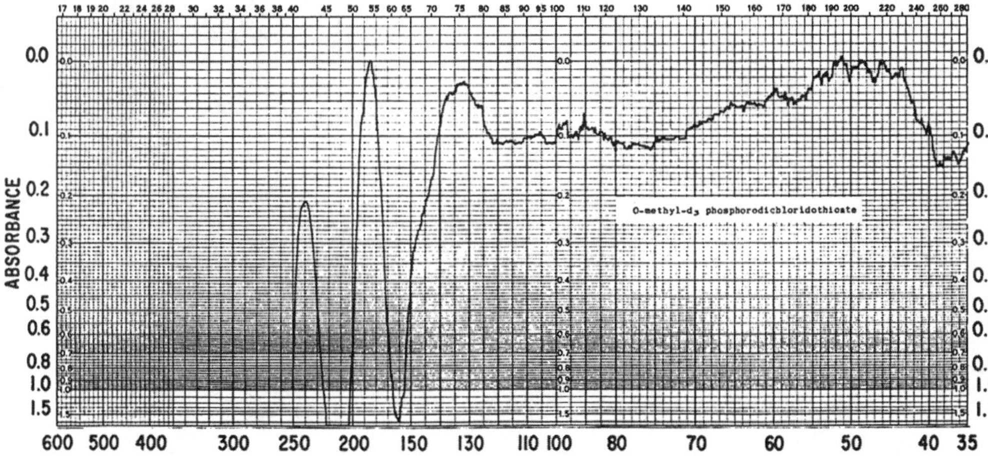

FIGURE 8.7 Top: Infrared spectrum of O-methyl-d_3 phosphorodichloridothioate in 10 wt./vol. % hexane solution in a 1-mm polyethylene cell. Bottom: Infrared spectrum of O-methyl-d_3 phosphorodichloridothioate in 25 wt./vol. % hexane solution in a 2-mm polyethylene cell and compensated with polyethylene (25).

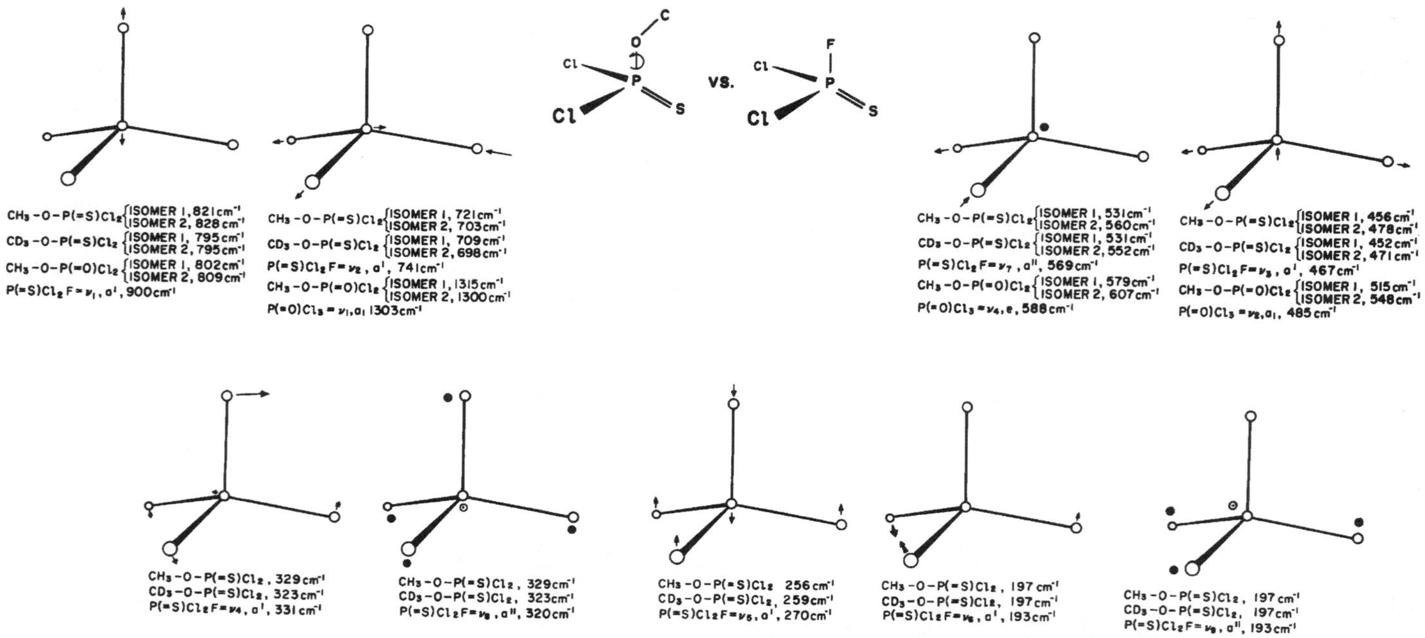

FIGURE 8.8 Assumed normal vibrations of organophosphorus and inorganophosphorus compounds (25).

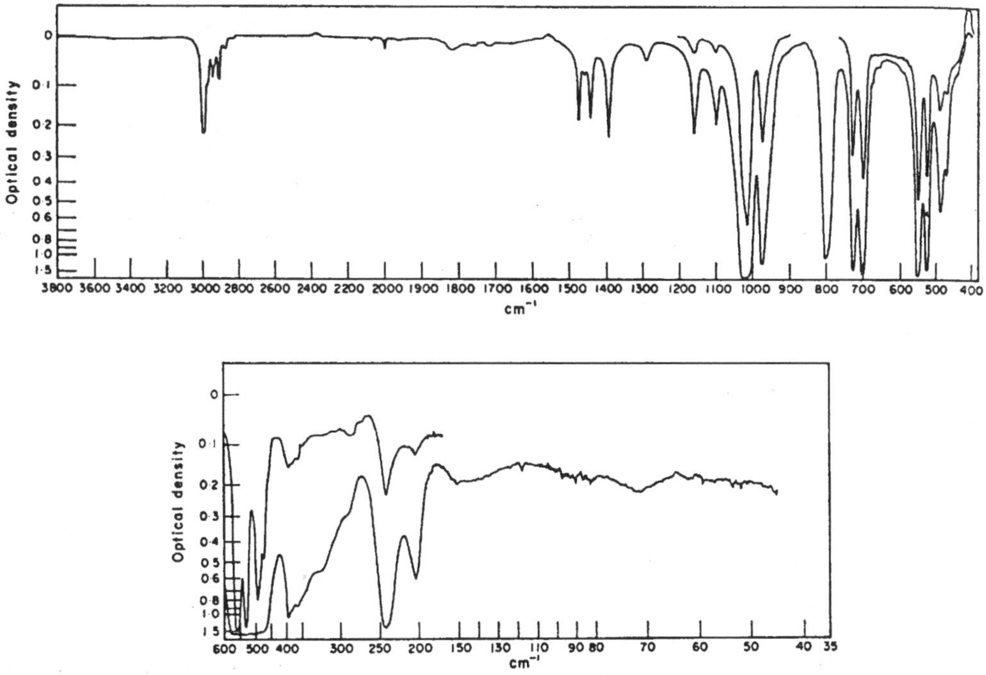

FIGURE 8.9 Top: Infrared spectrum of O-ethyl phosphorodichlorodiothioate 10 wt./vol. % in CCl_4 solution (3800–1333 cm^{-1}) and 10 and 2 wt./vol. % in CS_2 solution (1333–400 cm^{-1}) in 0.1-mm KBr cells. The solvents were compensated. Bottom: Infrared spectrum of O-ethyl phosphorodichloridothioate in 25 and 2 wt./vol. % in hexane solution (600–45 cm^{-1}) in a 1-mm polyethylene cell (22).

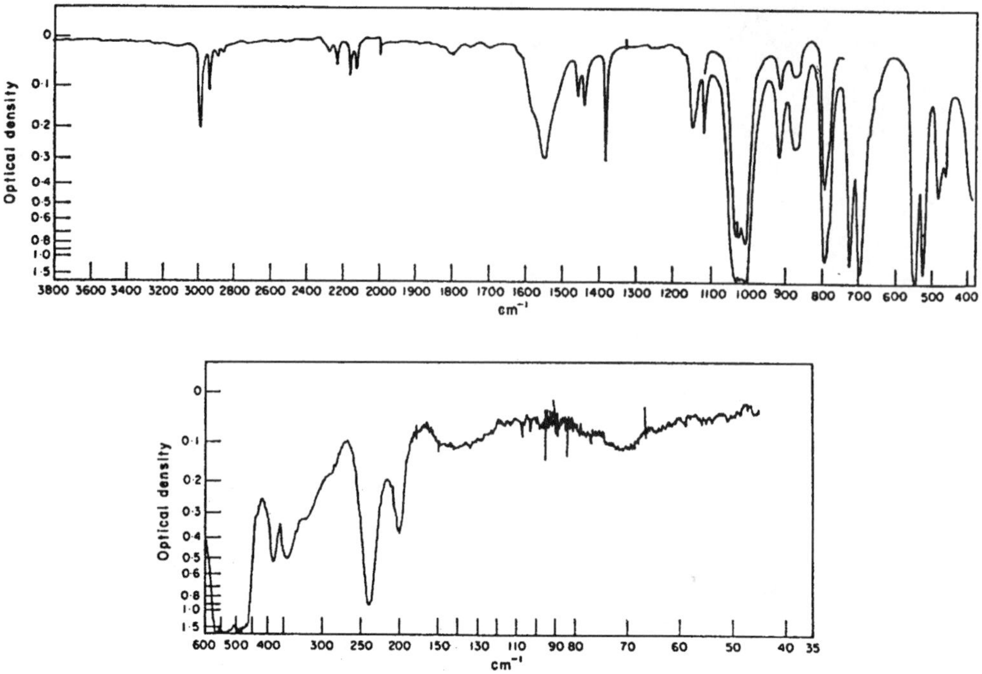

FIGURE 8.10 Top: Infrared spectrum of O-ethyl-1,1-d_2 phosphorodichloridothioate in 10 wt./vol. % in CCl_4 solution (3800–1333 cm^{-1}) and 10 and 2.5 wt./vol. % in CS_2 solution (1333–400 cm^{-1}) in 0.1-mm KBr cells. The solvents are compensated. Bottom: Infrared spectrum of O-ethyl-1,1-d_2 phosphorodichloridothioate in 10 wt./vol. hexane solution (600–45 cm^{-1}) in a 1-mm polyethylene cell (22).

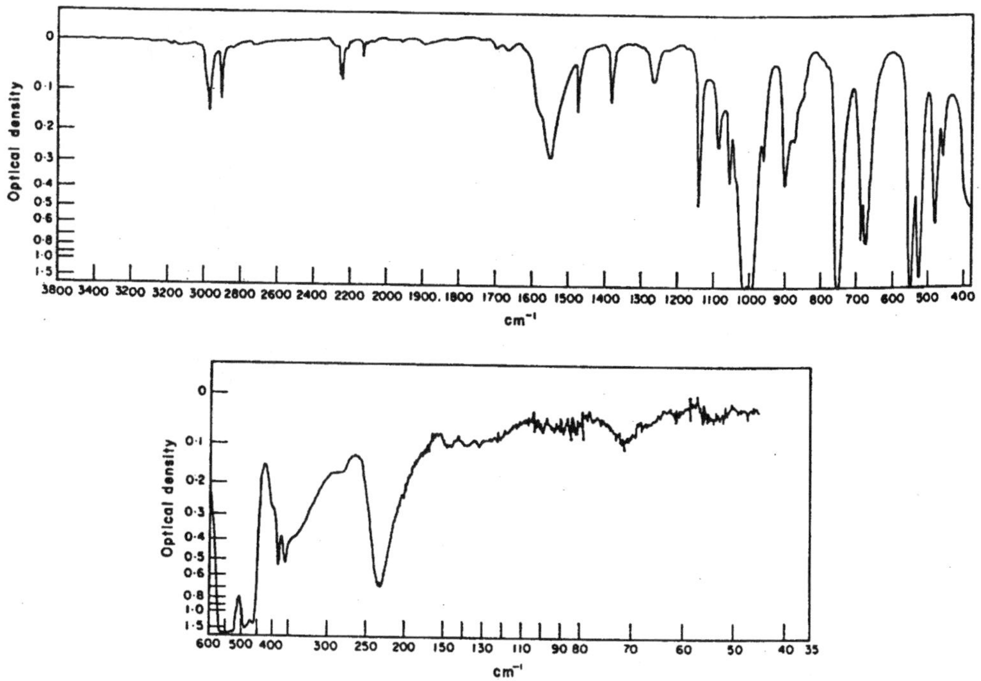

FIGURE 8.11 Top: Infrared spectrum of O-ethyl-2,2,2-d_3 phosphorodichloridothioate in 10 wt./vol. % CCl_4 solution (3800–1333 cm^{-1}) and 10 wt./vol. % CS_2 solution (1333–400 cm^{-1}) in 0.1-mm KBr cells. The solvents are not compensated. Bottom: Infrared spectrum of O-ethyl-2,2,2-d_3 phosphorodichloridothioate in 10 wt./vol. hexane solution in a 1-mm polyethylene cell (22).

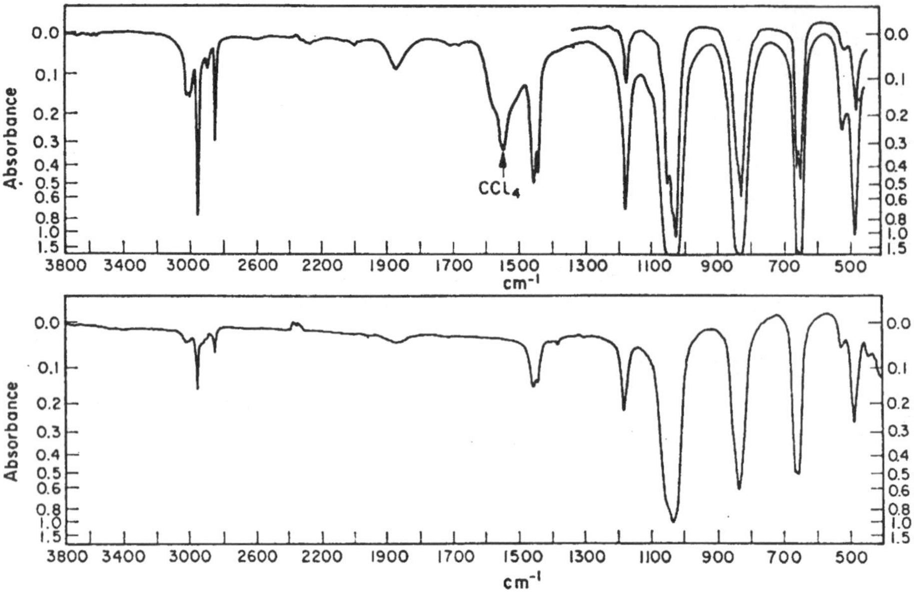

FIGURE 8.12 Top: Infrared spectrum of O,O-dimethyl phosphorochloridothioate in 10 wt./vol. % in CCl_4 solution (3800–1333 cm^{-1}) and 10 and 2 wt./vol. % in CS_2 solution (1333–450 cm^{-1}) in 0.1-mm KBr cells. The solvents have not been compensated. Bottom: Liquid-phase IR spectrum of O,O-dimethyl phosphorochloridothioate between KBr plates (27).

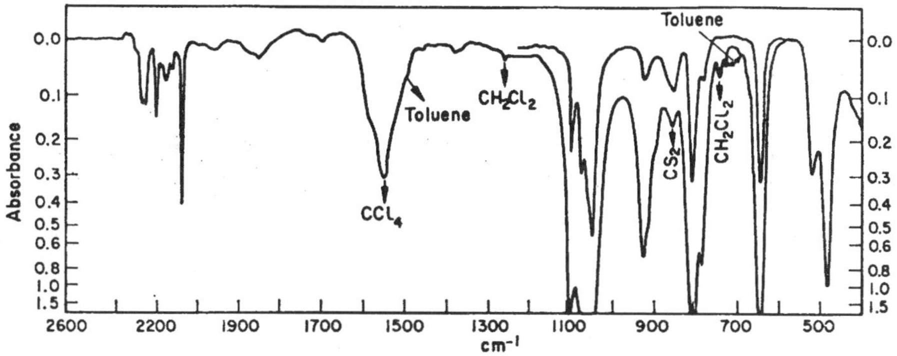

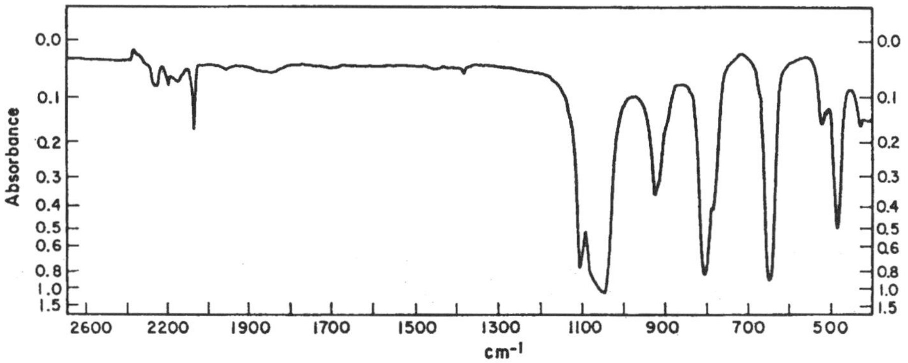

FIGURE 8.13 Top: Infrared spectrum of O,O-dimethyl-d_6 phosphorochloridothioate in 10 wt./vol. % in CCl_4 solution (3800–1333 cm^{-1}) and 10 and 1 wt./vol. % in CS_2 solution (1333–400 cm^{-1}) in 0.1-mm KBr cells. Traces of toluene and methylene chloride are present. Bottom: Liquid-phase IR spectrum of O,O-dimethyl-d_6 phosphorochloridothioate between KBr plates (27).

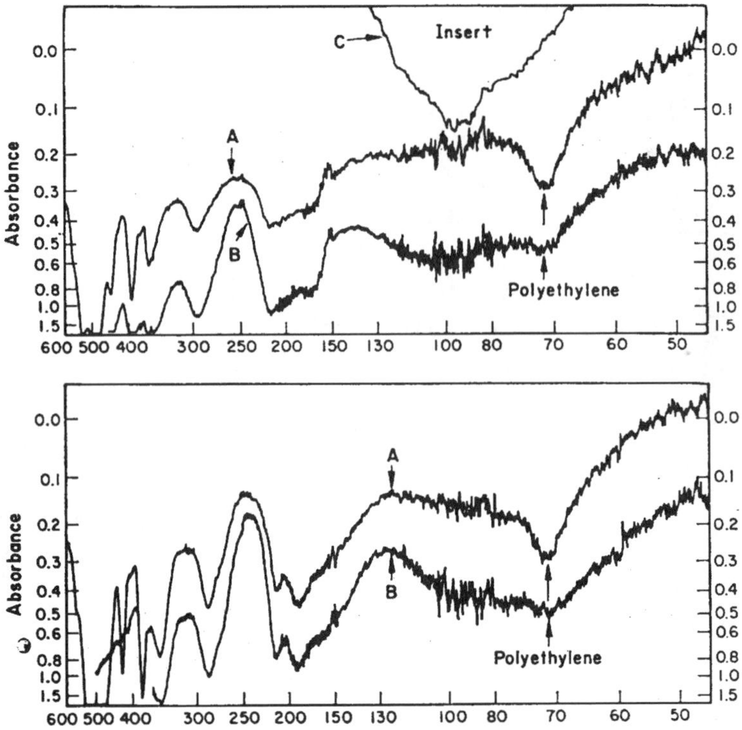

FIGURE 8.14 Top: Infrared spectrum of O,O-dimethyl phosphorochloridothioate in 2-mm polyethylene cells. The polyethylene has been compensated. Bottom: Infrared spectrum of O,O-dimethyl-d_6 phosphorochloridothioate in 10 wt./vol. % n-hexane in 1- and 2-mm polyethylene cells, respectively (27).

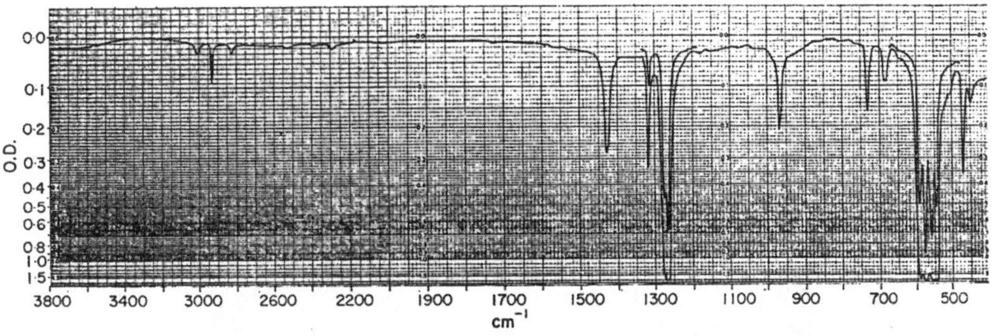

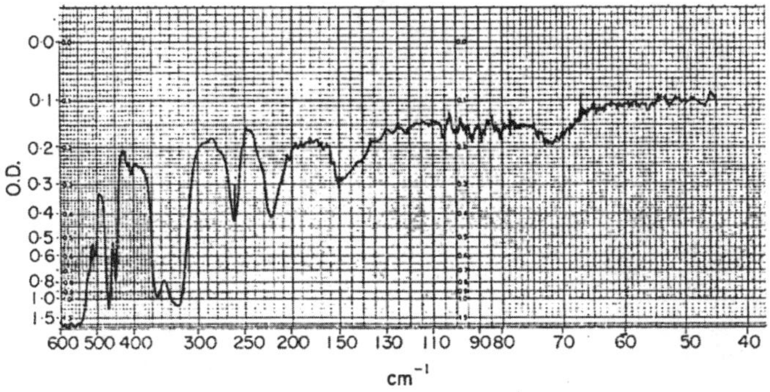

FIGURE 8.15 Top: Infrared spectrum of S-methyl phosphorodichloridothioate in 10 wt./vol. % in CCl_4 solution (3800–1333 cm^{-1}) and in 10 and 2 wt./vol. % in CS_2 solution (1333–400 cm^{-1}) in 0.1-mm KBr cells. Bottom: Infrared liquid phase spectrum of S-methyl phosphorodichloridothioate in 10 wt./vol. % hexane solution (600–45 cm^{-1}) in a 1-mm polyethylene cell (23).

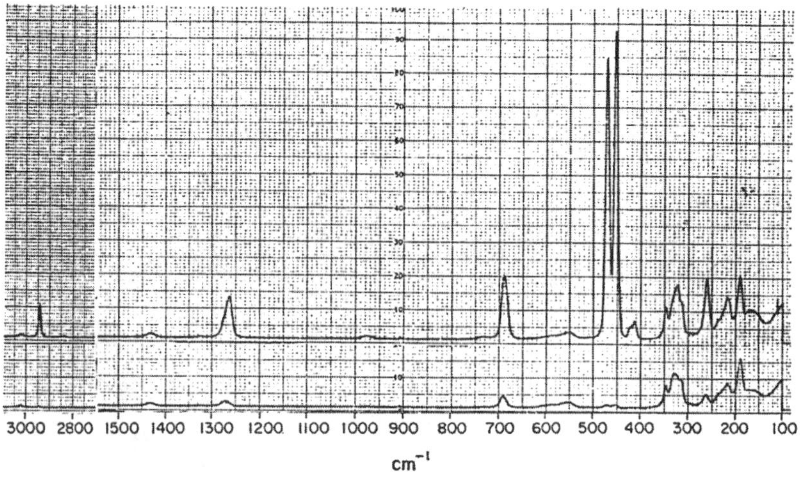

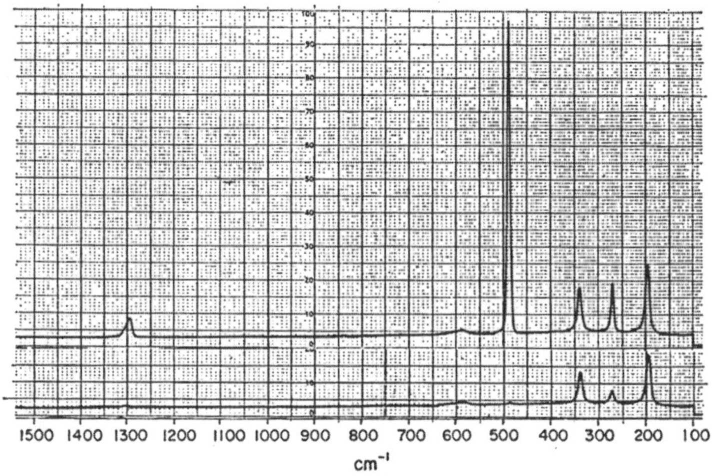

FIGURE 8.16 Top: upper: Raman liquid-phase spectrum of S-methyl phosphorodichloridothioate. Lower: Polarized Raman liquid-phase spectrum of S-methyl phosphorodichloridothioate (23). Bottom upper: Raman liquid-phase spectrum of phosphoryl chloride. Lower: Polarized Raman spectrum of phosphoryl chloride (23).

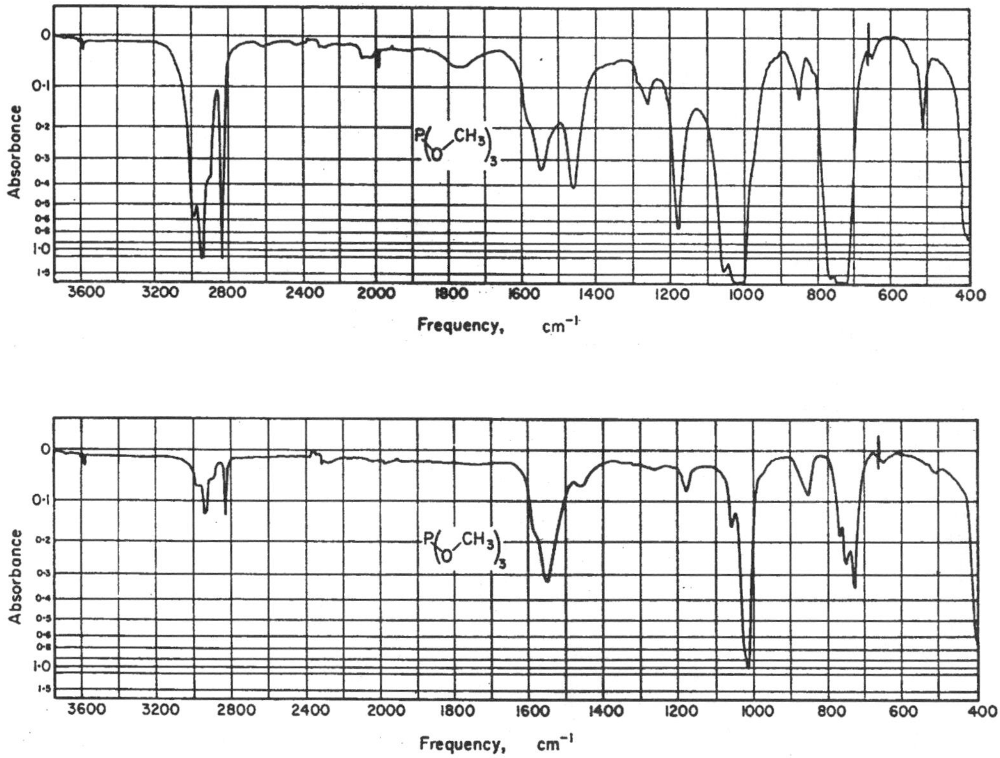

FIGURE 8.17 Top: Infrared spectrum of trimethyl phosphite 10 wt./vol. % in CCl_4 solution (3800–1333 cm^{-1}) and 10 wt./vol. % in CS_2 solution (1333–400 cm^{-1}) in 0.1-mm KBr cells. The solvents have not been compensated. Bottom: Infrared spectrum of trimethyl phosphite in 1 wt./vol. % CCl_4 solution (3800–1333 cm^{-1}) and in 1 wt./vol. % CS_2 solution (1333–400 cm^{-1}) using 0.1-mm KBr cells. The solvents have not been compensated (36).

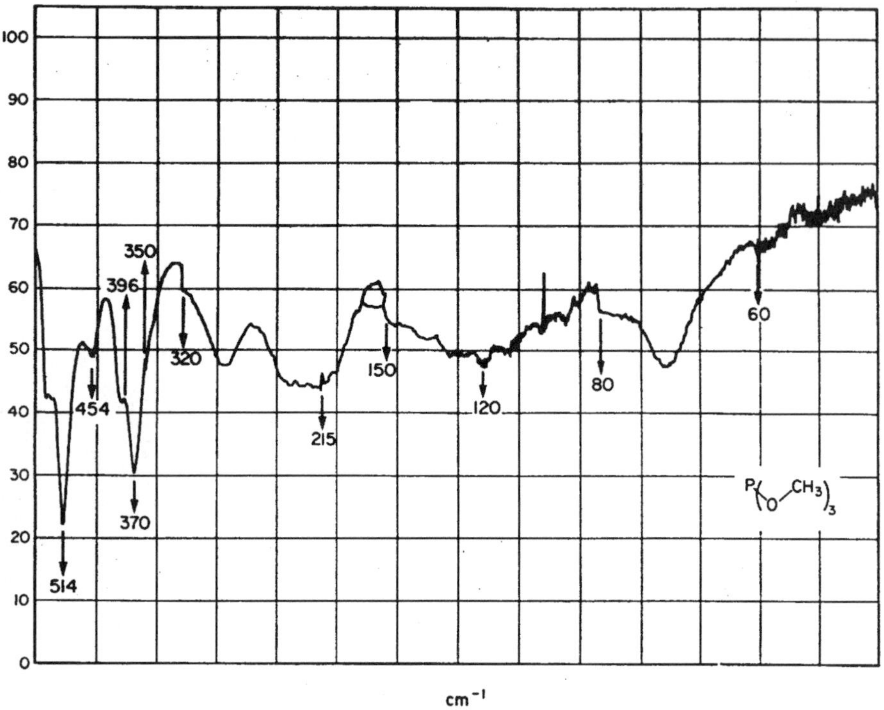

FIGURE 8.18 Infrared spectrum of trimethyl phosphite in 10 wt./vol. % hexane solution in a 0.1-mm polyethylene cell. The IR band at $\sim 70\,cm^{-1}$ in a lattice mode for polyethylene (36).

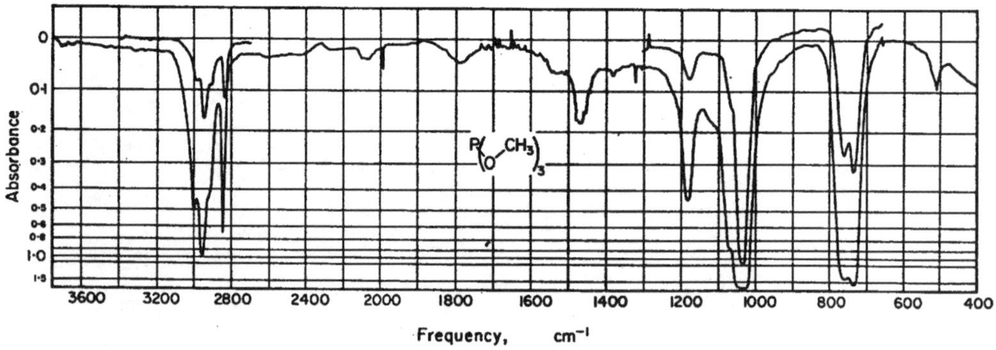

FIGURE 8.19 Vapor-phase IR spectrum of trimethyl phosphite in a 10-cm KBr cell (36).

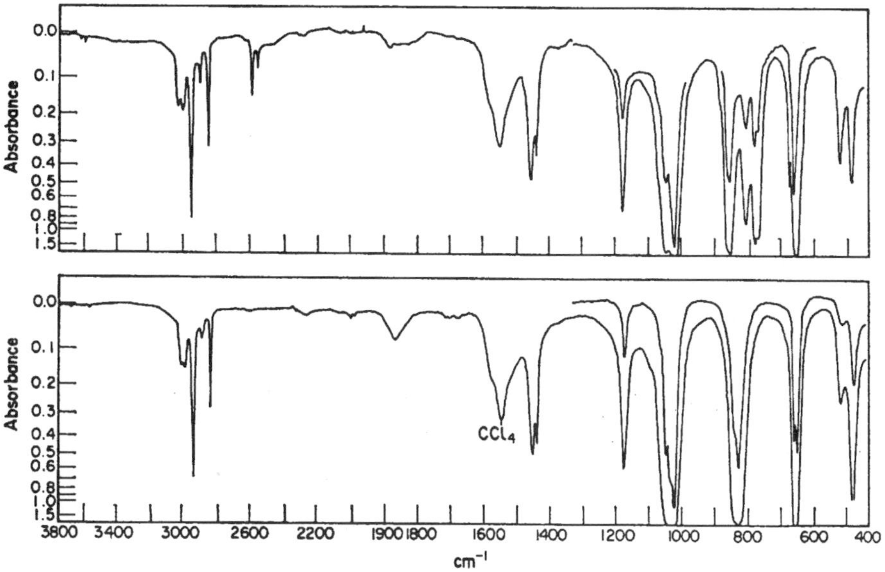

FIGURE 8.20 Top: Infrared spectrum of O,O-dimethyl phosphorodithioic acid in 10 wt./vol. % in CCl_4 solution (3800–1333 cm^{-1}) and 10 and 2 wt./vol. % in CS_2 solution (1333–450 cm^{-1}) in 0.1-mm KBr cells. Bottom: Infrared spectrum of O,O-dimethyl phosphorochloridothioate in CCl_4 solution (3800–1333 cm^{-1} and 10 and 2 wt./vol. % in CS_2 solutions (1333–450 cm^{-1}) in 0.1-mm KBr cells (32).

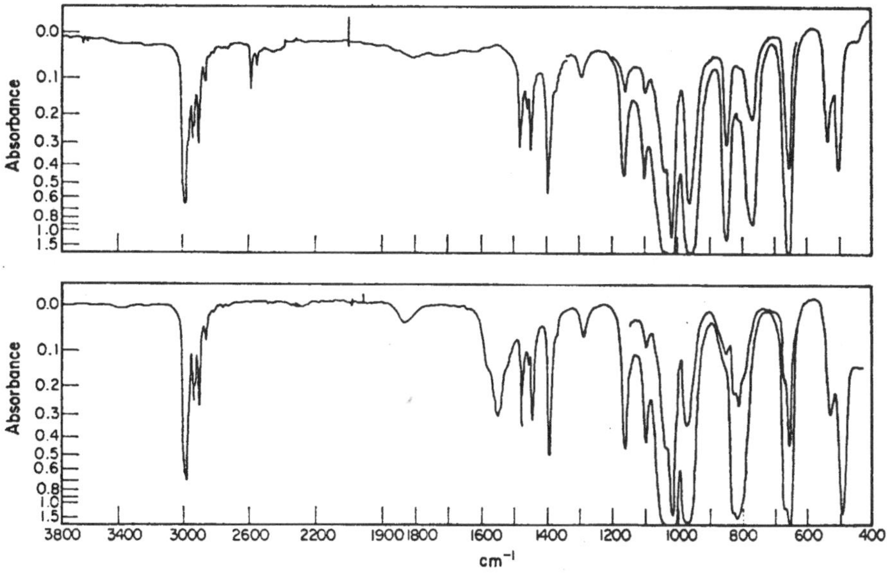

FIGURE 8.21 Top: Infrared spectrum of O,O-diethyl phosphorodithioic acid in 10 wt./vol. % in CCl_4 solution (3800–1333 cm^{-1}) and 10 and 2 wt./vol. % in CS_2 solutions (1333–400 cm^{-1}) in 0.1-mm KBr cells. The solvents have been compensated. Bottom: Infrared spectrum of O,O-diethyl phosphorochloridothioate in 10 wt./vol. % in CCl_4 solution and in 10 and 2 wt./vol. % CS_2 solutions (1333–450 cm^{-1}) in 0.1-mm KBr cells (32).

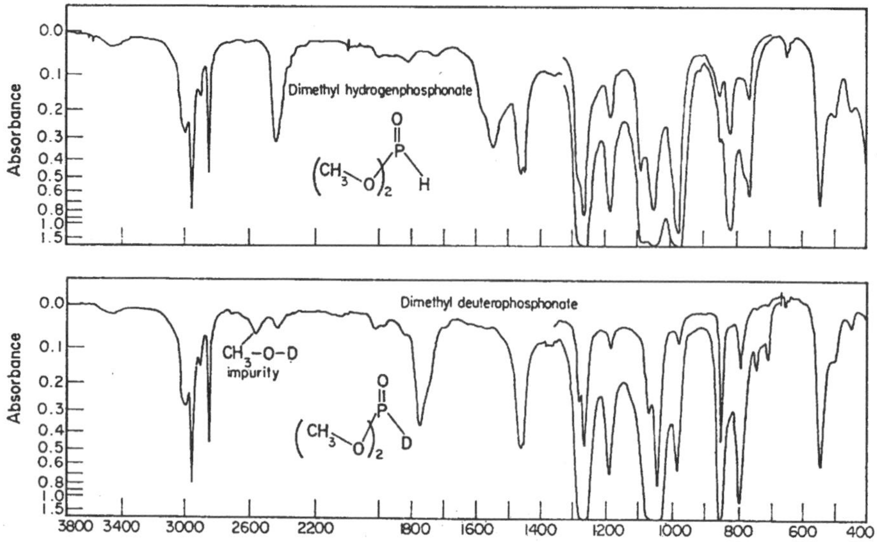

FIGURE 8.22 Top: Infrared spectrum of dimethyl hydrogenphosphonate in 10 wt./vol. % in CCl_4 solution (3800–1333 cm^{-1}) and in 10 and 2 wt./vol. % CS_2 solutions in 0.1-mm KBr cells. The solvents have not been compensated. Bottom: Infrared spectrum of dimethyl deuterophosphonate in 10% wt./vol. CCl_4 solution (3800–1333 cm^{-1}) and in 10 and 2 wt./vol. % solutions in CS_2 solution (1333–400 cm^{-1}). The solvents have been compensated (32).

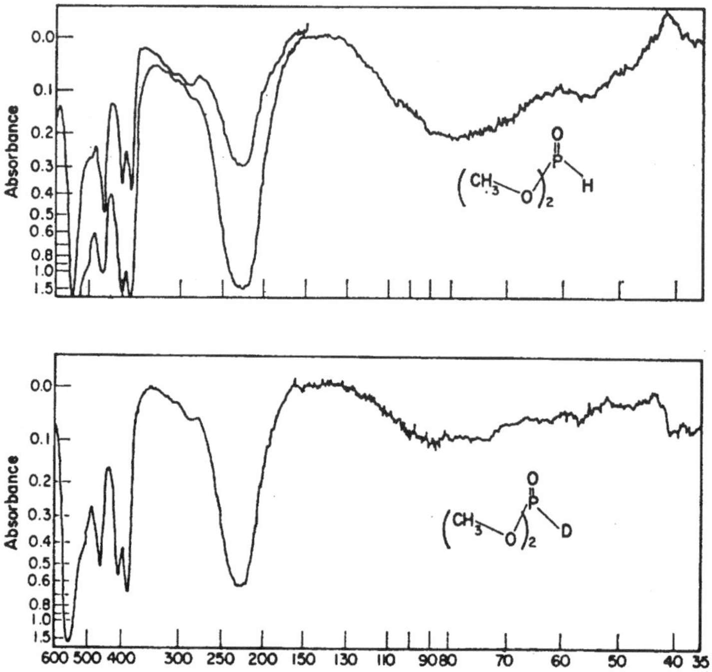

FIGURE 8.23 Top: Infrared spectrum of dimethyl hydrogenphosphonate saturated in hexane solution (600–35 cm^{-1}) in a 2-mm polyethylene cell. Bottom: Infrared spectrum of dimethyl deuterophosphonate saturated in hexane solution (600–35 cm^{-1}) in a 2-mm polyethylene cell (32).

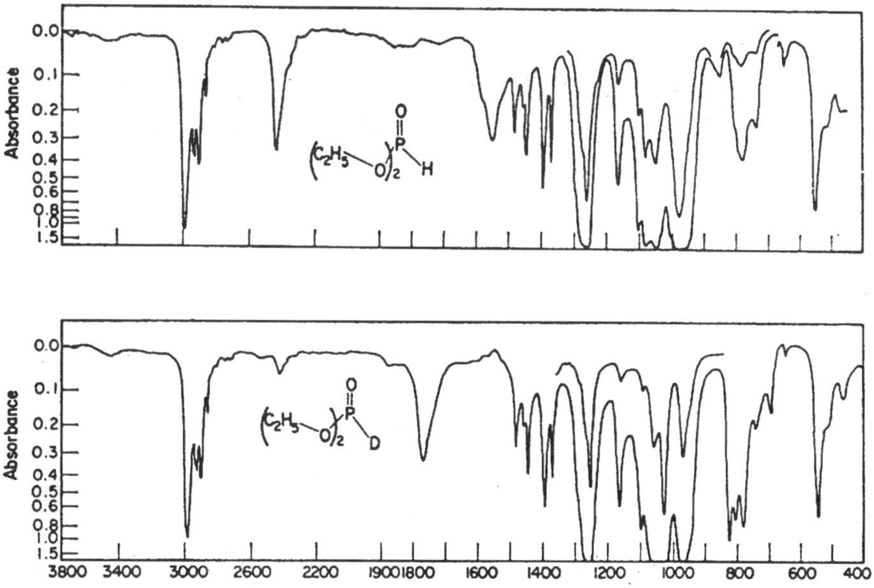

FIGURE 8.24 Top: Infrared spectrum of diethyl hydrogenphosphonate in 10 wt./vol. % in CCl_4 solution (3800–1333 cm^{-1}) and in 10 and 2 wt./vol. % in CS_2 solutions (1333–400 cm^{-1}) in 0.1-mm KBr cells. The solvents have not been compensated. Bottom: Infrared spectrum of diethyl deuterophosphonate in 10% wt./vol. % in CCl_4 solution (3800–1333 cm^{-1}) and 10 and 2 wt./vol. % CS_2 solutions (1333–400 cm^{-1}) in 0.1-mm KBr cells. The solvents have been compensated (32).

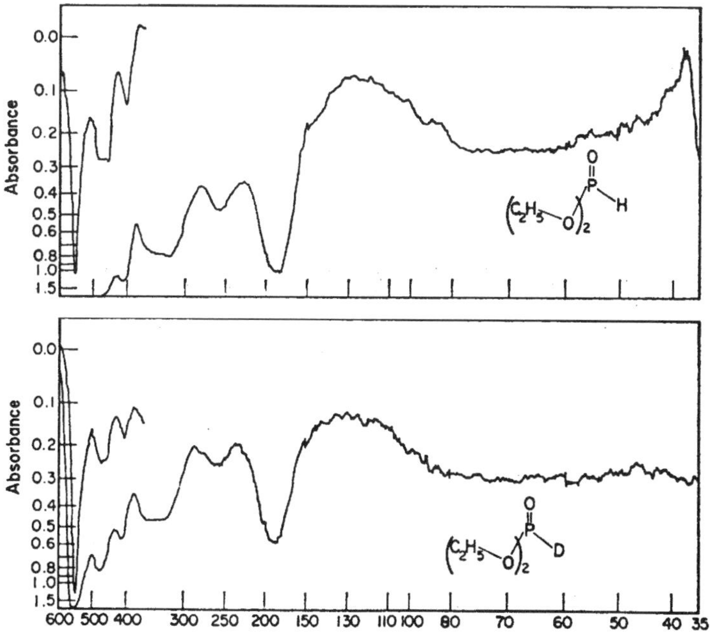

FIGURE 8.25 Top: Infrared spectrum for diethyl hydrogenphosphonate in 0.25 and 5 wt./vol. % hexane solutions (600–35 cm^{-1}) in a 2-mm polyethylene cell. Bottom: Infrared spectrum for diethyl deuterophosphonate in 10 wt./vol. % hexane solution (600–35 cm^{-1}) in a 1-mm polyethylene cell.

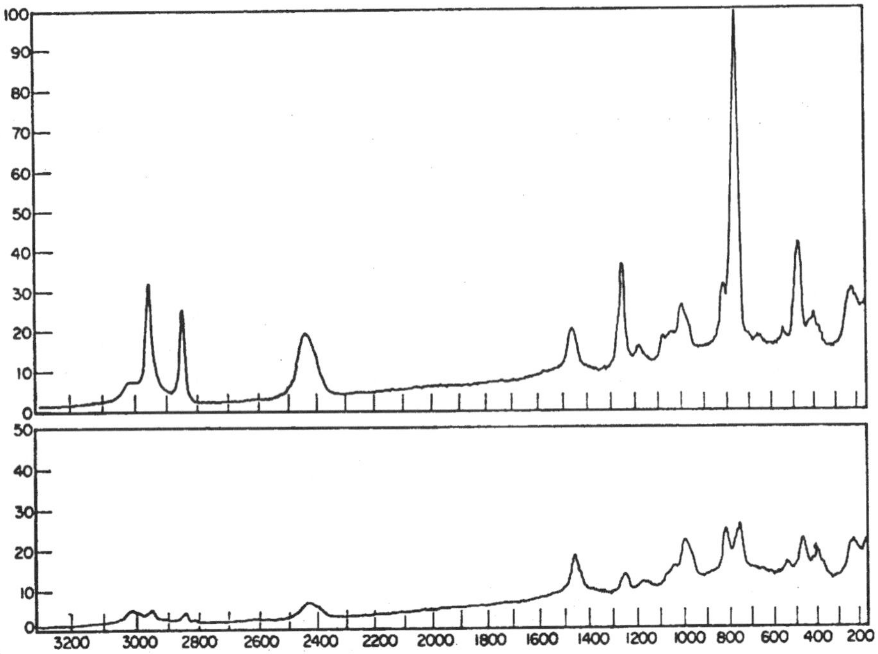

FIGURE 8.26 Top: Raman liquid-phase spectrum of dimethyl hydrogenphosphonate in a 2.5-ml multipass cell, gain 7, spectral slit-width 0.4 cm^{-1}. Bottom: Same as top but with the plane of polarization of the incident beam rotated through 90° (32).

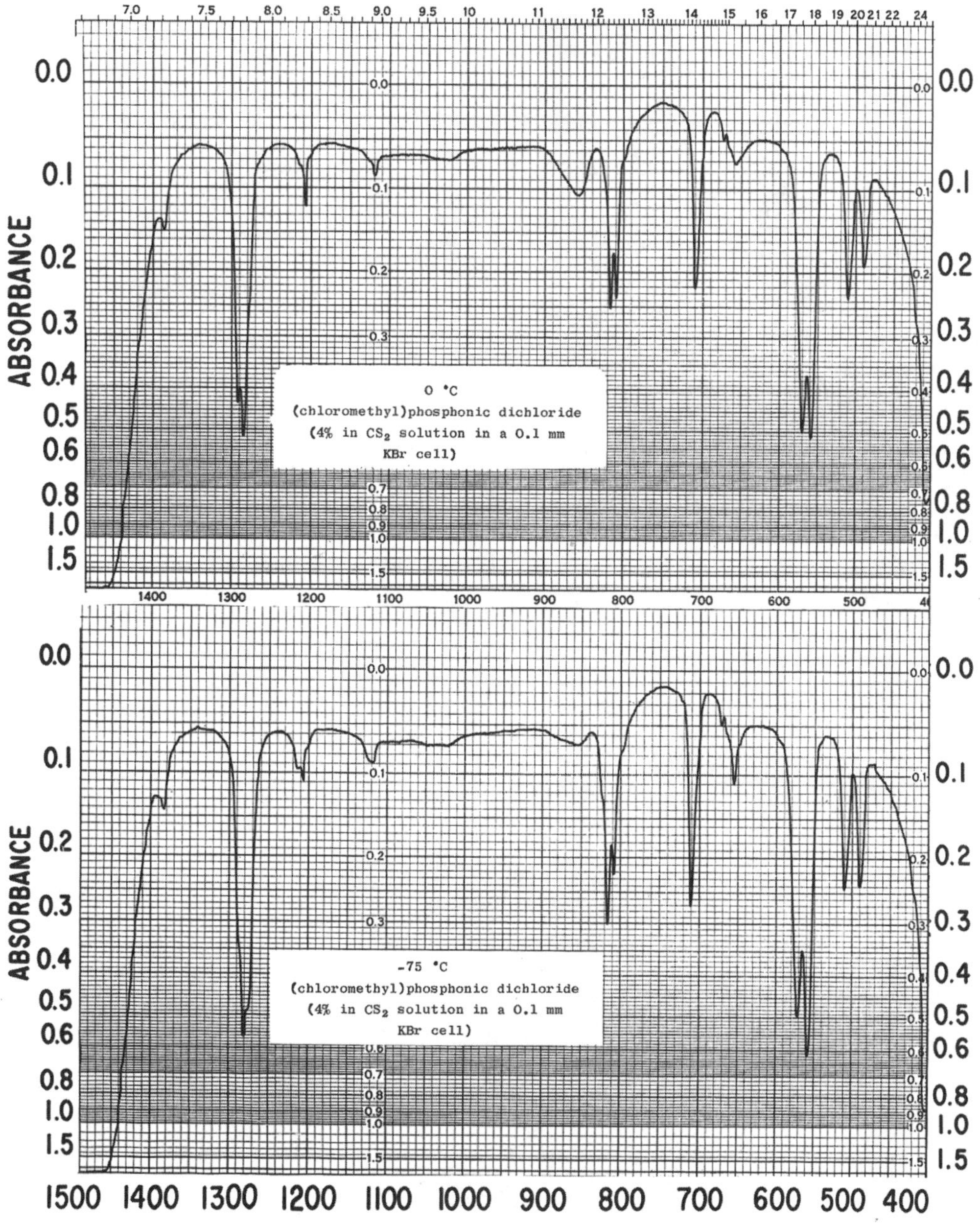

FIGURE 8.27 Top: Infrared spectrum of (chloromethyl) phosphonic dichloride in 4 wt./vol. % CS_2 solution (1500–400 cm^{-1}) at 0 °C in a 0.1-mm KBr cell. Bottom: Same as above except at −75 °C.

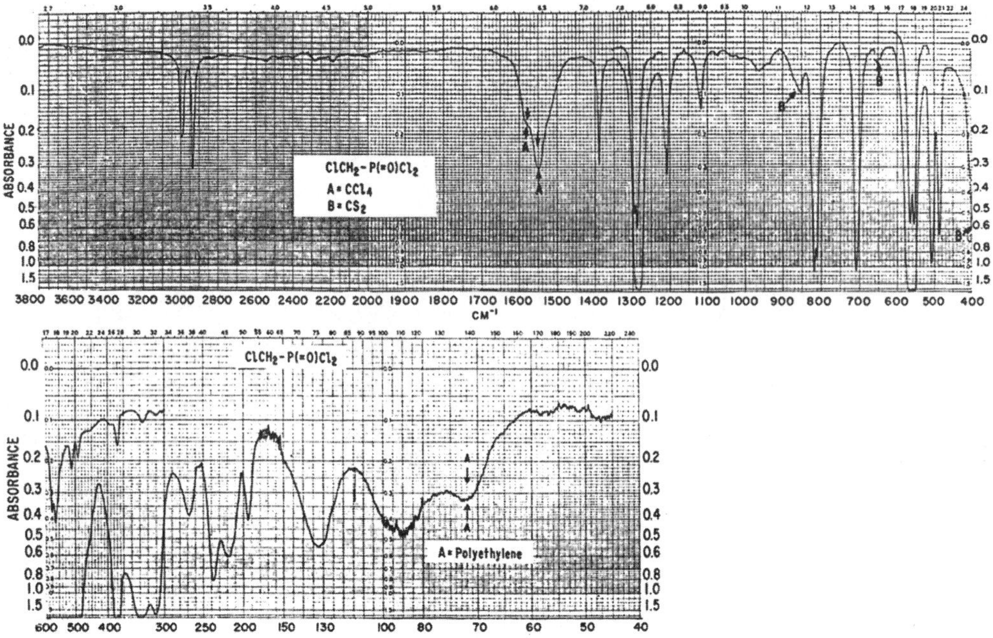

FIGURE 8.28 Top: Infrared spectrum of (chloromethyl) phosphonic dichloride in 10 wt./vol. % in CCl_4 solution (3800–1333 cm^{-1}) and 10 wt./vol. % in CS_2 solution (1333–400 cm^{-1}) in 0.1-mm KBr cells. Bottom: Infrared spectrum of (chloromethyl) phosphonic dichloride in 20 wt./vol. % hexane solution (600–45 cm^{-1}) in a 1-mm polyethylene cell (17).

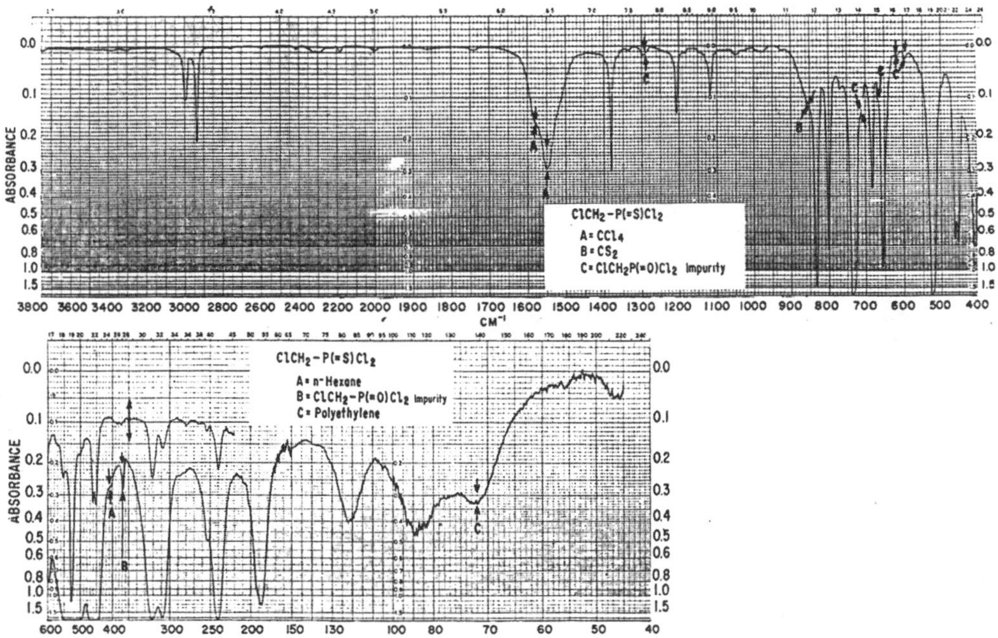

FIGURE 8.29 Top: Infrared spectrum of (chloromethyl) phosphonothioic dichloride in 10 wt./vol. % CCl_4 solution (3800–1333 cm^{-1}) and in 10 wt./vol. % CS_2 solution (1333–400 cm^{-1}) in 0.1-mm KBr cells. Bottom: Infrared spectrum of (chloromethyl) phosphonothioic dichloride in 20 wt./vol. % hexane solution (600–45 cm^{-1}) in a 1-mm polyethylene cell (17).

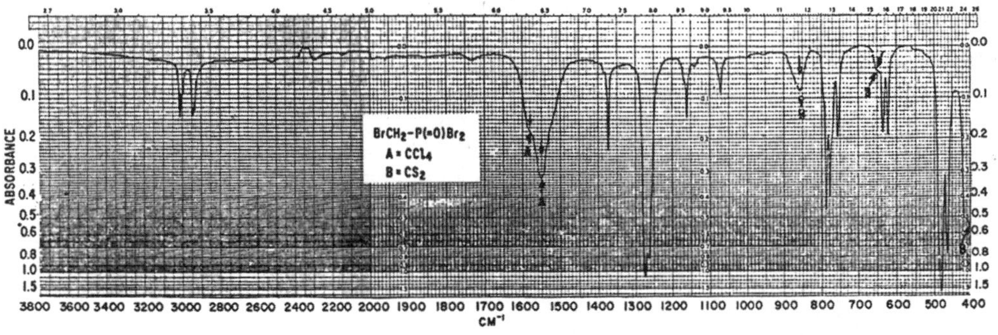

FIGURE 8.30 Infrared spectrum of (bromomethyl) phosphonic dibromide in 10 wt./vol. % CCl_4 solution (3800–1333 cm^{-1}) and 10 wt./vol. % in CS_2 solution (1333–400 cm^{-1}) in 0.1-mm KBr cells (17).

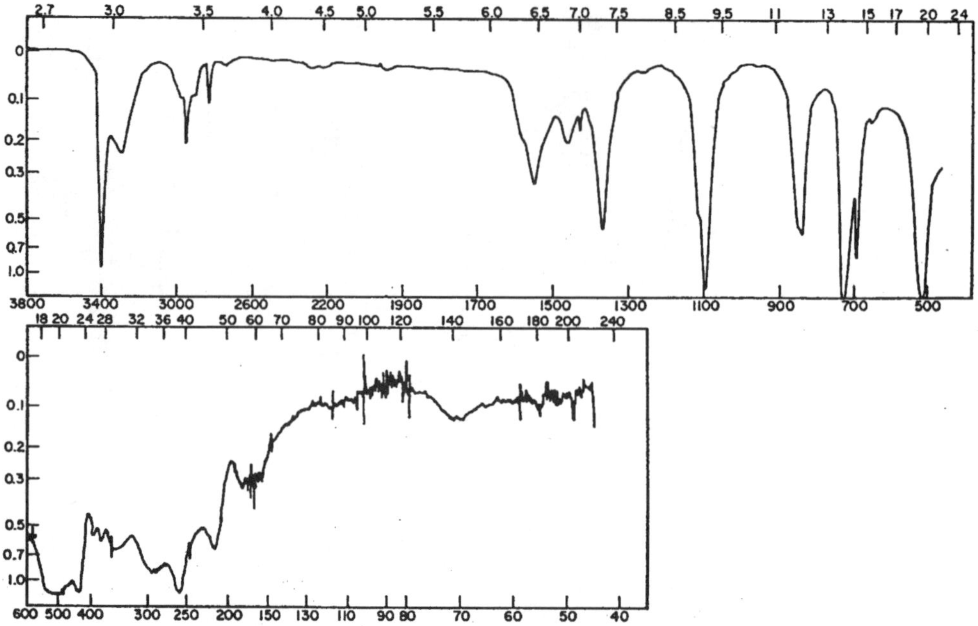

FIGURE 8.31 Top: Infrared spectrum of N-methyl phosphoramidodichloridothioate 10 wt./vol. % in CCl_4 solution (3800–1333 cm^{-1}) and 10 wt./vol. % in CS_2 solution (1333–45 cm^{-1}) in 0.1-mm KBr cells. The solvents have not been compensated. Bottom: Infrared spectrum of N-methyl phosphoramidodichloridothioate in 10 wt./vol. hexane solution (600–45 cm^{-1}) in a 1-mm polyethylene cell (24).

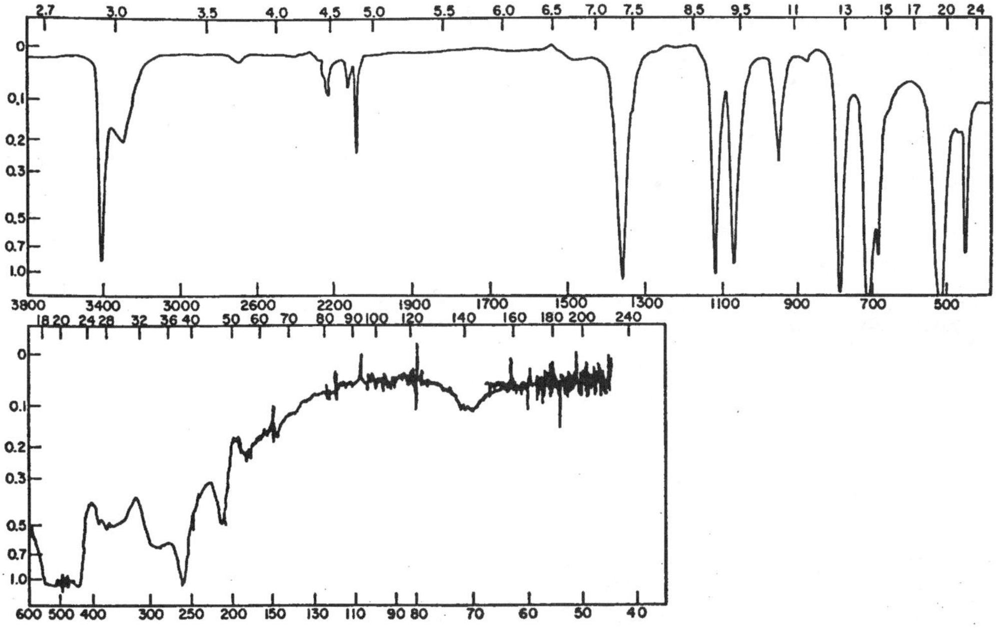

FIGURE 8.32 Top: Infrared spectrum of N-methyl-d_3 phosphoramidodichloridothioate in 10 wt./vol. % in CCl_4 (3800–1333 cm^{-1}) and 10 wt./vol. % in CS_2 solution (1333–45 cm^{-1}) in 0.1-mm KBr cells. Bottom:Infrared spectrum of N-methyl-d_3 phosphoramidodichloridothioate in 10 wt./vol. % in hexane solution (600–45 cm^{-1}) in a 1-mm polyethylene cell (24).

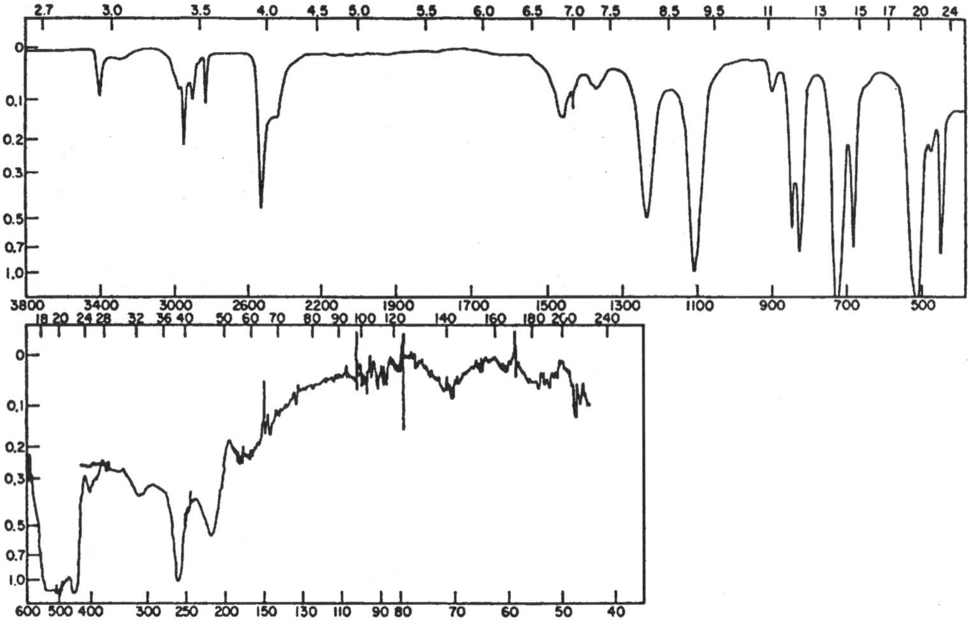

FIGURE 8.33 Top: Infrared spectrum of N—D, N-methyl phosphoramidodichloridothioate in 10 wt./vol. % CCl_4 solution (3800–1333 cm^{-1}) and in 10 wt./vol. % in CS_2 solution (1333–45 cm^{-1}) in 0.1-KBr cells. The solvents have been compensated. Bottom: Infrared spectrum of N—D, N-methyl phosphoramidodichloridothioate in 10 wt./vol. % hexane solution in a 1-mm polyethylene cell. Both samples contain N-methyl phosphoramidodichloridothioate as an impurity (24).

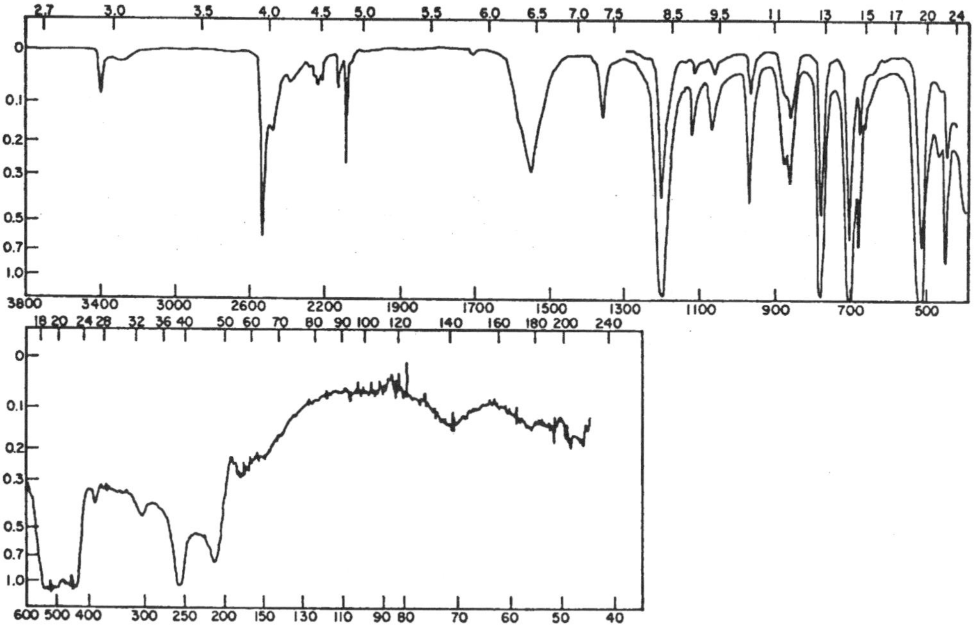

FIGURE 8.34 Top: Infrared spectrum for N—D, N-methyl-d$_3$ phosphoramidodichloridothioate in 10 wt./vol. % CCl_4 solution (3800–1333 cm^{-1}) and 10 and 2 wt./vol. % in CS_2 solutions (1333–400 cm^{-1}) in 0.1-mm KBr cells. The solvent bands are compensated. Bottom: Infrared spectrum of N—D, N-methyl-d$_3$ phosphoramidodichloridothioate in 10 wt./vol. % hexane solution in a 1-mm polyethylene cell. This sample contains N-methyl-d$_3$ phosphoramidodichloridothioate as an impurity (24).

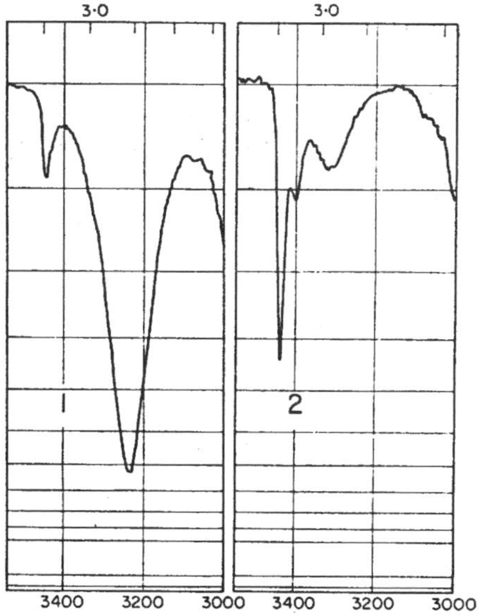

FIGURE 8.35 Infrared spectrum 1 is for O-alkyl O-aryl N-methylphosphoramidate in 10 wt./vol. % CCl_4 solution (3500–3000 cm^{-1}) in a 0.1-mm NaCl cell. Infrared spectrum 2 is for O-alkyl O-aryl N-methylphosphoramidothioate in 10 wt./vol. % in CCl_4 solution (3500–300 cm^{-1}) in a 0.1-mm NaCl cell (42).

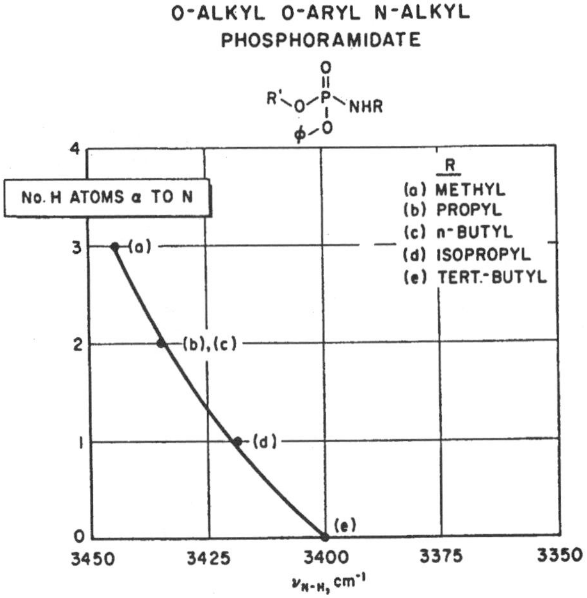

FIGURE 8.36 A plot of N—H stretching frequencies of O-alkyl O-aryl N-alkylphosphoramidates in the region 3450–3350 cm^{-1} vs an arbitrary assignment of one for each proton joined to the N-α-carbon atom (42).

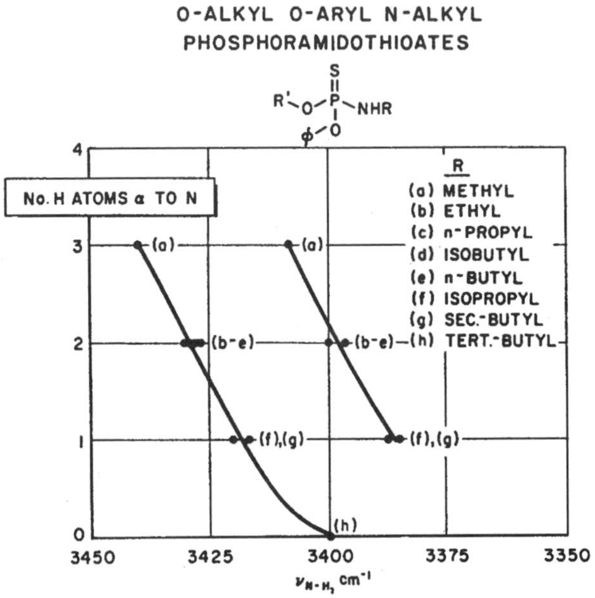

FIGURE 8.37 Plots of cis and trans N—H stretching frequencies of O-alkyl O-aryl N-alkylphosphoramidothioates in the region 3450–3350 cm^{-1} vs an arbitrary assignment of one for every proton joined to the N-α-carbon atom (42).

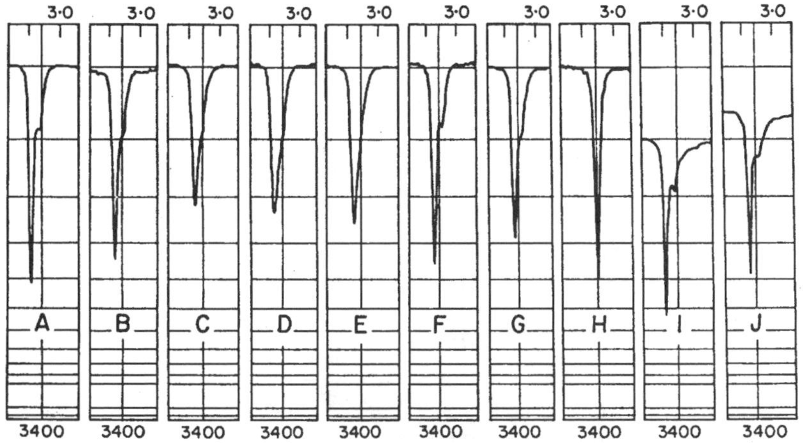

FIGURE 8.38 Infrared spectra of the cis and trans N—H stretching absorption bands of O-alkyl O-aryl N-alkylphosphoramidothioates in 0.01 molar or less in CCl_4 solutions (3450–3350 cm^{-1}) in a 14-mm NaCl cell. The N-alkyl group for A is methyl, B is ethyl, C is n-propyl, D is isobutyl, E is n-butyl, F is isopropyl, G is sec-butyl, and H is tert-butyl. Spectra I and J are O,O-dialkyl N-methylphosphoramidothioate and O,O-dialkyl N-propylphosphoramidothioate, respectively (42).

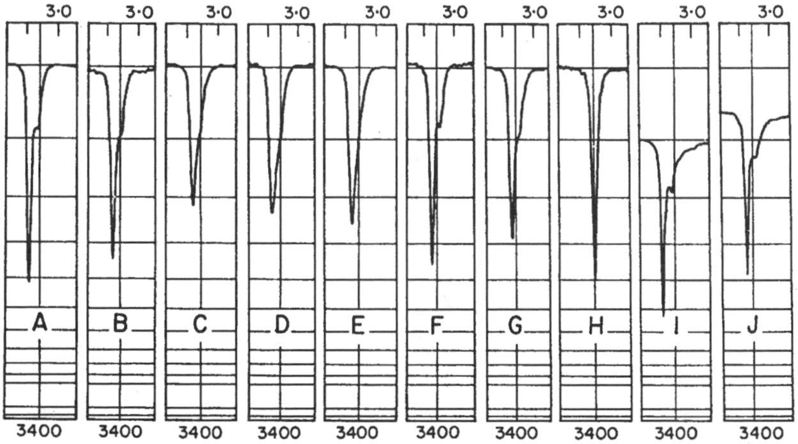

FIGURE 8.39 Infrared spectra of the N—H stretching absorption bands of O-alkyl N,N′-dialkyl phosphorodiamidothioates. The N,N′-dialkyl group for A is dimethyl, B is diethyl, C is dipropyl, D is dibutyl, E is diisopropyl, F is di-sec-butyl, G is dibenzyl. The H and I are IR spectra for the N—H stretching bands of O-aryl N,N′-diisopropylphosphordiamidothioate and O-aryl N-isopropyl, N-methylphosphorodiamidothioate, respectively. Spectra A through J were recorded for 0.01 molar or less in CCl_4 solutions using 3-mm KBr cells. Spectra D and J are for O-aryl N,N′-dibutylphosphorodiamidothioate in 0.01 molar CCl_4 solution in a 3-mm cell and 10 wt./vol. % in CCl_4 solution in a 0.1-mm KBr cell (42).

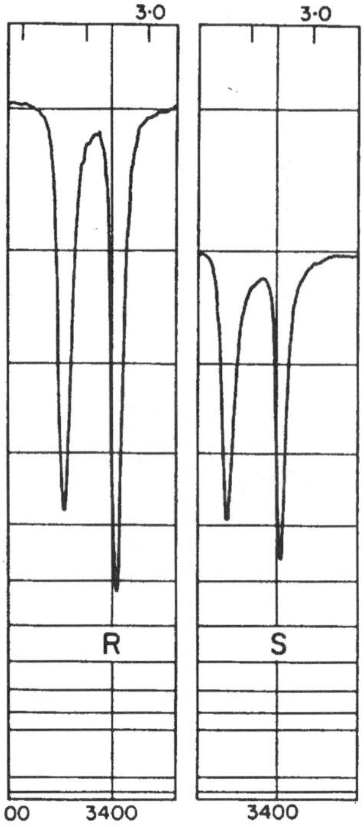

FIGURE 8.40 Infrared spectrum R gives the N—H stretching absorption bands for O-alkyl O-aryl phosphoramidothioate in 0.01 molar CCl_4 solution in a 3-mm NaCl cell. Infrared spectrum S gives O,O-dialkyl phosphoramidothioate in 0.01 molar CCl_4 solution in a 3-mm NaCl cell (42).

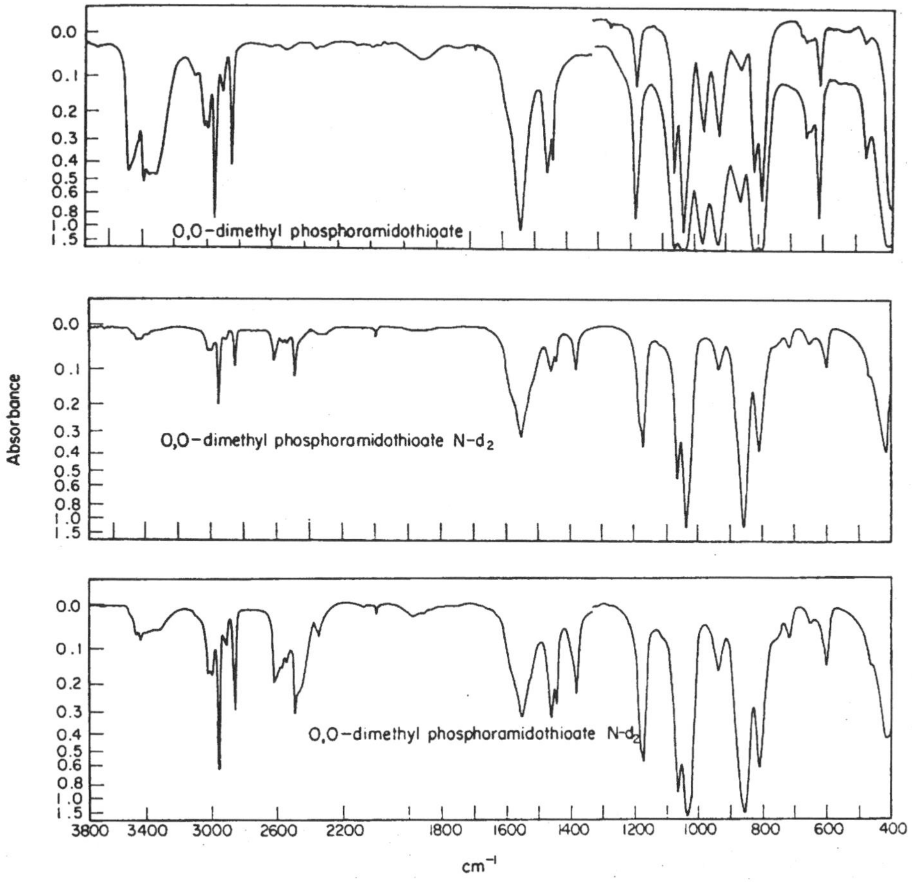

FIGURE 8.41 Top: Infrared spectrum of O,O-diethyl methylphosphoramidothioate in 10 wt./vol. % CCl_4 solution (3800–1333 cm^{-1}) and 10 and 2 wt./vol. % CS_2 solutions (1333–400 cm^{-1}) in 0.1-mm KBr cells. Middle: Infrared spectrum of O,O-diethyl N-methyl, N—D-phosphoramidothioate in 2 wt./vol. % in CCl_4 solution (3800–1333 cm^{-1}) and in 2 wt./vol. % in CS_2 solution (1333–400 cm^{-1}) in 0.1-mm KBr cells. Bottom: Infrared spectrum of O,O-diethyl N-methyl, N-D-phosphoramidothioate 10 wt./vol. % in CCl_4 solution (3800–1333 cm^{-1}) and in 10 wt./vol. % CS_2 solution (1333–400 cm^{-1}) in 0.1-mm KBr cells (37).

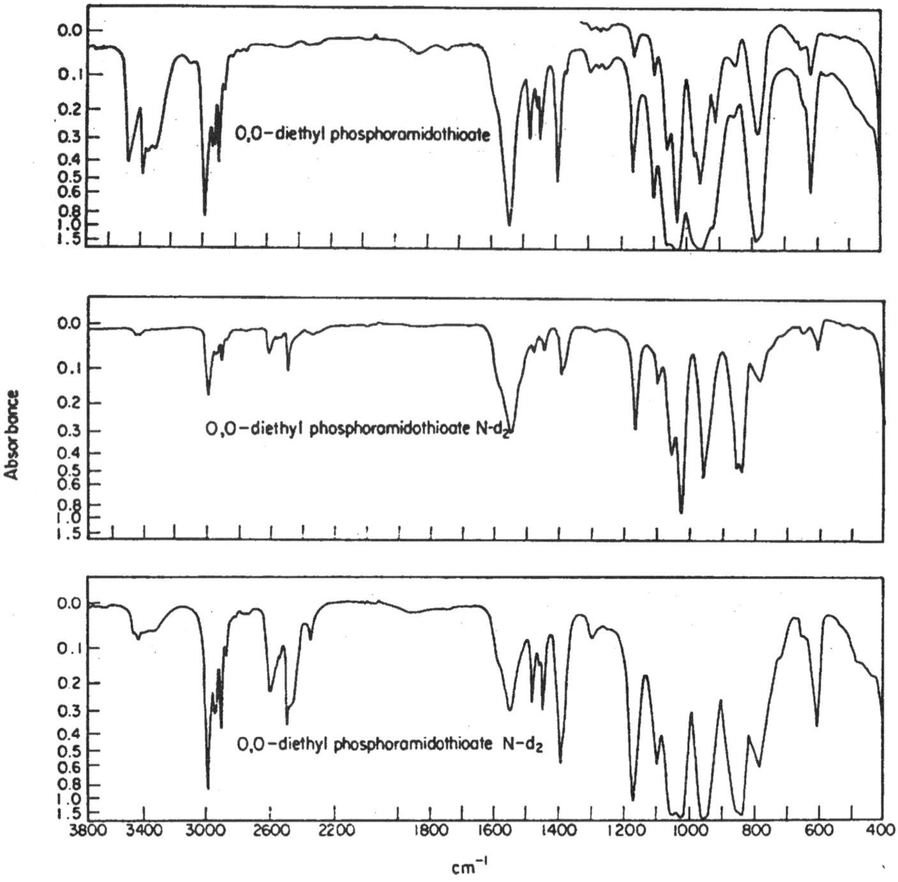

FIGURE 8.42 Top: Infrared spectrum of O,O-diethyl phosphoramidothioate in 10 wt./vol. % in CCl_4 solution (3800–1333 cm^{-1}) and 2 and 10 wt./vol. % in CS_2 solutions (1333–400 cm^{-1}) in 0.1-mm KBr cells. Middle: Infrared spectrum of O,O-diethyl phosphoramidothioate-N-D_2 in 2 wt./vol. % in CCl_4 solution (3800–1333 cm^{-1}) and in 2 wt./vol. % CS_2 solution (1333–400 cm^{-1}) in 0.1-mm KBr cells. Bottom: Infrared spectrum of O,O-diethyl phosphoramidothioate-N—D_2 10 wt./vol. % in CCl_4 solution (3800–1333 cm^{-1}) and 10 wt./vol. % in CS_2 solution (1333–400 cm^{-1}) in 0.1-mm KBr cells (37).

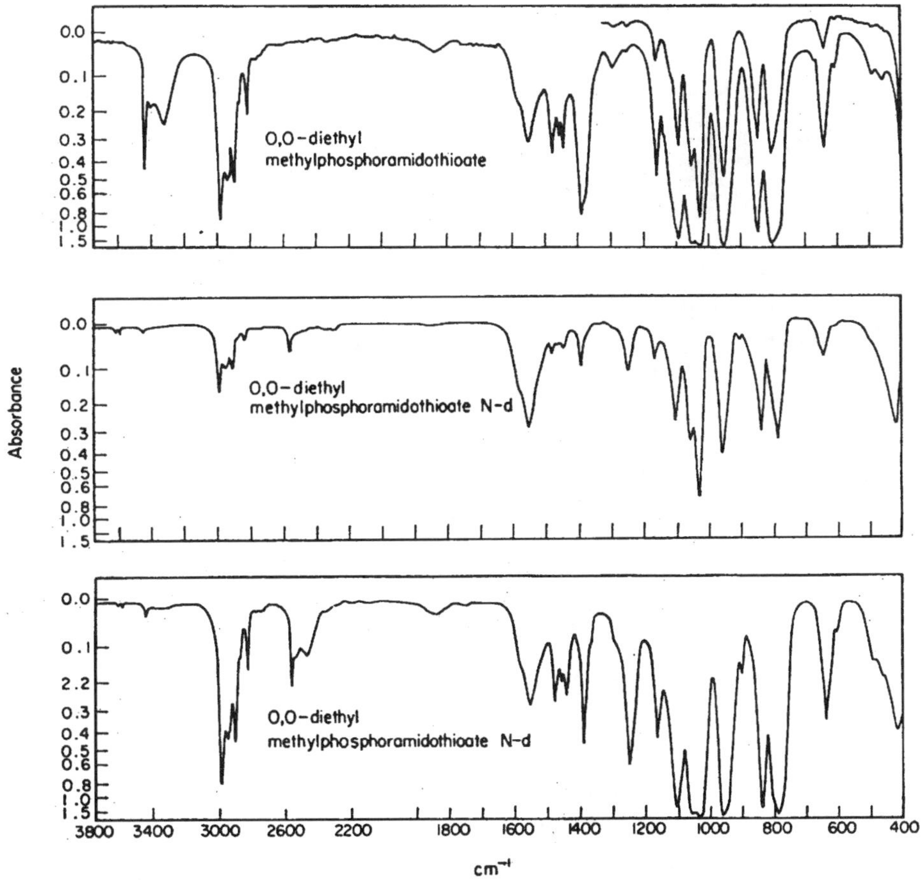

FIGURE 8.43 Top: Infrared spectrum of O,O-diethyl N-methylphosphoramidothioate 10 wt./vol. % in CCl_4 solution (3800–1333 cm^{-1}) and 10 and 2 wt./vol. % in CS_2 solutions (1333–400 cm^{-1}) in 0.1-mm KBr cells. Middle: Infrared spectrum of O,O-diethyl N-methyl, N–D phosphoramidothioate in 2 wt./vol. % CCl_4 solution (3800–1333 cm^{-1}) and 2 wt./vol. % in CS_2 solution (1333–400 cm^{-1}) in 0.1-mm KBr cells. Bottom: Infrared spectrum of O,O-diethyl N-methyl, N–D-phosphoramidothioate in 10 wt./vol. % CCl_4 solution (1333–400 cm^{-1}) in 0.1-mm KBr cells. The solvents are not compensated (37).

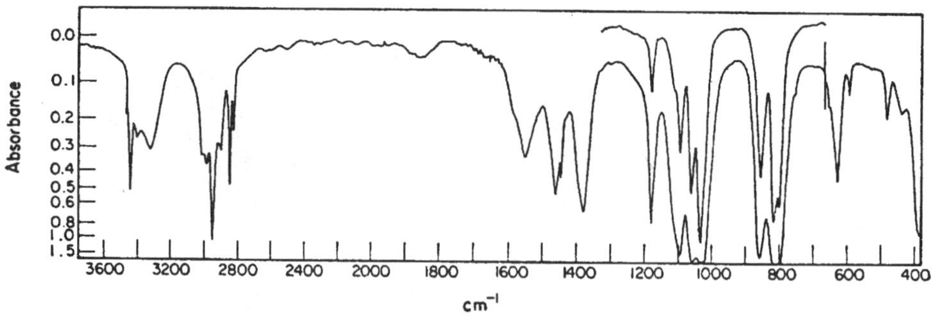

FIGURE 8.44 Infrared spectrum of O,O-dimethyl N-methylphosphoramidothioate in 10 wt./vol. % in CCl_4 solution (3800–1333 cm^{-1}) and 10 and 2 wt./vol. % in CS_2 solutions (1333–400 cm^{-1}) in 0.1-mm KBr cells (37).

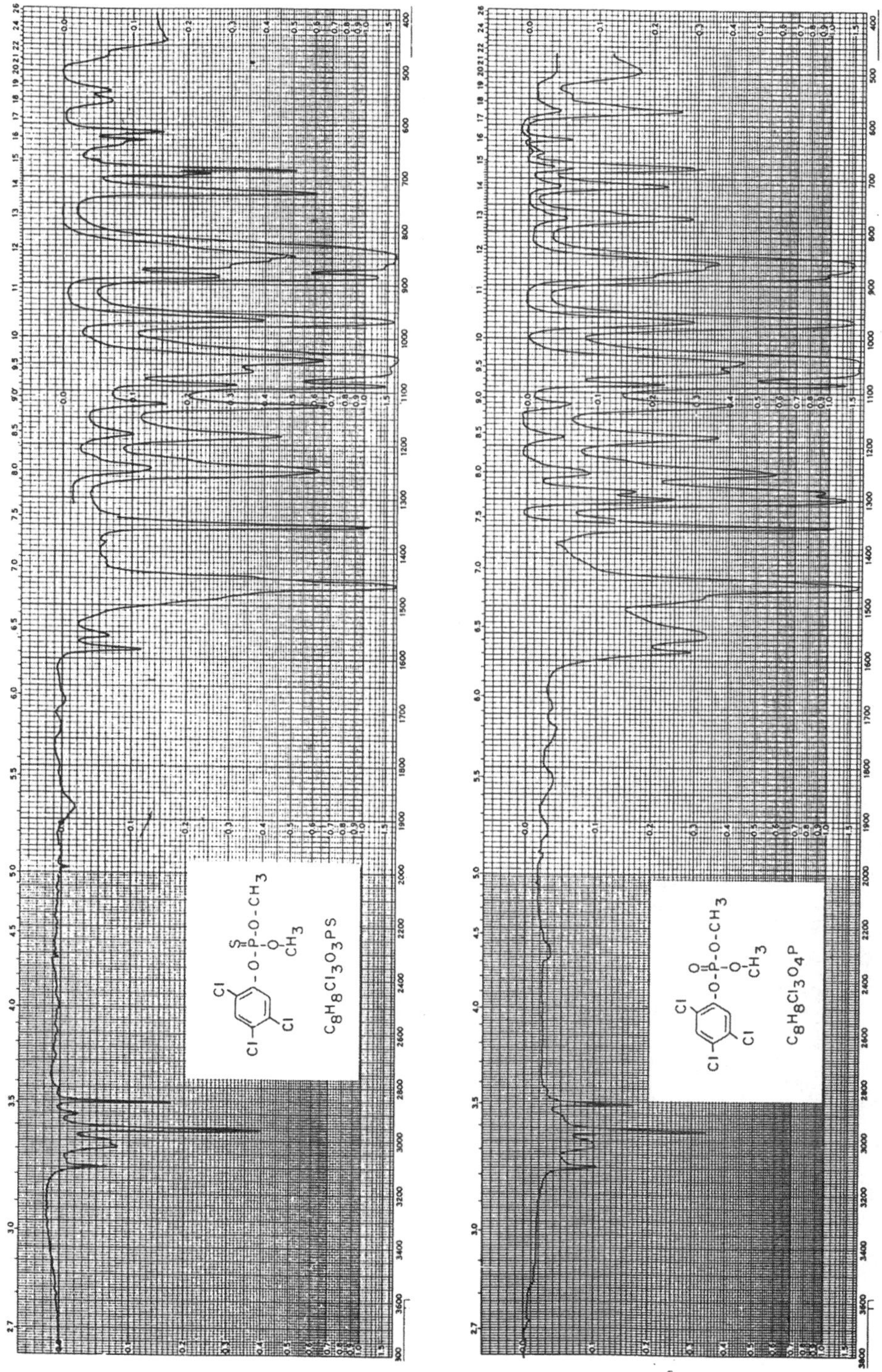

FIGURE 8.45 Top: Infrared spectrum of O,O-dimethyl O-(2,4,5-trichlorophenyl) phosphorothioate 10 wt./vol. % in CCl_4 solution (3800–1333 cm^{-1}) and 10 and 2 wt./vol. % in CS_2 solutions (1333–400 cm^{-1}) in 0.1-mm KBr cells. Bottom: Infrared spectrum of O,O-dimethyl O-(2,4,5-trichlorophenyl) phosphate in 10 wt./vol. % in CCl_4 solution (3800–1333 cm^{-1}) in 0.1-mm KBr cells. The solvents are not compensated (43).

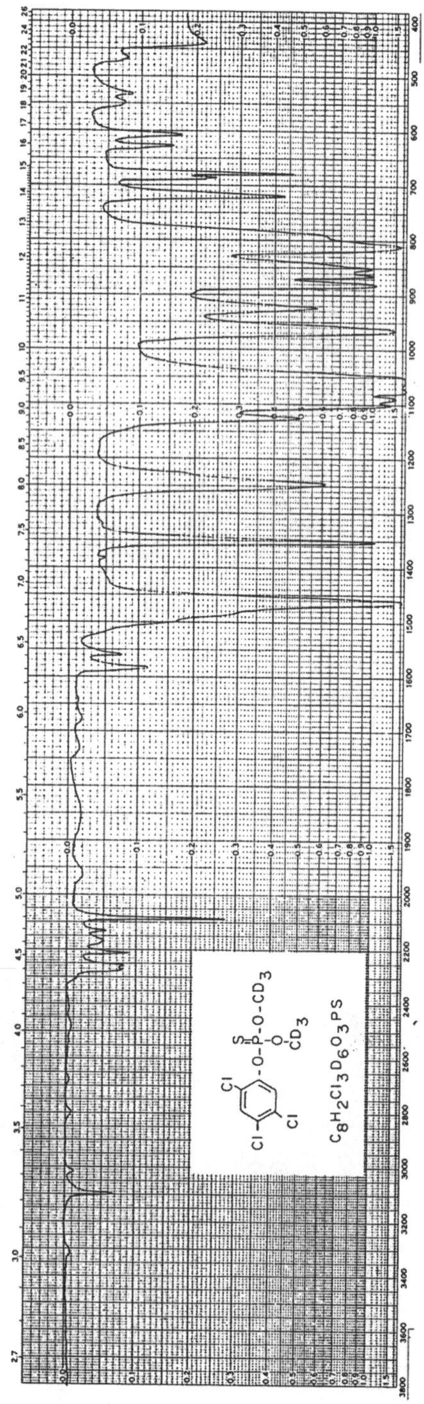

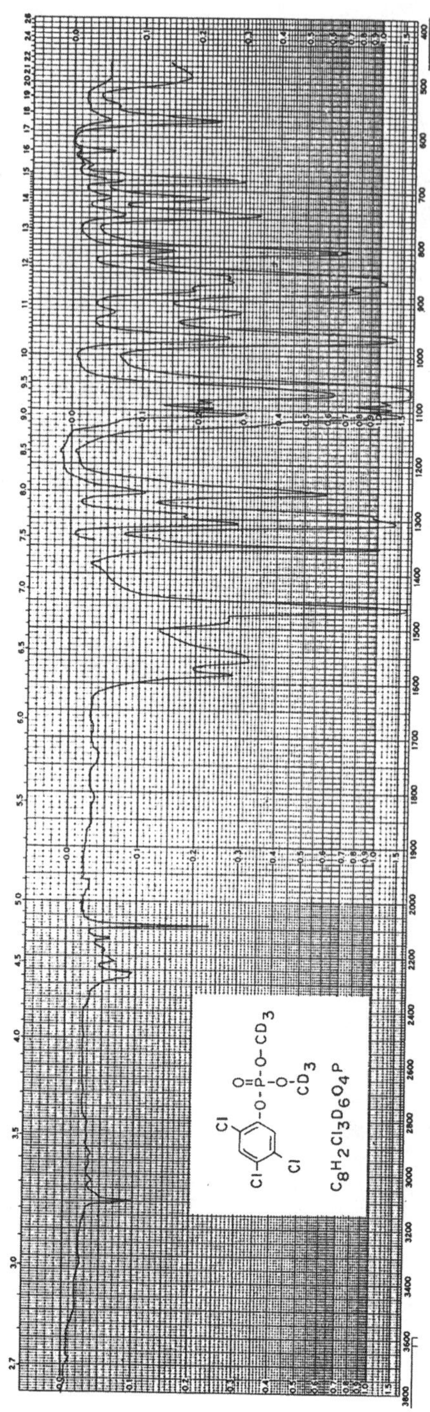

FIGURE 8.46 Top: Infrared spectrum for O,O-dimethyl-d_6 O-(2,4,5-trichlorophenyl) phosphorothioate in 10 wt./vol. % CCl_4 solution (3800–1333 cm^{-1}) and in 10 and 2 wt./vol. % CS_2 solution (1333–400 cm^{-1}) in 0.1-mm KBr cells. Bottom: Infrared spectrum for O,O-dimethyl-d_6 O-(2,4,5-trichlorophenyl) phosphate in 10 wt./vol. % CS_2 solutions (1333–450 cm^{-1}) in 0.1-mm KBr cells. The solvents have not been compensated (37).

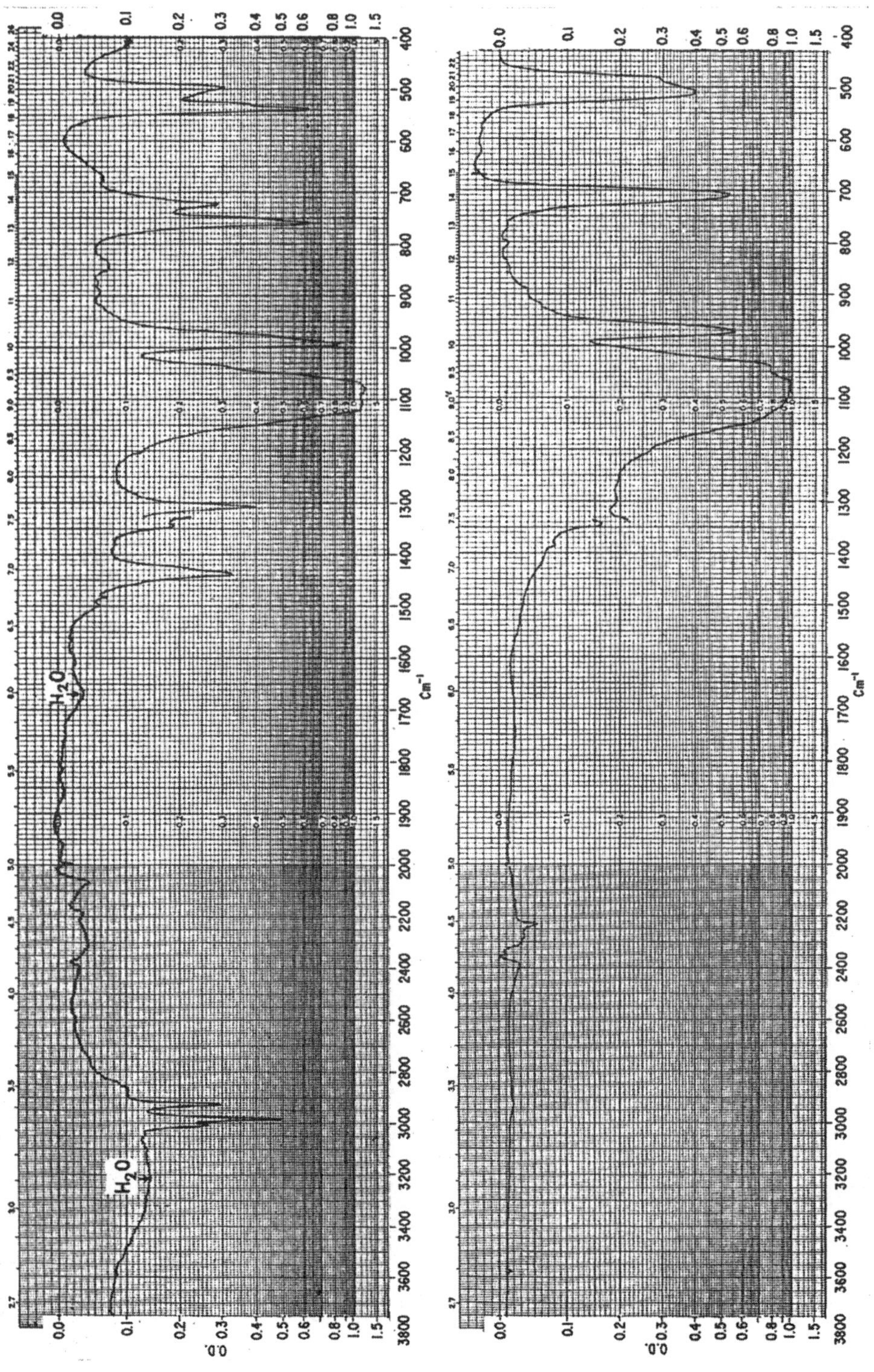

FIGURE 8.47 Top: Infrared spectrum of disodium methanephosphonate. Bottom: Infrared spectrum of disodium methane-d$_3$-phosphonate. In fluorolube oil mull (3800–1333 cm^{-1}) and in Nujol oil mull (1333–400 cm^{-1}) (40). These mulls were placed between KBr plates.

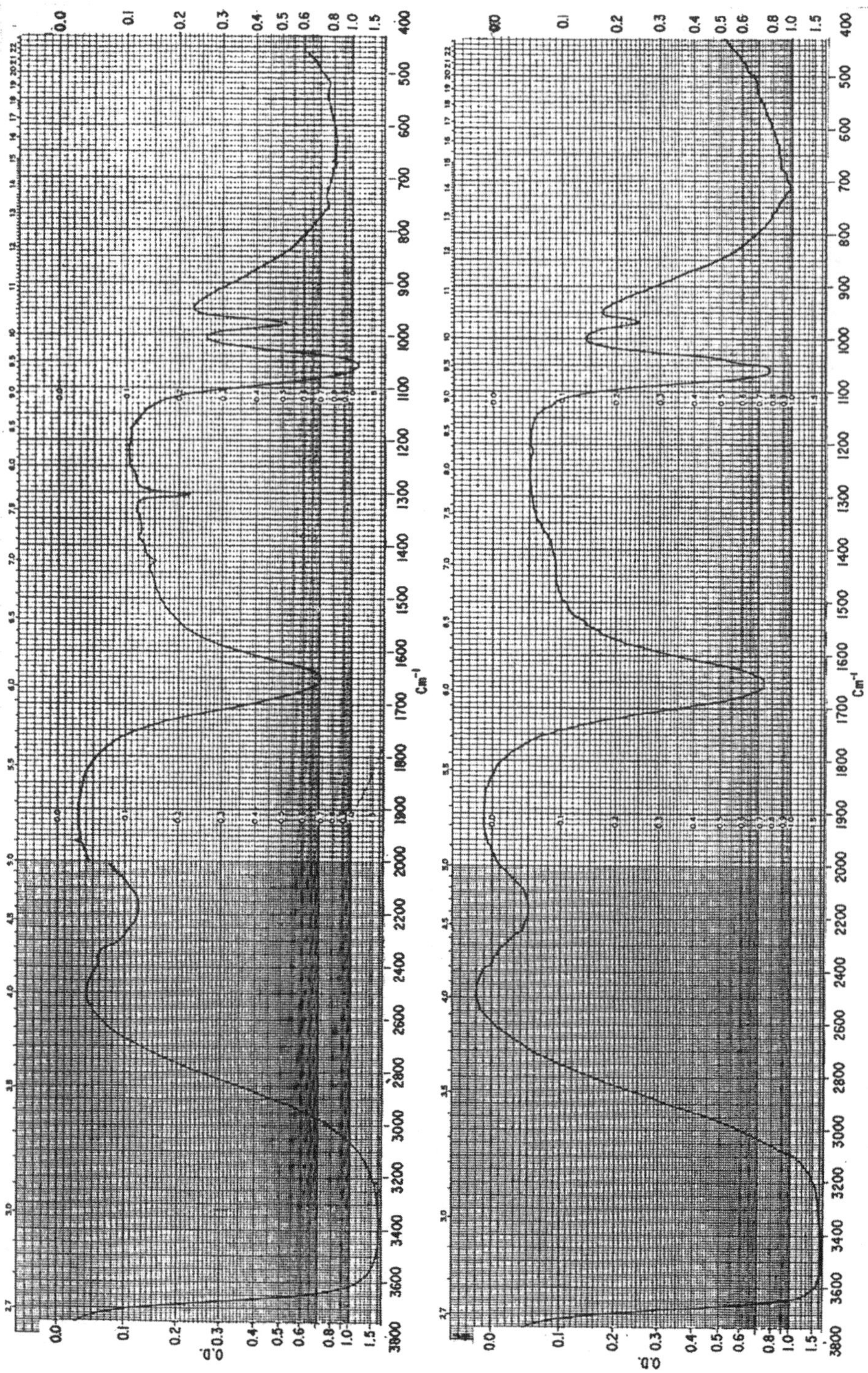

FIGURE 8.48 Top: Infrared spectrum of disodium methanephosphonate in water solution between AgCl plates. Bottom: Infrared spectrum of disodium methane-d_3-phosphonate in water solution between AgCl plates (40).

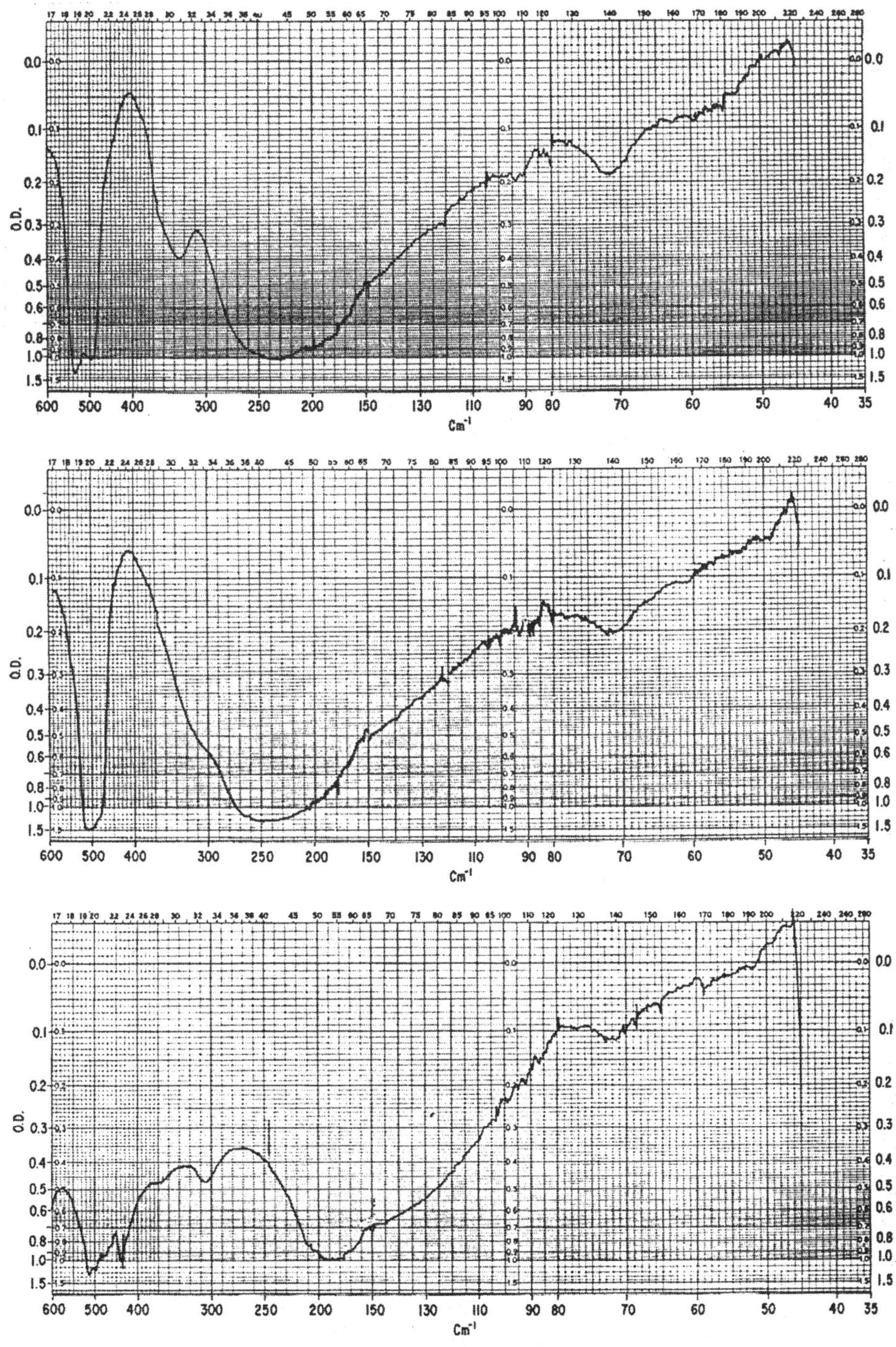

FIGURE 8.49 Top: Infrared spectrum of disodium methanephosphonate. Middle: Infrared spectrum of disodium methane-d_3-phosphonate. Bottom: Infrared spectrum of dipotassium methanephosphonate. These spectra were recorded from samples prepared as Nujol mulls between polyethylene film (600–45 cm^{-1}) (40).

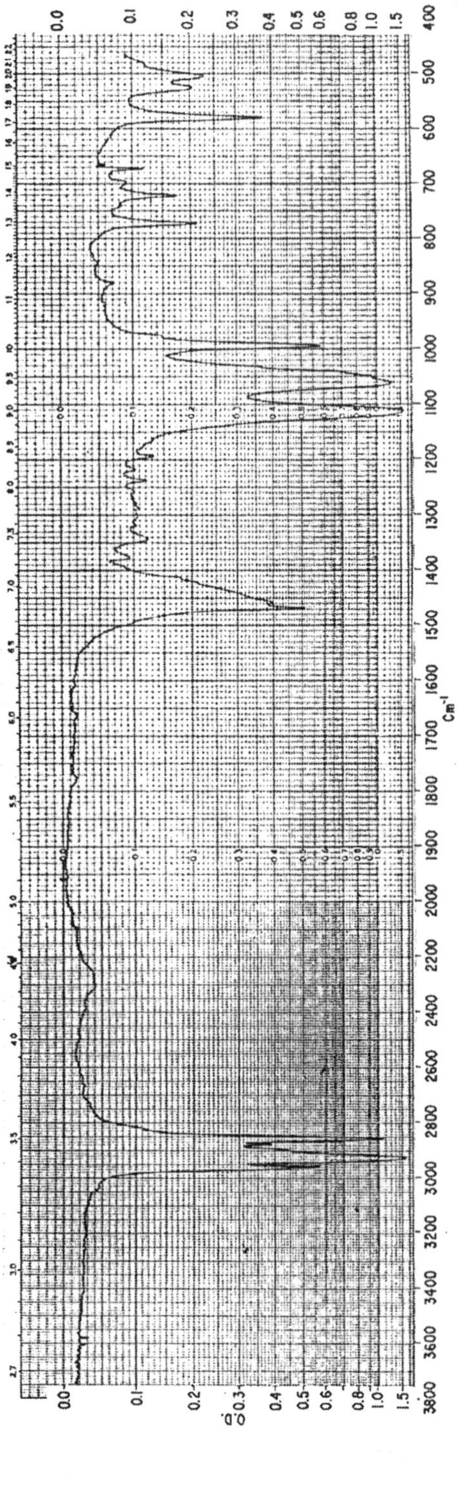

FIGURE 8.50 Infrared spectrum of disodium n-octadecanephosphonate prepared as a fluoroluble mull (3800–1333 cm^{-1}) prepared as a Nujol mull (1333–450 cm^{-1}) (40).

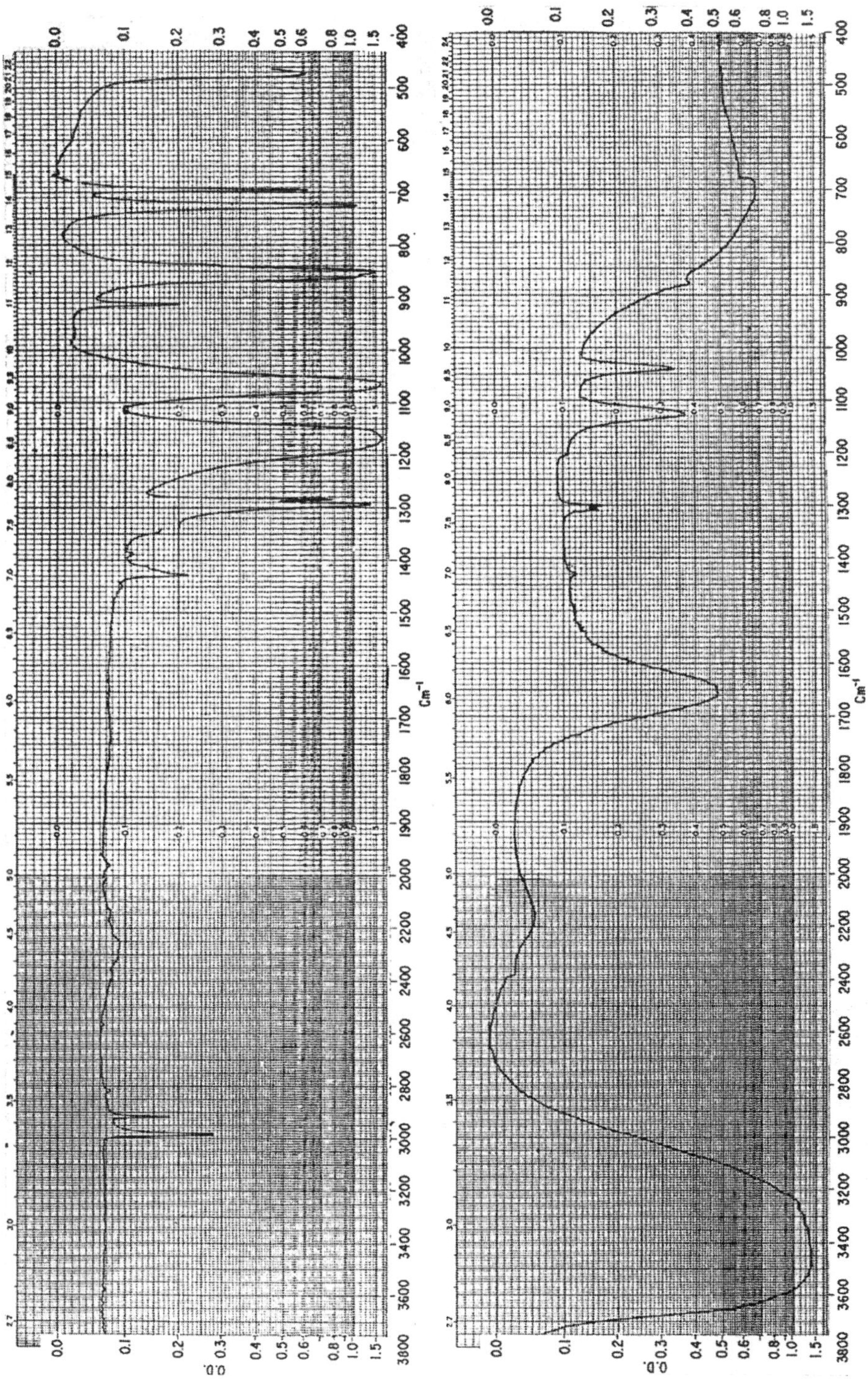

FIGURE 8.51 Top: Infrared spectrum of sodium dimethylphosphinate prepared as a fluorolube mull (3800–1333 cm^{-1}) and prepared as a Nujol mull (1333–400 cm^{-1}) between KBr plates. Bottom: Infrared spectrum of sodium dimethylphosphinate saturated in water solution between AgCl plates (41).

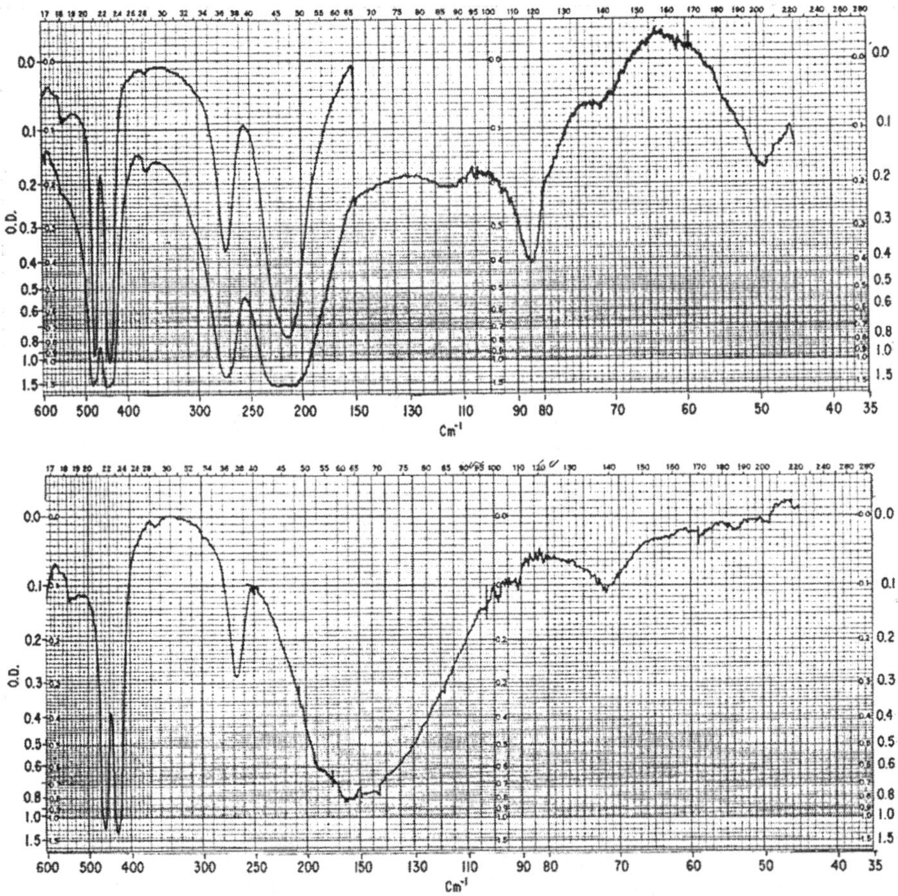

FIGURE 8.52 Top: Infrared spectrum of sodium dimethylphosphinate prepared as Nujol mull between polyethylene film (600–45 cm^{-1}). Bottom: Infrared spectrum of potassium dimethylphosphinate prepared as a Nujol mull between polyethylene film (600–45 cm^{-1}) (41).

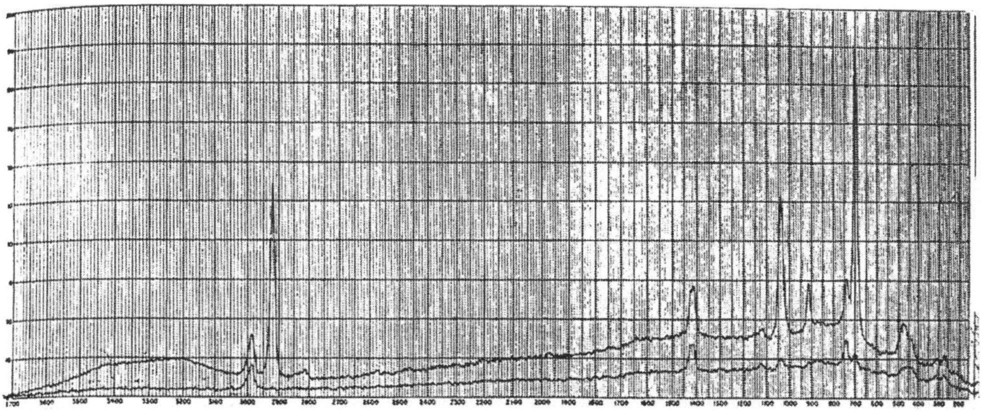

FIGURE 8.53 Top: Raman saturated water solution of sodium dimethylphosphinate using a 0.25-ml multipass cell, gain 13.4, spectral slit 10 cm^{-1}. Bottom: Same as above, except with the plane of polarization of the incident beam rotated 90° (41).

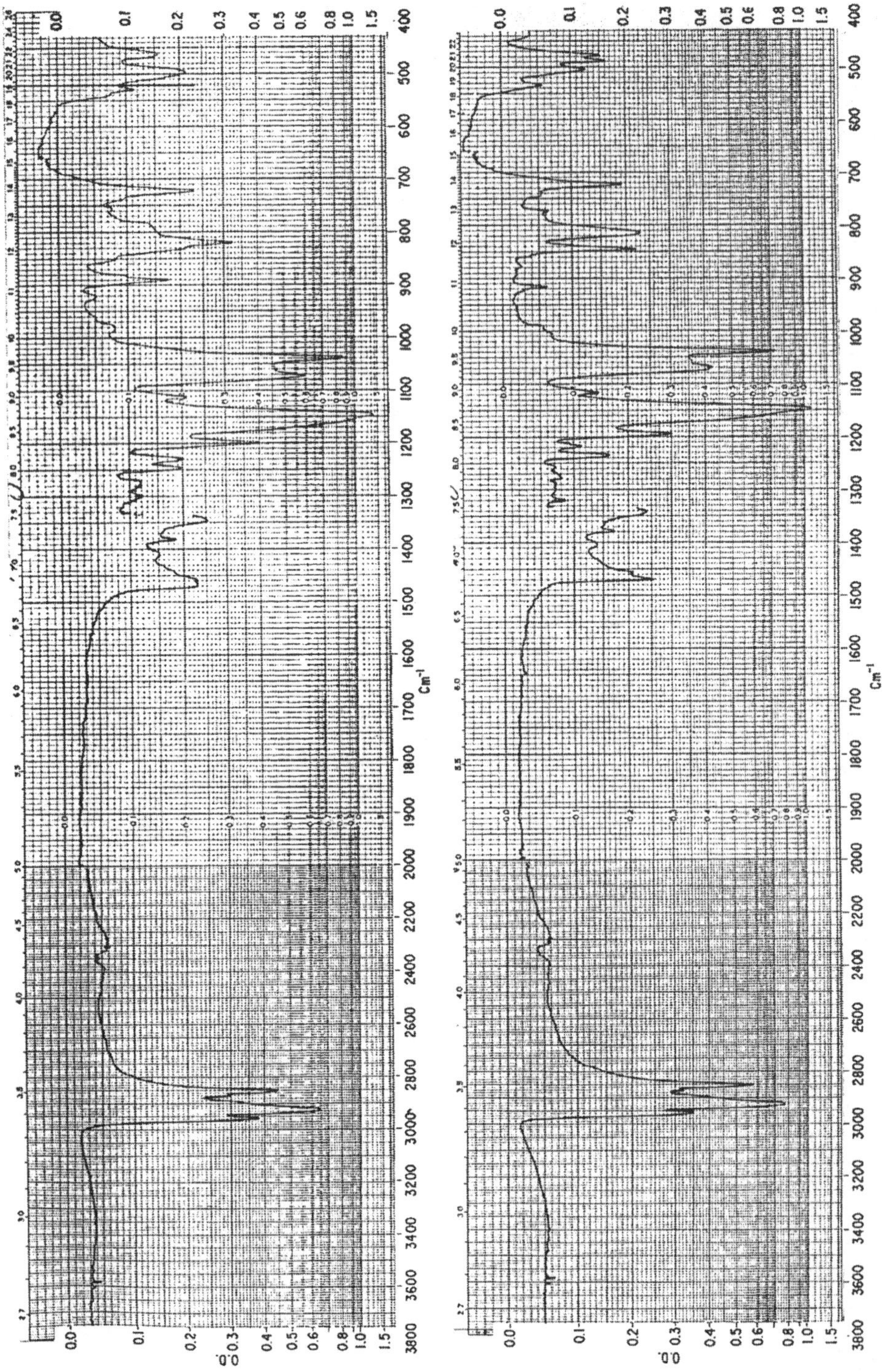

FIGURE 8.53a Top: IR spectrum for sodium diheptylphosphinate. (Bottom): IR spectrum for sodium dioctylphosphinate.

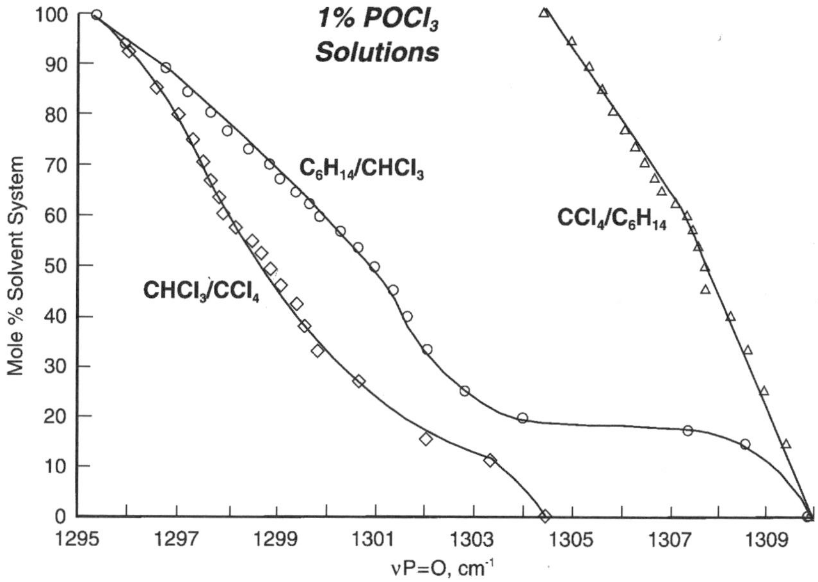

FIGURE 8.54 Plots of ν P=O for P(=O)Cl_3 vs mole % solvent system. The open triangles represent CCl_4/C_6H_{14} as the solvent system. The open circles represent $CHCl_3/C_6H_{14}$ as the solvent system; and the open diamonds represent $CHCl_3/CCl_4$ as the solvent system (26).

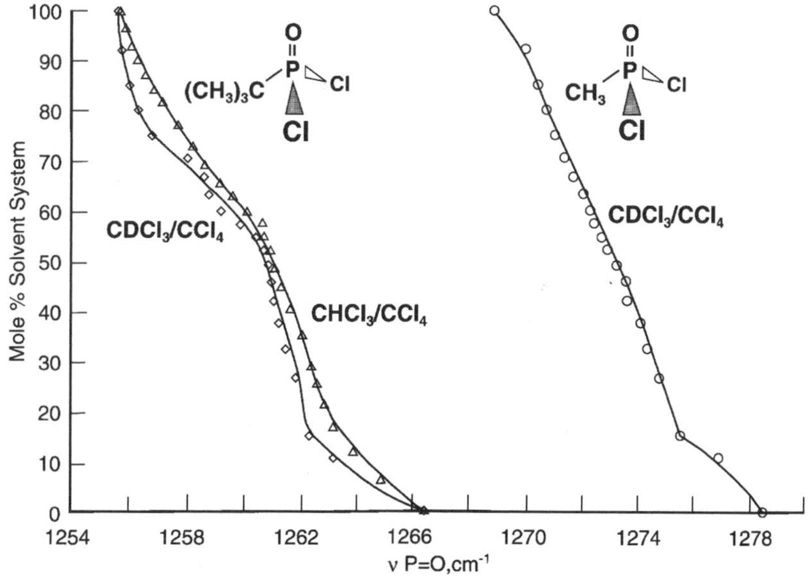

FIGURE 8.55 A plot of ν P=O for $CH_3P(=O)Cl_2$ vs mole % $CDCl_3/CCl_4$ as represented by open circles, and plots of ν P=O for $(CH_3)_3C$ P(=O)Cl_2 vs mole % $CHCl_3/CCl_4$ (open triangles) and mole % ($CDCl_3/CCl_4$ (open diamonds) (26).

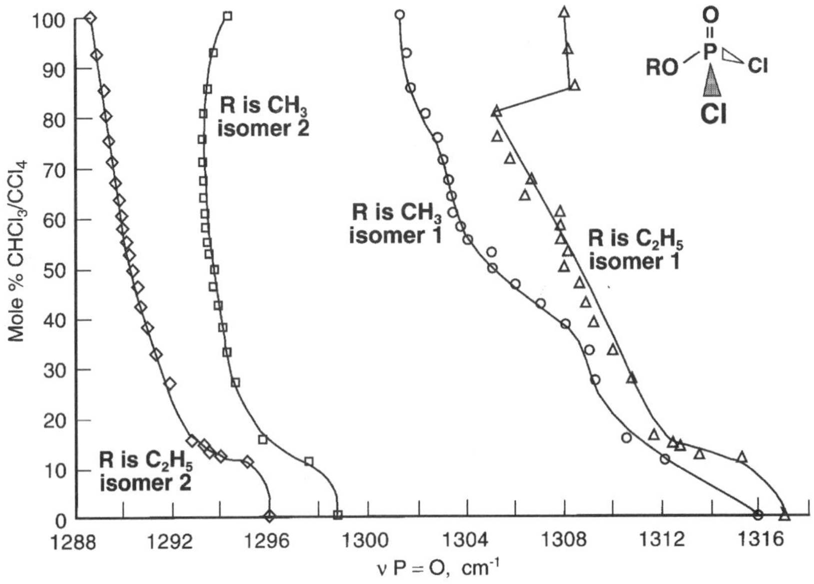

FIGURE 8.56 Plots of ν P=O rotational conformers 1 and 2 for CH_3O P(=O)Cl_2 (conformer 1, open circles; conformer 2, open squares), and for $C_2H_5OP(=O)Cl_2$ (conformer 1, open triangles; conformer 2, open diamonds) vs mole% $CHCl_3/CCl_4$ (26).

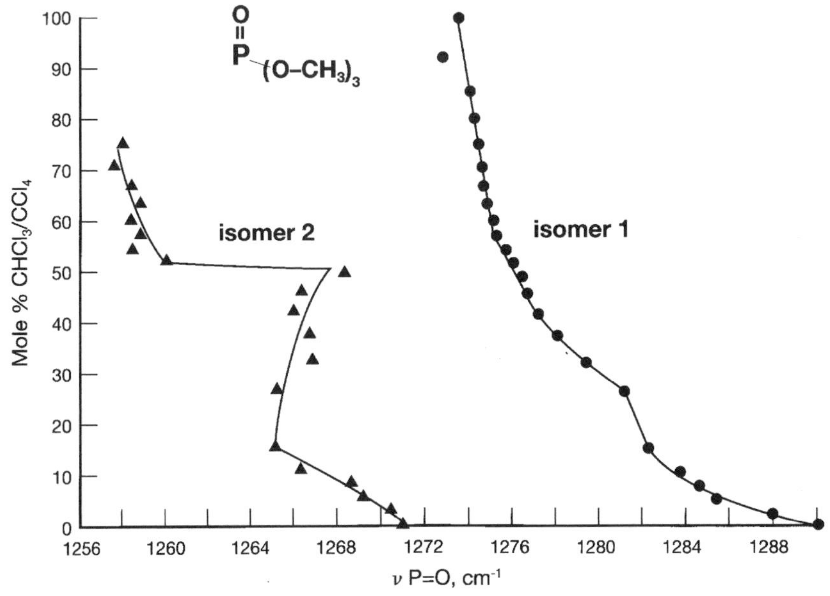

FIGURE 8.57 Plots of ν P=O rotational conformers 1 and 2 for $(CH_3O)_3P=O$ vs mole% $CHCl_3/CCl_4$ (conformer 1, solid circles, and conformer 2, solid triangles) (55).

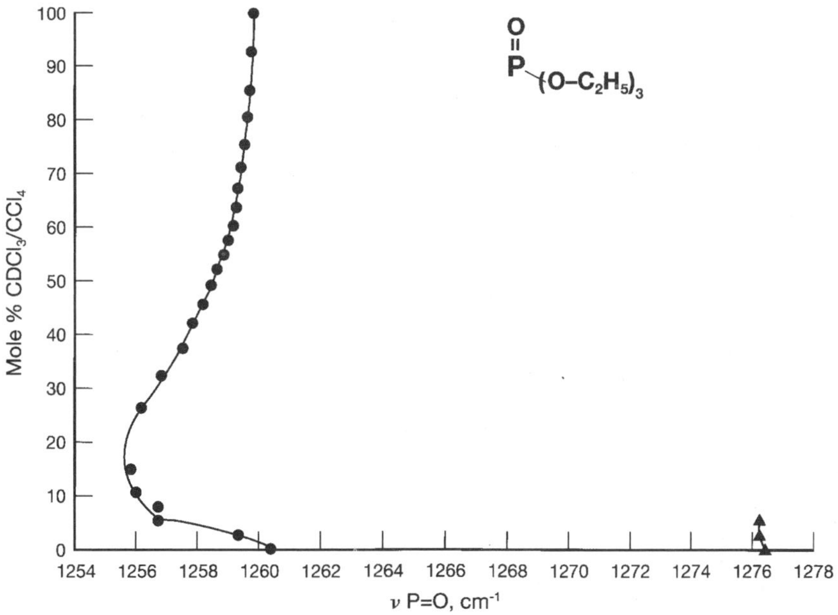

FIGURE 8.58 Plots of ν P=O rotational conformers 1 and 2 for $(C_2H_5O)_3P{=}O$ vs mole % $CDCl_3/CCl_4$ (conformer 1, solid triangles, conformer 2, closed circles) (55).

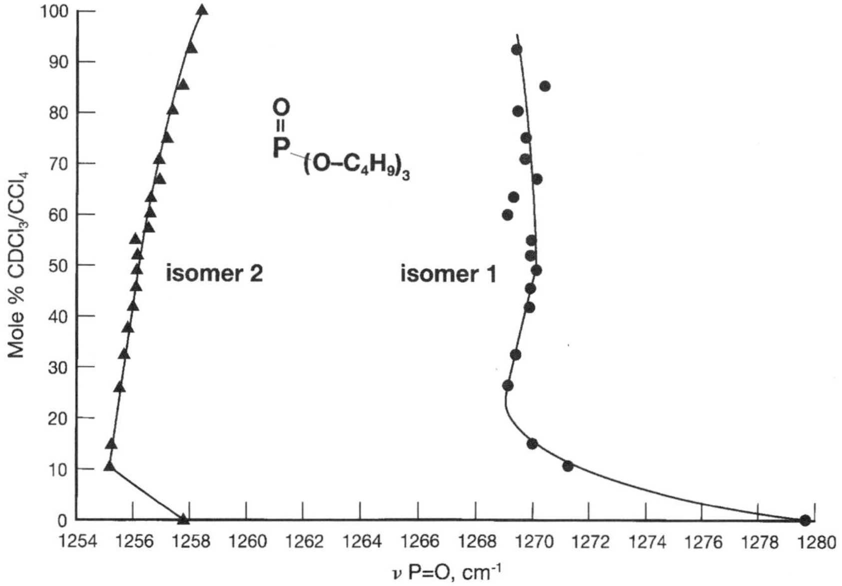

FIGURE 8.59 Plots of ν P=O rotational conformers 1 and 2 for $(C_4H_9O)_3P{=}O$ vs mole % $CDCl_3/CCl_4$ (conformer 1, solid circles, conformer 2, solid triangles) (55).

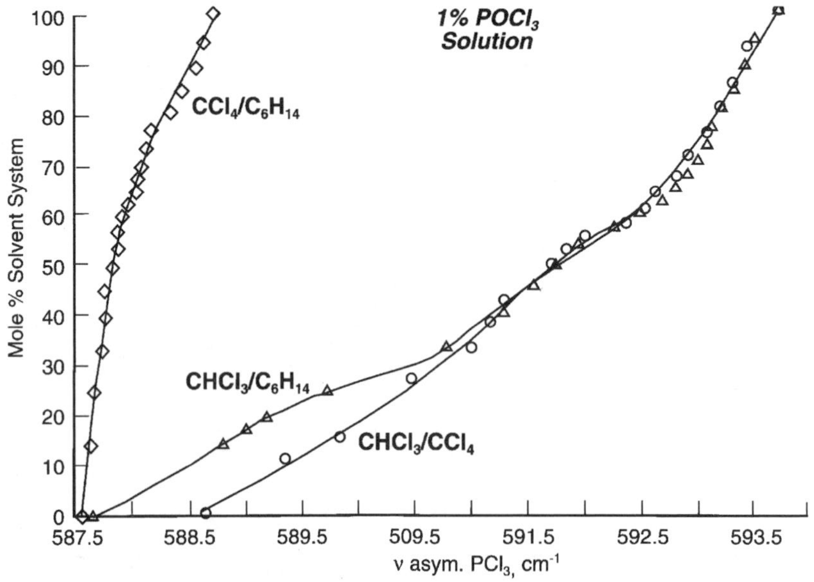

FIGURE 8.60 Plots of ν asym. PCl_3 for $P(=O)Cl_3$ vs mole % solvent system. The open circles represent the $CHCl_3/CCl_4$ solvent system; the open triangles represent the $CHCl_3/C_6H_{14}$ solvent system; and the open diamonds represent the CCl_4/C_6H_{14} solvent system (26).

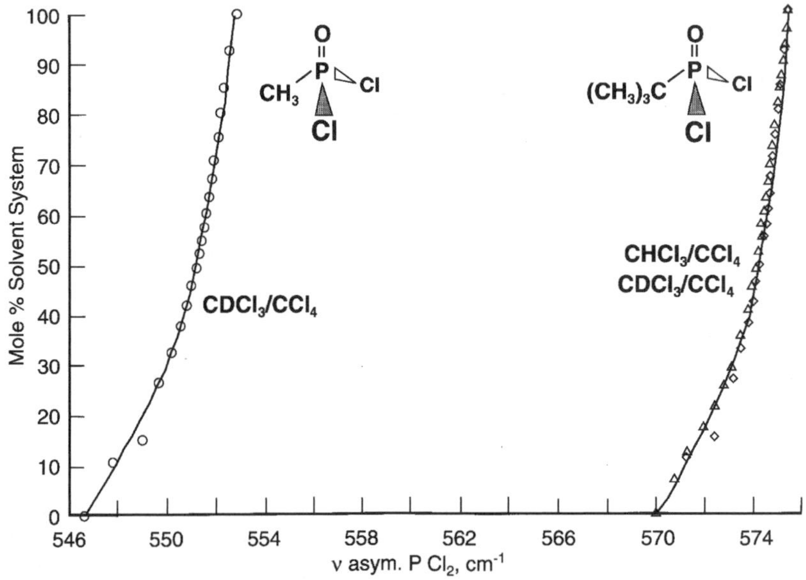

FIGURE 8.61 Plots of ν asym. PCl_2 frequencies for $CH_3P(=O)Cl_2$ (open circles) and for $(CH_3)_3CP(=O)Cl_2$ (open triangle and open diamonds) vs mole % $CDCl_3/CCl_4$ and $CHCl_3/CCl_4$ (26).

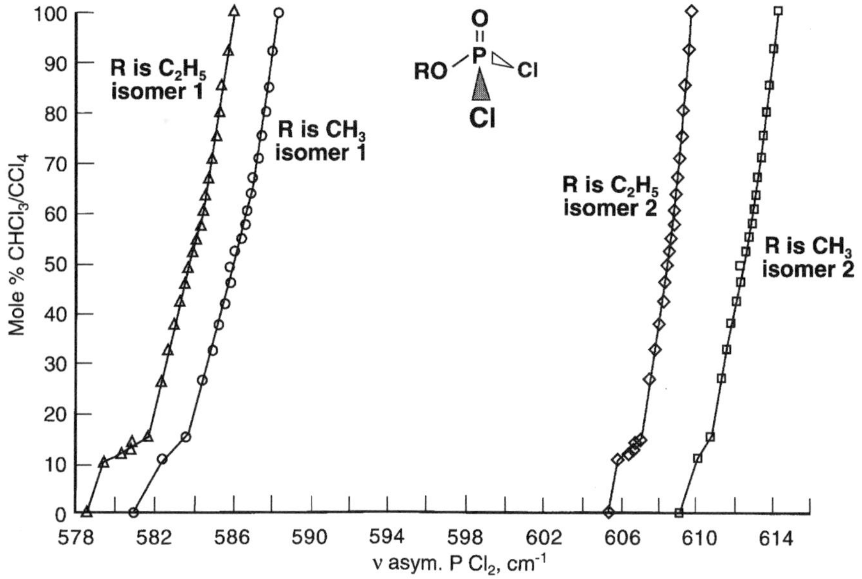

FIGURE 8.62 Plots of ν asym. PCl_2 rotational conformers 1 and 2 frequencies for $CH_3OP(=O)Cl_2$ (conformer 1, open circles; conformer 2, open squares) and for $C_2H_5OP(-O)Cl_2$ (conformer 1, open triangles, conformer 2, open diamonds) frequencies vs mole % $CHCl_3/CCl_4$ (26).

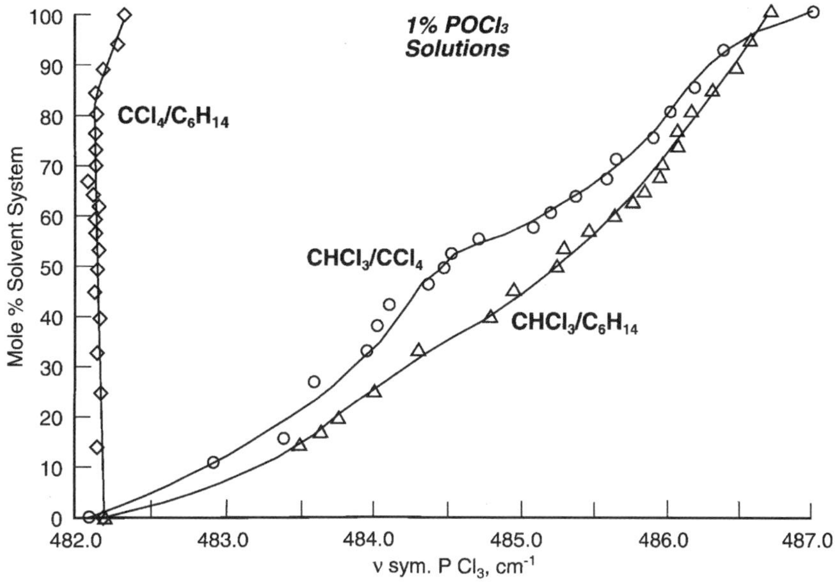

FIGURE 8.63 Plots of ν sym. PCl_3 frequencies for $P(=O)Cl_3$ vs mole % solvent system. The open circles represent $CHCl_3/CCl_4$ as the solvent system, the open triangles represent the $CHCl_3/C_6H_{14}$ solvent system, and the open diamonds represent the CCl_4/C_6H_{14} solvent system.

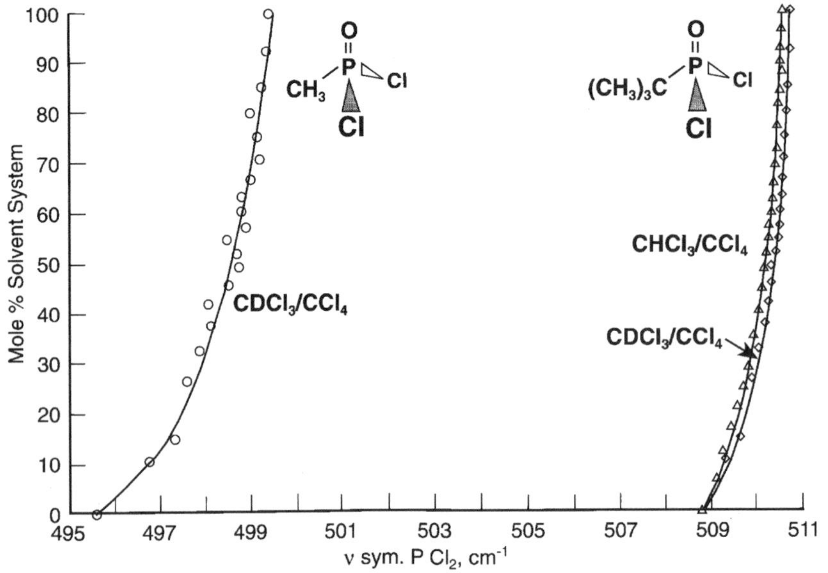

FIGURE 8.64 A plot of ν sym. PCl_2 frequencies for $CH_3P(=O)Cl_2$ vs mole % $CDCl_3/CCl_4$ (open circles) and plots of ν sym. PCl_2 frequencies for $(CH_3)_3CP(=O)Cl_2$ vs mole % $CHCl_3/CCl_4$ (open triangles) and mole % $CDCl_3/CCl_4$ (open diamonds) (26).

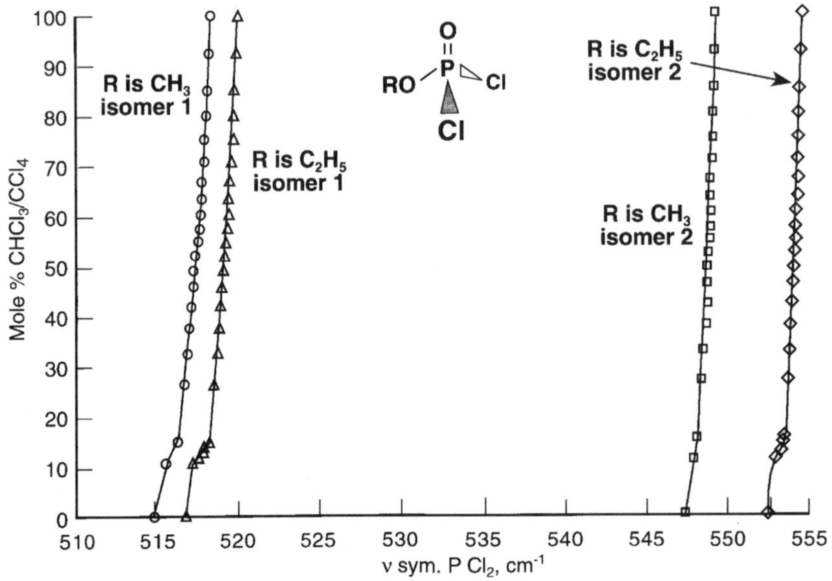

FIGURE 8.65 Plots of ν sym. PCl_2 rotational conformers 1 and 2 vs mole % $CHCl_3/CCl_4$. The open circles and open squares are for $CH_3OP(=O)Cl_2$ conformers 1 and 2, respectively. The open triangles and open diamonds are for $C_2H_5OP(=O)Cl_2$ conformers 1 and 2, respectively (26).

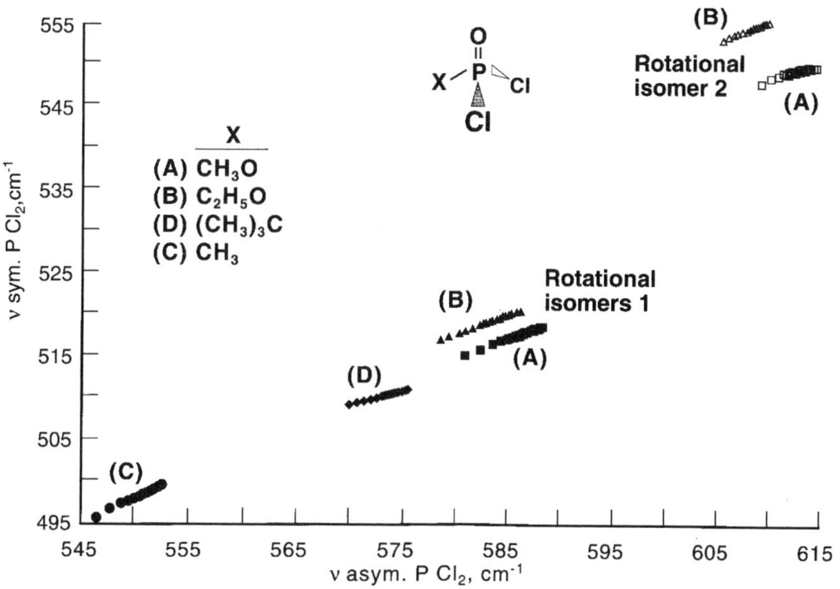

FIGURE 8.66 Plots of ν asym. PCl_2 frequencies vs ν sym. PCl_2 frequencies for $CH_3OP(=O)Cl_2$ (conformer 1, solid squares); $CH_3OP(=O)Cl_2$ (conformer 2, open squares); $C_2H_5OP(=O)Cl_2$ (conformer 1, solid triangles) $C_2H_5OP(=O)Cl_2$ (conformer 2, open triangles); $(CH_3)_3CP(=O)Cl_2$ (solid diamonds); and $CH_3P(=O)Cl_2$ (solid circles) (26).

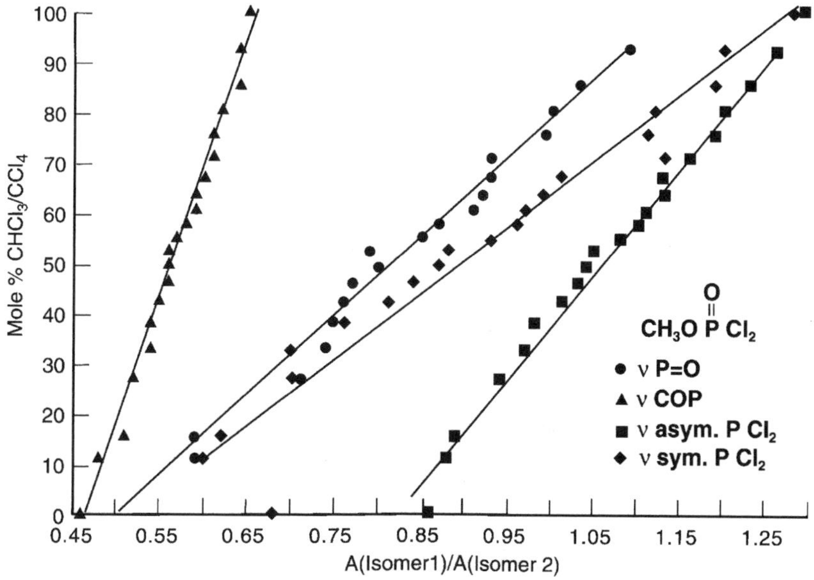

FIGURE 8.67 Plots of the absorbance (A) ratio of rotational conformer 1 and 2 band pairs for ν P=O (solid circles), ν COP (solid triangles), ν asym. PCl_2 (solid squares), and ν sym. PCl_2 (solid diamonds) for $CH_3OP(=O)Cl_2$ vs mole % $CHCl_3/CCl_4$ (26).

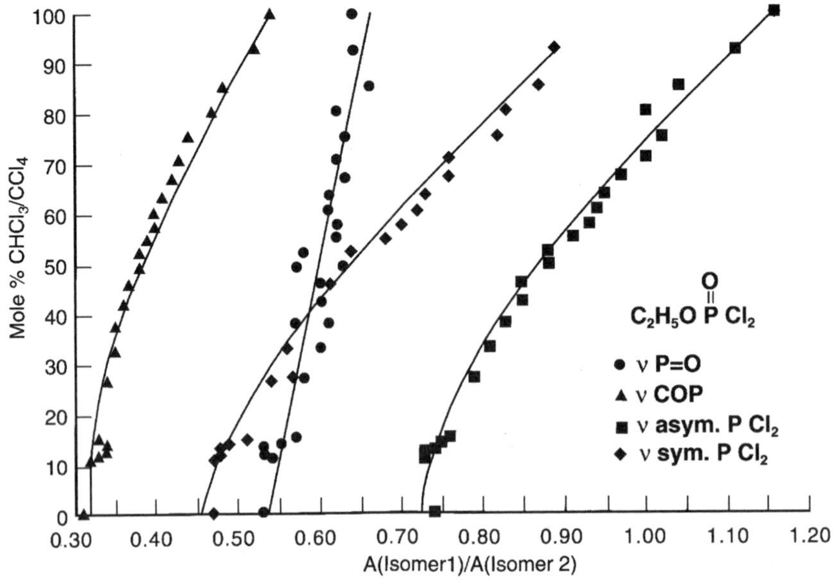

FIGURE 8.68 Plots of the absorbance (A) ratio rational conformer 1 and 2 band pairs for *ν* P=O (solid circles), *ν* COP (solid triangles), *ν* asym. PCl_2 (solid squares), and *ν* sym. PCl_2 (solid diamonds) for $C_2H_5OP(=O)Cl_2$ vs mole% $CHCl_3/CCl_4$ (26).

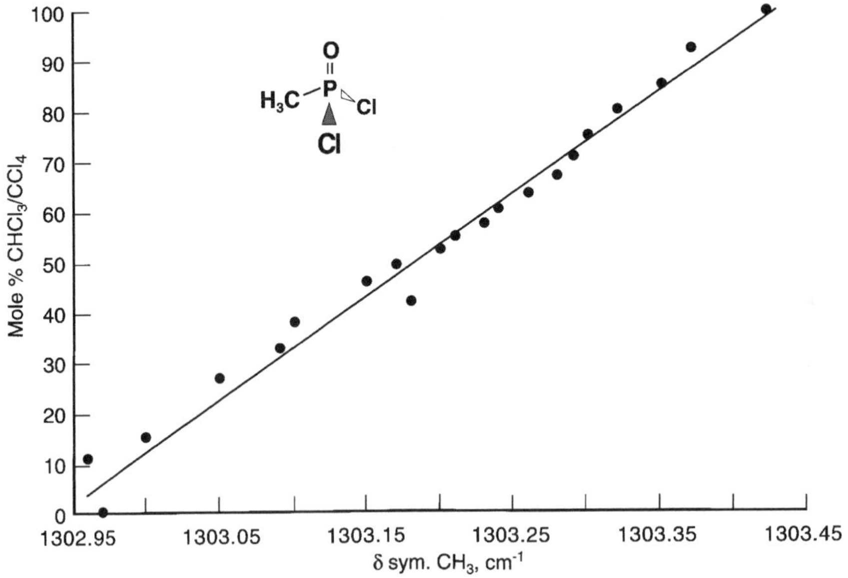

FIGURE 8.69 A plot of Δ sym. CH_3 frequencies for $CH_3P(=O)Cl_2$ vs mole% $CHCl_3/CCl_4$ (26).

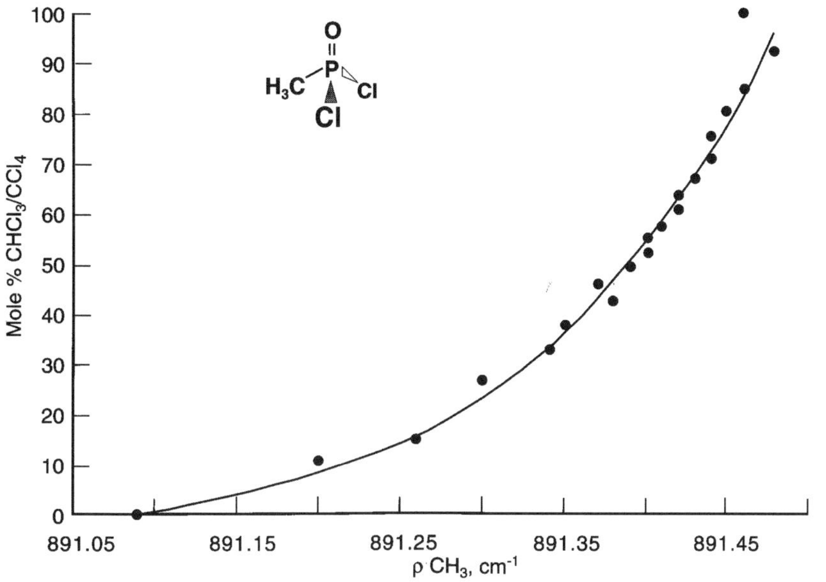

FIGURE 8.70 A plot of ρCH_3 frequencies for $CH_3P(=O)Cl_2$ vs mole % $CHCl_3/CCl_4$ (26).

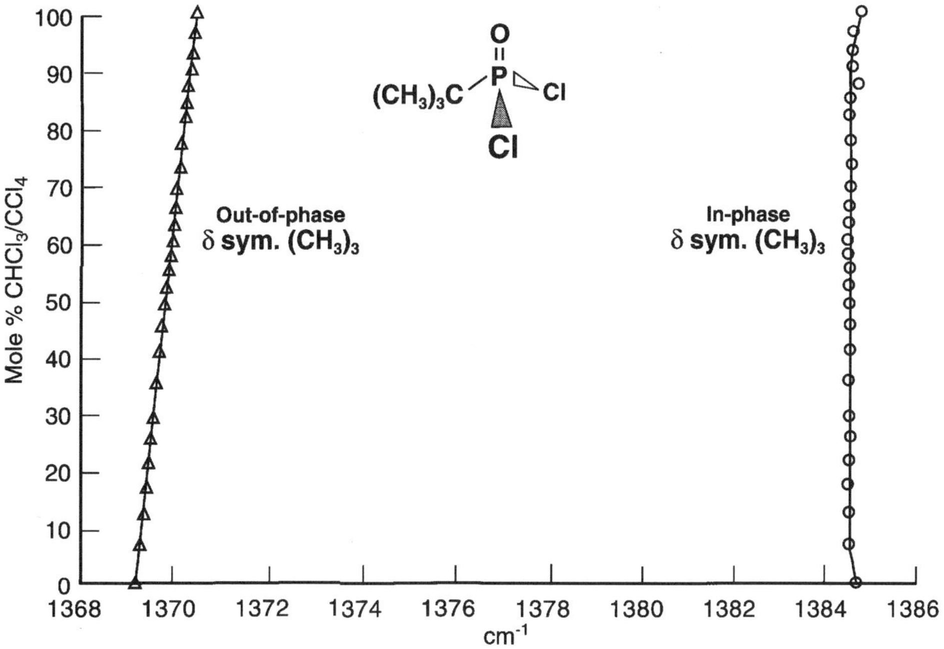

FIGURE 8.71 Plots of the in-phase δ sym. $(CH_3)_3$ frequencies (open circles) and the out-of-phase δ sym. $(CH_3)_3$ frequencies (open triangles) vs mole % $CHCl_3/CCl_4$ (26).

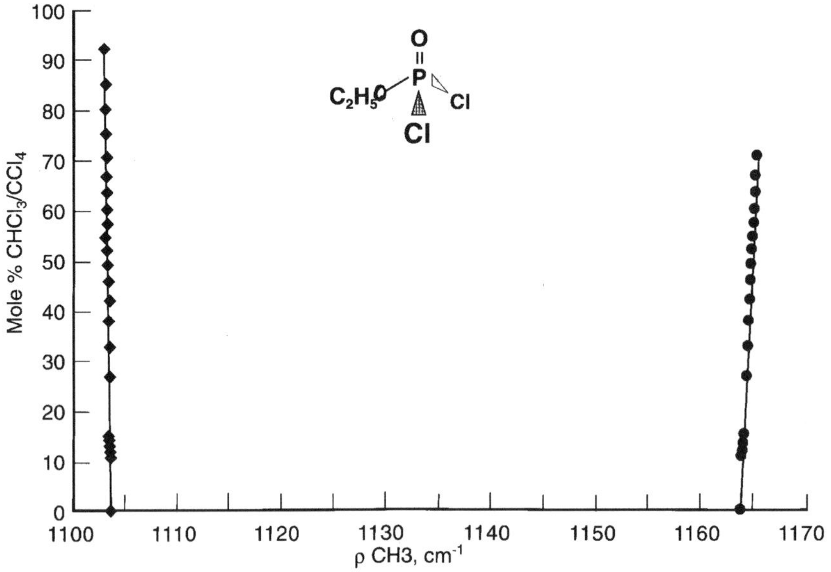

FIGURE 8.72 Plots of the a′ and a″ ρCH_3 frequencies for C_2H_5O P(=O)Cl_2 vs mole% $CHCl_3/CCl_4$ (26).

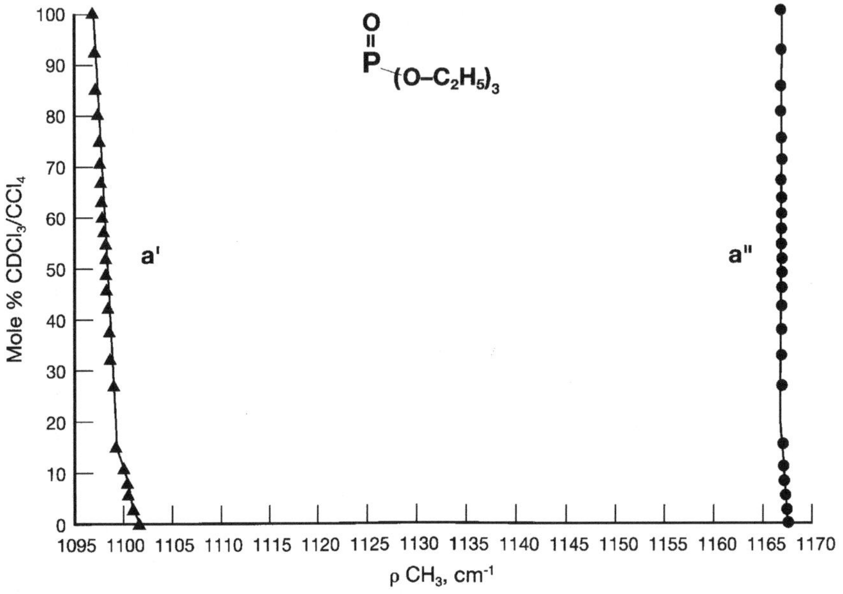

FIGURE 8.73 Plots of the a′ and a″ ρCH_3 modes of $(C_2H_5O)_3P{=}O$ vs mole% $CDCl_3/CCl_4$ (55).

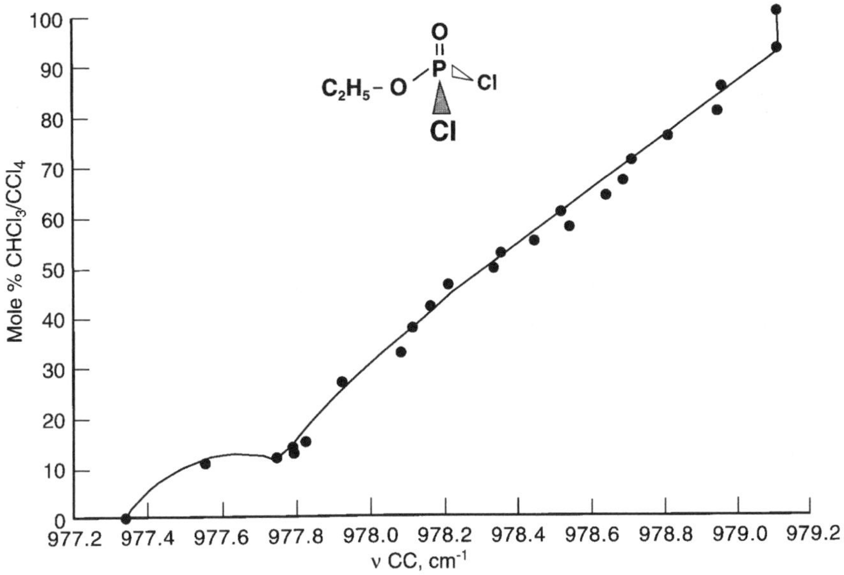

FIGURE 8.74 A plot of ν CC for $C_2H_5OP(=O)Cl_2$) vs mole % $CHCl_3/CCl_4$ (26).

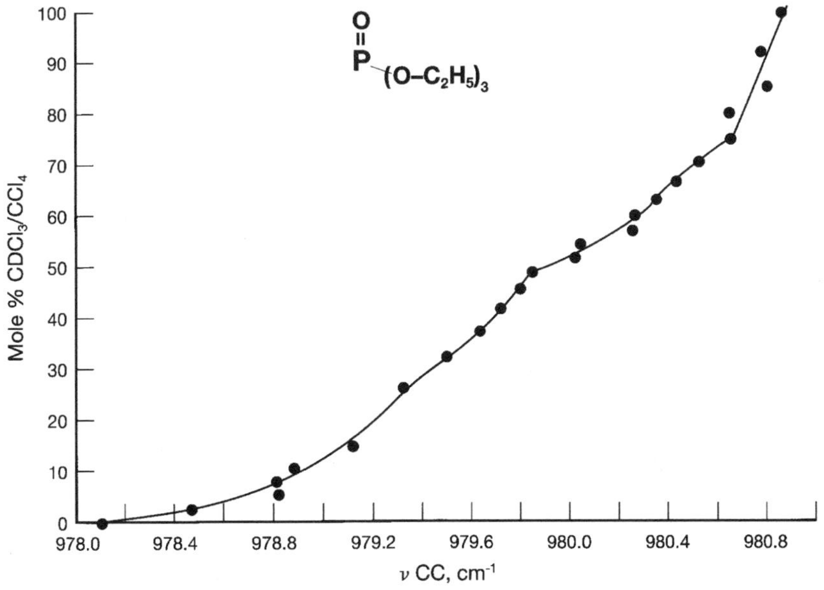

FIGURE 8.75 A plot of ν CC for $(C_2H_5O)_3$ P=O vs mole % $CDCl_3/CCl_4$ (55).

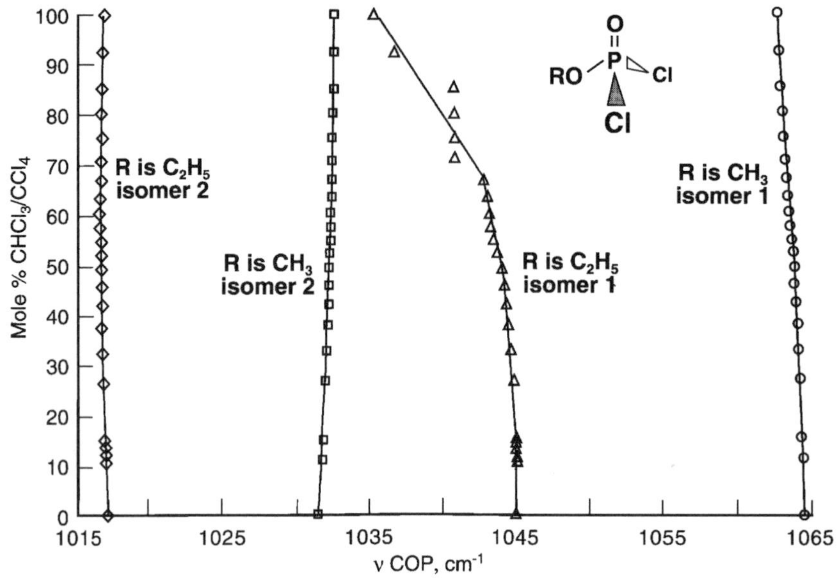

FIGURE 8.76 Plots of ν COP rotational conformers 1 and 2 for $CH_3O\ P(=O)Cl_2$ and $C_2H_5OP(=O)Cl_2$ vs mole % $CDCl_3/CCl_4$ (26).

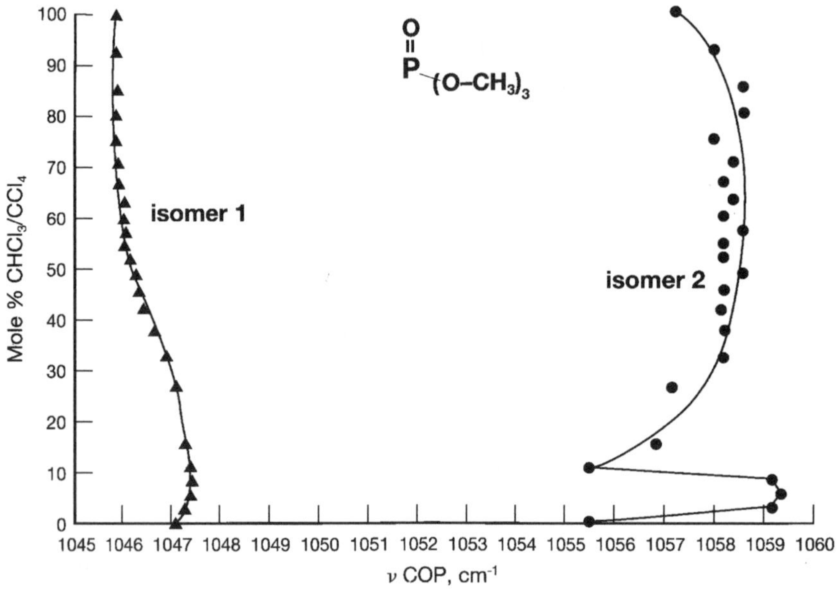

FIGURE 8.77 Plots of ν COP rotational conformers 1 and 2 frequencies for $(CH_3O)_3P{=}O$ vs mole % $CHCl_3/CCl_4$. The solid triangles represent conformer 1 and the solid circles represent conformer 2 (55).

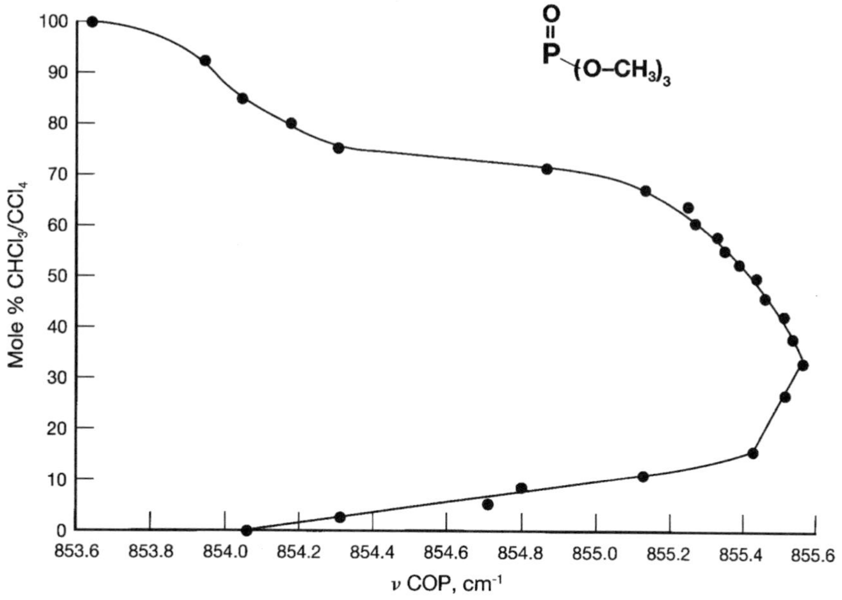

FIGURE 8.78 A plot of ν COP frequencies for $(CH_3O)_3P{=}O$ vs mole% $CHCl_3/CCl_4$ (55).

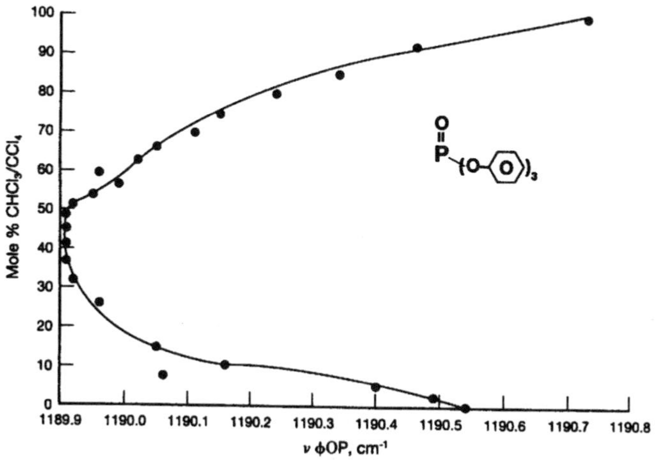

FIGURE 8.79 A plot of ν (φOP), "ν Φ-O", frequencies for triphenyl phosphate vs mole% $CHCl_3/CCl_4$ (55).

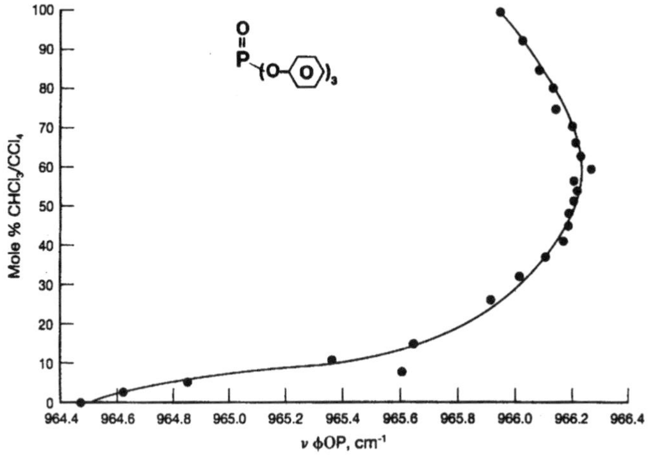

FIGURE 8.80 A plot of ν (ΦOP), "νP-O", frequencies for triphenyl phosphate vs mole% $CHCl_3/CCl_4$ (55).

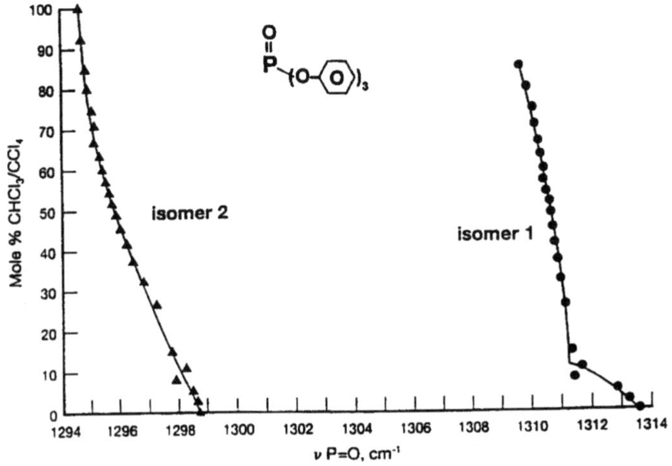

FIGURE 8.81 A plot of the in-plane hydrogen deformation frequencies for the phenyl groups of triphenyl phosphate vs mole% $CHCl_3/CCl_4$ (55).

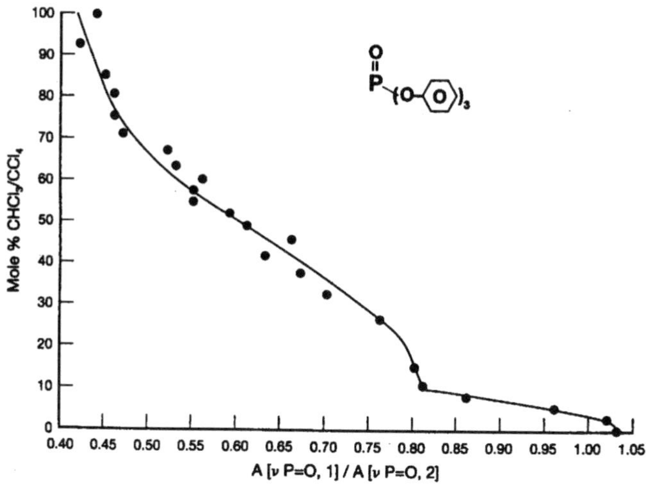

FIGURE 8.82 A plot of the out-of-plane ring deformation frequencies for the phenyl groups of triphenyl phosphate vs mole % $CHCl_3/CCl_4$ (55).

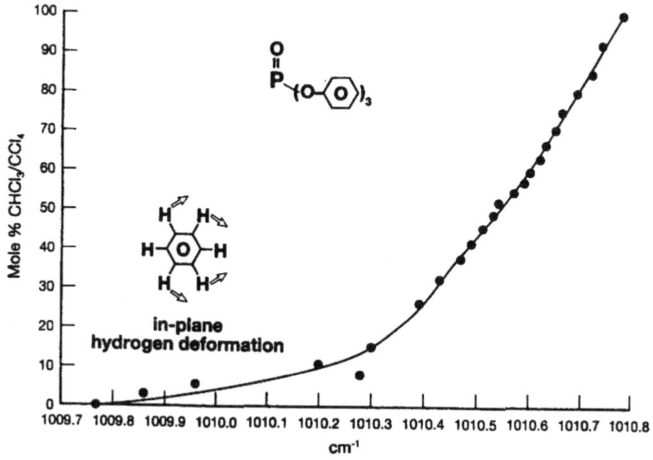

FIGURE 8.83 A plot of νCD frequencies (for the solvent system $CDCl_3/CCl_4$) vs mole % $CDCl_3/CCl_4$ (55).

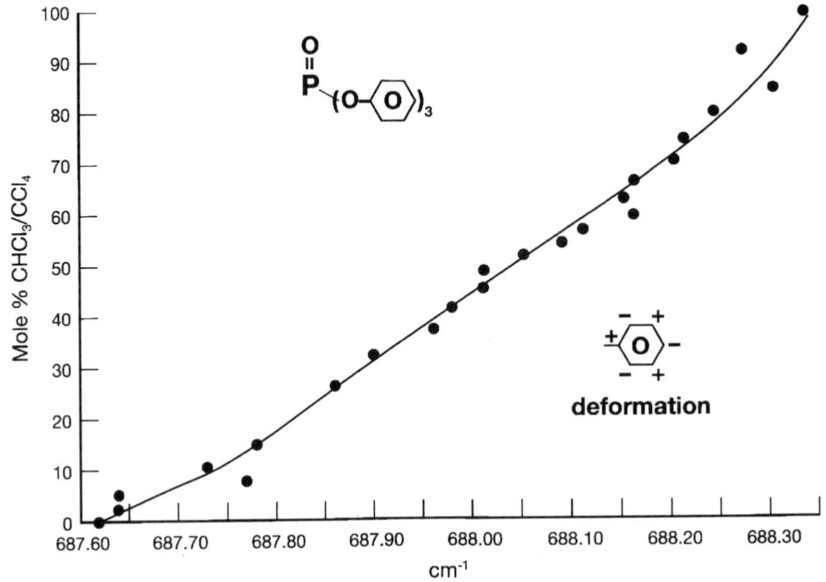

FIGURE 8.84 A plot of the ν CD frequencies for $CDCl_3$ vs mole% $CDCl_3/CCl_4$ for the $CDCl_3/CCl_4$ solvent system containing 1 wt./vol. % triphenylphosphate (55).

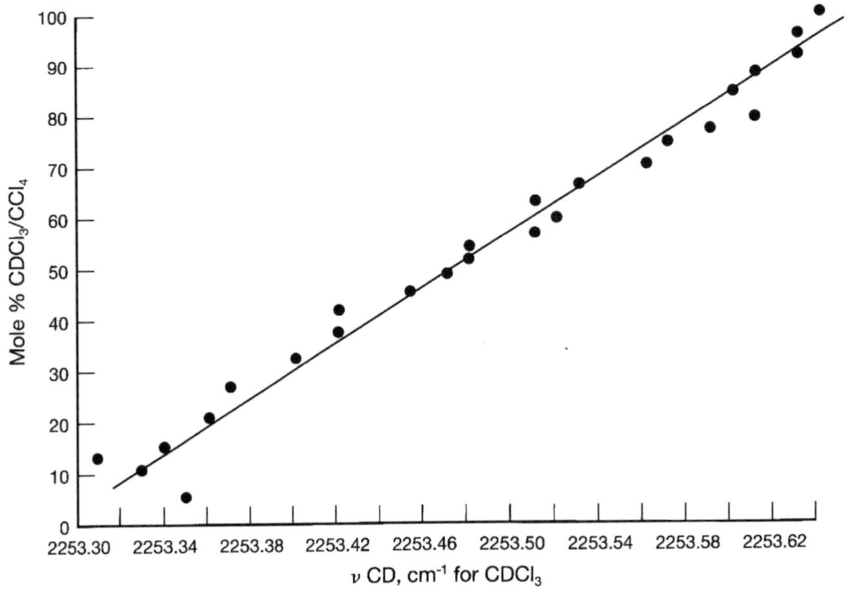

FIGURE 8.85 A plot of ν CD for $CDCl_3$ vs mole % $CDCl_3/CCl_4$.

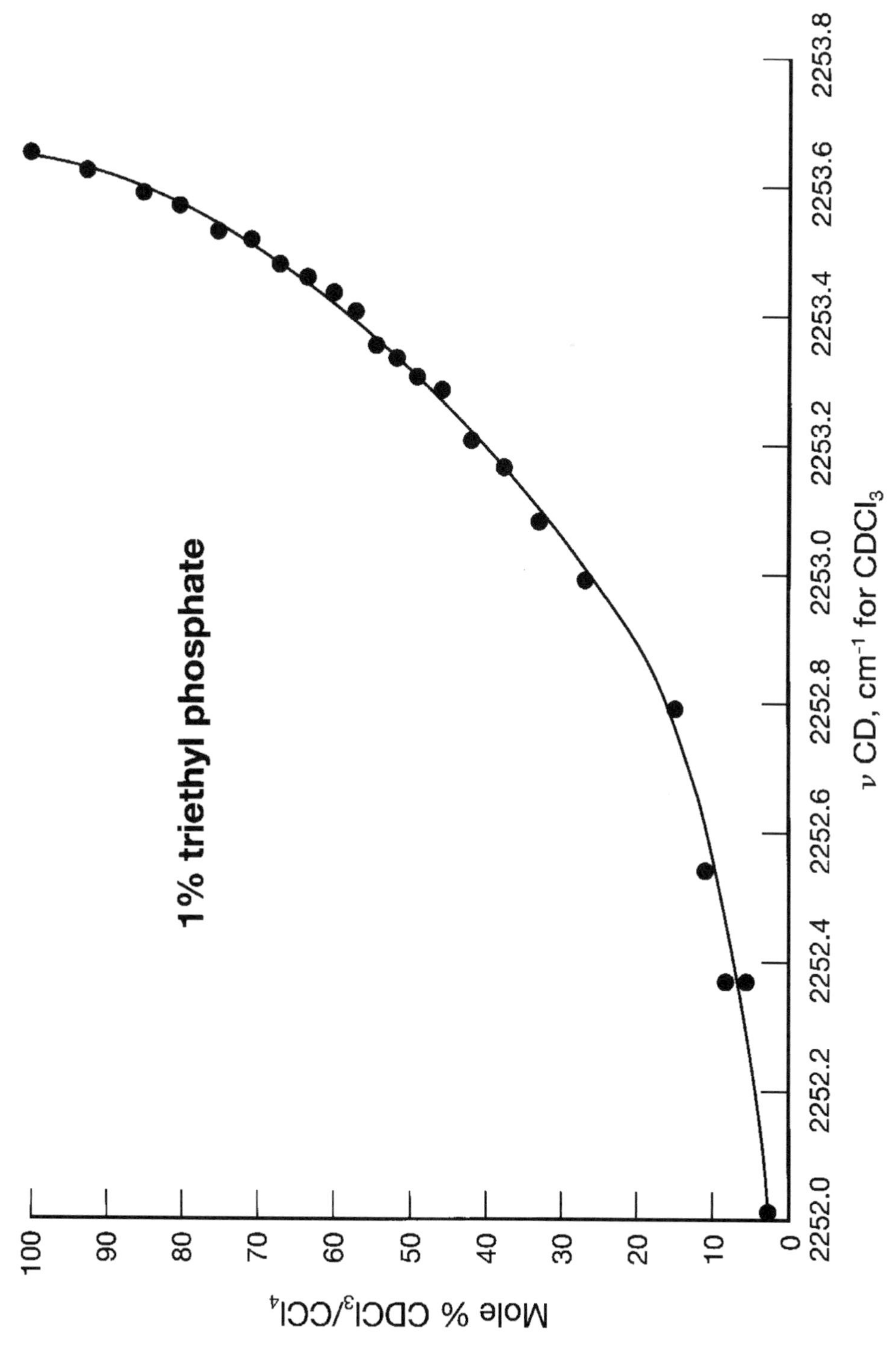

FIGURE 8.86 A plot of ν CD for $CDCl_3$ vs mole % $CDCl_3/CCl_4$ containing 1% triethylphosphate.

TABLE 8.1 Phosphorus halogen stretching frequencies for inorganic compounds*

Compound PX_3 X	F stretching cm^{-1}	Cl stretching cm^{-1}	Br stretching cm^{-1}	I stretching cm^{-1}	Assignment	References
F_3	890(s.)840(a.)				v1,a1;v3,e	1,2,3,4
FCl_2	827	524(s.);496(a.)			a′ ; a′ ; a″	5
FClBr	822	500	415		a ; a ; a	5
FBr_2	817		421(s.);393(a.)		a′ ; a′ ; a″	5
Cl_3		511(s.);484(a.)			v1,a1;v3,e	6
Cl_2Br		510(s.);480(a.)	~ 400		a′ ; a″ ; a′	5
$ClBr_2$		~ 480	400(s.);380(a.)		a′ ; a′ ; a″	5
Br_3			380(s.);400(a.)		v1,a1;v3,e	5
I_3				303(s.);325(a.)	v1,a1;v3,e	6
P(=S)X_3 X						
F_3	981(s.);945(a.)				v2,a1;v4,e	7
F_2Cl	946(s.);920(a.)	541			a′ ; a″ ; a′	7
F_2Br	930(s.);899(a.)		~ 468		a′ ; a″ ; a′	8
FCl_2	900	476(s.);567(a.)			a′ ; a″ ; a′	9
Cl_3		431(s.);547(a.)			v2,a1;v4,e	10,11
Br_3			299(s.);438(a.)		v2,a1;v4,e	12
P(=O)X_3 X						
F_3	875(s.);982(a.)				v2,a1;v4,e	1
F_2Cl	895(s.);948(a.)	618			a′ ; a″ ; a′	13
F_2Br	888(s.);940(a.)		554		a′ ; a″ ; a′	13
FCl_2	894	547(s.);620(a.)			a′ ; a′ ; a″	14
FClBr	890	590	495		a ; a ; a	13
FBr_2	880		466(s.);538(a.)		a′ ; a′ ; a″	13
Cl_3		486(s.);581(a.)			v2,a1;v4,e	15
Cl_2Br		545(s.);580(a.)	432		a′ ; a″ ; a′	13
$ClBr_2$		552	391(s.);492(a.)		a′ ; a′ ; a″	13
Br_3			340(s.);488(a.)		v2;a1;v4,e	8,15

*See Reference 16.

TABLE 8.2 The PX_3 bending frequencies for compounds of form PX_3, PXY_2, and PXYZ*[1]

Compounds & symmetry	Species & Assignment s. bending	Compound & symmetry	Species & Assignment	Compound & symmetry	Species & Assignment	Compound References
C3v	v2,a1 cm^{-1}	Cs	v3,a′ cm^{-1}	Ci	v4,a cm^{-1}	see Table 8.1
PF_3	486					
		PClF2	[390–400]*[2]			
		PBF2	[367–375]			
		PF2I	[327–340]			
		PCl2F	327			
				PBrClF	302	
		PBr2F	257			
				PClFI	[250–275]	
PCl_3	261					
				[PBrFI]	[220–252]	
		PBrCl2	230			
		PFI2	[190–220]			
		PBr2Cl	197			
		PCl2I	[175–202]			
PBr_3	165			PBrClI	[167–195]	
		PBr2I	[135–157]			
		PClI2	[133–155]			
		PbrI2	[120–135]			
PI_3	113					

*[1]See Reference 16.
*[2]All numbers in brackets are estimated frequencies.

TABLE 8.3 Vibrational data and assignments for PX_3, $P(=O)X_3$, and $P(=S)X_3$*[1]

PX(1–3)	P C3v,e v4,cm^{-1}	P=O C3v,e v6,cm^{-1}	P=S C3v,e v6,cm^{-1}	P Cs,a′ v6,cm^{-1}	P=O Cs,a′ v6,cm^{-1}	P=S Cs,a′ v6,cm^{-1}	P Ci,a v6,cm^{-1}	P=O Ci,a v9,cm^{-1}	P=S Ci,a v9,cm^{-1}	Compound References
F3	347	345	276							See Table 8.1
ClF				[245–262]*[2]	274	209				
BrF_2				[166–178]	[170–190]	175				
F_2I				[126–145]	[140–150]	[130–140]				
Cl_2F				200	207	191				
BrClF							161	173	[140–160]	
Br_2F				126	134	[125–135]				
ClFI							[126–136]	[130–140]	[125–135]	
Cl_3	191	193	172							
BrFI							[109–119]	[125–135]	[110–130]	
$BrCl_2$				149	161	[135–155]				
FI_2				[97–105]	[100–120]	[100–120]				
Br_2Cl				123	130	[115–130]				
Cl_2I				[119–129]	[125–135]	[115–130]				
Br_3	116	118	115							
BrClI							[105–116]	[110–125]	[100–120]	
Br_2I					[90–115]	[90–115]				
ClI_2				[94–101]	[90–115]	[90–115]				
BrI_2				[87–93]	[90–110]	[90–110]				
I_3	85	[70–90]	[70–90]							

*[1]See Reference 16.
*[2]All numbers in brackets are estimated frequencies.

TABLE 8.4 The asym. and sym. PCl_2 stretching frequencies for $XPCl_2$, $XP(=O)Cl_2$, and $XP(=S)Cl_2$ groups

Compound $X-PCl_2$ X	a.PCl_2 str. cm^{-1}	s.PCl_2 str. cm^{-1}	Ref.	[a.PCl_2 str.]−[s.PCl_2.str.] cm^{-1}	Compound	a.PCl_2 str. cm^{-1}	s.PCl_2 str cm^{-1}	Ref.	[a.PCl_2 str.]−[s.PCl_2 str.] cm^{-1}
CH_3	495	495	17,18	0					
$ClCH_2$	490	490	17	0					
CH_3-O	506	453	19	53					
CH_3-O	505	456	19	49					
$X-P(=O)Cl_2$	$[CCl_4]/[CHCl_3]$	$[CCl_4]/[CHCl_3]$			$X-P(=S)Cl_2$				
X									
	cm^{-1}	cm^{-1}							
CH_3	552	497	17,18	55	CH_3	520	464	17,18	56
CH_3	[546.6*1]/[552.8*2]	[495.6*1]/[499.4*2]	26	51*1, 53.4*2					
$(CH_3)_3C$	[570.1*1]/[575.5*2]	[508.8*1]/[510.6*2]	26	61.3*1; 64.9*2					
$ClCH_2$	556/566	487/508	17	69;61	$ClCH_2$	527	452/461	17	65; or 75
C_6H_5	565	542	20	23	C_6H_5	521	501	20	20
CH_3-O	579–607	515/548	21	64;59	CH_3-O	531/560	456/478	21,25	75;82
					CD_3-O	531/552	452/471	21,25	79;81
CH_2H_5-O	576/600	516/550	22	60;50	C_2H_5-O	528/558	472/490	22	56;68
					CH_3-CD_2-O	525/550	465/484	22	69;66
					CD_3-CH_2-O	531/552	462/483	22	69;69
					iso-C_3H_7-O	521/548	483/500	20	59;69
					iso-C_4H_9-O	528/553	473/493	20	55;60
					s.-C_4H_9-O	525/547	485/500	20	40;47
					iso-$C_5H_{11}-O$	528/552	455/465	20	73;87
					$H-CC-CH_2-O$	529/557	440/474	26	89;83
4-Cl−C_6H_4-O	585/600	?/555	20	[−; 45]	4-Cl−C_6H_4-O	560	?/471	20	89
					2,4-$Cl_2-C_6H_3-O$	?/551	?	20	[−]
					2,6-$Cl_2-C_6H_3-O$	555	472	20	83
2,4,5-$Cl_3-C_6H_2-O$	587/602	?	20	[−; −]	2, 4, 5-$Cl_3-C_6H_2-O$	?/565	?/448	20	117?
CH_3-S	547/560	449/471	23	98;89					
CH_3-NH	560	517	24	43	CH_3-NH	512	450/468	24	62; or 44
					CD_3-NH	515	446/463	24	69; or 52
					CH_3-ND	513	447/470	24	66; or 43
					CD_3-ND	510	443/461	24	67; or 49
					C_2H_5-NH	512	?463	24	[−; 49]
					iso-C_3H_7-NH	514	?/481	24	[−; 33]
$(CH_3)_2-N$	560	517	20	43	$(CH_3)_2-N$	512	450	20	62
$(C_2H_5)_2-N$	560	523	20	37	$(C_2H_5)_2-N$	522	481	20	41

*1 In CCl4 soln.

*2 In $CHCl_3$ soln. Other data are in CS_2 soln.

TABLE 8.5 Vibrational assignments for $F_2P(=S)Cl$, $(CH_3-O-)_2P(=S)Cl$, and $(CD_3-O-)_2P(=S)Cl$

Assignment	[11] $ClF_2P(=S)$ cm^{-1}	[27] $Cl(CH_3-O-)_2P(=S)$ cm^{-1}	[27] $Cl(CD_3-O-)_2P(=S)$ cm^{-1}	$[(CH_3O)_2]-[(CD_3O_2]$ cm^{-1}	Assignment
s.PF_2 str., a′	939	~ 845	783.2	61.8	s.p$(-O-)_2$ str.
a.PF_2 str., a″	913	836.2	807.3	28.9	a.P$(-O-)_2$ str.
P=S str. ,a′	727	666.1	649.1	17	P=S str.*1
		654.7	649.1	5.6	P=S str.*1
PCl str., a′	536	524.7	519.6	5.1	P–Cl str.*1
		485.7	482	3.7	P–Cl str.*1
PF_2 wag, a′	394	396	376	2	P$(-O-)_2$ wag*1
		375	348	27	P$(-O-)_2$ wag*1
PF_2 wag, a′	359	352	341	11	P$(-O-)_2$ bend
PF_2 twist, a″	314	291	288	3	p$(-O-)_2$ twist*1
		276	272	4	P$(-O-)_2$ twist*1
Cl–P(=S) out-of-plane bend, a″	251	233	215	18	Cl–P(=S) out-of-plane bend
Cl–P(=S) in-of-plane bend, a″	207	217	215	2	Cl–P(=S) in-of-plane bend
		1055.9	1076.4	[−20.5]	s.$(C-O-)_2$P str.*1
		1042	1059	[−17.0]	s.$(C-O-)_2$P str.*1
		1031.1	1050	[−18.9]	a$(C-O-)_2$P str.
		449	428	21	$(C-O-)_2$P bend, a′
		426	405	21	$(C-O-)_2$P bend, a″
		171	157	14	$(C-O-)_2$P,a′ & a″
				CH_3/CD_3	
		3025.7	2273	1.33	a.CH_3 str. or a.CD_3 str., a″
		3002.2	2255.3	1.33	a.CH_3 str. or a.CD_3 str., a′
		2952.3	2077.1	1.42	s.CH_3str. or CD_3 str., a′
		[2922.3]*2	[2109.3]*1	1.39	2(CH_3 bend) or 2(CD_3 bend). a′
		2848	2203	1.29	
		[2878]*1	[2170.9]*1	1.33	
		1456.9	1099.4	1.33	a.CH_3 or a.CD_3 bend a′ & a″
		1443.1	1099.4	1.31	s.CH_3 or CD_3 bend, a′
		1179.9	926.1	1.27	CH_3 or CD_3 rock, a′ (in-plane)
		1160	900	1.29	CH_3 or CD_3 rock, a″ (out-of-plane)
		200	189	1.06	CH_3 or CD_3 torsion a′ & a″

*1Rotational conformers.

TABLE 8.5a The PCl_2 stretching frequencies for methyl phosphorodichloridate and O-methyl phosphorodichlorothioate*

Mode	[A] $CH_3-O-P(=O)Cl_2$ cm^{-1}	[B] $CH_3-O-P(=S)Cl_2$ cm^{-1}	[A]-[B] cm^{-1}	[C] $CD_3-O-P(=S)Cl_2$ cm^{-1}	[A]-[C] cm^{-1}	[B]-[C] cm^{-1}
a.PCl_2 str.						
conformer 1	579	531	48	531	48	0
conformer 2	607	560	47	552	55	8
delta[C_2-C_1]	[28]	[29]	[1]	[21]	[7]	[8]
s.PCl_2 str.						
conformer 1	515	456	59	452	63	4
conformer 2	548	478	70	471	77	7
delta[C_2-C_1]	[33]	[22]	[11]	[19]	[14]	[3]

*See Reference 21.

TABLE 8.6 P–Cl stretching frequencies

Compound	P–Cl str.	Ref.	Frequency separation between rotational conformers	Compound	P–Cl str.	Ref.	Frequency separation between rotational conformers
X–P(=O)Cl				X–P(=S)Cl			
X				X			
$(CH_3-O)_2$	553/598	27	45	$(CH_3-O)_2$	486/525	27	39
				$(CD_3-O)_2$	482/519	27	37
$(C_2H_5-O)_2$	554/589	20	35	$(C_2H_5-O)_2$	502/539	32	37
$(n\text{-}C_3H_7-O)_2$	557/597	20	40	$(n\text{-}C_3H_7-O)_2$	505/538	20	33
$[(CH_3)_2N-]_2$	540	20		$[(CH_3)_2N-]^2$	493	20	
				$[(C_2H_5)_2N-]_2$	515	20	
(CH_3-O) (CH_3-NH)	532/583	20	51	(CH_3-O) (CH_3-NH)	479/530	20	51
				(CH_3-O) $[(CH_3)2N]$	470/528	20	58
				(CH_3-O) (CH_3-NH)	491/543	20	52
				(CH_3-O) $(iso\text{-}C_3H_7-NH)$	508/538	20	30
				(C_2H_5-O) (CH_3-NH)	479/532	20	53
				(C_2H_5-O) (C_2H_5-NH)	483/532	20	49
				(C_2H_5-O) $(isoC_3H_7-NH)$	501/536	20	35
				$(isoC_3H_7-O)$ (CH_3-NH)	483/531	20	48
				$(isoC_3H_7-O)$ (C_2H_5-NH)	485/532	20	47
				$(isoC_3H_7-O)$ $(isoC_3H_7-NH)$	498/532	20	34

TABLE 8.7. Vibrational assignments for the CH_3, CD_3, C_2H_5, CH_3CD_2 and CD_3CH_2 groups of R–O–R(=O)Cl_2 and R–O–P(=S)Cl_2 analogs

[A] Compound CH_3–O–P(=S)Cl_2 cm^{-1}	[B] Compound CD_3–O–P(=S)Cl_2 cm^{-1}	[C] Compound	[D] Compound CH_3–O–P(=O)Cl_2 cm^{-1}	Assignment [22]	[A]/[B]
3030	2279		3030	a.CH_3 str., a″ or CD_3, a″	1.33
3004	2250		3008	a.CH_3 str., a′ or CD_3, a′	1.34
2959	2198		2960	s.CH_3 str., a′ [in Fermi res.] or CD_3, a′	1.35
2848	2071		2852	2(CH_3 bend), a′[in Fermi res.] or CD_3, a′	1.38
1455	1100		1455	a.CH_3 bend, a′ and a″ or CD_3, a′	1.32
1442	1095		1448	s.CH_3 bendR, a′ or CD_3, a′	1.32
1175*[1]	913		1178	CH_3 rock, a′ and a″ or CD_3, a″	1.29
?	?		?	CH_3 torsion, a″ or CD_3, a″	[?]
C_2H_5–O–P(=S)Cl_2 cm^{-1}	CH_3 – CD_2–O–P(=S)Cl_2 cm^{-1}	CD_3 – CH_2–O–P(=S)Cl_2 cm^{-1}	C_2H_5–O–P(=O)Cl_2 cm^{-1}		[D]/[C]
3000	2988	2258	3000	a.CH_3 str., a″ or CD_3, a″	1.33
2990	2988	2244	2991	a.CH_3 str., a′ or CD_3, a′	1.33
2945	2938	2129	2944	s.CH_3 str., a′ or CD_3, a′	1.38
1458	1458	1086	1457	a.CH_3 bend, a″ or CD_3, a″	1.34
1443	1441	1143	1445	a.CH_3 bend, a′ or CD_3, a′	1.26
1394	1383	1055	1374	s.CH_3 bend, a′ or CD_3, a′	1.3
1159	1148	876	1160	CH_3 rock, a″ or CD_3, a″	1.32
1100	1148	962	1100	CH_3 rock, a′ or CD_3, a′	1.14
					[D]/[B]
325?	320?	masked ? (200–300)		CH_3 torsion, a″ or CD_3, a″	[–]
2990	2275	2980	2991	a.CH_2 str., a″ or CD_2, a″	1.31
2972	2168	2970	2970	s.CH_2 str., a′ or CD_2, a′	1.37
1475	1119	1472	1479	CH_2 bend, a′ or CD_2, a′	1.32
1394	1034	1383	1394	CH_2 wag, a′ or CD_2, a′	1.35
1290	917	1268	1394	CH_2 twist, a″ or CD_2, a″	1.52
masked near 800 951/876	masked near 550	785	810	CH_2 rock, a″ or CD_2, a″	[?]

TABLE 8.8 The P=O stretching frequencies for inorganic and organic phosphorus compounds

Compound P=O(X)$_3$ X	References	P=O str. Rotational conformers/ cm^{-1}	Frequency difference between rotational conformers cm^{-1}	P=O str. for the other compounds minus P=O str. for P(=O)Cl$_3$ at 1303 cm^{-1} cm^{-1}
F_3	1	1405		102
Cl_3	15	1303		0
Br_2	13	1277		[−26]
$Br_2(BrCH_2)$	17	1275/1264	11	[−28]/[−39]
$Cl_2(CH_3)$	17,18	1277		[−26]
$Cl_2(CH_3)$	26	[1278.5*^{1};1268.9*^{2}]		[−24.5*^{1}]/[−34.6*^{2}]
Cl_2(tert-C_4H_9)	26	[1266.4*^{1};1255.6*^{2}]		[−36.6*^{1}]/[−47.4*^{2}]
$Cl_2(ClCH_2)$	17	1295/1288	7	[−8]/[−15]
$Cl_2(C_6H_5)$	20	1280		[−23]
$cl_2(CH_3-O)$	21	1322/1300	22	[19]/[−3]
$Cl_2(C_2H_5-O)$	22	1317/1296	21	[14]/[−7]
Cl_2(n-C_3H_7−O)	20	1313/1294	19	[10]/[−9]
Cl_2(n-C_4H_9−O)	20	1313/1295	18	[10]/[−8]
$Cl_2(C_6H_5-O)$	20	1316/1305	11	[13]/[2]
Cl_2(4-Cl−C_6H_4−O)	20	1315/1306	9	[12]/[3]
$Cl_2(CH_3-NH)$	24	1292/1279*^{3}	13	[−11]/[−24]
	20	1267*^{4}		[−36]
$Cl_2[(CH_3)_2N]$	20	1270		[−33]
$Cl_2[(C_2H_5)_2N]$	20	1280		[−23]
$Cl_2(CH_3-S)$	23	1279/1271	8	[−24]/[−32]
$Cl(CH_3-O)_2$	27	1308/1293	15	[5]/[−10]
$Cl(C_2H_5-O)_2$	20	1298/1285	13	[−5]/[−18]
Cl(n-C_3H_7−O)$_2$	20	1303/1285	18	[0]/[−18]
$Cl(O-CH_2CH_2-O)$	20	1315		[12]
$Cl(C_6H_5-O)_2$	20	1313/1301	12	[10]/[−2]
$Cl[(CH_3)_2-N]$	20	1243		[−60]
$Cl[(C_2H_5)_2-N]$	20	1258		[−45]
$Cl(CH_3)[(CH_3)_2-N]$	20	1248		[−55]
$Cl(CH_3-O)(CH_3-NH)$	20	1260*^{4}		[−43]
$(CH_3-O)_3$	20,26(mortimer)	1291/1272	19	[−12]/[−31]
$(C_2H_5-O)_3$	20	1280/1263	17	[−23]/[−40]
(n-C_3H_7−O)$_3$	20	1279/1265	14	[−24]/[−38]
(n-C_4H_9−O)$_3$	20	1280/1265	15	[−23]/[−32]
(isoC_3H_7−O)$_3$	20	1272/1257	15	[−31]/[−46]
$CH_3-O)_2(C_6H_5-O)$	20	1305/1297(1281)	8;24	[3]/[−6]/[−22]
$(C_2H_5-O)_2(C_6H_5-O)$	20	1305/1292(1276)	13;29	[2]/[−11]/[−27]
$(CH_3-O)(C_6H_5-O)_2$	20	1307/1298(1286)	9;21	[4]/[−5]/[−17]
$(C_2H_5-O)(C_6H_5-O)_2$	20	1306/1292(1285)	14;21	[3]/[−11]/[−18]
$(C_6H_5-O)_3$	20	1312/1298	14	[9]/[−5]
$(CH_3-O)_2(CH_3)$	20	1268/1252	16	[−35]/[−51]
$(C_2H_5-O)(CH_3)$	20	1268/1248	20	[−35]/[−55]
$(CH_3-O)_2(C_6H_5)$	20	1258		[−45]
$(CH_3-O)[(CH_3)_2-N]$	20	1260		[−43]
$[(CH_3)_2-N]_3$	20	1209		[−94]
[(n-C_3H_7−O)(C_2H_5−NH)P−O[$_2$(−O−)	20	~ 1248*^{4}		[−55]
$[(CH_3)_2-N]_3P=O]_2(-O-)$	20	~ 1255*^{4}		[−48]

*^{1}[CCl_4 soln.]

*^{2}[$CHCl_3$ soln.]

*^{3}[dilute soln.]

*^{4}[concentrated soln.]

TABLE 8.8a A comparison of P=O stretching frequencies in different physical phases

Compound	P=O str. liquid cm^{-1}	P=O str. CS_2 soln. rotational conformers cm^{-1}	P=O str. vapor cm^{-1}	Frequency difference between rotational conformers cm^{-1}	P=O str. (CS_2) minus P=O str. (liquid) or (solid) cm^{-1}	P=O str. (vapor) minus P=O str. (liquid) cm^{-1}	P=O str. (vapor) minus P=O str. (CS_2 soln.) cm^{-1}
CH_3–O–P(=O)Cl_2	1296	1322/1300		22	26		
C_2H_5–O–P(=O)Cl_2	1303/1289	1317/1296		21	14/7		
C_6H_5–O–P(=O)Cl_2	1304	1316/1305	1324	11	14	20	8
(CH_3–O–)$_3$P=O	1290 (solid)	1291/1272		19	18		
	1275				16		
(C_2H_5–O)$_3$P=O	1270 (solid)	1280/1263	1300/1280	17/20	20/17		20/17
	1290/1270					[10/10]	
		1261(CDCl3)					
(C_3H_7–O–P)$_3$=O	1269	1279/1265	1300/1282	14/18	10	31	21/17
(C_4H_9–O–)$_3$P=O	1270	1280/1265	1300/1282	16/18	10	20/17	
($C_6H_5CH_2$–O–)$_3$P=O	1255						
(CH_3–O–)2(CH_6H_5–O–)P=O	1285	1305/1297		20			
(CH_3–O–)(C_6H_5–O–)$_2$P=O	1297sh/1291	1307/1298		[10/7]			
(C_6H_5–O–)$_3$P=O	1314/1298	1312/1298	1311/1300	16/14/11			

TABLE 8.9 Infrared data for the rotational conformer P=O stretching frequencies for O,O-dimethyl O-(2-chloro-4-X-phenyl) phosphate

O,O-Dimethyl O-(2-chloro-4-X-phenyl) phosphate X	σ_p(29)	P=O str. rotational conformers (20) CS_2 soln. cm^{-1}	Frequency separation between rotational conformers cm^{-1}
NO_2	0.78	1307.4/1291.1	16.3
CN	0.66	1308.5/1291.0	17
Cl	0.23	1304.5/1288.4	15.6
H	0	1303.6/1287.8	15.8
$C(CH_3)_3$	−0.27	1304.2/1285.8	18.4
CH_3–O	−0.27	1303.2/1286.7	16.5

TABLE 8.10 Infrarred observed and calculated P=O stretching frequencies compared*[1]

G − P(=O)Cl_2 Substituent Group G	π	P=O str. Calc. by Eq. 1 cm^{-1}	P=O str. observed cm^{-1}	δ P=O str. cm^{-1}	Average δ P=O str. cm^{-1}
R	2	1262	1277	15	
R_2N	2.4	1278	1280	2	
C_6H_5	2.4	1278	1280	2	
C=C	2.4	1278			
H	2.5	1282			
C=O	2.5	1282			
$ClCH_2$	2.7	1290	1295/1288	5/−2	1.5
R-O	2.8	1294	1322/1300	28/6	17
Cl_2CH	2.9	1298			
CCl_3	3	1302			
C_5H_6−O	3	1302	1316/1305	14/3	8.5
Br	3.1	1306	1295	−11	
C=C−O	3.1	1306			
CF_3	3.3	1314			
R(C=O)−O	3.4	1318			
Cl*[2]	3.4	1318	1300	−18	
CN	3.5	1322			
F	3.9	1338	1340	2	

*[1]The calculated frequencies were obtained using the equation developed by Thomas and Chittenden (30).

*[2][2(π) for $_2$Cl is 6.3 not 6.8].

$\nu P{=}O(cm^{-1}) = 930 + 40\sigma\pi$.

TABLE 8.11 The P=O and P=S stretching frequencies for phenoxarsine derivatives

Phenoxarsine	[31] P=S str. [CS_2] cm^{-1}
X=S−P=S(O−R)$_2$	
R	
methyl	648
isopropyl	645
X=OP=O(O−R)$_2$	P=O str.
R	
ethyl	1235
isopropyl	1228

TABLE 8.12 The P=O and P=S stretching frequencies for P(=O)X3 and P(=S)X3 type compounds

X_3	$P(=O)X_3$ P=O str. cm^{-1}	$P(=S)X_3$ [P=S str.] cm^{-1}	[P=O str.]– [P=S str.] cm^{-1}	[P=O.]/ [P=S str.]	References
F_3	1415	695	720	2.04	1,2,3,4
F_2Cl	1358				7
F_2Br	1350				7
FCl_2	1331	755	576	1.76	9
Cl_3	1326	771	555	1.72	8
FBr_2	1303				10,11
Cl_2Br	1285				5
Br_3	1277	730	547	1.75	5
$ClBr_2$	1275				4
					5

TABLE 8.13 The P=S stretching frequencies for inorganic and organic phosphorus compounds

Compound P=SX$_3$ X	P=S str. rotational conformers cm^{-1}	Ref.	Frequency separation between rotational conformers	Compound P=SX$_3$ X	P=S str. rotational conformers/ cm^{-1}	Ref.	Frequency separation between rotational conformers
F_3	696	33		$CH_3-O)_3$	620/600	20	20
F_2Cl	727	33		$(C_2H_5-O)_3$	635/614	20	21
F_2Br	711	34		$(isoC_3H_7-O)_3$	672/650	20	22
FCl_2	741	9,33		$(CH_3)(C_2H_5-O)_2$	~632	20	
Cl_3	752	33		$(C_6H_5)(CH_3-O)_2$	624	20	
Br_3	730	5		$(CH_3-O)_2[(CH_3)_2-N]$	588	20	
$Cl_2(CH_3)$	670	7,18		$(C_2H_5-O)2(N_3)$	640/627	20	13
$Cl_2(ClCH_2)$	683/657	17	26	$(CH_3-O)_2(CH_3-NH)$	630/597	20	33
$Cl_2(C_6H_5)$	690	20		$(C_2H_5-O)_2(CH_3-NH)$	640/608	20	32
$Cl_2(CH_3-O)$	721/703	21,25	18	$(CH_3-O)_2(isoC_3H_7-NH)$	645/611	20	34
$Cl_2(CD_3-O)$	709/698	21,25	11	$(C_2H_5-O)_2(isoC_3H_7-NH)$	650/617	20	33
$Cl_2(C_2H_5-O)$	728/701	22	27	$(CH_3-O)2(NH_2)$	635/611	20	24
$Cl_2(CH_3-CD_2-O)$	727/699	22	28	$(C_2H_5-O)_2(NH_2)$	641/618	20	23
$Cl_2(CD_3-CH_2-O)$	695/675	22	30	$(isoC_3H_7-O)_2(NH_2)$	643/621	20	22
$Cl_2(isoC_4H_9-O)$	737/715	20	22	$(n-C_4H_9-O)_2(NH_2)$	653/628	20	25
$Cl_2(H-CC-CH_2-O)$	742/716	35	26	$(isoC_4H_9-O)(NH_2)$	665/641	20	24
$Cl_2(4-Cl-C_6H_4\text{-}O)$	721	20		$[(CH_3)_2-N]_3$	565	20	
$Cl_2(CH_3-NH)$	725/690	22	35	$[(CH_3)_2-N]_2(N_3)$	607	20	
$Cl_2(CD_3-NH)$	711/683	22	28	$[(CH_3)_2-N]_2(NH_2)$	579	20	
$Cl_2(CH_3-ND)$	728/681	22	47	$(CH_3)[(CH_3)_2-N]_2$	569	20	
$Cl_2(CD_3-ND)$	701/673	22	28	$(C_2H_5-S)_3$	685	20	
$Cl_2(C_2H_5-NH)$	725/688	22	37	$(isoC_2H_7-S)_3$	685	20	
$Cl_2(isoC_3H_7-NH)$	727/690	22	37	$(C_2H_5-S)(CH_3-NH)$	693/663	20	30
$Cl_2[(CH_3)_2-N]$	673	20		$(CH_3)_2(P{=}S)(P{=}S)(CH_3)_2$	576	20	
$Cl_2[(C_2H_5)_2-N]$	667	20		$[(C_2H_5-O)_2(P{=}S)]_2(-O-)$	~636	20	
$Cl_2(C_2H_5-S)$	730			$[(CH_3-O)(CH_3-NH)P{=}S]_2(-O-)$	651/635	20	16
$Cl(CH_3-O)_2$	666/655	27	11	$[(CH_3-O)(C_2H_5-NH)P{=}S]_2(-O-)$	660/642	20	18
$Cl(CD_3-O)_2$	~650	27		$[(CH_3-O)(n\text{-}C_3H_7-NH)P{=}S]_2(-O-)$	660/642	20	18
$Cl(C_2H_5-O)_2$	673/660	32	13	$[(CH_3-O)(isoC_3H_7-NH)P{=}S]_2)-O-)$	666/647	20	19
$Cl(n\text{-}C_3H_7-O)_2$	675/661	20	14	$[(CH_3-O)(isoC_4H_9-NH)P{=}S]_2(-O-)$	660/645	20	15
$Cl[(CH_3)_2-N]_2$	613	20		$(C_2H_5-O)(CH_3-NH)P{=}S]_2(-O-)$	654/641	20	13

(continues)

TABLE 8.13 (*continued*)

Compound $P{=}SX_3$ X	P=S str. rotational conformers cm^{-1}	Ref.	Frequency separation between rotational conformers	Compound $P{=}SX_3$ X	P=S str. rotational conformers/ cm^{-1}	Ref.	Frequency separation between rotational conformers
$Cl[(C_2H_5)_2{-}N]_2$	611	20		$[(n{-}C_3H_7{-}O)(CH_3{-}NH)P{=}S]_2({-}O{-})$	660/645	20	15
$Cl[{-}N(CH_3)CH_2CH_2(CH_3)N{-}]$	603	20		$[(isoC_3H_7{-}O)(CH_3{-}NH)P{=}S]_2({-}O{-})$	~652/638	20	14
$Cl(CH_3{-}O)[(CH_3)_2{-}N]$	633	20		$[(isoC_3H_7{-}O)(C_2H_5{-}NH)P{=}S]_2({-}O{-})$	652/639	20	13
$Cl(CH_3{-}O)(CH_3{-}NH)$	660/639	20	21	$[(isoC_3H_7{-}O)(NH_2)P{=}S]_2({-}O{-})$	632	20	
$Cl(CH_3{-}O)(C_2H_5{-}NH)$	663/639	20	24	$[(CH_3)_2{-}N)_2P{=}S]_2({-}O{-})$	598	20	
$Cl(CH_3{-}O)(isoC_3H_7{-}NH)$	673/643	20	30	$[(C_2H_5{-}O)_2P{=}S]_2({-}S{-})$	~645	20	
$Cl(C_2H_5{-}O)(CH_3{-}NH)$	666/643	20	23	$[(C_2H_5{-}O)_2P{=}S({-}S{-})]_2$	662	20	
$Cl(C_2H_5{-}O)(C_2H_5{-}NH)$	663/643	20	20				
$Cl(C_2H_5{-}O)(isoC_3H_7{-}NH)$	673/647	20	26				
$Cl(isoC_3H_7{-}O)(CH_3{-}NH)$	660/639	20	21				
$Cl(isoC_3H_7{-}O)(C_2H_5{-}NH)$	663/642	20	21				
$Cl(isoC_3H_7{-}O)(isoC_3H_7{-}NH)$	665/645	20	20				
$Cl(C_2H_5{-}S)_2$	705	20					
$Cl(isoC_3H_7{-}S)_2$	705	20					

TABLE 8.14 Vibrational assignments for the skeletal modes of S-methyl phosphorothiodichloridate and phosphoryl chloride

$CH_3{-}S{-}P({=}O)Cl_2$ cm^{-1}	Assignment(23)	$P({=}O)Cl_3$ cm^{-1}	Species
1275	P=O str., a′, rotational conformer 1	1303	al
1266	P=O str., a′, rotational conformer 2		
450	s.PCl_2S str.,a′	483	al
471	s.PCl_2S str., a′		
235	s.PCl_2S bend, a′, rotational conformer 1	268	al
262	s.PCl_2S bend, a′, rotational conformer 2		
593	a.PCl_2S str., a′, rotational conformer 1	588	e
579	a.PCl_2S str., a′, rotational conformer 2		
547	a.PCl_2S str., a″, rotational confomer 1		
560	a.PCl_2S str., a″, rotational confomer 2		
323	P=O rock, a′	335	e
344	P=O rock, a″		
190	PCl_2S bend, a′ and a″	193	e
698	C–S str., rotational conformer 1		
732	C–S str., rotational conformer 2		
223	P–S–C bend, a′		
168 or 148	P–S–C torsion, a″		

TABLE 8.15 Vibrational assignments of the $P({-}O{-}C)_3$ group of trimethyl phosphite and trimethyl phosphate

[36] $(CH_3{-}O)_3P$ cm^{-1}	[26-mortimer] $(CH_3{-}O)_3P{=}O$ cm^{-1}	Assignment	Number of modes
1059*1	1045	s.$P({-}O{-}C)_3$ str.	1
1016	1080	a.$P({-}O{-}C)_3$ str.	2
769*1	750	s.$P({-}O{-})_3$ str.	1
750	865	a.$P({-}O{-})_3$ str.	2
732	858	a.$P({-}O{-})_3$ str.	
513*1	533*2	s.$P({-}O{-}C)_3$ bending	1
534	510*2	a.$P({-}O{-}C)_3$ bending	2
370*1	367(R)*3	s.$P({-}O{-})_3$ bending	1
280*1	239(R)	a.$P({-}O{-})_3$ bending	2
225		a.$P({-}O{-})_3$ bending	
190	184?(R)	s.P–O–C torsion	1
120(?)	(?)	a.P–O–C torsion	2

*1Raman polarized band.
*2Assignments were reversed by Mortimer (26a).
*3Abbreviations: R = Raman; s. = symmetric; and a. = antisymmetric.

TABLE 8.16 Vibrational assignments for the CH_3 groups of trimethyl phosphite

$(CH_3-O)_3P$ cm^{-1}	Number of normal modes	Assignment (36)
2990	6	a.CH_3 str.
2949		
2939	3	s.CH_3 str.
1459	6	a.CH_3 bend
1436	3	s.CH_3 bend
1179*3	3	CH_3 rock*1
1159(R)	3	CH_3 rock*2
*4	3	CH_3 torsion

*11 Out of P–O–C plane.
*2In P–O–C plane.
*3R = Raman.
*4Well below 300 cm^{-1}.

TABLE 8.17 The P=S, P–S, and S–H stretching frequencies for O,O-dialkyl phosphorodithioate and O,O-diaryl phosphorodithioate

$(CH_3-O-)_2P(=S)SH$ cm^{-1}	$(C_2H_5-O-)_2P(=S)SH$ cm^{-1}	$(2,4,5-Cl_3-C_6H_4-O-)_2$ P(=S)SH cm^{-1}	Assignment(32)
2588	2582	2575	S–H str., rotational conformer 1
2550	2550	2549	S–H str., rotational conformer 2
670sh	670sh		P=S str., rotational conformer 1
659	659		P=S str., rotational conformer 2
524	535		P–S str., rotational conformer 1
490	499		P–S str., rotational conformer 2

TABLE 8.18 The P=O, PH stretching and PH bending vibrations for O,O-dialkyl hydrogenphosphonates*

$(R-O-)_2P(=O)H$ R	P−H str. $[CCl_4]$ cm^{-1}	P=O str. $[CS_2]$ cm^{-1}	P−H bend $[CS_2]$ cm^{-1}	P−H str. neat cm^{-1}	P=O str. neat cm^{-1}	P−H bend neat cm^{-1}	P−H str. cm^{-1}	P−H bend cm^{-1}	delta P=O str. cm^{-1}
CH_3	2438	1283sh 1266	979	2415	1255	968	23	11	28;11
C_2H_5	2439	1275sh 1262	~980	2419	1250	970	24	10	25;12
$n\text{-}C_3H_7$	2441	1273sh 1261	973						
$isoC_4H_9$	2435	~1275 1257	~975						
$2\text{-}C_2H_5-C_6H1_3$	2431	1273 1257	~970	2410	1250	955	21	15	23;7
$2\text{-}(2-C_2H_5-C_6H_{12}-O-)$ C_2H_4	2445	1273 1261	968						

*See Reference 32.

TABLE 8.19 The "C–O" and "P–O" stretching frequencies for the C–O group

Compound	"C–O stretching" cm^{-1}	"P–O stretching" cm^{-1}	References
$(C-O)_3P$	1059(s.)	769(s.)	36
	1016(a.)	750,732(a.)	
$(C_2H_5-O-)_3P$	1060(s.)	765(s.)	20
	1035[1025](a.)	736(a.)	
$(C_6H_5-O-)_3P$	[1200]1190,1162	878,862	20
$CH_3-O-P(=O)Cl_2$	1060/1027	809/802	26 (mortimer)
$C_2H_5-O-P(=O)Cl_2$	1041/1013	779	22
n-$C_3H_7-O-P(=O)Cl_2$	1032,1013[995]	794	20
n-$C_4H_9-O-P(=O)Cl_2$	[1034,1032,1013,995]	794	20
$CH_3-O-P(=S)Cl_2$	1041/1023	828/821	20
$C_2H_5-O-P(=S)Cl_2$	1028/1014	790/804	20
n-$C_3H_7-O-P(=S)Cl_2$	1050,1036,1020[1000]	838	20
$H-CC-CH_2-OP(=S)Cl_2$	1015/984	857/822	20
$(CH_3-O=)(CH_3-NH-)P(=S)Cl$	~1043	~820	20
$(C_2H_5-O-)(CH_3-NH-)P(=S)Cl$	~1030	~808	20
(iso$C_3H_7-O-)(CH_3-NH-)P(=S)Cl$	~990	~780	20
$(CH_3-O-)(C_2H_5-NH-)P(=S)Cl$	~1041	~812	20
$(C_2H_5-O-)(C_2H_5-NH-)P=S)Cl$	~1025	~799	20
(iso$C_3H_7-O-)(C_2H_5-NH-)P(=S)Cl$	~990	~781	20
(CH_3-O-)(iso$C_3H_7-NH-)P(=S)Cl$	~1040	~835(?)	20
(C_2H_5-O-)(iso$C_3H_7-NH-)P(=S)Cl$	~1030	~787(?)	20
(iso$C_3H_7-O-)$(iso$C_3H_7-NH-)P(=S)Cl$	~990	~778	20
$CH_3-O)[(CH_3)_2N-]P(=S)Cl$	~1042	~828	20
[(iso$C_3H_7-O-)(NH_2-)P(=S)]2(-O-)$	1050	816	20
$[(C_2H_5-O-)(CH_3NH-)P(=S]_2(-O-)$	~1045	811	20
[(n-$C_3H_7-O-)(CH_3-NH-)P(=S)]_2)-O-)$	1060[1007]	~855	20
[(iso$C_3H_7-O-)(CH_3-NH-)P(=S)]_2(-O-)$	~1005	~795	20
$[(CH_3-O-)(C_2H_5-NH-)P(=S)]_2)-O-)$	1050	815	20
[(iso$C_3H_7-O-)(C_2H_5-NH-)P(=S)]_2(-O-)$	~1000	~780	20
$[(CH_3-O-)$(n-$C_3H_7-NH-)P(=S)]_2(-O-)$	1050	819	20
$[(CH_3-O-)$(iso$C_3H_7-NH-)P(=S)]_2(--)$	1040	805	20
$[(CH_3-O-)$(iso$C_4H_9-NH-)P(=S)]_2(-O-)$	~1045	810	20
$(CH_3-O-)_2P(=O)Cl$	1068,1044[1036]	[845]779,769	27
$(C_2H_5-O-)_2P(=O)Cl$	1057,1030[1022]	800,763	27
(n-$C_3H_7-O-)_2P(=O)Cl$	1060,1050[1025]	853,827	20
$(-O-CH_2-CH_2-O-)P(=O)Cl$	1031	861,826	20
$(CH_3-O-)_2P(=O)H$	1051(s.)	820(a.)	32
	1080(a.)	765(s.)	
$(CH_3-O-)_2P(=O)D$	1068(s.)	850(a.)	32
	1040(a.)	791(s.)	
$(C_2H_5-O-)_2P(=O)H$	1081(s.)	782(a.)	32
	1052(a.)	741(s.)	
$(C_2H_5-O-)_2P(=O)D$	1061(s.)	8216(a.)	32
	1030(a.)	781(s.)	
$(C_2H_5-O-)_2(CH_3-)P(=O)$	1061[1032]	770,713	20
$(CH_3-O-)_2(C_6H_5-)P=O)$	1059[1029]	828,788	20
$(CH_3-O-)_2(CH_3-NH-)P(=O)$	1063[1033]	[830]742	37
$(C_2H_5-O-)_2(CH_3-NH-)P(=O)$	1059[1032]	793,740	37

(*continues*)

TABLE 8.19. *(continued)*

$(CH_3-O-)_2[(CH_3)_2N-]P(=O)$	1065[1030]	[825]808	20
$(CH_3-O-)_2P(=S)Cl$	1056/1042(s.)	845(s.)	18
	1040(s.)	831(s.)	20
$(C_2H_5-O-)_2P(=S)Cl$	1017(a.)	819(a.)	20
$(C_2H_5-O-)_2(CH_3-)P(=S)$	1054[1048]	762.755	20
$(CH_3-O-)_2(C_6H_5-)P(=S)$	1058[1030]	817,794	20
$(CH_3-O-)_2(NH_2-)P(=S)$	1065[1034]	790(?)	20
$(C_2H_5-O-)_2(NH_2-)P(=S)$	1058[1028]	781	20
$(isoC_3H_7-O-)_2(NH_2-)P(=S)$	1011[982]	803[771]	20
$(n\text{-}C_4H_9-O-)_2(NH_2-)P(=S)$	1065,1048[1022][979]	839,823[798]	20
$(isoC_4H_9-O-)_2(NH_2-)P{=}S)$	1050[1110]	[859]839,820	20
$(CH_3-O-)_2(CH_3-NH-)P(=S)$	1061[1033]	[816]799	20
$(C_2H_5-O-)_2(CH_3-NH-)P(=S)$	1055[1029]	809,775	20
$(CH_3-O-)_2(isoC_3H_7-NH-)P(=S)$	1060[1033]	805,784	20
$(C_2H_5-O-)_2(iso(C_3H_7-NH-)P(=S)$	1058[1035]	780	20
$(CH_3-O-)_2[(CH_3)_2N-]P(=S)$	1060[1030]	821,803	20
$(C_2H_5-O)_2(N_3-)P(=S)$	1044[1019]	820,800	20
$(CH_3-O-)_2(HS-)P(=S)$	1050/1037(s.)	801(s.)	32
$(C_2H_5-O-)_2(HS-)P(=S)$	1040(s.)	849	32
	1017(a.)	810(?)	
$[(C_2H_5-O-)_2P(=S)]_2(-O-)$	1059[1030]	755/739(s.)	20

TABLE 8.20. The "C–O" and "P–O" stretching frequencies for compounds containing C–O–P=O, C–O–P=S, and C–O–P=Se groups

Compound	Ref.	"C–O stretching" R-O–P cm^{-1}	"P–O stretching" R-O–P cm^{-1}	"Aryl-O stretching" Aryl-O–P cm^{-1}	"P–O stretching" Aryl-O–P cm^{-1}
$(CH_3–O–)_3P=O$	26 (Mortimer)	1085(s.)	755/739(s.)		
		1-55/1040(a.)	835/844(a.)		
$(C_2H_5–O–)_3P=O$	20	1073(s.)	744(s.)		
		1033(a.)	892/798(a.)		
$(n\text{-}C_3H_7–O–)_3P=O$	20	1053,1039[1008]	888,863,756		
$(isoC_3H_7–O–)_3P=O$	20	1028[1002]	[791]752		
$(n\text{-}C_4H_9–O–)_3P=O$	20	1058[1030]1003,987	862,850,770		
$(CH_3–O–)_3P=S$	20	1071(s.)	812(s.)		
		1043(a.)	832(a.)		
$(C_2H_5–O–)_3P=S$	20	1063(s.)	792(s.)		
		1031(a.)	825(a.)		
$(isoC4H_9–O–)_3P=S$	20	1058[1018]	[865]822		
$(CH_3–O–)_3P=Se$	38	1028(s.)	780(s.)		
		1028(a.)	826(a.)		
$Cl_2(C_6H_5–O–)P=O$	20			[1183]1160	957[942]
$Cl_2(4–Cl–C_6H_4–O–)P=O$	20			[1185]1160	953,931
$Cl_2(C_6H_5–O–)P=S$	20			[1180]1157	964[932]
$Cl_2(_4–Cl–C_6H_4–O–)P=S$	20			[1187]1158	[937,920]

Table 8.20 (*continued*)

$Cl(C_6H_5{-}O{-})_2P{=}O$	20			1205[1180]1159	[962]940
$Cl(C_6H_5{-}O{-})_2P{=}S$	20			{1201,1182,1161}	~940
$(C_6H_5{-}O{-})_3P{=}O$	20			[1190]1160	~960
$(CH_3{-}O{-})_2(C_6H_5{-}O{-})P{=}O$	20	1069[1047]	[854]796(?)	[1213]1165	[951]937
$(C_2H_5{-}O{-})_2(C_6H_5{-}O{-})P{=}O$	20	1063[1036]	820,802[780]	[1215]1165	[960]940
$(CH_3{-}O{-})(C_6H_5{-}O{-})_2P{=}O$	20	1050	823,812	1220[1192]1161	951
$(C_2H_5{-}O{-})(C_6H_5{-}O{-})_2P{=}O$	20	1041	792,770	1230[1192]1162	950
$(CH_3{-}O{-})_2(C_6H_5{-}O{-})P{=}S$	20	1066[1041]	[834]818	1207[1165]	938
$(C_2H_5{-}O{-})_2(C_6H_5{-}O{-})P{=}S$	20	1062[1035]	[828]822	1212[1165]	948
$(C_2H_5{-}O{-})(C_6H_5{-}O{-})_2P{=}S$	20	1048	[837]816	1211[1187]1161	938
$(CH_3{-}O{-})_2(2\text{-}Cl_6H_4{-}O{-})P{=}O$	20	1073[1056]	859	[1234]	950,938
$(CH_3{-}O{-})_2(2\text{-}Cl{-}4{-}NO_2{-}C_6H_3{-}O{-})P{=}O$	20	1071[1050	861	[1261]	941
$(CH_3{-}O{-})_2(2\text{-}Cl,4{-}CN{-}C_6H_3{-}O{-})P{=}O$	20	1075[1056]	862	[1256]	941
$(CH_3{-}O{-})_2(2,4\text{-}Cl_2{-}C_6H_3{-}O{-})P{=}O$	20	1073[1075]	859	{1260,1238}	949,937
$(CH_3{-}O{-})_2(2\text{-}Cl,4{-}t{-}C_4H_9{-}C_6H_3{-}O{-})P{=}O$	20	1075p1053]	858	{1263,1243}	954,944
$(CH_3{-}O{-})_2(2\text{-}Cl,_4{-}CH_3{-}O{-}C_6H_3{-}O{-})P{=}O$	20	1071[1048]	857	{1263,1214}	951,941
$(CH_3{-}O{-})_2(2\text{-}Cl{-}C_6H_4{-}O{-})P{=}S$	20	1067,1052[1038]	838	1232	938,928
$(CH_3{-}O{-})_2(2\text{-}Cl,4{-}NO_2{-}C_6H_3{-}O)P{=}S$	20	1061[1040]	844	1260	928
$(CH_3{-}O{-})_2(2\text{-}Cl,4{-}CN{-}C_6H_3{-}O{-})P{=}S$	20	1065,1049[1035]	840	1248	922
$(CH_3{-}O{-})_2(2,4\text{-}Cl_2{-}C_6H_3{-}O{-})P{=}S$	20	1065,1052[1038]	845	{1259,1238}	952,940
$(CH_3{-}O{-})_2(2\text{-}Cl,4\text{-}t{-}C_4H_9{-}C_6H_3{-}O{-})P{=}S$	20	1066,1051[1040]	841	{1261,1239}	948,937

TABLE 8.21 The aryl-O stretching frequencies for O-methyl O-(X-phenyl) N-methylphosphoramidate

Compound	Aryl-O stretching
O-methyl O-(x-Phenyl) N-methylphosphoramidate X	[39] cm^{-1}
4-nitro	1237
2,4,5-trichloro	1259
2,4-dichloro	1245
2-chloro	1235
4-chloro	1228
hydrogen	1213
4-methoxy	1207
3-tert-butyl	1209
4-tert-butyl	1225
2-chloro-4-tert-butyl	1243

TABLE 8.22. The C–P stretching for organophosphorus compounds

Compound	Ref.	C–P str.	Assignment	Frequency separation between rotational conformers
$Cl_2(CH_3-)P$	17	699		
$Cl_2(CH_3-)P(=O)$	17	757		
$Cl_2(CH_3-)P(=S)$	17	810		
$Cl_2(ClCH_2-)P$	17	812	[not split by rotational conformers]	0
$Cl_2(ClCH_2-)P(=O)$	17	811,818	[rotational conformers]	7
$Cl_2(ClCH_2-)P(=S)$	17	801,833	[rotational conformers]	32
$Cl_2(BrCH_2-)P(=O)$	17	778,788	[rotational conformers]	10
$CH_3-PO_3Na_2$	40	753		
$(CH_3-)_2PO_2Na$.	41	737	a.PC2 str.	
	41	700	s.PC2 str.	
$(CH_3-O-)_2(CH_3-)P(=O)$	20	711		
$(CH_3-O-)_2(CH_3-)P(=S)$	20	793		
Range		699–833		

TABLE 8.23 The trans and/or cis N–H stretching frequencies for compounds containing the P–NH–R group

Compound	Ref.	NH str. cis CCl_4 soln. cm^{-1}	NH str. trans CCl_4 soln. cm^{-1}	Frequency separation [cis]-[trans] cm^{-1}
$Cl_2(CH_3{-}NH{-})P(=O)$	24	3408		
$Cl_2(CH_3{-}NH{-})P(=S)$	24,42	3408		
$Cl_2(CD_3{-}NH{-})P(=S)$	24,42	3409		
$Cl_2(C_2H_5{-}NH{-})P(=S)$	24,42	3397		
$Cl_2(isoC_3H_7{-}NH{-})P(=S)$	24,42	3388		
$Cl(CH_3{-}O{-})(CH_3{-}NH{-})P(=S)$	28	3425		
$Cl(CH_3{-}O{-})(C_2H_5NH{-})P(=S)$	28	3411		
$Cl(CH_3{-}O{-})(isoC_3H_7{-}NH{-})P(=S)$	28	3403	3385sh	18
$Cl(C_2H_5{-}O{-})(CH_3{-}NH{-})P(=S)$	28	3430	~3400sh	30
$Cl(C_2H_5{-}O{-})(C_2H_5{-}NH{-})P(=S)$	28	3402		
$Cl(C_2H_5{-}O{-})(isoC_3H_7{-}NH{-}(P(=S)$	28	3408	3390sh	18
$Cl(isoC_3H_7{-}O{-})(CH_3{-}NH{-})P(=S)$	28	3422		
$Cl(isoC_3H_7{-}O{-})(C_2H_5{-}NH{-})P(=S)$	28	3409		
$Cl(isoC_3H_7{-}O{-})(isoC_3H_7{-}NH{-})P(=S)$	28	3405	3390sh	15
$(CH_3{-}O{-})_2(CH_3{-}NH{-})P(=O)$	28	3444		
$(CH_3{-}O{-})_2(CH_3{-}NH{-})P(=S)$	28	3447	3403	44
$(C_2H_5{-}O{-})_2(CH_3{-}NH{-})P(=O)$	28	3439		
$(C_22H_5{-}O{-})_2(CH_3{-}NH{-})P(=S)$	28	3442	3404	38
$(CH_3{-}O{-})_2(isoC_3H_7{-}NH{-})P(=S)$	28	3429	3387	42
$(C_2H^5{-}O{-})2(isoC_3H_7{-}NH{-})P(=S)$	28	3419	3380	39
$(CH_3{-}O{-})(2,4,5\text{-}Cl_3{-}C_6H_2{-}O{-})(CH_3{-}NH{-})P(=O)$	28	3429*		
$(CH_3{-}O{-})(2,4,5\text{-}Cl_3{-}C_6H_2{-}O{-})(CH_3{-}NH{-})P(=S)$	28	3442	3409	31
$(cH_3{-}O{=})(2,4,5\text{-}Cl_3{-}C_6H_2{-}O{-})(C_2H_5{-}NH{-})P(=O)$	28	3425		
$(CH_3{-}O{-})(2,4,5\text{-}Cl_3{-}C_6H_2{-}O{-})(C_2H_5{-}NH{-})P(=S)$	28	3429	3398	31
$(CH_3{-}O{-})(2,4,5\text{-}Cl_3{-}C_6H_2{-}O{-})(iso\text{-}C_4H_9{-}NH{-})P(=O)$	28	3428		
$(CH_3{-}O{-})(2,4,5\text{-}Cl_3{-}C_6H_2{-}O{-})(iso\text{-}C_4H_9{-}NH{-})P(=S)$	28	3431	3400	31
$(CH_3{-}O{-})(2,4,5\text{-}Cl;_3\ {-}C_6H_2{-}O{-})(isoC_3H_7{-}NH{-})P(=O)$	28	3413		
$(CH_3{-}O{-})(2,4,5\text{-}Cl_3{-}C_6H_2{-}O{-})(isoC_3H_7{-}NH{-})PP{=}S)$	28	3419	3385	34
$(CH_3{-}O{-})(2,4,5\text{-}Cl_3{-}C_6H_2{-}O{-})(t\text{-}C_4H_9{-}NH{-})P(=O)$	28	3401		
$(CH_3{-}O{-})(2,4,5\text{-}Cl_3{-}C_6H_2{-}O{-})(t\text{-}C_4H_9{-}NH{-})P)(=S)$	37	3402		
$(CH_3{-}O{-})_2(HDN{-})P(=S)$	37	3461	3434	27
$(C_2H_5{-}O{-})_2(HDN{-})P(=S)$	37	3452	3427	25

*In CH_2Cl_2 solution.

TABLE 8.24 The NH_2, NHD, NH_2, NH and ND frequencies for O,O-dimethyl phosphoramidothioate, O,O-diethyl phosphoramidothioate, and N-methyl O,O-dimethylphosphoramidothhioate

O,O-dimethyl phosphoramidothioate $NH_2:ND_2$ cm^{-1}	$[NH_2], [ND_2]$	O,O-diethyl phosphoramidothioate $NH_2:ND_2$ cm^{-1}	$[NH_2]/[NH_2]$	O,O-diethyl N-methyl-phosphoramidothioate NH:ND cm^{-1}	[NH]/[ND]	Assignment(37)
3480:2611	1.33	3481:2600	1.34			a.NH_2 or a.ND_2 str.
3390:2490	1.36	3389:2491	1.36			s.NH_2 or s.ND_2 str.
3450:(2515)*1	1.37	~3450				NH_2 or ND_2 str. bonded
3350:(2410)*1	1.39	~3352:~2460	1.36			
3310:~2450	1.35	~3305				
NHD:NHD		NHD:NHD				
3461:2561	1.35	3452:2553	1.35	3442:2562	1.34	cis NH or ND str.
3434:2535	1.34	3427:2530	1.35	3404:2539	1.34	trans NH or ND str.
				3320:2470	1.34	NH or ND str. bonded
~3070		2342		~3070:2345	1.31	2(NH or ND bend)
1542:1171	1.32	1542:1170 (1179)*1	1.32			NH_2 or ND_2 bend
				1380:1250	1.1	NH or ND bend
NHD		NHD				
~1382		~1385				NHD bend
971:760	30:43:12	977:785	1.24			NH_2 or ND_2 wag*2
922:718	1.28	913:721	1.27			NH_2 or ND_2 twist*2
:938		:masked*2				NHD twist

*1 Liquid.
*2 See text.

TABLE 8.25 The N–H and N–D stretching and bending frequencies for O-alkyl O-(2,4,5-trichlorophenyl) N-alkylphosphoramidate and O-methyl O-(2,4,5-trichlorophenyl) N-alkylphosphoramidothioate

O-methyl O-(2,4,5-trichlorophenyl) N-alkylphosphoramidate	NH str. $[CCl_4]$ *2 cm^{-1}	ND str. $[CCl_4]$ *3 cm^{-1}	δ $[CCl_4]$ cm^{-1}	NH str. $[CH_2Cl_2]$ *2 cm^{-1}	ND str. $[CH_2Cl_2]$ *3 cm^{-1}	δ $[CH_2Cl_2]$ cm^{-1}	O-methyl O-(2,4,5-trichlorophenyl) N-alkylphosphoramidothioate	NH str. cis $[CCl_4]$ cm^{-1}	NH str. trans $[CCl_4]$ cm^{-1}	[NH str. cis]-[NH str. trans] cm^{-1}
N-alkyl							N-alkyl			
methyl	insol.	insol.		3429	3256	173		3442	3409	33
				[2550]*1	[2432]	118				
(NH/ND)				(~1.34)	(~1.34)					
ethyl	3425	3224	201	3415	3245	170		3429	3398	31
	[~2545	[~2394]								
(NH/ND)	(~1.34)	(~1.35)	151							
isobutyl	3428	3228	200	3417	3249	168		3431	3400	31
	[2545]	[2412]	154							
(NH/ND)	(~1.34)	(~1.34)								
isopropyl	3413	3213	200	3405	3230	175		3419	3385	34
	[2538]	[2384]	154							
(NH/ND)	−1.34	−1.35								
t-butyl	3401	3201	200	3395	3214	181		3402		
	[2528]	[2384]	144							
[NH/ND]	1.34	(~1.34)								
	δ NH $[CCl_4]$*2	δ ND $[CCl_4]$*3	δ NH $[CH_2Cl_2]$*2	δ ND $[CH_2Cl_2]$*3						
methyl			1397	1426				1397		
ethyl	1415	~1442	1415	1442				1403		
		[1214]								
(NH/ND)		(~ 1.19)								
isobutyl	~1418	~1440	1421					1406		
		~1442								
isopropyl	~1417	1440	1417					1404		
		[~1211]								
(NH/ND)		(~1.19)								
t-butyl	1396	1440 : 1408	1393	1425				1382		
		[1222?]		1410						
(NH/ND)		(~1.18 or ~1.15)								

*1 N–D frequencies.
*2 Monomer.
*3 Bonded.

TABLE 8.26. The cis and trans NH stretching frequencies for compoundsk containing O=P−NH−R or S=P−NH=R groups*[1]

Compound	N-Alkyl	N−H str. cis cm^{-1}	N−H str. trans cm^{-1}	[NH str. cis]-[N−H str. Trans.] cm^{-1}	N−H bending cm^{-1}
O-alkyl kO-aryl N-alkylphosphoramidate	CH_3	3444			1393
	C_2H_5	3428			1413
	n-C_3H_7	3430			1415
	n-C_4H_9	3430			1416
	isoC_3H_7	3418			1415
	t-C_4H_9	3400			1396
O-aryl N,N′-dialkyl phosphorodiamidate	CH_3, CH_3	3436			1390
O-alkyl O-aryl N-alkylphosphoramido-thioate	CH_3	3440	3408	32	1381
	C_2H_5	3430	3399	31	1400
	n-C_3H_7	3428	3401	27	1403
	iso-C_4H_9	3431	3410	21	1404
	nC_4H_9	3427	3402	25	1402
	iso-C_3H_7	3419	3385	34	1408
	s-C_4H_9	3417	3387	30	1406
	t-C_4H_9	3400			1383
O,O-dialkyl N-alkyl phosphoramidothioate	CH_3	3443	3408	35	1377
	isoC_3H_7	3421	3385	36	1402
N-alkylphosphoramidodichloridothioate	CH_3	3414			1372
	C_2H_5	3397			1393
	isoC_3H_7	3388			1403
O-alkyl N,N′-dialkylphos-phorodiamidothioate	CH_3	3448	3414	34	1378
	C_2H_5	3433	3404	29	1400
	n-C_3H_7	3430	3405	25	1400
	n-C_4H_9	3429	3401	28	1401
	isoC_3H_7	3419	3397	22	1402
	s-C_4H_9	3414	3398	16	1402
	$C_6H_5CH_2$	3421	3395	26	1399
O-aryl N,N′-dialkylphos-phorodiamido-thioate	isoC_3H_7	3415	3396	19	1403
	CH_3, isoC_3H_7	3438;3415	3415;3395		1387;1403
N-alkylphosphoramidodi-chloridate	CH_3	3408			1390
		3195(H-bonded)			1418(H-bonded)
		a.NH_2 str.	s.NH_2 str.		NH_2 bending
O-alkyl O-aryl phosphor-amidothioate	H	*[2]			1579*[3]
O,O-dialkylphosphoramido-thioate	H	3490	3390		1539*[3]
O-alkyl O-arylphosphor-amidothioate	H	3490	3390		1538*[3]

*[1]See Reference 37.

*[2]Not studied in dilute solution.

*[3]Solid phase.

TABLE 8.27. **Vibrational assignments for N-alkyl phosphoramidodichloridothioate and the CD_3NH, CD_3ND and CH_3ND analogs**

$CH_3-NH-P(=S)Cl_2$ cm^{-1}	$CD_3-NH-P(=S)Cl_2$ cm^{-1}	$C_2H_5-NH-P(=S)Cl_2$ cm^{-1}	iso$C_3H_7-NH-P(=S)Cl_2$ cm^{-1}	$CH_3-ND-P(=S)Cl_2$ cm^{-1}	$CD_3-ND-P(=S)Cl_2$ cm^{-1}	Assignment(24)
1094	1117	1112	1129	1107	1197 or 965	C−N str. or a.P−N−C str., a′
838	785	780	812	825	780	P−N str. or s.P−N−C str., a′
725	711	725	727	728	701	P=S str., rotational conformer 1, a′
690	683	681	690	681	673	P=S str., rotational conformer 2, a′
512	515	512	514	513	510	a.PCl_2 str., a″
468	463			470	461	s.PCl_2 str., rotational conformer 2, a′
450	446			447	443	s.PCl_2 str. rotational conformer 1, a′
395	384			404	386	C−N−P bend, rotational conformer 2, a′
370	360			382	362	C−N−P bend, rotational conformer 1, a′
masked	masked			310	305	P−N rock,a′ and a″
261	251			260	258	s.$PSCl_2$ bend, a′
216	213			218	213	a.$PSCl_2$ def., a′ and a″
180	183			180	180	P−N−C torsion,a″
3408	3409			2532	2532	NH or ND str., monomer, a′
3300	3301			2242	2477	NH or ND str., bonded
2745	2892			near 2442	2384	2(NH or ND bend), A′
1370	1359			1234	1197	NH or ND bend, a′
340	350			near 260	near 258	γ NH or ND, bonded
295	292			near 218	near 213	γ NH or ND, monomer, a″

TABLE 2.28 Vibrational assignments for O,O-dimethyl O-(2,4,5-trichlorophenyl) phosphorothioate and its P=O and $(CD_3-O-)_2$ analogs

$(CH_3-O-)_2P=S$ $(2,4,5-Cl_3-C_6H_2-O-)$ cm^{-1}	$(CD_3-O-)_2P=S$ $(2,4,5-Cl_3-C_6H_2-O-)$ cm^{-1}	$(CH_3-O-)_2P=O$ $(2,4,5-Cl_3-C_6H_2-O-)$ cm^{-1}	$(CD_3-O-)_2P=O$ $(2,4,5-Cl_3-C_6H_2-O-)$ cm^{-1}	Assignment(43)
1062	1076	1061	1107	s.$(C-O-)_2P$ str.
1041	1061	1043	1070	a.(C−O−)P str.
847	793	776	742	s.$P(-O-)_2$ str.
830	811	859	810	a.$P(-O-)_2$ str.
1248	1247	1251	1252	$(C_6H_2-O-)P$ str.
968	965	968	968	$(P-O-)C_6H_2$ str.
616	617	1284	1292	P=S str. or P=O*[1]
		1300	1306	P=O*[1]
3022	2272	3021	2275	a.CH_3 or a.CD_3 str., a″
3001	2258	3001	2260	a.CH_3 or a.CD_3 str., a′
2960	2088	2961	2089	s.CH_3 or s.CD_3 str., a′*[2]
2857	2211	2861	2219	2(CH_3 or CD_3 bend)*[2]
masked	1101	masked	1097	a.CH_3 or CD_3 bend, a′ and a″
1447	1101	masked	1097	s.CH_3 or s.CD_3 bend, a′
1183	923	1184	922	CH_3 or CD_3 rock, a′ (P−O−C plane)
1183	masked	1184	masked	CH_3 or CD_3 rock, a″(P−O−C plane)
				CH_3 or CD_3 torsion, a′ and a″

*[1]Rotational conformers.
*[2]In Fermi Resonance.

TABLE 8.29 Vibrational assignments for 1-fluoro-2,4,5-trichlorobenzene and the ring modes for O,O-dimethyl O-(2,4,5-trichlorophenyl) phosphorothioate and its P=O and $(CD_3-O-)_2$ analogs

[43] $(CH_3-O-)_2P=S$ (2,4,5-$Cl_3-C_6H_2-O-$) cm^{-1}	[43] $(CD_3-O-)_2P=S$ (2,4,5-$Cl_3-C_6H_2-O-$) cm^{-1}	[43] $(CH_3-O-)_2P=O$ (2,4,5-$Cl_3-C_6H_2-O-$) cm^{-1}	[43] $(CD_3-O-)_2P=O$ (2,4,5-$Cl_3-C_6H_2-O-$) cm^{-1}	[9] 1-F,2,4,5Cl_3-C_6H_2 cm^{-1}	Assignment
					a′ species
3099	3098	3085	3098	3098	v1
3075	3070	3075	3075	3061	v2
1582	1583	1583	1587	1589	v3
1557	1560	1559	1562	1571	v4
1463	1463	1463	1465	1461	v5
1352	1353	1352	1352	1357	v6
1260	1258	1257	1258	1263	v7
1248	1247	1251	1251	1251	v8
1230	1228	1239 or 1220	1228	1239	v9
1127	1125	1122	1122	1128	v10
1088	1076	1087	1084	1086	v11
?	?	911	911	934	v12
729	720	712	715	725	v13
686	679	688	681	677	v14
556	548	570	572	572	v15
537	532	544	545	528	v16
383	375	364	365	390	v17
330	322	335	335	373	v18
257	250	247	246	275	v19
210	200	203	202	211	v20
170	180?	172	170	196	v21
					a″ species
887	882	882	882	882	v22
872	852	866	869	863	v23
689	686	680 or 688	687	669	v24
629	627	627	629	622	v25
440	439	452	460	447	v26
360	355	347	354	373	v27
231	225	225	230	249	v28
157	153	152	152	154	v29
95?	~90?	~80?	88?	93?	v30

TABLE 8.30 Infrared data for O,O-dialkyl phosphorochloridothioate and O,O,O-trialkyl phosphorothioate in different physical phases

O,O-Dialkyl phosphorochlorido-thioate	C—O str. +P—O str. cm^{-1}	C—O str. cm^{-1}	P—O str. cm^{-1}	P=S str. cm^{-1}	P—Cl str. cm^{-1}	C—C str. cm^{-1}	Physical state
Dimethyl	1888	1045	845	665	490		vp, 200 °C
Dimethyl	1878.7bd	[1055.9,1042sh]	[836.2,845sh]	[666.1,654.9]	[485.7,524.7]		solution
Dimethyl	1868.9bd	[1050.1035.4]	[835.4,850sh]	[665,656.4]	[486.8,526.9]		liquid
Diethyl	1845	1030	821	662	499	966	vp, 200 °C
Diethyl	~1820	1010	809	668,652	~490,529	969	neat
	25	20	12	6,10	9,30		
Dipropyl	~1870	[1005,1055]	853	664	504		vp, 200 °C
Dipropyl	~1855	[1005,1055]	858	664	509		neat
	15	[0,0]	−5	0	−5		
O,O,O-trialkyl phosphorothioate							
trimethyl	1885	1054	834	605			vp, 200 °C
trimethyl	~1863	~1025	820	616,595			neat
	22	29	14	[−11,10]			
triethyl	~1850	~1040	828	614		960	vp, 200 °C
triethyl		~1018	~815	631,604		960	neat
		22	13	[−17,10]			

TABLE 8.31 **Infrared data for organophosphates and organohydrogenphosphonates in different physical phases**

Compound	P=O str. cm^{-1}	P=O str. cm^{-1}	PH bend cm^{-1}	POC str. cm^{-1}	POC str. cm^{-1}	PO_3 sym. str. cm^{-1}	Physical state
Phosphate							
Triethyl	1300	*1280		1045	810	748	Vapor, 216 °C
Triethyl	1277	1262		1033	821;797	744	CS_2 soln.
	23	18		12	[−11];[13]	4	vapor-CS_2 soln.
Propyl	1299	*1282		[1050],*[1009]	820	750	Vapor, 280 °C
Propyl	1270	1270		[1055],*[1004]	860	751	neat
	19	12		[−5],[5]	−40	−1	vapor-neat
Butyl	1300	*1280		[1060],*[1030]	815	736	Vapor, 289 °C
Butyl	1280	1280		[1055],[1021]	802	731	neat
	20	0		[5],[9]	13	5	vapor-neat
Phenyl	*1311	1300		1191	955	770	Vapor, 280 °C
Phenyl	1300	*1288		1171	950	765	neat
	11	12		20	5	5	vapor-neat
Diphenyl methyl	1316	1316		1198	[821],[952]	765	Vapor, 280 °C
Diphenyl methyl	1305	*1290		1182	[815],[950]	762	neat
	11	26		16	[6],[2]	3	vapor-neat
Phosphonate		PH str. CCl_4 soln. cm^{-1}	PH bend cm^{-1}				
Dimethyl hydrogen	1290	~2439	985	1065	815	770	Vapor, 200 °C
Dimethyl hydrogen	[1282],*[1264]	2435	988	1055	821	766	CS_2 soln.
	[8],[26]		−3	10	−6	4	vapor-CS_2 soln.
		4					vapor-CCl_4 soln.
Diethyl hydrogen	1281	~2435	~983	1061	780	749	Vapor.200 °C
Diethyl hydrogen	[1280],*[1262]	~2438	~978	[1084],*[1054]	784	741	CS_2 soln.
	[1],[19]	~ −3	5	[−23],[7]	−4	8	vapor-CS_2 soln.

*Strongest band in the doublet.

TABLE 8.32 Infrared data for O-alkyl phosphorodichloridothioates and S-alkyl phosphorodichloridothioates in different physical phases

O-alkyl Phosphorodichloridothioate	C—O str. isomer 1 cm^{-1}	C—O str. isomer 2 cm^{-1}	P—O str. isomer 1 cm^{-1}	P—O str. isomer 2 cm^{-1}	a.PCl_2 str. isomer 1 cm^{-1}	a.PCl_2 str. isomer 2 cm^{-1}	s.PCl_2 str. isomer 1 cm^{-1}	s.PCl_2 str. isomer 2 cm^{-1}	P=S str. isomer 1	P=S str. isomer 2	Physical phase
O-ethyl	1026	1014	804	790	528	552	472	490	728	701	CS_2 soln.
	~1027	977	789bd	798bd	533	555	473	490	728	701	liquid
	−1	37	15	−8	−5	−3	−1	0	0	0	CS_2-liquid
O-ethyl-1,1-d_2	1025	1008	783	796	525	550	465	484	727	699	CS_2 soln.
	1025	1010	782	797	529	551	466	474	726	700	liquid
	0	−2	1	−1	−4	−1	−1	−10	1	−1	CS_2-liquid
O-ethyl-2,2,3d_3	1013	995	750	750	526	551	460	481	688	691	CS_2-soln.
	1010	998	750	750	531	552	462	?	675	677	liquid
	3	−3	0	0	−5	−1	−2	?	13	14	CS_2-liquid
O-(2-Propynyl) [32]	1015	984	857	822	529	557	440	474	742	716	CS_2-soln.
	1014	982	853	821	533	556	440	473	739	714	liquid
	1	2	4	1	−4	1	0	1	3	2	CS_2-liquid
	C—S str. isomer 1 cm^{-1}	C—S str. isomer 2 cm^{-1}	a.$PSCl_2$ str. isomer 1	a.$PSCl_2$ str. isomer 2	a.$PSCl_2$ str. isomer 1 cm^{-1}	a.$PSCl_2$ str. isomer 2 cm^{-1}	s.$PSCl_2$ str. isomer 1 cm^{-1}	s.$PSCl_2$ str. isomer 2 cm^{-1}	P=O str. isomer 1 cm^{-1}	P=O str. isomer 2 cm^{-1}	
S-methyl Phosphorodichloridothioate	689	732	547a″	560a″	593a′	579a′	450	471	1275	1266	CS_2 soln.
	685	729	547a″	555a″	598a′	582a′	451	470	1261		liquid
	4	3	0	5	−5	−3	−1	1	14		CS_2-liquid

TABLE 8.33 Infrared data for O,O-diethyl N-alkylphosphoramidates in different physical phases

Alkylphosphoramidic acid, diethyl ester N-alkyl	P=O str. cm^{-1} vapor phase	P=O:H str. cm^{-1} neat phase	POC str. cm^{-1} vapor phase	POC str. cm^{-1} vapor phase	PO_2N str. cm^{-1} vapor phase	NH str. cm^{-1} vapor phase	NH:O=P str. cm^{-1} neat phase	Physical phase vapor phase
Methyl	1275	1225	1041	842	749	3460	3240	240 °C
Ethyl	1275	1230	1044	795	750	3440	~3235	240 °C
Hexyl	1275	1230	1044	798	750	3450	~3230	240 °C
Phenethyl	1280	1229	1045	803	749	3440	~3230	280 °C
Isopropyl	1274	1230	1037	794	755	3430	3220	240 °C
Cyclohexyl	1275	1240	1040	795	755	3422	3198	280 °C
Tert-butyl	1275	1237	1045	796	760	3420	3208	240 °C
Phenyl	1275	1210	1039	796	749	3435	3210	280 °C

	P=O str.- P=O:H str. cm^{-1}	NH str.- NH:O=P str. cm^{-1}	cm^{-1} neat phase	cm^{-1} neat phase	cm^{-1} neat phase
Methyl	50	220	1024	860	791
Ethyl	45	205	1030	790	740
Hexyl	45	220	1030	792	755
Phenethyl	51	210	1020	791	745
Isopropyl	34	210	1030	790	745
Cyclohexyl	35	224	1053	793	752
Tert-butyl	38	212	1024	789	760
Phenyl	65	225	1010	791	745

TABLE 8.34 Vibrational assignments for $[CH_3-PO_3]^{2-}$, $[CD_3-PO_3]^{2-}$, $[H-PO_3]^{2-}$, and $[PO_4]_3-$

[40] $[CH_3-PO_3]^{2-}$ cm^{-1}	[40] $[CD_3-PO_3]^{2-}$ cm^{-1}	C_3v symmetry	[44] $[H-PO_3]^{2-}$ cm^{-1}	C_3v symmtry	[45] $[PO_4]_3-$ cm^{-1}	Td symmetry
1057	1060	a.PO_3 str.,e	1085	a.PO_3 str.,e	1082	a.PO_4 str., f2
753	720	C−P str.,a1				
977	972	s.PO_3 str.,a1	979	s.PO_3 str.,a1	980	s.PO_4 str., a1
510	510	e	465	a.PO_3 bend,e	515	a.PO_4 bend, f2
485	495	s.PO_3 bend,a1	567	s.PO_3 bend,a1		
335	~320	e			363	PO_4 def., e

		$CH_3(cm^{-1}/CD_3$ (cm^{-1})	Assignment
2983	~2225	1.34	a.CH_3 str. or a.CD_3 str.
2921	2142	1.36	s.CH_3 str. or s.CD_3 str.
1425	?	?	a.CH_3 bend or a.CD_3 bend
842	~622	1.35	s.CH_3 rock or s.CD_3 rock
?	?	?	CH_3 torsion or CD_3 torsion

TABLE 8.35 Infrared data and assignments for the $(CH_3)_2PO_2-$ anion

$(CH_3)_2P(O)_2Na$ cm^{-1}	Assignment for C_2v symmetry	$H_2P(O)_2K$ cm^{-1}	Assignment for C_2v symmetry
1120	a.PO_2 str., b1	1180	a.PO_2 str., b2
1040	s.PO_2 str., a1	1042	s.PO_2 str., a1
479	PO_2 bend,a1	469	PO_2 bend,a1
		[PO_4]3−	for Td symmetry
1120	a.PO_2 str., b1		
738	a.$P(C)_2$ str., b2	1082	a.PO_4 str., f2
700	s.$P(C)_2$ str., a1		
1040	s.PO_2 str., a1	980	s.PO_4 str., a1
479	s.$(C)_2P(O)_2$ bend,a1		
440	a.$(C)_2P(O)_2$ bend, b1 and b2	515	a.PO_4 bend,f2
352			
315	$(C)_2P(O)_2$ twist,a2	363	PO_4 defor., e
272	$(C)_2P(O)_2$ defor., a1		
		$(CH_3)_3P$	for C_3v symmetry
738	a.$P(C)_2$ str., b2	708	a.$P(C)_3$ str., e
700	s.$P(C)_2$ str., a1	653	s.$P(C)_3$ str., a1
		305	a.$P(C)_3$ bend,a1
		263	s.$P(C)_3$ bend,a1
		$(CH_3)_3PO$	for C_3v symmetry
735	a.$P(C)_3$ str., b2	756	a.$P(C)_3$ str., e
700	s.$P(C)_3$ str., a1	671	s.$P(C)_3$ str., a1
		311	$P(C)_3$ bend, e
		256	$P(C)_3$ bend, a1

CHAPTER 9

Benzene and Its Derivatives

Introduction	353
Polychlorobiphenyls	353
An A_1 Fundamental for Toluene and Related Analogs	355
Styrene-4-Methylstyrene Copolymers	355
Styrene-Acrylic Acid Coplymer	358
Styrene Acrylamide Copolymer	359
Styrene/2-Isopropenyl-2-Oxazoline Copolymer (SIPO Copolymer)	359
Ethynylbenzene and Ethynylbenzene-d	359
Bromodichlorobenzenes	360
Raman Data for 1,2-Disubstituted Benzenes	360
Vibrational Data for 1,3-Dichlorobenzene and Raman Data for 1,3-Disubstituted Benzenes	361
Raman Data and Assignments for Some In-Plane Ring Modes 1,4-Distributed Benzenes	361
Raman Data for Decabromobiphenyl and Bis (Pentabromophenyl) Ether	361
Infrared Data and Assignments for Benzene, Benzene-d6, Benzyl Alcohol, and Benzyl $-2, 3, 4, 5, 6-d5$ Alcohol	361
Out-of-Plane Deformations and Their Combination and Overtones for Substituted Benzenes	362
Polystyrene and Styrene Copolymers	362
Ethynylbenzene	363
1,2-Disubstituted Benzenes	363
1,3-Dibsubstituted Benfizenes	364
1,4-Disubstituted Benzenes	364
1,3,5-Trisubstituted Benzenes	364
1,2,3-Trisubstituted Benzenes	364
1,2,4-Trisubstituted Benzenes	365
1,2,3,4-Tetrasubstituted Benzenes	365
1,2,3,5-Tetrasubstituted Benzenes	365
1,2,4,5-Tetrasubstituted Benzenes	365
1,2,3,4,5-Pentasubstituted Benzenes	366
Summary of the Out-of-Plane Hydrogen Deformations for Substituted Benzenes and the Out-of-Plane Ring Deformation for Mono-substituted Benzenes, and Their Combination and Overtones	366
Correlation Chart	366
2,3,4,5,6-Pentachlorobiphenyl	366
References	367

Figures

Figure 9-1	368 (354)	Figure 9-25	383 (357)
Figure 9-2	368 (354)	Figure 9-26	384 (358)
Figure 9-3	369 (554)	Figure 9-27	385 (358, 362)
Figure 9-4	369 (354)	Figure 9-28	386 (358)
Figure 9-5	370 (354)	Figure 9-29	387 (359, 362)
Figure 9-6	370 (354)	Figure 9-30	388 (359, 362)
Figure 9-7	371 (354)	Figure 9-31	388 (359)
Figure 9-8	371 (354)	Figure 9-32	389 (359, 363)
Figure 9-9	372 (354)	Figure 9-33	390 (359)
Figure 9-10	372 (354)	Figure 9-34	391 (359)
Figure 9-11	373 (354)	Figure 9-35	392 (359)
Figure 9-12	373 (354)	Figure 9-36	393 (362)
Figure 9-13	374 (354)	Figure 9-37	394 (363)
Figure 9-14	374 (354)	Figure 9-38	395 (363)
Figure 9-15	375 (354)	Figure 9-39	396 (364)
Figure 9-16	375 (354)	Figure 9-40	397 (364)
Figure 9-17	376 (355)	Figure 9-41	398 (364)
Figure 9-18	377 (355)	Figure 9-42	399 (364)
Figure 9-19	378 (355, 362)	Figure 9-43	399 (365)
Figure 9-20	378 (355, 362)	Figure 9-44	400 (365)
Figure 9-21	379 (355, 362)	Figure 9-45	400 (365)
Figure 9-22	380 (355, 362)	Figure 9-46	401 (365)
Figure 9-23	381 (355, 362)	Figure 9-47	401 (366)
Figure 9-24	382 (356, 362)	Figure 9-48	402 (366)

Tables

Table 9-1	403 (357)	Table 9-13	417 (354)
Table 9-2	404 (354)	Table 9-14	418 (355)
Table 9-3	405 (354)	Table 9-15	418 (360)
Table 9-4	406 (354)	Table 9-16	419 (360)
Table 9-5	407 (354)	Table 9-17	419 (361)
Table 9-6	408 (354)	Table 9-18	420 (361)
Table 9-7	409 (354)	Table 9-19	420 (361)
Table 9-8	411 (354)	Table 9-20	421 (361)
Table 9-9	413 (354)	Table 9-21	422 (362)
Table 9-10	414 (354)	Table 9-22	422 (362)
Table 9-11	415 (354)	Table 9-23	423 (362)
Table 9-12	416 (354)	Table 9-24	423 (366)

*Numbers in parentheses indicate in-text page reference.

INTRODUCTION

Benzene has 30 fundamental vibrations, and benzene substituted with atoms such as halogen also has 30 fundamental vibrations. Twenty-one fundamentals are in-plane vibrations, and nine fundamentals are out-of-plane vibrations. Moreover, there are 30 fundamental benzene ring vibrations for all of its derivatives. Thus, for a compound such as toluene, which has 39 fundamental vibrations, 30 fundamentals result from C_6H_5C vibrations, and 9 fundamentals result from CH_3 vibrations.

Benzene and derivatives of benzene, which have a center of symmetry, have IR vibrations that are IR active (allowed in the IR) and vibrations that are Raman active (allowed in the Raman). Therefore, it is most helpful in the spectra-structure identification of benzene and its derivatives to obtain both IR and Raman data. Of course, a standard IR or Raman spectrum, if available for comparison, is sufficient to identify a chemical compound when the spectrum of the sample and standard reference are identical.

In the case of IR, it is helpful to obtain a spectrum in the vapor phase. This is because the IR band shapes help in determining which vibrations are in-plane vibrations and which vibrations are out-of-plane vibrations. In the case of Raman, it is helpful to obtain polarized and depolarized data to help distinguish between in-plane and out-of-plane vibrations. In certain cases, a fundamental is both IR and Raman inactive. Moreover, compounds with no molecular symmetry, but with only its identity, have molecular vibrations that are both IR and Raman active. Another helpful feature in interpreting IR and Raman data is that normal vibrations with strong IR band intensity have weak Raman band intensity and vice-versa.

Several publications are available to assist one in the spectra-structure identification of benzene and its derivatives (1–6). Reference 6 discusses in detail both in-plane and out-of-plane normal vibrations for substituted benzenes together with schematics of their approximated normal vibrations. This reference is recommended for those who are unfamiliar with this topic.

Some of the topics covered in this section are the result of chemical problems submitted to us for spectra-structure identification by The Dow Chemical Company.

POLYCHLOROBIPHENYLS

Polychlorobiphenyls have been utilized in the electrical power industry. These chemicals are alleged to be carcinogenic materials, and their possible presence in the environment requires methods for their detection and identification.

There are 66 possible pentachlorobiphenyl isomers, and 16 of these isomers are included here. These spectra were recorded utilizing the DRIFT technique (diffuse reflectance infrared Fourier transform), because it is especially useful in obtaining IR spectra of liquid chromatograph (LC) fractions where the amount of sample available is limited (7). All 16 spectra were recorded by using 1–5 μg of sample deposited from hexane solution via a syringe on approximately 100 mg of KBr powder placed in the sample cup. The solvent was evaporated using an IR heat lamp.

Biphenyl and each polychlorobiphenyl isomer have 60 fundamental vibrations; therefore, it is twice as difficult to determine ring substitution for the polychlorobiphenyl isomers as it is, for example, benzene and any of the chlorinated benzenes. Moreover, there is no indication which ring modes belong to one substituted phenyl group from the other substituted phenyl group in the case of polychlorobiphenyl isomers. In order to be able to predict which bands belong to each substituted phenyl group, it is necessary to know the potential energy distribution for each of the normal modes for both 1,2,4-trichlorobenzene and 1,2,4,5-tetrachlorobenzene to determine whether the structure of a pentachlorobiphenyl isomer is 2,2′,4,4′,5-pentachlorobiphenyl or 2,3′,4,4′,5-pentachlorobiphenyl, for example. This is because normal vibrations, including a significant potential energy contribution from carbon-chlorine stretching, are expected to be affected significantly by substitution of a 2,4-dichlorophenyl group or a 3,4-dichlorophenyl group from 1,2,4-trichlorophenyl for a chlorine atom of 1,2,4,5-tetrachlorobenzene. Scherer (3) has performed normal coordinate calculations for chlorobenzenes and deuterated chlorobenzenes, and the resulting potential energy distribution data were used to enable the development of IR group frequency verification of these 16 polychlorobiphenyl isomers. The IR spectra and assignments are reported in the literature (7).

The diffuse reflectance FT-IR spectra are presented in Figs. 9.1 through 9.16, and vibrational assignments for the 16 polychlorobiphenyl isomers are presented in Tables 9.1 through 9.12 (7); these assignments are compared to those for correspondingly substituted chlorobenzenes.

Out-of-plane ring hydrogen deformations are also useful in characterizing substituted benzenes, and these spectra-structure correlations for polychlorobiphenyls are discussed later in this chapter.

Table 9.13 lists Raman data for a variety of compounds containing the phenyl group (8). It has been reported that two planar polarized Raman bands and one planar depolarized Raman band yield characteristic frequencies and intensities regardless of the nature of the substituent group (9). These three planar ring vibrations together with a fourth planar ring mode are shown here:

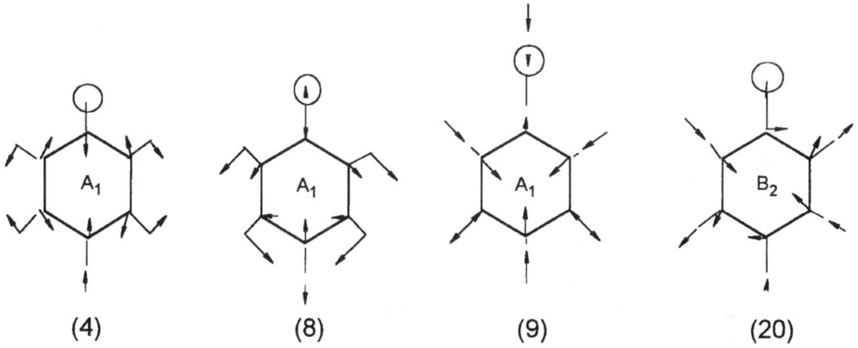

Ring mode 9 is essentially a ring breathing vibration, and its Raman band occurs in the region 994–1010 cm^{-1}. This Raman band is usually the most intense band in the spectrum, and it is polarized. Notable exceptions are exhibited by cinnamic acid, glycidyl cinnamate, and 1,6-diphenyl-1,3,5-hexatriene.

Ring mode 8 occurs in the region 1010–1032cm^{-1}, and this polarized Raman band is relatively weak. Ring mode 20 occurs in the region 603–625 cm^{-1}, and the Raman band is depolarized and has weak intensity. Ring mode 4 occurs in the region 1585–1610 cm^{-1}, and the Raman band has variable intensity.

The relative Raman band intensity ratio for Ring 4/Ring 9 varies between 0.11 and 1.14, for Ring 8/Ring 9 it varies between 0.06 and 0.33, and for Ring 20/Ring 9 it varies between 0.06 and 0.25. All of these Raman data help in the spectra-structure identification of the phenyl group in organic materials.

AN A_1 FUNDAMENTAL FOR TOLUENE AND RELATED ANALOGS

Lau and Snyder performed a normal coordinate analysis for toluene and related compounds (10). They determined that the potential energy distribution of the 790 cm^{-1} planar A_1 fundamental for toluene is 32% ring CCC bend, 31% C—C ring stretch, 30% C—C stretch for methyl-to-ring bond. The 790 cm^{-1} Raman band for toluene is strong, and it is also polarized. Strong Raman bands that are also polarized are observed near 760, 740, and 705 cm^{-1} for ethylbenzene, isopropylbenzene, and tert-butylbenzene, respectively (8).

Figure 9.17 shows a plot of the planar A1 fundamental for toluene, ethylbenzene, isopropylbenzene, and tert-butylbenzene in the region 700–790 cm^{-1} vs the number of protons on the ring α-carbon atom (11), and Fig. 9.18 shows a plot of the planar A1 fundamental for toluene, ethylbenzene, isopropylbenzene, and tert-butylbenzene in the region 700–790 cm^{-1} vs Tafts σ^* values of 0, −0.01, −0.19, and −0.30 for methyl, ethyl, isopropyl, and tert-butyl groups, respectively (11, 12). Both plots are essentially linear.

Table 9.14 lists IR and Raman data for this A_1 fundamental. These Raman band correlations for this A_1 fundamental were used to assign the comparable Raman band at 773.4 cm^{-1} for α-syndiotactic polystrene and at 785 cm^{-1} for isotactic polystyrene (13). Copies of the syndiotactic polystyrene uniaxially stretched film (perpendicular) and (parallel) polarized IR spectra are presented in Figs. 9.19 and 9.20, respectively (13).

Variable temperature studies of isotactic polystyrene have shown that most of the IR bands shift to lower frequency by 1 to 5 cm^{-1} as the temperature is raised from 30 °C to 180 °C (14). A variable-temperature study of syndiotactic polystyrene film cast from a solution of 1,2-dichlorobenzene shows that it changes crystalline form at temperatures of 190 °C and above (14). Figure 9.21 shows the IR spectra of syndiotactic polystyrene of the cast film from boiling 1,2-dichlorobenzene, and Fig. 9.22 is an ambient temperature IR spectrum of the same syndiotactic polystyrene film used to record Fig. 9.21 except that the film was heated to 290 °C and then allowed to cool to ambient temperature (15).

Spectral differences between Figs. 9.21 and 9.22 will be discussed under the section for the out-of-plane vibrations for the phenyl group.

STYRENE-4-METHYLSTYRENE COPOLYMERS

Figure 9.23 is an IR spectrum of a syndiotactic styrene (98%) −4-methylstyrene (2%) copolymer. This film was cast on a cesuim iodide plate from boiling 1,2-dichlorobenzene (16). A comparison of Fig. 9.23 with Fig. 9.21 shows that the syndiotactic copolymer has the same crystal structure as the syndiotactic polystyrene. The IR band at 1511 cm^{-1} is due to an in-plane ring mode and the 816 cm^{-1} band is due to the out-of-plane hydrogen deformation for the 4-methylphenyl group in the copolymer (16). Figure 9.24 is an IR spectrum of syndiotactic styrene (93%) −4-methylstyrene (7%) copolymer. This film was cast on a cesuim iodide plate from boiling 1,2-dichlorobenzene. A comparison of Fig. 9.24 with Fig. 9.22 shows that the syndiotactic copolymer has the same crystal structure as syndiotactic polystyrene. The IR band near 1511 cm^{-1} is due to an in-plane ring deformation, and the band near 816 cm^{-1} is due to an out-of-plane hydrogen deformation for the 4-methylphenyl group in the copolymer (16).

It is possible to utilize IR spectroscopy in the quantitative analyses of copolymer films where the film thickness is unknown. This method requires measurement of an absorbance band for each of the components in the copolymer. This analysis can be performed using one of the following methods.

A sample containing known concentrations of the monomer units is required in all cases. A film of the copolymer is then cast on a suitable IR plate such as preheated potassium bromide (KBr) placed under an IR heat lamp in a nitrogen (N_2) atmosphere.

The absorbance (A) is proportional to the concentration of each component in the copolymer. Thus,

$$C_a = kA_a;\ C_b = kA_b$$

$$\frac{C_a}{C_b} = K\frac{A_a}{A_b} \qquad K = \frac{C_a}{C_b}\cdot\frac{A_b}{A_a} \tag{1}$$

As an example, component (*a*) of the copolymer is *70%* and component (*b*) of the copolymer is *30%*. The absorbance of a band at 1730 cm^{-1} is *0.753* for component (*a*) and the absorbance of a band at 1001 cm-01- for component (*b*) is *0.542*. These numbers are used to determine the value of K.

$$K = \frac{70}{30}\cdot\frac{0.542}{0.753} = 1.680$$

The copolymer containing components (a) and (b) of unknown concentrations is submitted for analysis. The absorbance (A) for component (a) is measured at 1730 cm^{-1} and for component (b) is measured at 1001 cm^{-1}. The A for (a) is found to be 0.664 and for (b) A is 0.557.

Equation 1 is then utilized using the value for K of 1.680

$$\frac{C_a}{C_b} = 1.680\frac{(0.664)}{(0.557)} = 2.-3;\ C_a = 2.003\ C_b$$

Then $C_a + C_b = 100\%$.

$$2.003\ C_2 + C_b = 100\%$$

$$C_b = \mathit{100\%} = 33.3\%$$

$$3.003$$

Then $C_a = 66.7\%$.

Quantitation of a copolymer consisting of three different components can also be analyzed by application of IR spectroscopy. A copolymer of known composition is required. The measured values of the IR spectrum for the known concentration of the copolymer are:

C_a is 42%; A_a (at 2230 cm^{-1}) is 0.452

C_b is 26%; A_b (at 1730 cm^{-1}) is 0.847

C_c is 32%; A_c (at 700 cm^{-1}) is 0.269

$$K_1 = \frac{C_a}{C_b} \cdot \frac{A_b}{A_a} \qquad K_1 = \frac{42}{26} \cdot \frac{0.847}{0.542} = 2.524$$

$$K_1 = \frac{C_c}{C_b} \cdot \frac{A_b}{A_c} \qquad K_2 = \frac{32}{26} \cdot \frac{0.847}{0.269} = 3.875$$

The spectrum of the unknown copolymer containing components (a), (b), and (c) is obtained, and the following absorbance bands are measured:

The IR band at 2230 cm^{-1} has an absorbance of 0.757

The IR band at 1730 cm^{-1} has an absorbance of 0.798

The IR band at 700 cm^{-1} has an absorbance of 0.345.

Then

$$\frac{C_{(a)}}{C_{(b)}} = K_1 \frac{A_{(a)}}{A_{(b)}}; \frac{C_{(a)}}{C_{(b)}} = 2.524 \times \frac{0.757}{0.798} = 2.394; C_{(a)} = 2.394\ C_{(b)}$$

$$\frac{C_{(c)}}{C_{(b)}} = K_2 \frac{A_{(c)}}{C_{(b)}}; \frac{C_{(c)}}{C_{(b)}} = 3.875 \times \frac{0.345}{0.798} = 1.675 \quad C_{(c)} = 1.675\ C_{(b)}$$

Then

$$C_{(a)} + C_{(b)} + C_{(c)} = 100$$

$$2.394\ C_{(b)} + C_{(b)} + 1.675\ C_{(b)} = 100\%$$

$$C_{(b)} = 100/5.069 = 19.73\%$$

$$C_{(a)} = 2.394\ C_{(b)} = 2.394\ (19.73\%) = 47.23\%$$

$$C_{(c)} = 1.675\ C_{(b)} = 1.675\ (19.73\%) = 33.05\%$$

$$C_{(a)} = 47.23 + C_{(b)} = 19.73\%; C_{(b)} = 33.05\% = 100.01\%$$

Another way to set up a quantitative method for the analysis of a copolymer is to record IR spectra of known quantitative composition over a range of concentrations that are to be manufactured. The IR spectra of a series of styrene-p-methylstyrene copolymers were prepared and the % p-methylstyrene in the copolymer is: 1,2,3,5 and 7%. Absorbance (A) for styrene was measured near 900 cm^{-1}, and the absorbance (A) also was measured for p-methylstyrene at 1511 cm^{-1} and near 816 cm^{-1}. Figure 9.25 shows a plot of the absorbance ratio A(1511 cm^{-1})/A(901–904 cm^{-1}) vs % p-methylstyrene in the styrene-p-methylstyrene copolymer; Fig. 9.26 shows the plot of the absorbance ratio A(815–817 cm^{-1})/A(900–904 cm^{-1}) vs % p-methylstyrene in the styrene-p-methylstyrene copolymer. Both plots are linear in this concentration range, and either method is suitable for a routine analytical method for this copolymer.

A matrix method for the quantitative IR multicomponent analysis for films with indeterminate path-length has been reported (17).

STYRENE-ACRYLIC ACID COPLYMER

Figure 9.27 (top) is an IR spectrum of styrene (92%)—acrylic acid (8%) copolymer recorded at 35 °C, and Fig. 9.27 (bottom) is an IR spectrum of the same copolymer at 300 °C (18). There are significant differences between these two IR spectra of the same copolymer. The ratio of the IR band intensities at 1742 to 1700 cm^{-1} is low in the upper spectrum and is high in the lower spectrum. The 1700 cm^{-1} IR band results from acrylic acid intermolecular hydrogen bonded dimers, and it is actually an out-of-phase $(CO_2H)_2$ vibration as depicted here:

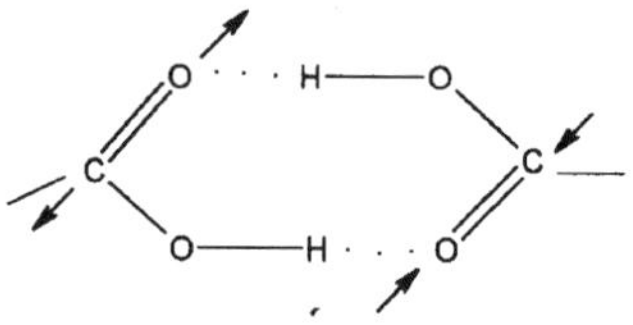

out-of-phase $\nu(CO_2H)_2$

The broad IR band in the region 3500–2000 cm^{-1} with subsidiary maxima results for $(OH)_2$ stretching of the $(CO_2H)_2$ groups in Fermi resonance with combinations and overtones. Note how there is less absorption in this region at high temperature than there is at 35 °C. The three bands in the region 3000–3100 cm^{-1} result from phenyl hydrogen stretching vibrations, and the two bands in the region 2950–2800 cm^{-1} result from CH_2 stretching vibrations. The IR bands near 3520 and 1125 cm^{-1} in the lower spectrum result from ν OH and ν C—O for the CO_2H group not existing in the $(CO_2H)_2$ dimer form. The IR band near 3470 cm^{-1} is assigned to ν O—H intermolecularly hydrogen bonded to the π system of the phenyl groups in the copolymer (18).

It is interesting to study Fig. 9.28, which shows five plots of absorbance ratios vs °C from 30 to ~310 °C. The plots are all parallel below ~150 °C, and then they increase or decrease essentially in a linear manner. These changes do not take place until the copolymer becomes molten, and the $(CO_2H)_2$ groups are free to form two CO_2H groups. The 1600 cm^{-1} IR band results from a phenyl ring bend-stretching vibration, and the 752 cm^{-1} IR band results from the five hydrogen atoms vibrating in-phase out-of-the plane of the phenyl groups. The 1742 cm^{-1} and 1700 cm^{-1} IR bands have been assigned previously; R_1 is the ratio of $(A)CO_2H/A(CO_2H)_2$, and it increases as the sample temperature is increased; R_2 is the ratio of (A) $CO_2H/A(\phi)$, and it increases as the sample temperature is increased; R_3 is the ratio of $A(CO_2H)_2/A(\phi)$, and it decreases as the sample temperature is increased; R_4 is the ratio $A(CO_2H)/A(\phi)$, and it increases as the sample temperature is increased; and R_5 is the ratio $A(CO_2H)_2/A(\phi)$, and it decreases as the sample temperature is increased. All of these absorbance ratios show that after ~150 °C the $(CO_2H)_2$ groups split into two CO_2H groups. Upon cooling to ambient temperature the CO_2H groups reform $(CO_2H)_2$ groups (18).

STYRENE ACRYLAMIDE COPOLYMER

Figure 9.29 (top) is an IR spectrum of styrene acrylamide recorded at 27 °C, and Fig. 9.29 (bottom) is an IR spectrum of the same copolymer recorded at 275 °C. The ν C=O vibration for acrylamide in the copolymer occurs at 1682 cm^{-1} at 27 °C and at 1690 cm^{-1} at 275 °C. Moreover, this change in the ν C=O frequency occurs between 125 to 167 °C. In addition, ν asym. NH_2 occurs at 3480 cm^{-1} and ν sym. NH_2 occurs at 3380 cm^{-1} at temperature below 125 °C, and ν asym. NH_2 occurs at 3499 cm^{-1} and ν sym. NH_2 occurs near 3380 cm^{-1} in the temperature range 27 to 275 °C. In addition, bands in the region 3200–3370 cm^{-1} in the 27 °C spectrum are not present in the 275 °C spectrum. These data show that the NH_2 groups are not hydrogen bonded to C=O groups in temperature above 167 °C. Upon cooling the intermolecular hydrogen, bonds reform (18).

STYRENE/2-ISOPROPENYL-2-OXAZOLINE COPOLYMER (SIPO COPOLYMER)

Figure 9.30 is an IR spectrum of a styrene (93.8%)/2-isopropenyl-2-oxazoline (6.2%) copolymer film cast from methylene chloride onto a KBr plate (19). The IR band at 1656 cm^{-1} is assigned to ν C=N of the oxazoline ring. The IR band at 1600 cm^{-1} results from and in-plane bend-stretching vibration of the phenyl ring. Figure 9.31 is a plot of the wt % IPO in the SIPO copolymer vs the absorbance ratio A(1656 cm^{-1})/A(1600 cm^{-1}) for 10 SIPO copolymers. This plot is linear over the % SIPO copolymers studied, and the calculated and % IPO values agree within 0.15% (19).

ETHYNYLBENZENE AND ETHYNYLBENZENE-D

Figure 9.32 (top) is a vapor-phase IR spectrum of ethynylbenzene (phenylacetyline) in a 4-m cell with the vapor pressure of the liquid sample at 25 °C. Figure 9.32 (bottom) is a vapor-phase IR spectrum of ethynylbenzene-d run under the same conditions as the top spectrum (20).

Figure 9.33 (top) is a solution-phase IR spectrum of ethynylbenzene, and Fig. 9.33 (bottom) is a solution-phase IR spectrum of ethynylbenzene-d (20).

Figure 9.34 (top) is an IR spectrum of ethynylbenzene in the liquid phase, and Fig. 9.34 (bottom) is an IR spectrum of ethynylbenzene-d in the liquid phase (20).

Figure 9.35 (top) is a Raman spectrum of ethynylbenzene, and Fig. 9.35 (bottom) is a Raman spectrum of ethynylbenzene-d (20).

The $\nu\equiv$C—H mode occurs at 3340; 3320 sh cm^{-1} in the vapor, at 3315; 3305 sh cm^{-1} in CCl_4 solution, and at 3291; 3310 sh cm^{-1} in the liquid phase. The $\nu\equiv$C—D mode occurs at 2608 cm^{-1}

in the vapor, at 2596 cm^{-1}; 2560 sh cm^{-1} in CCl_4 solution, and at 2550 cm^{-1} in the liquid phase (20). Both $\nu{\equiv}C{-}H$ and $\nu{\equiv}C{-}D$ shift to lower frequency in the order: vapor, CCl_4 solution, and liquid phase. The ν C≡C mode for the C≡C−H group occurs at 2122 cm^{-1} in the vapor, at 2119 cm^{-1} in CCl_4 solution, and at 2118 cm^{-1} in the liquid phase while the ν C≡C mode for the C≡C−D group occurs at 1989 cm^{-1} in CCl_4 solution and at 1983 cm^{-1} in the liquid phase. The in-plane ≡C−H deformation occurs at 648 cm^{-1} in the vapor, at 648 cm^{-1} in CS_2 solution, and at 653 cm^{-1} in the vapor phase. By contrast, the out-of-plane deformation occurs at 612 cm^{-1} in the vapor, at 610 cm^{-1} in CS_2 solution, and at 619 cm^{-1} in the liquid phase. The in-plane ≡C−D deformation occurs at 481 cm^{-1} in the vapor, 482 cm^{-1} in CS_2 solution, and at 486 cm^{-1} in the liquid phase. The out-of-plane ≡C−D deformation appears to be coincident with the in-plane ≡C−D deformation in the case due to coupling with a C−C≡C bending mode (20). As discussed in Volume 1, Chapter 5, $\nu{\equiv}C{-}D$ and ν C≡C are coupled.

BROMODICHLOROBENZENES

There are six isomers of bromodichlorobenzene, and these are: 1-bromo,3,5-dichlorobenzene; 1-bromo, 2,6-dichlorobenzene; 1-bromo,2,3-dichlorobenzene; 1-bromo,2,4-dichlorobenzene; 1-bromo,2,5-dichlorobenzene; and 1-bromo,3,4-dichlorobenzene. There is no problem in distinguishing between the 1,3,5-; 1,2,3-; and 1,2,4-positions by application of IR spectroscopy; however, the problem arises in the unambiguous identity of 1-Br,2,6-$Cl_2\phi$ from 1-Br,2,3-$Cl_2\phi$3-$Cl_2\phi$, and between 1-Br, 2,4-$Cl_2\phi$, 1-Br,2,5-$Cl_2\phi$, and 1-Br,3,4-$Cl_2\phi$. Both IR and Raman spectroscopy were used to correctly identify each of the six bromodichlorobenzenes (21). These assignments were made possible by knowing the potential energy distributions for the corresponding trichlorobenzene isomers for planar vibrations involving carbon-halogen stretching (3). Table 9.15 lists the vibrational assignments for the six bromodichlorobenzene isomers as well as the molecular symmetry of each isomer (21).

RAMAN DATA FOR 1,2-DISUBSTITUTED BENZENES

Raman spectra for 1,2-disubstituted benzenes exhibit characteristic group frequencies (see Table 9.16). A strong Raman band occurs in the region 1020–1044 cm^{-1}, and a weaker Raman band occurs in the region 642–668 cm^{-1}. Raman bands also occur in the region 1595–1610 cm^{-1} and in the region 1577–1586 cm^{-1}, and the higher frequency band is more intense than the lower frequency band (8). The approximate normal vibrations for ring modes 3, 14, 18, and 9 are presented here.

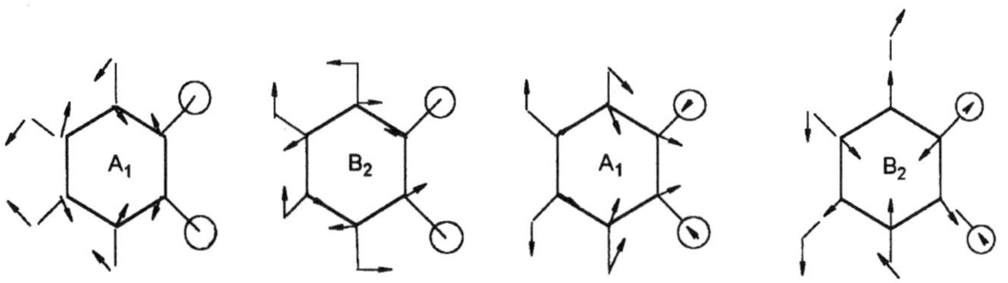

(3) (14) (9) (18)

VIBRATIONAL DATA FOR 1,3-DICHLOROBENZENE AND RAMAN DATA FOR 1,3-DISUBSTITUTED BENZENES

Vibrational assignments for the in-plane modes of 1,3-dichlorobenzene have been reported (3). Table 9.17 lists Raman data for 1,3-disubstituted benzenes (8). A strong Raman band occurs in the region 995–1005 cm^{-1}, and a weak to medium Raman band occurs in the region 1574–1618 cm^{-1}.

RAMAN DATA AND ASSIGNMENTS FOR SOME IN-PLANE RING MODES FOR 1,4-DISUBSTITUTED BENZENES

Table 9.18 lists Raman data and assignments for some in-plane ring modes for 1,4-disubstituted benzenes (8). Weak to strong Raman bands occur in the regions 1591–1615 cm^{-1}; 1268–1308 cm^{-1}; 819–881 cm^{-1}; 801–866 cm^{-1}, and 625–644 cm^{-1}.

RAMAN DATA FOR DECABROMOBIPHENYL AND BIS (PENTABROMOPHENYL) ETHER

Raman data for decabromobiphenyl and bis-(pentabromophenyl) ether are listed in Table 9.19 (8). Ring mode assignments are also listed for the pentabromophenyl group.

INFRARED DATA AND ASSIGNMENTS FOR BENZENE, BENZENE-d6, BENZYL ALCOHOL, AND BENZYL −2,3,4,5,6-d5

The task of assigning the infrared and Raman data of a molecule is aided by obtaining deuterium analogs. In Table 9.20, vapor-phase infrared data for benzene and benzyl alcohol are compared to the vapor-phase infrared data for benzene-d6 and benzyl-2,3,4,5,6-d5 alcohol, respectively. Benzene has a center of symmetry, and overtones are not allowed in the infrared spectrum. Molecular vibrations involving primarily motion of the H atoms joined to the benzene carbon atoms are expected to decrease in frequency by a factor of the square root of 2 or approximately 1.41. Thus, for benzene and benzene-d6 the assignments where corresponding vibrations decrease in frequency by a factor of over 1.3 involve primarily motion of 6H or 6D atoms. For example, benzene exhibits a type-C band whose strong Q-branch occurs at 670 cm^{-1} (with P- and R- branches and 688 cm^{-1}, respectively). Benzene-d6 exhibits the corresponding type-C band at 505 cm^{-1}, 491 cm^{-1}, and 470 cm^{-1}. The frequency ratio of 670/491 is 1.365. These bands are due to a vibration where either the 6 H atoms or the 6 D atoms bend in-phase out of the plane of the benzene ring.

Benzyl alcohol and benzyl-2,3,4,5,6-d5 alcohol exhibit a vapor-phase infrared band at 738 cm^{-1} and 534 respectively. The frequency ratio 738/534 is 1.382. This molecular vibration is where the 5 H atoms or 5 C atoms based in-phase out of the plane of the phenyl group.

OUT-OF-PLANE DEFORMATIONS AND THEIR COMBINATION AND OVERTONES FOR SUBSTITUTED BENZENES

Young *et al.* have shown the substitution pattern for mono-through hexa-substituted benzenes, in the region 5–6μ 2000–1666.7 cm^{-1}), and their correlation presentation is shown in Fig. 9.36 (2). The substitution patterns are less complex with increased benzene substitution or with fewer protons joined directly to the benzene ring. In fact, hexa-substituted benzenes do not contain protons directly joined to the benzene ring. However, extensive studies of benzene derivatives have shown that most of the IR bands in this region of the spectrum result from combination and overtones of the out-of-plane hydrogen deformations (see Reference 6).

Table 9.21 lists a number of mono-substituted benzenes in the order of increasing σ_ρ values (6,22). These vibrational modes are listed I through V for the five out-of-plane hydrogen deformations for a mono-substituted benzene in the order of decreasing frequency. The number VI normal vibration is the out-of-plane ring deformation.

A study of Table 9.21 shows that modes I through III generally increase in frequency as the σ_ρ value is increased. These out-of-plane hydrogen deformations are also dependent upon the physical phase. This phase dependence is the result of the reaction field between the mono-substituted benzene molecules. This reaction field could also affect these spectra-structure conditions. However, a linear correlation is not obtained for data recorded in the vapor phase vs σ_ρ values (6).

POLYSTYRENE AND STYRENE COPOLYMERS

Tables 9.22 and 9.23 summarize the out-of-plane hydrogen deformation and the out-of-plane ring deformation for substituted benzenes and their combination and overtones, respectively (6).

Figures 9.19–9.24, 9.27, 9.29 and 9.30 are IR spectra for either polystyrene or a copolymer containing styrene units. Compare Figs. 9.19–9.30 with the mono-pattern shown in Fig. 9.36. The pattern for the styrene units is the same as that shown in Fig. 9.36.

In the IR spectra of α- and β-syndiotactic polystyrene the IR bands in the region 2000–1666.7 cm^{-1} are assigned as presented here (13):

α-Syndiotactic Polystyrene

Parallel Polarization cm^{-1}	Perpendicular Polarization cm^{-1}	Assignment (13)
1955	—	2 (977.5) = 1955; 2 I
—	1960	2 (980) = 1960; 2 I
1941.7	—	977.5 + 964.4 = 1941.9; I + II
—	1943.6	980 + 964 = 1944; I + II
1869.1	—	964.4 + 903 = 1867.4; II + III
—	1871.3	964 + 904.8 = 1869; II + III
1800.9	—	964.4 + 841 = 1805.4; II + IV

—	1802.8	$964 + 840.7 = 1805$; II + IV
1744.2	—	$903 + 841 = 1744$; III + IV
—	1746.3	$904.8 + 840.7 = 1745.5$; III + IV
1704	—	$2\,(852.75) = 1705.5$; 2 IV
1673.3	—	$977.5 + 695.3 = 1672.8$; I + VI
—	1675.2	$980 + 695.9 = 1676$

ETHYNYLBENZENE

In the case of ethynylbenzene (Fig. 9.33 in CCl_4 solution) the combination and overtone assignments are as presented here:

Ethynylbenzene	Assignment
1964	$2\nu_{29}(A_1)$; $2(983) = 1966$; 2 I
1947	$\nu_{29} + \nu_{26}(B_1) = 983 + 967 = 1950$; I + II
1895	$\nu_{29} + \nu_{30}(A_1) = 983 + 917 = 1900$; I + III
1871	$\nu_{26} + \nu_{30}(B_1) = 967 + 917 = 1884$; II + III
1820	$\nu_{29} + \nu_{27}(B_1) = 983 + 842 = 1825$; I + IV
1802	$\nu_{26} + \nu_{27}\,(A_1) = 967 + 842 + 1809$; II + IV
1751	$\nu_{30} + \nu_{27}\,(B_1) = 917 + 842 = 1759$; II + IV

These assignments are in good agreement with the summary for out-of-plane hydrogen deformation and their combination and overtones for monosubstituted benzenes (see Fig. 9.37).

In the case of mono-substituted benzenes, modes I, II, and V belong to the B_1 symmetry species for molecules with C_{2v} symmetry, and modes I, II and V belong to the A'' symmetry species for molecules with C_s symmetry. Moreover, modes II and IV belong to the A_2 symmetry species for molecules with C_{2v} symmetry, and modes II and IV for molecules that belong to the A'' species for molecules with C_s symmetry. Thus, the author finds it most convenient to term these vibrations I through V in the order of their decreasing frequency.

It is only possible for vibrations belonging to the same symmetry species to couple. Thus, there is more probability of coupling between the A'' vibrations in the case of mono-substituted benzenes with C_s symmetry than there is between the B_1 fundamentals and between the A_2 fundamentals for mono-substituted benzenes with C_{2v} symmetry. This may be one of the reasons that modes I through V do not correlate on a one-to-one basis with σ_ρ. Another factor is that not all spectra were recorded in the same physical phase.

1,2-DISUBSTITUTED BENZENES

Figure 9.38 is a summary of the out-of-hydrogen deformations and their combination and overtones for 1,2-disubstituted benzenes (6). For 1,2-disubstituted benzenes with C_{2v} symmetry modes I and II belong to the A_2 species and modes II and IV belong to the B_1 species. For 1,2-disubstituted benzenes with C_s symmetry, modes I through IV belong to the A'' species. Modes I through IV occur in the regions 956–989 cm^{-1}; 915–970 cm^{-1}, 833–886 cm^{-1}, and 725–

791 cm^{-1}, respectively. The two overtone and five combination tones of modes I through IV are presented on Fig. 9.38.

1,3-DIBSUBSTITUTED BENZENES

Figure 9.39 is a summary of the out-of-plane hydrogen deformations and their combination and overtones for 1,3-disubstituted benzenes (6). For 1,3-disubstituted benzenes with C_{2v} symmetry, modes I, III and IV belong to the B_1 symmetry species and mode II belongs to the A2 symmetry species. For 1,3-disubstituted benzenes with C_s symmetry, modes I through IV belong to the A'' species. Modes I through IV occur in the regions 947–999 cm^{-1}, 863–941 cm^{-1}, 831–910 cm^{-1}, and 761–815 cm^{-1}, respectively. The three overtones and four combination tones of modes I through IV are presented in Fig. 9.39.

1,4-DISUBSTITUTED BENZENES

Figure 9.40 is a summary of the out-of-plane hydrogen deformations and their combination and overtones for 1,4-disubstituted benzenes (6). For 1,4-disubstituted benzenes with V_h symmetry, modes I through IV belong to the A_u, B_{2g}, B_{1g}, and B_{3u} species, respectively. For molecules with C_{2v} symmetry, modes I and III belong to the A_2 species and modes II and IV belong to the B_1 species. For molecules with C_s symmetry, modes I through IV belong to the A'' species. For molecules with a center of symmetry such as 1,4-disubstituted benzenes with V_h symmetry, overtones of fundamentals are not allowed in the IR. A study of Fig. 9.40 shows that overtones are not present in the region 2000–1600 cm^{-1} for any of the 1,4-disubstituted benzenes. Modes I through IV occur in the regions 1852–1957 cm^{-1}, 916–971 cm^{-1}, 790–852 cm^{-1}, and 794–870 cm^{-1}, respectively. Four combinations tones of modes I through IV are listed in Fig. 9.40.

1,3,5-TRISUBSTITUTED BENZENES

Figure 9.41 is a summary of the out-of-plane deformations and their combination and overtones for 1,3,5-trisubstituted benzenes (6). For 1,3,5-trisubstituted benzenes with D_{3h} symmetry modes, the Ia and Ib modes are degenerate. For molecules with C_{2v} symmetry, the degeneracy is split into two B_1 modes and the A_2'' mode for molecules with D_{3h} symmetry becomes an A_2 fundamental. In the case of 1,3,5-trisubstituted benzenes with C_s symmetry, all three modes belong to the A'' species. The Ia and Ib modes occur in the region 819–920 cm^{-1} and mode II occurs in the region 786–910 cm^{-1}. The 2I overtone occurs in the regions 1635–1840 cm^{-1}, and the mode I + II combination occurs in the region 1605–1840 cm^{-1}.

1,2,3-TRISUBSTITUTED BENZENES

Figure 9.42 is a summary of the out-of-plane hydrogen deformations and their combination and overtones for 1,2,3-trisubstituted benzenes (6). For molecules with C_{2v} symmetry, modes I and

III belong to the B_1 species and mode II belongs to the A_2 species. For molecules with C_s symmetry, modes I through III belong to the A″ species. Modes I through III occur in the ranges 930–989 cm^{-1}, 848–930 cm^{-1}, and 731–810 cm^{-1}, respectively; 2I and 2II occur in the regions 1860–1975 and 1700–1860 cm^{-1}, respectively. The combination I + II occurs in the region 1848–1930 cm^{-1}.

1,2,4-TRISUBSTITUTED BENZENES

Figure 9.43 is a summary of the out-of-plane hydrogen deformations and their combination and overtones for 1,2,4-trisubstituted benzenes (6). As long as these molecules have a plane of symmetry their molecular symmetry is C_s. Modes I through III belong to the A″ species and occur in the regions 926–982 cm^{-1}, 841–923 cm^{-1}, and 790–852 cm^{-1}, respectively. Their three overtones and two combinations are shown in Fig. 9.43.

1,2,3,4-TETRASUBSTITUTED BENZENES

Figure 9.44 is a summary of the out-of-plane hydrogen deformations and their combination and overtones for 1,2,3,4-tetrasubstituted benzenes (6). For molecules with C_{2v} symmetry, mode I and mode II belong to the A_2 and B_1 species, and for molecules with C_s symmetry, modes I and II belong to the A″ species. Modes I and II occur in the regions 919–949 cm^{-1} and 770–821 cm^{-1}, respectively. Their two overtones and one combination are shown in Fig. 9.44.

1,2,3,5-TETRASUBSTITUTED BENZENES

Figure 9.45 is a summary of the out-of-plane hydrogen deformations and their combination and overtones for 1,2,3,5-tetrasubstituted benzenes (6). For molecules with C_{2v} symmetry, modes I and II belong to the A_2 and B_1 species, and occur in the regions 821–942 cm^{-1} and 839–920 cm^{-1}, respectively. The overtones I and II occur in the regions 1635–1860 cm^{-1} and 1638–1841 cm^{-1}, respectively. The combination I + II occurs in the region 1659–1862 cm^{-1}.

1,2,4,5-TETRASUBSTITUTED BENZENES

Figure 9.46 is a summary of the out-of-plane hydrogen deformations and their combination and overtones for 1,2,4,5-tetrasubstituted benzenes (6). For molecules with v_h symmetry, modes I and II belong to the B_{3u} and B_{2g} species, respectively. For molecules with C_{2v} symmetry, modes I and II belong to the B_1 and A_2 species, respectively, and for molecules with C_s symmetry both modes belong to the A″ species. Modes I and II occur in the regions 861–911 cm^{-1} and 828–980 cm^{-1}, respectively. These two overtones, if allowed, occur in the regions 1725–1830 cm^{-1} and 1651–1770 cm^{-1}, respectively, and the combination occurs in the region 1695–1799 cm^{-1}.

1,2,3,4,5-PENTASUBSTITUTED BENZENES

Figure 9.47 is a summary of the out-of-plane hydrogen deformation and its first overtone for pentasubstituted benzenes (6). For molecules with C_{2v} symmetry, mode I belongs to the B_1 species, and for molecules with C_s symmetry mode I belongs to the A″ species. Its first overtone occurs in the region 1655–1848 cm^{-1}.

SUMMARY OF THE OUT-OF-PLANE HYDROGEN DEFORMATIONS FOR SUBSTITUTED BENZENES AND THE OUT-OF-PLANE RING DEFORMATION FOR MONO-SUBSTITUTED BENZENES, AND THEIR COMBINATION AND OVERTONES

Table 9.22 is a summary of the frequency ranges for 11 types of substituted benzenes (6). The number of ranges for each class of substituted benzenes decreases as the number of protons joined to the ring decreases.

Table 9.23 is a summary of the frequency ranges for the combination and overtones of the out-of-plane hydrogen deforms for the 11 types of substitute benzenes (6). The number of ranges within each class of substituted benzenes decreases as the number of protons joined to the ring decreases.

CORRELATION CHART

Figure 9.48 is a correlation chart for the out-of-plane hydrogen deformations and their combination and overtones for substituted benzenes. The thickness of each bar graph for modes I through V and their combination and overtones indicates the relative band intensities exhibited within each class of substituted benzenes for most of the compounds within each class. The band intensities for the combination and overtones are generally 10 to 100 times less intense than the most intense IR band for the fundamental out-of-plane hydrogen deformations exhibited by each class of substituted benzenes.

It should be remembered that other overtones such as those for vinyl wag and vinylidene CH_2 wag also occur in the region 2000–1666 cm^{-1} (see Volume 1, Chapter 4). Of course, the vast majority of compounds containing a carbonyl group exhibit ν C=O in the region 2000–1666 cm^{-1}, and these ν C=O modes have strong IR band intensity that mask many of the combination and overtone bands (see Volume 1, Chapters 10–16).

2,3,4,5,6-PENTACHLOROBIPHENYL

Table 9.24 compares vibrational data for hexachlorobenzene vs 2,3,4,5,6-pentachlorobiphenyl. Eight fundamentals of the pentachlorophenyl group are assigned (7).

REFERENCES

1. Varsanyi, G. (1969). *Vibrational Spectra of Benzene Derivatives*. New York/London: Academic Press.
2. Young, C. W., Du Vall, R. B., and Wright, N. (1951). *Anal. Chem.* **23**: 709.
3. Scherer, J. R. (1964). *Planar Vibrations of Chlorinated Benzenes*. Midland, MI: The Dow Chemical Company.
4. Scherer, J. R., and Evans, J. C. (1963). *Spectrochim Acta* **19**: 1763.
5. Scherer, J. R., Evans, J. C., Muelder, W. W., and Overend, J. (1962). *Spectrochim. Acta* **18**: 57.
6. Nyquist, R. A. (1984). *The Interpretation of Vapor-Phase Infrared Spectra: Group Frequency Data*. Philadelphia: Sadtler.
7. Nyquist, R. A., Putzig, C. L., and Peterson, D. P. *Appl. Spectros.* **37**: 140.
8. *Sadtler Standard Raman Spectral Collection*. Philadelphia: Sadtler Research Laboratories.
9. Nyquist, R.A., and Kagel, R. O. (1977). Organic Materials. Chapter 6, in *Infrared and Raman Spectroscopy*, Part B, E. G. Frame, Jr. and J. G. Grassellie, eds., New York: Marcel Dekker, Inc., p. 476.
10. Lau, C. L., and Snyder, R. G. (1971) *Spectrochim. Acta* **27A**: 2073.
11. Nyquist, R. A. (1988). *Appl. Spectrosc.* 1314.
12. Taft, Jr., R. W. (1956) *Steric Effects in Organic Chemistry*. M. S. Newman, ed., New York: Wiley & Sons, p. 592.
13. Nyquist, R. A., Putzig, C. L., Leugers, M. A., McLachlan, R. D., and Thill, B. (1992). *Appl. Spectrosc.* **46**: 981.
14. Nyquist, R. A. (1984). *Appl. Spectrosc.* **38**: 264.
15. Nyquist, R. A. (1989). *Appl. Spectrosc.* **43**: 440.
16. Nyquist, R. A., and Malanga, M. (1989) *Appl. Spectrosc.* **43**: 442.
17. Loy, B. R., Chrisman, R. W., Nyquist, R. A., and Putzig, C. L. (1979). *Appl. Spectrosc.* **33**: 638.
18. Nyquist, R. A., Platt, A. E., and Priddy, D. B. (1982). *Appl. Spectrosc.* **36**: 417.
19. Nyquist, R. A., and Schuetz, J. E. (1985). *Appl. Spectrosc.* **39**: 595.
20. Evans, J. C., and Nyquist, R. A. (1960). *Spectrochim. Acta* **16**: 918.
21. Nyquist, R. A., Loy, B. R., and Chrisman, R. W. (1981). *Spectrochim. Acta* **37A: 319.**
22. Taft, R. W. (1976) *Progress in Physical Organic Chemistry*. vol. 12, New York: Interscience Publication, J. Wiley, p. 74.

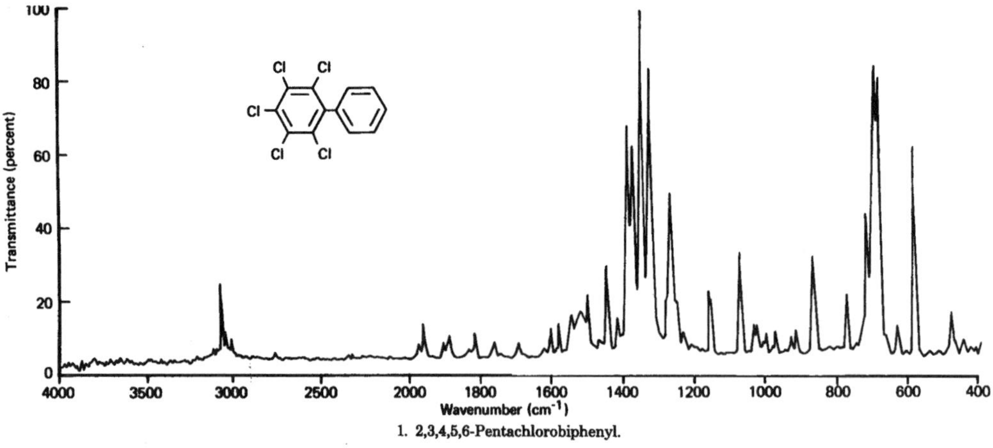

FIGURE 9.1 Infrared spectrum for 2,3,4,5,6-pentachlorobiphenyl.

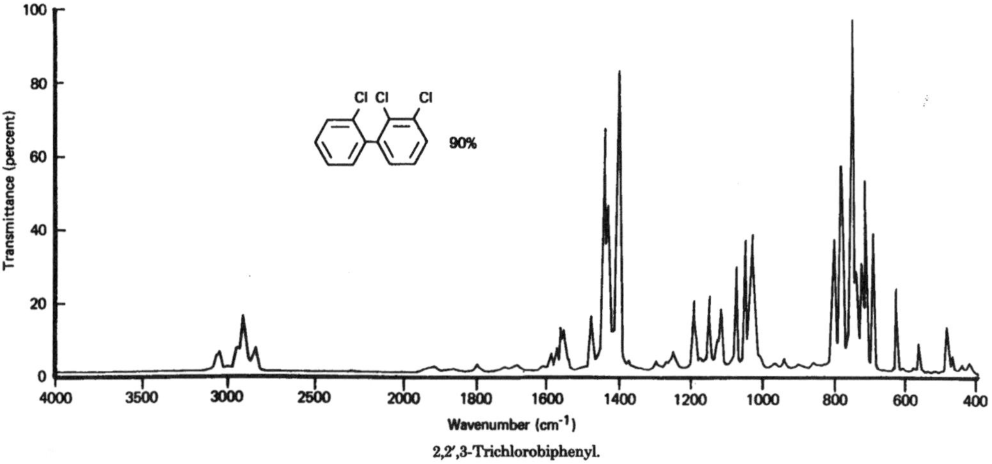

FIGURE 9.2 Infrared spectrum for 2,2′,3-trichlorobiphenyl.

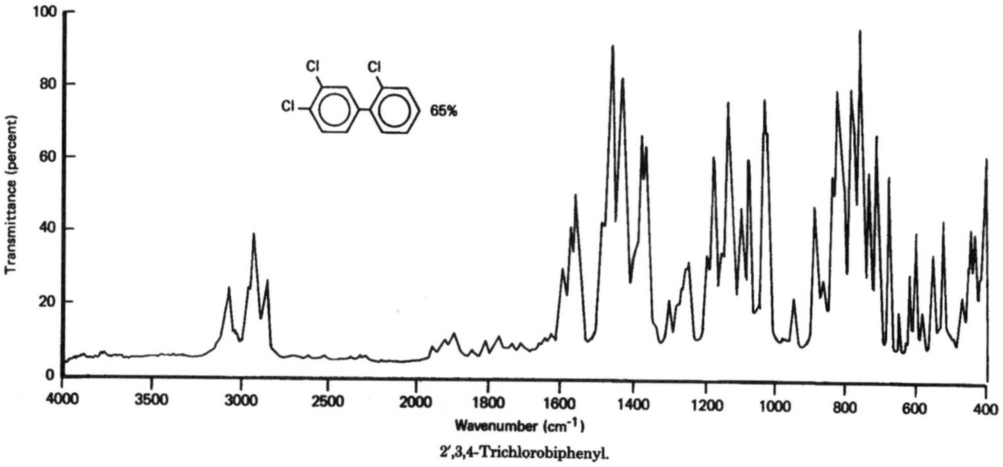

FIGURE 9.3 Infrared spectrum for 2′,3,4-trichlorobiphenyl.

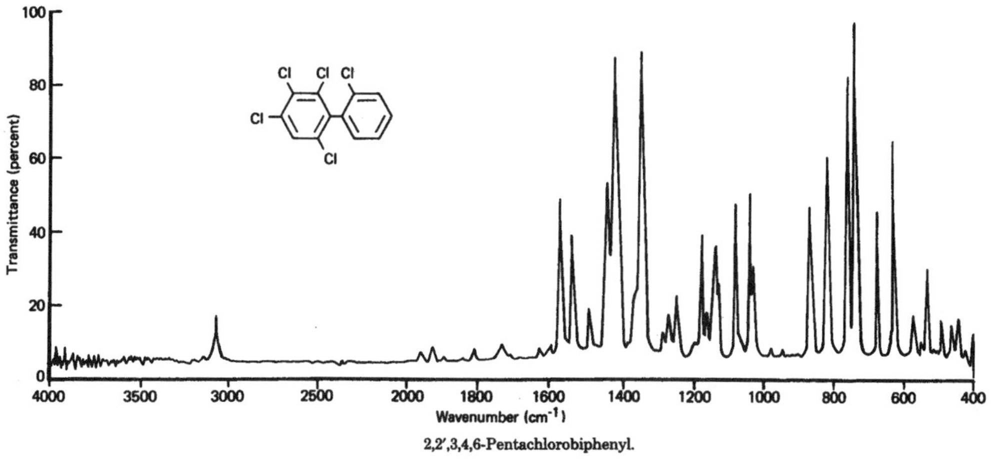

FIGURE 9.4 Infrared spectrum for 2,2′,3,4,6-pentachlorobiphenyl.

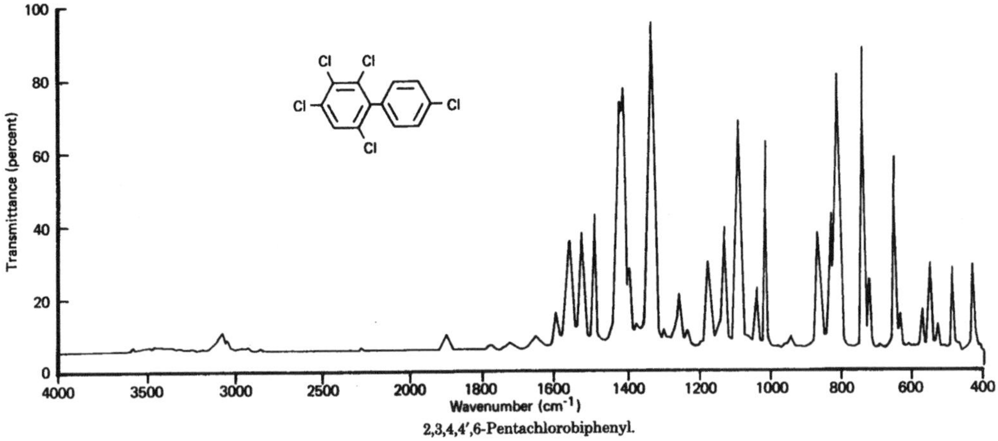

FIGURE 9.5 Infrared spectrum for 2,3,4,4′,6-pentachlorobiphenyl.

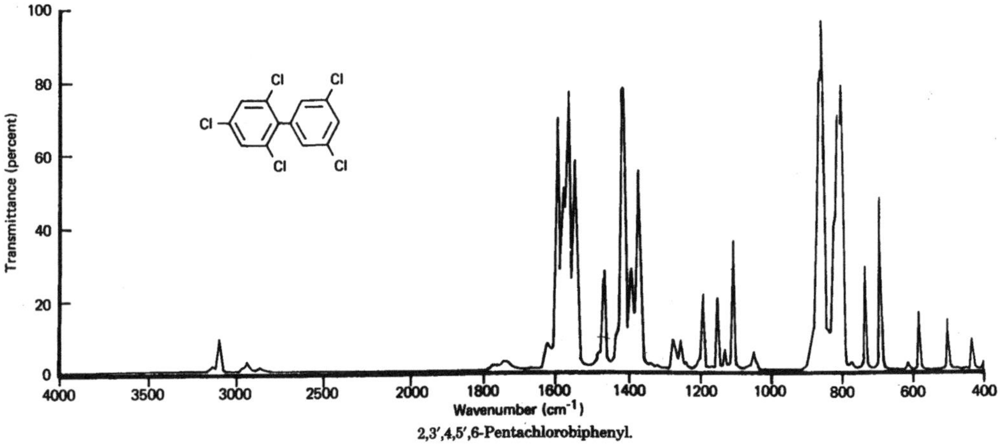

FIGURE 9.6 Infrared spectrum for 2,3′,4,5′,6-pentachlorobiphenyl.

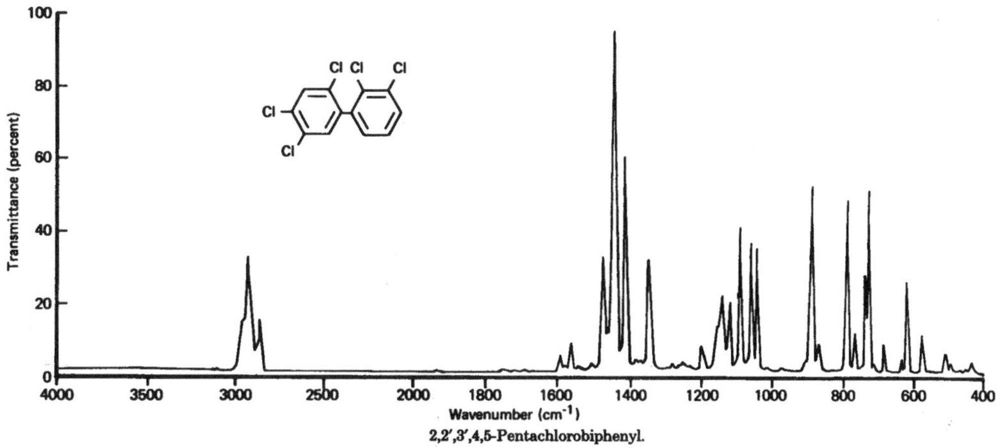

FIGURE 9.7 Infrared spectrum for 2,2′,3′,4,5-pentachlorobiphenyl.

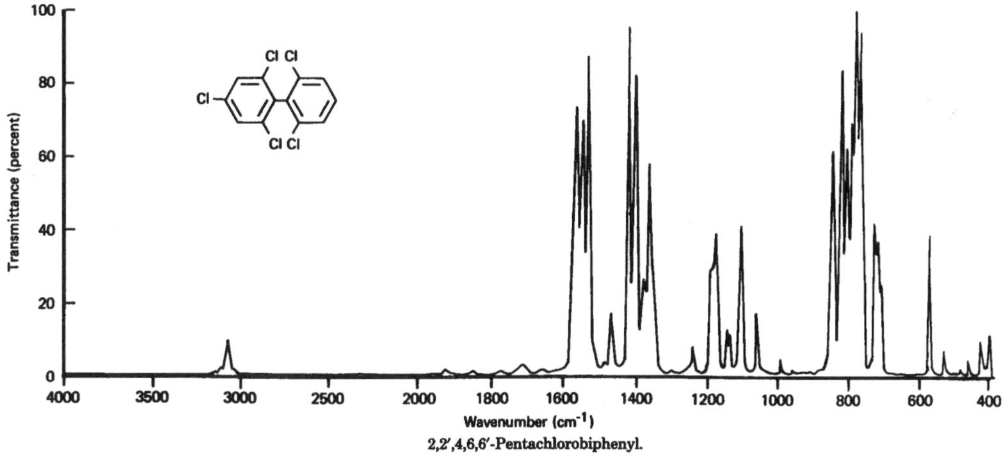

FIGURE 9.8 Infrared spectrum for 2,2′,4,6,6′-pentachlorobiphenyl.

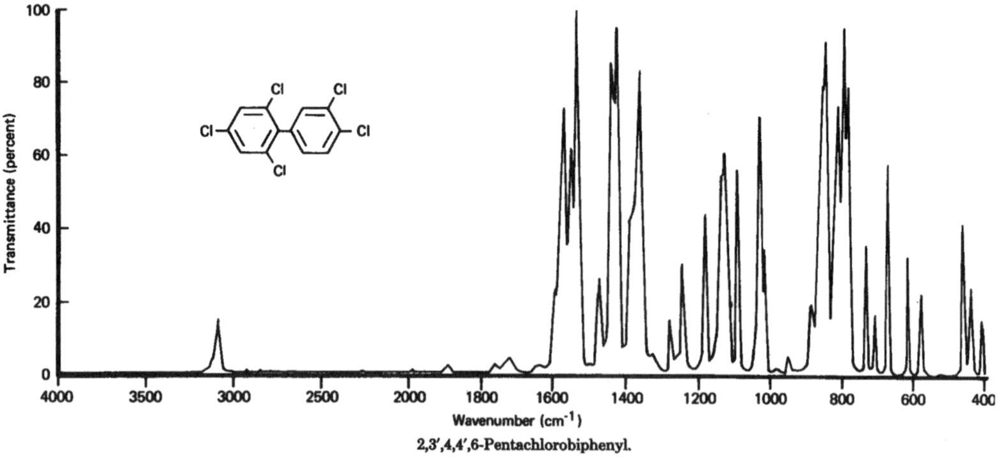

FIGURE 9.9 Infrared spectrum for 2,3′,4,4′,6-pentachlorobiphenyl.

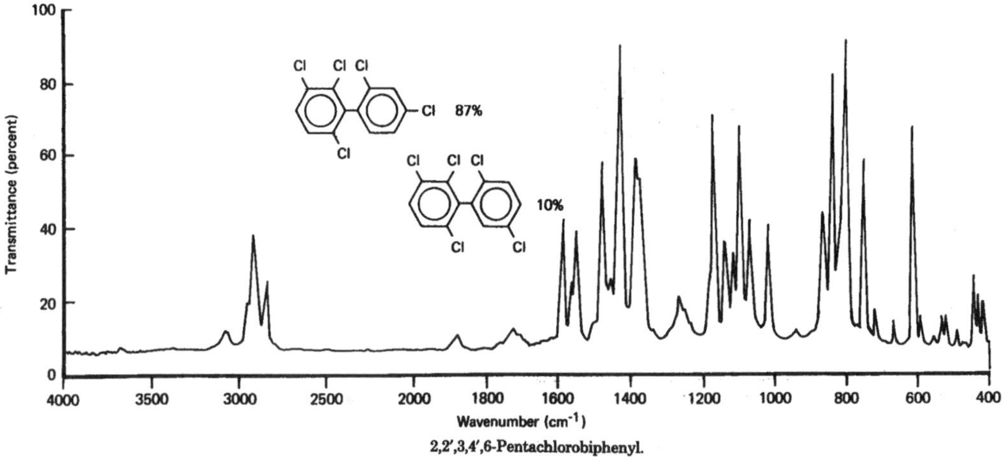

FIGURE 9.10 Infrared spectrum for 2,2′,3,4′,6-pentachlorobiphenyl.

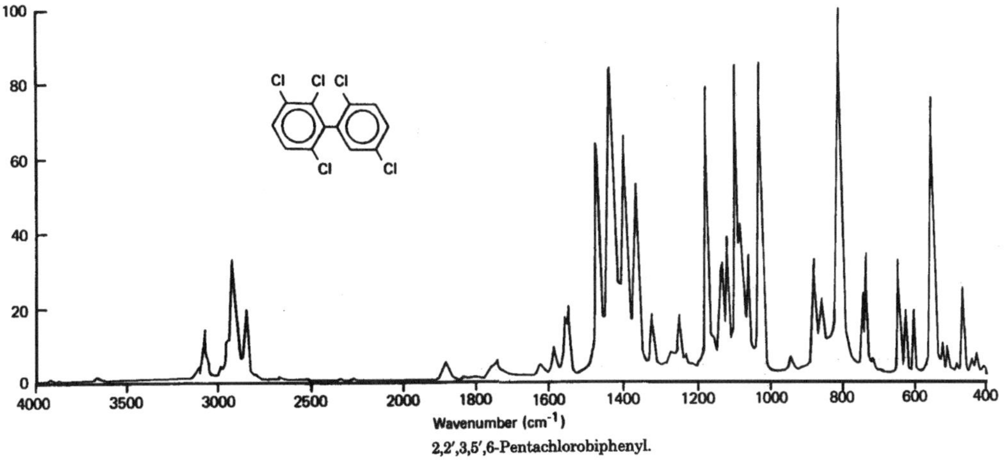

FIGURE 9.11 Infrared spectrum for 2,2′,3,5′,6-pentachlorobiphenyl.

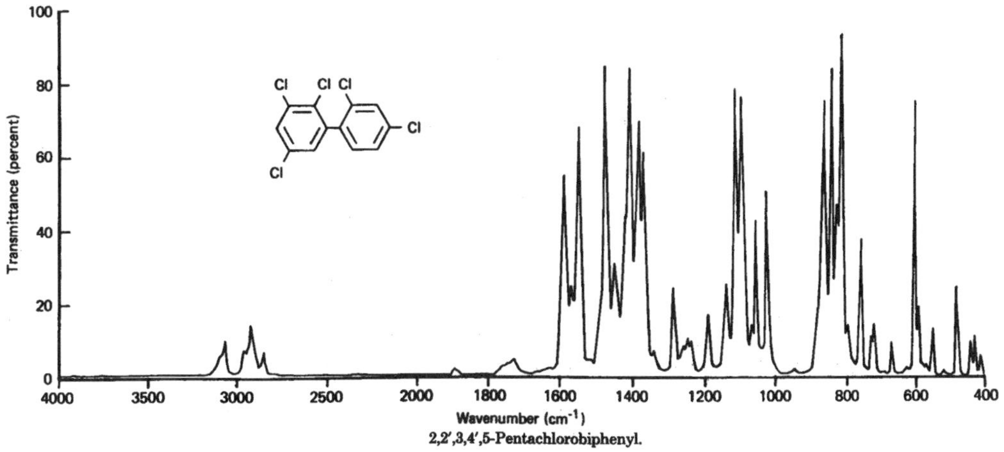

FIGURE 9.12 Infrared spectrum for 2,2′,3,4′,5-pentachlorobiphenyl.

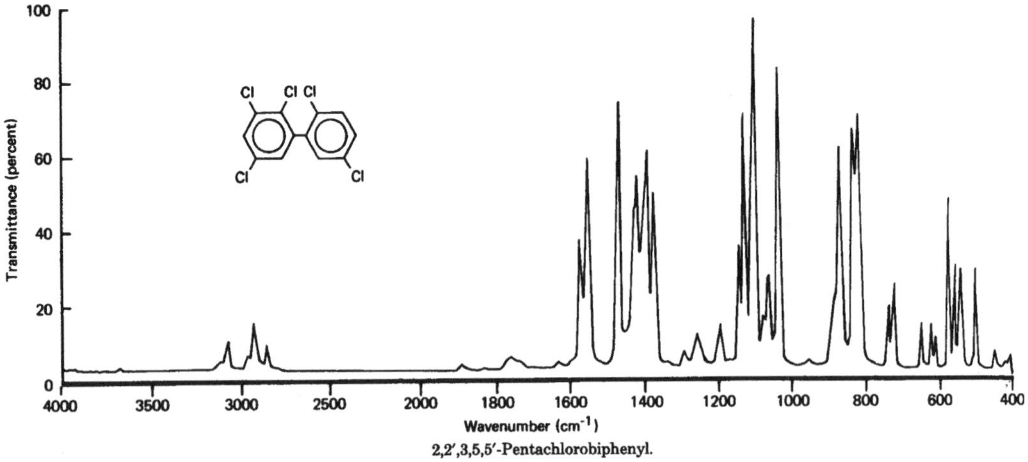

FIGURE 9.13 Infrared spectrum for 2,2′,3,5,5′-pentachlorobiphenyl.

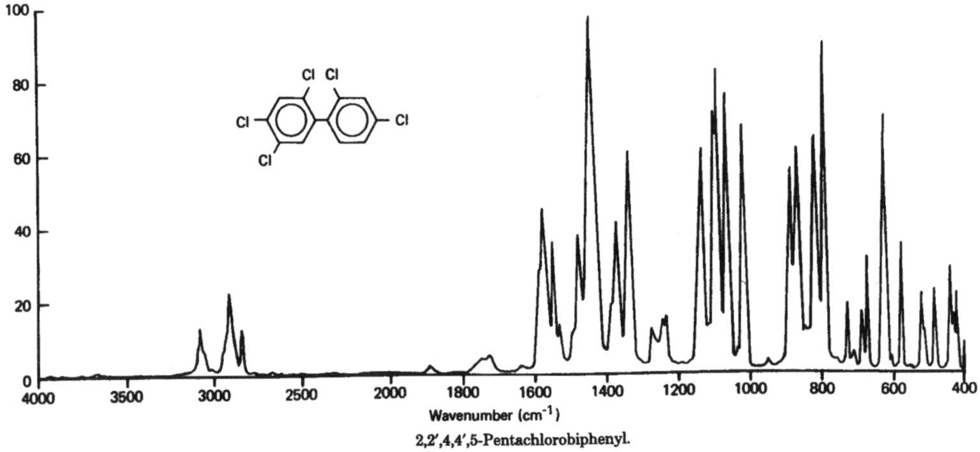

FIGURE 9.14 Infrared spectrum for 2,2′,4,4′,5-pentachlorobiphenyl.

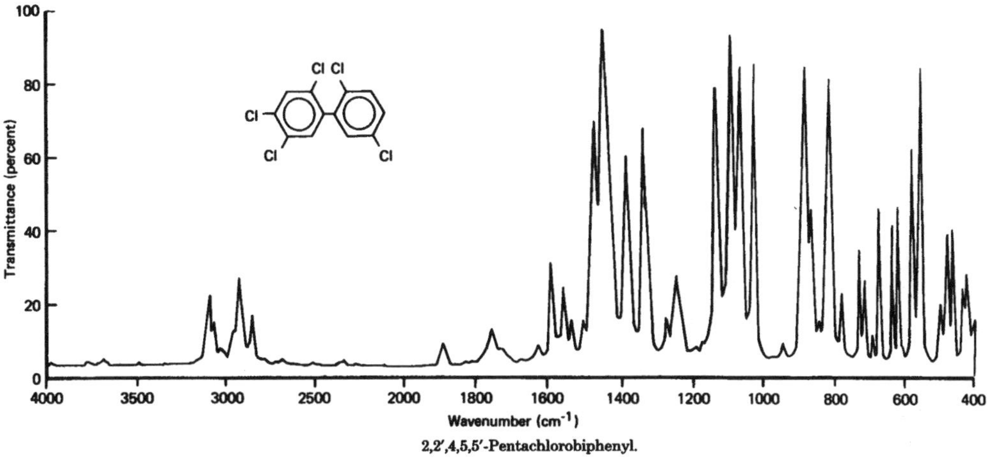

FIGURE 9.15 Infrared spectrum for 2,2′,4,5,5′-pentachlorobiphenyl.

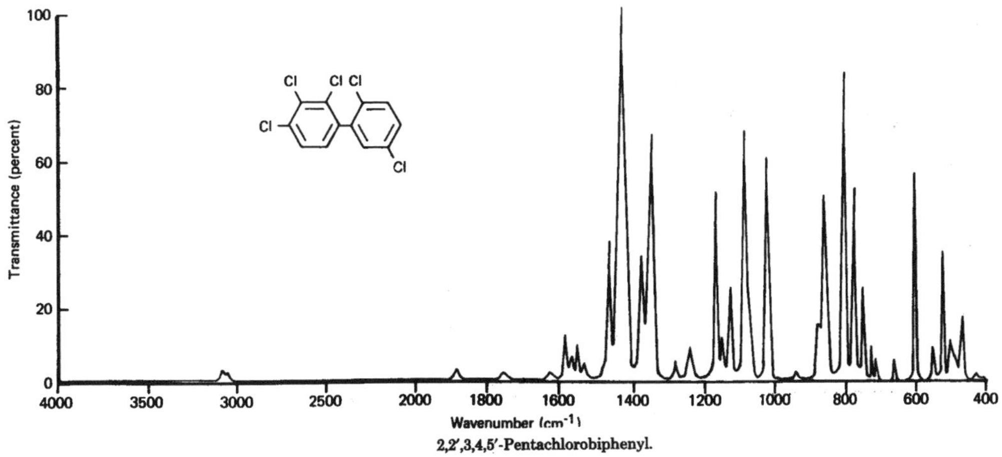

FIGURE 9.16 Infrared spectrum for 2,2′,3,4,5′-pentachlorobiphenyl.

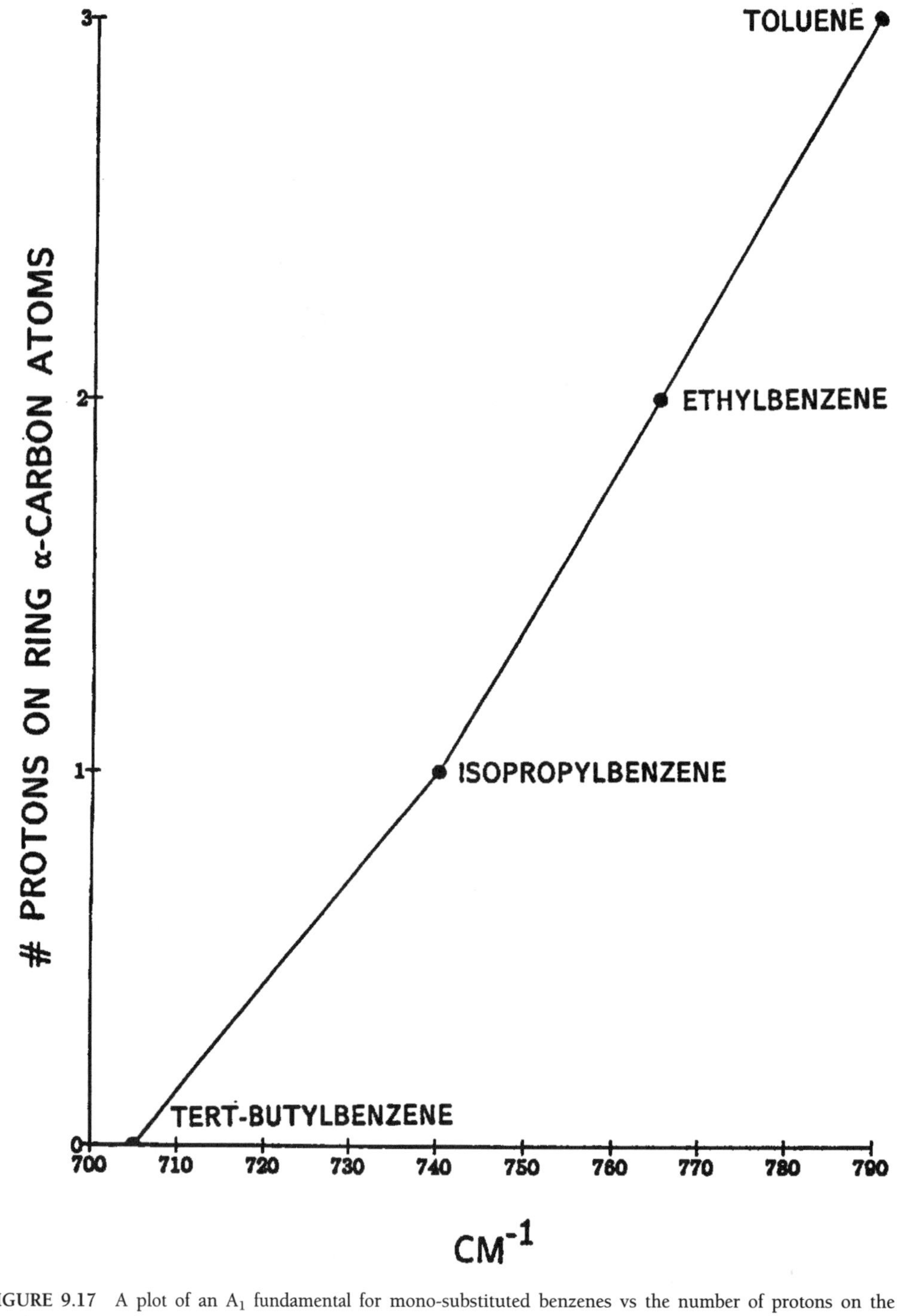

FIGURE 9.17 A plot of an A_1 fundamental for mono-substituted benzenes vs the number of protons on the ring α-carbon atom.

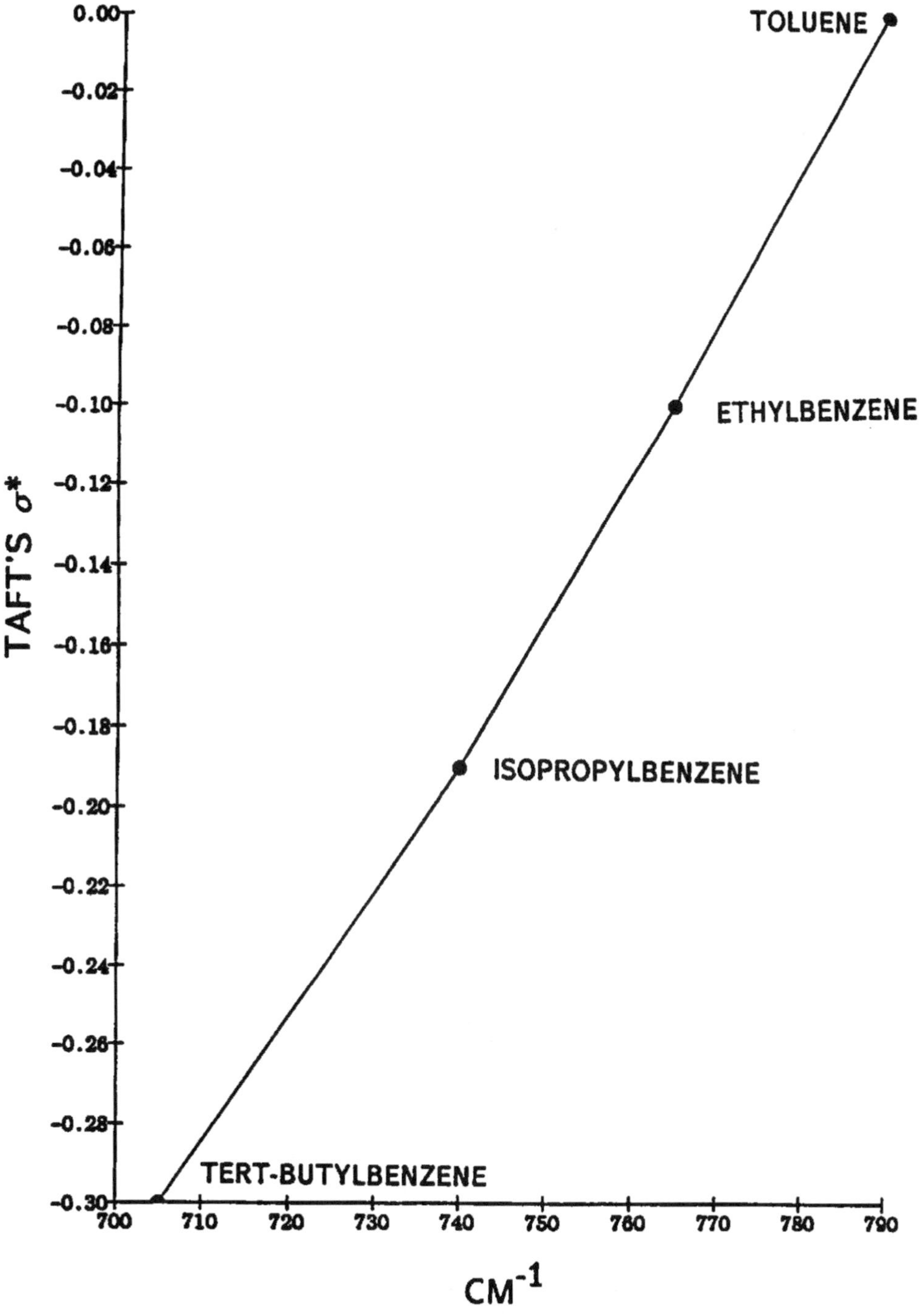

FIGURE 9.18 A plot of an A_1 fundamental for mono-substituted benzenes vs Tafts σ^*.

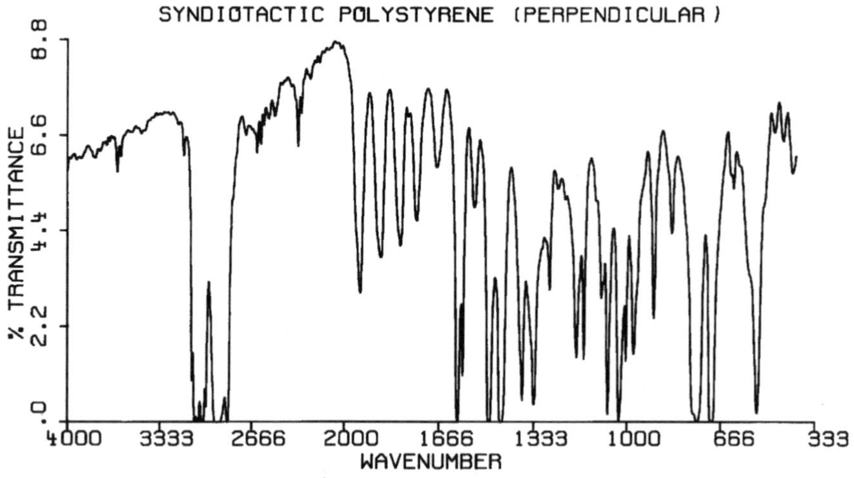

FIGURE 9.19 An IR spectrum for a uniaxially stretched syndiotactic polystyrene film (perpendicular polarization).

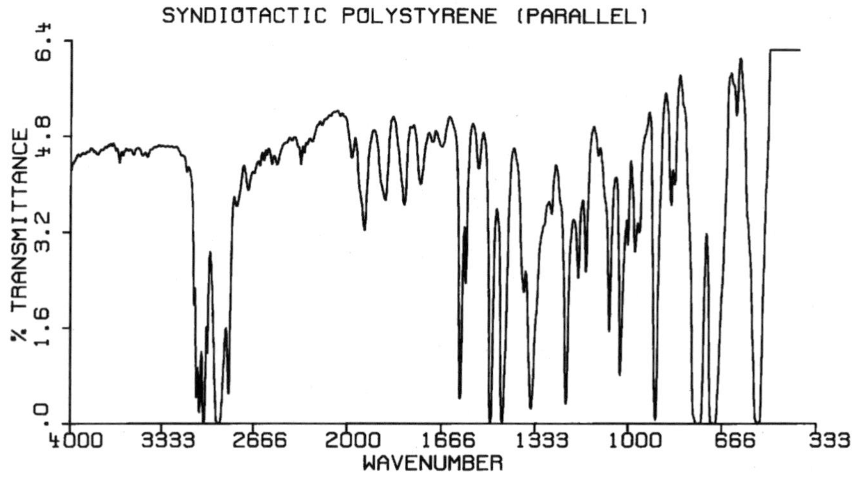

FIGURE 9.20 An IR spectrum for a uniaxially stretched syndiotactic polystyrene film (parallel polarization).

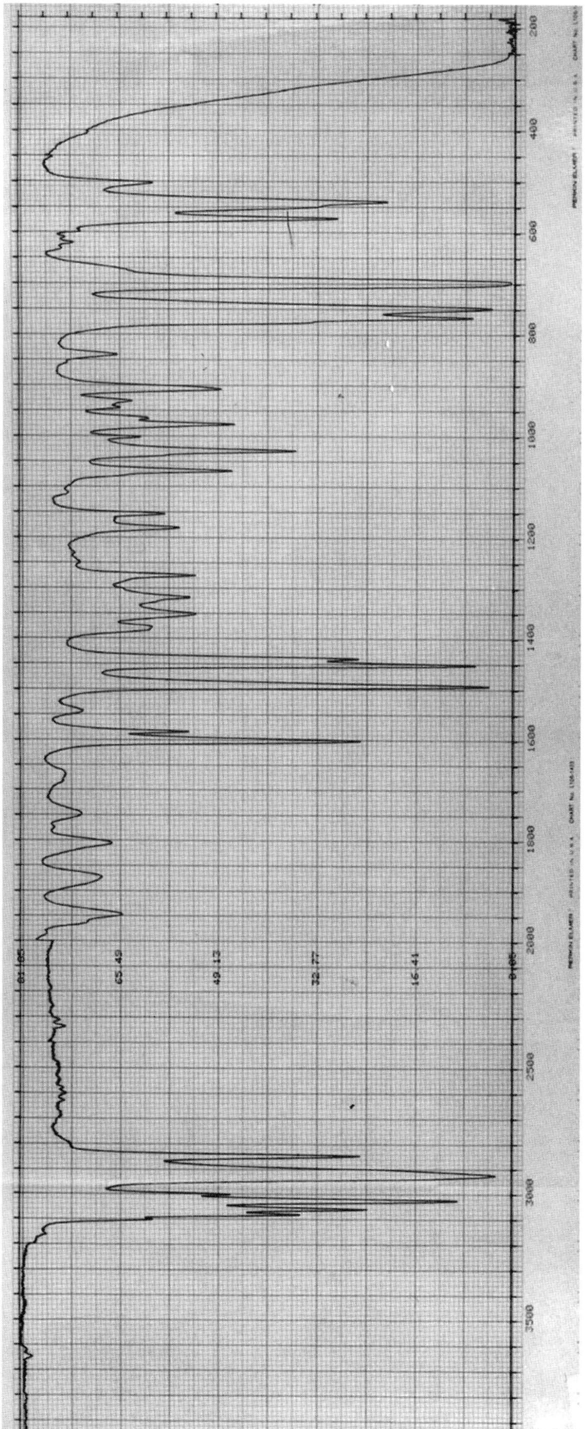

FIGURE 9.21 An IR spectrum of syndiotactic polystyrene film cast from boiling 1,2-dichlorobenzene onto a KBr plate.

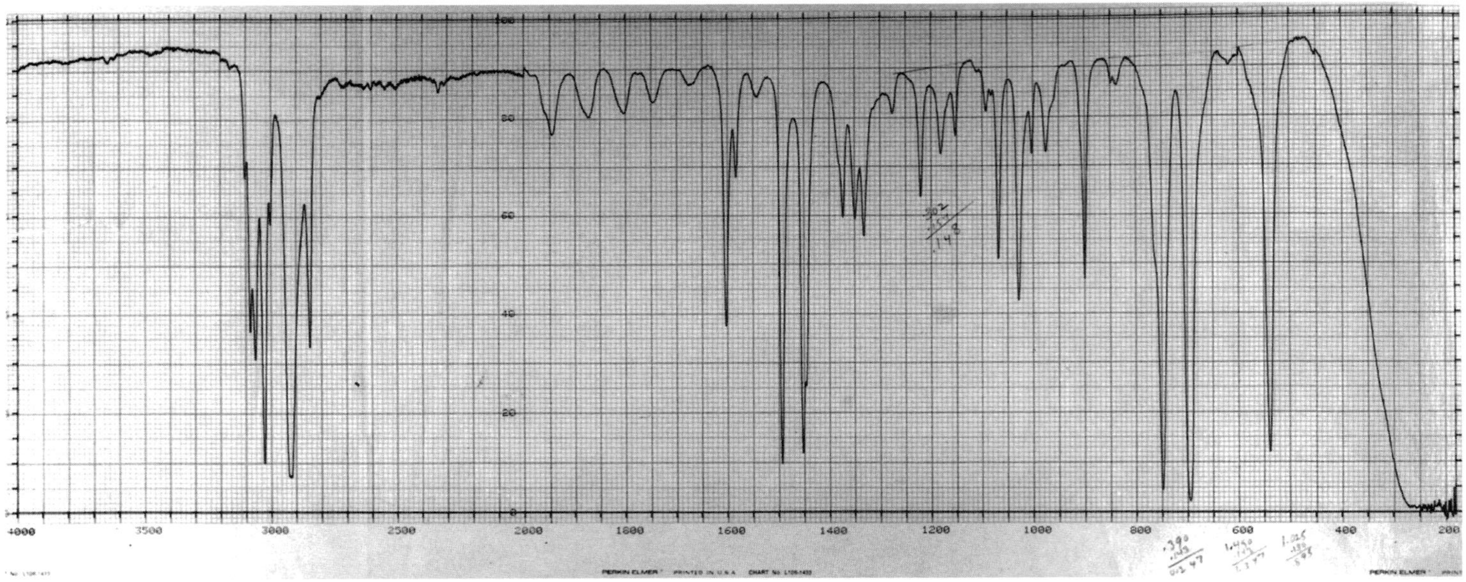

FIGURE 9.22 An IR spectrum of the same film used to record the IR spectrum shown in Fig. 9.21 except that the film was heated to 290 °C then allowed to cool to ambient temperature before recording the IR spectrum of syndiotactic polystyrene.

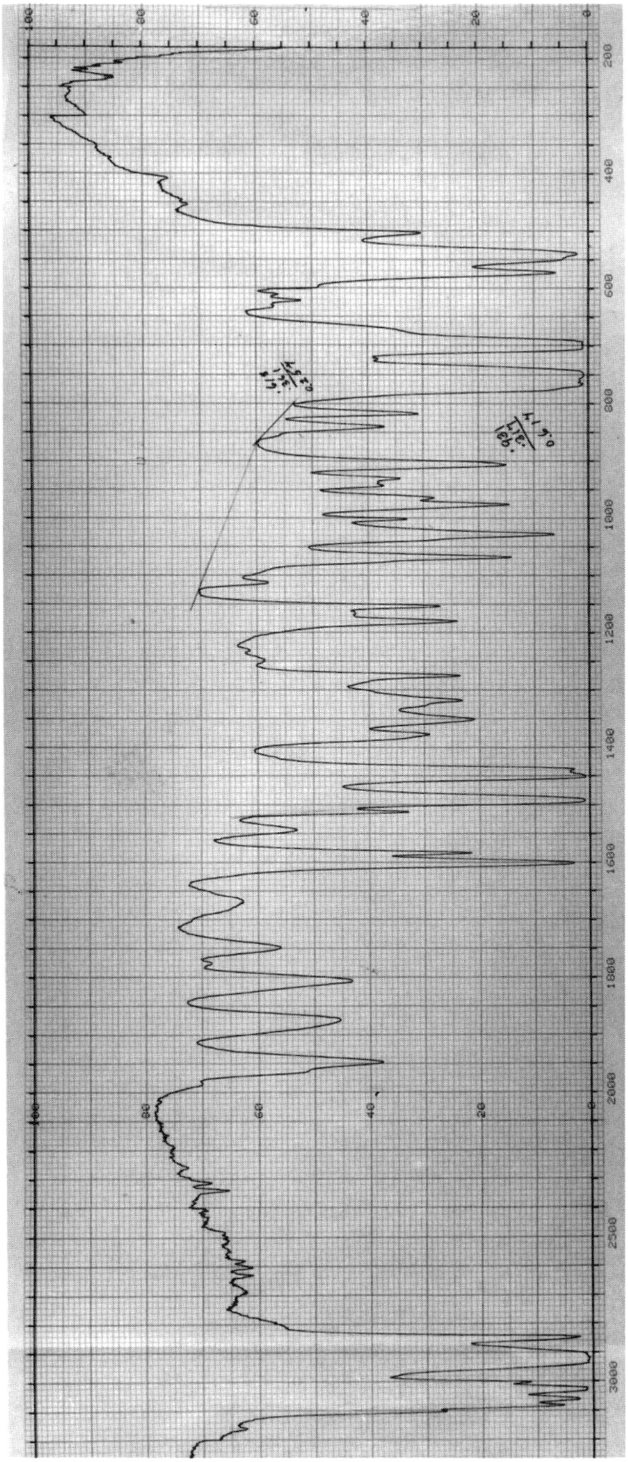

FIGURE 9.23 An IR spectrum syndiotactic styrene (98%)-4-methyl-styrene (2%) copolymer cast from boiling 1,2-dichlorobenzene onto a C_sI plate.

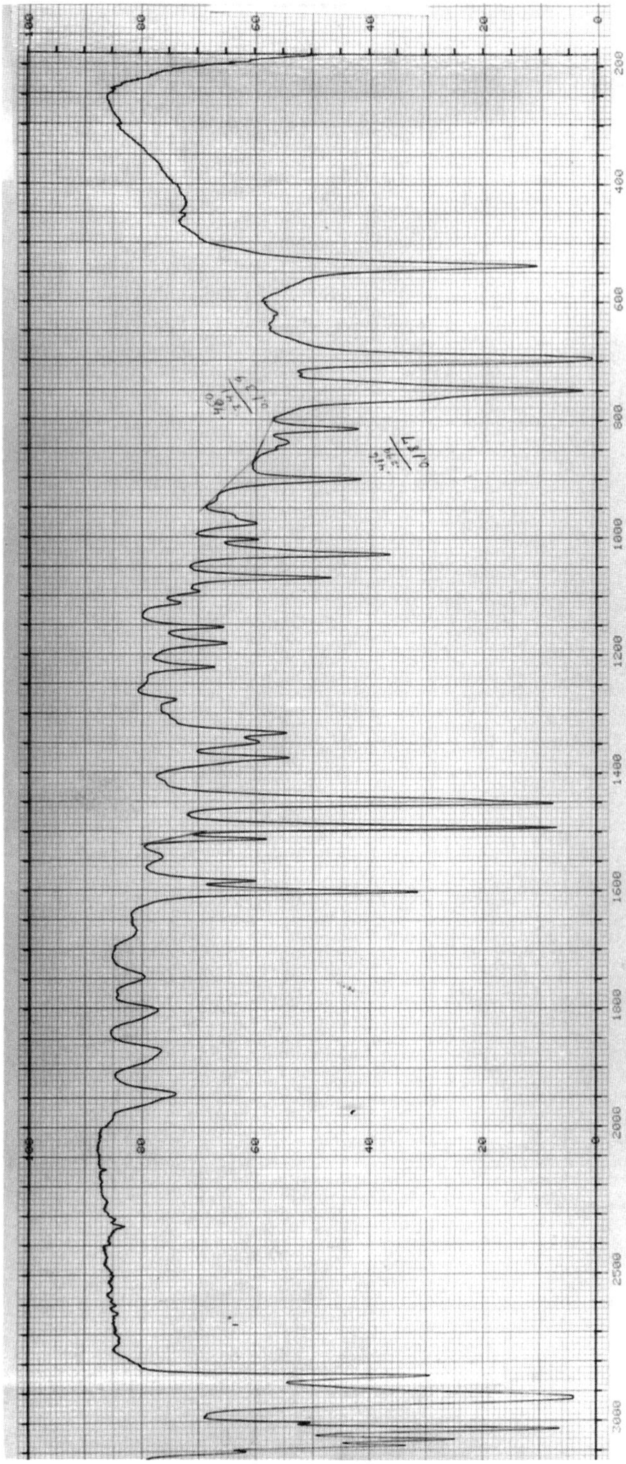

FIGURE 9.24 An IR spectrum of a styrene (93%)-4-methylstyrene (7%) copolymer cast from boiling 1,2-dichlorobenzene onto a C_sI plate.

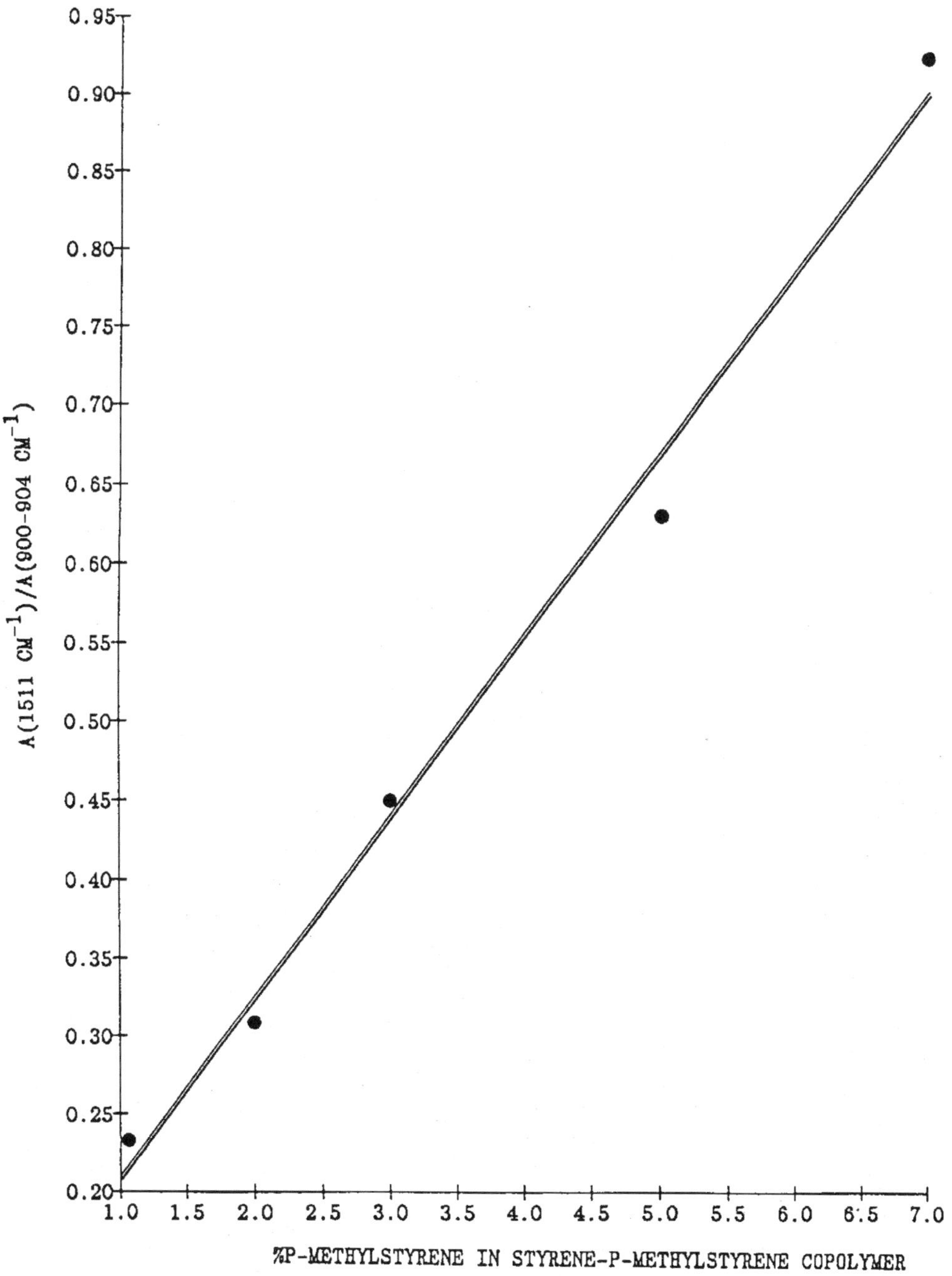

FIGURE 9.25 A plot of the IR band intensity ratio A(1511 cm^{-1})/A (900–904 cm^{-1}) vs the weight % 4-methylstyrene in styrene –4-methylstyrene copolymers.

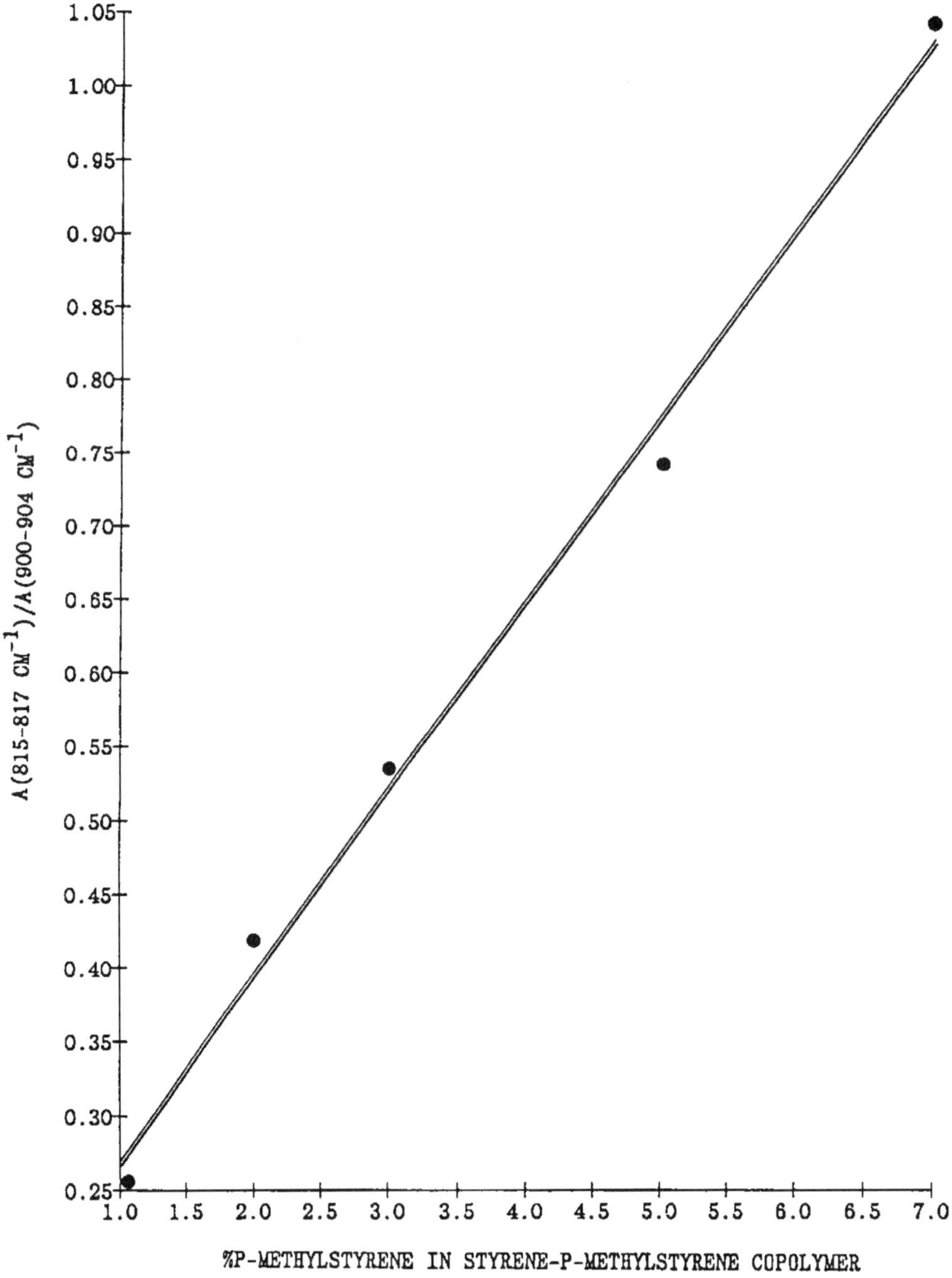

FIGURE 9.26 A plot of the IR band intensity ratio A(815–817 cm^{-1})/A(900–904 cm^{-1}) vs the weight % 4-methylstyrene in styrene –4-methylstyrene copolymers.

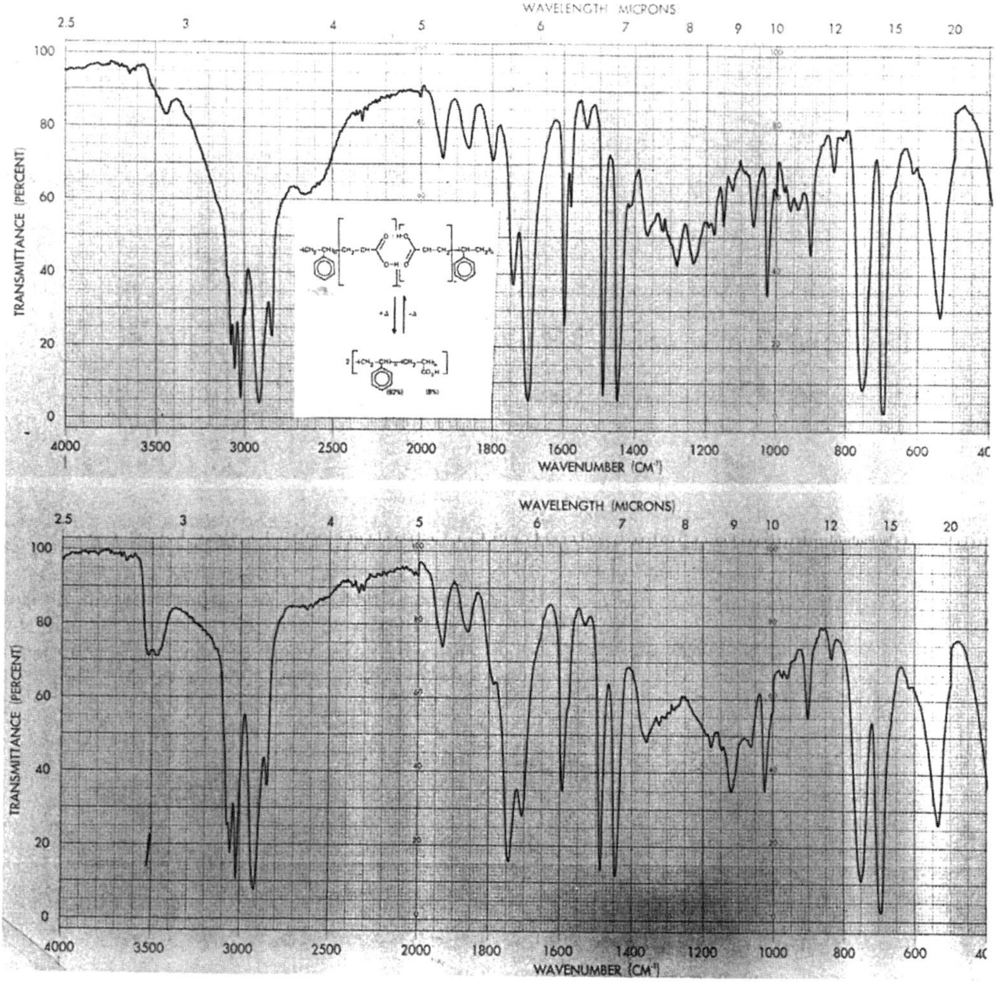

FIGURE 9.27 Top: Infrared spectrum of styrene (92%) -acrylic acid (8 %) copolymer recorded at 35 °C. Bottom: Infrared spectrum of styrene (92%) –acrylic acid (8%) copolymer recorded at 300 °C.

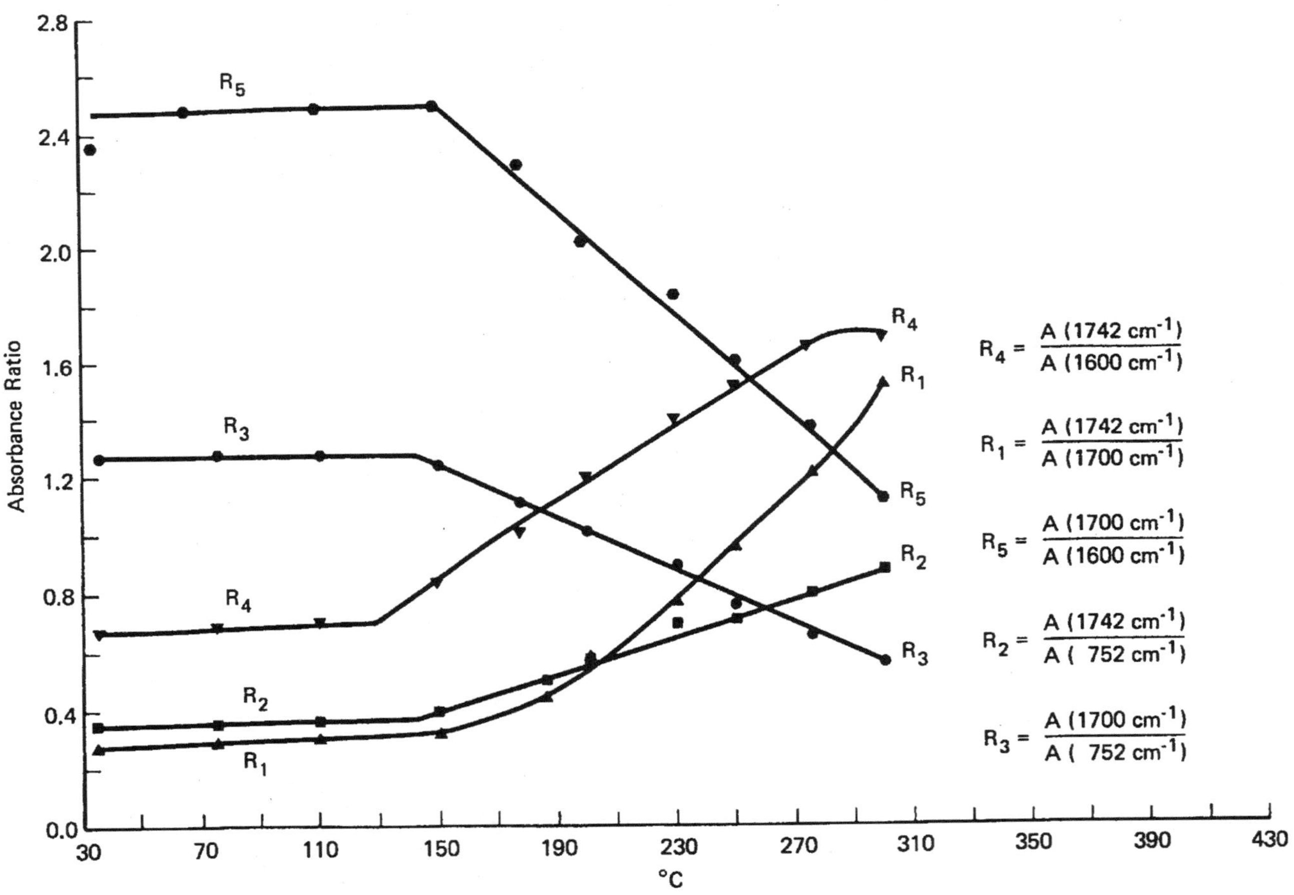

FIGURE 9.28 Styrene (92%) – acrylic acid (8%) copolymer absorbance ratios at the indicated frequencies vs copolymer film temperatures in °C; R_1, R_2, and R_4 indicate an increase in CO_2H concentrations at temperatures >150 °C; R_3 and R_5 indicate a decrease in $(CO_2H)_2$ concentrations at >150 °C. Base line tangents were drawn from 1630–1780 cm^{-1}, 1560–1630 cm^{-1} and 720–800 cm^{-1} in order to measure the absorbance values at 1746 cm^{-1} and 1700 cm^{-1}, 1600 cm^{-1} and 752 cm^{-1}, respectively.

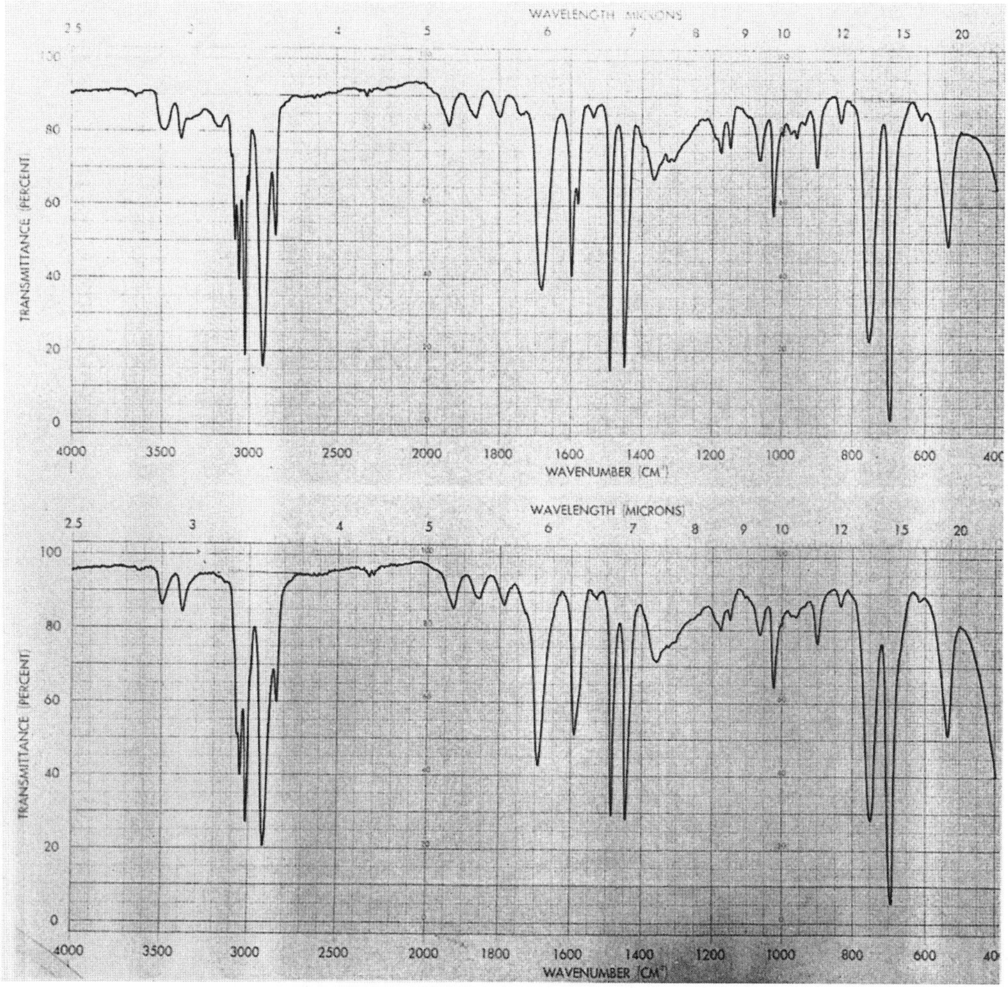

FIGURE 9.29 Top: Infrared spectrum of styrene–acrylamide copolymer recorded at 27 °C. Bottom: Infrared spectrum of styrene–acrylamide copolymer recorded at 275 °C.

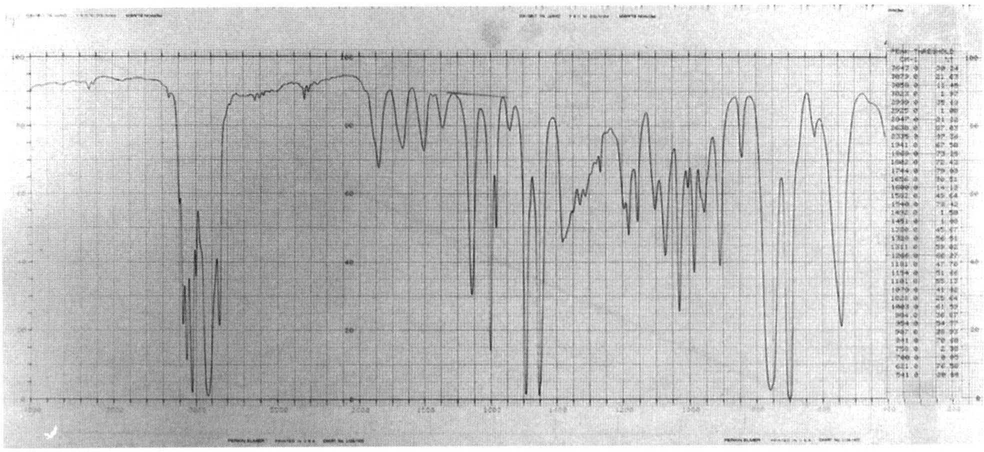

FIGURE 9.30 An IR spectrum for a styrene–2-isopropenyl-2-oxazoline copolymer (SIPO) cast from methylene chloride onto a KBr plate.

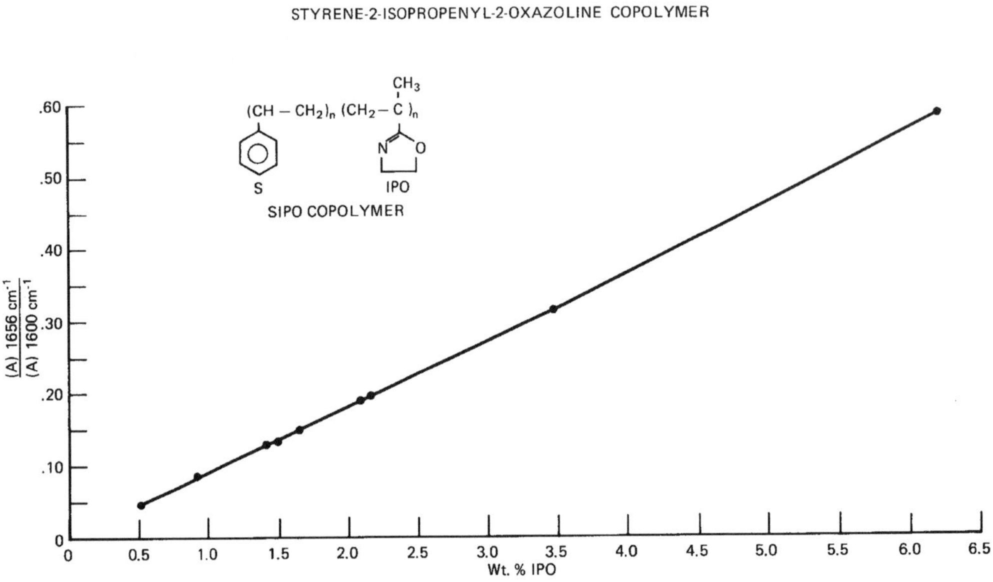

FIGURE 9.31 A plot of the weight % IPO in the SIPO copolymer vs the absorbance ratio (A)(1656 cm^{-1})/(A)(1600 cm^{-1}).

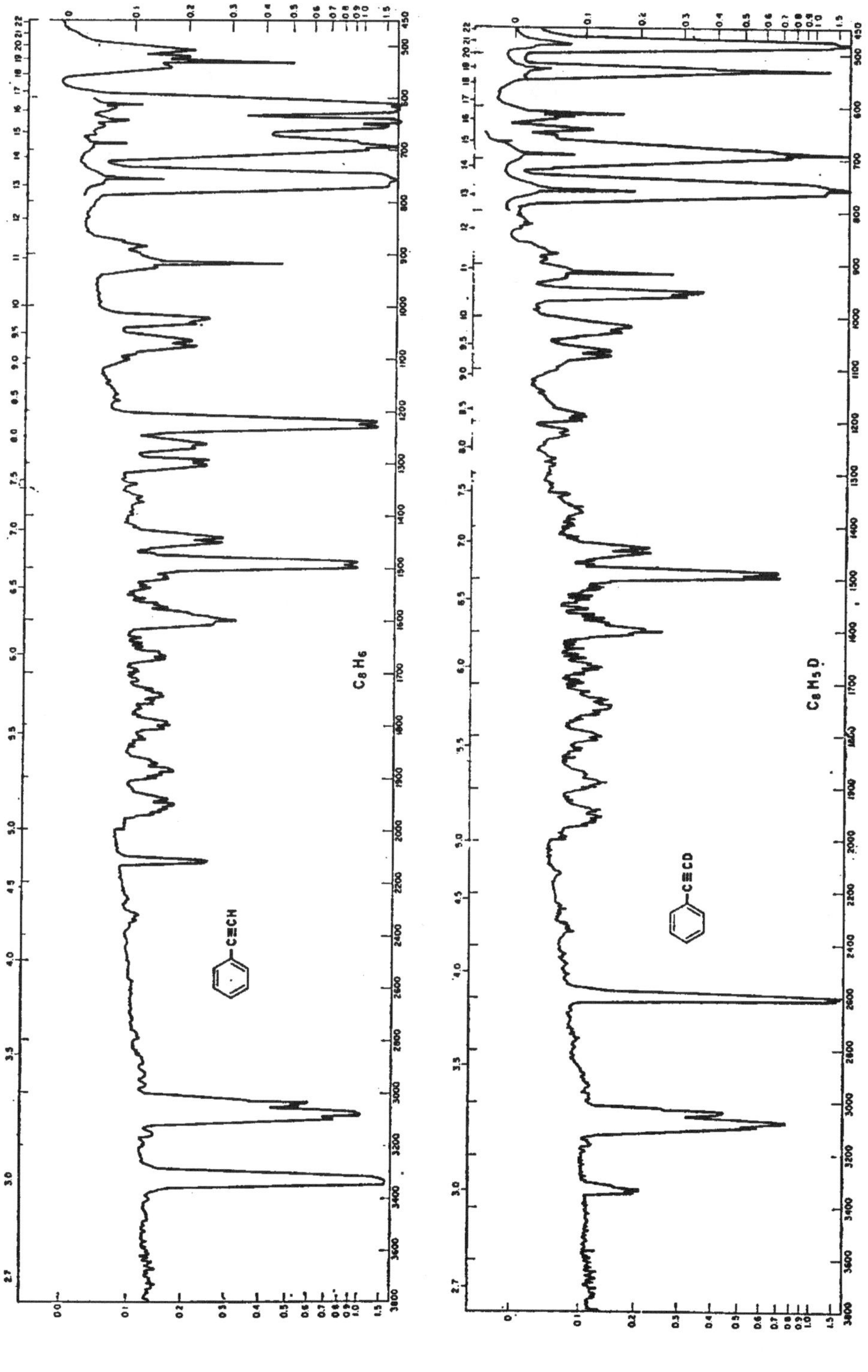

FIGURE 9.32 Top: Vapor-phase IR spectrum of ethynylbenzene in a 4-m cell (vapor pressure is in an equilibrium with the liquid at 25 °C). Bottom: Vapor-phase IR spectrum of ethynylbenzene-d in a 4-m cell (vapor pressure is in an equilibrium with the liquid at 25 °C).

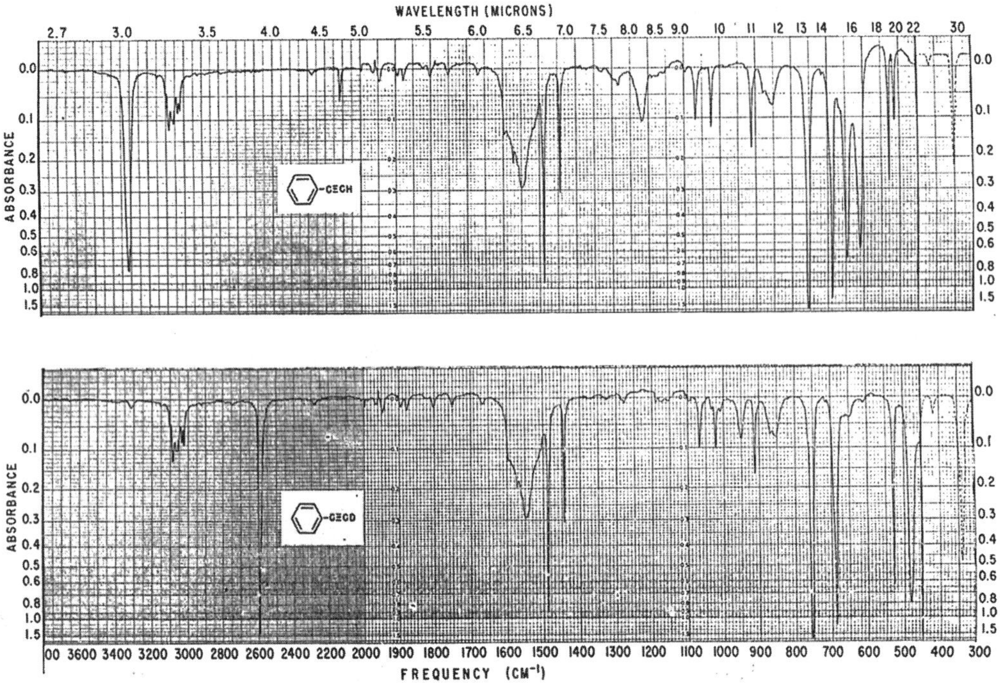

FIGURE 9.33 Top: An IR solution spectrum for ethynylbenzene (3800–1333 cm^{-1} in CCl_4 (0.5 M) solution in a 0.1 mm NaCl cell), (1333–450 cm^{-1} in CS_2 (0.5 M) solution in a 0.1 mm KBr cell), and in hexane (0.5 M) solution using a 2 mm cis I cell. The IR band at 1546 and 853 cm^{-1} is due to the solvents. Bottom: An IR solution spectrum for ethynylbenzene-d recorded under the same conditions used to record the top spectrum.

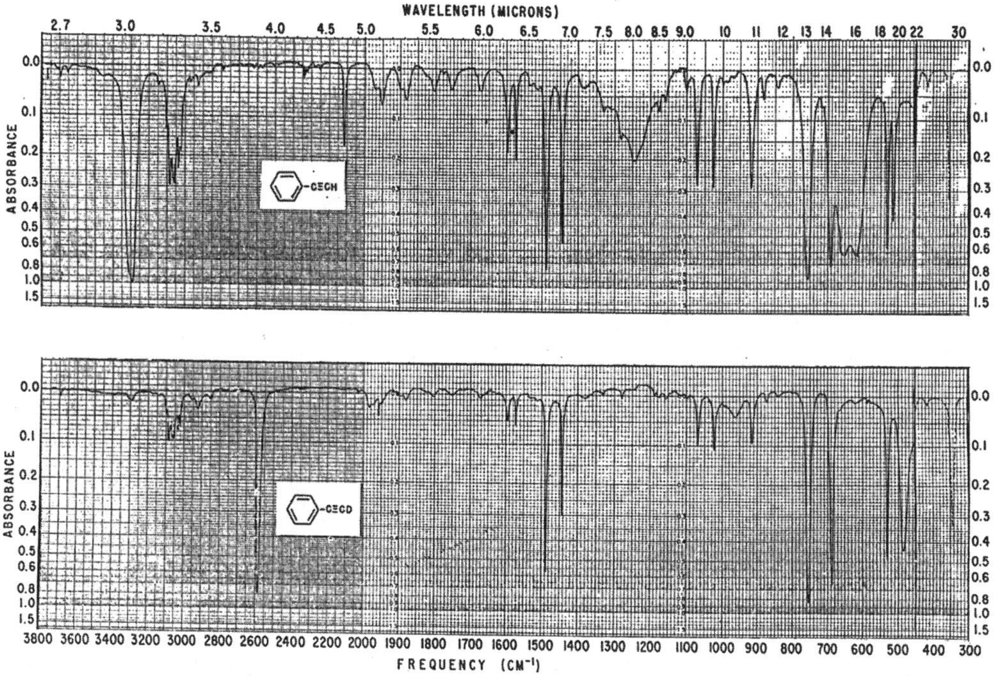

FIGURE 9.34 Top: Liquid-phase IR spectrum ethynylbenzene between KBr plates in the region 3800–450 cm^{-1}, and between C_SI plates in the region 450–300 cm^{-1}. Bottom: Liquid-phase IR spectrum of ethynylbenzene-d recorded under the same conditions as the top spectrum.

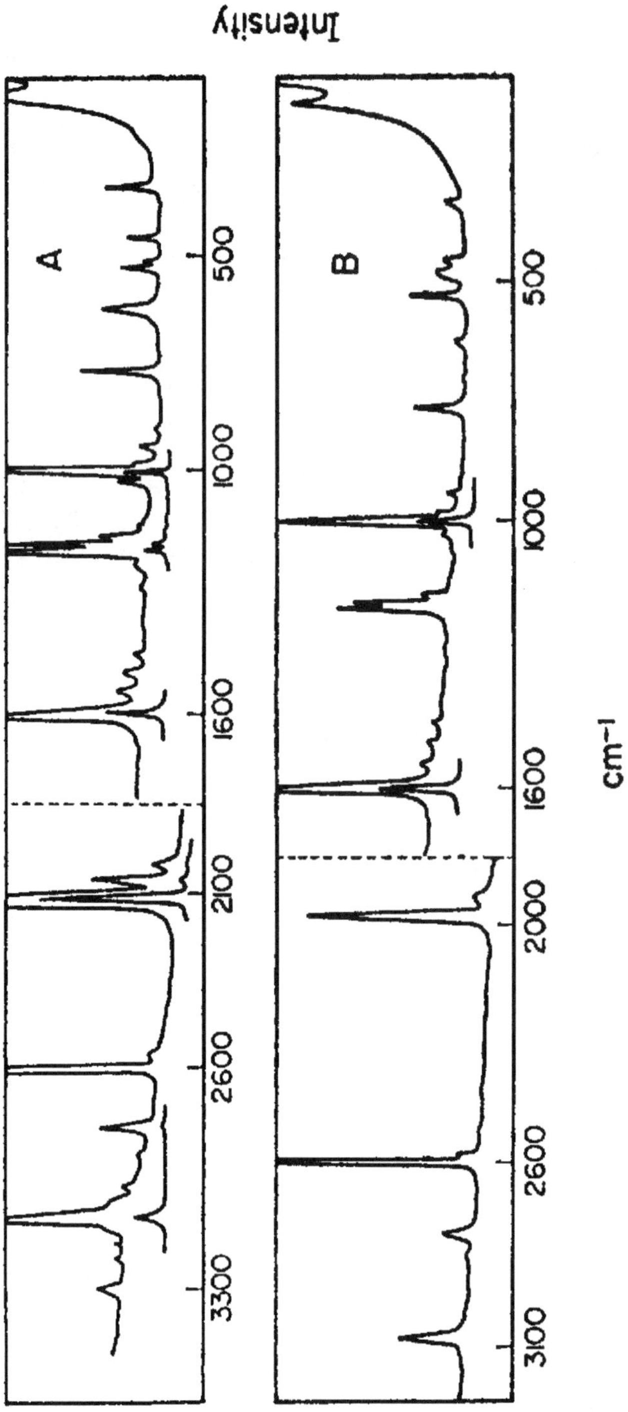

FIGURE 9.35 Top: Raman spectra for ethynylbenzene in the liquid phase. Bottom: Raman spectra for ethynylbenzene-d in the liquid phase.

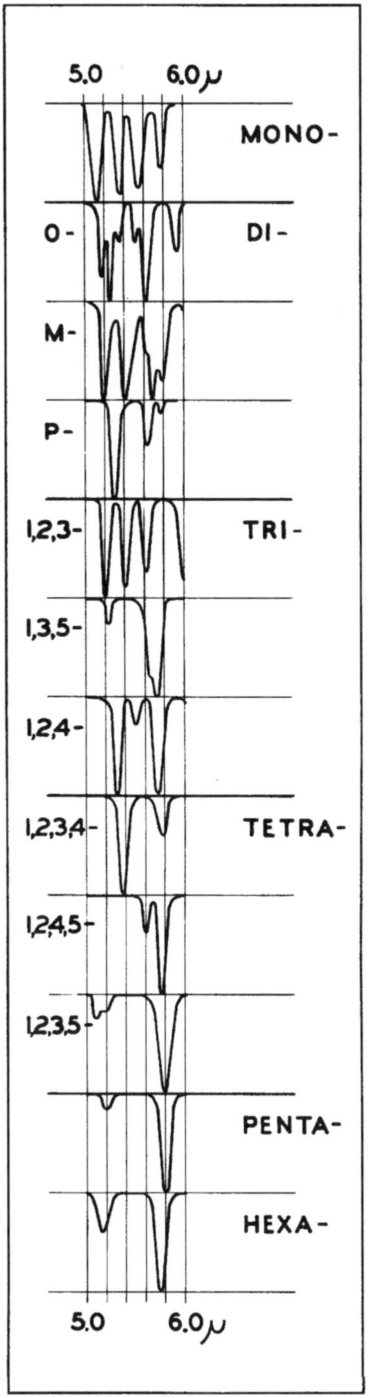

FIGURE 9.36 A correlation chart for substituted benzenes in the region 5–6 μ (after Young, DuVall, and Wright).

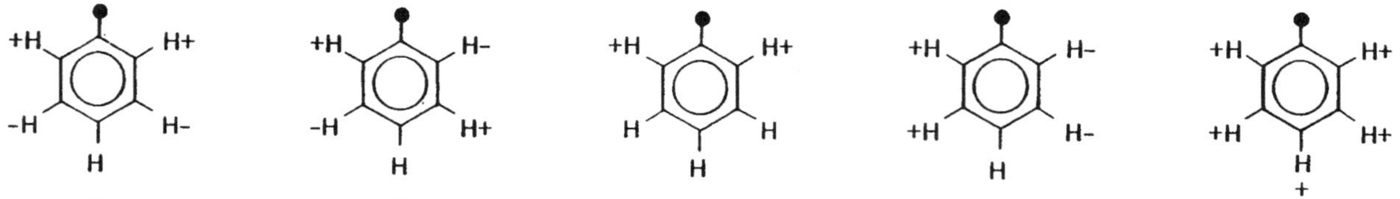

Symmetry	I	II	III	IV	V
C_{2v}	B_I	A_2	B_I	A_2	B_I
C_s planar	A^{II}	A^{II}	A^{II}	A^{II}	A^{II}
C_s perpendicular plane	A^I	A^{II}	A^I	A^{II}	A^I
C_I range	965.5-996 cm^{-1}	945-981 cm^{-1}	862-940 cm^{-1}	789–852 cm^{-1}	(725-810)

Symmetry			
C_{2v}	$2 \times I = 2 \times B_I = A_I$	$I + II = B_I \times A_2 = B_2$	$I + III = B_I \times B_I = A_I$
C_s planar	$2 \times I = 2A^{II} = A^I$	$I + II = A^{II} \times A^{II} = A^I$	$I + III = A^{II} \times A^{II} = A^I$
C_s perpendicular plane	$2 \times I = 2 \times A^I = A^I$	$I + II = A^I \times A^{II} = A^{II}$	$I + III = A^I \times A^I = A^I$
range	1931–1992 cm^{-1}	1915–1976 cm^{-1}	1815–1936 cm^{-1}

Symmetry			
C_{2v}	$II + III = A_2 \times B_I = B_2$	$II + IV = A_2 \times A_2 = A_1$	$III + IV = B_I \times A_2 = B_2$
C_s planar	$II + III = A^{II} \times A^{II} = A^I$	$II + IV = A^{II} \times A^{II} = A^I$	$III + IV = A^{II} \times A^{II} = A^I$
C_s perpendicular plane	$II + III = A^{II} \times A^I = A^{II}$	$II + IV = A^{II} \times A^{II} = A^I$	$III + IV = A^I \times A^{II} = A^{II}$
range	1815-1914 cm^{-1}	1737-1829 cm^{-1}	1680–1787 cm^{-1}

Symmetry	
C_{2v}	$III + V = B_I \times B_I = A_I$
C_s planar	$III + V = A^{II} \times A^{II} = A^I$
C_s perpendicular plane	$III + V = A^I \times A^I = A^I$
range	1611–1748 cm^{-1}

FIGURE 9.37 Summary of out-of-plane hydrogen deformations and their combination and overtones for mono-substituted benzenes.

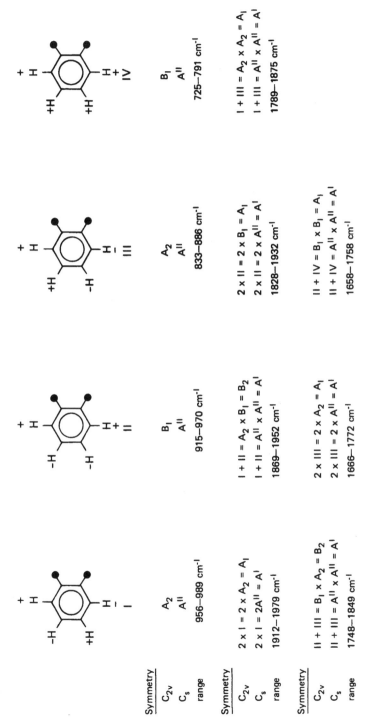

	I	II	III	IV
Symmetry				
C_{2v}	A_2	B_1	A_2	B_1
C_s	A^{II}	A^{II}	A^{II}	A^{II}
range	956–989 cm^{-1}	915–970 cm^{-1}	833–886 cm^{-1}	725–791 cm^{-1}
Symmetry				
C_{2v}	$2 \times I = 2 \times A_2 = A_1$	$I + II = A_2 \times B_1 = B_2$	$2 \times II = 2 \times B_1 = A_1$	$I + III = A_2 \times A_2 = A_1$
C_s	$2 \times I = 2A^{II} = A^{I}$	$I + II = A^{II} \times A^{II} = A^{I}$	$2 \times II = 2 \times A^{II} = A^{I}$	$I + III = A^{II} \times A^{II} = A^{I}$
range	1912–1979 cm^{-1}	1869–1952 cm^{-1}	1828–1932 cm^{-1}	1789–1875 cm^{-1}
Symmetry				
C_{2v}	$II + III = B_1 \times A_2 = B_2$	$2 \times III = 2 \times A_2 = A_1$	$II + IV = B_1 \times B_1 = A_1$	
C_s	$II + III = A^{II} \times A^{II} = A^{I}$	$2 \times III = 2 \times A^{II} = A^{I}$	$II + IV = A^{II} \times A^{II} = A^{I}$	
range	1748–1849 cm^{-1}	1666–1772 cm^{-1}	1658–1758 cm^{-1}	

FIGURE 9.38 Summary of out-of-plane hydrogen deformations and their combination and overtones for 1,2-disubstituted benzenes.

	I	II	III	IV
Symmetry				
C_{2v}	B_1	A_2	B_1	B_1
C_s	A^{II}	A^{II}	A^{II}	A^{II}
range	947–999 cm^{-1}	863–941 cm^{-1}	831–910 cm^{-1}	761–815 cm^{-1}
Symmetry				
C_{2v}	2 x I = 2 x B_1 = A_1	I + II = B_1 x A_2 = B_2	2 x II = 2 x A_2 = A_1	II + III = A_2 x B_1 = B_2
C_s	2 x I = 2 x A^{II} = A^{I}	I + II = A^{II} x A^{II} = A^{I}	2 x II = 2 x A^{II} = A^{I}	II + III = A^{II} x A^{II} = A^{I}
range	1910–1991 cm^{-1}	1820–1940 cm^{-1}	1724–1885 cm^{-1}	1709–1839 cm^{-1}
Symmetry				
C_{2v}	I + IV = B_1 x B_1 = A_1	2 x III = 2 x B_1 = A_1	II + IV = A_2 x B_1 = B_2	
C_s	I + IV = A^{II} x A^{II} = A^{I}	2 x III = 2 x A^{II} = A^{I}	II + IV = A^{II} x A^{II} = A^{I}	
range	1710–1810 cm^{-1}	1680–1820 cm^{-1}	1637–1760 cm^{-1}	

FIGURE 9.39 Summary of out-of-plane hydrogen deformations and their combination and overtones for 1,3-disubstituted benzenes.

Symmetry	I	II	III	IV
V_h	A_u	B_{2g}	B_{1g}	B_{3u}
C_{2v}	A_2	B_I	A_2	B_I
C_s	A^{II}	A^{II}	A^{II}	A^{II}
range	933–985 cm^{-1}	916–971 cm^{-1}	790-852 cm^{-1}	794-870 cm^{-1}

Symmetry		
V_h	$I + II = A_u \times B_{2g} = B_{2u}$	$I + III = A_u \times B_{1g} = B_{Iu}$
C_{2v}	$I + II = A_2 \times B_I = B_2$	$I + III = A_2 \times A_2 = A_I$
C_s	$I + II = A^{II} \times A^{II} = A^I$	$I + III = A^{II} \times A^{II} = A^I$
range	1852–1957 cm^{-1}	1729–1835 cm^{-1}

Symmetry		
V_h	$II + IV = B_{2g} \times B_{3u} = B_{Iu}$	$III + IV = B_{1g} \times B_{3u} = B_{2u}$
C_{2v}	$II + IV = B_I \times B_I = A_I$	$III + IV = A_2 \times B_I = B_2$
C_s	$II + IV = A^{II} \times A^{II} = A^I$	$III + IV = A^{II} \times A^{II} = A^I$
range	1727–1850 cm^{-1}	1619–1702 cm^{-1}

FIGURE 9.40 Summary of out-of-plane hydrogen deformations and their combination and overtones for 1,4-disubstituted benzenes.

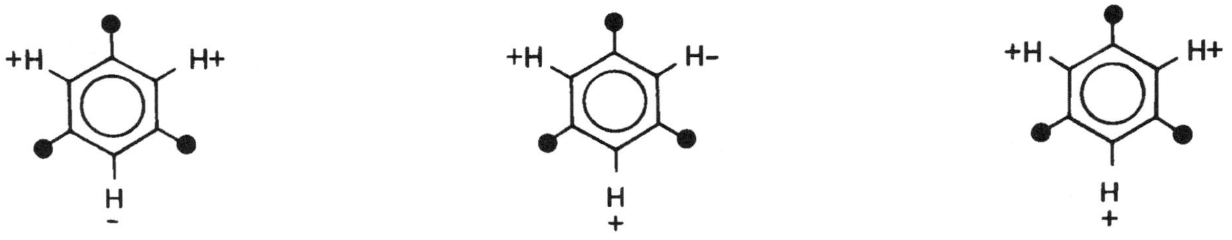

Ia   Ib   II

Symmetry	Ia	Ib	II
D_{3h}	E''		A''_2
C_{2v}	B_1	B_1	A_2
C_s	A''	A''	A''
range	819–920 cm^{-1}		786–910 cm^{-1}

Symmetry		
D_{3h}	$2 \times I = 2 \times E'' = A'_1, E'$	$I + II = E'' \times A''_2 = E'$
C_{2v}	$2 \times I = 2 \times B_1 = A_1$	$I + II = B_1 \times A_2 = B_2$
C_s	$2 \times I = 2 \times A'' = A'$	$I + II = A'' \times A'' = A'$
range	1635–1840 cm^{-1}	1605–1840 cm^{-1}

FIGURE 9.41 Summary of out-of-plane hydrogen deformations and their combination and overtones for 1,3,5-trisubstituted benzenes.

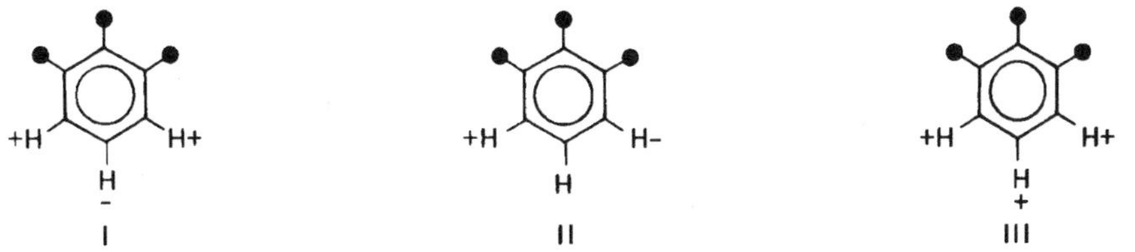

Symmetry	I	II	III
C_{2v}	B_1	A_2	B_1
C_s	A^{II}	A^{II}	A^{II}
range	930–989 cm^{-1}	848–930 cm^{-1}	731–810 cm

Symmetry			
C_{2v}	$2 \times I = 2 \times B_1 = A_1$	$I + II = B_1 \times A_2 = B_2$	$2 \times II = 2 \times B_1 =$
C_s	$2 \times I = 2A^{II} = A^I$	$I + II = A^{II} \times A^{II} = A^I$	$2 \times II = 2 \times A^{II} =$
range	1860–1975 cm^{-1}	1795–1899 cm^{-1}	1700–1860 cm^{-1}

FIGURE 9.42 Summary of out-of-plane hydrogen deformations and their combination and overtones for 1,2,3-trisubstituted benzenes.

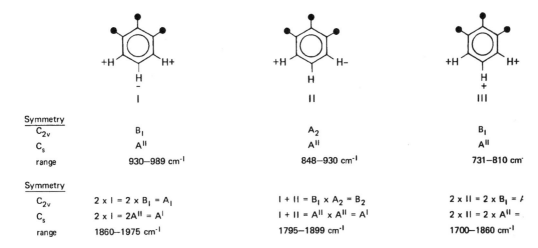

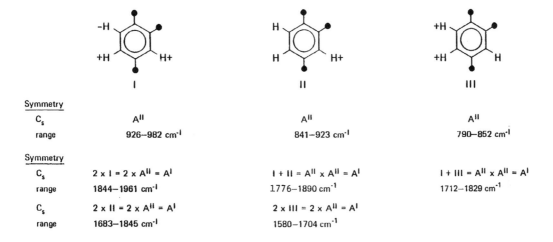

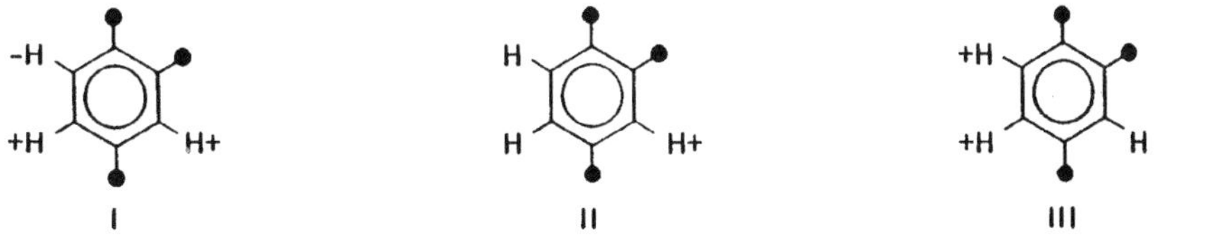

Symmetry	I	II	III
C_s	A^{II}	A^{II}	A^{II}
range	926–982 cm^{-1}	841–923 cm^{-1}	790–852 cm^{-1}

Symmetry			
C_s	$2 \times I = 2 \times A^{II} = A^I$	$I + II = A^{II} \times A^{II} = A^I$	$I + III = A^{II} \times A^{II} = A^I$
range	1844–1961 cm^{-1}	1776–1890 cm^{-1}	1712–1829 cm^{-1}
C_s	$2 \times II = 2 \times A^{II} = A^I$	$2 \times III = 2 \times A^{II} = A^I$	
range	1683–1845 cm^{-1}	1580–1704 cm^{-1}	

FIGURE 9.43 Summary of out-of-plane hydrogen deformations and their combination and overtones for 1,2,4-trisubstituted benzenes.

Symmetry	I		II
C_{2v}	A_2		B_1
C_s	A''		A''
range	919–949 cm^{-1}		770–821 cm^{-1}

Symmetry			
C_{2v}	2 x I = 2 x A_2 = A_1	I + II = A_2 x B_1 = B_2	2 x II = 2 x B_1 = A_1
C_s	2 x I = 2 x A'' = A'	I + II = A'' x A'' = A'	2 x II = 2 x A'' = A'
range	1838–1902 cm^{-1}	1715–1770 cm^{-1}	1610–1650 cm^{-1}

FIGURE 9.44 Summary of out-of-plane hydrogen deformations and their combination and overtones for 1,2,3,4-tetrasubstituted benzenes.

Symmetry	I		II
C_{2v}	A_2		B_1
C_s	A''		A''
range	821–942 cm^{-1}		838–920 cm^{-1}

Symmetry			
C_{2v}	2 x I = 2 x A_2 = A_1	I + II = A_2 x B_1 = B_2	2 x II = 2 x B_1 = A_1
C_s	2 x I = 2 x A'' = A'	I + II = A'' x A'' = A'	2 x II = 2 x A'' = A'
range	1635–1860 cm^{-1}	1659–1862 cm^{-1}	1638–1841 cm^{-1}

FIGURE 9.45 Summary for out-of-plane hydrogen deformations and their combination and overtones for 1,2,3,5-tetrasubstituted benzenes.

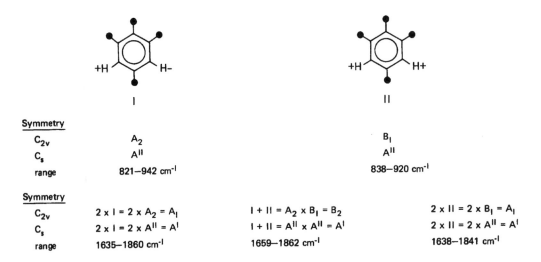

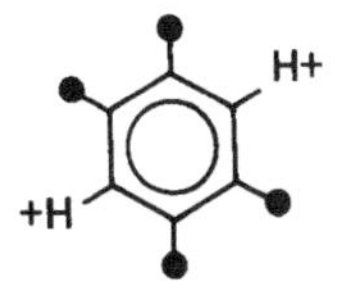

I

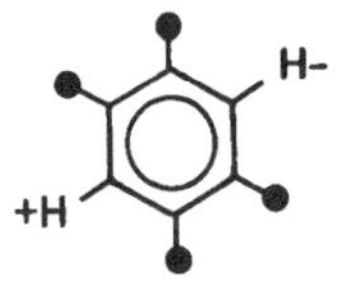

II

Symmetry	I	II
V_h	B_{3u}	B_{2g}
C_{2v}	B_1	A_2
C_s	A^{II}	A^{II}

Symmetry			
V_h	$2 \times I = 2 \times B_{3u} = A_g$	$I + II = B_{3u} \times B_{2g} = B_{1u}$	$2 \times II = 2 \times B_{2g} = A_g$
C_{2v}	$2 \times I = 2B_1 = A_1$	$I + II = B_1 \times A_2 = B_2$	$2 \times II = 2A_2 = A_1$
C_s	$2 \times I = 2 \times A^{II} = A^I$	$I + II = A^{II} \times A^{II} = A^I$	$2 \times II = 2 \times A^{II} = A^I$
range	1725–1830 cm^{-1}	1695–1799 cm^{-1}	1651–1770 cm^{-1}

FIGURE 9.46 Summary of out-of-plane hydrogen deformations and their combination and overtones for 1,2,4,5-tetrasubstituted benzenes.

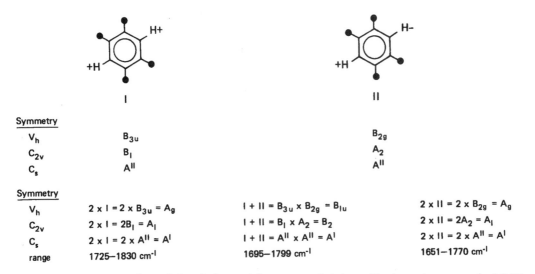

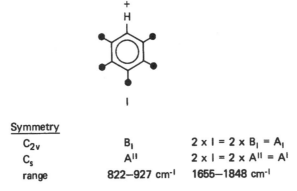

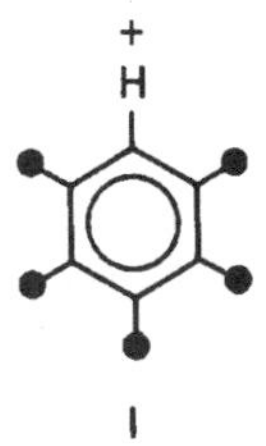

I

Symmetry		
C_{2v}	B_1	$2 \times I = 2 \times B_1 = A_1$
C_s	A^{II}	$2 \times I = 2 \times A^{II} = A^I$
range	822–927 cm^{-1}	1655–1848 cm^{-1}

FIGURE 9.47 Summary of out-of-plane hydrogen deformation and its first overtone for 1,2,3,4,5-pentasubstituted benzenes.

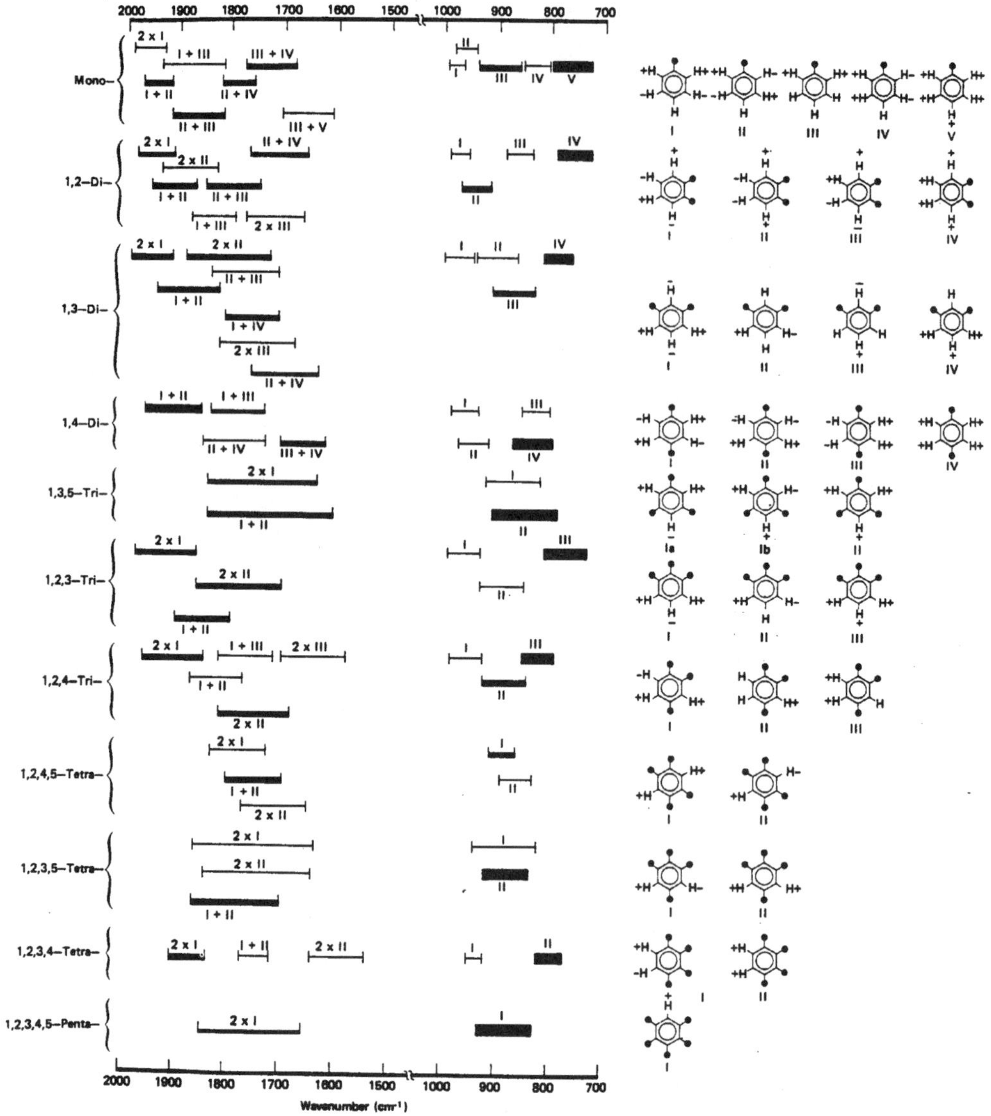

FIGURE 9.48 Infrared correlation chart for out-of-plane hydrogen deformations and their combination and overtones for substituted benzenes.

TABLE 9.1 Vibrational data for chlorobenzene vs chlorinated biphenyls

Chlorobenzene vs phenyl In plane modes	Chlorobenzene cm^{-1}	2,3,4,5,6-Pentachlorobiphenyl cm^{-1}
A1		
	3087	3068
	3072	3030
	3067	3000
	1584	1602
	1480	1497
	1174	1155
	1084	1151?
	1023	1023
	1002	1000
	703	719
	418	449
B2		
	3087	3058
	3059	3030
	1584	1580
	1447	1442
	1325	?
	1272	1262
	1157	1157
	1068	1070
	613	610
	297	?
A2		
	957	974
	825	[845]
		414?
B1		
	976	982
	897	918
	734	773
	679	686
	?	472
	?	?

? tentative assignment or not assigned.

TABLE 9.2 Vibrational data for 1,2-dichlorobenzene vs chlorinated biphenyls

1,2-Dichlorobenzene cm^{-1}	2,2′,3-Trichlorobiphenyl cm^{-1}	2,3′4′-Trichlorobiphenyl cm^{-1}	2,2′,3,4,6-Pentachlorophenyl cm^{-1}
A1 species			
3070	3068	3060	3062
3070	3068	3060	3062
1575	1592	1589	1592
1458	1479	1481	1481
1276	1278	1274	1279
1162	1178	1175	1175 or 1182
1130	1135	1151?	1125 or 1132
1041	1029	1021	1032 or 1025
660	715	710	?
480	627	648	628
203	?	?	?
B2 species			
3072	3068	3060	3062
3072	3068	3060	3062
1575	1578	1570	1562
1438	1421	1429	1417
1252	1250	1242	1040
1130	1129	1131	1132
1038	1029	1029	1025
740	735	730?	748sh
429	422	430	423
334	?	?	?
Out-of-plane modes			
A2 species			
975	977	978	974
850	[861]	[863]	[862.5]
695	725	730?	728sh
564	580	581	570
154	?	?	?
B1 species			
940	941	944	941
749	751	751	754
435	434	431	438

? not assigned.

TABLE 9.3 Vibrational data for 1,3-dichlorobenzene vs 2,3,3′,5,6-pentachlorobiphenyl

1,3-Dichlorobenzene In-plane modes	Assignment cm^{-1}	2,3,3′,5,6-Cl5 biphenyl cm^{-1}
A1		
	3093.8	
	3085.8	
	3064.6	
	1574.6	1575.8
	1392.6	1390.4
	1126.2	
	1075.2	1083.9
	995.6	
	662.6	658
	401.4	
	201.4	
B2		
	3083.6	
	1586.2	
	1474.5	1475.4
	1326.7	
	1238.9	1247.4
	1169	1165.1
	1069.2	1063.6
	786.3	792.1
	422	
	370.6	

TABLE 9.4 Vibrational data for 1,4-dichlorobenzene vs chlorinated biphenyls

Species	1,4-Dichlorobenzene cm^{-1}	2,3,4,4′,6-Pentachlorobiphenyl cm^{-1}
In-plane modes		
Ag[R]		
	3087	3078
	1574	1599
	1169	1176
	1096	1131
	747	?
	328	?
B3g		
	3065	3042
	1574	[1560]
	1290	1300
	626	638
	350	?
B1u		
	0	3078
	390	
	1477	1492
	1090	1091
	1015	1013
	550	550
B2u		
	3090	3078
	1394	1397
	1221	1234
	1107	1102sh
	226	?
Out-of-plane modes		
B1g(R)		
	815	[817]
B2g(R)		
	934	941
	687	721
	298	?
Au		
	951	959
	407	410?
B3u		
	819	815
	485	488
	125	?

? not assigned.

TABLE 9.5 Vibrational data for 1,3,5-trichlorobenzene vs chlorinated biphenyls

Species	1,3,5-Trichlorobenzene cm^{-1}	2,3′,4,5′,6-Pentachlorobiphenyl cm^{-1}
In-plane modes		
A1′		
	3084	3080
	1149	1131
	997	997
	379	?
A2′		
	1379	1386
	1249	1233
	471	449
E′		
	3089	3080
	1570	1590 & 1575
	1420	1459
	1098	1099
	816	805
	429	[430]
	191	?
Out-of-plane modes		
A2″		
	853	849
	662	689
	148.5	?
E″		
	868.5	[870]
	530	[525]
	215	?

? not assigned.

TABLE 9.6 Vibrational data for 1,2,3-trichlorobenzene vs 2,2′,3-trichlorobiphenyl, 2,2′,3′,4,5-pentachlorobiphenyl, and 2,2′,4,6,6′-pentachlorobiphenyl

Species	1,2,3-Trichloro-benzene cm^{-1}	2,2′,3-Trichloro-biphenyl cm^{-1}	2,2′,3′,4,5-Pentachloro-biphenyl cm^{-1}	2,2′,4,6,6′-Pentachloro-biphenyl cm^{-1}
A1				
	3090			
	1566	1591	1590	1559 or masked
	1416	1401	1411	1411
	1161	1150	1139	1146
	1087	1072	1088	1113
	1049	1050	1040	1069
	737	740	731	733
	513	562	575	580
	352[R]	?	?	?
	212[R]	?	?	?
B2				
	3060			
	1566	1557	1560	1559
	1436	1431	1440	1430
	1260	1250	1249	1248
	1196	1190	1195	1195sh or 1184
	1156[R]	1174?	1150?	1151?
	791	803	762	787
	486	483	484	491
	398	?	?	?
Out-of-plane modes				
A2				
	896	900	masked	892
	524	527?	510?	490?
	212[R]	?	?	?
B1				
	963	967	969	968
	773	782	784	772
	697	692	680 or 721	680 or 721
	500	?	?	?
	242[R]	?	?	?
	90[R]	?	?	?

? not assigned.

TABLE 9.7 Vibrational data for 1,2,4-trichlorobenzene vs chlorinated biphenyls

Species	1,2,4-Trichlorobenzene cm^{-1}	2′,3,4-Trichlorobiphenyl cm^{-1}	2,3′,4,4′,6-Pentachlorobiphenyl cm^{-1}	2,2,3′,4′,6-Pentachlorobiphenyl cm^{-1}	2,2′,3,5′,6-Pentachlorobiphenyl cm^{-1}
In-plane modes					
A′					
	3094	3060	3079	?	?
	3072	3060	3079	?	?
	1571	1570	1573	1589	?
	1562	1557	1552	1565	?
	1461	1466	1472	1479	?
	1377	1372	1347sh	1382	?
	1267	1272	1278	1280	?
	1245	1249	1244	1250	?
	1156	1151	1152sh		?
	1132	1131	1130	1119	1119
	1096	1094	1093	1100	1091
	1036	1035	1031	1021	1026
	817	885	884	?	?
	697	710	708	?	?
	576	615	615	612	?
	456	469	459	471	463
	396	?	405	?	?
	328	?	?	?	?
	211	?	?	?	?
	1578	1590	1586	1572	1571
	1551	1556	1568	1550	?
	1469	1472	1472	1469	1471
	1270	1272	1285	1285	1288
	1244	1245	1246	1249	1248

(*continues*)

TABLE 9.7 *(continued)*

Species	1,2,4-Trichlorobenzene cm^{-1}	2′,3,4-Trichlorobiphenyl cm^{-1}	2,3′,4,4′,6-Pentachlorobiphenyl cm^{-1}	2,2,3′,4′,6-Pentachlorobiphenyl cm^{-1}	2,2′,3,5′,6-Pentachlorobiphenyl cm^{-1}
	1163	1162sh	?	?	?
	1098	?	?	?	1135sh
	1089	1093	1100	1094	1096
	1020	1028	1024	1029	1031
	621	552?	602	?	612
	481	461	480	?	474
Out-of-plane modes					
A′					
	942	944	951	943	943
	869	863	855	863	859
	811	820	812	809	
	688	672	672	668	695
	551	550	[572]	534	551
	305	?	?	?	?
	183	?	?	?	?
	117	?	?	?	?
	947	946	947	946	[947]
	862	863	860	863	866
	818	812	811	813	812
	670	674	?	644?	668
	575	577	549	551	557
	?	?	439	442	435

? not assigned.

TABLE 9.8 Vibrational data for 1,2,4,5-tetrachlorobenzene vs chlorinated biphenyls

Species	1,2,3,4,5-Tetrachlorobenzene cm^{-1}	2,2′,4,4′,5-Pentachlorobiphenyl cm^{-1}	2,2′,4,5,5′-Pentachlorobiphenyl cm^{-1}	2,2′,3′,4,5-Pentachlorobiphenyl cm^{-1}
In-plane modes				
Ag				
	3070			
	1549	1550	1557	1560
	1165	1167	masked	1150?
	684	723	723	731
	352			
	190			
B3g				
	1566	1533	1536	1540
	1240	1245	1244	1248
	868			
	511	481	495	494
	312			
B1u				
	3094			
	1327	1337	1334	1342
	1063	1063	1067	1057
	510	510 or 518	510 or 524	510
	218			
B2u				
	1473 in FR	1479 in FR	1472 in FR	1472 in FR
	[860 + 600]	[862 + 603]	[605 + 863]	[865 + ?]
	1448 in FR	1443 in FR	1446 in FR	1440 in FR

(continues)

TABLE 9.8 (*continued*)

Species	1,2,3,4,5-Tetrachlorobenzene cm^{-1}	2,2′,4,4′,5-Pentachlorobiphenyl cm^{-1}	2,2′,4,5,5′-Pentachlorobiphenyl cm^{-1}	2,2′,3′,4,5-Pentachlorobiphenyl cm^{-1}
	1226	1233	~1235sh	1229
	1118	1133	1138	1139
	645	670	668	679
	209			
Out-of-plane modes				
B1g				
	348			
B2g				
	860	862 or masked	863	865
	681	684	679	680
	225			
Au				
	600	603?	605?	618 or masked
	80			
B3u	878	882	880	882
	442	429	430	432
	140			

? tentative assignment.

TABLE 9.9 Vibrational data for 1,2,3,5-tetrachlorobenzene vs chlorinated biphenyls

Species	1,2,3,5-Tetra-chloro-benzene cm^{-1}	2,3′,4,4′,6-Penta-chloro-biphenyl cm^{-1}	2,3′,4,5′,6-Penta-chloro-biphenyl cm^{-1}	2,2′,4,6,6′-Penta-chloro-biphenyl cm^{-1}	2,2′,3,4′,5-Penta-chloro-biphenyl cm^{-1}	2,2′,3,5,5′-Penta-chloro-biphenyl cm^{-1}
In-plane modes						
A1						
	3078	3079	3080		3065	3065
	1566	1573	1560	1573	1568	1571
	1412	1425	1411	1411	1407	1417
	1170	1182	1183	1184	1188	1190
	1120	1138 or 1130	1142	1115	1119	1121
	1049	1032	1040	1069	1056	1056
	836	812 masked	815sh	830	841	827
	598 stg	580sh	610?wk	580	601 stg	601 stg
	326					
	206					
B2						
	3078		3080		3065	3065
	1550	1539	1542	1542	1549	1550
	1378	1362	1368	1371	1369	1371
	1254	1260	1269	1248	1246	1250
	1191	1182	1182	1183	1188	1190
	810[FR]	797	797	800	819	829
	521	[572]	579	580	550	551
	431	[435]	[430]	431	439	442
	215					
	192					
Out-of-plane modes						
A2						
	871	884	[870]	[870]	[869]	[867]
	520	521?	[525?]	525?	520	520
	215					
B1						
	859	855	855	855	861	863
	692	732	729	733	716	719
	560	572	579	580	569	570
	316					
	147					
	80					

TABLE 9.10 Infrared data for 1,2,3,4-tetrachlorobenzene vs chlorinated biphenyls

Species	1,2,3,4-Tetra-chlorobenzene cm^{-1}	2,2′,3,4′,6-Penta-chlorophenyl cm^{-1}	2,2′,3,5,6-Penta-chlorobiphenyl cm^{-1}	2,2′,3′,4,5′-Penta-chlorobiphenyl cm^{-1}
In-plane modes				
A1				
	3074			
	1560	1565	1557	
	1364	1372?	1362?	1358
	1248	1250	1249	1248
	1178	1173	1175	1176
	1132	1140	1132	1135
	836	865	870	882
	515	590	602	612
B2				
	3074			
	1560	1551	1550	1559
	1428	1431	1433	1438
	1168			
	1077	1100	1091	1096
	775	808sh	806(masked)?	781
	609	[612]	602	[612]
	482	490	463	473
	356			
	202			
Out-of-plane modes				
A2				
	940			[947]
	756[R]			
	307[R]			
B1				
	808	800	806	813
	557	557	[549]	532
	240[R]			
	116[R]			

TABLE 9.11 Vibrational data for pentachlorobenzene vs 2,2′3,4,6-pentachlorobiphenyl and 2,3,4,4′,6-pentachlorobiphenyl

Species	Pentachlorobenzene cm^{-1}	2,2′,3,4,6-Pentachlorobiphenyl cm^{-1}	2,3,4,4′,6-Pentachlorobiphenyl cm^{-1}
In-plane modes			
A1			
	3069	3064	3078
	1528	1529	1529
	1338	1338	1337
	1198[R]	1191	masked
	1087	1076	masked
	823	811	820
	563	570	573
B2			
	1559	1562	[1560]
	1398	1417	1421
	1235[R]	1230	1235
	1169	1171	1177
	[863]calc.	[862.5]calc.	[862.5]calc.
	683	671	650
	556	548	549
Out-of-plane modes			
A2			
	597		
B1			
	863	862	863
	701	735sh?	741?
	526	529	529
		1725 = 2[862.5]	

? tentative assignment.

TABLE 9.12 Vibrational data for 1,2,3,4,5,6-hexachlorobenzene vs 1,2,3,4,5-pentachlorobiphenyl

Species	Hexachlorobenzene cm^{-1}	2,3,4,5,6-Pentachlorobiphenyl cm^{-1}
A1g		
	1210[R]	1206
	372[R]	
B1u		
	1108	
	369	
A2g		632
	630 est.	
B2u		
	1224 Comb.	1230
	230 est.	
E2g		
	1512	1516 & 1497
	870[R]	868
	323[R]	
	219[R]	
E1u		
	1350	1344 & 1320
	699	685
	218	
A2u		
	209	
B2g		
	704 Comb.	750?
	97 Comb.	
E1g		
	340[R]	
E2u		
	594	586
	80 Comb.	

TABLE 9.13 Raman data for mono-substituted benzenes

Compound	Ring 4	Ring 8	Ring 9	Ring 20	R.I.[Ring 4]/ R.I.[Ring 9]	R.I.[Ring 8]/ R.I.[Ring 9]	R.I.[Ring 20]/ R.I.[Ring 9]
Allylbenzene	1604(1)	1031(2)	1004(9)	621(0)	0.11	0.22	0.06
N-Benzyl acrylamide		1031(1)	1004(9)	622(1)		0.11	0.11
Benzyl acrylate	1609(1)	1031(2)	1004(9)	622(1)	0.11	0.22	0.11
N-Benzyl methacrylamide	1604(2)	1030(2)	1003(9)	621(1)	0.22	0.22	0.11
Cinnamyl acrylate		1031(2)	1002(9)	620(1)		0.22	0.11
Cinnamic acid	1600(8)	1027(2)	1001(7)	620(0)	1.14	0.29	0.07
Benzil	1595(4)	1020(2)	1000(9)	616(1)	0.44	0.22	0.11
Benzoin ethyl ether	1599(7)	1033(3)	1005(9)	617(2)	0.77	0.33	0.22
Diphenyl divinyl silane	1590(2)	1030(2)	999(9)	620(0)	0.22	0.22	0.06
1,6-Diphenyl-1,3,5-hexatriene	1592(9)		1000(1)		0.11		
Glycidyl cinnamate	1601(6)	1030(1)	1001(5)		1.2	0.2	
2-Hydroxy-4-acryoxyethoxy benzophenone	1602(5)	1027(3)	1001(9)	618(1)	0.56	0.33	0.11
Hydroxy-4-dodecyloxy benzophenone	1601(9)	1030(2)	1001(8)	618(2)	1.13	0.25	0.25
2Hydroxy-4-methoxy-5-sulfonic acid	1598(7)	1029(3)	1003(9)	617(1)	0.77	0.33	0.11
4-Methacryloxy-2-hydroxy benzophenone	1600(7)	1030(3)	1002(9)	621(1)	0.77	0.33	0.11
α-Methylstyrene	1602(4)	1029(2)	1001(9)		0.44	0.22	
Phenyl acrylate	1594(2)	1025(3)	1007(9)	615(1)	0.22	0.33	0.11
2-Phenylethyl acrylate	1606(2)	1032(3)	1004(9)	622(1)	0.22	0.33	0.11
2-Phenylethyl methacrylate	1606(2)	1032(3)	1004(9)	622(1)	0.22	0.33	0.11
Phenyl glycidyl ether	1601(1)	1026(2)	997(9)	615(1)	0.11	0.22	0.11
N-Phenyl methacrylamide	1598(4)	1031(1)	1002(9)	616(1)	0.44	0.11	0.11
Phenyl methacrylate	1594(2)	1026(3)	1002(9)	615(2)	0.22	0.33	0.22
Phenyl vinyl sulfone	1585(4)	1025(1)	1002(9)	616(1)	0.44	0.11	0.11
Vinyl benzoate	1603(3)	1028(1)	1004(9)	618(1)	0.33	0.11	0.11
Diphenyl terephthalate	1610(3)	1026(1)	1001(9)	614(0)	0.33	0.11	0.06
Diphenyl isophthalate	1607(1)	1022(0)	1004(9)	615(1)	0.11	0.06	0.11

TABLE 9.14 Infrared and Raman data for an A_1 fundamental for mono-X-benzenes

Compound	Raman(liquid) A_1 Fundamental cm^{-1} strong	IR(liquid) A_1 Fundamental cm^{-1} weak	# Protons on α carbon	Taft σ*	Es
Toluene	790	781	3	0	0
Ethylbenzene	760	770	2	−0.1	−0.07
Isopropylbenzene	740	741	1	−0.19	−0.47
s-Butylbenzene		731	1	−0.21	
t-Butylbenzene	705	masked	0	−0.3	−1.54
α-Syndiotactic polystyrene	773.4[solid]				
Isotactic polystyrene	785[solid]				

TABLE 9.15 Vibrational assignments for the bromodichlorobenzene

	1-Br,3,5-Cl_2 benzene C_2v symmetry	1-Br,2,6-Cl_2 benzene C_2v symmetry	1-Br,2,3-Cl_2 benzene C_2v symmetry	1-Br,2,4-Cl_2 benzene C_2v symmetry	1-Br,2,5-Cl_2 benzene C_2v symmetry	1-Br,3,4-Cl_2 benzene C_2v symmetry
In-plane vibrations	3090	3060	3059	3087	3088	3089
	3090	3060	3059	3075	3070	3065
	3090	3060	3059	3075	3070	3065
	1570	1578	1580	1562	1569	1561
	1559	1561	1558	1553	1555	1554
	1417	1431	1431	1450	1454	1452
	1417	1403	1411	1368	1369	1369
	1365	1261	1252	1258	1259	1260
	1230	1189	1190	1240	1241	1244
	1142	1150	1155	1141	1115	1137
	1097	1144	1144	1135	1124	1119
	1097	1080	1078	1088	1097	1075
	999	1032	1040	1019	1927	1031
	801	791	749	811	792	791
	766	710	728	678	661	684
	470	460	500	518	570	547
	435	470	461	449	429	449
	419	401	382	402	349	381
	311	305	287	270	329	266
	187	199	209	202	200	207
	160	171	170	171	161	170
Out-of-plane vibrations	867.5	965	963	942	943	941
	867.5	884	891.5	865	870	869
	850	774	770	805	811	809
	660	691	693	667	681	675
	527.5	522	512	544	540	541
	527.5	480	492	430	431	432
	218	249	250	300	313	304
	218	213	215	189	171	182
	140	91	91	108	118	110

TABLE 9.16 Raman data and assignments for some in-plane ring modes of 1,2-disubstituted benzenes

Compound	3 cm^{-1} (R.I.)	14 cm^{-1} (R.I.)	18 cm^{-1} (R.I.)	9 cm^{-1} (R.I.)
Bis(1-acryloxy-2-hydroxypropyl) phthalate	1600(6)	1581(2)	1042(9)	653(3)
Mono(2-methacryloxy-ethyl) phthalate	1602(6)	1582(4)	1042(9)	646(1)
Isooctadecyl 2-sulfobenzoate NA salt	1595(4)	1577(1)	1044(7)	642(2)
Poly(vinyl hydrogen phthalate)	1601(8)	1582(3)	1042(9)	645(3)
Diallyl phthalate	1602(6)		1041(9)	
Poly (diallyl phthalate)	1601(8)	1580(3)	1041(9)	657(2)
2-Ethylhexyl salicylate	1615(3)	1586(2)	1034(9)	668(4)
N-vinylphthalimide	1610(2)		1020(2)	666(0)
Poly(hexamethylene terephthalamide)	1614(9)	1566(1)		634(1)

TABLE 9.17 Raman data and assignments for some in-plane ring modes for 1,3-disubstituted benzenes

Compound	4 cm^{-1} (R.I.)	13 cm^{-1} (R.I.)	8 cm^{-1} (R.I.)	9 cm^{-1} (R.I.)
1,3-Dichlorobenzene	1574	1586	995	662
N,N′-(m-phenylene) Bis-maleimide	1608(2)	1588(3)	1005(5)	683(9)
Bis-maleimide Diallyl phthalate	1609(4)		1004(9)	658(2)
Poly(diallyl isophthalate)	1608(4)	1590(2)	1004(9)	657(2)
1,3-Diisopropenyl benzene	1599(4)	1577(2)	1000(9)	692(4)
3-Fluorostyrene	1603(4)		1002(9)	
3-Chlorostyrene	1596(3)		999(9)	
3-Nitrostyrene	1618(1)		1001(7)	

TABLE 9.18 Raman data and assignments for some in-plane ring modes for 1,4-disubstituted benzenes

Compound	2 cm^{-1} (R.I.)	9 cm^{-1} (R.I.)	5? cm^{-1} (R.I.)	 cm^{-1} (R.I.)	10 cm^{-1} (R.I.)
Diethyl terephthalate	1615(9)	1276(3)	853(3)	836(2)	634(2)
Bis(2-hydroxyethyl) terephthalate	1615(9)	1284(4)	850(3)		634(2)
4-Ethylstyrene	1612(5)	1308(1)	819(1)	805(2)	641(1)
4-Fluorostyrene	1603(4)	1295(1)	842(4)		637(1)
4-Nitrophenyl acrylate	1591(4)		866(3)	848(1)	625(1)
4-Nitrophenyl methacrylate	1591(4)		866(3)	866(3)	632(1)
4-Nitrostyrene	1599(5)		859(2)	859(2)	
4-Nonylphenyl methacrylate	1606(4)	1296(1)	881(2)	812(4)	639(2)
4-Phenoxystyrene	1609(4)	1296(0)		801(2)	
1,4-Phenylene diacrylate	1597(3)	1285(3)	859(5)	803(0)	636(3)
α,α-Dihydroxy	1615(2)				639(2)
Poly(4-methylstyrene)	1614(9)		830(9)	809(9)	644(8)
Poly(4-vinylphenol)	1612(6)		843(9)	826(7)	644(5)
Poly(4-tert-butylstyrene)	1612(9)		845(1)		643(4)
Poly(ethylene terephthalate)	1615(9)	1288(2)	857(2)		633(3)
Poly(1,4-butylene terephthalate)	1614(9)	1277(9)	857(3)		633(3)
1,4-Bis(hydroxymethyl) benzene	1616(2)		839(9)		639(4)
1,4-Dimethoxybenzene	1615(1)	1268(3)		821(9)	638(3)
4-Vinylbenzoic acid	1609(8)	1288(2)		826(3)	638(1)
4-Na vinylbenzenesulfonate	1600(8)	1300(1)		808(1)	637(1)
Bisphenol A/epichlorohydrin liquid epoxy	1610(8)	1298(3)		823(9)	639(8)
4-Vinylphenyl acetonitrile	1607(4)	1302(1)		825(1)	

TABLE 9.19 Raman data and tentative assignments for decabromobiphenyl and bis-(pentabromophenyl) ether

Decabromobiphenyl cm^{-1} (R.I.)	Assignment Ring modes	Bis(pentabromophenyl) ether cm^{-1} (R.I.)	Assignment Ring modes
1559(1)	3	1524(1)	3
1521(1)	9	1514(1)	9
1256(2)	1		
392(3)	16,20?	431(0)	16,20?
359(9)	2	322(1)	8
345(4)	16,20?	237(7)	5,11
327(4)	8	226(9)	6,12,&14?
250(4)	5,11	208(3)	17,21
217(4)	6,12,&14?		

TABLE 9.20 Vibrational assignments for benzene, benzene d_6, benzyl alcohol and benzyl-2,3,4,5,6-d_5 alcohol

Compound	v12(e1u) in F.R. cm^{-1} (A)	v12 + v16(E1u) in F.R. cm^{-1} (A)	cm^{-1} (A)	v2 + v4(A2u) cm^{-1} (A)		v18 + v19(A2u) cm^{-1} (A)	cm^{-1} (A)	v4 + v11(E1u) cm^{-1} (A)	v12(e1u) cm^{-1} (A)	cm^{-1} (A)	v14(e1u) cm^{-1} (A)	v4(a2u) cm^{-1} (A)	
				1964		1829			1498		1051	686	
				(0.065)		(0.080)			(0.190)		(0.082)	(0.950)	
Benzene	3099	3050		1960		1810		1520	1485	1385	1038	670	
	(0.510)	(0.600)		(0.067)		(0.080)		(0.040)	(0.240)	(0.029)	(0.180)	(1.250)	
				1948		1801			1472		1022	651	
				(0.070)		(0.100)			(0.185)		(0.100)	(0.860)	
						1628		1164	1345		825(0.070)	505	
						(0.025)		(0.019)	(0.025)			(0.360)	
Benzene-d_6	2380	2282		1675		1619		1154	1330		811	491	
	(0.039)	(0.600)		(0.020)		(0.023)		(0.015)	(0.040)		(0.120)	(1.250)	
						1610		1146	1315		800	470	
						(0.026)		(0.020)	(0.025)		(0.080)	(0.439)	
Benzene/Benzene-d_6	1.302	1.337		1.169		1.118		1.317	1.117		1.279	1.365	
	vC–H	vC–H	4,14	5	7	8							
Benzyl alcohol	3082	3048	1605	1499	[1080 CP]	[1035 CP]	905					738	697
	(0.351)	(0.380)	(0.020)	(0.089)			(0.055)					(0.600)	(0.630)
	vC–D	vC–D	4,14	5	7	8							
Benzyl-2,3,4,5,6-d_5 alcohol	2395	2282	1228	1145	782	735	634					534	
	(0.010)	(0.320)	(0.120)	(0.200	(0.0156))	(0.030)	(0.030)					(0.529)	
Benzyl alcohol/Benzyl-2,3,4,5,6-d_5 alcohol	1,287	1.336	1,309	1.309	1.381	1.408	1.427					1.382	
	OH str.	a.CH_2 str.	s.CH_2 str.	CH_2 bend	COH bend	CH_2 wag	CH_2 twist	C–O str.	CH_2 rock				
				1462									
				(0.139)									
Benzyl alcohol	3658	2935	2895	1460	1385	1204	masked	1020	800				
	(0.161)	(0.240)	(0.318)	(0.110)	(0.375)	(0.195)		(1.220)	(0.030)				
				1451									
				(0.141)									
									824				
									(0.080)				
Benzyl-2,3,4,5,6-d_5 alcohol	3645	2921	2842	1467	1378	1184	masked	1018	820				
	(0.165)	(0.219)	(0.300)	(0.025)	(0.310)	(0.129)		(1.250)	(0.058)				
									812				
									(0.062)				
	2(I)	I + II	I + III	II + III	III + IV								
Benzyl alcohol	1958	1945	1882	1808	1750								
	(0.039)	(0.020)	(0.020)	(0.025)	(0.010)								
Benzyl-2,3,4,5,6-d_5 alcohol	1740	1605	1565	1555	1505								
	(0.020)	(0.0250)	(0.011)	(0.011)	(0.011)								
Benzyl alcohol/Benzyl-2,3,4,5,6-d_5 alcohol	1.125	1.212	1.203	1.162	1.162								

TABLE 9.21 Infrared data for the out-of-plane deformations for mono-x-benzenes

Group	I cm^{-1}	II cm^{-1}	III cm^{-1}	IV cm^{-1}	V cm^{-1}	VI cm^{-1}	σp
$N(CH_3)_2$	966	945	862	823	749	689	[—]
$NHCH_3$	966	947	867	812	749	691	[—]
NH_2	968	957	874	823	747	689	−0.66
OH	971	952	882	825	750	688	−0.37
OCH_3	970	953	881	814	753	691	−0.27
$C(CH_3)_3$	979	964	905	841	763	698	−0.2
CH_3	973	963	895	789	730	697	−0.17
C_2H_5	978	963	903	841	748	699	−0.15
C_3H_7-iso	979	961	905	840	762	701	−0.15
F	972	952	890	827	750	682	0.06
I	986	959	901	825	730	685	0.18
Cl	976	957	897	825	734	679	0.23
Br	980	960	902	830	738	685	0.23
CHO	989	970	919	809	753	689	0.42
CO_2H	995	973	940	852	810	707	0.45
$S(=O)CH_3$	988	969	913	843	744	690	0.49
CF_3	984	965	920	830;847	769	693	0.54
CN	987	967	920	843	756	685	0.66
SO_2CH_3	994	972	928	849	741	690	0.72
NO_2	990	969	932	840	793	705	0.78

TABLE 9.22 Summary of the out-of-plane hydrogen deformations for substituted benzenes, and the out-of-plane ring deformation for mono-substituted benzenes

Benzenes	I cm^{-1}	II cm^{-1}	III cm^{-1}	IV cm^{-1}	V cm^{-1}	VI cm^{-1}
Mono-substituted	965.5–996	945–981	862–940	789–852	725–810	679–720
1,2-Disubstituted	956–989	915–970	833–886	715–791	683–729	
1,3-Disubstituted	947–999	863–949	831–910	761–815	639–698	
1,4-Disubstituted	933–985	916–971	790–852	794–870	641–735	
1,3,5-Trisubstituted	819–927	786–910	642–739			
1,2,3-Trisubstituted	930–989	848–930	731–810	684–744		
1,2,4-Trisubstituted	926–982	841–923	790–852	652–750		
1,2,4,5-Tetrasubstituted	861–911	828–890				
1,2,3,5-Tetrasubstituted	821–942	838–920				
1,2,3,4-Tetrasubstituted	919–949	770–821				
Pentasubstituted	822–927					

TABLE 9.23 Combination and overtones of the out-of-plane hydrogen deformations for substituted benzenes

Benzenes	2xI cm^{-1}	I + II cm^{-1}	I + III cm^{-1}	II + III cm^{-1}	II + IV cm^{-1}	III + IV cm^{-1}	III + V cm^{-1}
Mono-substituted	1931–1992	1915–1976	1815–1936	1815–1914	1737–1829	1680–1787	1611–1748
						2xII cm^{-1}	2xIII cm^{-1}
1,2-Disubstituted	1912–1997	1869–1952	1789–1875	1748–1849	1655–1758	1828–1932	1666–1772
			I + IV cm^{-1}				
1,3-Disubstituted	1910–1991	1820–1940	1709–1820	1709–1852	1637–1760	1724–1890	1680–1820
1,4-Disubstituted		1852–1957	1729–1835		1727–1850	1619–1702	
1,3,5-Trisubstituted	1635–1849	1605–1840					
1,2,3-Trisubstituted	1860–1975	1792–1914				1700–1860	
1,2,4-Trisubstituted	1844–1961	1776–1890	1712–1829			1683–1845	1580–1704
1,2,4,5-Tetrasubstituted	1725–1830	1695–1799				1651–1770	
1,2,3,5-Tetrasubstituted	1635–1876	1659–1862				1675–1841	
1,2,3,4-Tetrasubstituted	1838–1902	1715–1770				1610–1650	
Pentasubstituted	1655–1848						

TABLE 9.24 Hexachlorobenzene vs Pentachlorobiphenyl

Species	Hexachloro-benzene cm^{-1}	2,3,4,5,6-Cl5 biphenyl cm^{-1}
A1g	1210(R)	1205
	372(R)	[—]
B1u	1108	[—]
	369	[—]
A2g	630(est)	632
B2u	1224Comb.	1230
	230(est)	[—]
E2g	1512	1516&1497
	870(R)	868
	323(R)	[—]
	219(R)	[—]
E1u	1350	685
	699	[—]
	218	[—]
A2u	209	[—]
B2g	704Comb.	750?
	340(R)	[—]
E2u	594	586
	80Comb.	[—]

CHAPTER 10

The Nyquist Vibrational Group Frequency Rule

Aliphatic Hydrocarbons — The Nyquist Vibrational Group Frequency Rule 426
Anilines — The Nyquist Vibrational Group Frequency Rule 427
Anhydrides, Imides and 1,4-Benzoquinones — The Nyquist Vibrational Group Frequency Rule 428
Substituted Hydantoins — The Nyquist Vibrational Group Frequency Rule 428
1,4-Diphenylbutadiyne and 1-Halopropadiene — The Nyquist Vibrational Group Frequency Rule 429
3-Nitrobenzenes and 4-Nitrobenzenes — The Nyquist Vibrational Group Frequency Rule 430
Organic Sulfates, Sulfonate, Sulfonyl Chloride and Sulfones — The Nyquist Vibrational Group Frequency Rule 430
References 431

Tables

Table 10-1	423 (426)
Table 10-2	433 (427)
Table 10-3	434 (428)
Table 10-4	435 (428)
Table 10-5	435 (429)
Table 10-6	436 (430)
Table 10-7	437 (430)

*Numbers in parentheses indicate in-text page reference.

The *Nyquist Vibrational Group Frequency Rule* is distinctly different from the *Nyquist Frequency*. The sampling rate, called the *Nyquist Frequency*, is defined here. A signal is made up of various frequency components. If the signal is sampled at regular intervals, then there will be no information lost if the rate at which it is sampled is twice the frequency of the highest frequency component of the signal. The author of this text has been asked many times if he developed the *Nyquist Frequency*. The answer is NO! This answer always causes a look of disappointment. When this question arises, I respond with this reply, "No, but I developed the *Nyquist Vibrational Group Frequency Rule*."

The Nyquist Vibrational Group Frequency Rule is defined here. With change in the physical phase or environment of a chemical compound, the group frequency shift for the antisymmetric stretching vibration is larger than it is for the symmetric vibration (or out-of-phase vs in-phase stretching vibrations) for a chemical group within the molecular structure of a chemical compound. This relative change between the group frequency shifts is caused by the difference in energy required in order for the particular chemical group to vibrate in the antisymmetric and symmetric (or out-of-phase and in-phase) modes in two physical phases. The Nyquist Vibrational Group Frequency Rule was developed from studies of alkanes, anhydrides, anilines, imides, nitro-containing compounds, and SO_2-containing compounds in different solutions (CCl_4 and $CHCl_3$), in the vapor phase vs solution or neat phase. This energy differential is believed to result from the difference in the dipolar interaction of the chemical group within the solute and the surrounding change in its reaction field. This change can also result from intermolecular hydrogen bonding between solvent molecules and the chemical group within the solute, or from intermolecular hydrogen bonding between molecules in the vapor phase compared to the molecules in the vapor phase that are not intermolecularly hydrogen-bonded (1).

ALIPHATIC HYDROCARBONS — THE NYQUIST VIBRATIONAL GROUP FREQUENCY RULE

Table 10.1 lists IR data for the ν asym. CH_3, ν sym. CH_3, ν asym. CH_2 and ν sym. CH_2 vibrations for *n*-alkanes in CCl_4 and $CDCl_3$ solutions, and the frequency difference between each of these vibrations in CCl_4 and $CDCl_3$ solutions (2). In all cases, the ν asym. CH_3 and ν asym. CH_2 vibrations occur at higher frequency in $CDCl_3$ solution than in CCl_4 solution. In all cases, the ν sym. CH_3 and ν sym. CH_2 vibrations occur at lower frequency in $CDCl_3$ solution than in CCl_4 solution.

The Cl atoms in CCl_4 are more basic in the case of CCl_4 than they are in $CDCl_3$ or $CHCl_3$. If there is intermolecular hydrogen bonding between aliphatic hydrocarbon protons and the Cl atoms in CCl_4 and $CDCl_3$, one would expect the strongest intermolecular association to be formed between the protons of the aliphatic hydrocarbon and the Cl atoms of CCl_4.

This appears to be the case, because both ν asym. CH_3 and ν asym. CH_2 occur at lower frequency by 0.08 to 0.22 cm^{-1} in CCl_4 solution than in $CDCl_3$ solution. If so, another factor must be considered in the case of ν sym. CH_3 and ν sym. CH_2, because these vibrations occur at higher frequency in $CDCl_3$ solution than in CCl_4 solution. During the ν asym. CH_3 and ν asym. CH_2 vibration there is a trade-off in energy arising from simultaneous expansion and contraction of the $CH_3\cdots Cl{-}C$ and $CH_2\cdots Cl{-}C$ bonds while it requires more energy to expand and contract these same $CH_3\cdots Cl{-}C$ and $CH_2\cdots Cl{-}C$ bonds in phase. These factors more than offset the expected decrease in ν sym. CH_3 and ν sym. CH_2 frequencies due to the strength of the intermolecular hydrogen bond formed between the solute and solvent. Thus, the ν CH_3 and ν CH_2 vibrations conform to the Nyquist Vibrational Group Frequency Rule.

Progressing in the series pentane through octadecane, the ν asym. CH_3 vibration decreases in frequency by a factor of 2 to 3 times that exhibited by ν sym. CH_3. In the case of the CH_2 vibrations, ν asym. CH_2 decreases in frequency by a factor of ~1/6 of that exhibited by ν sym. CH_2. Progressing in the series pentane through octadecane, the frequency separation between ν asym. CH_3 and ν asym. CH_2 decreases progressively from 31.82 cm^{-1} to 30.65 cm^{-1} in CCl_4 solution and from 31.52 cm^{-1} to 30.56 cm^{-1} in $CDCl_3$ solution. Progressing in the series pentane through octadecane, the frequency separation between ν sym. CH_3 and ν sym. CH_2

increases progressively from 11.30 cm^{-1} to 17.76 cm^{-1} in $CDCl_3$ solution. These data indicate that as n increases in the series $CH_3(CH_2)_n$ CH_3 there is less coupling between ν asym. CH_3 and ν asym. CH_2 and there is more coupling between ν sym. CH_3 and ν sym. CH_2.

ANILINES — THE NYQUIST VIBRATIONAL GROUP FREQUENCY RULE

Table 10.2 list IR data for the ν asym. NH_2 and ν sym. NH_2 frequencies for 4-X and 3-X anilines in both CCl_4 and $CHCl_3$ solutions and the frequency separation between ν asym. NH_2 in CCl_4 and $CHCl_3$ solutions and the frequency separation between ν sym. NH_2 in CCl_4 and $CHCl_3$ solutions (3). The ν asym. NH_2 and ν sym. NH_2 frequencies for these anilines in CCl_4 solution are suggested to result from complexes such as A:

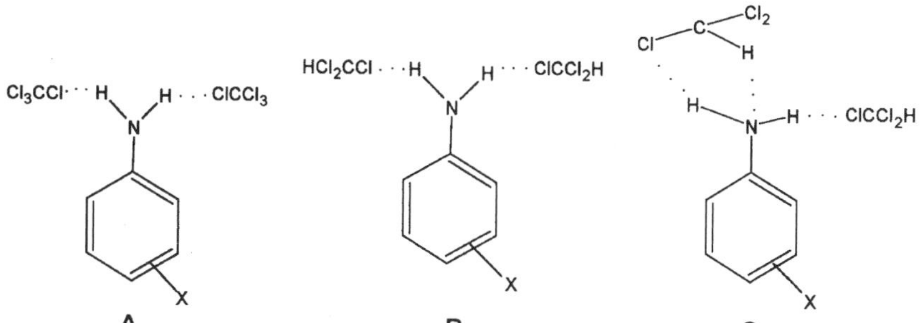

A B C

ν asym. NH_2 : 3453.8-3508.6 cm^{-1}
ν sym. NH_2 : 3376.8-3415.0 cm^{-1}

ν asym. NH_2 : 3459.1-3510.3 cm^{-1}
ν sym. NH_2 : 3377.0-3416.7 cm^{-1}

ν asym. NH_2 : 3436.4-3478.7 cm^{-1}
ν sym. NH_2 : 3362.0-3386.8 cm^{-1}

Two sets of ν asym. NH_2 and ν sym. NH_2 are noted for these anilines in $CHCl_3$ solution. The higher frequency set is suggested to result from a complex such as B, and the lower frequency set is suggested to result from a complex such as C (3).

The intermolecular hydrogen-bonded complexes formed for A are stronger than those complexes for B, because ν asym. NH_2 in CCl_4 solution occurs at lower frequency by 1.16 to 9.19 cm^{-1} than those for the corresponding aniline in $CHCl_3$ solution.

However, in the case of complex C, ν asym. NH_2 occurs at lower frequency than in complex A by 17.4 to 29.7 cm^{-1}. The CCl_3H proton intermolecularly hydrogen bonded to the free pair of electrons on the nitrogen atom in complex B weakens the NH_2 bonds, causing their ν asym. NH_2 vibration to occur at even lower frequency than in the case of complex B. The ν sym. NH_2 vibrations shift to lower frequency by a factor of 2 to 8.5 times more than the ν asym. NH_2 vibrations for the correspond 3-X and 4-X-aniline. This frequency difference between the ν asym. NH_2 and ν sym. NH_2 vibrational in complexes B and C is attributed to the Nyquist Vibrational Group Frequency Rule. During the ν sym. NH_2 vibration in complexes B and C, the NH_2 vibrations expand and contract in phase from the chlorine atoms of the CCl_3H molecules. The amount of energy required during the compression of the $NH_2(\cdots ClCCl_2H)_2$ intermolecular hydrogen bonds is larger in the case of ν sym. NH_2 than the amount of energy required for ν asym. NH_2 where the compression of one $NH \cdots ClCCl_2H$ intermolecular hydrogen bond is offset by the expansion of the other intermolecular $NH \cdots ClCCl_2H$ hydrogen bond in complexes such as B and C.

ANHYDRIDES, IMIDES AND 1,4-BENZOQUINONES — THE NYQUIST VIBRATIONAL GROUP FREQUENCY RULE

Table 10.3 lists IR data for anhydrides, imides, and 1,4-benzoquinones (4–8). The out-of-phase and in-phase $(C{=}O)_2$ stretching frequencies for these compounds in CCl_4 and $CHCl_3$ solution are listed as well as the frequency difference between $\nu_{op}(C{=}O)_2$ in CCl_4 solution and $\nu_{op}(C{=}O)_2$ in $CHCl_3$ solution, and the frequency difference between $\nu_{ip}(C{=}O)_2$ in CCl_4 solution and $\nu_{ip}(C{=}O)_2$ in $CHCl_3$ solution.

In all cases of anhydrides and imides $\nu_{op}(C{=}O)_2$ and $\nu_{ip}(C{=}O)_2$ occur at lower frequency in $CHCl_3$ solution than in CCl_4 solution. These lower ν $(C{=}O)_2$ frequencies in CCl_3H solution are attributed to complexes such as $(C{=}O \cdots HCCl_3)_2$. In the 1,4-benzoquinone series the $\nu_{op}(C{=}O)_2$ mode occurs at lower frequency in CCl_3H solution than in CCl_4 solution while $\nu_{ip}(C{=}O)_2$ (with one exception) increases in frequency in $CHCl_3$ solution compared to the corresponding compounds in CCl_4 solution.

The carbonyl-containing compounds differ from those for the aliphatic hydrocarbons and aniline in that the former compounds involve stretching of the $(C{=}O)_2$ bond while the latter compounds involve stretching of CH_3, CH_2 and NH_2 bonds. Intermolecular hydrogen bonding of two CCl_3H protons with two oxygen atoms $(C{=}O \cdots HCCl_3)_2$ lowers $\nu_{op}(C{=}O)_2$ from 1.7 to 5.2 cm^{-1} for the anhydrides, by 5.7 to 10.6 cm^{-1} for the imide type compounds, and by 0.9 to 2.9 cm^{-1} for the 1,4-benzoquinones. Intermolecular hydrogen bonding in the case of $\nu_{ip}(C{=}O)_2$ does not lower this vibrational mode as much in frequency as the $\nu_{op}(C{=}O)_2$ vibration, because it requires more energy to compress two $C{=}O \cdots HCCl_3$ intermolecular hydrogen bonds in the case of $\nu_{ip}(C{=}O)_2$ than it requires to expand one $C{=}O \cdots HCCl_3$ bond while the other $C{=}O \cdots HCCl_3$ contracts during the $\nu_{op}(C{=}O)_2$ vibration. Thus, the ν $(C{=}O)_2$ frequency behavior for these compounds conforms to the Nyquist Vibrational Group Frequency Rule.

SUBSTITUTED HYDANTOINS — THE NYQUIST VIBRATIONAL GROUP FREQUENCY RULE

Table 10.4 lists IR data for the $\nu_{op}(C{=}O)_2$ and $\nu_{ip}(C{=}O)_2$ vibrations for substituted hydantoins in the vapor- and solid phases (7). The in-phase mode occurs in the region 1809-1825 cm^{-1} in the vapor phase and in the region 1761–1783 cm^{-1} in the solid phase. These $\nu_{ip}(C{=}O)_2$ vibrations occur at lower frequency by 29–48 cm^{-1} in the solid phase than they occur in the vapor phase. The out-of-phase $(C{=}O)_2$ mode occurs in the region 1774–1785 cm^{-1} in the vapor phase, and in the region 1707–1718 cm^{-1} in the solid phase. These $\nu_{op}(C{=}O)_2$ vibrations occur at lower frequency by 56–68 cm^{-1} in the solid phase than in the vapor phase.

The hydantoins listed in Table 10.4 exists as intermolecular hydrogen bonded dimers in the solid phase as depicted here:

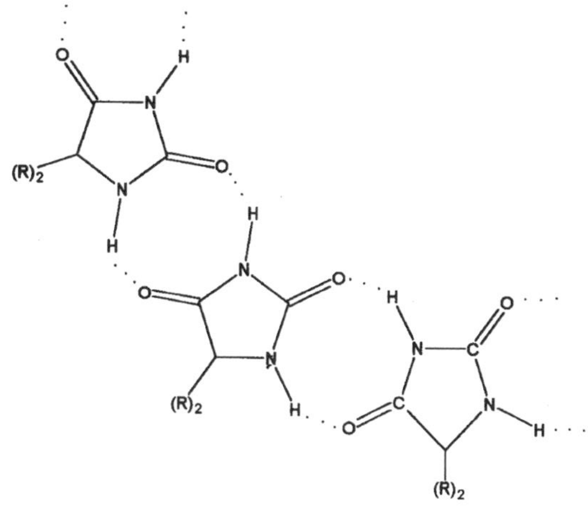

However, in the vapor phase these 5,5-dialkyl hydantoins are not intermolecularly hydrogen bonded. The $\nu_{ip}(C{=}O)_2$ modes shift to lower frequency in the solid phase because of the intermolecular hydrogen-bond formation between the N–H···O=C groups, but the $\nu_{ip}(C{=}O)_2$ mode does not shift as much to lower frequency due to the fact that it requires more energy to compress two intermolecular N–H···O=C bonds than is required to compress one intermolecular N–H···O=C bond and expand one N–H···O=C intermolecular bond in the case of $\nu_{op}(C{=}O)_2$. Thus, these group frequencies fulfill the requirement of the Nyquist Vibrational Group Frequency Rule.

1,4-DIPHENYLBUTADIYNE AND 1-HALOPROPADIENE— THE NYQUIST VIBRATIONAL GROUP FREQUENCY RULE

Table 10.5 lists the $\nu_{op}(CC)_2$, $\nu_{ip}(CC)_2$, $\nu_{op}C{=}C{=}C$, and $\nu_{ip}C{=}C{=}C$ vibrations for 1,4-diphenylbutadiyne and the 1-halopropadiene in CCl_4 solution and in $CHCl_3$ solution, and their frequency differences between each of these vibrations in CCl_4 and $CHCl_3$ solutions (9 11).

In the case of 1,4-diphenylbutadiyne, both $\nu_{op}(CC)_2$ and $\nu_{ip}(CC)_2$ occur at lower frequency in CCl_3H solution than in CCl_4 solution. These frequency decreases are attributed to intermolecular hydrogen bonding between the π systems of the $(C{\equiv}C)_2$ groups and CCl_3 H protons (eg. –C≡C–C≡C–). Most likely the CCl_3 protons also intermolecularly hydrogen

$HCCl_3$ $HCCl_3$

bond to the π system of the two phenyl groups. The C≡C–C≡C group is a linear rod and the CCl_3H molecules would be surrounding and intermolecularly hydrogen bonding with the two C≡C groups. The $\nu_{ip}(CC)_2$ vibration takes relatively more energy to vibrate due to the compression and expansion of the intermolecularly hydrogen-bonded CCl_3H molecules than the energy required during $\nu_{op}(CC)_2$ where one intermolecularly hydrogen-bonded C≡C is

$HCCl_3$

compressed while the other C≡C group is expanded. In the $\nu_{op}(CC)_2$ case there is less

⋮
$HCCl_3$

interaction between the CCl_3H molecules than in the case of $\nu_{ip}(CC)$ interactions with CCl_3H molecules. Thus, $\nu_{op}(CC)_2$ decreases twice as much as $\nu_{ip}(CC)_2$ in going from solution in CCl_4 to solution in $CHCl_3$.

The ν asym. C=C=C vibration decreases in frequency in going from the vapor phase to the liquid phase while under the same physical conditions ν sym. C=C=C does not decrease as much in frequency as the ν asym. C=C=C vibration or else actually increases in frequency (9–11). These data fulfill the Nyquist Vibrational Group Frequency Rule.

3-NITROBENZENES AND 4-NITROBENZENES — THE NYQUIST VIBRATIONAL GROUP FREQUENCY RULE

Table 10.6 lists IR data for ν asym. NO_2 and ν sym. NO_2 in both CCl_4 and CCl_3H solutions and their frequency difference in CCl_4 and CCl_3H solutions (5,12). The ν asym. NO_2 mode decreases in frequency by 1.7 to 8.6 cm^{-1} in going from CCl_4 to CCl_3H solution while the ν sym. NO_2 mode for the same compound decreases less in frequency or in many cases actually increases in frequency. In CCl_3H solution the NO_2 group is intermolecularly hydrogen-bonded to the oxygen atoms of the NO_2 group.

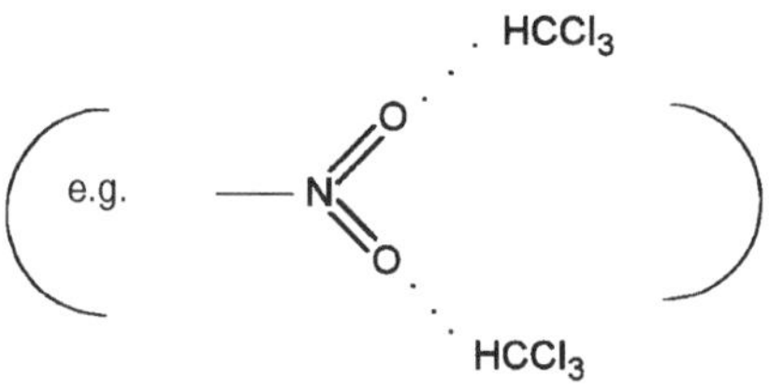

It takes more energy to compress and expand the intermolecular hydrogen-bonded NO_2 groups during the ν sym. NO_2 vibration than it does to expand one N=O···$HCCl_3$ intermolecular hydrogen bond and compress one N=O···HCl_3 during ν_{op} NO_2. Thus, the ν NO_2 most conform to the Nyquist Vibrational Group Frequency Rule.

ORGANIC SULFATES, SULFONATE, SULFONYL CHLORIDE AND SULFONES — THE NYQUIST VIBRATIONAL GROUP FREQUENCY RULE

Table 10.7 lists ν asym. SO_2 and ν sym. SO_2 frequencies in both CCl_4 and CCl_3H solutions and their frequency differences in each solvent. In all cases the ν SO_2 vibrations decrease in frequency in going from solution in CCl_4 to solution in CCl_3H, and this decrease in frequency is larger for

ν asym. SO_2 than for ν sym. SO_2 by a factor of ~2 to 10. In CCl_3H solution, the SO_2 oxygen atoms are intermolecularly hydrogen bonded as presented here:

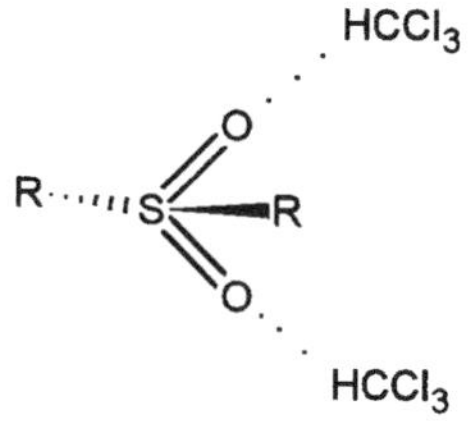

The energy to expand and contract the $(S{=}O \cdots HCCl_3)_2$ intermolecular hydrogen bonds during ν sym. SO_2 is greater than that needed to compress one $S{=}O \cdots HCCl_3$ intermolecular hydrogen bond and expand one $S{=}O \cdots HCCl_3$ intermolecular hydrogen bond during ν asym. SO_2.

REFERENCES

1. Nyquist, R. A. (1994). *Structural Information From Infrared Studies on Solute-Solvent Interactions*. Ph. D. Thesis, Utrecht University, The Netherlands, ISBN 90-393-0743-1, p. 260.
2. Nyquist, R. A. and Fiedler, S. L. (1993). *Appl. Spectrosc.* 47: 1670.
3. Nyquist, R. A., Luoma, D. A., and Puehl, C. W. (1992). *Appl. Spectrosc.* 46: 1273.
4. Nyquist, R. A. (1990). *Appl. Spectrosc.* 44: 438.
5. Nyquist, R. A. (1990). *Appl. Spectrosc.* 44: 594.
6. Nyquist, R. A. (1990). *Appl. Spectrosco.* 44: 783.
7. Nyquist, R. A. and Fielder, S. L. (1995). *Vib Spectrosc.* 8: 365.
8. Nyquist, R. A., Luoma, D. A., and Putzig, C. L. (1992). *Vib. Spectrosc.* 3: 181.
9. Nyquist, R. A. and Putzig, C. L. (1992). *Vib. Spectrosc.* 3: 35.
10. Nyquist, R. A., Lo, Y-S., and Evans, J. C. (1964). *Spectroschim. Acta* 20: 619.
11. Nyquist, R. A., Reder, T. L., Stec, F. F., and Kallos, G. J., (1971). *Spectrochem. Acta* 27A: 897.
12. Nyquist, R. A. and Settineri, S. E. (1990). *Appl. Spectrosc.* 44: 1552.

TABLE 10.1 Aliphatic hydrocarbons—the Nyquist Rule

Compound	a.CH_3 str. CCl_4 soln. cm^{-1}	a.CH_3 str. $CDCl_3$ soln. cm^{-1}	s.CH_3 str. CCl_4 soln. cm^{-1}	s.CH_3 str. $CDCl_3$ soln. cm^{-1}	a.CH_3 str. [CCl_4]–[$CDCl_3$] cm^{-1}	s.CH_3 str. [CCl_4]–[$CDCl_3$] cm^{-1}
Pentane	2959.55	2959.66	2873.12	2872.67	−0.11	0.45
Hexane	2958.69	2958.8	2872.92	2872.34	−0.11	0.58
Hexane	2959.23	2959.31	2873.05	2872.54	−0.08	0.51
Heptane	2958.86	2958.98	2872.81	2872.23	−0.12	0.58
Octane	2958.62	2958.73	2872.82	2872.2	−0.11	0.62
Nonane	2958.46	2958.58	2872.79	2872.17	−0.12	0.62
Decane	2958.35	2958.49	2872.78	2872.14	−0.14	0.64
Undecane	2958.14	2859.33	2872.7	2872.05	−0.19	0.65
Dodecane	2958.03	2958.23	2872.66	2871.96	−0.2	0.7
Tridecane	2957.94	2958.1	2872.65	2871.91	−0.16	0.74
Tetradecane	2957.79	2957.99	2872.6	2871.81	−0.2	0.79
Octadecane	2957.26	2957.48	2872.35	2871.5	−0.22	0.85
Δ cm^{-1}	2.29	2.18	0.77	1.17		

	a.CH_2 str. CCl_4 soln. cm^{-1}	a.CH_2 str. $CDCl_3$ soln. cm^{-1}	s.CH_2 str. CCl_4 soln. cm^{-1}	s.CH_2 str. $CDCl_3$ soln. cm^{-1}	a.CH_2 str. [CCl_4]–[$CDCl_3$] cm^{-1}	s.CH_2 str. [CCl_4]–[$CDCl_3$] cm^{-1}
Pentane	2927.73	2928.14	2861.82	2861.21	−0.41	0.61
Hexane	2928	2928.25	2859.4	2859.14	−0.25	0.26
Hexane	2927.89	2928.12	2859.12	2858.74	−0.23	0.38
Heptane	2927.29	2927.58	2857.81	2857.37	−0.29	0.44
Octane	2926.95	2927.26	2856.55	2856.22	−0.31	0.33
Nonane	2926.79	2927.07	2855.68	2855.58	−0.28	0.1
Decane	2926.83	2927.13	2855.31	2855.25	−0.3	0.06
Undecane	2926.76	2927.03	2855.09	2855.05	−0.27	0.04
Dodecane	2926.72	2927.01	2854.92	2854.88	−0.29	0.04
Tridecane	2926.7	2926.98	2854.83	2854.79	−0.28	0.04
Tetradecane	2926.66	2926.96	2854.77	2854.71	−0.3	0.06
Octadecane	2926.61	2926.92	2854.59	2854.55	−0.31	0.06
Δ cm^{-1}	1.12	1.22	7.23	6.66		

	a.CH_3 str.–a.CH_2 str. CCl_4 soln. cm^{-1}	a.CH_3 str.–a.CH_2 str. $CDCl_3$ soln. cm^{-1}	s.CH_3 str.–s.CH_2 str. CCl_4 soln. cm^{-1}	s.CH_3 str.–s.CH_2 str. $CDCl_3$ cm^{-1}
Pentane	31.82	31.52	11.3	11.46
Hexane	30.69	30.55	13.52	13.2
Hexane	31.34	31.19	13.93	13.8
Heptane	31.57	31.4	15	14.86
Octane	31.67	31.15	16.27	15.98
Nonane	31.67	31.51	17.11	16.59
Decane	31.52	31.36	17.47	16.89
Undecane	31.38	32.3	17.61	17
Dodecane	31.31	31.22	17.82	17.08
Tridecane	31.24	31.12	17.82	17.12
Tetradecane	31.13	31.03	17.83	17.1
Octadecane	30.65	30.56	17.76	16.95
Δ cm^{-1}	0.69	0.96	6.46	5.49

TABLE 10.2 Anilines — the Nuquist Rule

4-X-Aniline	a.NH_2 str. CCl_4 soln. cm^{-1}	a.NH_2 str. $CHCl_3$ soln. cm^{-1}	a.NH_2 str. $CHCl_3$ soln. cm^{-1}	s.NH_2 str. CCl_4 soln. cm^{-1}	s.NH_2 str. $CHCl_3$ soln. cm^{-1}	s.NH_2 str. $CHCl_3$ soln. cm^{-1}	a.NH_2 str. [CCl_4]–[$CHCl_3$] cm^{-1}	s.NH_2 str. [CCl_4]–[$CHCl_3$] cm^{-1}	a.NH_2 str. [CCl_4]–[$CHCl_3$] cm^{-1}	s.NH_2 str. [CCl_4]–[$CHCl_3$] cm^{-1}
4-X										
OH	[–]	3468.9	3440.31	[–]	3386.8	3367.89	[–]	[–]	[–]	[–]
OCH_3	3459.71	3468.9	3441.09	3381.41	3388.75	3366.62	−9.19	−7.34	18.62	14.79
$N(CH_3)_2$	3454.12	3459.13	3435.4	3377.35	3377.02	3362.02	−5.01	0.33	18.72	15.33
NH_2	3453.81	[–]	3436.41	3376.79	3388.75	3362.43	[–]	−11.96	17.4	14.63
OC_6H_5	3468.78	3474.77	3445.62	3387.3	3389.28	3373.49	−5.99	−1.98	23.16	13.81
F	3472.02	3476.72	3449.2	3390.56	3390.71	3374.57	−4.7	−0.15	22.82	15.99
Cl	3484.52	3486.89	3457.74	3398.35	3400.11	3380.93	−2.37	−1.76	26.78	17.42
Br	3486.23	3488.96	3459.32	3399.48	3401.15	3380.93	−2.73	−1.67	26.91	18.55
H	3480.29	3481.75	3453.95	3394.77	3396.43	3378.03	−1.46	−1.66	26.34	16.74
H	3480.53	3481.57	3453.78	3394.88	3396.75	3377.76	−1.04	−1.87	26.75	17.12
C_2H_5	3471.32	3478.68	3446.47	3389.44	3390.77	3377.91	−7.63	−1.33	24.85	11.53
$C(CH_3)_3$	3472.72	3480.63	3447.29	3390.13	3391.79	3373.96	−7.91	−1.66	25.43	16.17
C_6H_5	3483.8	3486.75	3454.79	3396.84	3399.33	3378.79	−2.95	−2.49	29.01	18.05
$C(=O)OC_2H_5$	3500.08	3504.25	3478.68	3408.59	3411.06	[–]	−4.17	−2.47	21.4	[–]
CN	3504.31	3507.43		3412.05	3413.73	[–]	−3.12	−1.68	[–]	[–]
CF_3	3498.73	3501.31	3476.72	3408.35	3409.82	3386.8	−2.59	−1.47	22.01	21.55
$C(=O)CH_3$	3501.51	3505.77	[–]	3409.6	3412.09	[–]	−4.26	−2.49	[–]	[–]
NO_2	3508.61	3510.27	[–]	3414.96	3416.69	[–]	−1.66	−1.73	[–]	[–]
3-X										
OCH_3	3483.54	3485.08	3457.05	3397.02	3399.07	3380.93	−1.54	−2.05	26.49	16.09
F	3491.35	3492.65	3464.99	3402.99	3404.16	3380.93	−1.3	−1.17	26.36	22.06
Cl	3490.49	3492.02	3464.99	3401.94	3403.23	3380.93	−1.53	−1.29	25.5	21.01
Br	3490.31	3491.94	3463.79	3401.31	3403.05	3378.98	−1.63	−1.74	26.52	22.33
CH_3	3479.05	3482.59	3453.48	3393.8	3395.42	3377.62	−3.54	−1.62	25.57	16.18
$C(=O)OC_2H_5$	3486.72	3488.89	3459.56	3399.26	3401.39	3380.93	−2.17	−2.13	27.16	18.33
CN	3494.7	3496.96	3474.77	3405.49	3406.75	3382.89	−2.26	−1.26	19.93	22.6
CF_3	3492.74	3493.9	3463.04	3403.81	3404.53	3378.98	−1.16	−0.73	29.7	24.83

TABLE 10.3 Andydrides imides, and 1,4-benzoquinones—the Nyquist Rule

Anhydride	op.(C=O)$_2$ str. CCl_4 soln. cm^{-1}	op(C=O)$_2$ str. $CHCl_3$ soln. cm^{-1}	ip(C=O)$_2$ str. CCl_4 soln. cm^{-1}	ip(C=O)$_2$ str. $CHCl_3$ soln. cm^{-1}	op(C=O)$_2$ str. [CCl_4]–[$CHCl_3$] cm^{-1}	ip(C=O)$_2$ str. [CCl_4]–[$CHCl_3$] cm^{-1}
Hexahydrophthalic	1792.7	1787.5	1861.4	1859.1	5.2	2.3
Phthalic	1784.6	1779.4	1856	1853	5.2	3
Tetrachlorophthalic	1790.7	1786.8	1845.9	1845.5	3.9	0.4
Tetrabromophthalic	1795.6	1790.3	1867.2	1863.4	5.3	3.8
Maleic	1787.1	1785.4	1851.74	1851.68	1.7	0.06
Dichloromaleic	1799	1796.4	1876.7	1876.2	2.6	0.5
Phthalimide						
N-(4-bromobutyl)	1718.7	1713	1773.8	1772.2	5.7	1.6
Caffeine	1667.5	1658.4	1710.6	1707.9	9.1	2.7
Isocaffeine*[1]	1673.2	1664	1716.7	1711.3	9.2	5.4
1,3,5-Trimethyluracil	1663.4*[2]	1652.4*[2]	1706.6	1700	11	6.6
1,3,6-Trimethyluracil	1668.8*[2]	1658.2*[2]	1710.8	1702	10.6	8.8
1,3-Dimethyl-2,4-(1H,3H)-quinazolinedione	1667*[2]	1657.4*[2]	1711.6	1703.7	9.6	7.9
1,4-Benzoquinone						
Tetrafluoro-	1702.71	1701.42	1696.8	1697.7	1.29	−0.9
Tetrachloro-	1694.68	1693.37			1.31	
Trichloro-	[1694]*[3]		[1657]*[3]			
Tetrabromo-	1687.48	1685.37			2.11	
2,5-Dichloro-	1682.18	1681.25	1672.5	1673.6	0.9	−1.1
Chloro-	1681.89	1680.95	1659.3	1659.3	0.94	0
Unsubstituted	1663.51*[2]	1662	1661.4	1661.8	1.51	−0.4
Methyl-	1662.47	1659.57	1665.7	1665.9	2.9	−0.2
2,6-Dimethyl-	1657					

*[1] 10.74 mol % $CHCl_3/CCl_4$.
*[2] corrected for Fermi resonance.
*[3] In CS_2 solution.

TABLE 10.4 Substituted hydantoins—the Nyquist Rule

Hydantion 1,3,5,5-	op(C=O)$_2$ str. vapor cm^{-1}	op(C=O)$_2$ str. solid cm^{-1}	ip(C=O)$_2$ str. vapor cm^{-1}	ip(C=O)$_2$ str. solid cm^{-1}	op(C=O)$_2$ str. [vapor]–[solid] cm^{-1}	ip(C=O)$_2$ str. [vapor]–[solid] cm^{-1}
H,H,CH_3,C_3H_7	1774	1707	1809	1761	67	48
H,H,CH_3,isoC_3H_7	1774	1712	1808	1770	62	38
H,H,H,H	1785	1717	1825	1783	68	42
H,H,CH_3,isoC_4H_9	1774	1718	1809	1780	56	29

TABLE 10.5 1,4-Diphenylbutadiyne and 1-halopropadiene—the Nyquist Rule

Compound	op(CC)$_2$ str. CCl_4 cm^{-1}	op(CC)$_2$ str. $CHCl_3$ cm^{-1}	ip(CC)$_2$ str. CCl_4 cm^{-1}	ip(CC)$_2$ str. $CHCl_3$ cm^{-1}	op(CC)$_2$ str. [CCl_4]–[$CHCl_3$] cm^{-1}	ip(CC)$_2$ str. [CCl_4]–[$CHCl_3$] cm^{-1}
1,4-Diphenylbutadiyne	2152.23	2150.22	2220.51	2219.45	2.01	1.06
Propadiene	**a.C=C=C str. CCl_4 cm^{-1}**	**a.C=C=C str. liquid cm^{-1}**	**s.C=C=C str. CS_2 cm^{-1}**	**s.C=C=C str. liquid cm^{-1}**	**a.C=C=C str. [CCl_4]–[liquid] cm^{-1}**	**s.C=C=C str. [CS_2]–[liquid] cm^{-1}**
1-Chloro-	1958	1951	1101	1095	7	6
1-Bromo-	1957	1954	1082	1086	3	−4
1-Bromo,1-d-	1940	1943	857	861	−3	−4
1-Iodo-	1947	1947	1076	1076	0	0
Propadiene	**a.C=C=C str. vapor cm^{-1}**	**a.C=C=C str. CCl_4 cm^{-1}**	**s.C=C=C str. vapor cm^{-1}**	**s.C=C=C str. CCl_4 cm^{-1}**	**a.C=C=C str. [vapor]–[CCl_4] cm^{-1}**	**s.C=C=C str. [vapor]–[CCl_4] cm^{-1}**
1-Chloro-	1971	1958	1108	1101	13	7
	1956		1094		−2	−7
1Bromo-	1969	1957	1081	1082	12	−1
	1954		1076		−3	−6
1-Bromo,1-d-	1943	1940	861	857	3	−4
	1929				−11	
1-Iodo-	1958	1947	1081	1076	11	5
	1947		1071		0	−5
Propadiene	**a.C=C=C str. vapor cm^{-1}**	**a.C=C=C str. liquid cm^{-1}**	**s.C=C=C str. vapor cm^{-1}**	**s.C=C=C liquid cm^{-1}**	**a.C=C=C str. [vapor]–[liquid] cm^{-1}**	**s.C=C=C str. [vapor]–[liquid] cm^{-1}**
1-Chloro-	1971	1951	1108	1095	20	13
	1956		1094		−1	−1
1-Bromo-	1969	1954	1081	1086	15	−5
	1954		1076		0	−1
1-Bromo,1-d-	1943	1940	861	861	3	0
	1929				−11	
1-Iodo-	1958	1947	1081	1076	11	5
	1947		1071		0	−5

TABLE 10.6 3=Nitrobenzenes and 4-nitrobenzenes—the Nyquist Rule

4-Nitrobenzene [1wt./vol. % or less]	a.NO_2 str. CCl_4 soln.$CHCl_3$ soln. cm^{-1}	a.NO_2 str. $HCCl_4$ soln. cm^{-1}	s.NO_2 str. $CHCl_3$ soln. cm^{-1}	s.NO_2 str. [CCl_4]–[$CHCl_3$] cm^{-1}	a.NO_2 str. [CCl_4]–[$CHCl_3$] cm^{-1}	s.NO_2 str. [CCl_4]–[$CHCl_3$] cm^{-1}
4-X						
NH_2	1513.94	1505.34	1338.08	1335.15	8.6	2.93
CH_3O	1519.32	1514.12	1343.07	1342.76	5.2	0.31
C_6H_5O	1522.43	1517.85	1344.98	1344.88	4.58	0.1
C_6H_5	1524.26	1519.47	1346.67	1347.74	4.79	−1.07
CH_3	1524.8	1520.1	1346.18	1346.93	4.7	−0.75
HO	1526.74	1521.41	1344.45	1342.46	5.33	1.99
Cl	1527.01	1523.64	1344.3	1345.39	3.37	−1.09
isoC_3H_7	1527.68	1519.8	1347.11	1348.08	7.8	−0.97
CH_2Cl	1529.17	1525.44	1348.01	1350.09	3.73	−2.08
I	1529.78	1526.08	1348.7	1350.3	3.7	−1.6
C_6H_5–C=O	1531.06	1525.77	1353.19	1355.23	5.29	−2.04
H	1531.53	1527.37	1348.24	1349.76	4.16	−1.52
Br	1531.95	1524.88	1350.1	1351.46	7.07	−1.36
F	1532.31	1528.48	1346.41	1349.26	3.83	−1.85
CH_3CO_2	1532.87	1530.26	1348	1350.03	2.62	−2.03
CH_3C=O	1533.76	1531.06	1345.71	1347.46	2.7	−1.75
CN	1535.33	1533.21	1345.37	1347.56	2.12	−2.19
CH_3SO_2	1537.97	1535.9	1348.2	1350.6	2.07	−2.4
CF_3	1538.91	1535.44	1354.35	1356.34	3.47	−1.99
SO_2Cl	1541.25	1539.26	1346.06	147.86	1.99	−1.8
NO_2	1556.41	1554.37	1338.68	1340.6	2.04	−1.92
Δ cm^{-1}	42.47	49.03	16.27	20.08		
3-X-Nitrobenzene						
3–X						
$N(CH_3)_2$	1535.29	1531.64	1349.47	1349.81	3.65	−0.34
CH_3	1532.78	1529.16	1350.88	1352.04	3.62	−1.16
Br	1536.44	1533.09	1348.28	1349.97	3.35	−1.69
I	1533.69	1530.42	1346.97	1348.62	3.27	−1.65
NO_2	1544.05	1541.63	1345.29	1347.44	2.42	−2.15
Δ cm^{-1}	11.27	12.47	5.59	4.6		
Nitromethane	1564.5	1562.4	1374.4	1376.2	2.1	−1.8
Trichloronitromethane	1608.2	1606.5	1307.5	1309.6	1.7	−2.1
Tribromonitromethane	1595.1	1593.3	1305.9	1308.7	1.8	−2.8
Ethyl nitrate	1637.3	1631.7	1281.5	1282.9	5.6	−1.4

TABLE 10.7 Organic sulfate, sulfonate, sulfonyl chloride, and sulfones—the Nyquist Rule

Compound	a.SO_2 str. CCl_4 soln. cm^{-1}	a.SO_2 str. $CHCl_3$ soln. cm^{-1}	s.SO_2 str. CCl_4 soln. cm^{-1}	s.SO_2 str. $CHCl_3$ soln. cm^{-1}	a.SO_2 str. $[CCl_4]-[CHCl_3]$ cm^{-1}	s.SO_2 str. $[CCl_4]-[CHCl_3]$ cm^{-1}
Dimethyl sulfate	1405.7	1395.7	1203.8	1201.8	−10	−2
4′-Chlorophenyl 4-chlorophenyl benzenesulfonate	1392.4	1382.9	1178.9	1176.8	−9.5	−2.1
Benzenesulfonyl chloride	1387.4	1380.3	1177.8	1177.1	−7.1	−0.7
Diphenyl sulfone	1326.8	1319.1	1161.4	1157.6	−7.7	−3.8
Methyl phenyl sulfone	1325.7	1317.7	1157.4	1153.2	−8	−4.2

CHAPTER 11

Infrared, Raman, and Nuclear Magnetic Resonance Spectra-Structure Correlations for Organic Compounds

Introduction	441
4-x-Anilines	441
3-x and 4-x Benzoic Acids	442
4-x-Acetanilides	442
4-x-Benzaldehydes	443
4-x-Acetophenones	443
4-x and 4,4′-x,x-Benzophenones	444
Alkyl 3-x and 4-x-Benzoates	444
νC=O vs δ ^{13}C=O	444
Acetone In $CHCl_3/CCl_4$ Solutions	445
N,N′-Dimethylacetamide in $CHCl_3/CCl_4$ Solutions	445
Maleic Anhydride (MA) in $CHCl_3/CCl_4$ Solutions	446
3-x and 4-x-Benzonitriles	446
Organonitriles	446
Alkyl Isonitriles	447
Nitroalkanes	447
Alkyl Isocyanates	448
Alkylamines	448
A Summary of the Correlations for the Nitrogen Compounds	449
Organophosphorus Compounds	450
A Summary of Correlations for the Phosphorus Compounds	452
Anisoles	453
Mono-Substituted Benzenes	454
4-x and 4,4′-x,x′-Biphenyls in $CHCl_3$ Solutions	454
Tetramethylurea (TMU)	454
Dialkylketones	455
Acetates	456
Acrylates and Methacrylates	456
References	457

Figures

Figure 11-1	459 (441)	Figure 11-3	461 (441)
Figure 11-2	460 (441)	Figure 11-4	462 (441)

Figure 11-5	463 (442)
Figure 11-6	464 (442)
Figure 11-7	465 (443)
Figure 11-8	466 (442)
Figure 11-9	467 (442)
Figure 11-10	468 (442)
Figure 11-11	469 (472)
Figure 11-12	470 (442)
Figure 11-13	471 (442)
Figure 11-14	472 (442)
Figure 11-15	473 (442)
Figure 11-16	474 (442)
Figure 11-17	475 (442)
Figure 11-18	476 (443)
Figure 11-19	477 (443)
Figure 11-20	478 (443)
Figure 11-21	479 (448)
Figure 11-22	480 (443)
Figure 11-23	481 (443)
Figure 11-24	482 (443)
Figure 11-25	483 (444)
Figure 11-26	484 (444)
Figure 11-27	485 (444)
Figure 11-28	486 (444)
Figure 11-29	487 (444)
Figure 11-30	488 (444)
Figure 11-31	489 (444)
Figure 11-32	490 (444)
Figure 11-33	491 (445)
Figure 11-34	492 (445)
Figure 11-35	493 (445)
Figure 11-35a	493 (445)
Figure 11-35b	494 (446)
Figure 11-35c	494 (446)
Figure 11-36	495 (446)
Figure 11-37	496 (446)
Figure 11-38	497 (446)
Figure 11-39	498 (446)
Figure 11-40	498 (447)
Figure 11-41	499 (447)
Figure 11-42	500 (447)
Figure 11-43	501 (447)
Figure 11-44	502 (447)
Figure 11-45	503 (447)
Figure 11-46	504 (448)
Figure 11-47	505 (448)
Figure 11-48	506 (448)
Figure 11-49	507 (448)
Figure 11-50	508 (448)
Figure 11-51	509 (448)
Figure 11-52	509 (448)
Figure 11-53	510 (448)
Figure 11-54	511 (448)
Figure 11-55	512 (449)
Figure 11-56	513 (449)
Figure 11-57	514 (449)
Figure 11-58	515 (450)
Figure 11-59	516 (450)
Figure 11-60	517 (450)
Figure 11-61	518 (450)
Figure 11-62	519 (450)
Figure 11-63	520 (450)
Figure 11-64	521 (450)
Figure 11-65	522 (450)
Figure 11-66	523 (450)
Figure 11-67	524 (450)
Figure 11-68	525 (450)
Figure 11-69	526 (451)
Figure 11-70	527 (451)
Figure 11-71	528 (451)
Figure 11-72	529 (451)
Figure 11-73	530 (451)
Figure 11-74	531 (451)
Figure 11-75	532 (451)
Figure 11-76	533 (452)
Figure 11-77	534 (452)
Figure 11-78	535 (452)
Figure 11-79	536 (452)
Figure 11-80	537 (452)
Figure 11-81	538 (452)
Figure 11-82	539 (452)
Figure 11-83	540 (452)
Figure 11-84	541 (452)
Figure 11-85	542 (453)
Figure 11-86	543 (453)
Figure 11-87	544 (453)
Figure 11-88	544 (453)
Figure 11-89	545 (453)
Figure 11-90	545 (454)
Figure 11-91	546 (454)
Figure 11-92	547 (454)
Figure 11-93	548 (457)
Figure 11-94	549 (454)
Figure 11-95	550 (454)
Figure 11-96	551 (454)
Figure 11-97	552 (454)
Figure 11-98	552 (454)
Figure 11-99	553 (454)
Figure 11-100	554 (454)
Figure 11-101	555 (455)

Figure 11-102	556 (455)	Figure 11-104	558 (456)
Figure 11-103	557 (455)	Figure 11-105	559 (456)

Tables

Table 11-1	560 (441)	Table 11-14	571 (448)
Table 11-2	561 (442)	Table 11-15	571 (450)
Table 11-3	562 (443)	Table 11-16	572 (450)
Table 11-4	563 (443)	Table 11-17	574 (453)
Table 11-5	564 (444)	Table 11-17a	575 (453)
Table 11-6	565 (444)	Table 11-18	576 (454)
Table 11-7	566 (444)	Table 11-19	577 (454)
Table 11-8	567 (444)	Table 11-20	577 (454)
Table 11-9	567 (445)	Table 11-21	578 (455)
Table 11-10	568 (445)	Table 11-22	578 (455)
Table 11-11	569 (446)	Table 11-23	579 (456)
Table 11-12	570 (446)	Table 11-24	579 (456)
Table 11-13	570 (446)		

*Numbers in parentheses indicate in-text page reference.

INTRODUCTION

Spectra-structure correlations exist for IR, Raman and nuclear magnetic resonance (NMR) data. Spectra-structure correlations have been developed for the carbonyl stretching frequencies and the carbon-13 NMR chemical shift data for over 600 compounds (1), and this work has aided us in the solution of many complex chemical problems. Development of other spectra-structure cross correlations between IR, Raman and NMR data would be most helpful in the solution of many complex problems arising in both academia and industry.

4-x-ANILINES

Table 11.1 lists NMR ^{13}C chemical shift data for anilines in CCl_3H solution (2,3). Figure 11.1 is a plot of the δC-1 chemical shift for 4-x-anilines in $CHCl_3$ solution vs Hammett σ_p values for the 4-x atom or group, and Fig. 11.2 is a plot of the NMR δ ^{13}C-1 chemical shift data for 4-x-anilines in $CHCl_3$ solution vs Taft σ_{R° values for the 4-x atom or group. Both plots show a pseudolinear relationship (2).

Figure 11.3 is a plot of ν asym. NH_2 frequencies for 4-x-anilines in the vapor phase (4) vs the δ ^{13}C-1 chemical shift data for 4-x-anilines in $CHCl_3$ solution (2,3).

Compound 24 (4-aminoacetophenone) does not correlate well with the other aniline derivatives, and this most likely results from the intermolecular hydrogen bond formed between the C=O group and the CCl_3H proton (C=O$\cdots$$HCCl_3$). Otherwise, this IR–NMR spectra-structure correlation shows a pseudolinear relationship.

Figure 11.4 is a plot of ν asym. NH_2 frequencies for 3-x- and 4-x-anilines in hexane solution vs the δ ^{13}C-1 chemical shift data of 3-x- and 4-x-aniline in $CHCl_3$ solution (2,3). This IR–NMR spectra-structure correlation shows a pseudolinear relationship.

Figure 11.5 is a plot of ν asym. NH_2 frequencies for 3-x- and 4-x-anilines in CCl_4 solution vs the δ ^{13}C-1 chemical shift data for 3-x- and 4-x-anilines in $CHCl_3$ solution. Compounds 17 (4-amino-acetophenone) and 18 (4-nitroaniline) do not fit as well to this pseudolinear correlation between IR and NMR data, and this may be due to the fact that both the oxygen atoms of the C=O and NO_2 groups are intermolecularly hydrogen bonded in CCl_3H solution [C=O$\cdots$$HCCl_3$ and NO_2 ($\cdots$$HCCl_3$)$_2$] (2,3).

Figure 11.6 is a plot of the ν asym. NH_2 frequencies for 3-x- and 4-x-anilines in $CHCl_3$ solution vs δ ^{13}C-1 chemical shift data for 3-x- and 4-x-anilines in $CHCl_3$ solution. Both plots show a pseudolinear relationship between IR and NMR data (2,3).

Figure 11.7 is a plot of the ν sym. NH_2 frequencies for 4-x-anilines in the vapor phase vs δ ^{13}C-1 chemical shift data for 3-x- and 4-x-anilines in $CHCl_3$ solution. This plot shows a pseudolinear relationship between IR and NMR data.

Figure 11.8 is a plot of ν sym. NH_2 frequencies for 3-x- and 4-x-anilines in hexane solution vs the δ ^{13}C-1 chemical shift data for 3-x- and 4-x-anilines in $CHCl_3$ solution. This plot shows a pseudolinear relationship between IR and NMR data (2,3).

Figure 11.9 is a plot of the ν sym. NH_2 frequencies for 3-x- and 4-x-anilines in CCl_4 solution vs the δ ^{13}C-1 chemical shift data for 3-x- and 4-x-anilines in $CHCl_3$ solution. This plot shows a pseudolinear relationship between IR and NMR data (2,3).

Figure 11.10 is a plot of the νsym. NH_2 frequencies for 3-x- and 4-x-anilines in $CHCl_3$ solution vs the δ ^{13}C-1 chemical shift data for 3-x- and 4-x-anilines in $CHCl_3$ solution. Both plots show a pseudolinear relationship between IR and NMR data (2,3).

Figure 11.11 is a plot of the frequency difference between ν asym. NH_2 and ν sym. NH_2 for 3-x- and 4-x-anilines in hexane solution vs δ ^{13}C-1 chemical shift data for 3-x- and 4-x-anilines in $CHCl_3$ solution. This plot shows a pseudolinear relationship between IR and NMR data (2,3).

Figure 11.12 is a plot of the frequency difference between ν asym. NH_2 and ν sym. NH_2 for 3-x- and 4-x-anilines is CCl_4 solution vs the δ ^{13}C-1 chemical shift data for 3-x- and 4-x-anilines in $CHCl_3$ solution. This plot shows a pseudolinear relationship between IR and NMR data (2,3).

Figure 11.13 is a plot of the frequency difference between ν asym. NH_2 and ν sym. NH_2 for 3-x- and 4-x-anilines in $CHCl_3$ solution vs the δ ^{13}C-1 chemical shift data of 3-x- and 4-x-anilines in $CHCl_3$ solution (2,3).

Figure 11.14 is a plot of the absorbance ratio A(ν asym. NH_2)/A(ν sym. NH_2) for 3-x- and 4-x-anilines in CCl_4 solution vs δ ^{13}C-1 chemical shift data for 3-x- and 4-x-anilines in $CHCl_3$ solution (2,3). This plot shows that a pseudolinear relation exists between IR and NMR data.

Figure 11.15 is a plot of δ ^{13}C-1 chemical shift data for 4-x-anilines in $CHCl_3$ solution vs Tafts (σ_{R° values for the 4-x atom or group. This plot shows that there is a pseudolinear relationship between δ ^{13}C-1 and the resonance parameter of the 4-x group.

3-X AND 4-X BENZOIC ACIDS

Table 11.2 lists IR and NMR data for 3-x- and 4-x-benzoic acids (1,3,4). Figure 11.16 is a plot of δ ^{13}C-1 for 3-x- and 4-x-benzoic acids vs Hammett σ values for the x group. The Hammett σ values are from Kalinowski *et al.* (5). This plot shows that δ ^{13}C-1 generally increases as the σ_m and σ_p values increase (4).

Figure 11.17 show plots of ν C=O vs Hammett σ values for 3-x- and 4-x-benzoic acids. The solid circles are for IR data in the vapor phase. The open circles are for IR dilute CCl_4 solution data for unassociated benzoic acids (4). Both sets of data are for unassociated benzoic acids (3-x- and 4-x-$C_6H_4CO_2H$). Both plots increase in frequency as the value of σ_p and σ_m values increases.

Figure 11.18 show plots of ν C=O vs δ ^{13}C-1 for 3-x- and 4-x-benzoic acids. The NMR data are for $CDCl_3$ solutions. The open circles are for IR CCl_4 solution data, and the closed circles include vapor-phase IR data (4). This cross correlation shows that in general ν C=O increases in frequency as δ ^{13}C-1 increases in frequency.

4-X-ACETANILIDES

Table 11.3 lists NMR data in $CHCl_3$ solution and IR data in CCl_4 and $CHCl_3$ solution for 4-x-acetanilides (3,6). Figure 11.19 is a plot of δ ^{13}C-1 vs Hammett σ_p for 4-x-acetanilides. This plot shows a pseudolinear relationship between Hammett σ_p values and δ ^{13}C-1 for 4-x-acetanilides.

Figure 11.20 is a plot of ν C=O vs Hammett σ_p for 4-x-acetanilides (4). The solid circles represent IR CCl_4 solution data and the solid squares represent IR $CHCl_3$ solution data. The IR solution data are for dilute solutions of the 4-x-acetanilides where the acetanilide molecules are not intermolecularly hydrogen bonded via C=O···H–N. In this case ν C=O occurs at lower frequency in $CHCl_3$ solution as the result of C=O···$HCCl_3$ intermolecular hydrogen bonding plus the difference in the refraction field between $CHCl_3$ and CCl_4.

4-X-BENZALDEHYDES

Table 11.4 lists IR and NMR data for 4-x-benzaldehydes in $CHCl_3$ and/or CCl_4 solution (3,4,7). Figure 11.21 is a plot of δ ^{13}C-1 vs Taft $\sigma_{R°}$ for 4-x-benzaldehydes, and there is a pseudolinear correlation between these two parameters. This correlation shows that as $\sigma_{R°}$ increases in value the δ^{13} C-1 value increases in frequency. The chemical significance of increasingly more negative values is that there is an increasing amount of π electron contribution to the C-1 atom. Thus, with an increasing contribution of π electrons to the C-1 atom from the atom or group joined to the C-4 atom there is more π overlap with the π system of the C=O group. Consequently, the ν C=O mode should decrease in frequency as the $\sigma_{R°}$ value decreases. Figure 11.22 is a plot of ν C=O for 4-x-benzaldehydes vs Taft $\sigma_{R°}$. The solid circles represent ν C=O frequencies in CCl_4 solution and the solid triangles represent ν C=O frequencies in $CHCl_3$ solution. In general, the pseudolinear relationships noted for the two sets of solution data support the preceding hypothesis. The ν C=O frequencies are lower in $CHCl_3$ solution than they are in CCl_4 solution, and this is due in part to intermolecular hydrogen bonding (e.g., C=O···$HCCl_3$) together with an increased field effect in going from solution in CCl_4 to solution in $CHCl_3$.

Figure 11.23 is a plot of ν C=O for 4-x-benzaldehydes vs Hammett σ_p. The solid circles represent ν C=O frequencies in CCl_4 solution and the solid triangles represent ν C=O frequencies in $CHCl_3$ solution. The σ_p values include both resonance and inductive effects. These plots also exhibit pseudolinear relationships, which indicates that there must be some contribution from the σ electrons to the shift of ν C=O with change in the 4-x atom or group.

Figure 11.24 gives plots of ν C=O vs δ ^{13}C-1 for 4-x-benzaldehydes in CCl_4 solution and in $CHCl_3$ solution (4). The filled-in circles represent ν C=O frequencies in CCl_4 solution and the filled-in triangles represent ν C=O frequencies in $CHCl_3$ solution. The δ ^{13}C-1 data are for $CDCl_3$ solutions. Both plots show a pseudolinear relationship.

4-X-ACETOPHENONES

Table 11.5 lists IR and NMR data for 4-x-acetophenones. Figure 11.25 is a plot of δ ^{13}C-1 chemical shifts data in $CDCl_3$ solution vs Taft σ_{R° for 4-x-acetophenones (4). This plot shows a pseudolinear relationship. Figure 11.26 is a plot of ν C=O frequencies for 4-x-acetophenones in $CHCl_3$ solution vs Taft σ_{R°, and this plot shows a pseudolinear relationship. Figure 11.27 is a plot of ν C=O frequencies for 4-x-acetophenones in $CHCl_3$ solution vs δ ^{13}C-1 chemical shifts for 4-x-acetophenones in $CDCl_3$ solution. This plot shows that, in general, ν C=O increases in frequency as δ ^{13}C-1 increases in frequency.

4-X AND 4,4′-X,X-BENZOPHENONES

Table 11.6 lists vapor-phase IR and NMR $CHCl_3$ solution-phase data for 4-x- and 4,4′-x,x-benzophenones (4,8). Figure 11.28 is a plot of δ ^{13}C-1 vs Taft σ_{R° for the 4-x and 4,x,x atoms or groups for 4-x- and 4,4′-x,x-benzophenones in $CHCl_3$ solution (4). The first group in parentheses is plotted vs its σ_{R° value. This relationship between δ ^{13}C-1 and σ_{R° appears to be pseudolinear (4). Figure 11.29 is a plot of the ν C=O frequencies vs δ ^{13}C-1 chemical shift frequencies for 4-x-benzophenone and vs δ ^{13}C-1 for 4,4′-x,x-benzophenones. Figure 11.30 is a plot of ν C=O frequencies vs Hammett σ_p for 4-x-benzophenones and the sum of σ_p, σ_p for 4,4′-x,x-benzophenones. With the exception of the 4-methoxy analog, the 4-x and 4,4′-x,x-benzophenones yield separate relationships where ν C=O increases in frequency as the sum of the σ_p values increases.

ALKYL 3-X AND 4-X-BENZOATES

Table 11.7 lists IR and NMR data for methyl and/or ethyl 3-x or 4-x-benzoates (4,8). Figure 11.31 shows plots of δ ^{13}C-1 vs Hammett σ values for methyl and ethyl 3-x- and/or 4-x-benzoates (4). These plots show that, in general, δ 13 C-1 increases in frequency as the Hammett σ value increases, and that the δ ^{13}C-1 chemical shift for the ethyl analogs generally occurs at higher frequency than for the methyl analogs.

ν C=O VS δ ^{13}C=O

Table 11.8 is a summary of some of the IR and NMR data for C=O containing compounds (4).

Figure 11.32 shows plots of the range of ν C=O frequencies for the compounds studied in different physical phases vs the range of δ ^{13}C=O for the compounds studied in $CDCl_3$ solutions. The IR data for plots 1,4,6, and 7 have been recorded in CCl_4 solution, and plots 1,4,6, and 7 are for 4-x-acetophenones, 4-x-benzaldehydes, 4-x-acetonilides, and 3-x and 4-x-benzoic acids, respectively. Infrared plots for 3 and 5 have been recorded in $CHCl_3$ solution, and these plots are for 4-x-benzaldehydes and 4-x-acetanilides, respectively. The IR data for 2,8,9, and 10 have been recorded in the vapor phase. Plots 2, 8, 9, and 10 are for 4-x and 4,4′-x,x-benzophenones, ethyl 3-x and 4-x-benzoates, methyl 3-x and 4-x-benzoates, and 3-x and 4-x-benzoic acids, respectively. These data show that ν C=O is affected by change of phase.

ACETONE IN $CHCl_3/CCl_4$ SOLUTIONS

Table 11.9 lists IR data for ν C=O and δ ^{13}C=O for acetone in mol % $CHCl_3/CCl_4$ solutions. Figure 11.33 is a plot of δ ^{13}C=O in ppm for acetone vs mol % $CHCl_3/CCl_4$, and δ ^{13}C=O increases in frequency as the mol % $CHCl_3/CCl_4$ is increased.

Figure 11.34 shows a plot of ν C=O vs δ ^{13}C=O for acetone, and this plot is essentially a linear relationship. In Volume I the deviation from linearity at low mol % $CHCl_3/CCl_4$ was attributed to intermolecular hydrogen bonding between solute and solvent (viz. $(CH_3)_2$ C=O$\cdots$$HCCl_3$). The steady decrease in the ν C=O frequency and the steady decrease in δ ^{13}C=O with increase in the mol % $CHCl_3CCl_4$ is attributed to the field effect of the solvent system (9).

N,N′-DIMETHYLACETAMIDE IN $CHCl_3/CCl_4$ SOLUTIONS

Table 11.10 lists IR and NMR data for N,N′-dimethylacetamide in $CHCl_3/CCl_4$ solutions (10). Figure 11.35 is a plot of ν C=O for N,N′-dimethylacetamide vs mol % $CHCl_3/CCl_4$ (10). The plot shows that ν C=O decreases in frequency by 26.27 cm^{-1} in going from 1650.50 cm^{-1} in CCl_4 solution to 1634.23 cm^{-1} in $CHCl_3$ solution. At 5.68 mol % $CHCl_3/CCl_4$ ν C=O has decreased 9.20 cm^{-1} or 35% of the entire ν C=O decrease in going from solution in CCl_4 to solution in $CHCl_3$. The principal cause for this 35% ν C=O frequency decrease is attributed to intermolecular hydrogen bonding (viz. C=O$\cdots$$HCCl_3$). Between 5.68 and 41.93 mol % $CHCl_3/CCl_4$ there is a linear relationship in the ν C=O frequency decrease, and this is attributed to the increase of the field effect of the solvent system. The second break suggests that a second intermolecular hydrogen bond is formed,

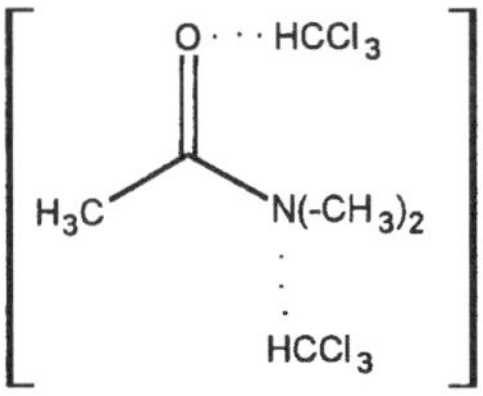

and the continued ν C=O frequency decrease is attributed to a further continued increase of the solvent field effect.

Figure 11.35a is a plot of δ ^{13}C=O for dimethylacetamide vs mol % $CHCl_3/CCl_4$. The δ ^{13}C=O frequency increases as the mol % $CHCl_3/CCl_4$ is increased.

The data in Table 11.10 show that as the mol % $CHCl_3/CCl_4$ is increased from 0 to 100 the δ ^{13}C=O increases in frequency from 168.4 to 170.8 ppm while ν C=O decreases in frequency from 1660.5 to 1634.2 cm^{-1}.

MALEIC ANHYDRIDE (MA) IN $CHCl_3/CCl_4$ SOLUTIONS

Table 11.11 lists IR and NMR data for maleic anhydride (MA) in $CHCl_3/CCl_4$ solutions (11,12). The out-of-phase ν $(C{=}O)_2$ mode is in FR with a combination tone (560 cm^{-1}, A and B_1 fundamental at 1235 cm^{-1}, and an A_1 fundamental yielding at 1795 cm^{-1}, B_1 combination tone). The out-of-phase ν $(C{=}O)_2$ mode for maleic anhydride listed in Table 11.11 has been corrected for FR. The δ $^{13}C{=}O$ chemical shift increases from 163.2 ppm in CCl_4 solution to 164.1 ppm in $CHCl_3$ solution while ν out-of-phase $(C{=}O)_2$ corrected for FR decreases in frequency from 1787.1 cm^{-1} in CCl_4 solution to 1785.4 cm^{-1} in $CHCl_3$. The ν in-phase $(C{=}O)_2$ mode decreases 0.06 cm^{-1} in going from solution in CCl_4 to solution in $CHCl_3$.

Figure 11.35b is a plot of δ $^{13}C{=}O$ for maleic anhydride vs mol % $CHCl_3/CCl_4$. This plot shows two linear segments with a break point near 37 mol % $CHCl_3/CCl_4$. The δ $^{13}C{=}O$ chemical shift increases in frequency as the mol % $CHCl_3/CCl_4$ is increased. These data suggest that at <37 mol % $CHCl_3/CCl_4$, the $CHCl_3/CCl_4$ complex with (MA) is most likely forming $C{=}O\cdots(HCCl_3)_x(CCl_4)_x$ type complexes and > ~37 mol % $HCCl_3/CCl_4$ solution $CHCl_3/CCl_4$ is forming $(C{=}O)_2\cdots(HCCl_3)_x(CCl_4)$ type complexes. Complexes with the C=C π system are also likely (see in what follows).

Figure 11.35c is a plot of δ $^{13}C{=}O$ vs δ $^{13}C{=}C$ for maleic anhydrides. This plot shows a linear relationship between these two parameters, and both chemical shifts increase in frequency as the mol % $CHCl_3/CCl_4$ increases from 0 to 100.

3-X AND 4-X-BENZONITRILES

Table 11.12 lists IR and NMR data for 3-x- and 4-x-benzonitriles (4). Figure 11.36 is a plot of δ ^{13}C-1 vs Taft σ_{R° values for 3-x- and 4-x-benzonitriles, and Fig. 11.37 is a plot of δ ^{13}C-1 vs Hammett σ values for x for 3-x and 4-x-benzonitriles. Both plots show that δ ^{13}C-1 increases in frequency as σ_p, σ_m and σ_{R° increase in value. However, the plot of δ ^{13}C-1 vs σ_{R° has a better pseudolinear relationship than the plot of δ ^{13}C-1 vs σ_p and σ_m.

Figure 11.38 is a plot of δ ^{13}CN vs Hammett σ values for 3-x and 4-x-benzonitriles. The δ ^{13}CN values are in the range 116.3 ppm and 120.6 ppm. The literature assignments for δ ^{13}CN and δ ^{13}C-1 for 4-cyanoacetophenone are 118.0 ppm and 116.3 ppm, respectively (3). However, Fig. 11.38 suggests that the δ ^{13}CN chemical shift should be 116.3 ppm rather than 118.0 ppm (3). The δ ^{13}CN decreases in frequency as the σ_p and σ_m values increase. It has been shown that ν CN frequencies and intensities correlate with Hammett σ_p and σ_m values (13–16). Electron-donating groups such as OCH_3 and NH_2 increase the IR intensity and decrease the ν CN frequency, while electron-withdrawing groups such as NO_2 and $CH_3{-}C{=}O$ have the opposite effect (increase the ν CN frequency and decrease its IR band intensity). The ν CN mode for benzonitriles occurs in the range 2220–2241 cm^{-1}.

ORGANONITRILES

Table 11.13 lists IR, Raman and NMR data for some organonitriles (17). Figure 11.39 is a plot of ν CN vs δ ^{13}CN for some organonitriles. In the series CH_3CN through $(CH_3)_3$ CCN, ν CN decreases in frequency while the δ ^{13}CN frequency increases in essentially a linear manner with increased branching on the α-carbon atom. Substitution of an α-Cl atom raises the ν CN frequency and lowers the δ ^{13}CN frequency. In addition, conjugation lowers both ν CN frequencies.

Figure 11.40 is a plot of δ ^{13}CN vs the number of protons on the α-carbon atom (17). This plot shows that δ ^{13}CN increases in frequency with increased branching on the α-carbon atom. The σ electron donation to the CN group also increases with increased branching on the α-carbon atom. Therefore, δ ^{13}CN increases in frequency with increased electron donation to the nitrile group, and ν CN decreases in frequency with increased electron donation to the nitrile group.

ALKYL ISONITRILES

The ν NC frequencies for methyl isonitrile (2183 cm^{-1}), ethyl isonitrile (2160 cm^{-1}), isopropyl isonitrile (2140 cm^{-1}), and tert-butyl isonitrile (2134 cm^{-1}) decrease in frequency in the order of increased branching on the alkyl α-carbon atom (18). The δ ^{15}N chemical shift for methyl isonitrile (−219.6 ppm), ethyl isonitrile (−205.1 ppm), isopropyl isonitrile (−193.4 ppm), and tert-butyl isonitrile (−184.9 ppm) increases in frequency with increased branching in the alkyl α-carbon atom. The δ ^{15}N frequencies reported for these isonitriles are based upon nitromethane as the δ ^{15}N reference point (19).

Figure 11.41 is a plot of δ ^{15}N vs Taft σ^* for alkyl isonitriles (18). With increased electron contribution to the isonitrile group, the δ ^{15}N chemical shift increases in frequency. Figure 11.42 is a plot of ν NC vs δ ^{15}N for alkyl isonitriles. This plot shows that as the ν NC vibration decreases in frequency the δ ^{15}N increases in frequency. With increased electron contribution to the isonitrile group the ν NC vibration decreases in frequency while the δ^{15} N chemical shift increases in frequency.

NITROALKANES

In the vapor phase both ν asym. NO_2 and ν sym. NO_2 decrease in frequency with increased branching on the alkyl α-carbon atom. For example, the ν asym. NO_2 and ν sym. NO_2 frequencies for the alkyl group are CH_3 (1582 cm^{-1}, 1397 cm^{-1}), C_2H_5 (1575 cm^{-1}, 1380 cm^{-1}), iso-C_3H_7 (1568 cm^{-1}, 1360 cm^{-1}), and tert-butyl (1555 cm^{-1}, 1349 cm^{-1}) (8). On the other hand, the δ ^{13}C (20) and δ ^{15}N (21) chemical shifts for the alkyl α-carbon atom and the NO_2 nitrogen atom both increase in frequency with increased branching on the alkyl α-carbon atom. For example δ ^{13}C and δ ^{15}N are, respectively, CH_3(63.0 ppm, −0.77 ppm), C_2H_5(70.9 ppm, 9.37 ppm), iso-C_3H_7(79.2 ppm, 19.40 ppm), and tert-butyl (84.3 ppm, 25.95 ppm). [The δ ^{13}C values in $CDCl_3$ solution referred to TMS (20), and the δ ^{15}N values in neat nitromethane; samples in 0.3 M in acetone (21)].

Figure 11.43 shows plots of ν asym. NO_2 and ν sym. NO_2 vs the number of protons on the α-carbon atom of the nitroalkanes. Figure 11.44 shows plots of ν asym. NO_2 and ν sym. NO_2 vs δ ^{13}C for the nitroalkanes, and Fig. 11.45 shows plots of ν asym. NO_2 and ν sym. NO_2 vs δ ^{15}N for the nitroalkanes (18).

These plots show that with increased electron donation of the alkyl group to the NO_2 group, both ν NO_2 vibrations decrease in frequency while the δ ^{13}C and δ ^{15}N chemical shifts increase in frequency.

Figures 11.46, 11.47, and 11.48 are plots of δ ^{13}C vs Taft σ^*, δ ^{15}N vs Taft σ^*, and δ ^{13}C vs δ ^{15}N for the nitroalkanes (18), respectively. Both δ ^{13}C and δ ^{15}N vs σ^* show a linear relationship for the methyl through isopropyl analogs. However, the plot of δ ^{13}C vs δ ^{15}N shows a linear relationship for methyl through the tert-butyl analogs.

ALKYL ISOCYANATES

The ν asym. N=C=O vibrations for alkyl isocyanates decrease in frequency with an increase in branching on the alkyl α-carbon atom. The ν asym. N=C=O frequencies for the alkyl isocyanates are CH_3 (2288 cm^{-1}), C_2H_5 (2280 cm^{-1}), $(CH_3)_2CH$ (2270 cm^{-1}), and $(CH_3)_3C$ (2252 cm^{-1}) (22). The δ ^{15}N (neat liquid, referred to neat nitromethane) values are CH_3 (−365.4 ppm), C_2H_5 (−348.6 ppm), $(CH_3)_2CH$ (−335.5 ppm), and $(CH_3)_3C$ (−326.0 ppm) (19).

Figure 11.49 is a plot of ν asym. N=C=O for alkyl isocyanates vs the δ ^{15}N for the N=C=O group. The ν asym. N=C=O vibrations decrease in frequency as the δ ^{15}N chemical shifts increase in frequency (18).

Figure 11.50 shows a plot of δ ^{15}N vs Taft's σ^* for both alkyl isocyanates and alkyl isothiocyanates. The δ ^{15}N chemical shifts decrease in frequency as the Taft σ^* values increase in value. [Taft σ^* values are CH_3(O), C_2H_5(−0.10), $(CH_3)_2CH$(−0.19), and $(CH_3)_3C$ −0.30] (23). The CH_3 and C_2H_5 analogs of alkyl isothiocyanate exhibit δ ^{15}N for alkyl isothiocyanates at higher frequency by 75.6 ppm and 71.6 ppm for the CH_3 and C_2H_5 analogs than for the corresponding methyl and ethyl analogs of alkyl isocyanates (18).

Figure 11.51 shows plots of δ ^{13}C for the N=C=O group of alkyl isocyanates or diisocyanates vs the number of protons on the alkyl α-carbon atom (24). These plots show that, in general, the chemical shift δ ^{13}C for the isocyanate group increases as the number of protons on the alkyl α-carbon atom decreases. Therefore, the δ ^{13}C for the isocyanate group increases in frequency as the electron donation of the alkyl group to the isocyanate group is increased.

Figure 11.52 shows plots of ν asym. N=C=O vs δ ^{13}C for the N=C=O group for alkyl isocyanates or diisocyanates. These data show that as the ν asym. N=C=O vibration decreases in frequency the δ ^{13}C chemical shift increases in frequency. Thus, as the electron contribution of the alkyl group to the isocyanate group is increased, ν asym. N=C=O decreases in frequency while δ ^{13}C increases in frequency (24).

ALKYLAMINES

Table 11.14 compares δ ^{15}N data for primary, secondary and tertiary amines (25). Figure 11.53 is a plot of δ ^{15}N vs ν C–N for four primary alkylamines. This plot shows that both ν C–N and δ ^{15}N increase in frequency with increased branching on the alkyl α-carbon atom (8,18,25).

Figure 11.54 is a plot of δ ^{15}N vs ωNH for dialkylamines (8,18,25). This plot shows that as δ ^{15}N decreases in frequency the ωNH vibration for dialkylamines increases in frequency.

The ωNH vibration decreases in frequency as more electrons are donated to the NH group by the two alkyl groups while the δ ^{15}N chemical shift increases in frequency.

Figures 11.55 and 11.56 are plots of ν C—N and δ ^{15}N for primary alkylamines vs Taft's σ^* values, respectively (18). Both plots increase in frequency as the σ^* values decrease. Thus, the ν C—N and δ ^{15}N increase in frequency as the electron contribution of the alkyl group to the nitrogen atom increases.

Figure 11.57 shows plots of δ ^{15}N vs the number of protons on the alkyl group(s) α-carbon atom(s) for alkylamines (18). These plots show a series of relationships. With methyl and ethyl groups, the correlations are more nearly linear. Deviation from linearity is apparent for the isopropyl and tert-butyl groups. These deviations suggest that steric factors are also important in determining the δ ^{15}N values, because a steric factor would alter the C—N—C bond angles (18).

A SUMMARY OF THE CORRELATIONS FOR THE NITROGEN COMPOUNDS

1. The cis and trans ν N=O frequencies for alkyl nitrites decrease with increased branching on the alkyl α-carbon atom. A plot of ν N=O cis ν N=O trans is linear for the alkyl nitrites with the exception of tert-butyl nitrite. This exception may be due to a steric factor of the $C(CH_3)_3$ group, which causes a change in the R—O—N=O bond angle.
2. Both ν sym. NO_2 and ν asym. NO_2 decrease in frequency with increased branching on the α-C—NO_2 carbon atom. A plot of ν sym. NO_2 vs ν asym. NO_2 is linear for nitroalkanes, with the exception of 1,1-dimethylnitroethane. This exception is attributed to a steric factor of the $C(CH_3)_3$ group.
3. Both ν sym. NO_2 and ν asym. NO_2 correlate with the number of protons on the α-C—NO_2 carbon atom, with δ ^{13}C of α-C—NO_2, with δ ^{15}N of C—NO_2, and with Taft σ^* values. Thus, ν sym. NO_2 and ν asym. NO_2 decrease in frequency with increased branching on the α-C—NO_2 carbon atom, and this frequency decrease is in the order of electron release to the alkyl group to the NO_2 group.
4. A nearly linear correlation exists between δ ^{13}C for the α-C—NO_2 carbon atom and both the ν sym. NO_2 and ν asym. NO_2 vibrations for nitroalkanes.
5. A nearly linear correlation exists between δ ^{15}N and ν sym. NO_2 and ν asym. NO_2 for nitroalkanes.
6. A linear correlation exists between the chemical shifts δ ^{15}N and δ ^{13}C for the α-C—NO_2 carbon atom for nitroalkanes.
7. The δ ^{15}N chemical shift increases in frequency while the ν asym. N=C=O decreases in frequency with increased branching on the alkyl group α-C—N carbon atom for alkyl isocyanates.
8. The ν asym. N=C=O mode decreases in frequency and δ ^{15}N increases in frequency with increasing electron release of the alkyl group in alkyl isocyanates. The chemical shift δ ^{15} N for alkyl isothiocyanates occurs at higher frequency than does δ ^{15}N for the corresponding alkyl isocyanate analogs.
9. The ν NC mode decreases in frequency and the δ ^{15}N chemical shift increases in frequency with increased branching on the α-C—NC carbon atom for alkyl isonitriles R—NC. With the exception of the tert-butyl analog a linear correlation exists between δ

^{15}N and ν NC for alkyl isonitriles. Comparable linear correlations exist between the electron release of the alkyl group and ν NC, and between electron release of the alkyl group and δ ^{15}N for alkyl isonitriles.

10. The chemical shift δ ^{15}N increases in frequency with increased branching on the α-C—N atom(s), progressing in and within the series, alkylamines, dialkylamines, and trialkyl amines.
11. A linear correlation exists between δ ^{15}N for alkylamines and the corresponding alkyl analogs of dialkylamines.
12. A correlation exists between ν C—N and δ ^{15}N for alkylamines. Both δ ^{15}N and ν C—N increase in frequency with increased electron release of the alkyl group to the N atom.
13. A correlation between ω NH and δ ^{15}N exists for dialkylamines. In this case ω NH decreases in frequency as δ ^{15}N increases in frequency.

ORGANOPHOSPHORUS COMPOUNDS

Table 11.15 lists NMR data for organophosphorus and organonitrogen compounds. Figure 11.58 is a plot of δ ^{31}P for trialkylphosphines vs the sum of the protons on the α-C—P carbon atoms. Figure 11.59 is a plot of δ ^{31}P for trialkylphosphines vs the sum of Taft σ^* values for the three alkyl groups. Both plots exhibit a linear relationship, and the δ ^{31}P chemical shift increases in frequency with increased branching on the α-carbon atom and with increased electron contribution to the phosphorus atom (26).

Figure 11.60 shows plots of δ ^{31}P vs δ ^{15}N for R—PH_2 vs R—NH_2, $(R—)_2PH$ vs (R—)NH, and $(R—)_3P$ vs $(R—)_3N$. These plots show that both δ ^{31}P and δ ^{15}N increase in frequency with increased alkyl substitution on the P or N atoms and with increased electron contribution from the number of alkyl groups to the P and N atoms (26).

Figure 11.61 shows plots of the number of protons on the α-C—P atom(s) vs δ ^{31}P for PH_3, RPH_2, and $(R—)_2PH$ (26). The plots show that δ ^{31}P increases in frequency in the order PH_3, R—PH_2, and $(R—)_2PH$ and in the order of increased branching on the α-C—P atom(s).

Table 11.16 lists IR and NMR data for organophosphorus compounds (26). Figure 11.62 is a plot of the sum of Taft σ^* values for the alkyl groups of trialkylphosphine oxides vs δ ^{31}P. This plot shows that δ ^{31}P increases in frequency with increased electron contribution to the P atom. Figure 11.63 is a plot of the sum of the number of protons on the α-C—P carbon atom(s) vs δ ^{31}P for trialkylphosphine oxides. The δ ^{31}P chemical shift increases in frequency with increased branching on the α-C—P carbon atom (26).

Figure 11.64 shows plots of ν P=O rotational conformers vs the sum of the protons on the α-C—O—P carbon atoms for trialkyl phosphates (26). Both plots show that ν P=O decreases in frequency as the branching is increased on the α-C—O—P carbon atom. Figure 11.65 shows plots of ν P=O rotational conformers vs δ ^{31}P for trialkyl phosphates. Both ν P=O and ν ^{31}P decrease in frequency with increased branching on the α-C—O—P carbon atom. Trialkyl phosphates exist as rotational conformers and exhibit ν P=O rotational conformers I and II at different frequencies; therefore, the P=O frequencies are affected by the molecular geometric configuration of the molecule because the electronic contribution of the O-alkyl groups to the P=O group is independent of the rotational molecular configuration of the trialkyl phosphate molecules.

Figures 11.66, 11.67, and 11.68 show plots of δ ^{31}P vs the sum of the number of protons on the α-C—O—P carbon atoms for trialkyl phosphates, O,O,O-trialkyl phosphorothioates, and

O,O,O-trialkyl phosphoroselenates (26). In all cases $\delta\ ^{31}P$ decreases in frequency with increased branching on the α-C—O—P carbon atom. In Fig. 11.66 the plot is essentially linear, but the plots in Figs. 11.67 and 11.68 show that the tert-butyl analogs deviate considerably from linearity. The electronegativity of oxygen, sulfur, and selenium decreases in the order 3.5; 2.5; and 2.4, respectively (32). The tetrahedral covalent radii of oxygen, sulfur, and selenium increases in the order 0.66; 1.04, and 1.14 Å, respectively (32). The $\delta\ ^{31}P$ frequency difference between the isopropyl and tert-butyl analogs of the P=O, P=S and P=Se analogs are 7.3 ppm, 21.3 ppm, and 36.6 ppm, respectively. The difference in the electronegativity of S and Se is only 0.1, and the difference in the tetrahedral covalent radii between S and Se is 0.1 Å. These differences do affect the $\delta\ ^{31}P$ chemical shifts of the tert-butyl analogs for O,O,O-tri-tert-butyl phosphorothioate and O,O,O-tri-tert-butyl phosphoroselenate. It is suggested that this difference in the $\delta\ ^{31}P$ chemical shifts is caused by a steric effect of the tert-butyl groups. The steric effect most likely changes the PO_3 bond angles in both the P=S and P=Se tri-tert-butyl analogs.

Figure 11.69 shows plots of $\delta\ ^{31}P$ vs the sum of the number of protons on the α-C—P carbon atoms and the α-C—O—P carbon atoms for $RP(=O)(OC_2H_5)_2$ and $(R)_2P{=}O(OR)$. These plots show that $\delta\ ^{31}P$ decreases in frequency with increased branching on the α-C—O—P carbon atoms, and increases in frequency with increased branching in the α C—P carbon atom and by substitution of $(CH3)_2P$ for $(CH_3)_1P$.

Figure 11.70 shows plots of $\delta\ ^{31}P$ for $RP(=O)(OC_2H_5)_2$ vs $\delta\ ^{13}C{=}O$ for $RC{=}(O)(OCH_3)$ and $\delta\ ^{31}P$ for $CH_3P(=O)(OR)_2$ vs $\delta\ ^{13}C{=}O$ for $CH_3C(=O)(OR)$. Both $\delta\ ^{13}C{=}O$ and $\delta\ ^{31}P$ decrease in frequency with decreased branching on the α-C—P carbon atom or α-C—C=O carbon atom and with increased branching on the α-C—O—P carbon atom or the O=C—O—C α-carbon atom.

Figure 11.71 shows plots of $\delta\ ^{31}P$ vs the sum of the protons on the α-C—O—P carbon atoms. These plots show that $\delta\ ^{31}P$ for $(RO)_2P(=S)H$ analogs occurs at higher frequency (≈60 ppm) than it does for the corresponding analogs. Moreover, $\delta\ ^{31}P$ decrease in frequency with increased branching in the α-C—O—P carbon atom (26).

The sulfur atom is less electronegative than the oxygen atom (2.5 vs 3.5), and, consequently, the P=O analogs occur at lower frequency than the P=S analogs.

Figure 11.72 is a plot of $\delta\ ^{31}P$ for $(RO)_2P(=O)H$ analogs vs $\delta\ ^{31}P$ for the corresponding $(RO)_2P(=S)H$ alkyl analogs. The two plots are for different $\delta\ ^{31}P$ literature values (27,28). This plot shows that the $\delta\ ^{31}P$ chemical shift values decrease in frequency with increased branching on the α-C—O—P carbon atom (26).

Figure 11.73 shows plots of the number of protons on the α-C—P carbon atoms vs $\delta\ ^{31}P$ for $RP(=O)Cl_2$ and $RP(=S)Cl_2$ (26). The $\delta\ ^{31}P$ chemical shift increases in frequency with increased branching on the α-C—P carbon atom. The $\delta\ ^{31}P$ chemical shifts for the $RP(=S)Cl_2$ analogs are larger than they are for the corresponding $RP(=O)Cl_2$ analogs (26). Changes in the bond angles caused by steric factors of the alkyl group could be the cause of the increased $\Delta\delta\ ^{31}P$ values with increased branching on the α-C—P carbon atom (P=O vs P=S analogs: CH_3, 34.9 ppm; C_2H_5, 42.4 ppm; and iso C_3H_7, 46.6 ppm).

Figure 11.74 is a plot of $\delta\ ^{31}P$ for $RP(=S)Cl_2$ vs $\delta\ ^{31}P$ for $RP(=O)Cl_2$, and the plot shows an essentially linear relationship (26). Both $\delta\ ^{31}P$ chemical shifts increase in frequency with increased branching on the α-C—P carbon atom.

Figure 11.75 is a plot of $\delta\ ^{31}P$ for $RP(=O)F_2$ vs the number of protons on the α-C—P carbon atom. The $\delta\ ^{31}P$ chemical shift increases in frequency with increased branching on the α-C—P carbon atom.

Figure 11.76 is a plot of α ^{31}P for $RP(=S)Br(OC_3H_{7-iso})$ vs the number of protons on the α-C—P carbon atom and the number of protons on the α-C—O—P carbon atom. The δ ^{31}P increases in frequency with increased branching on the α-carbon atoms.

Figure 11.77 shows plots of δ ^{31}P vs the number of chlorine atoms joined to phosphorus for $PCl_{3-x}(OR)_x$ (26). These plots show that δ ^{31}P decreases in frequency with a decrease in the number of atoms joining to the phosphorus atom, and that the ethyl analogs occur at lower frequency than the corresponding methyl analogs (26).

Figure 11.78 shows plots of δ ^{31}P for PX_{3-n} R_n vs the number of protons on the α-C—P carbon atom(s) (26). These plots show that δ ^{31}P decreases in frequency with decreased branching on the α-C—P carbon atom(s), and that δ ^{31}P for the Br analog occurs at lower frequency than the corresponding δ ^{31}P chemical shift for the corresponding Cl analog.

Figure 11.79 is a plot of δ ^{31}P vs the number of protons on the α-C—O—P carbon atoms for $(RO)_2P(=O)Cl_2$ (26). This plot shows that δ ^{31}P decreases in frequency with increased branching on the δ-C—O—P carbon atom.

Figure 11.80 shows plots of the number of Cl atoms joined to the P atom vs P=O frequencies for $P(=O)Cl_3$ through $P(=O)Cl_{3-x}(OR)_x$ where R is CH_3 or C_2H_5 (26). In all cases ν P=O for rotational conformer I occurs at higher frequency than ν P=O for rotational conformer II. The ν P=O frequencies decrease in frequency as the number of Cl atoms joined to P are decreased. The ν P=O frequencies for the CH_3O analogs occur at higher frequency than those exhibited by the C_2H_5O analogs.

Figure 11.81 is a plot of the number of Cl atoms joined to P vs ν P=O for $P(=O)Cl_{3-x}[N(CH_3)_2]_x$. The ν P=O frequency decreases with the decrease in the number of Cl atoms joined to the P atom.

Figure 11.82 is a plot of δ ^{31}P for $P(=S)(OR)_3$ vs δ ^{31}P for the corresponding alkyl analogs of $P(=O)(OR)_3$ (26). The δ ^{31}P chemical shifts decrease in frequency with increased branching on the α-C—O—P carbon atoms. The deviation from linearity for the tert-butyl analogs suggests that this results from a steric factor causing a change in the PO_3 bond angles.

Figure 11.83 is a plot of δ ^{31}P for $P(=Se)(OR)_3$ vs δ ^{31}P for the corresponding alkyl analogs of $P(=O)(OR)_3$ (26). The δ ^{31}P chemical shifts decrease in frequency with increased branching on the α-C—O—P carbon atoms. Deviation from linearity for the tert-butyl analogs supports the analysis that this is the result of a steric factor causing a change in the PO_3 bond angles.

Figure 11.84 shows plots of δ ^{31}P for $RP(=O)Cl_2$ vs δ ^{31}P for $RP(=O)Cl_2$ vs δ ^{31}P for corresponding alkyl analogs of $RP(=O)F_2$ The double plot presents the δ ^{31}P range given in the references for the $RP(=O)Cl_2$ analogs. The δ ^{31}P chemical shift decreases in frequency with decreased branching on the α-C—O—P carbon atom (26).

A SUMMARY OF CORRELATIONS FOR THE PHOSPHORUS COMPOUNDS

1. Increased branching on the α-C—P carbon atom causes δ ^{31}P to occur at increasingly higher frequency in a systematic manner. The δ ^{31}P chemical shift increases in frequency with increased electron release of the alkyl groups.

2. Correlations exist between $\delta\ ^{31}P$ and $\delta\ ^{15}N$ for corresponding alkyl analogs of RPH_2 vs RNH_2, $(R)_2PH$ vs $(R)_2NH$, and $(R)_3P$ vs $(R)_3N$.
3. The ν P=O frequency and $\delta\ ^{31}P$ frequency decrease with increased branching on the α-C—O—P carbon atoms for P(=O)(OR) analogs, and a correlation exists between ν P=O and $\delta\ ^{31}P$.
4. The chemical shift $\delta\ ^{31}P$ decreases in frequency with increased branching on the α-C—O—P carbon atoms for $P(=S)(OR)_3$ and $P(=Se)(OR)_3$.
5. The chemical shift $\delta\ ^{31}P$ increases in the order P=O, P=S, and P=Se for trialkyl esters, with the exception of the tert-butyl analog. It is suggested that the tert-butyl analogs sterically cause the $(RO)_3P$=S and $(RO)_3P$=Se bond angles to change so that $\delta^{31}P$ does not change in a linear manner, as in the case of the lesser branched alkyl groups.
6. Correlations exist between $\delta\ ^{31}P$ for comparable $(RO)_3P$=O, $(RO)_3P$=S and $(RO)_3P$=Se alkyl analogs.
7. The chemical shift value for $\delta\ ^{31}P$ is higher for R—P than for R—O—P.
8. The ν P=O frequency decreases with increased substitution of RO for Cl in the series $P(=O)Cl_{3-x}(OR)_x$. The decrease in ν P=O frequency is in the order of decreasing electronegativity of the sum of the Cl atoms and RO groups.
9. For RPX_2 analogs, $\delta\ ^{31}P$ decreases in frequency in the order Cl, Br, and I, which is in the order of decreasing electronegativity. However, $\delta\ ^{31}P$ for $RP(=O)Cl_2$ occurs at higher frequency than $\delta\ ^{31}P$ for corresponding $RP(=O)F_2$ analogs.
10. Correlations exist between $\delta\ ^{31}P$ for $RP(=O)(OC_2H_5)_2$ vs $\delta\ ^{13}C$=O for $RC(=O)(OCH_3)$ and $CH_3P(=O)(OR)_2$ vs $\delta\ ^{13}C$=O for $CH_3C(=O)(OR)$.

ANISOLES

Goldman *et al.* have reported that ν asym. ϕ-O—C for *p*-substituted anisoles increases in frequency with the electron withdrawing power of the *p*-substituent (30). Brown and Okamoto have reported a linear relationship between ν asym. ϕ-O—C and $\sigma+$ values (33).

Tables 11.17 and 11.17a list NMR and IR data for substituted anisoles, respectively (29). Figure 11.85 is a plot of ν asym. ϕ-O—C vs $\delta\ ^{13}C$-1 for 4-x-anisoles. This plot shows that, in general, both ν asym. ϕ-O—C and $\delta\ ^{13}C$-1 increase in frequency, progressing in the series 4-methoxy anisole through 4-nitroanisole (29).

Figure 11.86 is a plot of σ_{R° values vs $\delta\ ^{13}C$-1 for 3-x- and 4-x-anisoles (29). With the exception of point 22 (4-phenoxyanisole), there is a pseudolinear relationship between $\delta\ ^{13}C$-1 and Taft σ_{R°. The plot in Fig. 11.86 using $\delta\ ^{13}C$-1 of 150.4 ppm indicates that $\delta\ ^{13}C$-1 is misassigned. Assignment of $\delta\ ^{13}C$-1 at 156.0 ppm is indicated, and thus, we then assign the $\delta\ ^{13}C$-4 at 150.4 ppm for 4-phenoxyanisole.

Figure 11.87 is a plot of $\delta\ ^{13}C$-1 vs Hammett σ values for 3-x and 4-x-anisoles. The point 22 is for 4-phenoxyanisole, and it appears that the correct assignment for $\delta\ ^{13}C$-1 for this compound is 156.0 ppm rather than 150.4 ppm. The 150.4 ppm apparently results from $\delta\ ^{13}C$-4 (29).

Figure 11.88 is a plot of Taft σ_{R° values vs $\delta\ ^{13}C$-1 for 4-x-anisoles (29), and this plot exhibits a pseudolinear relationship between these two parameters. Figure 11.89 is a plot of Taft σ_{R° values vs ν asym. ϕ-O—C for 4-x-anisoles (26), and this plot shows that in general ν asym ϕ-O—C increases in frequency as the Taft σ_{R° increases in value.

Figure 11.90 is a plot of δ ^{13}C-1 vs Hammett σ_p values for 4-x-anisoles, and this plot exhibits a pseudolinear relationship between these two parameters. The δ ^{13}C-1 chemical shift increases in frequency as Hammett σ_p values increase.

Figure 11.91 is a plot of δ ^{13}C-1 vs δ ^{13}C-5 for 2-x-anisoles (29), and this plot shows a pseudolinear relationship between these two values. This plot includes 2-nitroanisole, the literature assignment for δ ^{13}C-5 was given as 134.4 ppm, and δ ^{13}C-3 was given as 125.3 ppm. The plot in Fig. 11.91 indicates that the δ ^{13}C-3 and δ ^{13}C-5 assignments should be reversed, and we have done so in the plot shown here.

MONO-SUBSTITUTED BENZENES

Table 11.18 lists IR and NMR data for mono-x-benzenes (4,34,35). Figure 11.92 is a plot of δ ^{13}C-4 for mono-substituted benzenes vs Taft σ_{R° values, and these parameters correlate in a pseudolinear manner. The significance of Taft σ_{R° is that there is a significant contribution of π electrons to the C-4 atom from the atom or group joined to the C-1 atom (4).

Figure 11.93 is a plot of the out-of-plane hydrogen deformation mode I vs Taft σ_{R° values, and these parameters correlate in a pseudolinear manner (4).

Figure 11.94 is a plot of the out-of-plane hydrogen deformation mode III vs Taft σ_{R° values, and these parameters correlate in a pseudolinear manner (4).

Figures 11.95 and 11.96 plot the out-of-plane hydrogen deformation modes I and III vs δ ^{13}C-4, respectively, and these parameters show that modes I and III increase in frequency as δ ^{13}C-4 increases in frequency (4).

4-X AND 4,4′-X,X′-BIPHENYLS IN $CHCL_3$ SOLUTIONS

Table 11.19 lists NMR data for 4-x and 4,4′-x,x-biphenyls in $CHCl_3$ solution (3). Figure 11.97 is a plot of δ ^{13}C-1 vs Taft σ_{R° for 4-x, and 4,4′-x,x-biphenyls and, in general, δ ^{13}C-1 increases in frequency as the Taft σ_{R° values increase (4).

TETRAMETHYLUREA (TMU)

Table 11.20 lists NMR data for 1 wt./vol. % tetramethylurea (TMU) in mol % $CHCl_3/CCl_4$ solutions. Figure 11.98 is a plot of δ ^{13}C=O for TMU vs mol % $CHCl_3$ CCl_4 (36). Figure 11.98 is a plot of δ ^{13}C=O for TMU vs mol % $CHCl_3/CCl_4$. This plot shows that δ ^{13}C=O increases in frequency (shifts down field) in a nonlinear manner as the mol % is increased from 0 to 100. These data indicate that there is changing solute-solvent interaction between TMU and $CHCl_3/CCl_4$ as the mol % $CHCl_3/CCl_4$ is increased from 0 to 100.

Figure 11.99 is a plot of ν C=O (37) vs δ ^{13}C=O for TMU in 0–100 mol % $CHCl_3/CCl_4$ solutions. This plot shows that ν C=O decreases in frequency while δ ^{13}C=O increases in frequency as the mol % $CHCl_3/CCl_4$ goes from 0–100 mol % $CHCl_3/CCl_4$ (36).

Figure 11.100 is a plot of δ ^{13}C=O for TMU vs δ $^{13}CH_3$ in 0–100 mol % $CHCl_3/CCl_4$ solutions. This plot shows that δ ^{13}C=O increases in frequency (shifts down-field) while

$\delta\ ^{13}CH_3$ decreases in frequency (shifts up-field) as the mol % $CHCl_3/CCl_4$ is increased from 0 to 100 (34). Because the $\delta\ ^{13}C{=}O$ chemical shift increases in frequency in going from CCl_4 solution to $CHCl_3$ solution, the CCl_4 deshields the carbonyl carbon atom more than $CHCl_3$ does. Simultaneously, $CHCl_3$ shields the CH_3 carbon atoms of atoms of TMU more than CCl_4 does, because $\delta\ ^{13}CH_3$ decreases in frequency as the mol % $CHCl_3/CCl_4$ is increased.

Table 11.21 lists NMR data for 1 wt./vol. % TMU in various solvents (36). Figure 11.101 is a plot of $\delta\ ^{13}C{=}O$ for TMU vs the solvent acceptor number (AN) for each of the 21 solvents. The solvents are presented in order, 1 through 21, in Table 11.21. This plot shows that in general there is a pseudolinear relationship between $\delta\ ^{13}C{=}O$ and AN, but the scatter of data points shows that this is not a precise indicator of the $\delta\ ^{13}C{=}O$ chemical shifts. If the AN values were a precise indicator of $\delta\ ^{13}C{=}O$ frequencies, the $\delta\ ^{13}C{=}O$ chemical shift for TMU in hexane would be expected to occur at lower frequency by approximately 0.8 ppm by extrapolation from the pseudolinear relationship exhibited by most of the solvents, AN values, and the corresponding $\delta\ ^{13}C{=}O$ frequencies.

These results indicate that the AN values do not take into account the steric effect of the solute and solvent, which alters the distance between the sites of solute-solvent interaction. Thus, there is not a comparable solute-solvent interaction between different but similar carbonyl compounds in the same solvent due to steric differences of the solute and solvent.

Figure 11.102 is a plot of $\delta\ ^{13}C{=}O$ for TMU in each of the solvents vs the difference between $\delta\ ^{13}C{=}O$ for TMU in acetic acid solution and each of the other solvents (see Table 11.21). There is no chemical significance in this linear plot, because this type of mathematical treatment of any data set always produces a linear relationship. It does help show that the AN values are not a precise indicator of $\delta\ ^{13}C{=}O$ chemical shifts. As examples: (A) The AN values for solvent 7 (tert-butyl alcohol) and solvent 14 (acetone) are 29.1 and 12.5, respectively; (B) The AN values for solvent 6 (isopropyl alcohol) and solvent 11 (acetonitrile) are 33.5 and 18.9, respectively; and (C) The AN values for solvent 13 (nitrobenzene), solvent 17 (benzene), and solvent 18 (methyl tert-butyl ether) are 14.8, 8.2, and 5.0, respectively. The $\delta\ ^{13}C{=}O$ frequencies for (A), (B), and (C) are 166.0, 166.1, and 165.1 ppm, respectively, yet the AN values for the solvents vary considerably. These data clearly show that the AN values are not a precise indicator of solute-solvent interactions (36).

DIALKYLKETONES

Table 11.22 lists NMR data for dialkylketones in various solvents at 1 wt./vol. % concentrations (36). Figure 11.103 is a plot of $\delta\ ^{13}C{=}O$ for dialkylketones at 1 wt./vol. % in different solvents vs the AN value for each of the solvents. Plots 1–6 show a pseudolinear relationship for each of the dialkylketones and $\delta\ ^{13}C{=}O$ increases in frequency with increased branching on the α-carbon atoms. In addition, plots 1–6 and 1′–6′ yield separate relationships. Plots 1′–6′ are $\delta\ ^{13}C{=}O$ frequencies for these dialkylketones in each of the four alcohols included in this study, and these plots are distinctly different from plots 1–6. An IR study of these same dialkylketones in solution with each of these four alcohols has shown that these ketones exist in alcoholic solution in two forms. These are: an intermolecularly hydrogen-bonded form $[(R)_2C{=}O\cdots HOR']$, and a form in which $(R)_2C{=}O$ is not intermolecularly hydrogen bonded, but surrounded by intermolecularly hydrogen-bonded alcohol molecules (38). This is why dialkylketones in alkyl alcohols differ from those in nonalcoholic solution.

Plots 1–6 also show that δ $^{13}C{=}O$ for these dialkylketones increases in frequency as the electron contribution of the alkyl groups increases, progressing in the series dimethyl ketone through di-tert-butyl ketone.

Figure 11.104 shows plots of δ $^{13}C{=}O$ (solvent) vs the δ $^{13}C{=}O$ chemical shift difference between each dialkylketone in methanol and the same ketone in each of the other solvents (36). It is noted that the δ $^{13}C{=}O$ chemical shifts for these dialkylketones increase in the order dimethyl ketone through ethyl isopropyl ketone, di-tert-butyl ketone, and di-isopropyl ketone. The inductive donation of electrons to the carbonyl group increases in the order dimethyl ketone through di-tert-butyl ketone. The di-tert-butyl ketone is out of order in this sequence because the steric effects of the tert-butyl group prevent a closer solute-solvent interaction with the carbonyl group as in the case of di-isopropyl ketone. Further support for this supposition is gained by study of these ketones in the four alcohols. One would expect the strongest intermolecular hydrogen bond (C=O···HO) to be formed between methanol and di-tert-butyl ketone. However, as methanol is the reference point in each plot, it does not provide support for the steric factor supposition. Comparison of the points on the plots for alcohols 2, 3, and 6 show that change for δ $^{13}C{=}O$ (methanol) minus δ $^{13}C{=}O$ (ethyl alcohol) is the least for di-tert-butyl ketone in going from solution in methanol to solution in ethanol, and the change is more in going from solution in isopropyl alcohol and tert-butyl alcohol. With the exception of diethyl ketone in benzonitrile, in going from solution in methanol to solution in any one of the other solvents, the δ $^{13}C{=}O$ frequency changes least in the case of di-tert-butyl ketone. These δ $^{13}C{=}O$ frequency shifts are what is expected when steric factors of the solute and solvent affect the distance between the sites of the solute-solvent interaction. The larger the steric factor, the less solute-solvent interaction regardless of the electrophilicity of the solvent and/or the acidity of the OH protons (36).

ACETATES

Table 11.23 lists NMR data for alkyl acetates and phenyl acetate 1 wt./vol. % in various solvents (36). Figure 11.105 is a plot of δ $^{13}C{=}O$ for alkyl acetates and phenyl acetate vs the difference of δ $^{13}C{=}O$ in methanol and δ $^{13}C{=}O$ in each of the other solvents (36). The numbers in each plot correspond to a particular solvent, and it can be seen that with the exception of methyl alcohol, the difference values for δ $^{13}C{=}O$ are not the same for each of the alkyl acetates and phenyl acetate. If the AN value for a solvent is constant, some other factor must be causing the change in the difference values for δ $^{13}C{=}O$ for each of the other acetates. This other factor is a steric factor, which causes the distance in space between sites of solute and solvent interaction to vary in the acetates studied.

ACRYLATES AND METHACRYLATES

Table 11.24 lists IR and NMR data for alkyl acrylates and alkyl methacrylates in CCl_4 and $CHCl_3$ solution (39). The ν C=O for alkyl acrylates in mol % $CHCl_3/CCl_4$ solutions continuously decreases, from 1722.9–1734.1 cm^{-1} in CCl_4 to 1713.8–1724.5 cm^{-1} in $CHCl_3$ solution while δ $^{13}C{=}O$ continually increases in frequency from 164.2–165.2 ppm in CCl_4 solution to 165.7–166.7 ppm in $CHCl_3$ solution (39). The ν C=O frequency for alkyl methacrylates is lower than the frequencies for the corresponding alkyl acrylates. This decrease in ν C=O frequency is the result of the electron contribution of the CH_3 group to the C=O group, which weakens the C=O bond via $\overset{+}{C}-\overset{-}{O}$ and thus causes the ν C=O mode to vibrate at a lower frequency. The ν C=O for alkyl methacrylates in mol % $CHCl_3/CCl_4$ solutions continuously decrease in frequency from 1719.5–1726.0 cm^{-1} in CCl_4 solution to 1709.5–1718.0 cm^{-1} in $CHCl_3$ solution, while δ $^{13}C{=}O$ continuously increase in frequency from 165.6–166.7 ppm in CCl_4 solution to 167.1–167.8 ppm in $CHCl_3$ solution. Moreover, the δ $^{13}C{=}O$ frequencies for the alkyl methacrylates are higher than those for the corresponding alkyl acrylate analog. Methyl acrylate exhibits the highest ν C=O frequency and highest δ $^{13}C{=}O$ chemical shift, while tert-butyl acrylate exhibits the lowest δ $^{13}C{=}O$ chemical shift for the seven alkyl acrylates included in the study. This difference is caused by the larger inductive-mesomeric contribution of the tert-butyl group to the carbonyl group relative to the contribution of the methyl group to the carbonyl group:

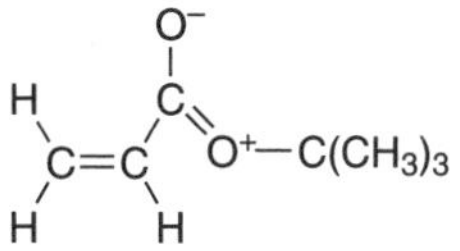

Both ν C=O and δ $^{13}C{=}O$ for alkyl methacrylates shift in a manner comparable to that shown for comparable alkyl acrylates. The behavior of ν C=O and δ $^{13}C{=}O$ is a function of the electron density on the carbonyl carbon atom. As the electron density on the carbonyl carbon atom increases, both ν C=O and δ $^{13}C{=}O$ decrease in frequency.

The C=O group in $CHCl_3$ solution is intermolecularly hydrogen bonded ($C{=}O\cdots HCCl_3$) and surrounded by other $CHCl_3$ molecules and the C=O groups are less shielded by Cl atoms when in CCl_4 solutions. Consequently, the δ $^{13}C{=}O$ chemical shift occurs at higher frequency in $CHCl_3$ than in CCl_4 solution. Intermolecular hydrogen bonding ($C{=}O\cdots HCCl_3$) causes the C=O mode to vibrate at lower frequency in $CHCl_3$ solution than in CCl_4 solution. Other factors that affect both ν C=O and δ $^{13}C{=}O$ are attributed to the reaction field of the solvent system and steric factors of the solute. As the reaction field is increased in going from 0 to 100 mol % $CHCl_3/CCl_4$, the ν C=O frequency is steadily decreased and δ $^{13}C{=}O$ is steadily increased (39).

REFERENCES

1. Nyquist, R. A. (1986). *IR and NMR Spectral Data-Structure Correlations for the Carbonyl Group*. Philadelphia: Sadtler.
2. Nyquist, R. A. (1993). *Appl. Spectrosc.* 47: 411.
3. *The Sadtler Standard Collection of NMR Carbon-13 NMR Spectra*. Philadelphia: Sadtler Research Laboratories,
4. Nyquist, R. A. and Hasha, D. L. (1991). *Appl. Spectrosc.* 45: 849.
5. Kalinowski, H. O., Berger, S., and Braun, S. (1988). *Carbon-13 NMR Spectroscopy.* New York: John Wiley & Sons, p. 313.
6. Nyquist, R. A. (1963). *Spectrochim. Acta* 19: 1595.
7. Nyquist, R. A., Settineri, S. E., and Luoma, D. A. (1991). *Appl. Spectrosc.* 45: 1641.

8. Nyquist, R. A. (1984). *The Interpretation of Vapor-Phase Infrared Spectra: Group Frequency Data*. Vol. 1, Philadelphia: Sadtler Res. Labs.
9. Nyquist, R. A., Putzig, C. L., and Hasha, D. L. (1989). *Appl. Spectrosc.* **43**: 1049.
10. Nyquist, R. A. and Luoma, D. A. (1991). *Appl. Spectrosc.* **45**: 1501.
11. Nyquist, R. A. (1990). *Appl. Spectrosc.* **44**: 438.
12. Nyquist, R. A., Streck, R., and Jeschek, G. (1996). *J. Mol. Struct.* **377**: 113.
13. Bellamy, L. J. (1975). *The Infrared Spectra of Complex Molecules*. New York: John Wiley & Sons; p. 295.
14. Rao, C. N. R. and Venkataraghavan, R. (1961). *Can. J. Chem.* **39**: 1757.
15. Deady, L., Katritzky, A. R., Shanks, R. A., and Topsom, R. D. (1973). *Spectrochim. Acta* **29A**: 115.
16. Saito, T., Yamakawa, M., and Takasuka, M. (1981). *J. Mol. Spectrosc.* **90**: 359.
17. Nyquist, R. A. (1987). *Appl. Spectrosc.* **41**: 904.
18. Nyquist, R. A. (1988). *Appl. Spectrosc.* **42**: 624.
19. Witanowski, M., Steaniak, L., and Webb, G. A. (1981). *Annual Reports on NMR Spectroscopy*. Vol. 11B, *Nitrogen NMR Spectroscopy*. New York: Academic Press, p. 301.
20. *Sadtler Research Laboratories Standard Carbon-13 NMR Spectra*.
21. Witanowski, M., Steaniak, L., and Webb, G. A. (1981). *Annual Reports on NMR Spectroscopy*. Vol. 11b, *Nitrogen NMR Spectroscopy*. New York: Academic Press, p. 385.
22. Hirschmann, R. P., Kniseley, R. N., and Fassel, V. A. (1965). *Spectrochim. Acta* **21**: 2125.
23. Taft, Jr., R. W. (1956). in *Steric Effects in Organic Chemistry*. M. S. Neuman, ed., New York: J. Wiley & Sons, Inc., p. 590.
24. Nyquist, R. A. and Jewett, G. L. (1992). *Appl. Spectrosc.* **46**: 841.
25. Witanowski, M., Steaniak, L., and Webb, G. A. (1981). *Annual Reports on NMR Spectroscopy*, Volume 11B, *Nitrogen NMR Spectroscopy*. New York: Academic Press, p. 305.
26. Nyquist, R. A. (1988). *Appl. Spectrosc.* **42**: 854.
27. Mavel, G. (1973). *Annual Reports on NMR Spectroscopy, NMR Studies of Phosphorus Compounds* (1965–1969), Vol. 5B, E. F. Mooney, ed., London/New York: Academic Press, pp. 103–288.
28. Mark, V., Dungan, C. H., Crutchfield, M. M., and Van Wazer, J. R. (1967). Compilation of ^{31}P NMR Data. in *Topics in Phosphorus Chemistry*, Vol. 5, Chap. 4, M. Grayson and E. J. Griffiths, eds., New York: Wiley Interscience, pp. 227–447.
29. Nyquist, R. A. (1991). *Appl. Spectrosc.* **45**: 1649.
30. Goldman, G. K., Lehman, H., and Rao, C. N. R. (1960). *Can. J. Chem.* **38**: 171.
31. Nyquist, R. A. and Potts, W. J. (1972). Vibrational Spectroscopy of Phosphorus Compounds. in *Chemical Analysis Series, Analytical Chemistry of Phosphorus Compounds*. M. Holman, ed., New York: John Wiley Interscience, pp. 189–293.
32. L. Pauling (1948). *The Nature of the Chemical Bond*. Ithaca, NY: Cornell University Press, pp. 60, 179.
33. Brown, H. C. and Okamoto, Y. J. (1958). *J. Am. Chem. Soc.* **80**: 4979.
34. Nyquist, R. A. (1984). *The Interpretation of Vapor-Phase Infrared Spectra: Group Frequency Data*. Vol. 1, Philadelphia: Sadtler Res. Labs., p. A-17.
35. Breitmaier, E. and Voelter, W. (1987). *Carbon-13 NMR Spectroscopy*. Federal Republic of Germany: VCH, Weinheim, p. 256.
36. Nyquist, R. A., Steck, R., and Jeschek, G. (1996). *J. Mol. Struct.* **377**: 113.
37. Nyquist, R. A. and Luoma, D. A. (1991). *Appl. Spectrosc.* **45**: 1491.
38. Nyquist, R. A. (1994). *Vib. Spectrosc.* **7**: 1.
39. Nyquist, R. A. and Streck, R. (1995). *Spectrochim. Acta* **51A**: 475.

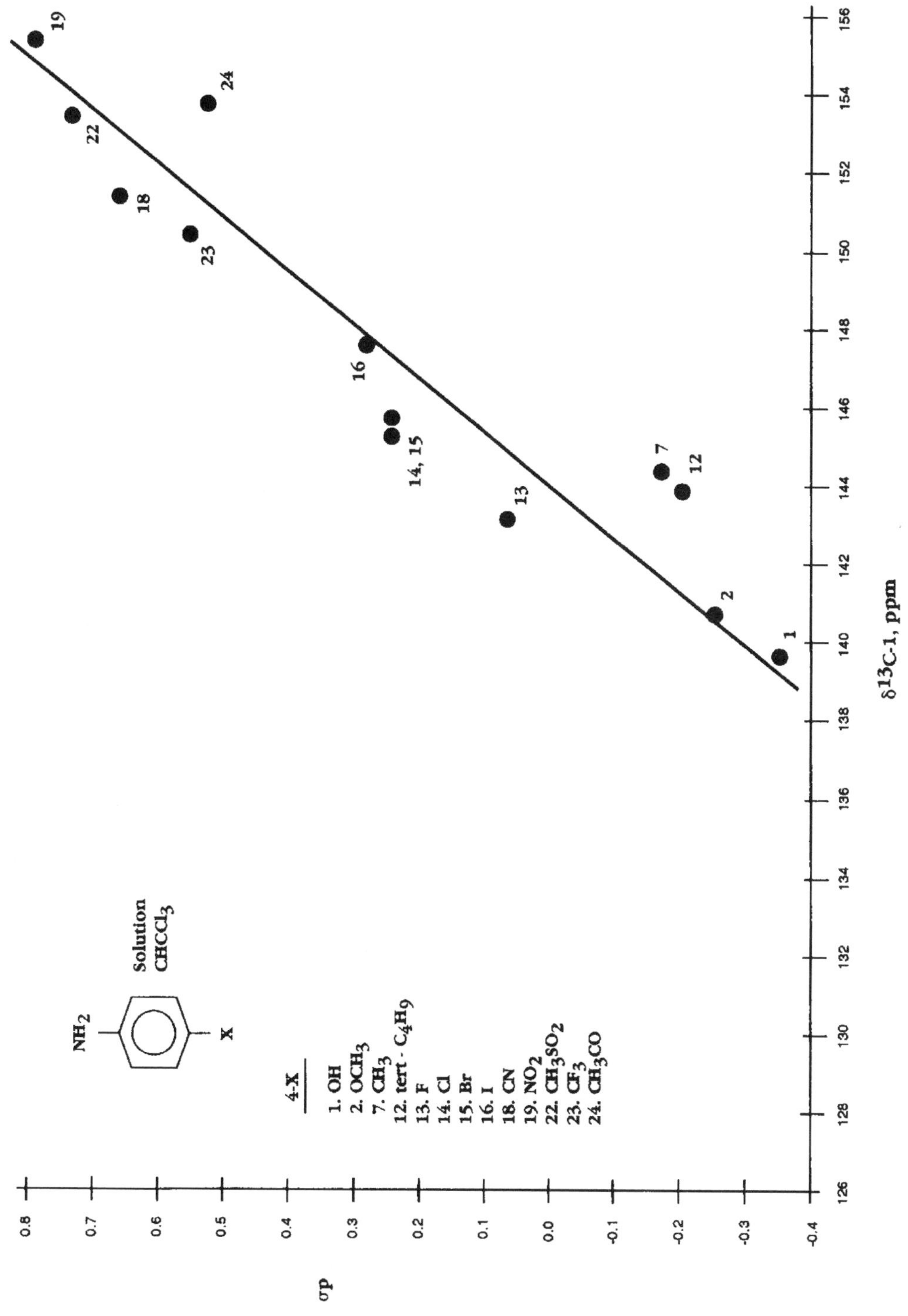

FIGURE 11.1 A plot of the NMR δ ^{13}C-1 chemical shift data for 4-x-anilines in $CHCl_3$ solution vs Hammett σ_p values for the 4-x atom or group.

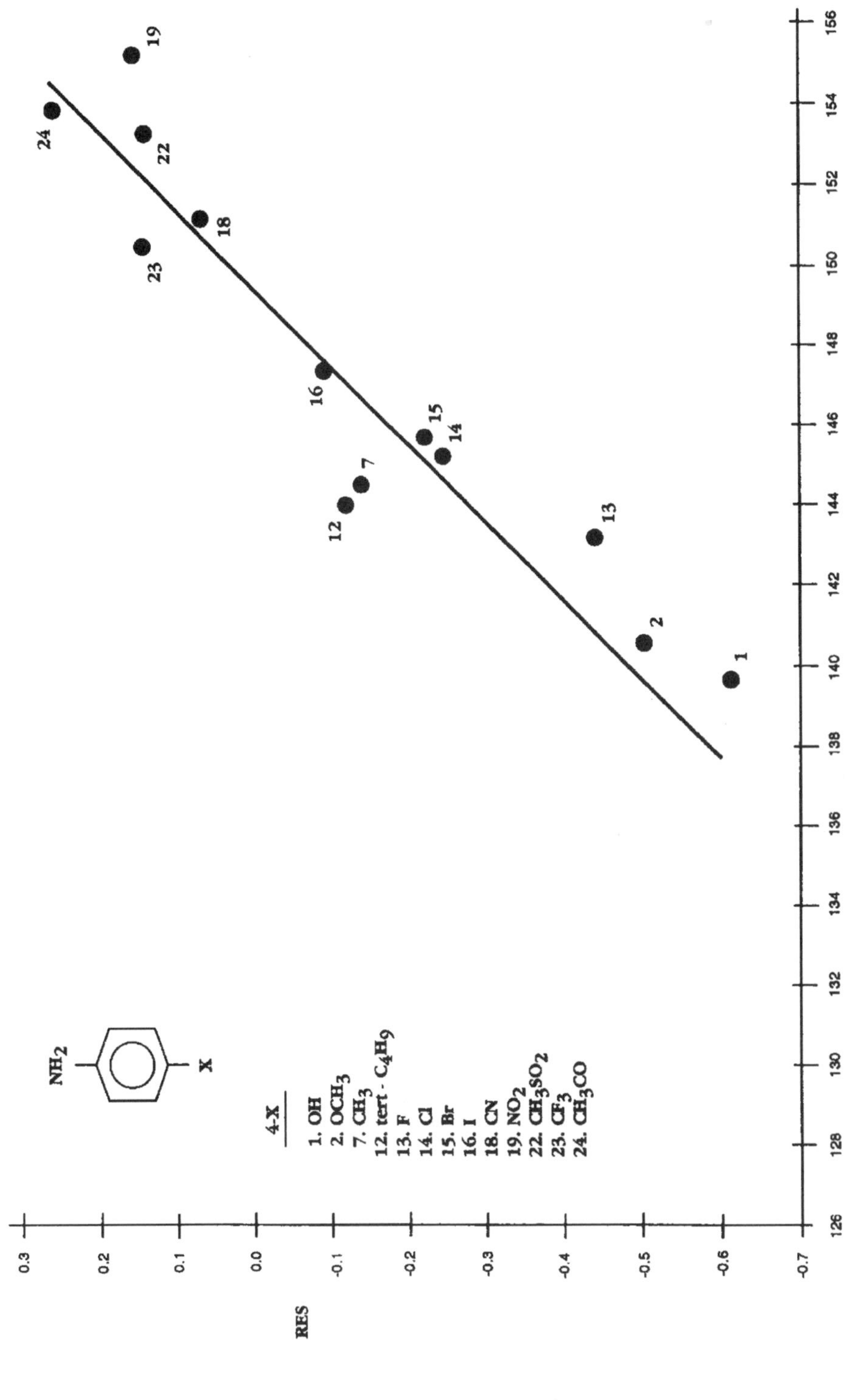

FIGURE 11.2 A plot of the NMR δ ^{13}C-1 chemical shift data for 4-x-anilines in $CHCl_3$ solution vs Taft σ_{R° values for the 4-x atom or group.

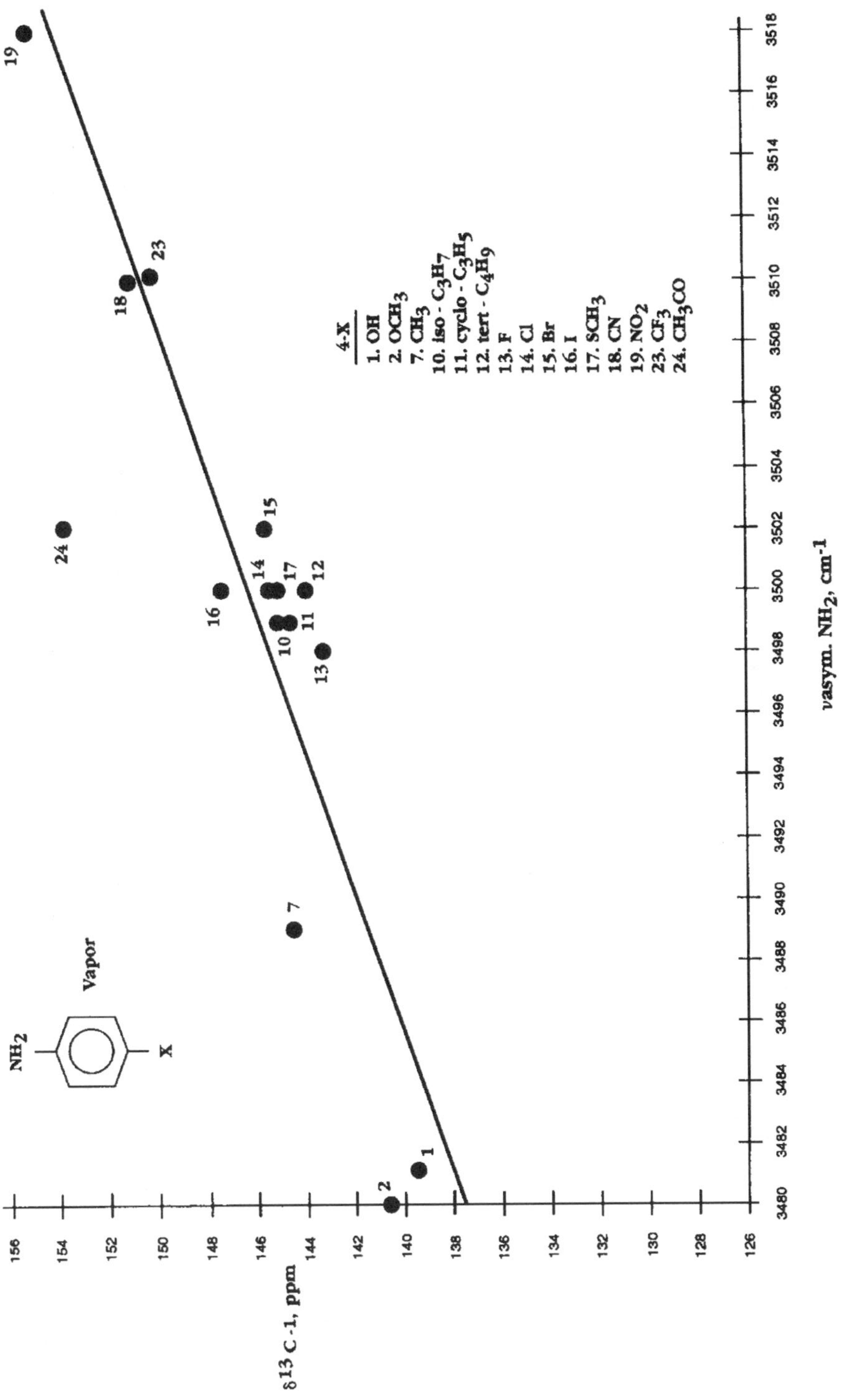

FIGURE 11.3 A plot of ν asym. NH_2 frequencies for 4-x-anilines in the vapor phase vs δ ^{13}C-1 chemical shift data for 4-x-anilines in $CHCl_3$ solution.

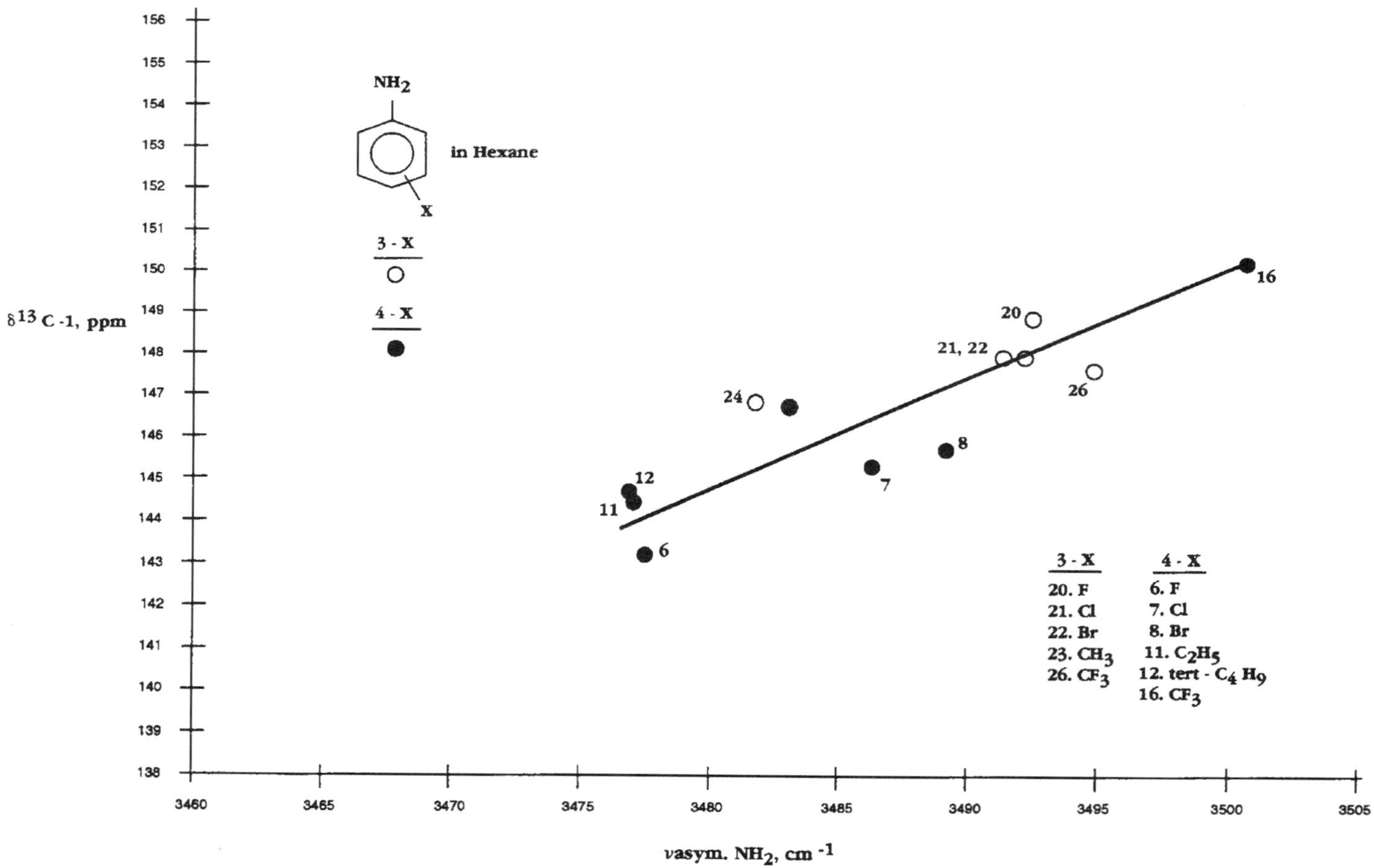

FIGURE 11.4 A plot of ν asym. NH_2 frequencies for 3-x- and 4-x-anilines in hexane solution vs δ ^{13}C-1 chemical shift data for 3-x- and 4-x-anilines in $CHCl_3$ solution.

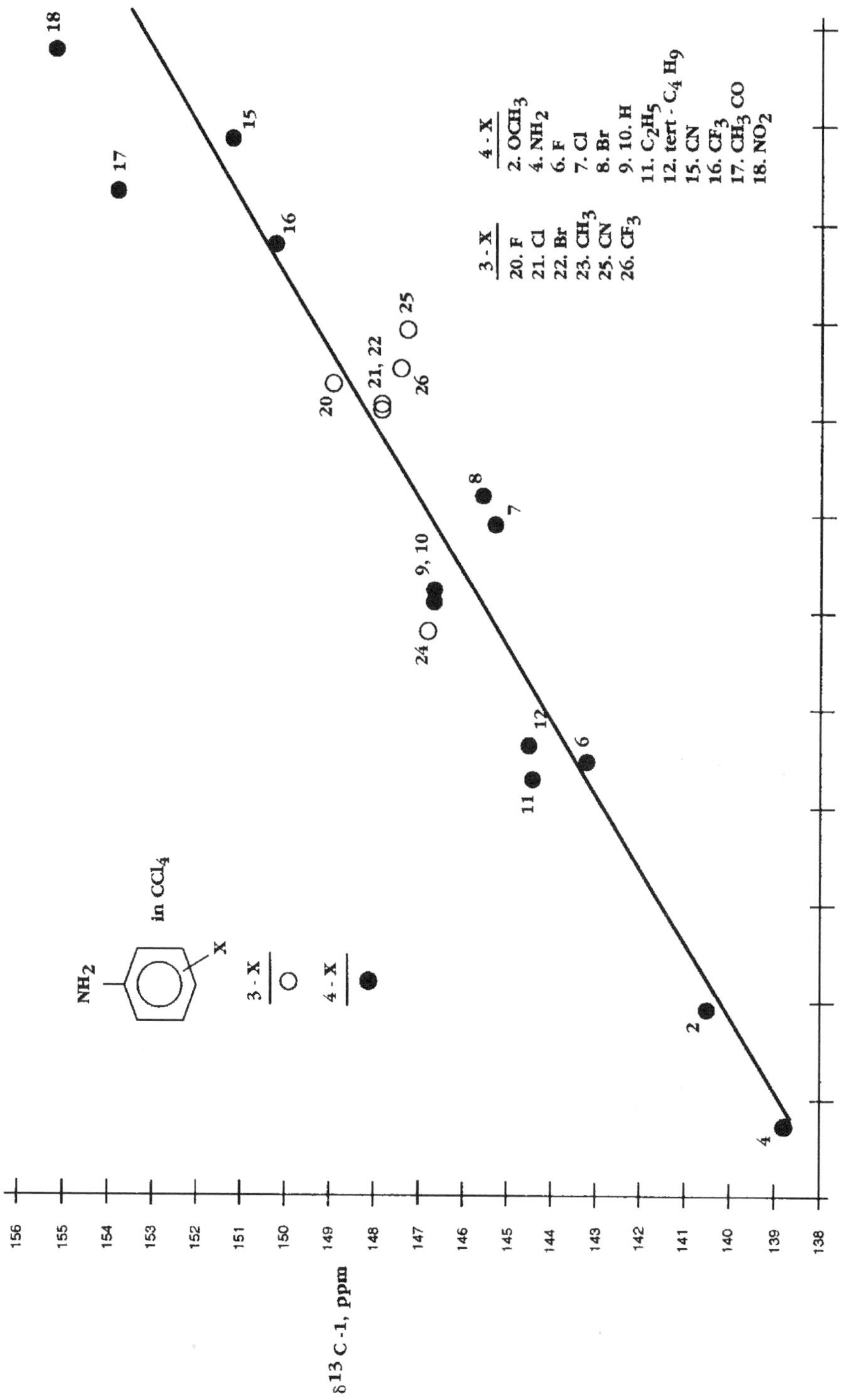

FIGURE 11.5 A plot of ν asym. NH_2 frequencies for 3-x- and 4-x-anilines in CCl_4 solution vs δ ^{13} C-1 chemical shift data for 3-x- and 4-x-anlines in $CHCl_3$ solution.

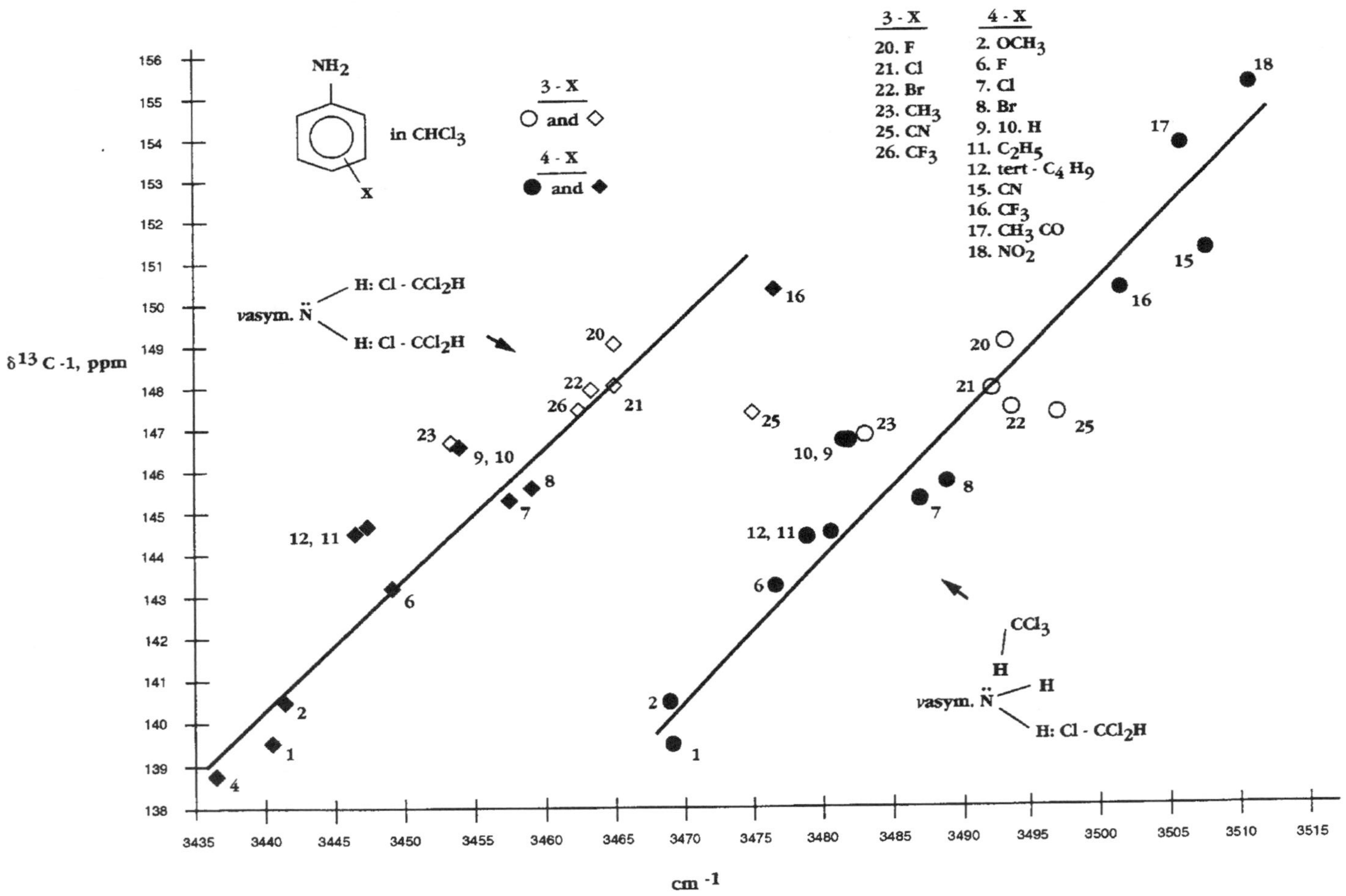

FIGURE 11.6 A plot of the ν asym. NH_2 frequencies for 3-x- and 4-x-anilines in $CHCl_3$ solution vs δ ^{13}C-1 chemical shift data for 3-x- and 4-x-anilines in $CHCl_3$ solution.

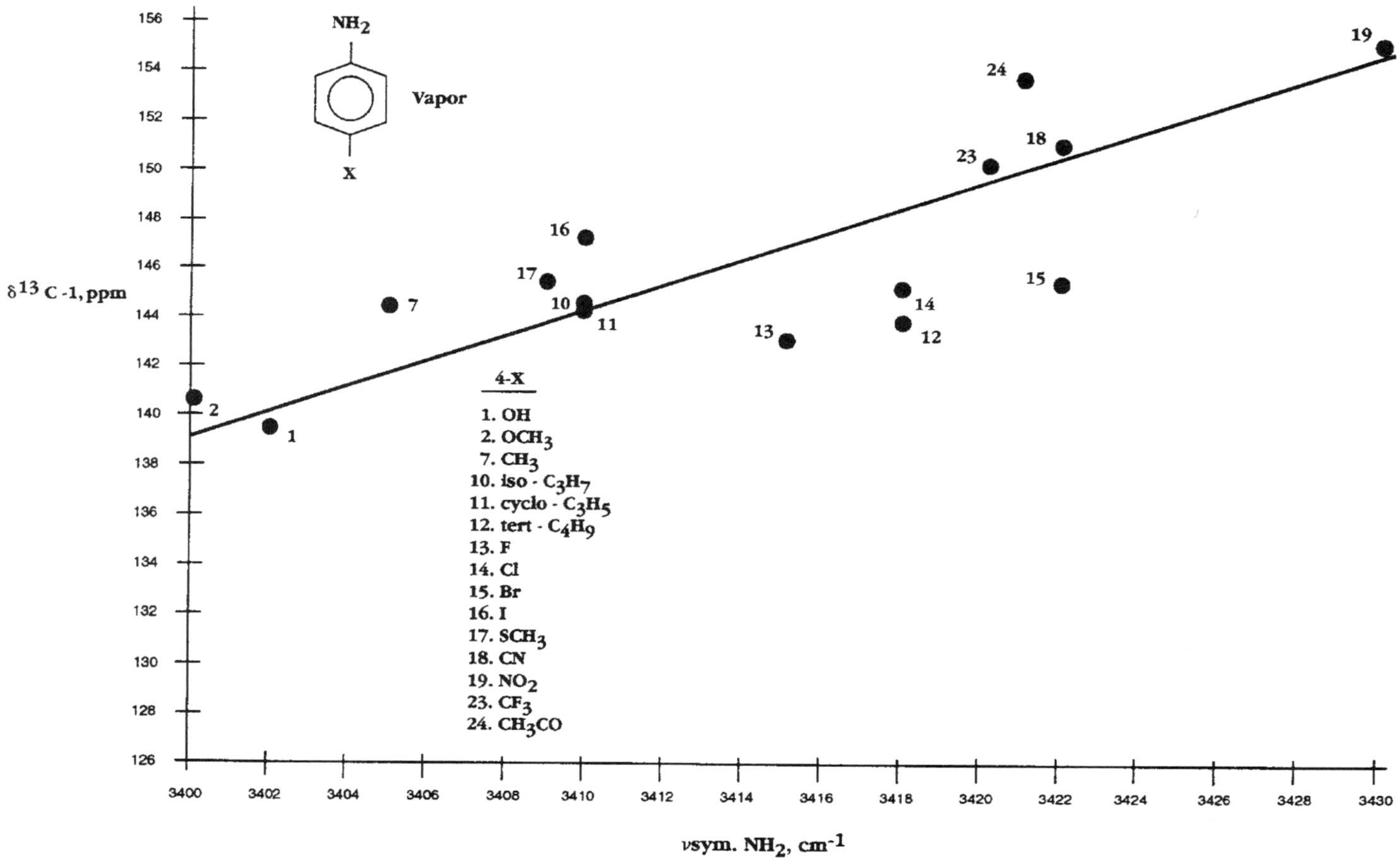

FIGURE 11.7 A plot of the ν sym. NH_2 frequencies for 4-x-anilines in the vapor phase vs δ ^{13}C-1 chemical shift data for 3-x- and 4-x-anilines.

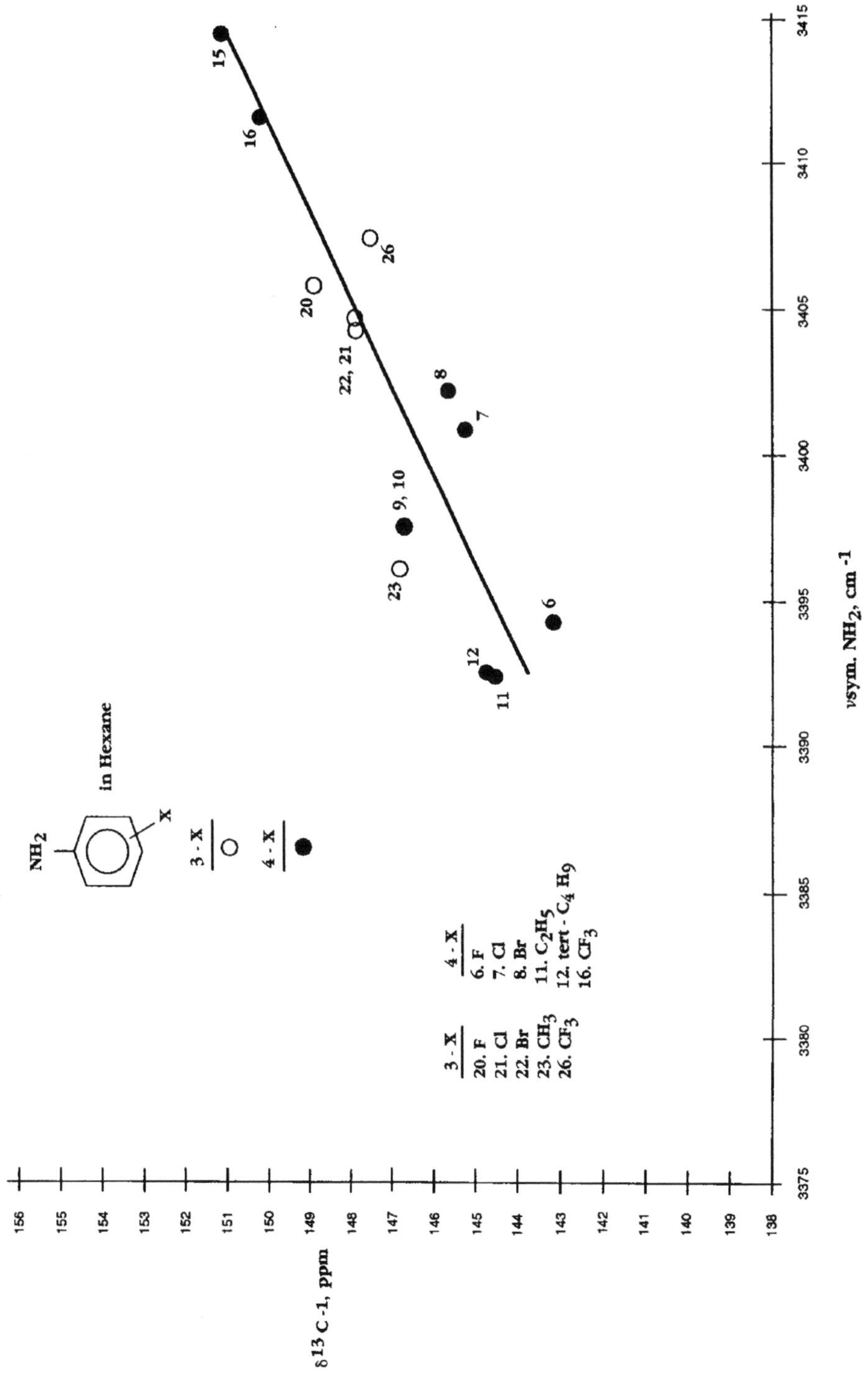

FIGURE 11.8 A plot of ν sym. NH_2 frequencies for 3-x- and 4-x-anilines in hexane solution vs δ ^{13}C-1 chemical shift data for 3-x- and 4-x-anilines in $CHCl_3$ solution.

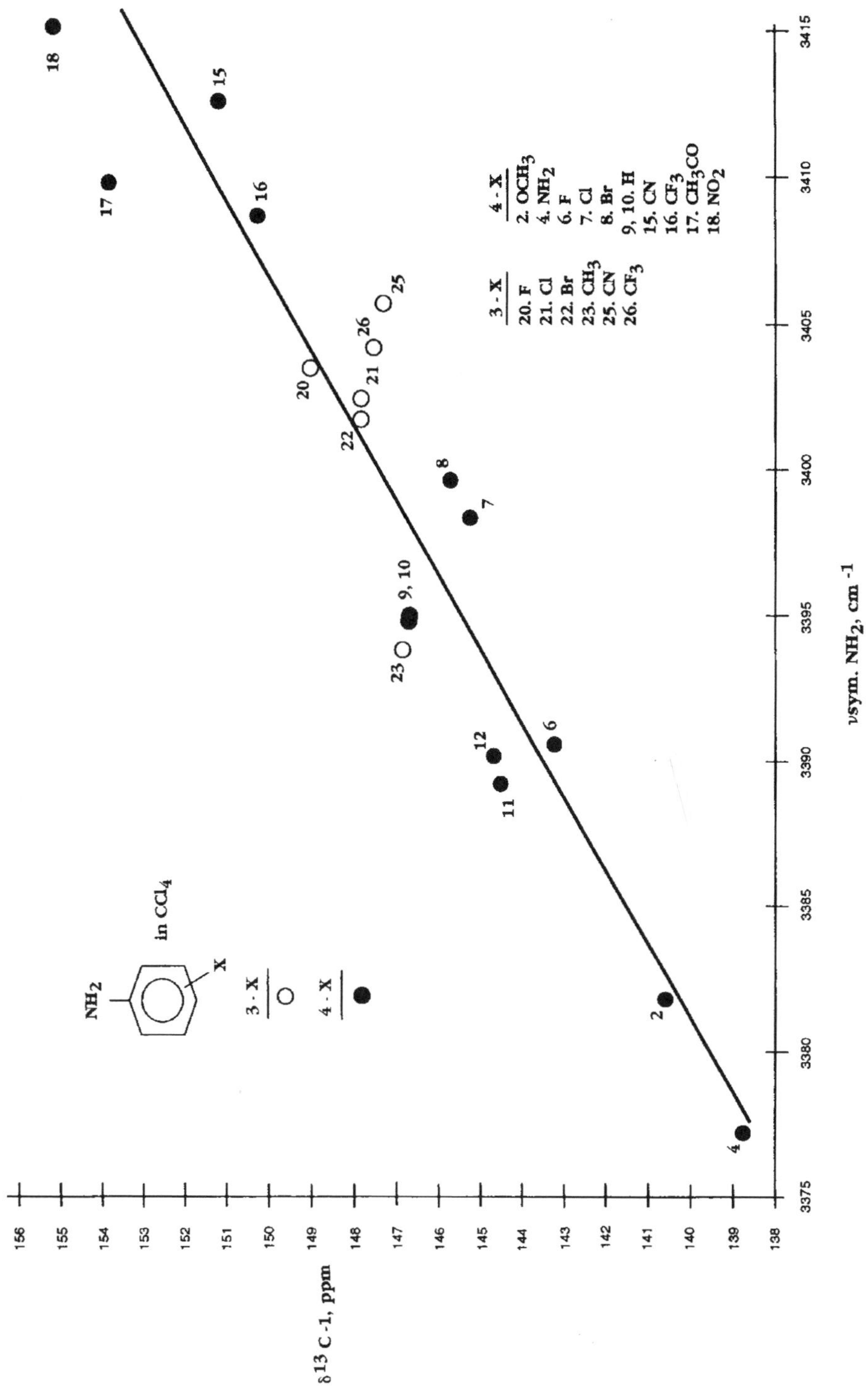

FIGURE 11.9 A plot of ν sym. NH_2 frequencies for 3-x- and 4-x-anilines in CCl_4 solution vs δ ^{13}C-1 chemical shift data for 3-x- and 4-x-anilines vs the δ ^{13}C-1 chemical shift data for 3-x- and 4-x-anilines in $CHCl_3$ solution.

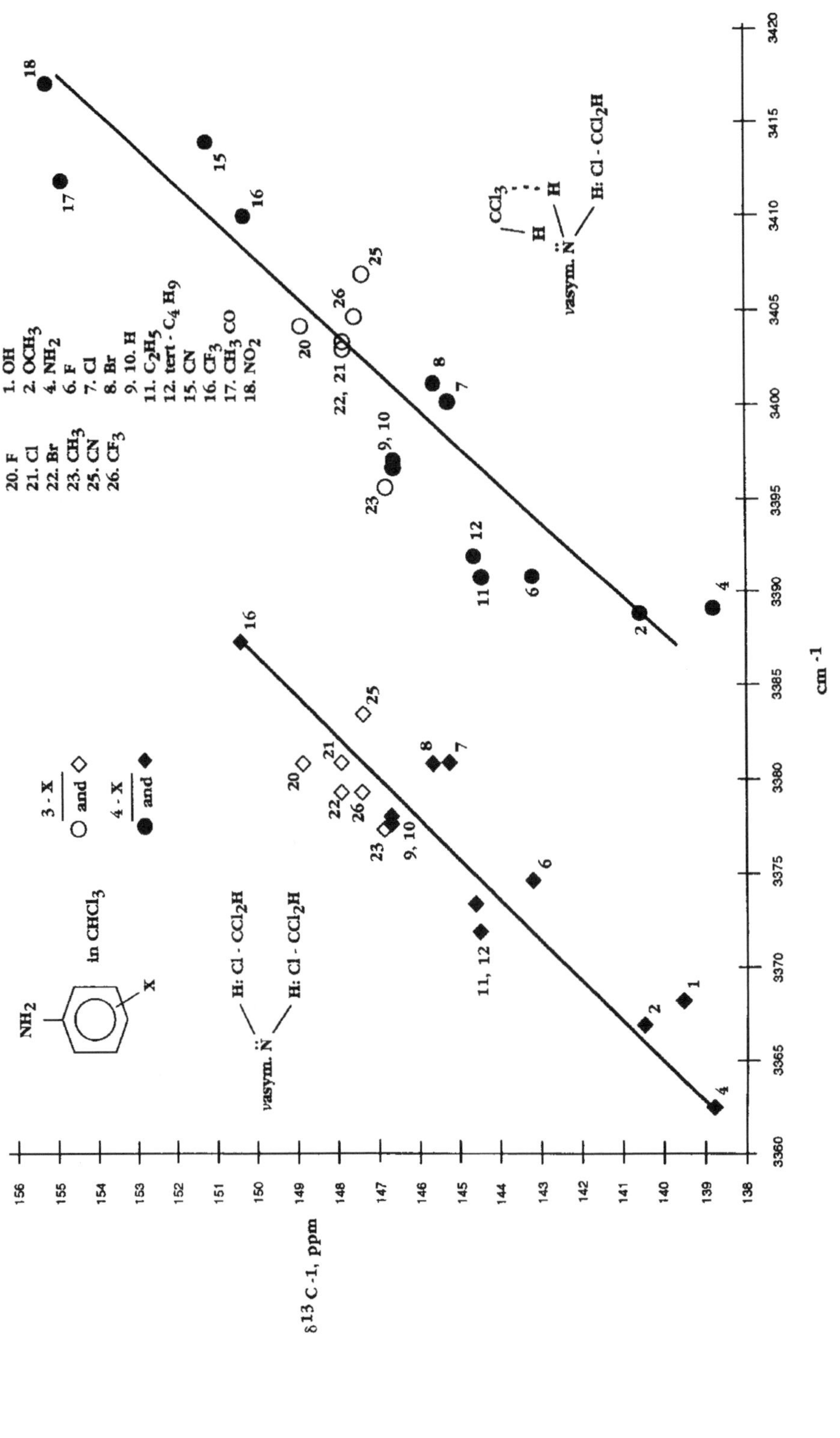

FIGURE 11.10 A plot of the ν sym. NH_2 frequencies for 3-x- and 4-x-anilines in $CHCl_3$ solution vs δ ^{13}C-1 chemical shift data for 3-x- and 4-x-anilnies in $CHCl_3$ solution.

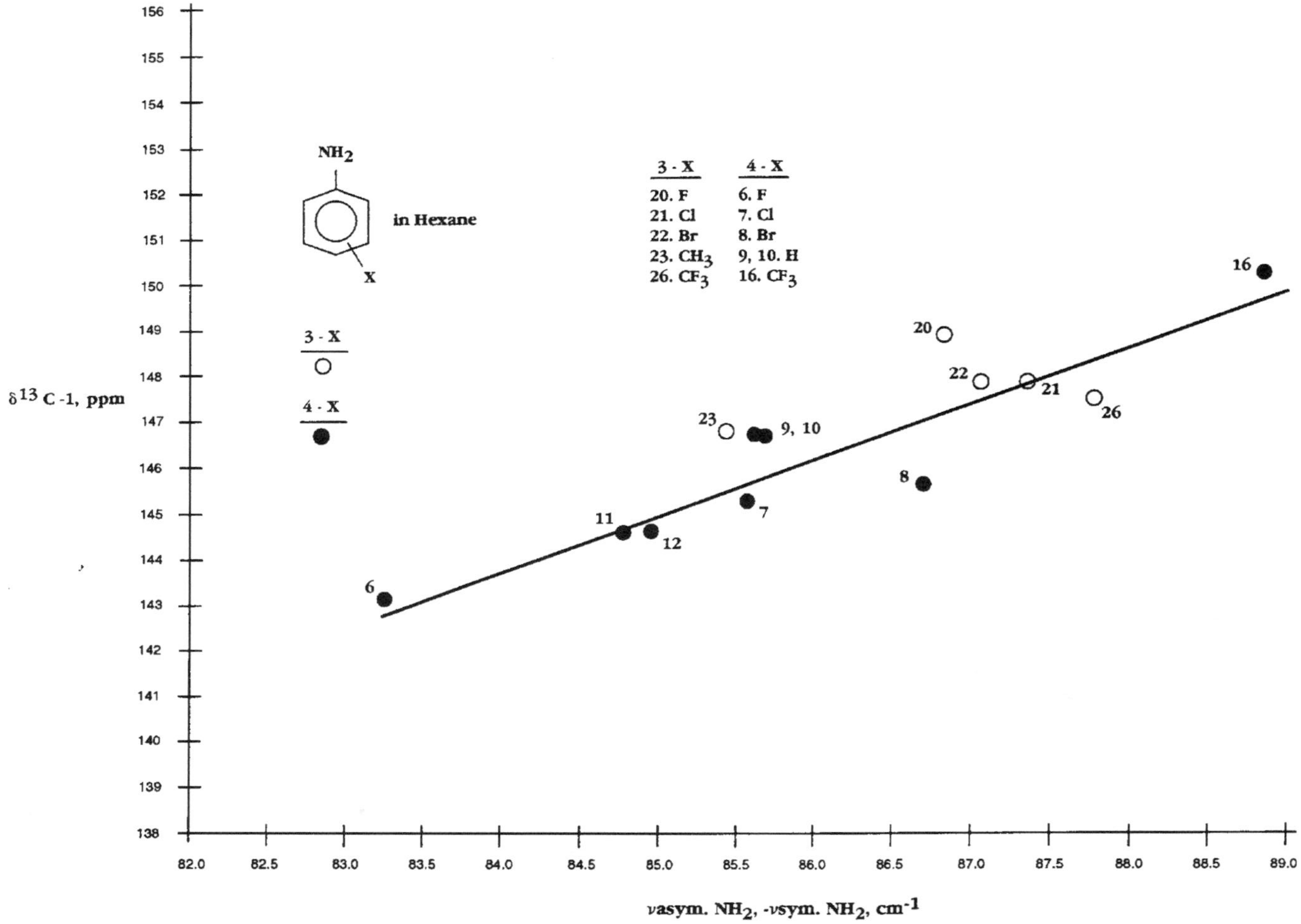

FIGURE 11.11 A plot of the frequency difference between ν asym. NH_2 and ν sym. NH_2 for 3-x- and 4-x-anilines in hexane solution vs δ ^{13}C-1 chemical shift data for 3-x- and 4-x-anilines in $CHCl_3$ solution.

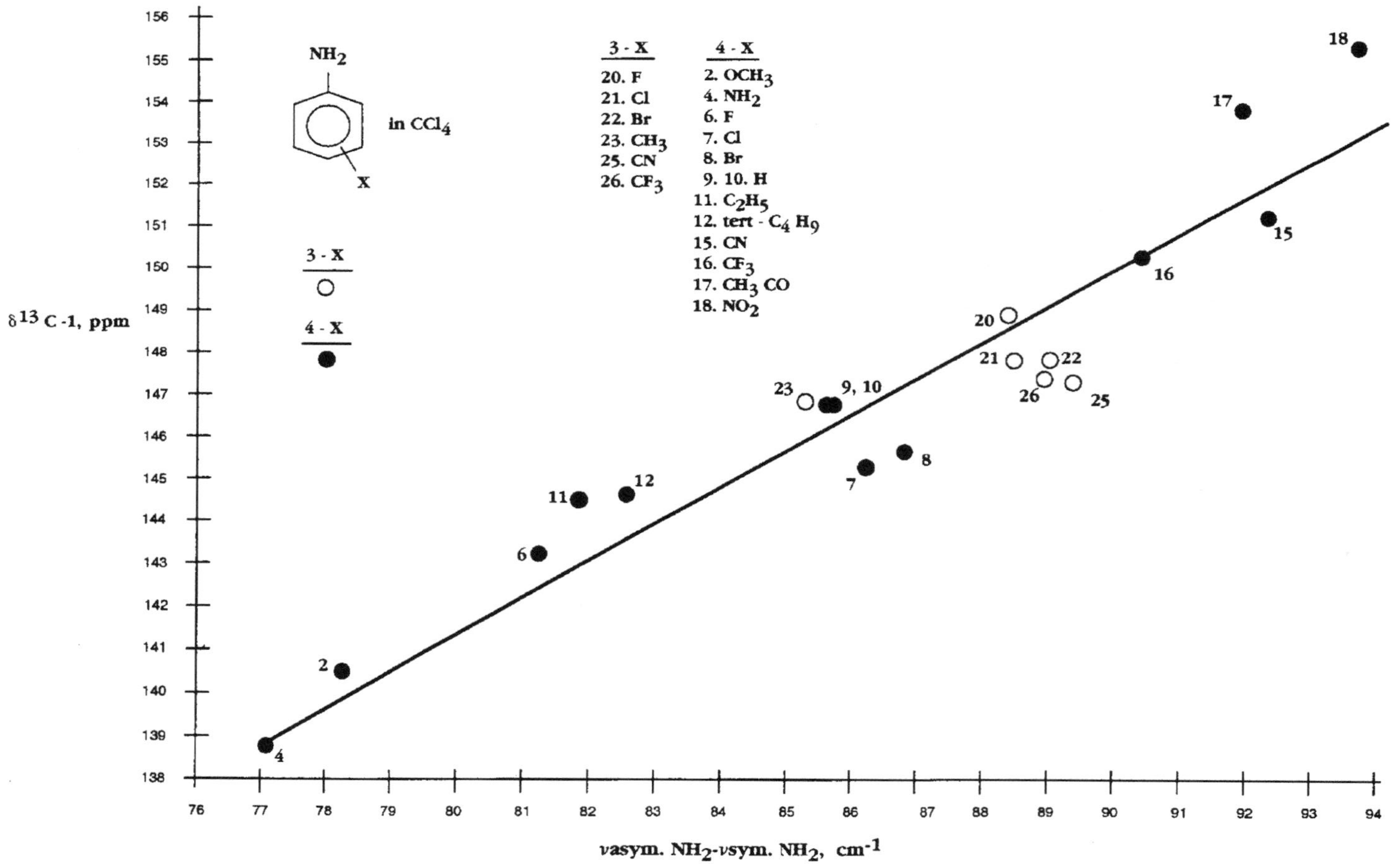

FIGURE 11.12 A plot of the frequency difference between ν asym. NH_2 and ν sym. NH_2 for 3-x- and 4-x-anilines in CCl_4 solution vs δ ^{13}C-1 chemical shift data for 3-x- and 4-x-anilines in $CHCl_3$ solution.

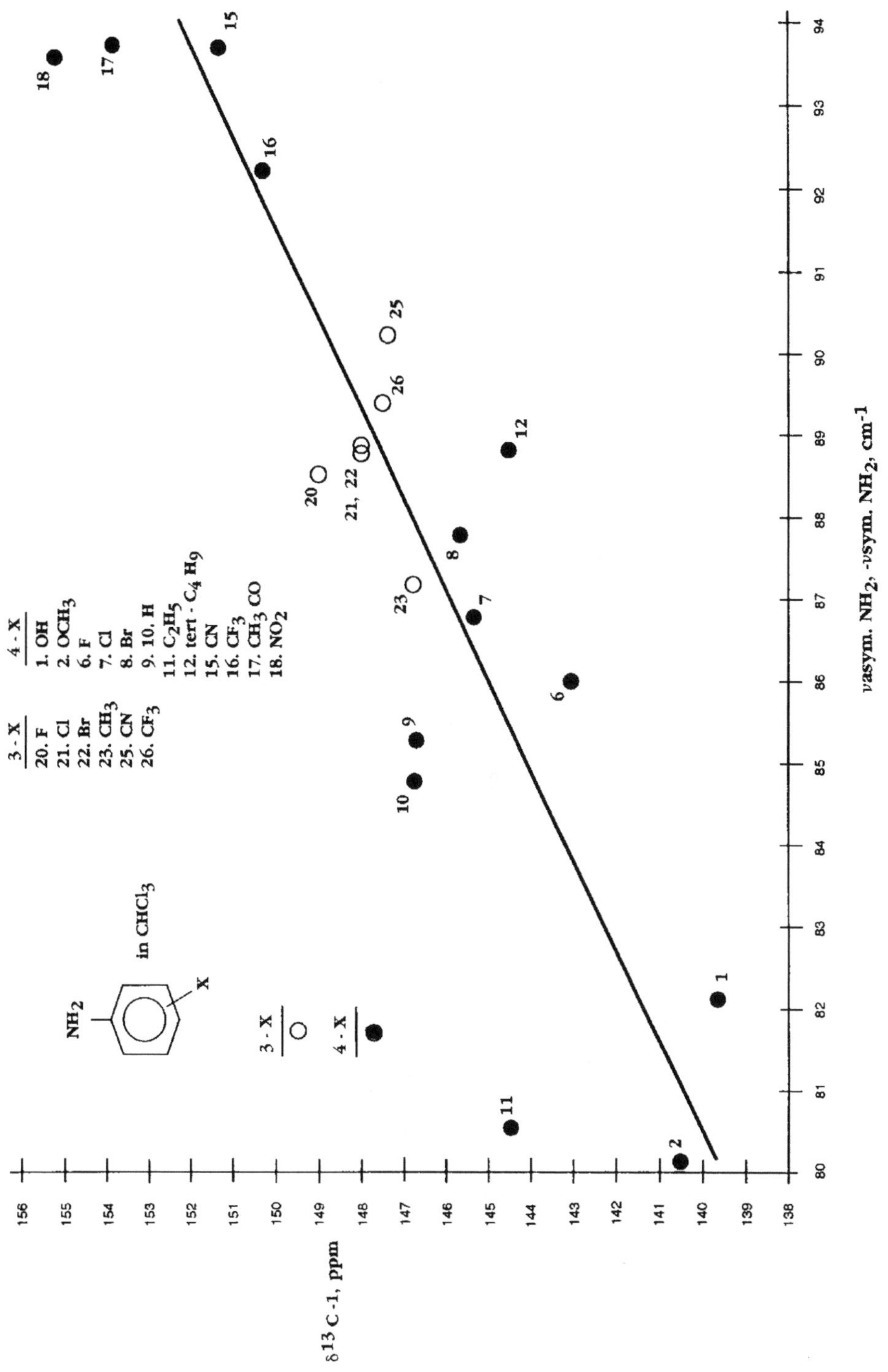

FIGURE 11.13 A plot of the frequency difference between ν asym. NH_2 and ν sym. NH_2 for 3-x- and 4-x-anilines in $CHCl_4$ solution vs δ ^{13}C-1 chemical shift data for 3-x- and 4-x-anilines in $CHCl_3$ solution.

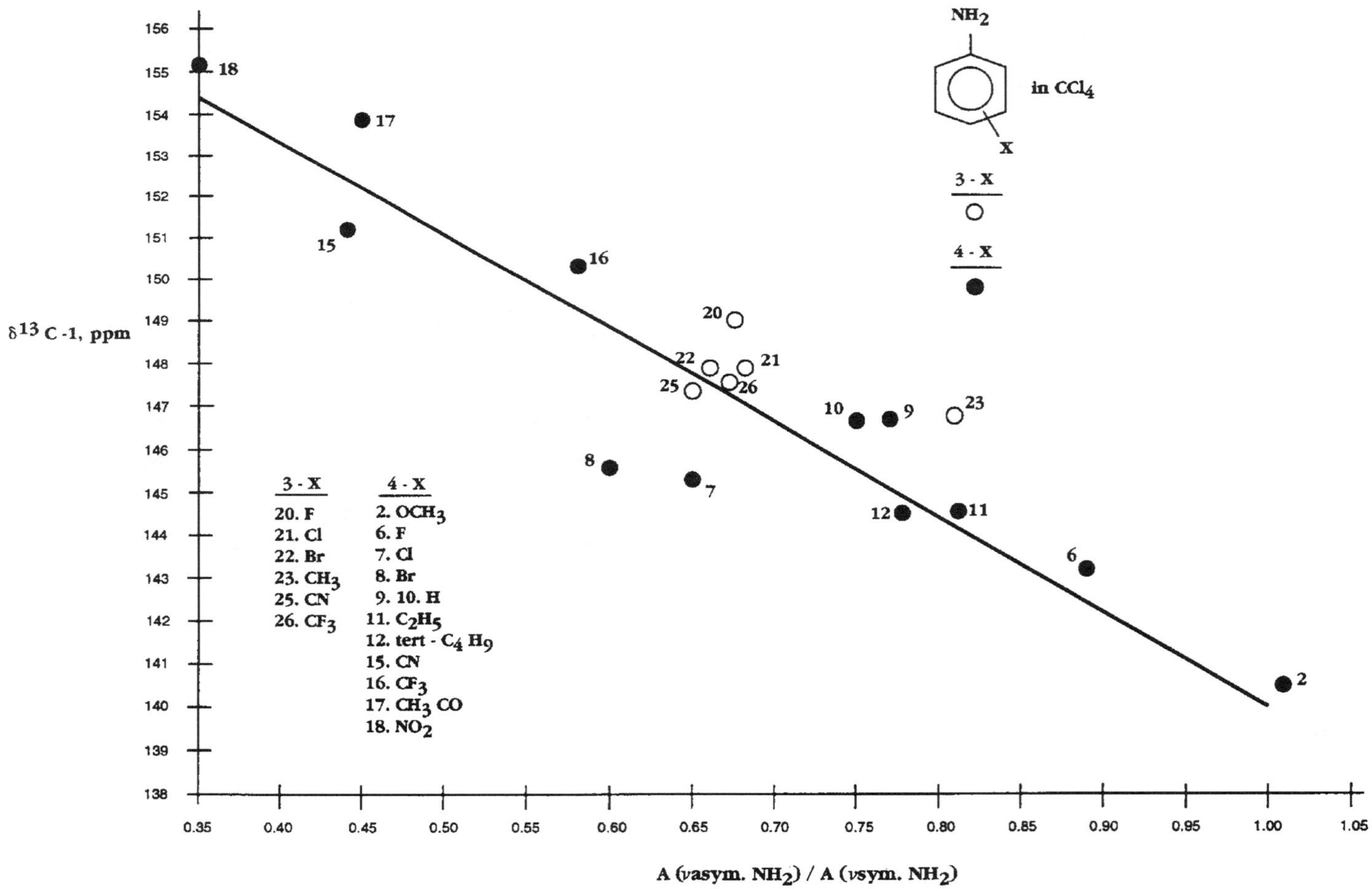

FIGURE 11.14 A plot of the absorbance ratio A(ν asym. NH_2)/A(ν sym. NH_2) for 3-x- and 4-x-anilines in CCl_4 solution vs δ ^{13}C-1 chemical shift data for 3-x- and 4-x-anilines in $CHCl_3$ solution.

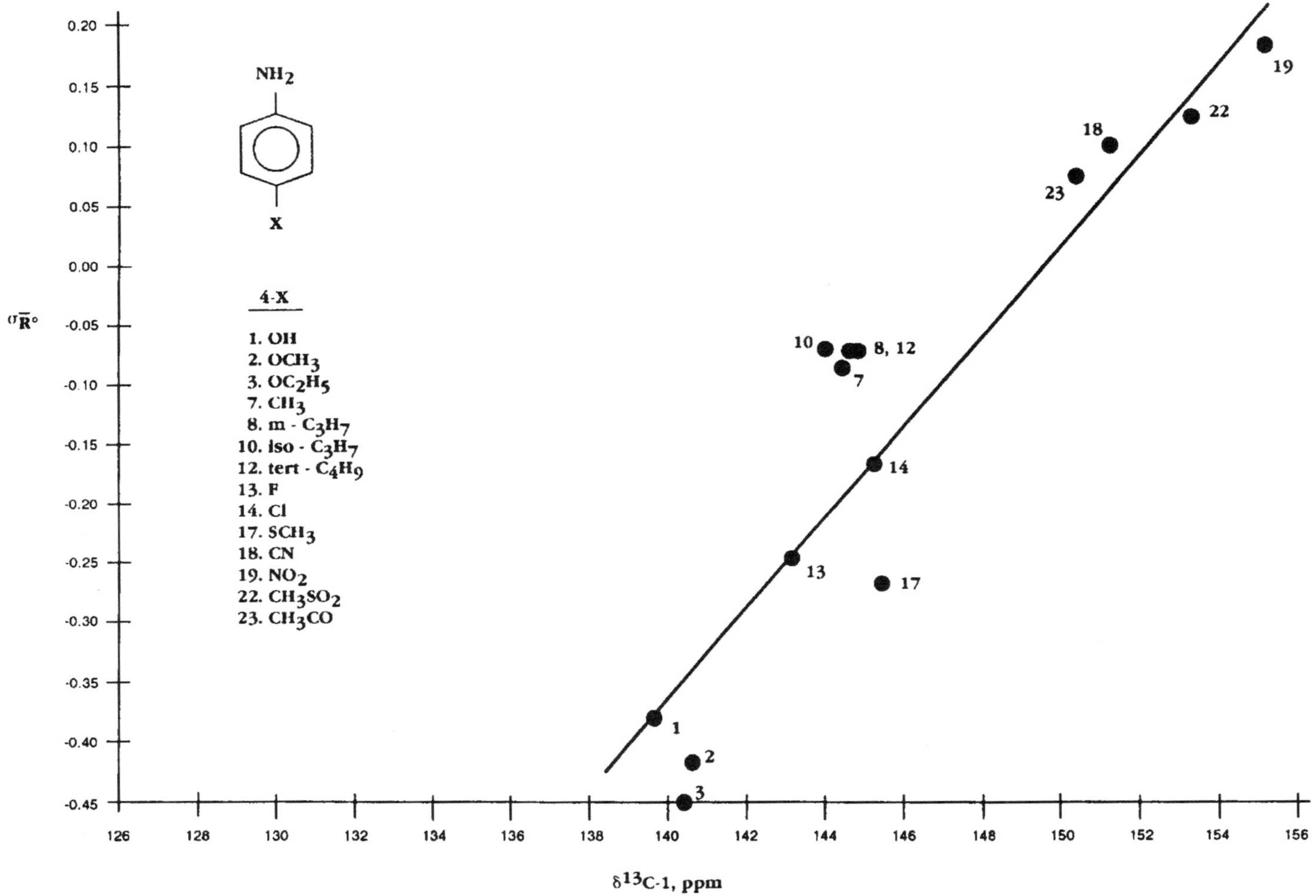

FIGURE 11.15 A plot of δ ^{13}C-1 chemical shift data for 4-x-anilines in $CHCl_3$ solution vs Taft σ_{R° values for the 4-x atom or group.

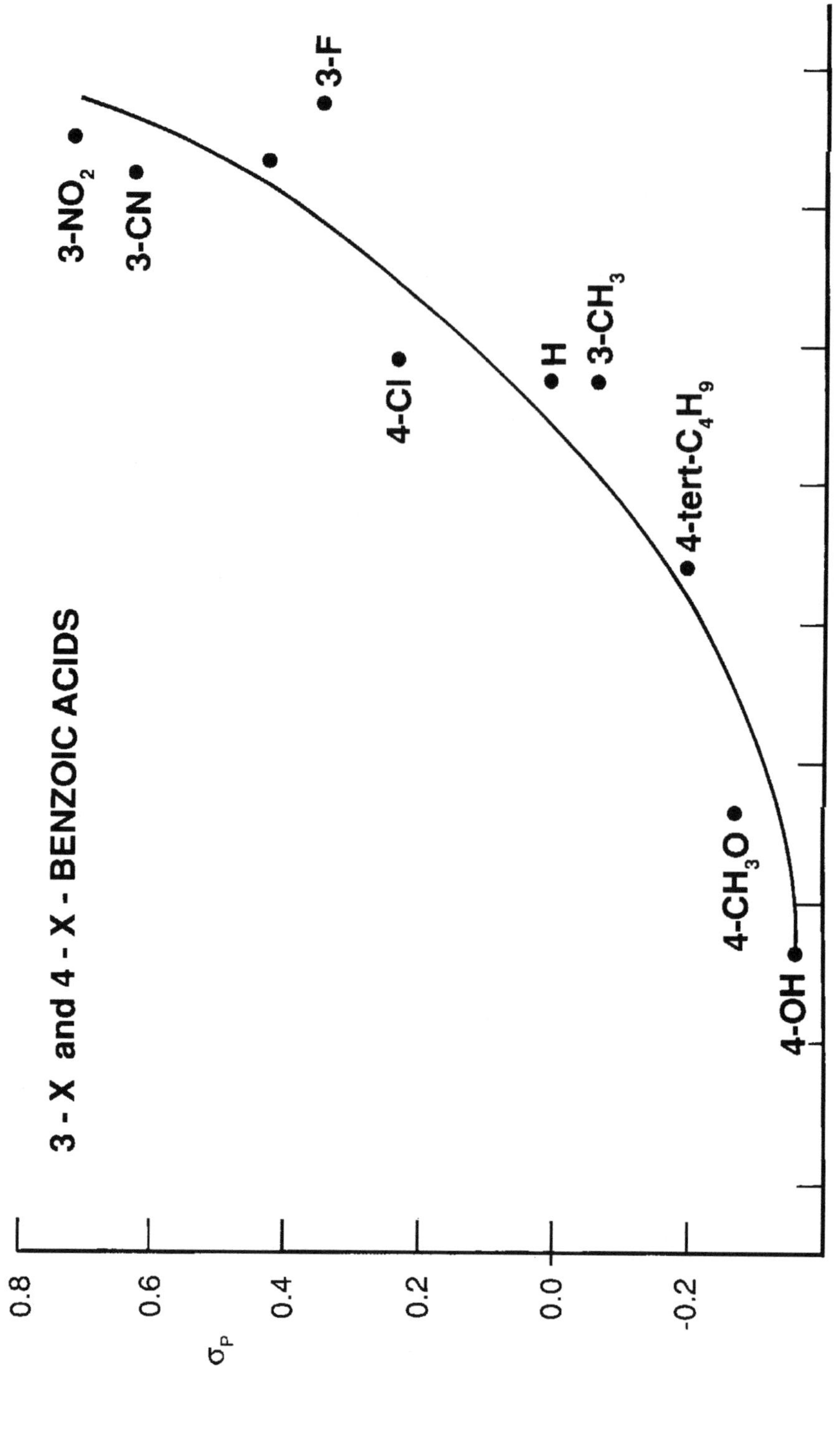

FIGURE 11.16 A plot of δ ^{13}C-1 for 3-x- and 4-x-benzoic acids vs Hammett σ values for the x atom or group.

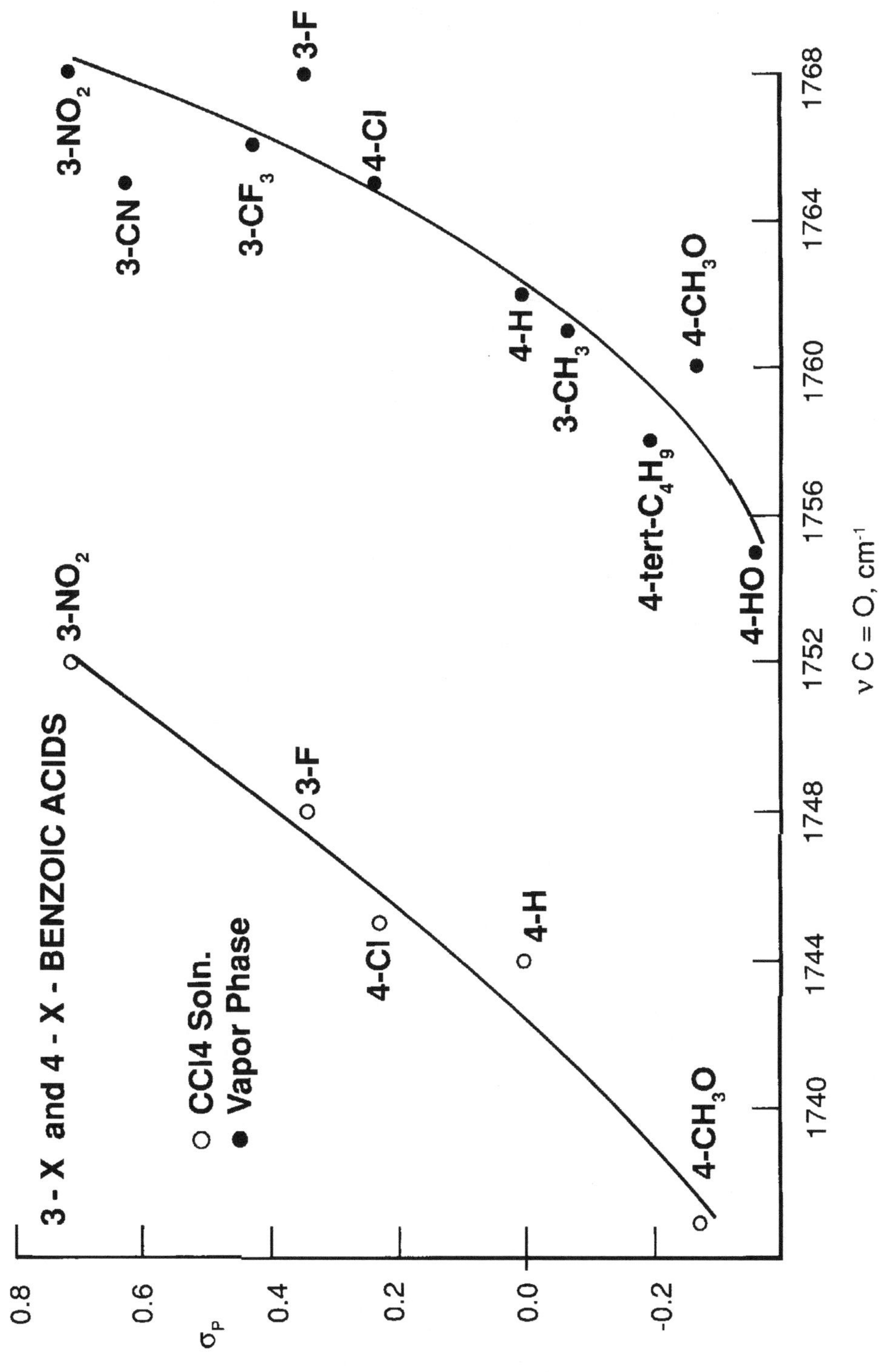

FIGURE 11.17 Plots of ν C=O vs Hammett σ values of 3-x- and 4-x-benzoic acids. The solid circles are for IR data in the vapor phase. The open circles are for IR dilute solution data for unassociated 3-x- and 4-x-benzoic acids.

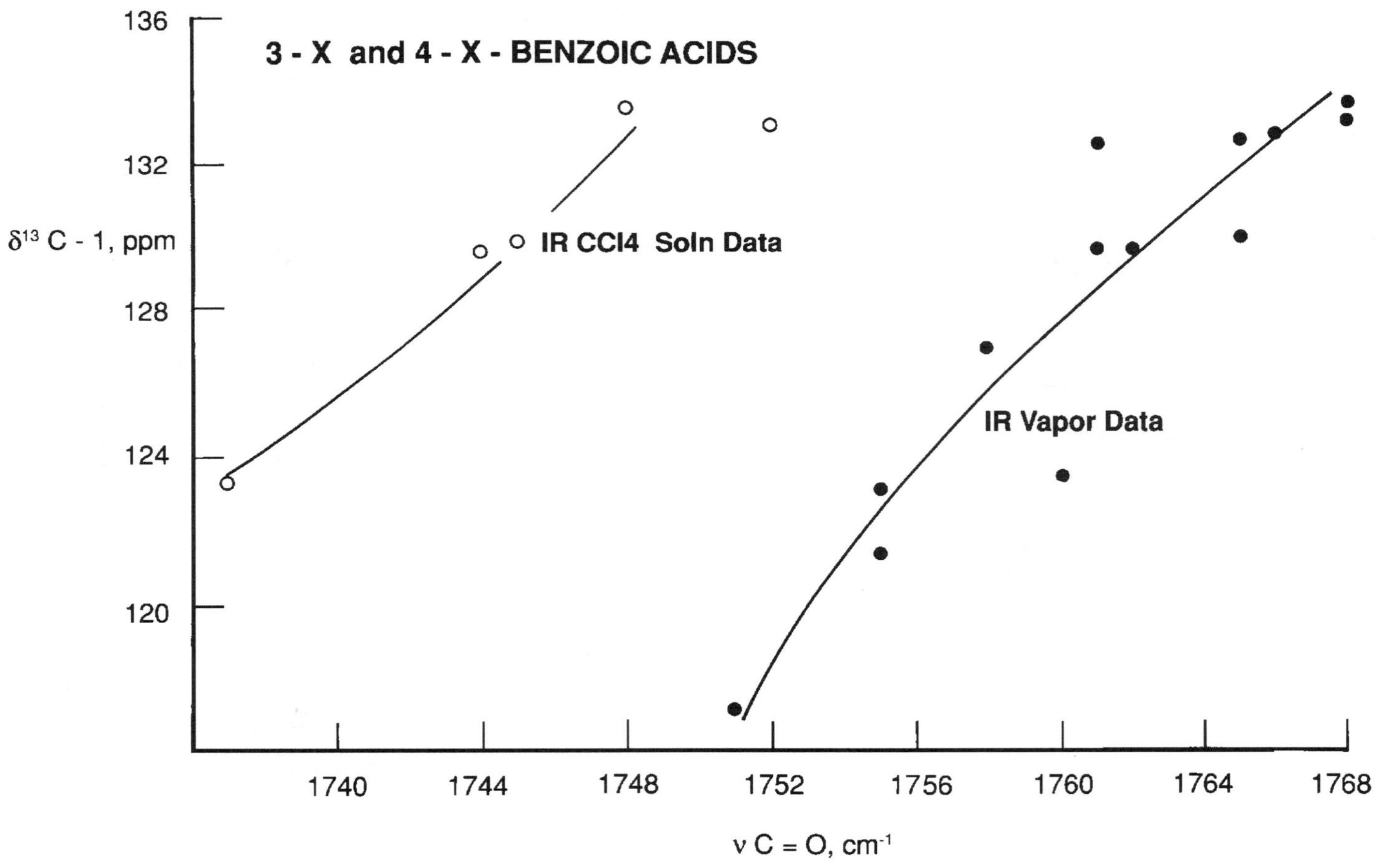

FIGURE 11.18 Plots of ν C=O vs δ ^{13}C-1 for 3-x- and 4-x-benzoic acids. The NMR data are for $CDCl_3$ solutions. The plot with closed circles includes vapor-phase IR data. The plot with open circles includes IR CCl_4-solution data.

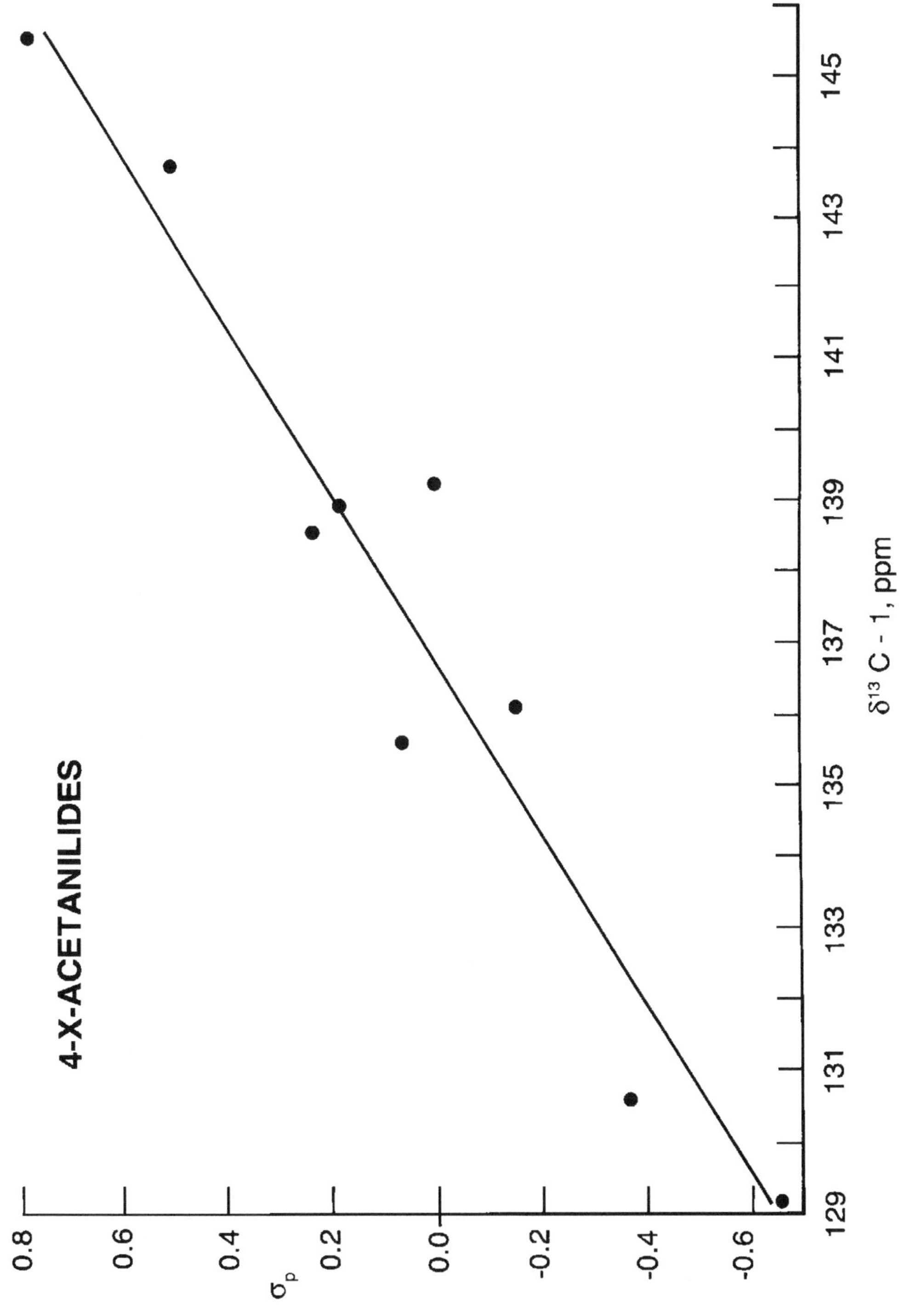

FIGURE 11.19 A plot of δ ^{13}C-1 for 4-x-acetanilides in $CDCl_3$ solution vs Hammett σ_p values for the x-atom or group.

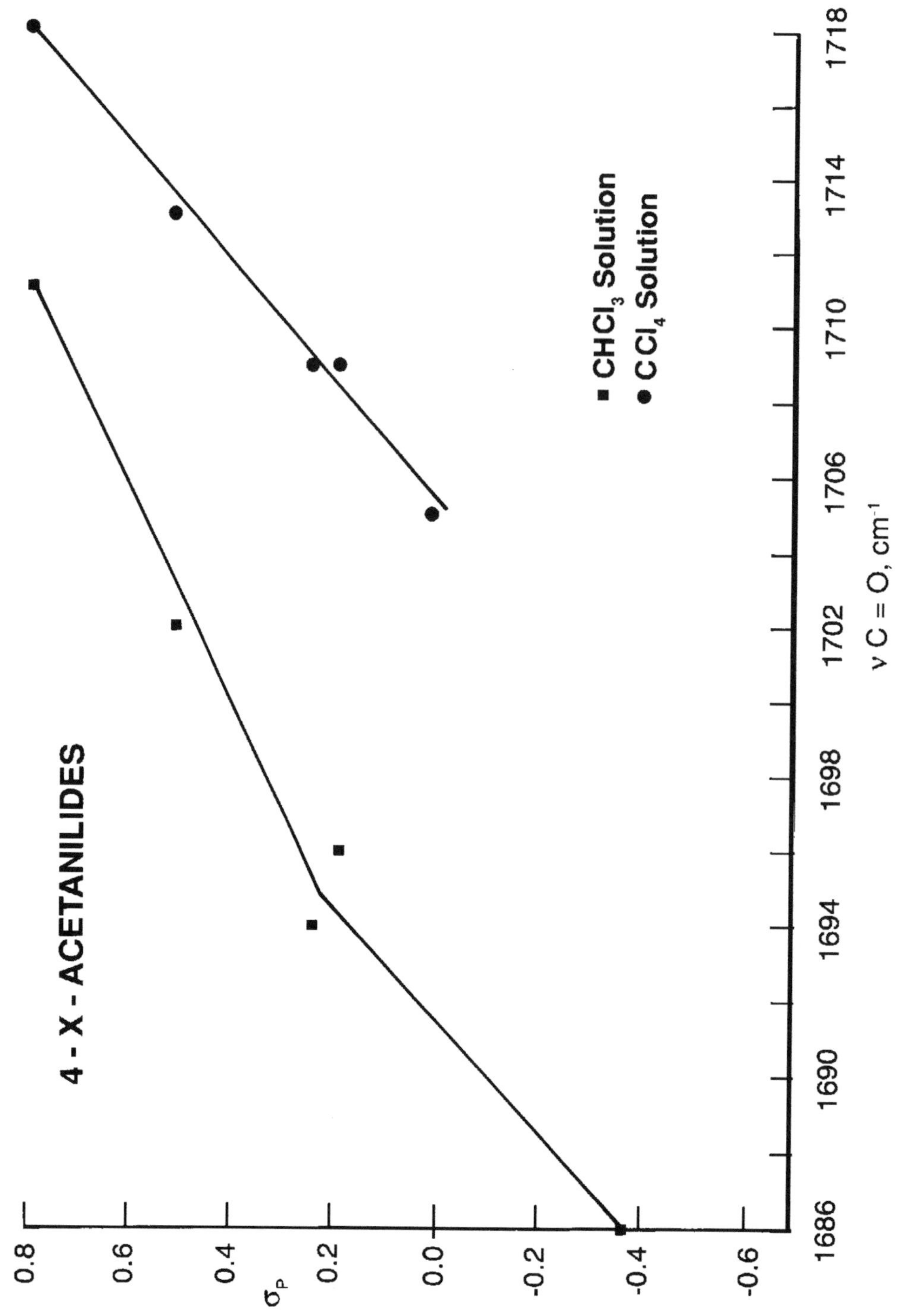

FIGURE 11.20 A plot of ν C=O for 4-x-acetanilides vs Hammett σ_p values. The solid circles represent IR CCl_4 solution data and the solid squares represent IR $CHCl_3$ solution data.

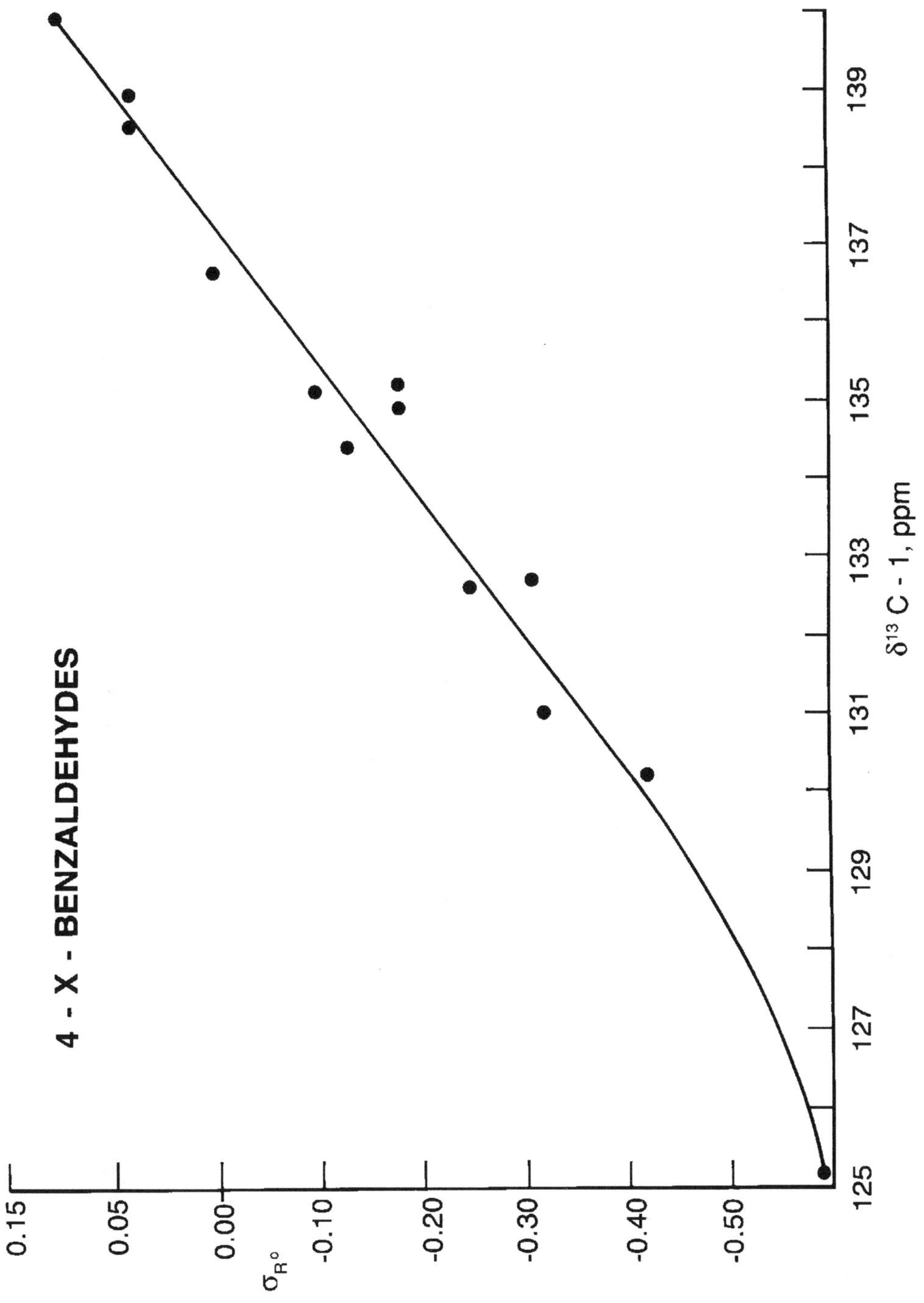

FIGURE 11.21 A plot of δ ^{13}C-1 for 4-x-benzaldehydes vs Taft σ_{R° values for the 4-x atom or group.

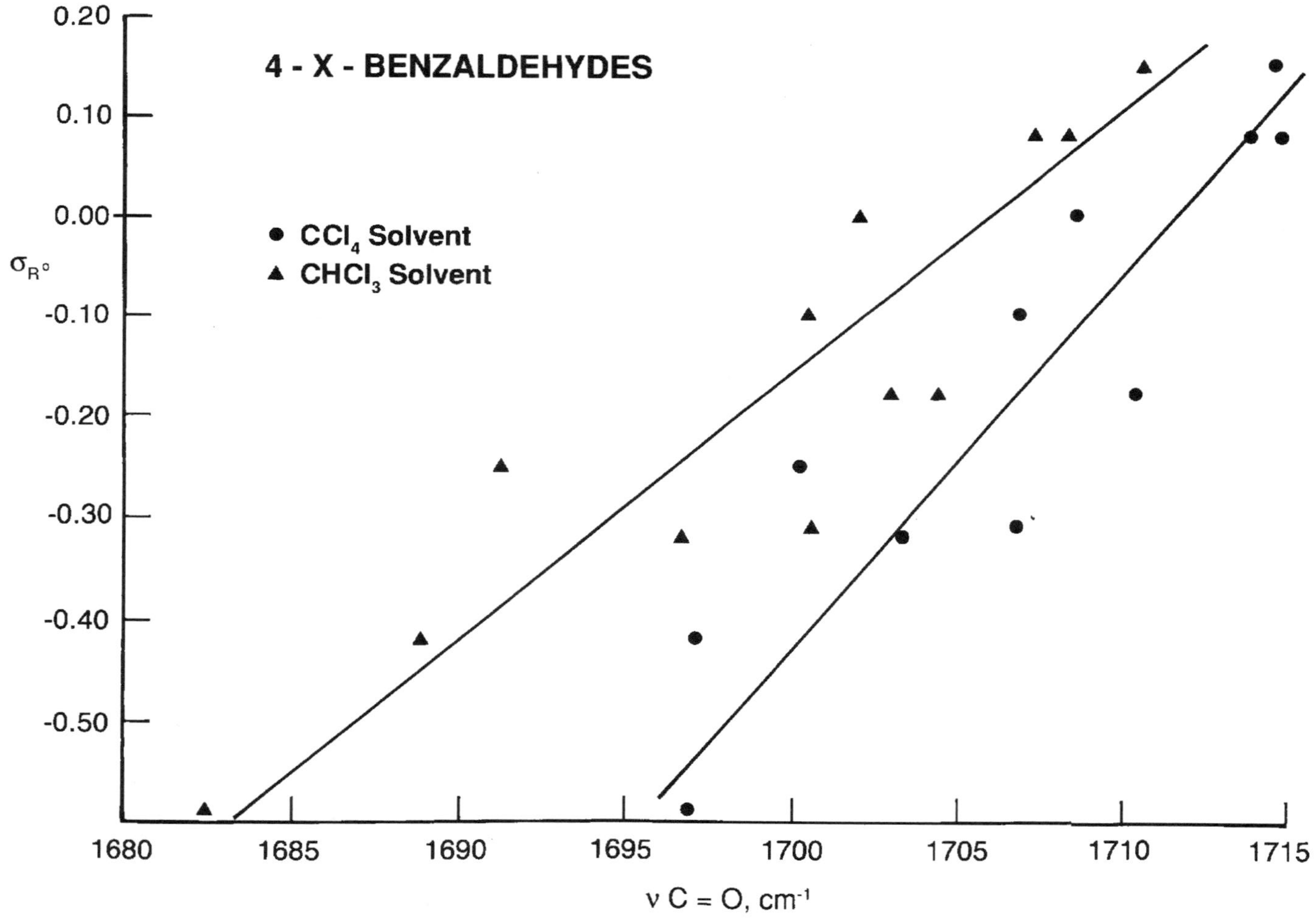

FIGURE 11.22 A plot of ν C=O for 4-x-benzaldehydes vs Taft σ_{R° values for the 4-x atom or group. The solid circles represent IR data in CCl_4 solutions and the solid triangles represent IR data in $CHCl_3$ solutions.

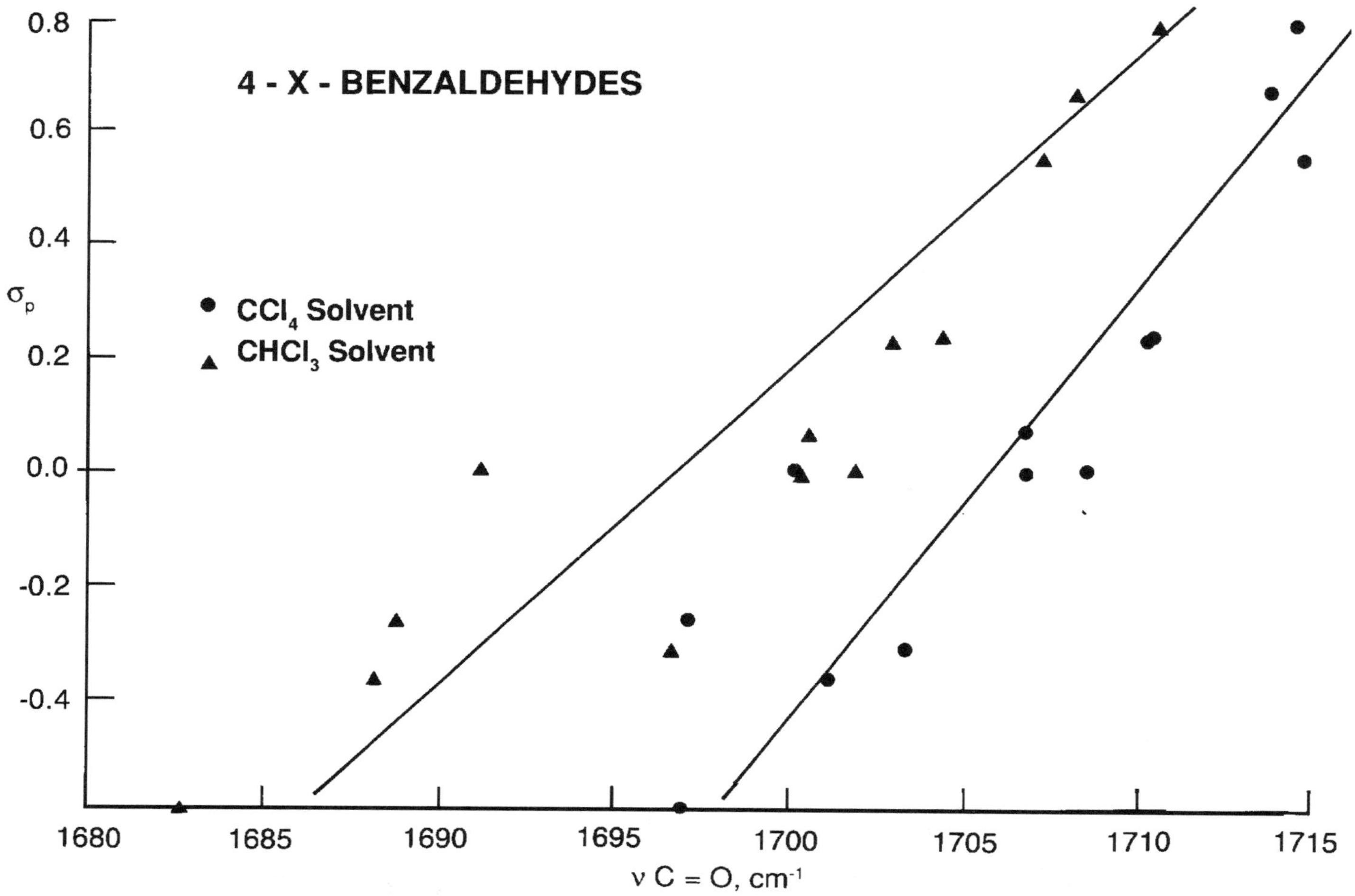

FIGURE 11.23 A plot of ν C=O for 4-x-benzaldehydes vs Hammett σ_p values for the 4-x atom or group. The solid circles represent IR data in CCl_4 solution and the solid triangles represent IR data in $CHCl_3$ solutions.

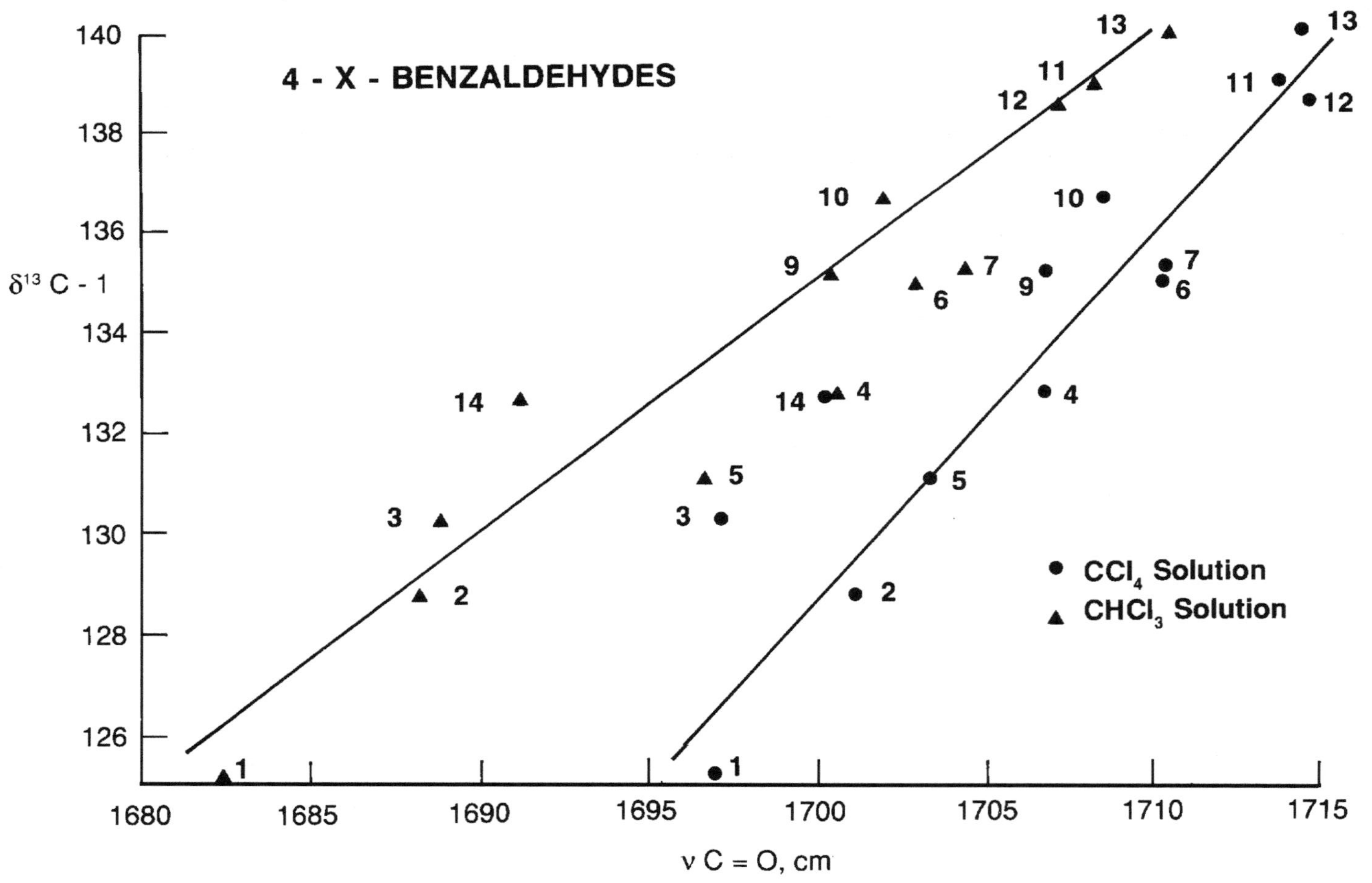

FIGURE 11.24 Plots of ν C=O vs δ ^{13}C-1 for 4-x-benzaldehydes. The δ ^{13}C-1 data are for $CDCl_3$ solutions. The plot with filled-in circles includes IR CCl_4 solution data and the plot with the filled-in triangles includes IR $CHCl_3$ solution data.

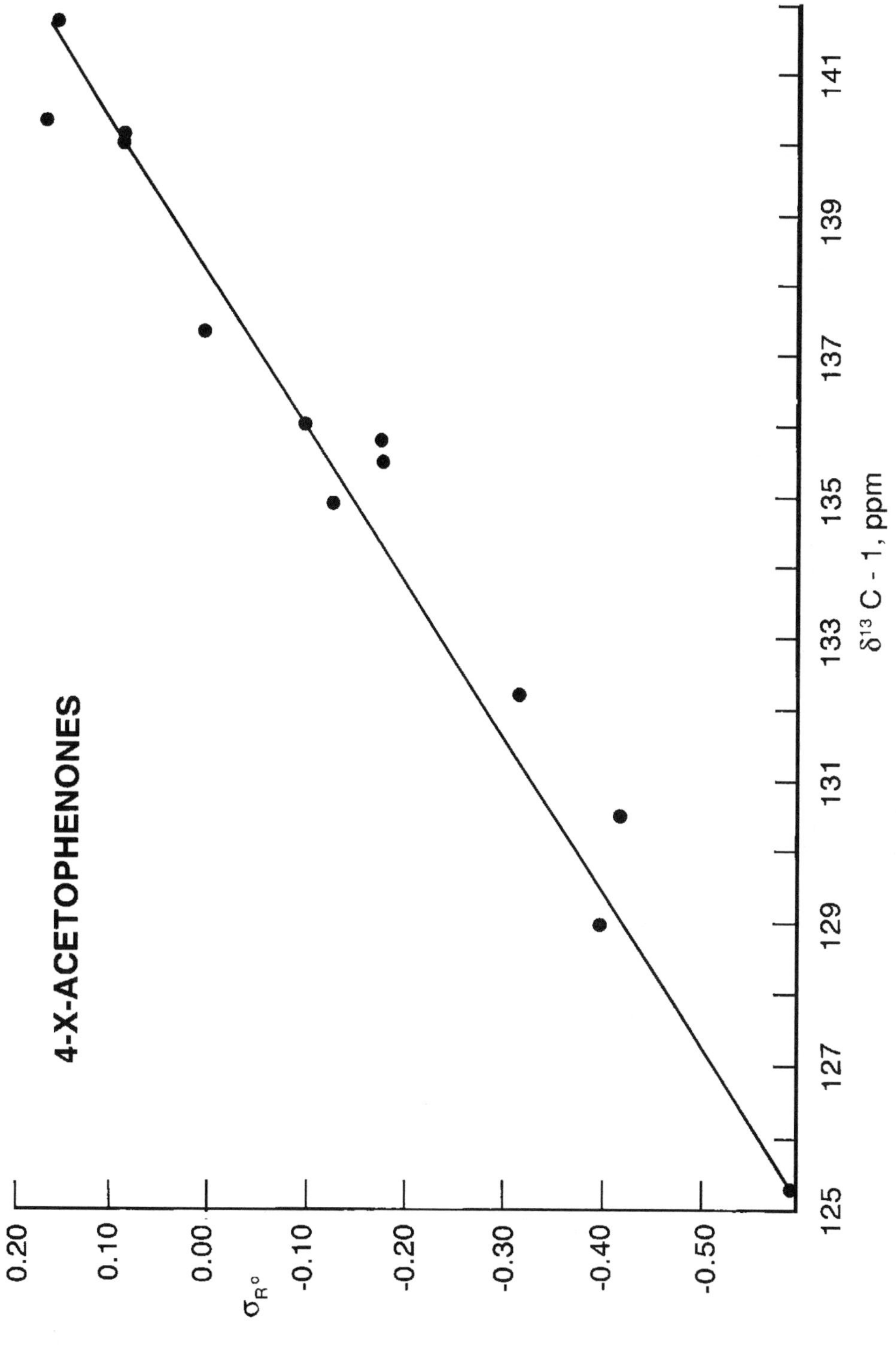

FIGURE 11.25 A plot of δ ^{13}C-1 for 4-x-acetophenones in $CDCl_3$ solutions vs Taft σ_{R° values for the 4-x atom or groups.

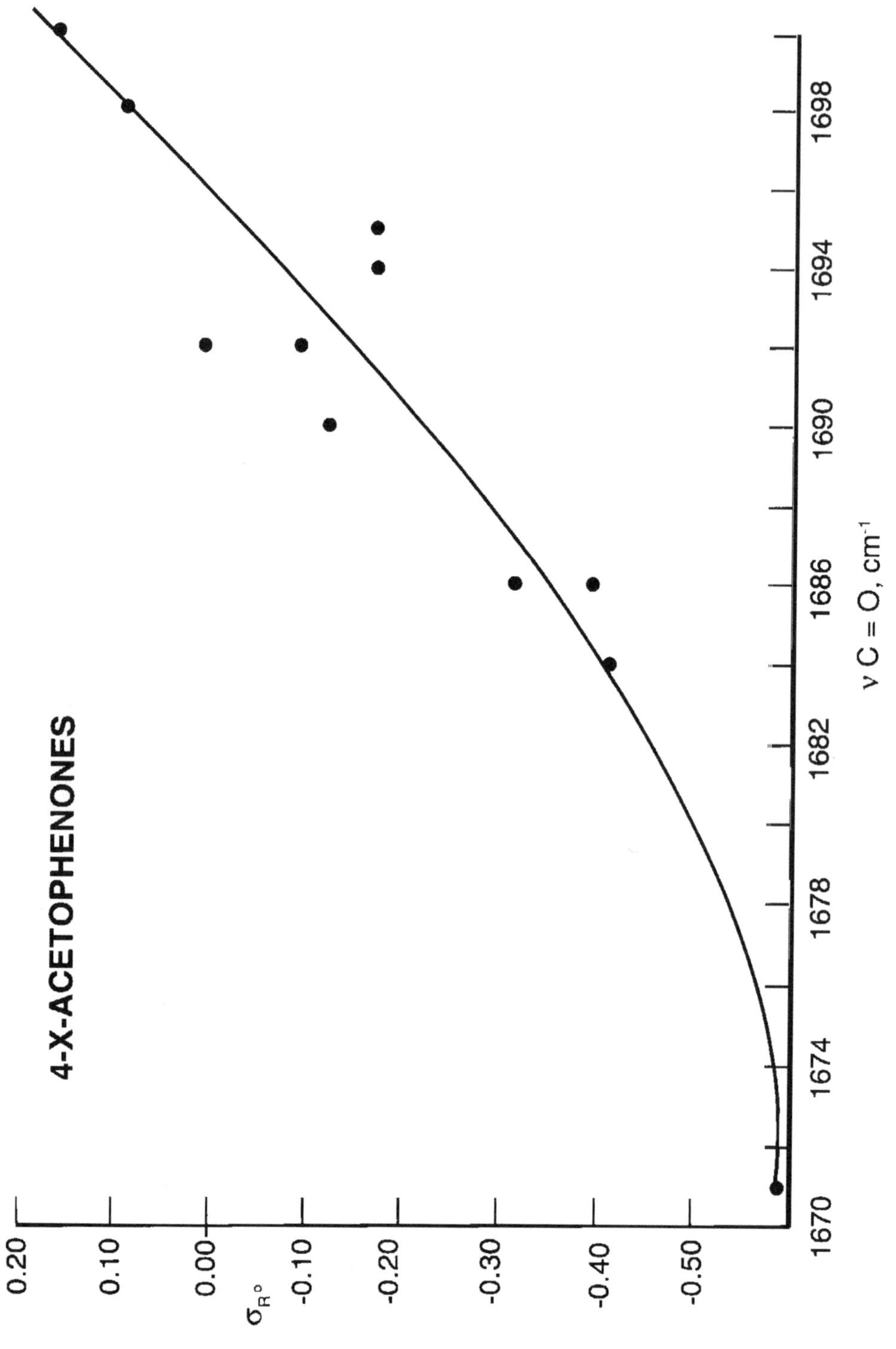

FIGURE 11.26 A plot of ν C=O for 4-x-acetophenones vs Taft $\sigma_{R°}$ values for the 4-x atom or group. The IR data is that for the 4-x-acetophenones in $CHCl_3$ solution.

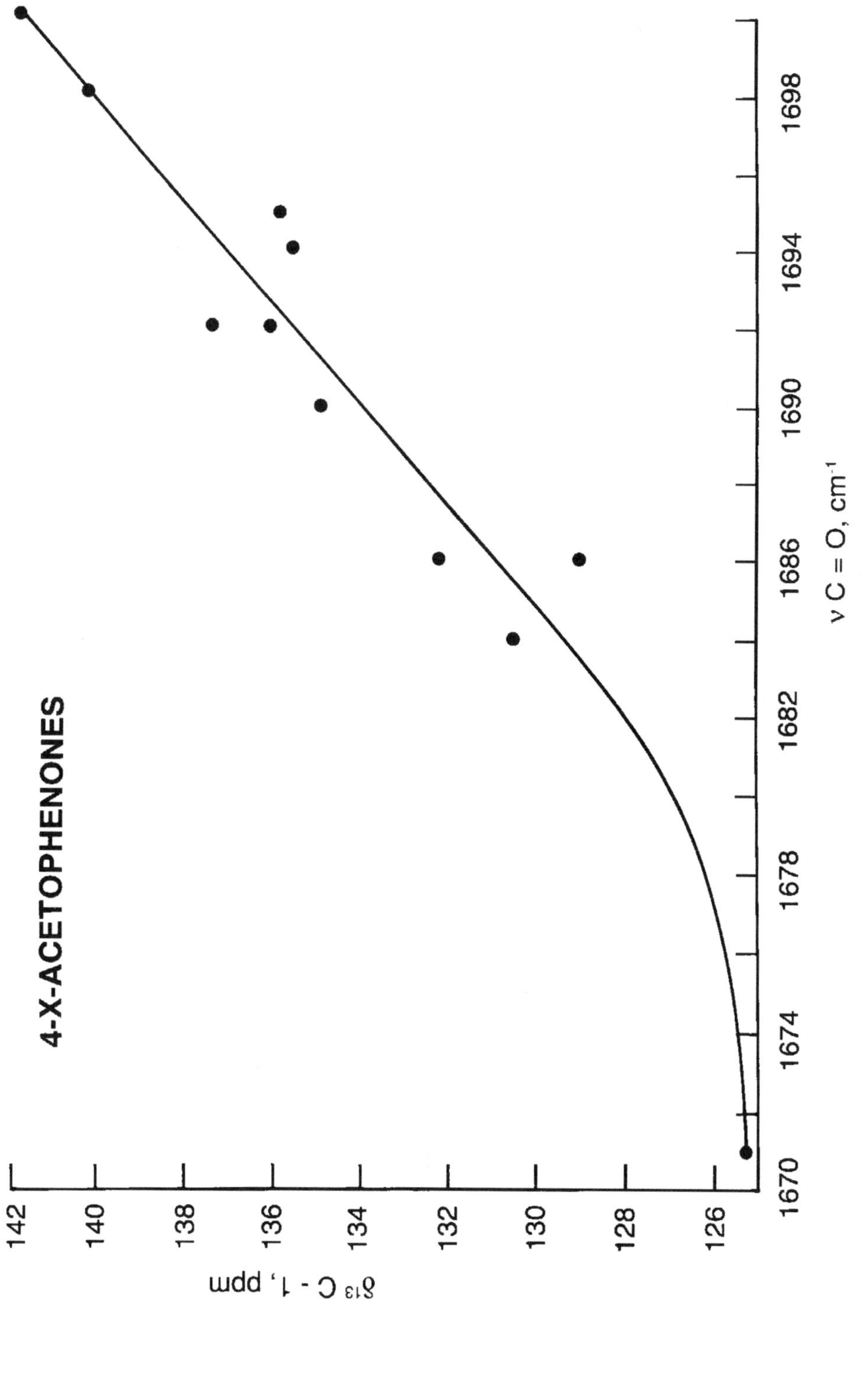

FIGURE 11.27 A plot of ν C=O ($CHCl_3$ solution) vs δ ^{13}C-1 ($CDCl_3$ solution) for 4-x-acetophenones.

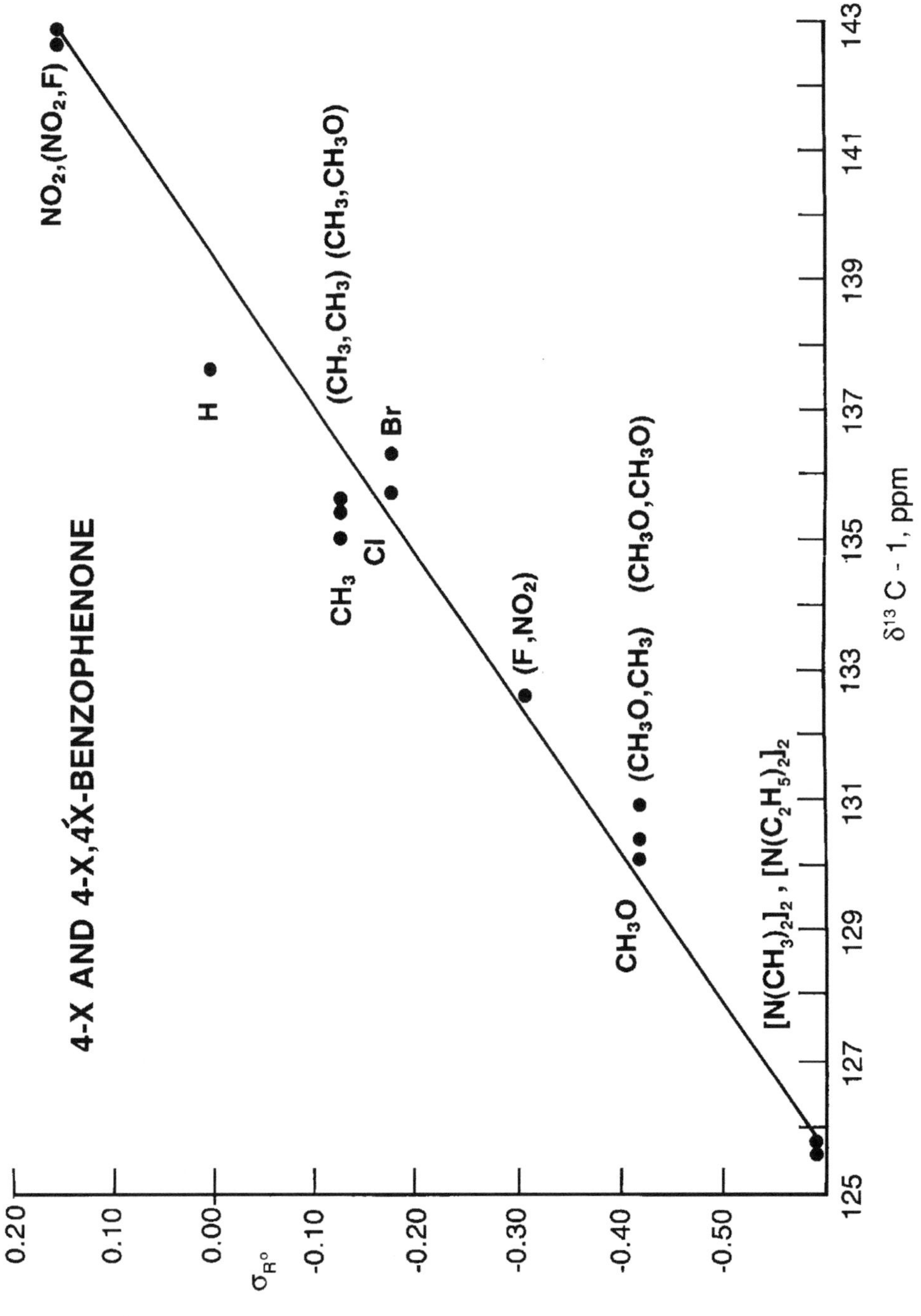

FIGURE 11.28 A plot of δ ^{13}C-1 vs Taft σ_{R° values for the 4-x and 4,4′-x,x atoms or groups for 4-x and 4,4′-x,x-benophenones. These data are for $CHCl_3$ solutions.

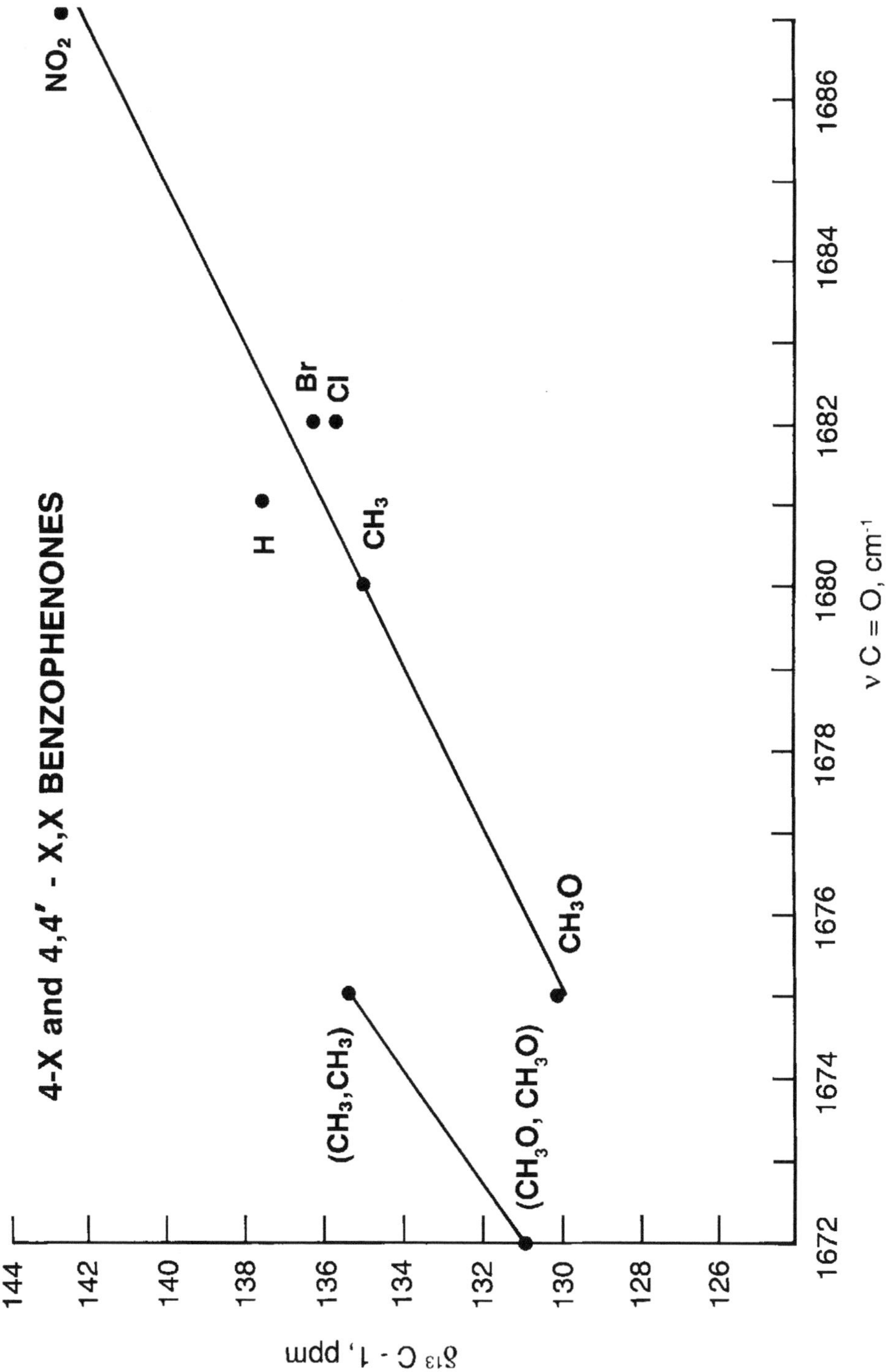

FIGURE 11.29 A plot of ν C=O in the vapor phase vs δ ^{13}C-1 in CCl_3 solution for 4-x and 4,4′-x,x-benzophenones.

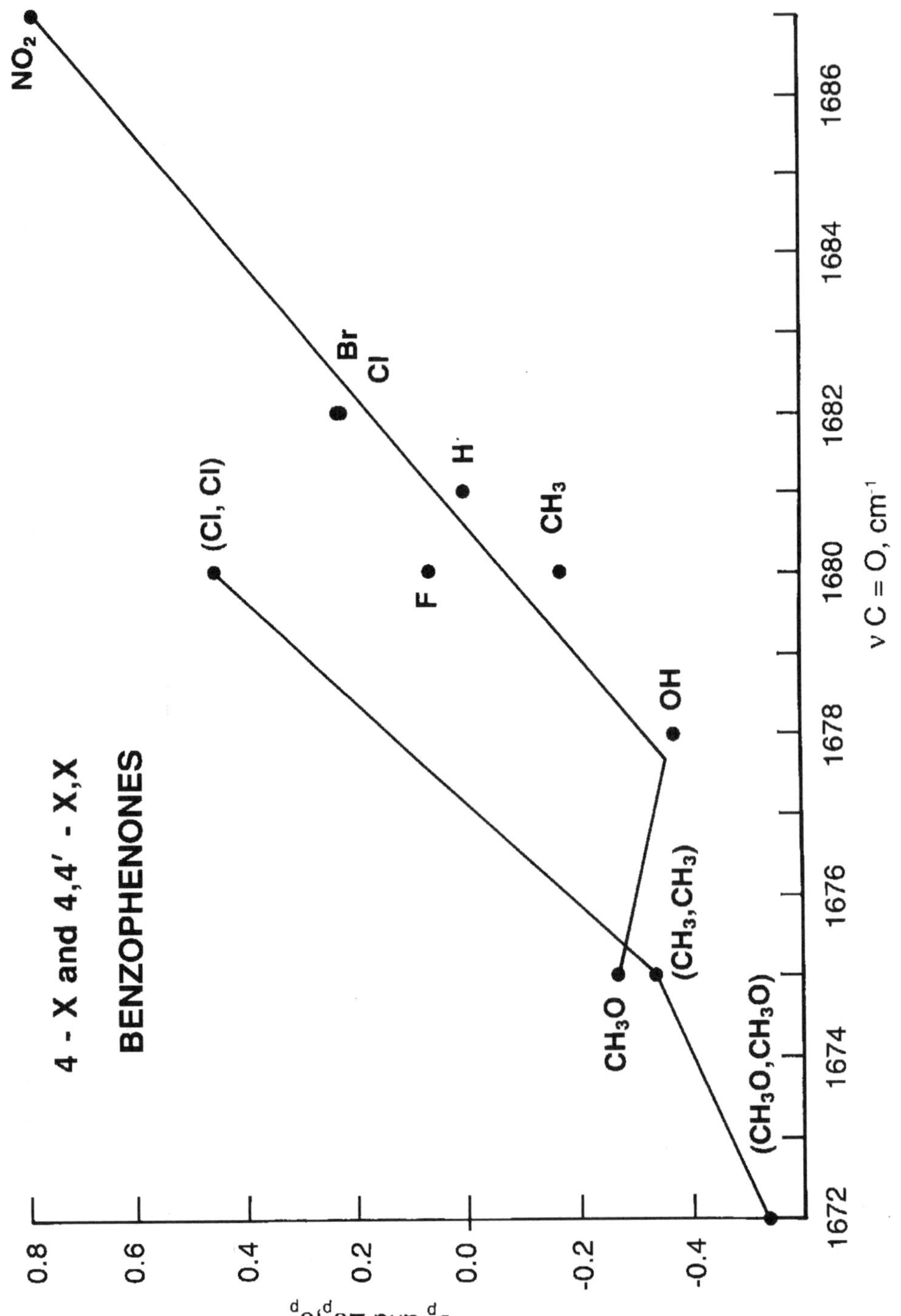

FIGURE 11.30 A plot of ν C=O in the vapor phase vs the sum of Hammett σ_p for the x-atom or group for 4-x- and 4,4′-x,x-benzophenones.

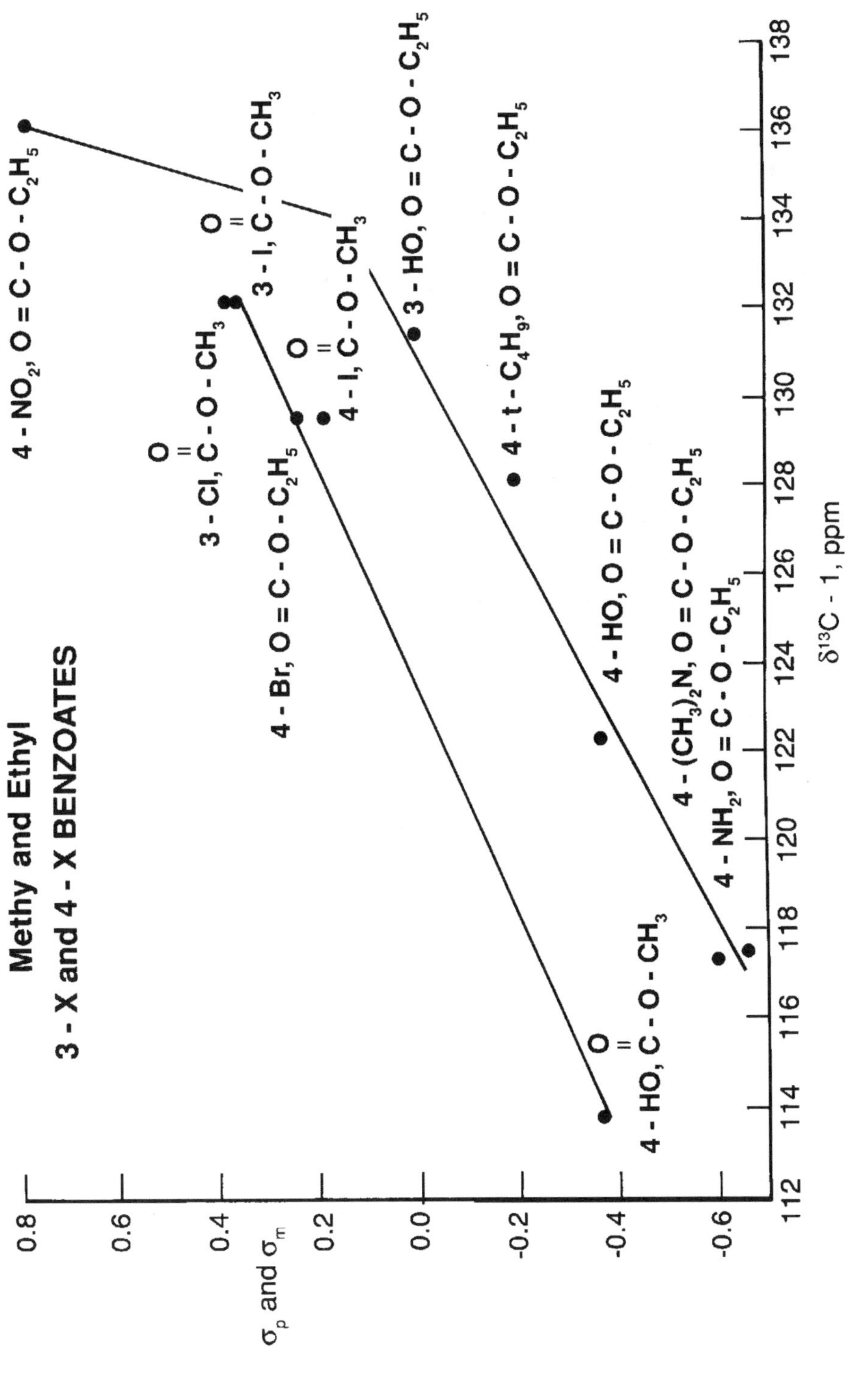

FIGURE 11.31 A plot of δ ^{13}C-1 in CHCl$_3$ solution vs Hammett σ values for the x atom or group for methyl and ethyl 3-x- and 4-x-benzoates.

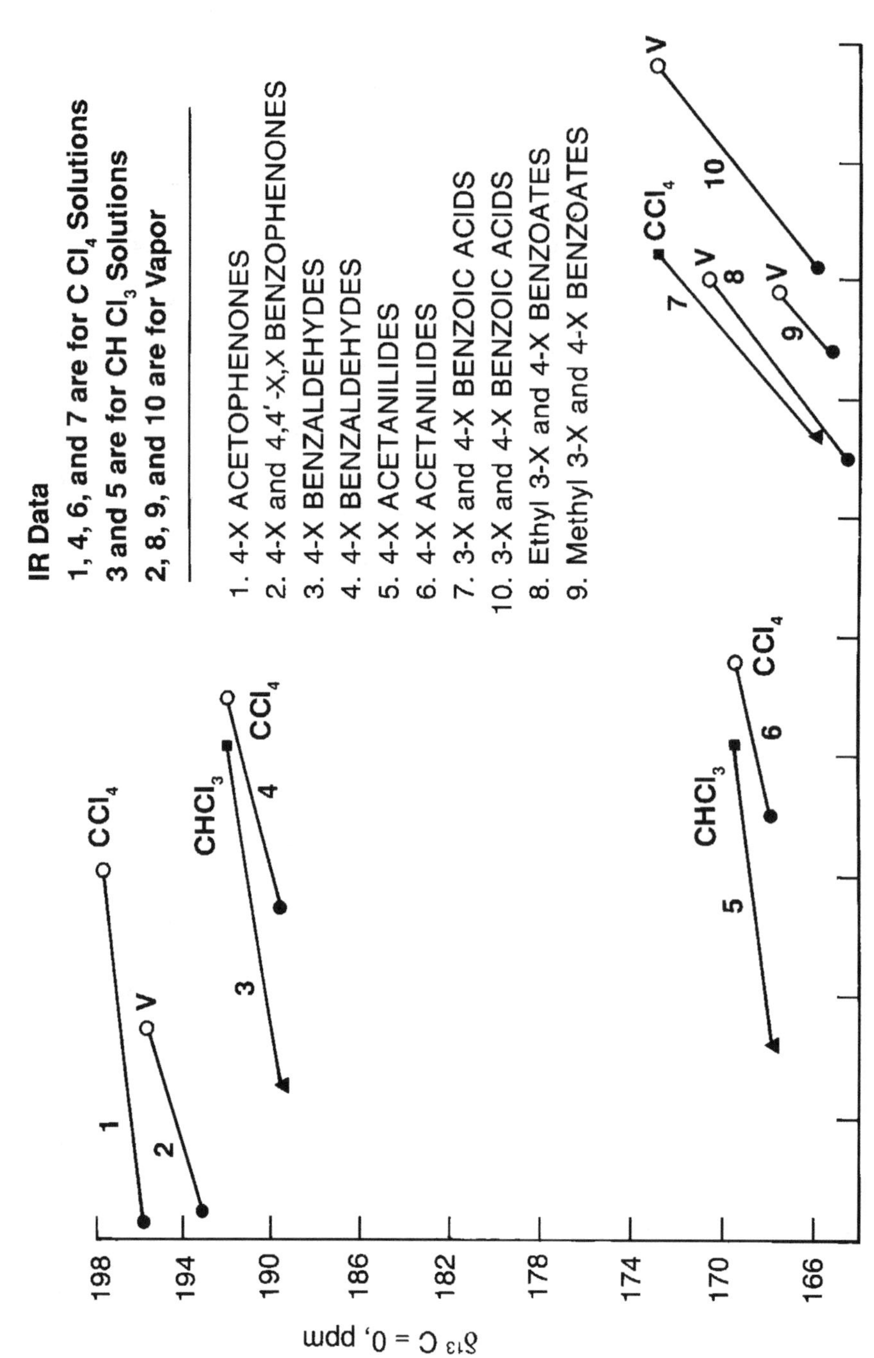

FIGURE 11.32 Plots of the range of ν C=O for the compounds studied in different physical phases vs the range of δ ^{13}C=O for the compounds studied in $CHCl_3$ solutions.

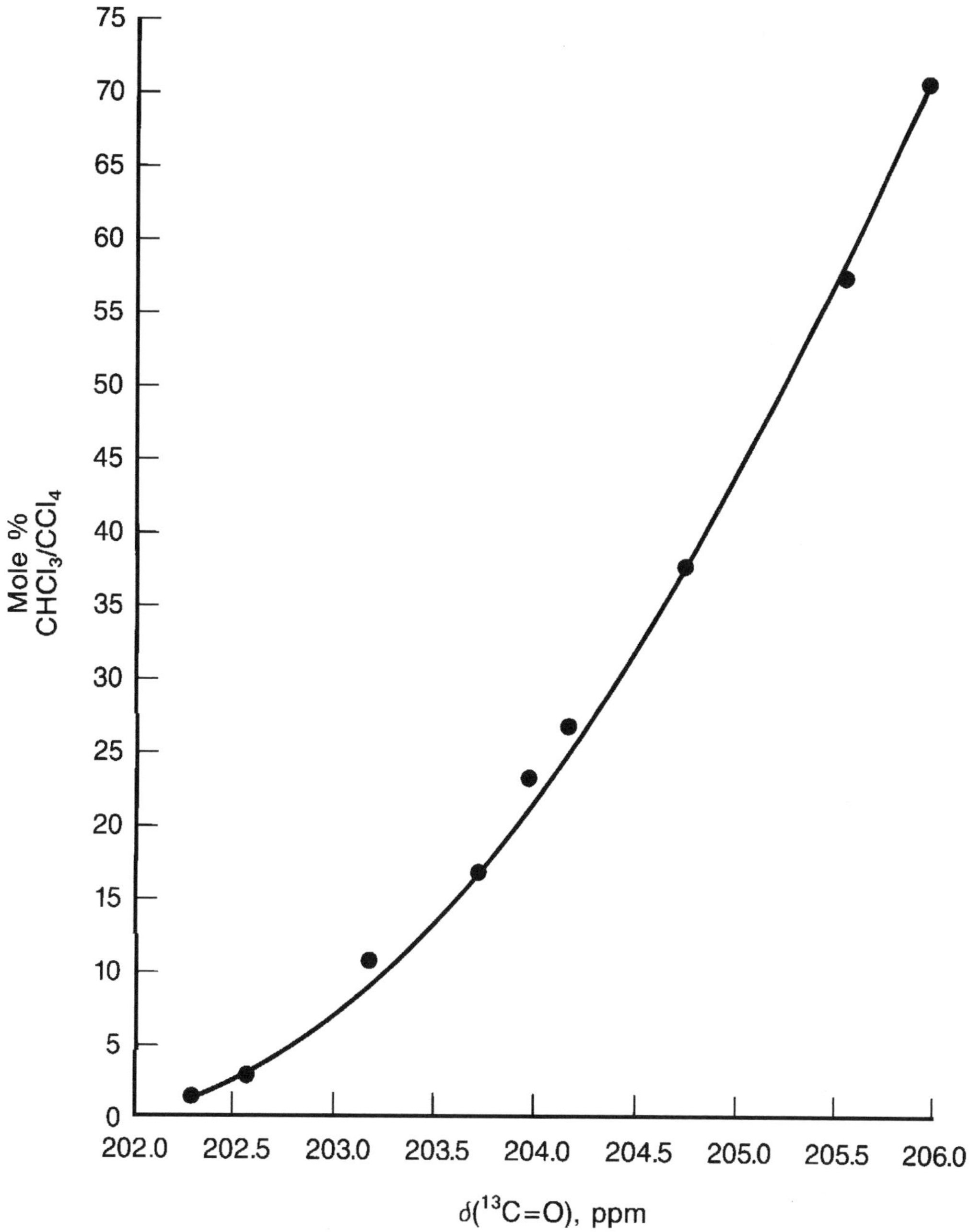

FIGURE 11.33 A plot of δ $^{13}C{=}O$ for acetone vs mole % $CHCl_3/CCl_4$ solutions.

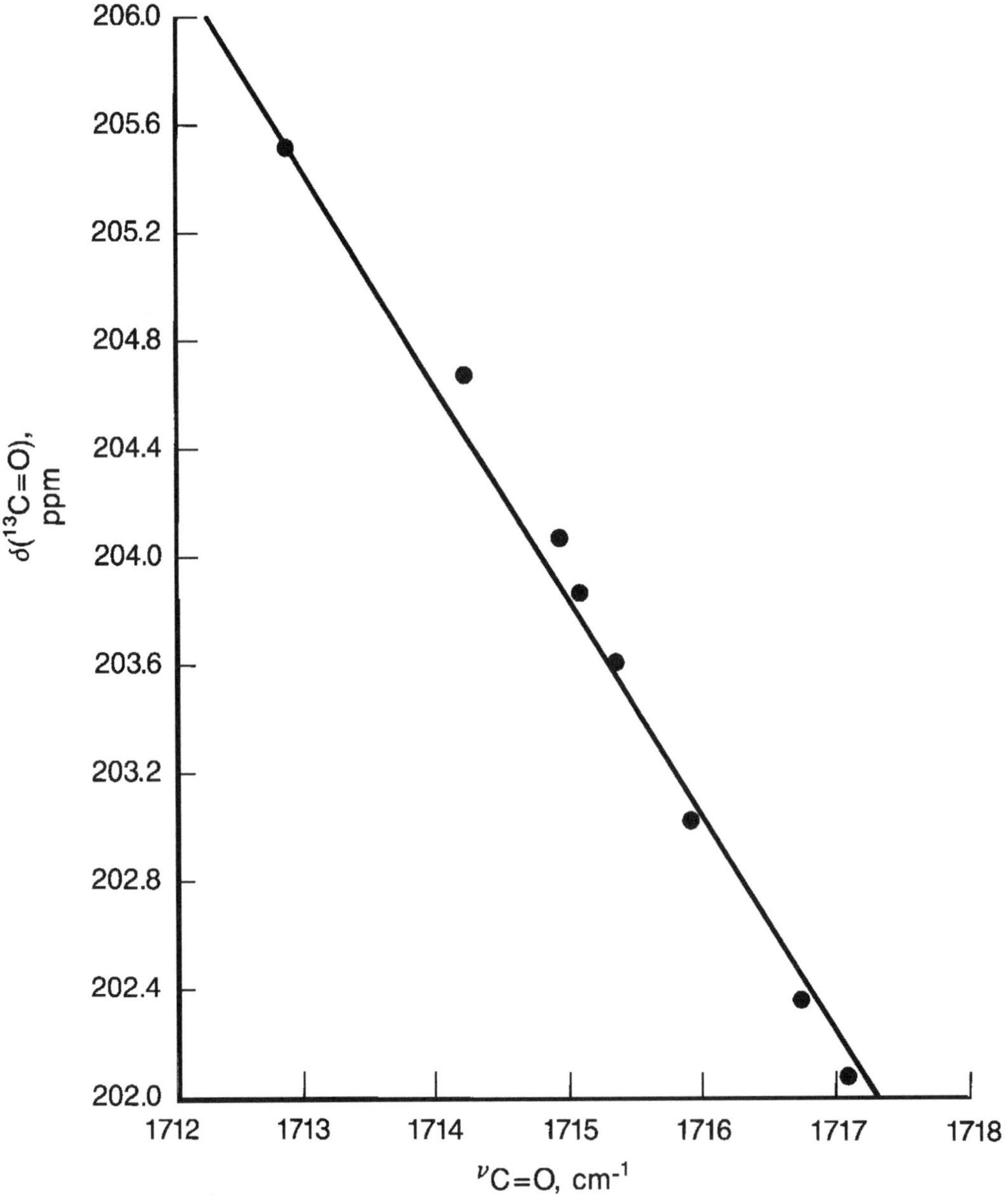

FIGURE 11.34 A plot of ν C=O vs δ ^{13}C=O for acetone in mole % $CHCl_3/CCl_4$ solutions.

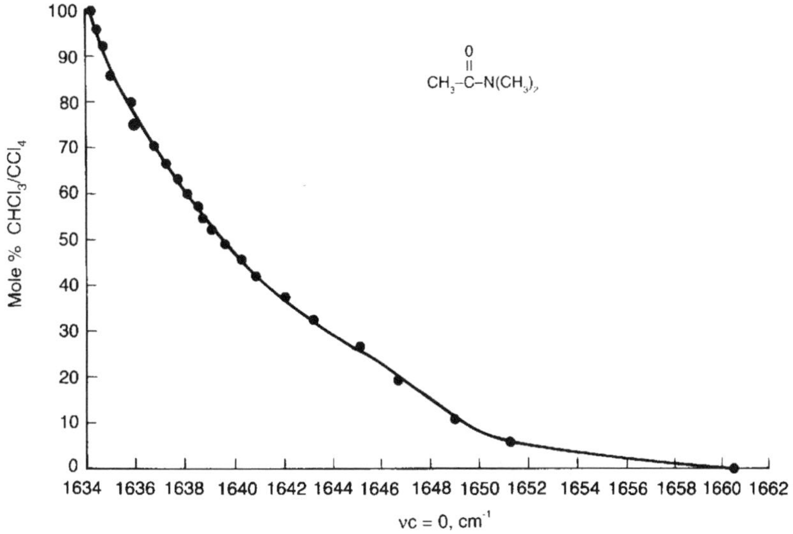

FIGURE 11.35 A plot of ν C=O for dimethylacetamide in CCl_4 and/or $CHCl_3$ solution vs mole % $CHCl_3/CCl_4$.

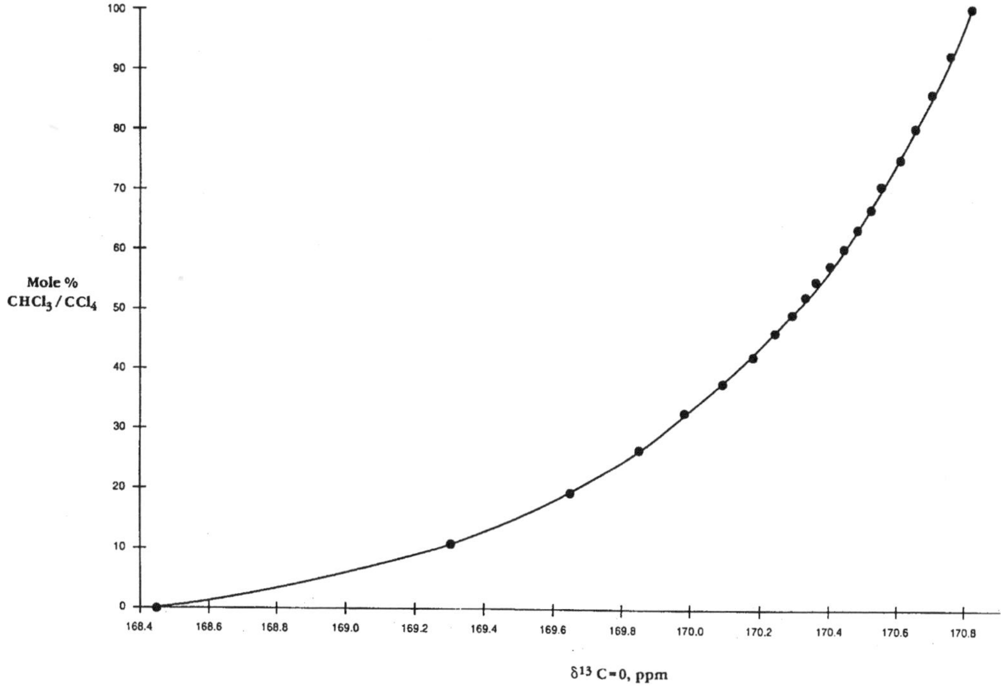

FIGURE 11.35a A plot of δ ^{13}C=O for dimethylacetamide in CCl_4 and/or $CHCl_3$ solution vs mole % $CHCl_3/CCl_4$.

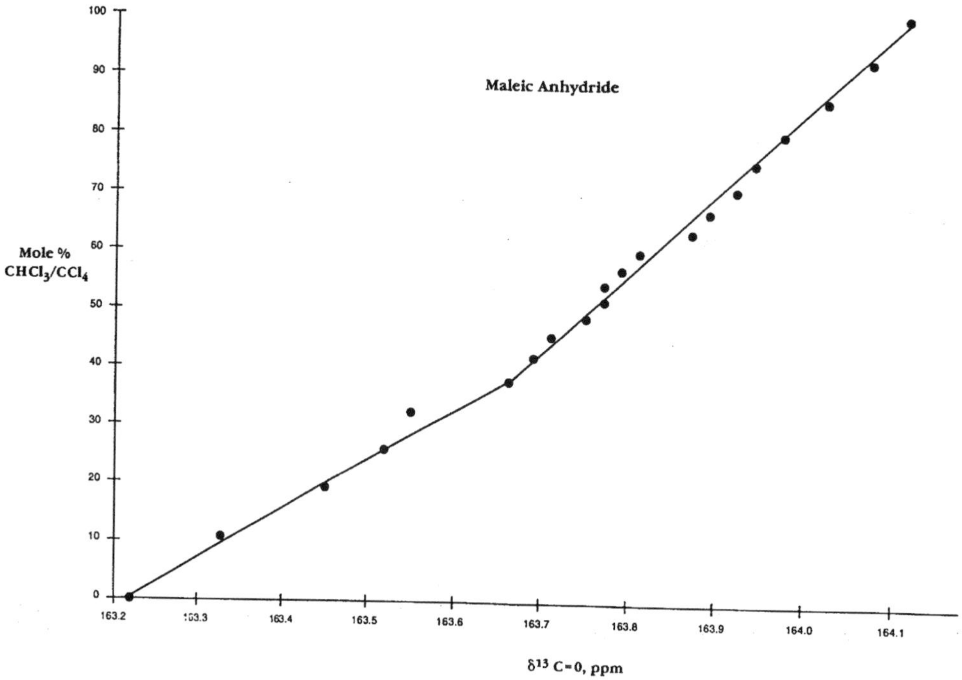

FIGURE 11.35b A plot of δ $^{13}C{=}O$ for maleic anhydride vs mole % $CHCl_3/CCl_4$.

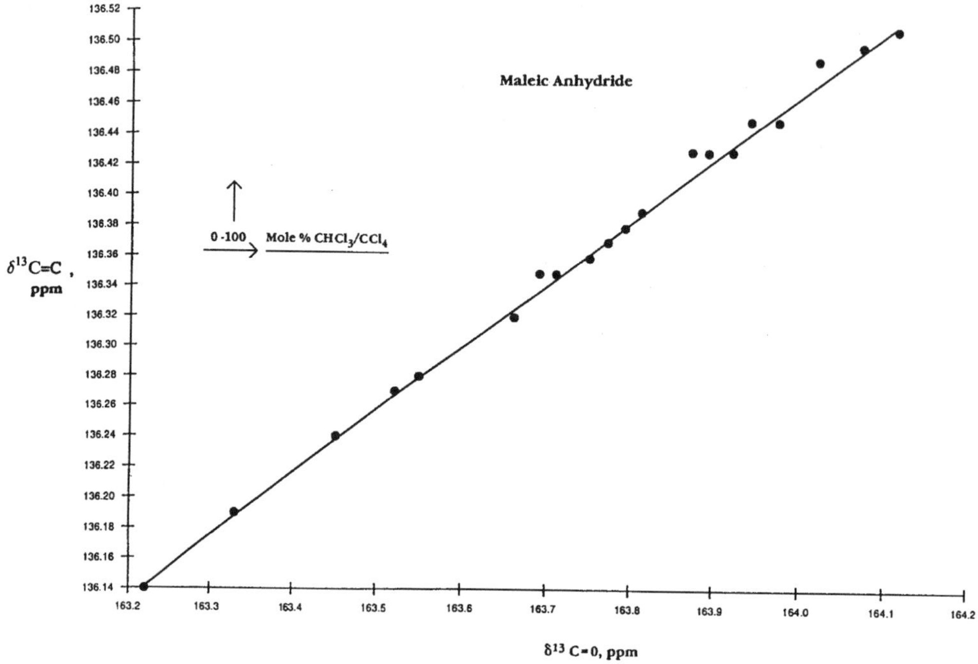

FIGURE 11.35c A plot of δ $^{13}C{=}O$ vs δ $^{13}C{=}C$ for maleic anhydride in 0 to 100 mol% $CHCl_3/CCl_4$ solutions.

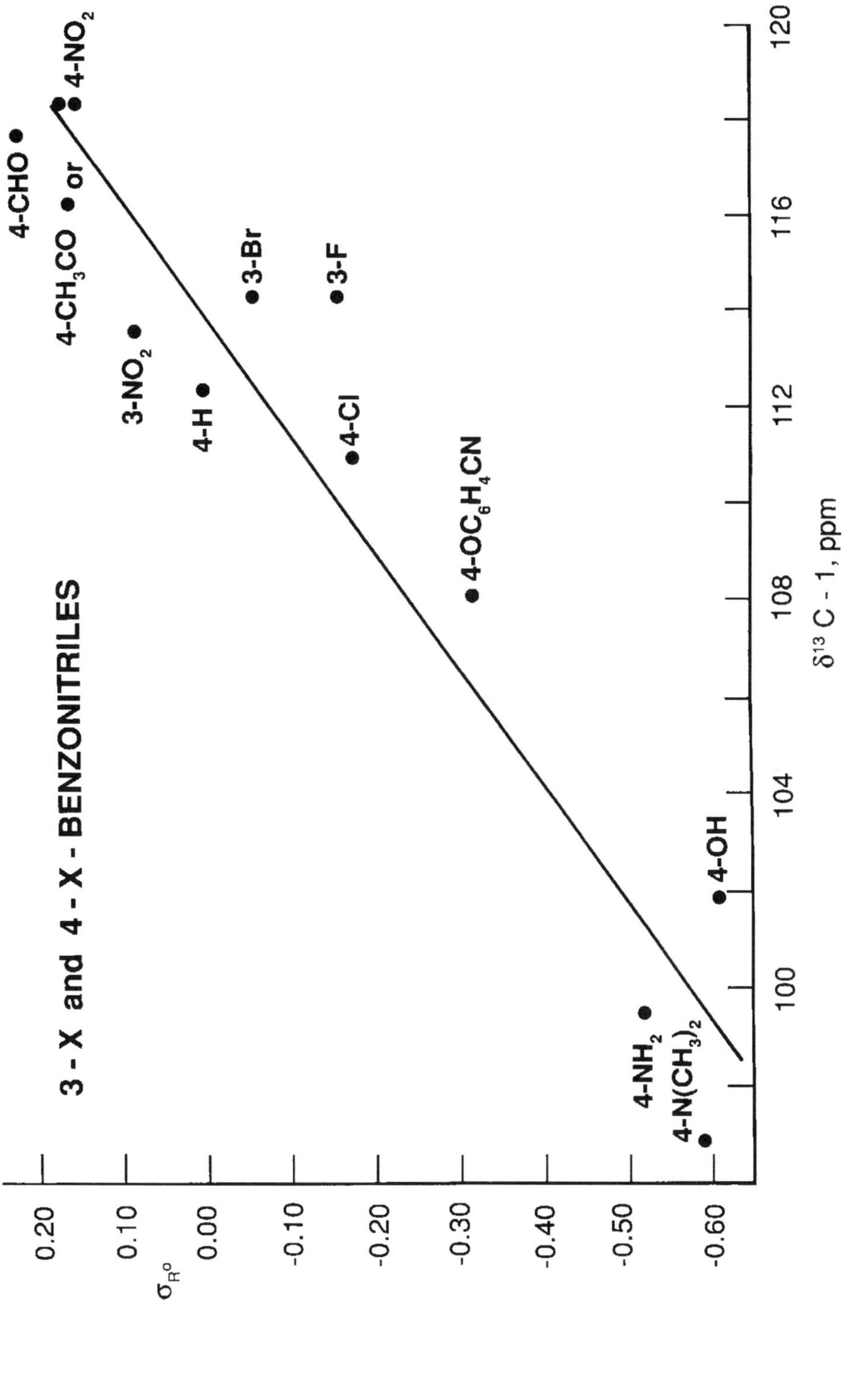

FIGURE 11.36 A plot of δ ^{13}C-1 vs Taft $\sigma_{R°}$ values for the 3-x and 4-x atom or group for 3-x- and 4-x-benzonitriles in $CDCl_3$ solution.

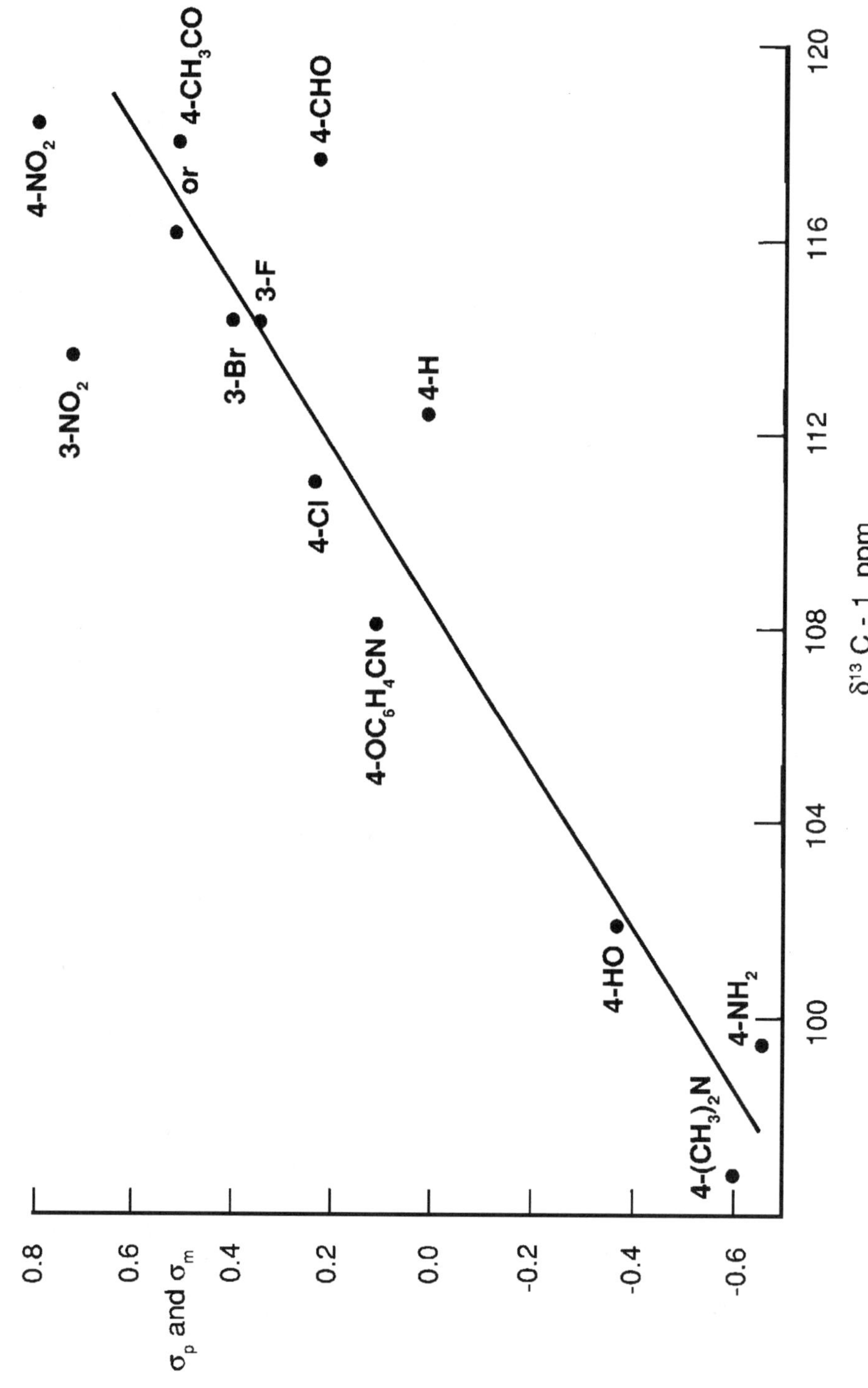

FIGURE 11.37 A plot of δ ^{13}C-1 vs Hammett σ values for the 3-x or 4-x atom or group for 3-x- and 4-x-benzonitriles in $CHCl_3$ solution.

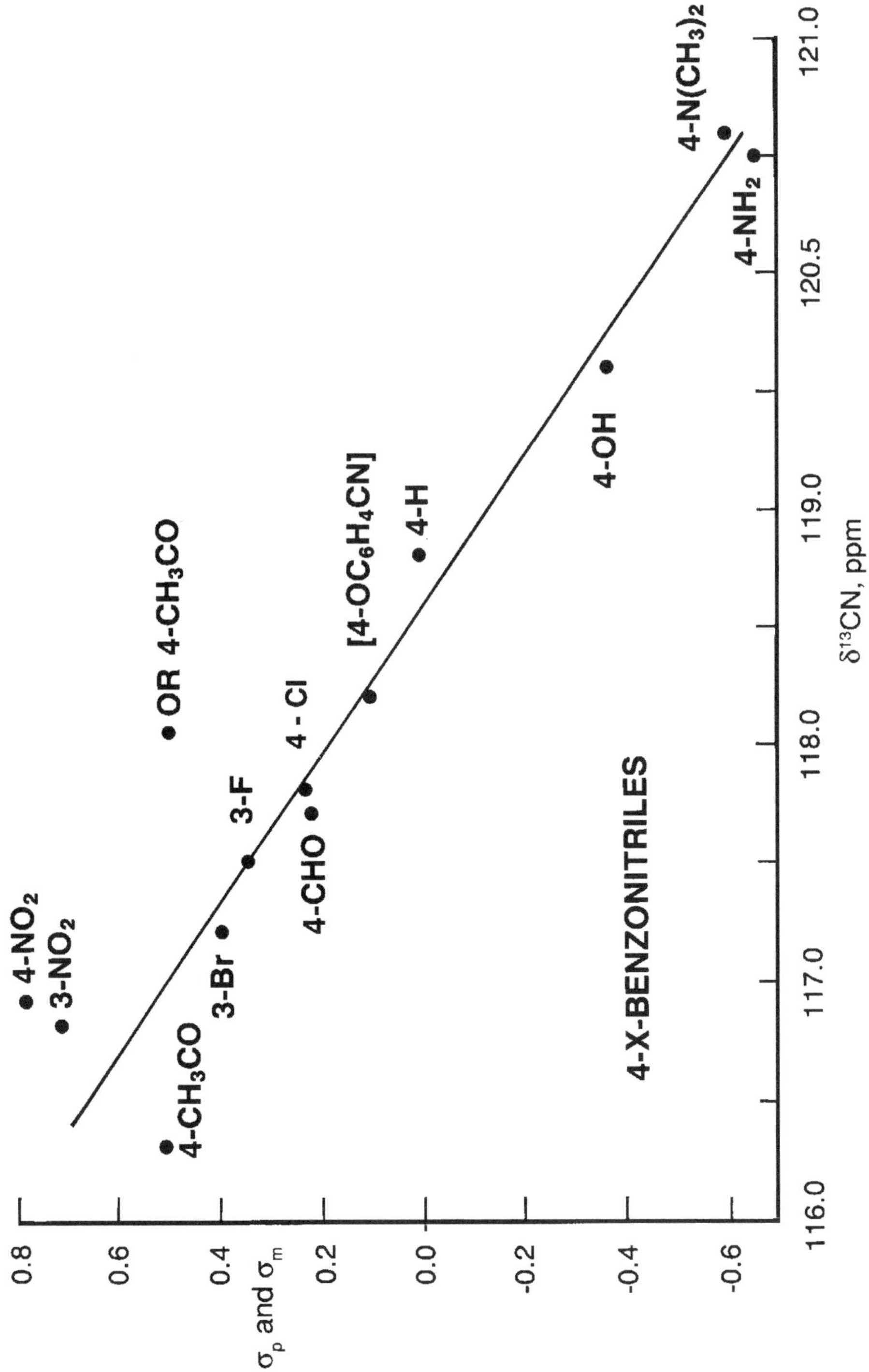

FIGURE 11.38 A plot of δ ^{13}CN vs Hammett σ values for the 3-x or 4-x atom or group for 3-x- and 4-x-benzonitriles in $CHCl_3$ solution.

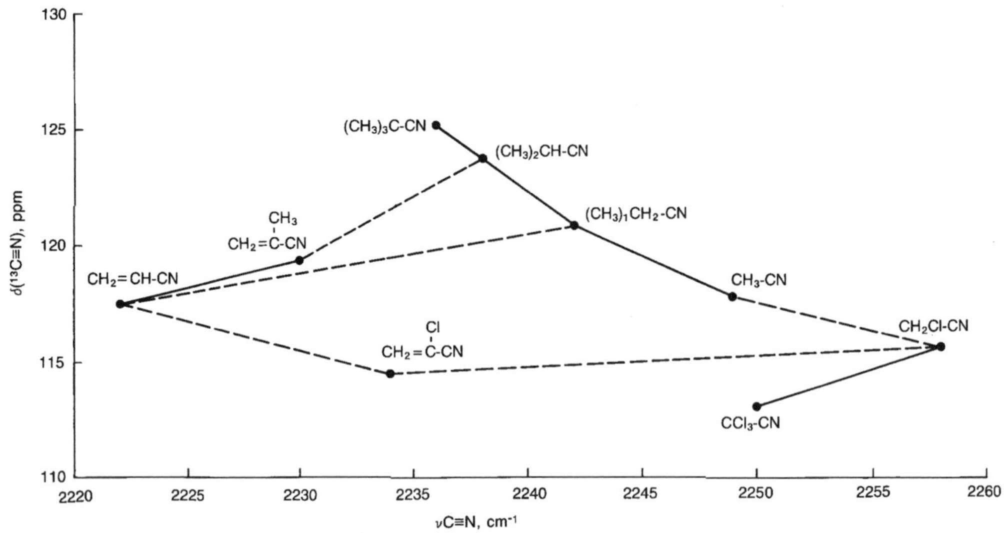

FIGURE 11.39 A plot of ν CN vs δ ^{13}CN for organonitriles.

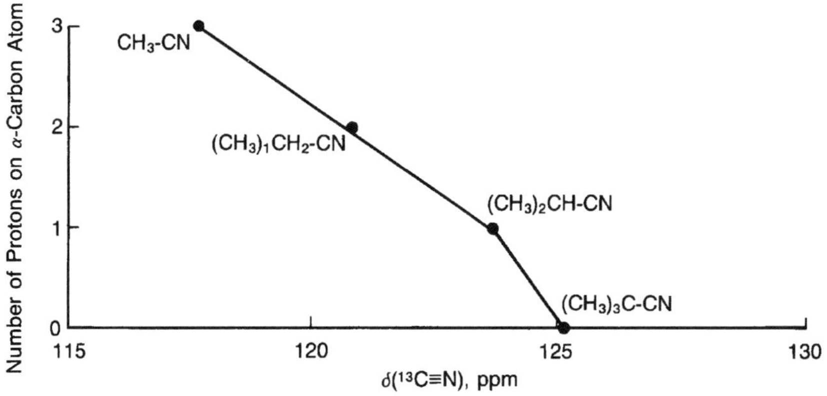

FIGURE 11.40 A plot of δ ^{13}CN vs the number of protons on the α-carbon atom of alkyl nitriles.

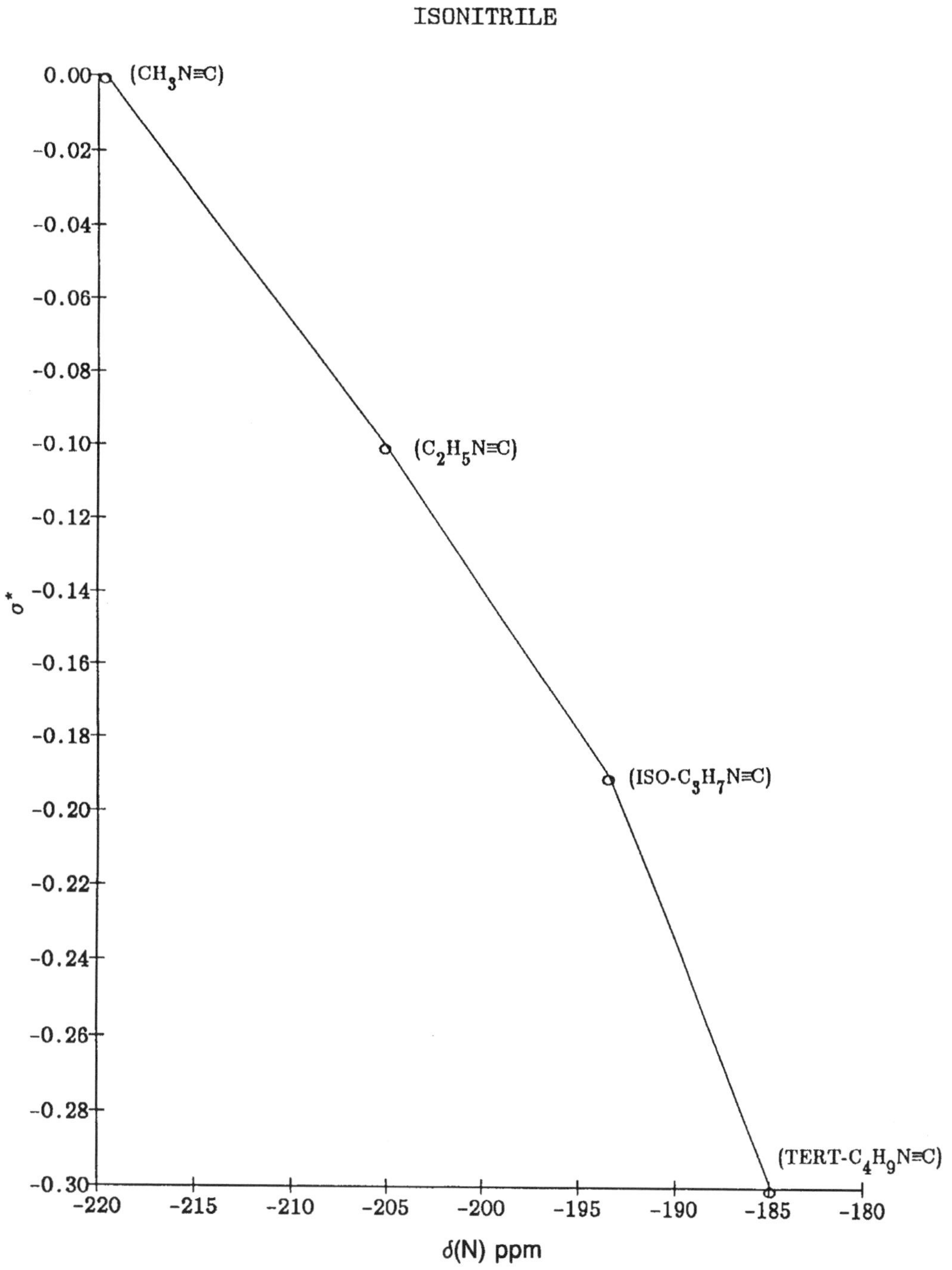

FIGURE 11.41 A plot of Taft σ^* values for the alkyl group vs δ ^{15}N for alkyl isonitriles.

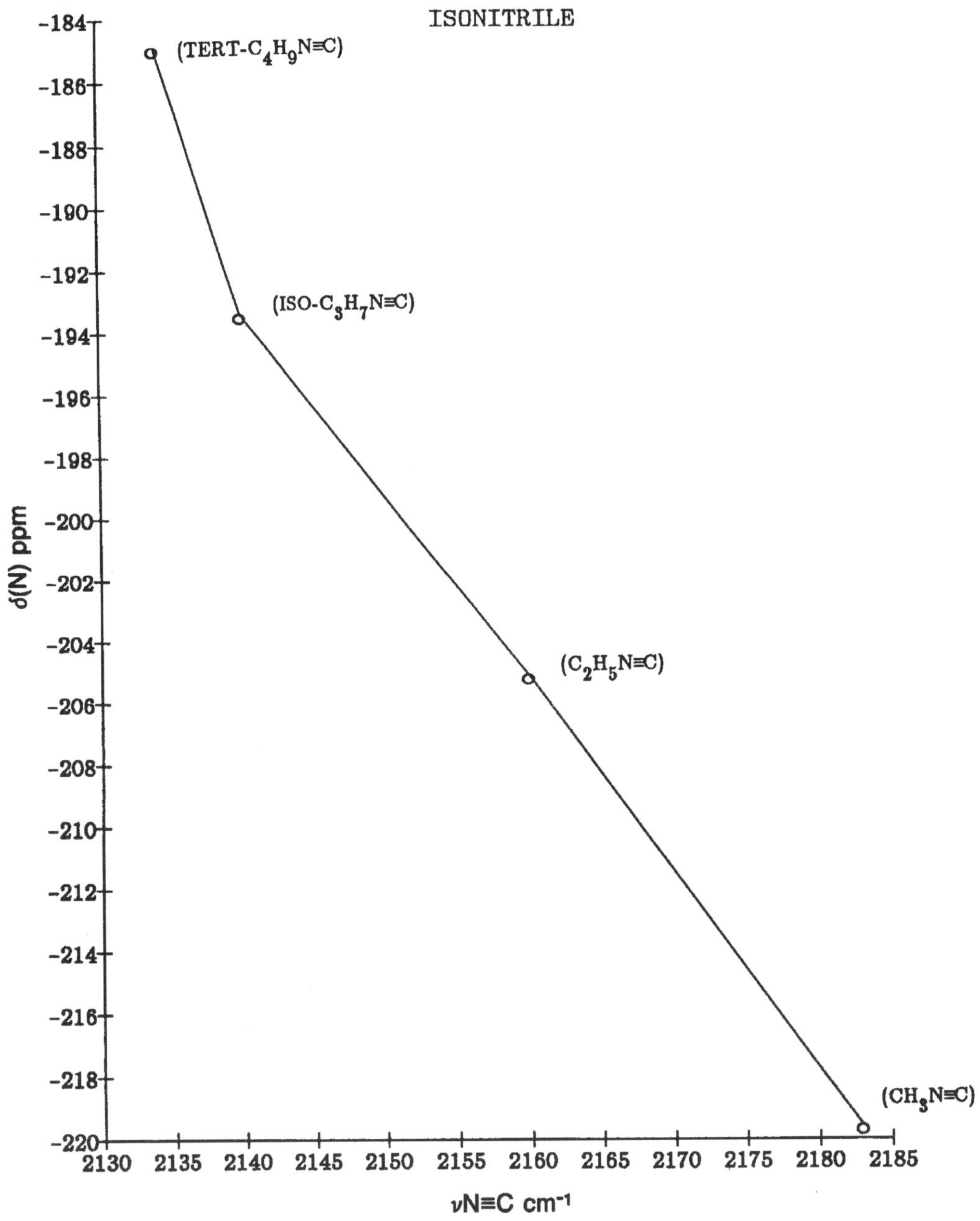

FIGURE 11.42 A plot of δ ^{15}N vs ν NC for alkyl isonitriles.

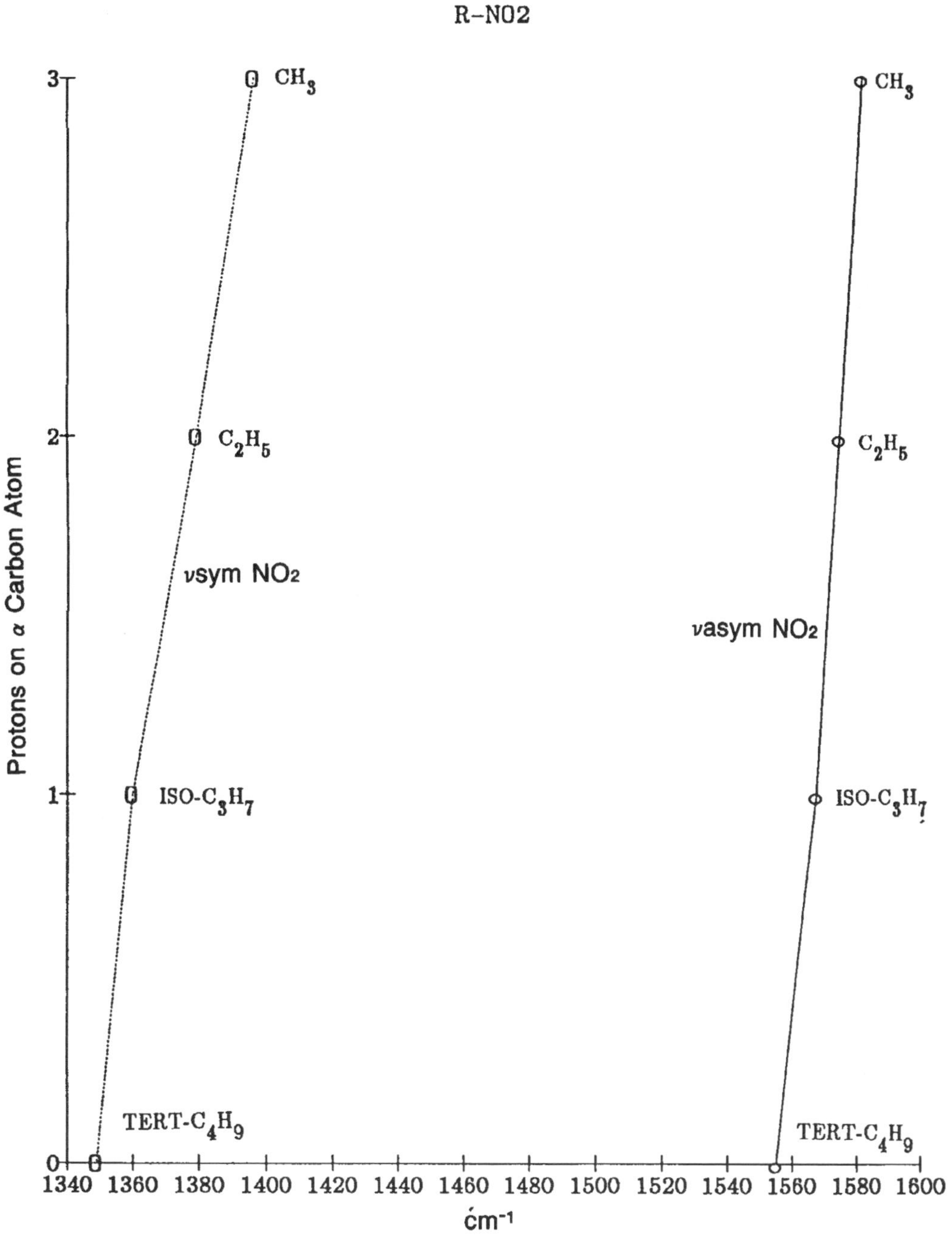

FIGURE 11.43 Plots of the number of protons in the alkyl α-carbon atom vs ν sym. NO_2 and ν asym. NO_2 for nitroalkanes in the vapor phase.

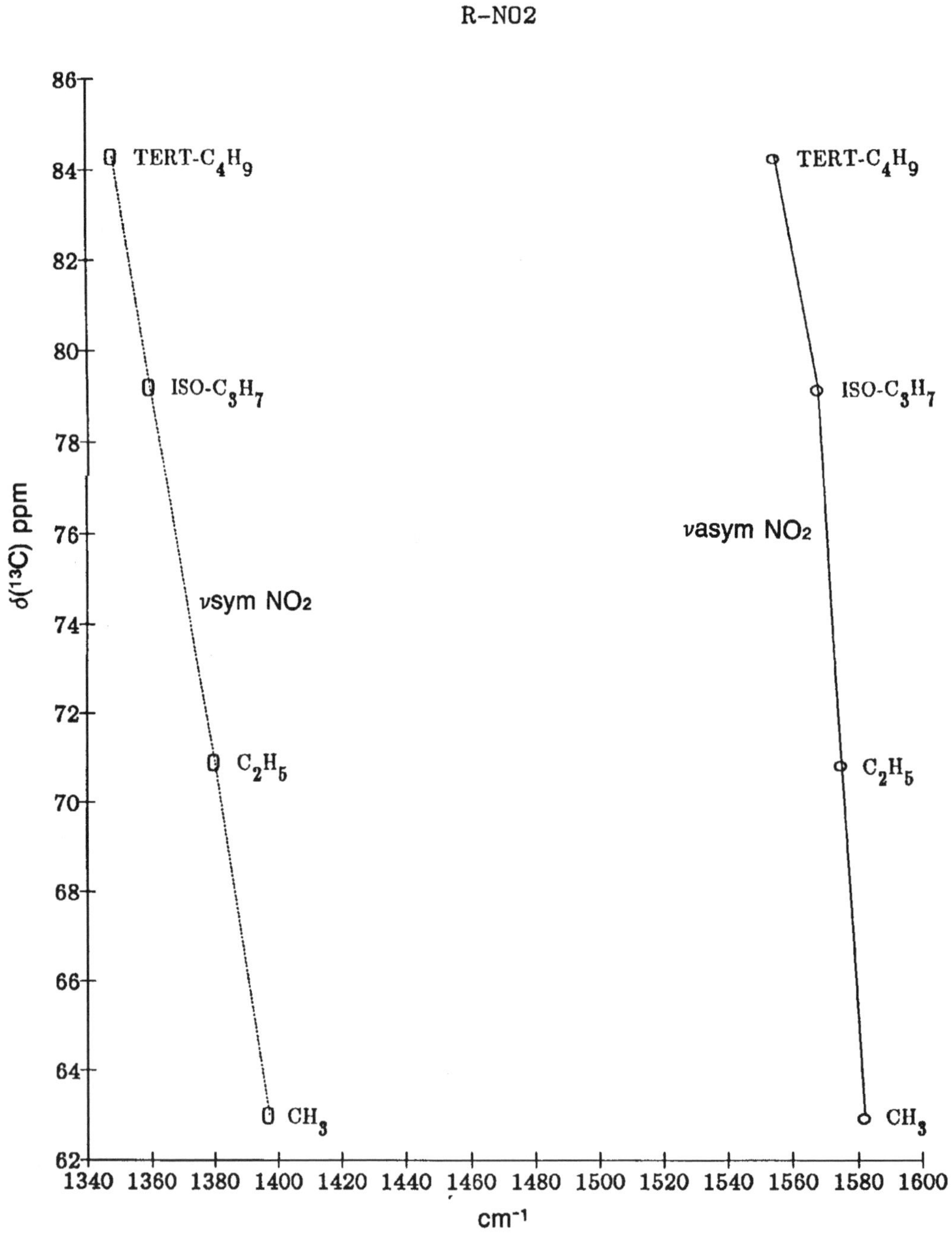

FIGURE 11.44 Plots of δ ^{13}C ($CHCl_3$ solution) for the alkyl α-carbon atom vs ν sym. NO_2 and ν asym. NO_2 (vapor phase) for nitroalkanes.

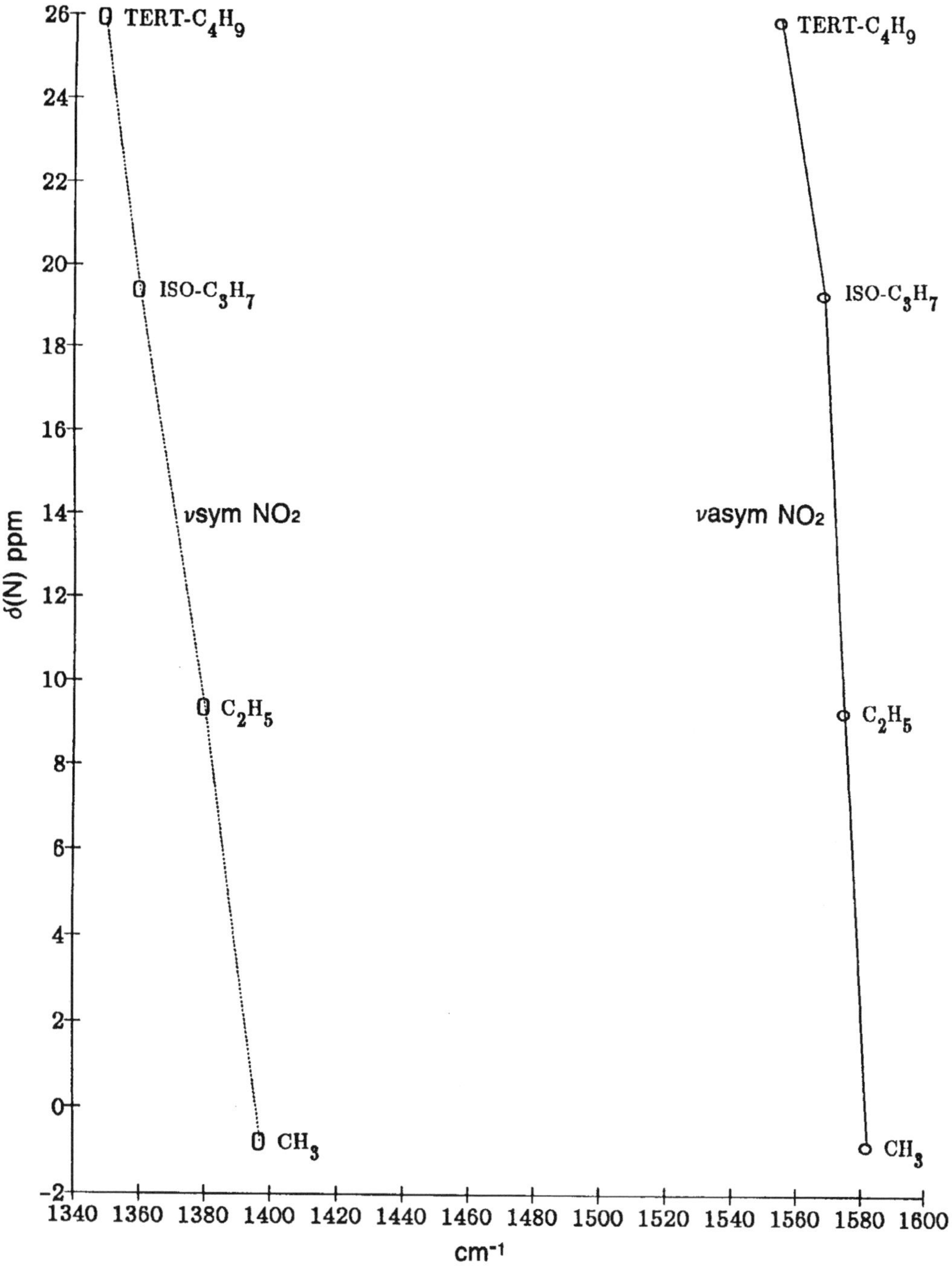

FIGURE 11.45 Plots of δ ^{15}N (in 0.3 M acetone) vs ν sym. NO_2 and ν asym. NO_2 (vapor phase) for nitroalkanes.

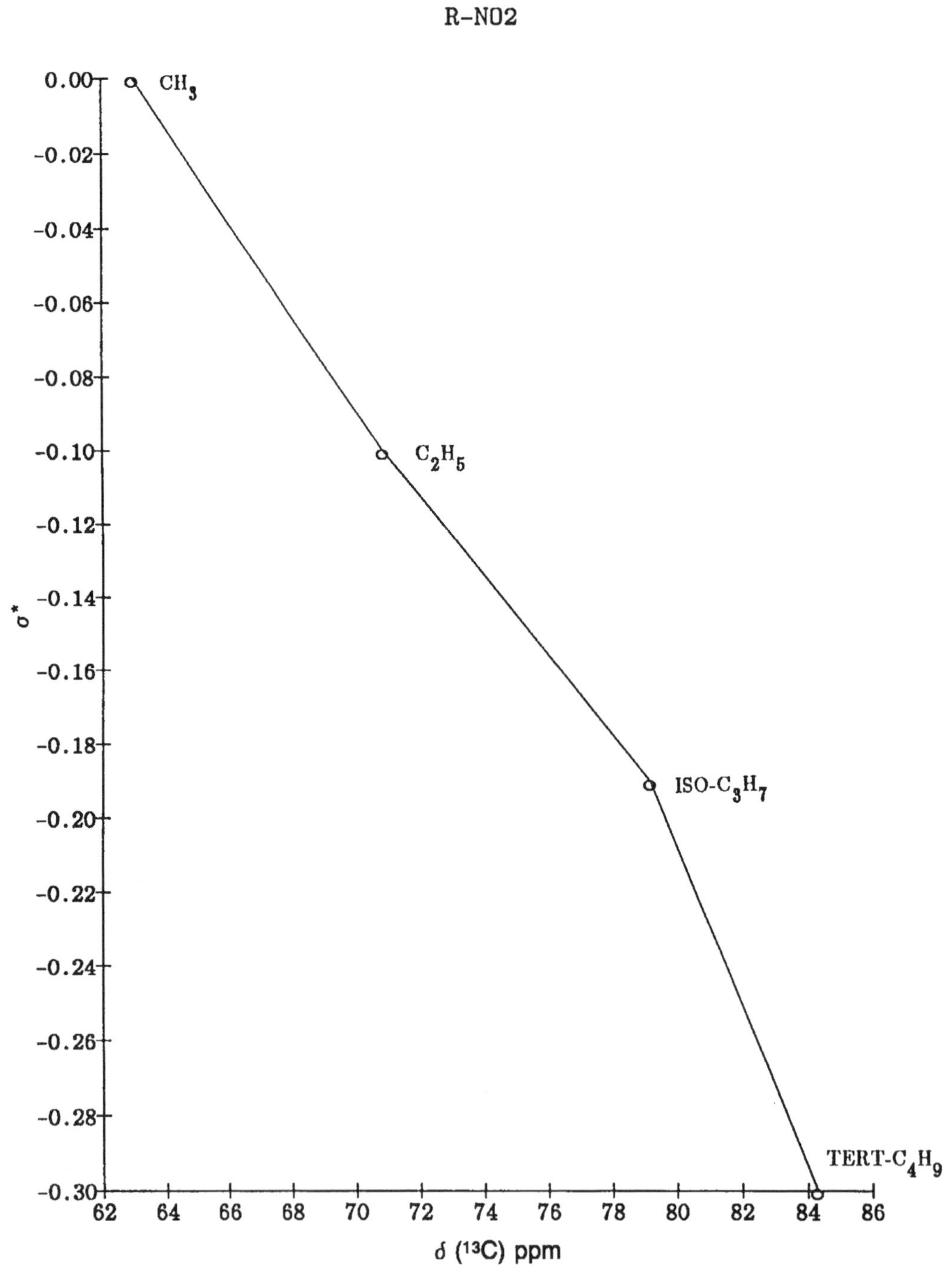

FIGURE 11.46 A plot of Taft σ^* values for the alkyl α-carbon atom group vs δ ^{13}C for nitroalkanes.

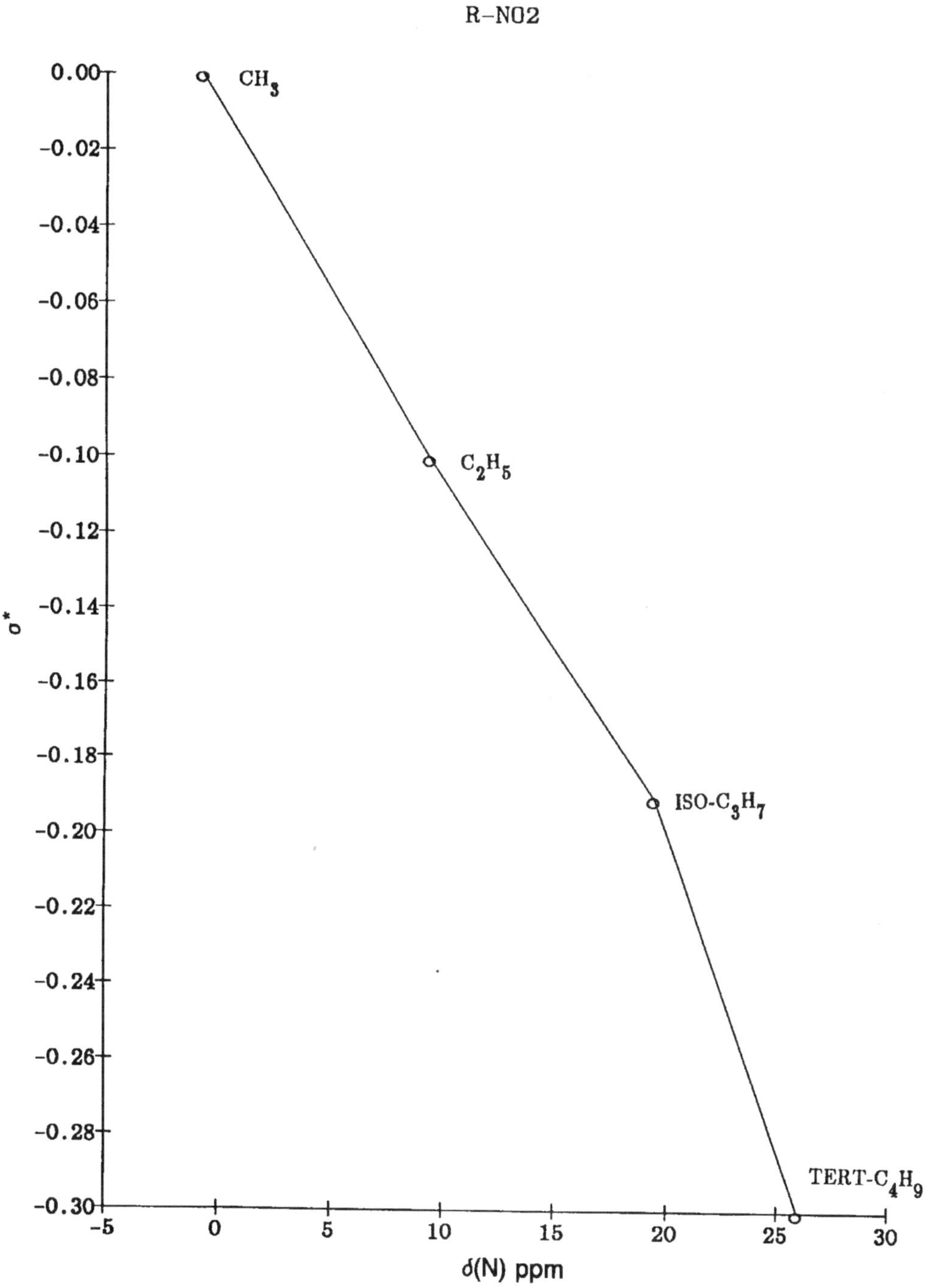

FIGURE 11.47 A plot of Taft σ^* values for the alkyl α-carbon atom vs δ ^{15}N for nitroalkanes.

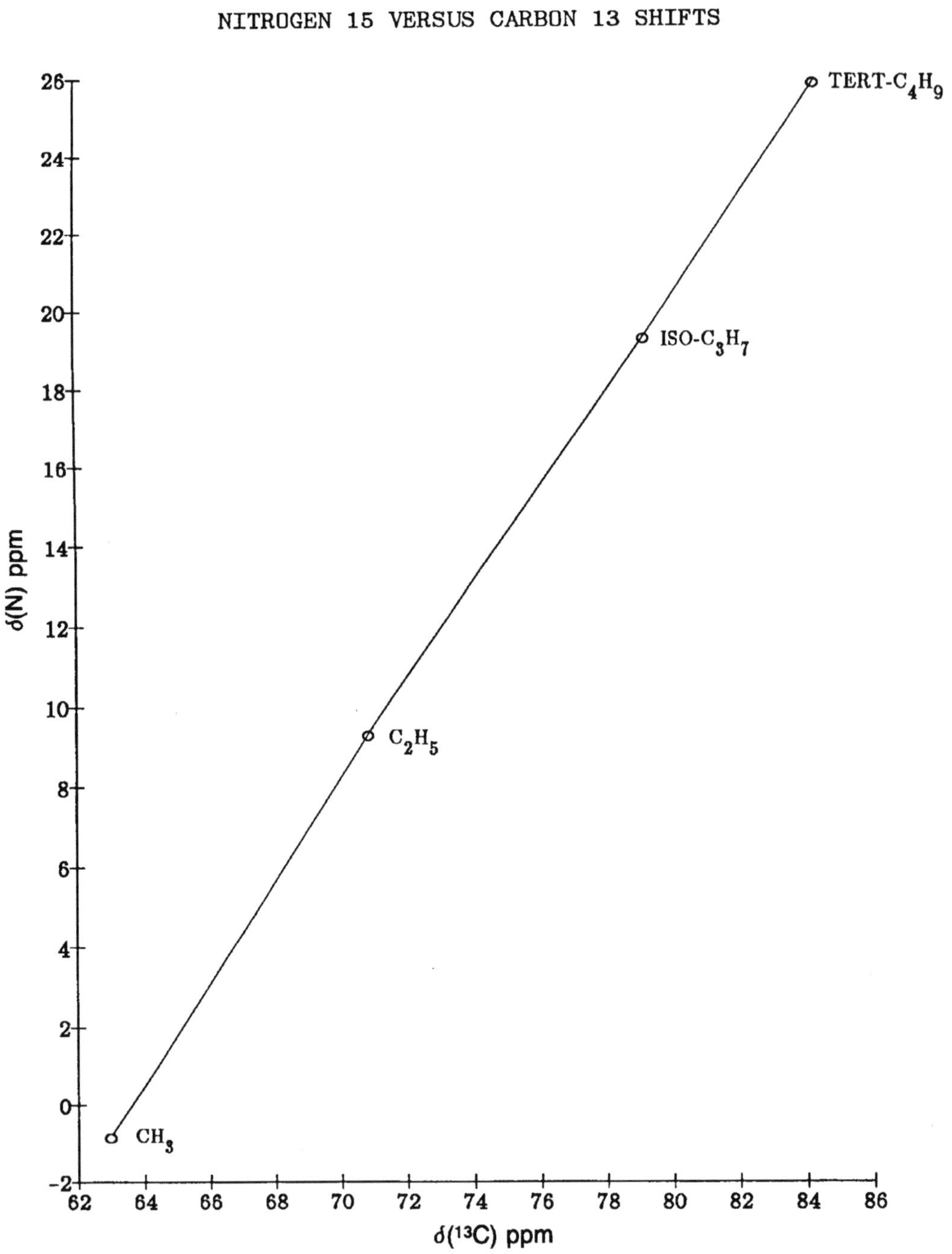

FIGURE 11.48 A plot of δ ^{15}N vs δ ^{13}C for the alkyl α-carbon atom for nitroalkanes.

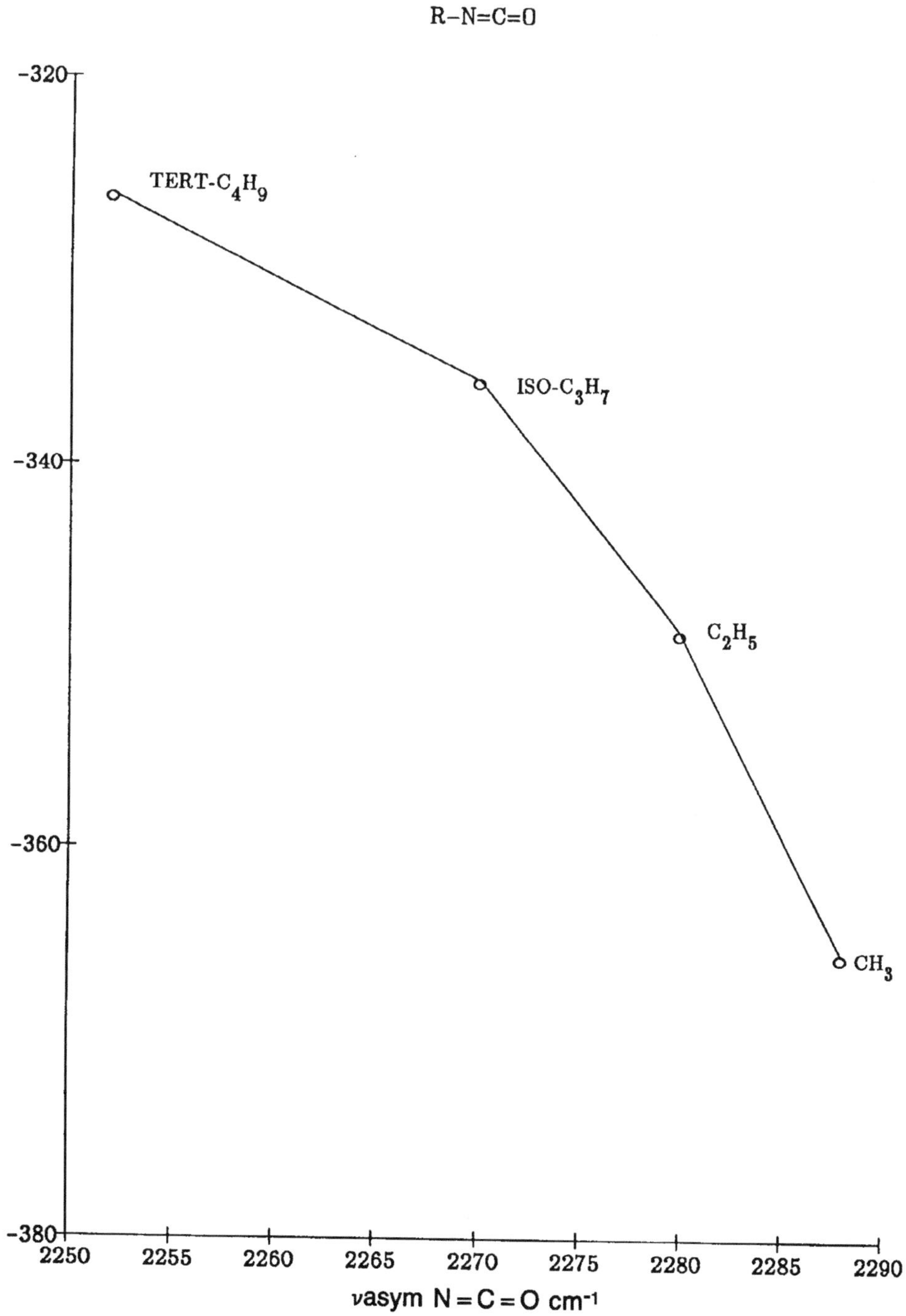

FIGURE 11.49 A plot of δ ^{15}N vs ν asym. N=C=O for alkyl isocyanates.

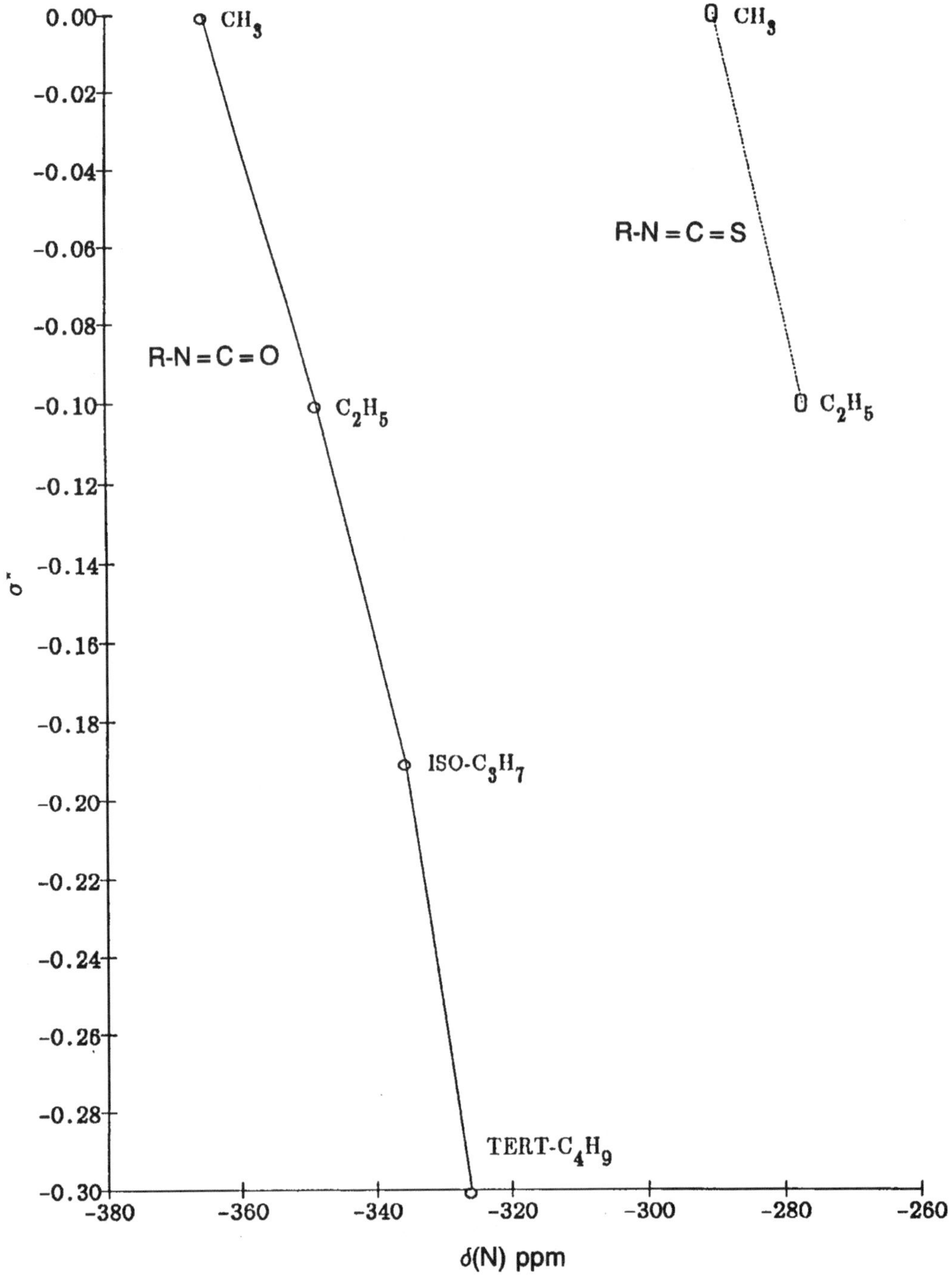

FIGURE 11.50 Plots of Taft σ^* values for the alkyl groups for alkyl isocyanates and alkyl isothiocyanates.

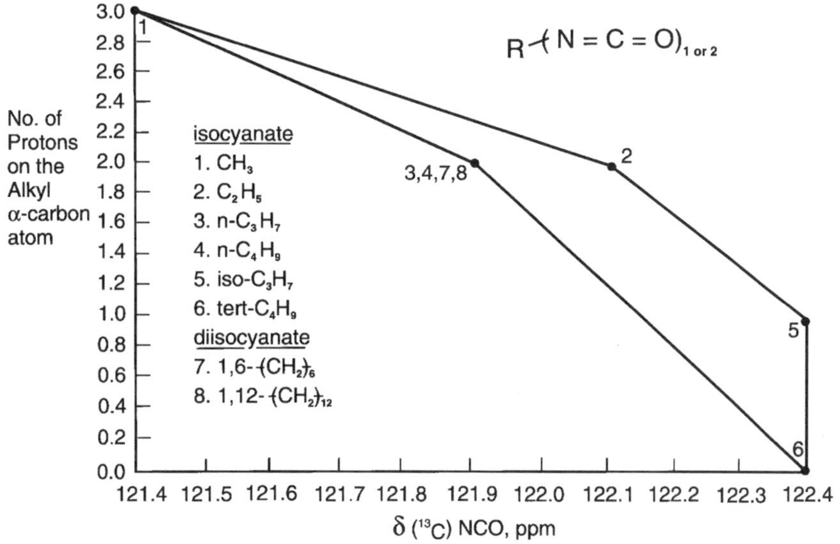

FIGURE 11.51 Plots of δ ^{13}C for the N=C=O group vs the number of protons on the alkyl α-carbon atom for alkyl isocyanates and alkyl diisocyanates.

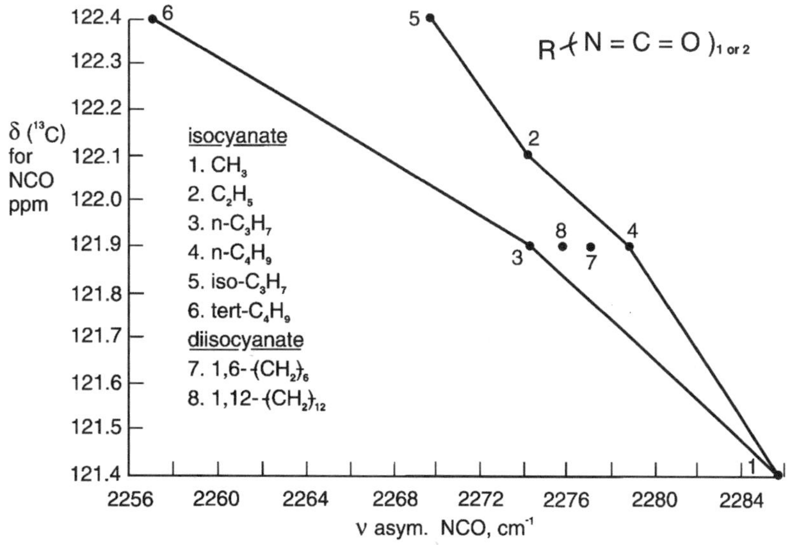

FIGURE 11.52 Plots of ν asym. N=C=O vs δ ^{13}C for the N=C=O group of alkyl isocyanates and alkyl diisocyanates.

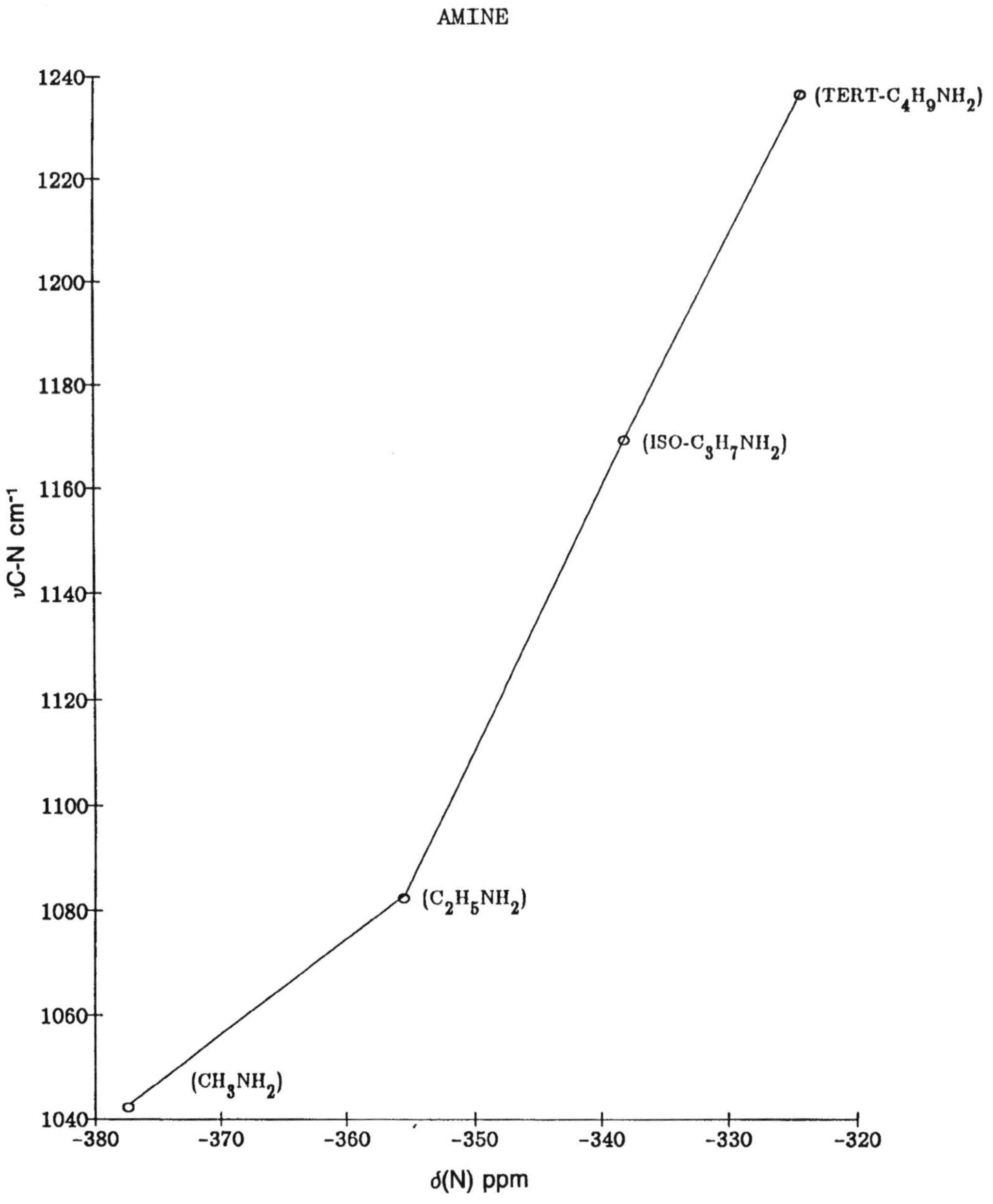

FIGURE 11.53 A plot of δ ^{15}N vs ν C—N for primary alkylamines.

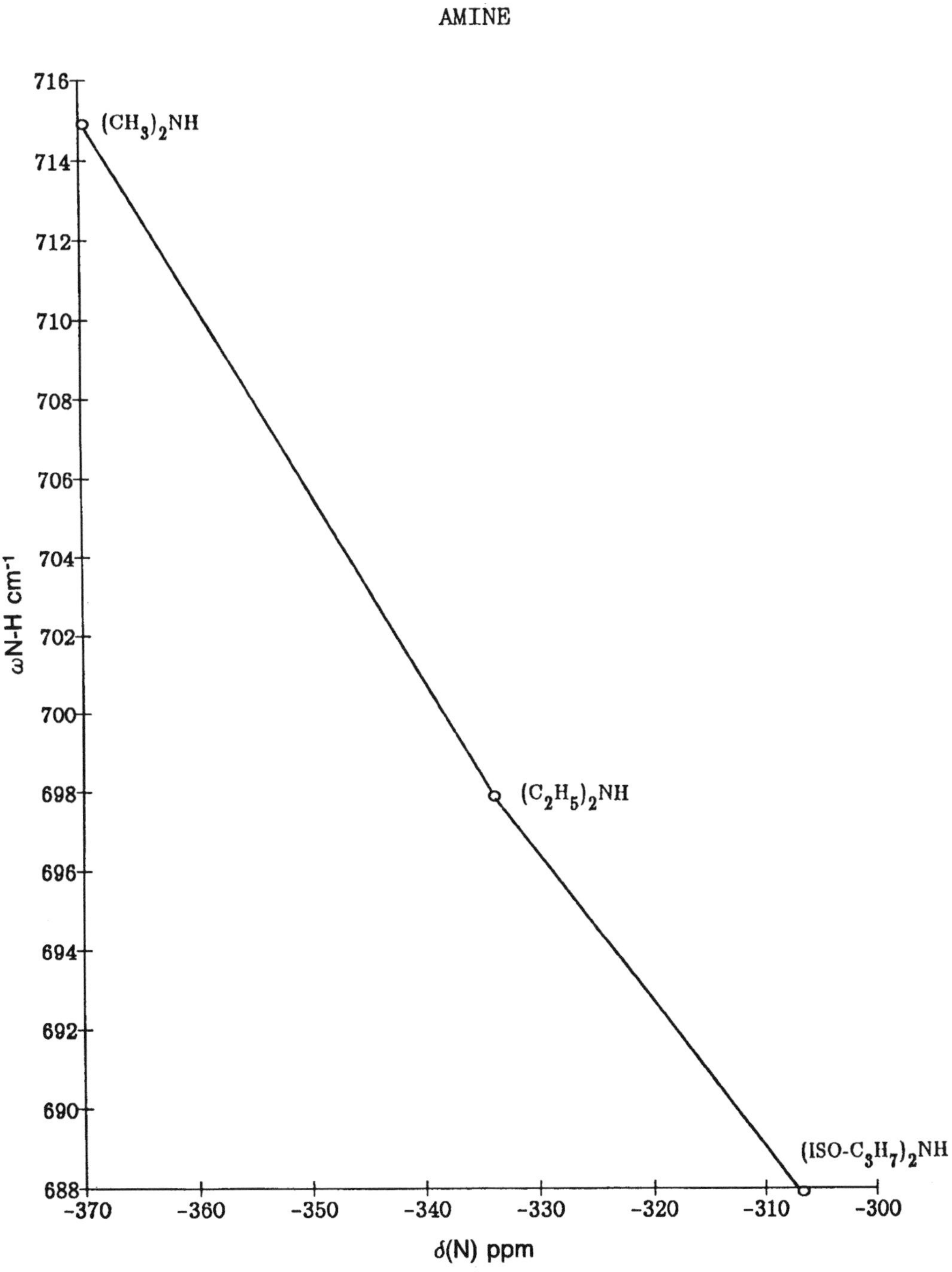

FIGURE 11.54 A plot of δ ^{15}N vs ΩN—H for dialkylamines.

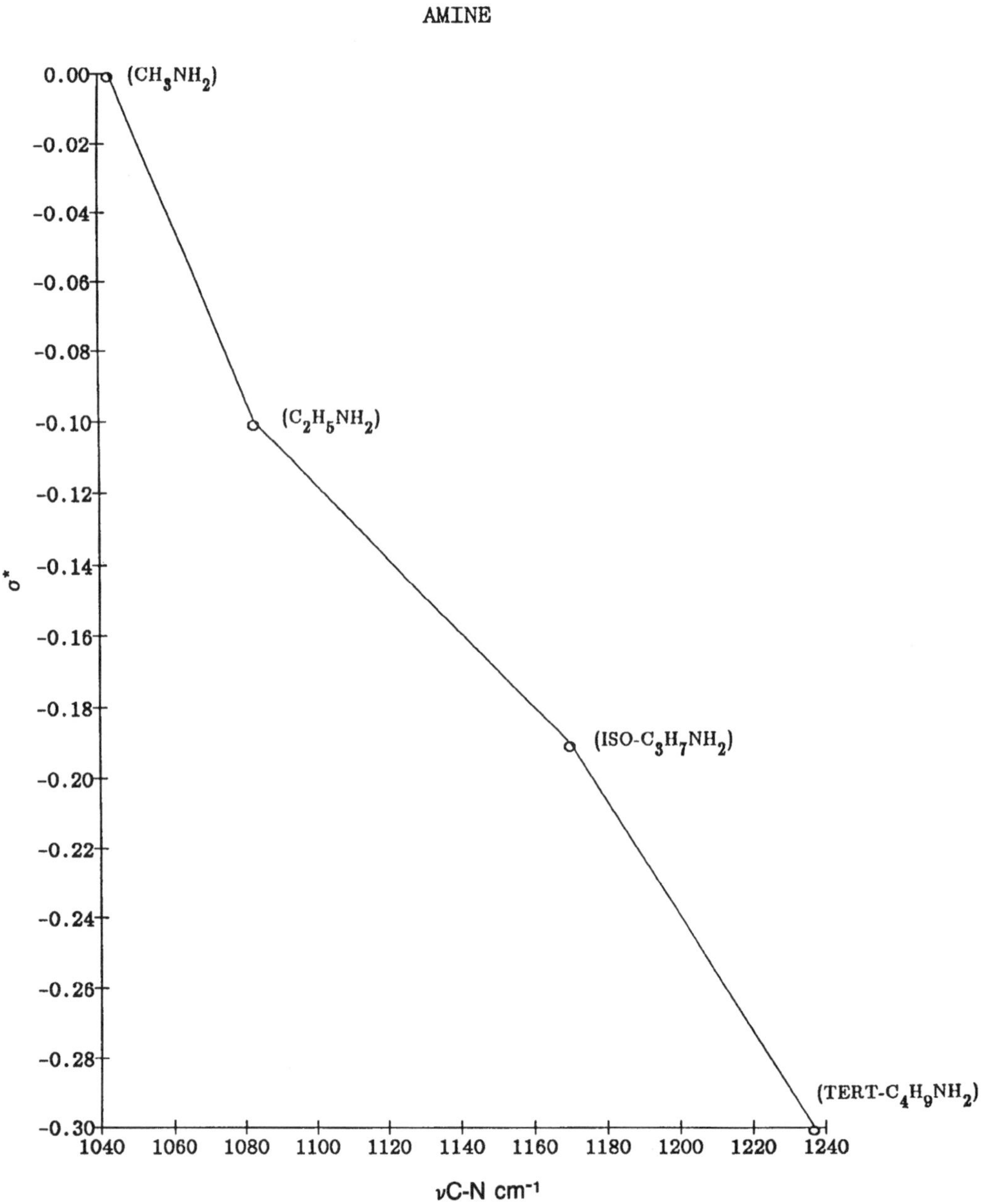

FIGURE 11.55 A plot of ν C—N vs Taft σ^* for the alkyl group of primary alkylamines.

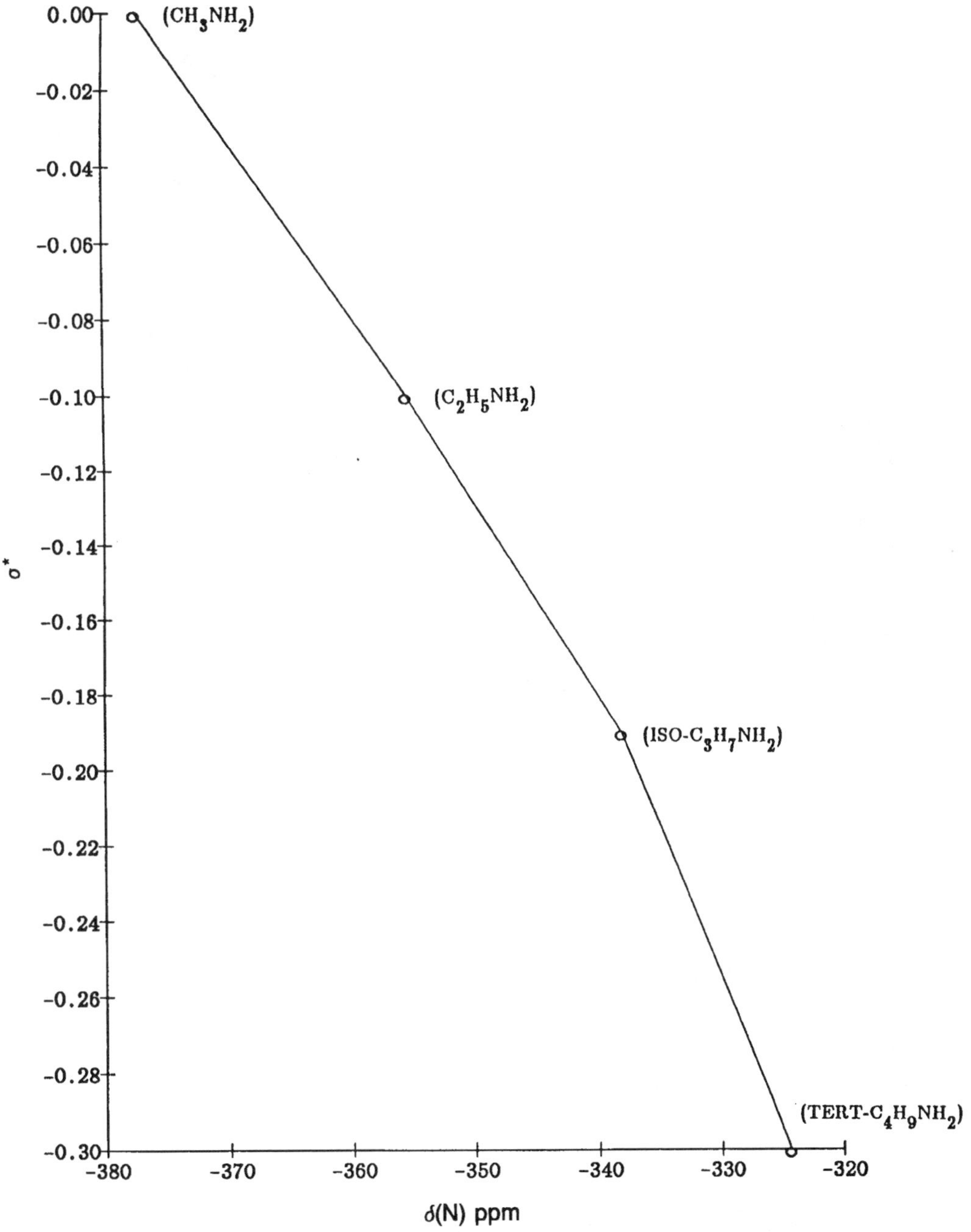

FIGURE 11.56 A plot of δ ^{15}N vs Taft σ^* for the alkyl group of primary alkylamines.

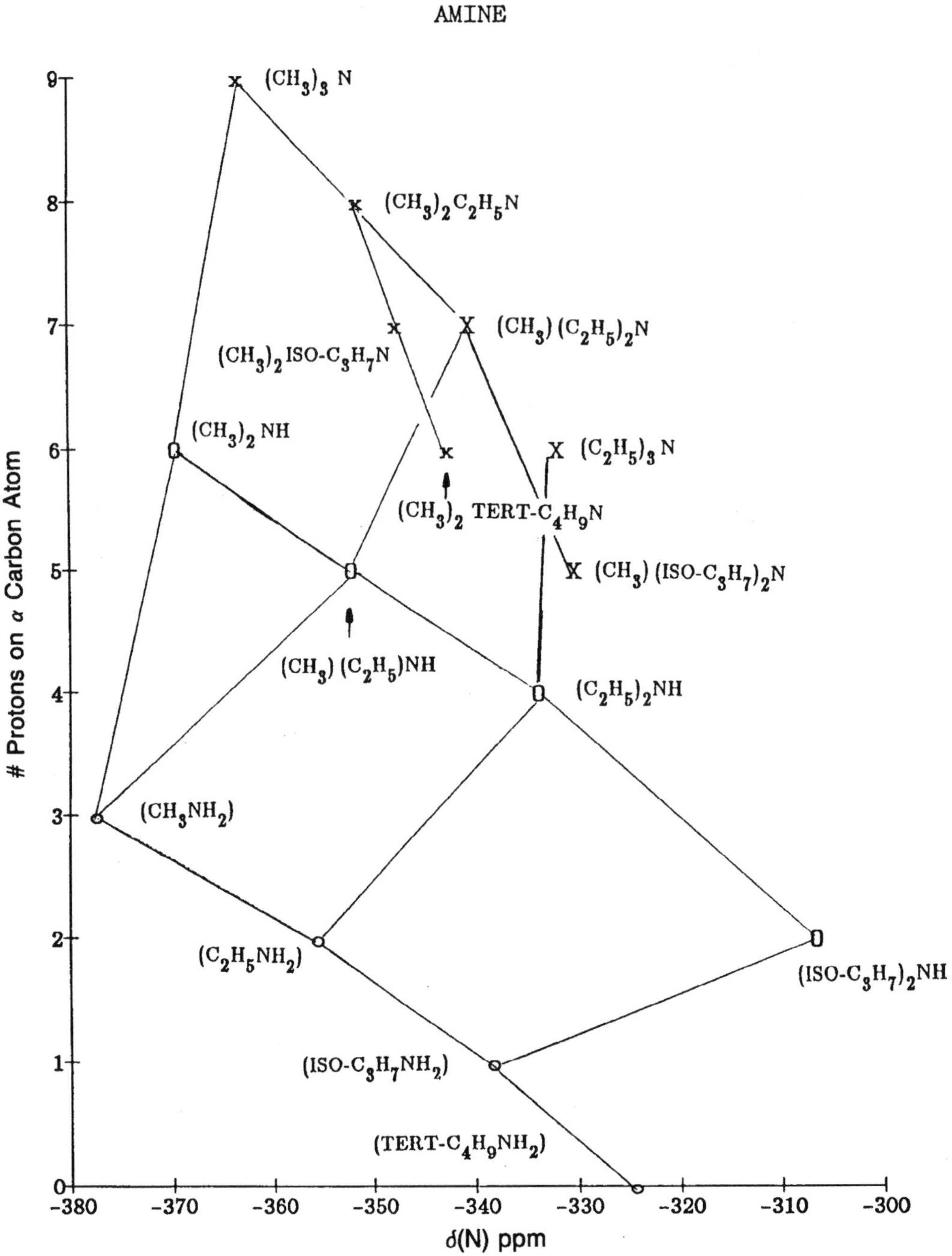

FIGURE 11.57 Plots of δ ^{15}N vs the number of protons on the alkyl α-carbon atom(s) for mono-, di-, and trialkylamines.

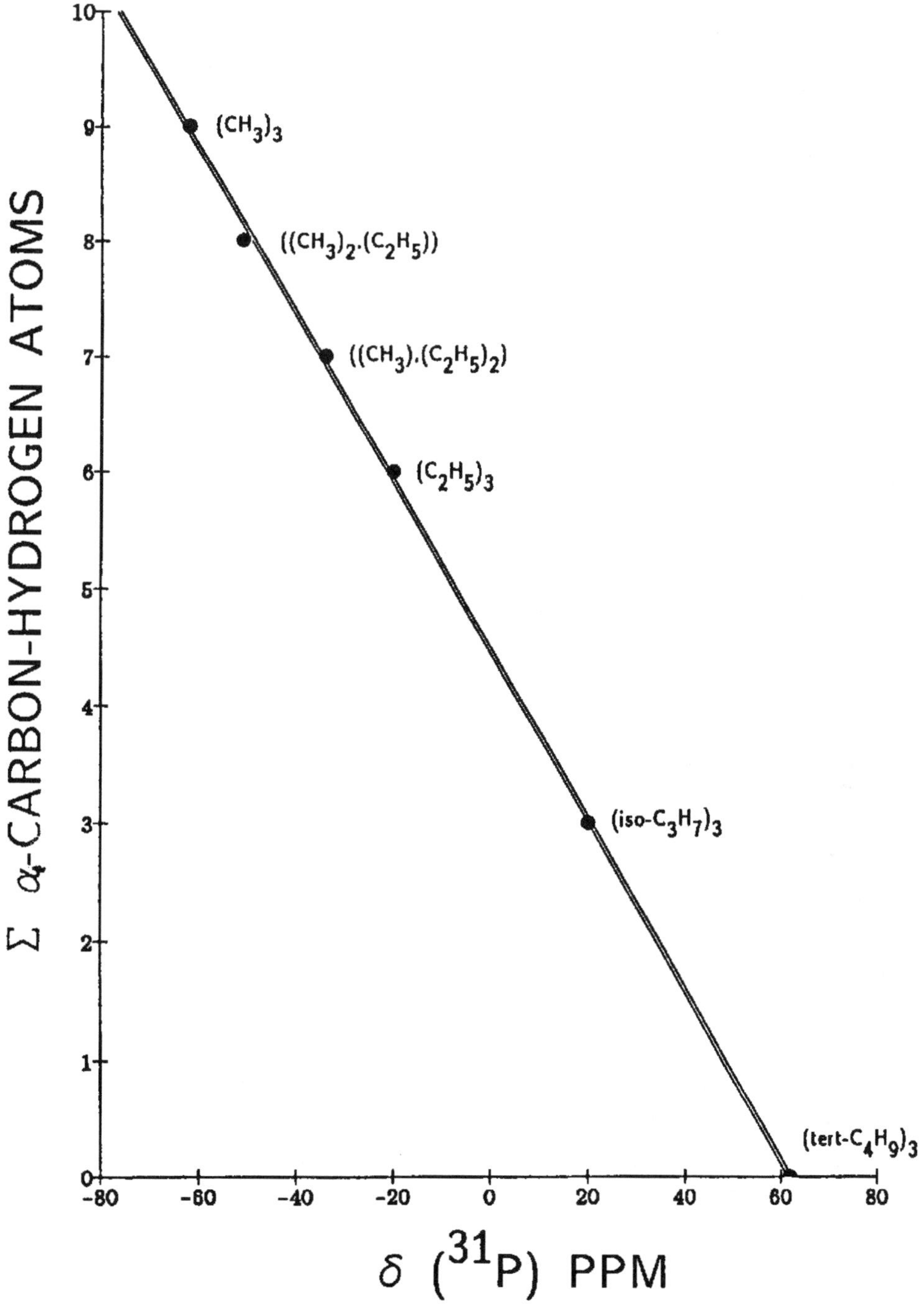

FIGURE 11.58 A plot of δ ^{31}P for trialkylphosphines vs the sum of the number of protons on the alkyl α-C—P carbon atoms.

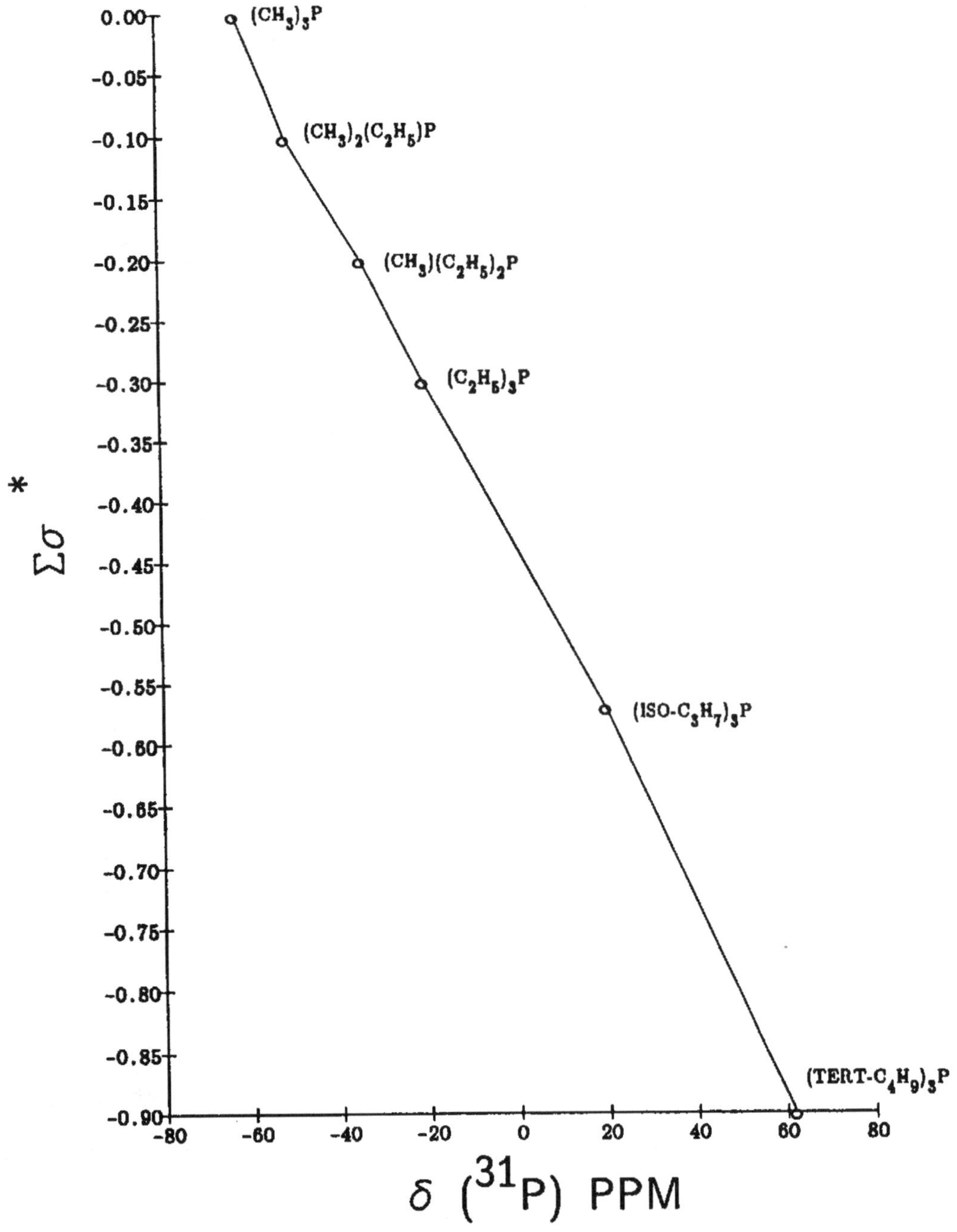

FIGURE 11.59 A plot of δ ^{31}P for trialkylphosphines vs Taft σ^* values for the alkyl groups.

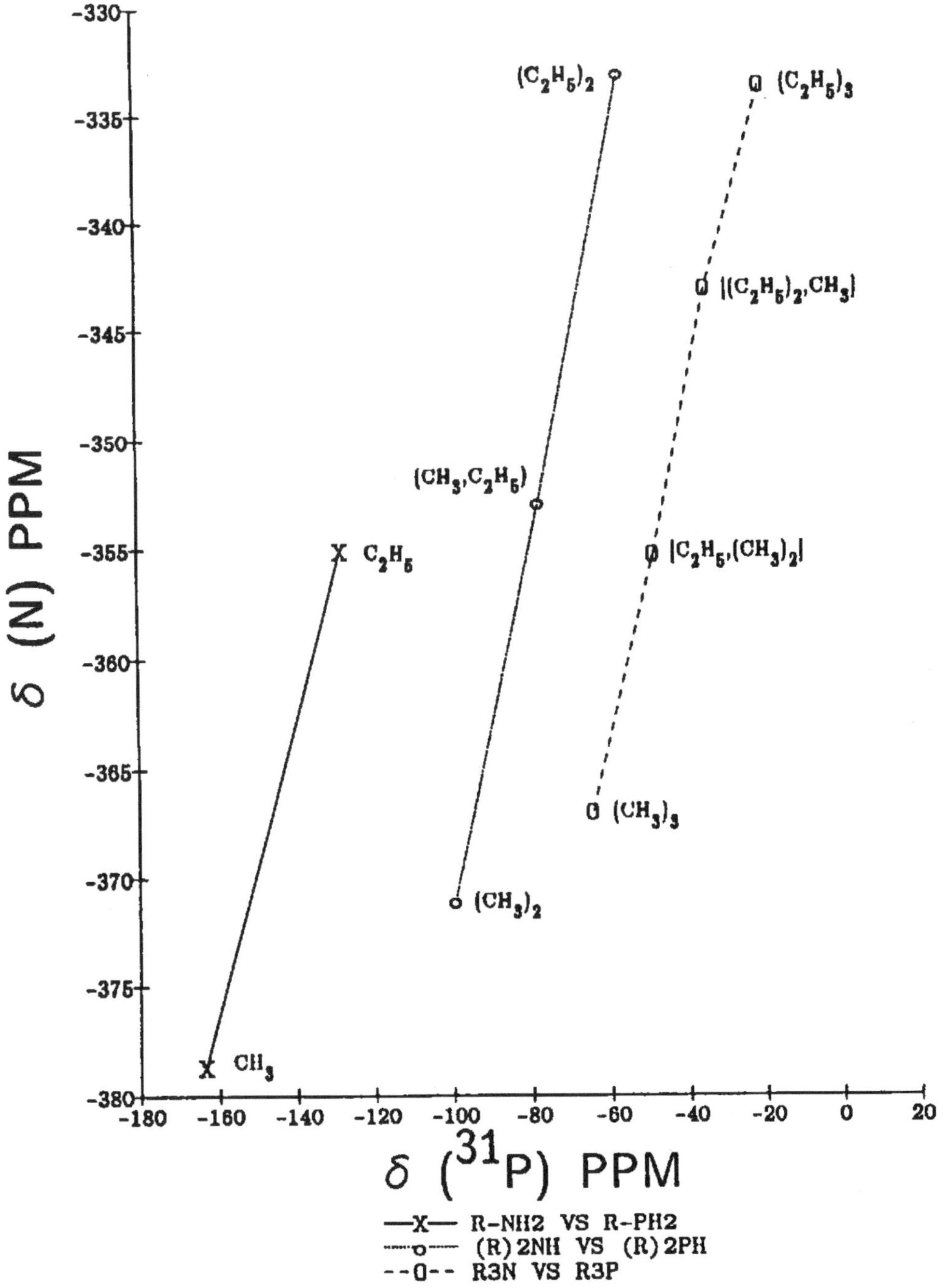

FIGURE 11.60 Plots of δ ^{31}P for R—PH$_2$ vs δ ^{15}N for R—NH$_2$, δ ^{31}P for (R—)$_2$PH vs δ ^{15}N for (R—)$_2$NH, and δ ^{31}P for (R—)$_3$P vs δ ^{15}N for (R—)$_3$N.

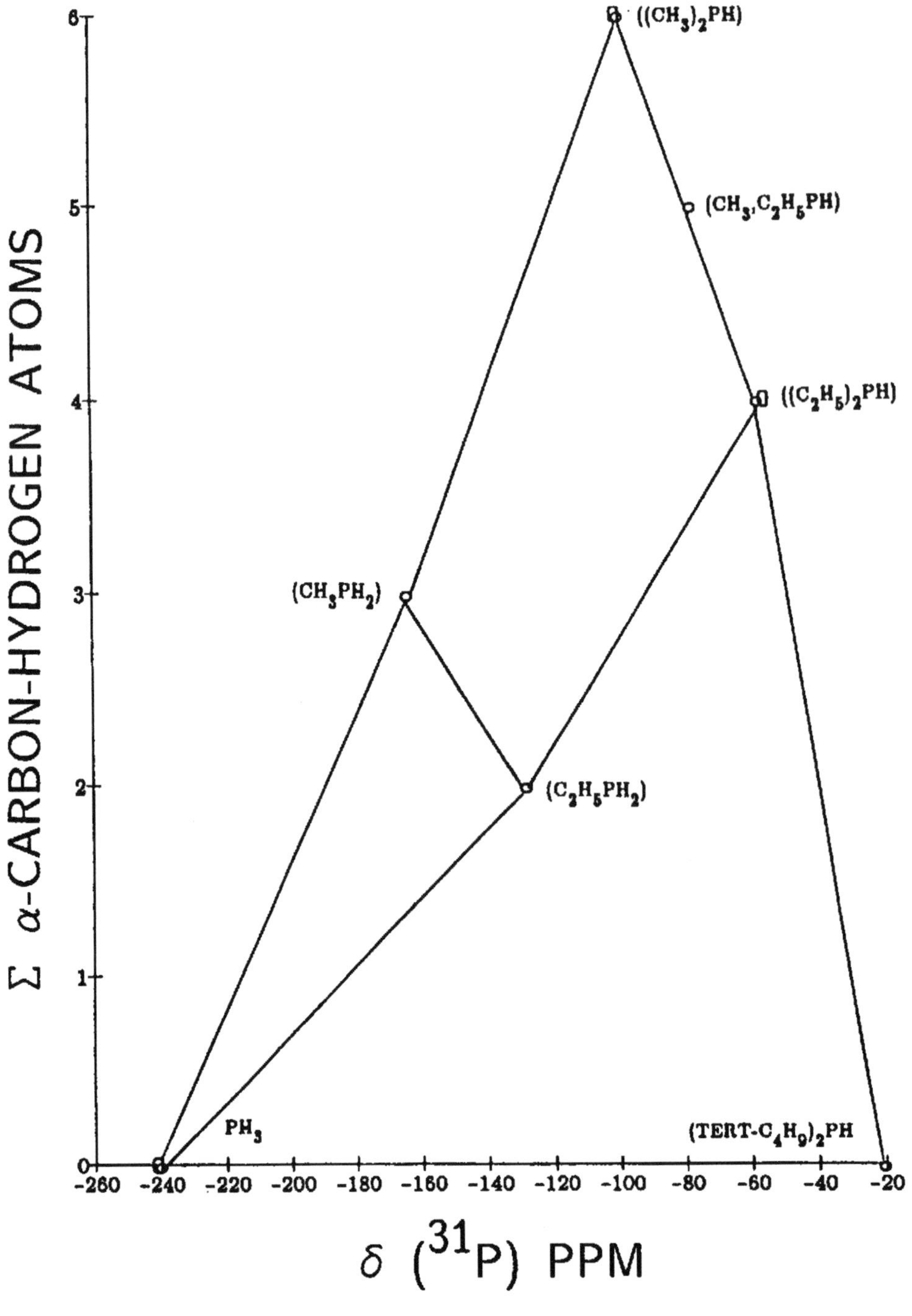

FIGURE 11.61 Plots of the number of protons on the α-C—P atom(s) vs δ ^{31}P for PH_3, R—PH_2, (R—)$_2$PH.

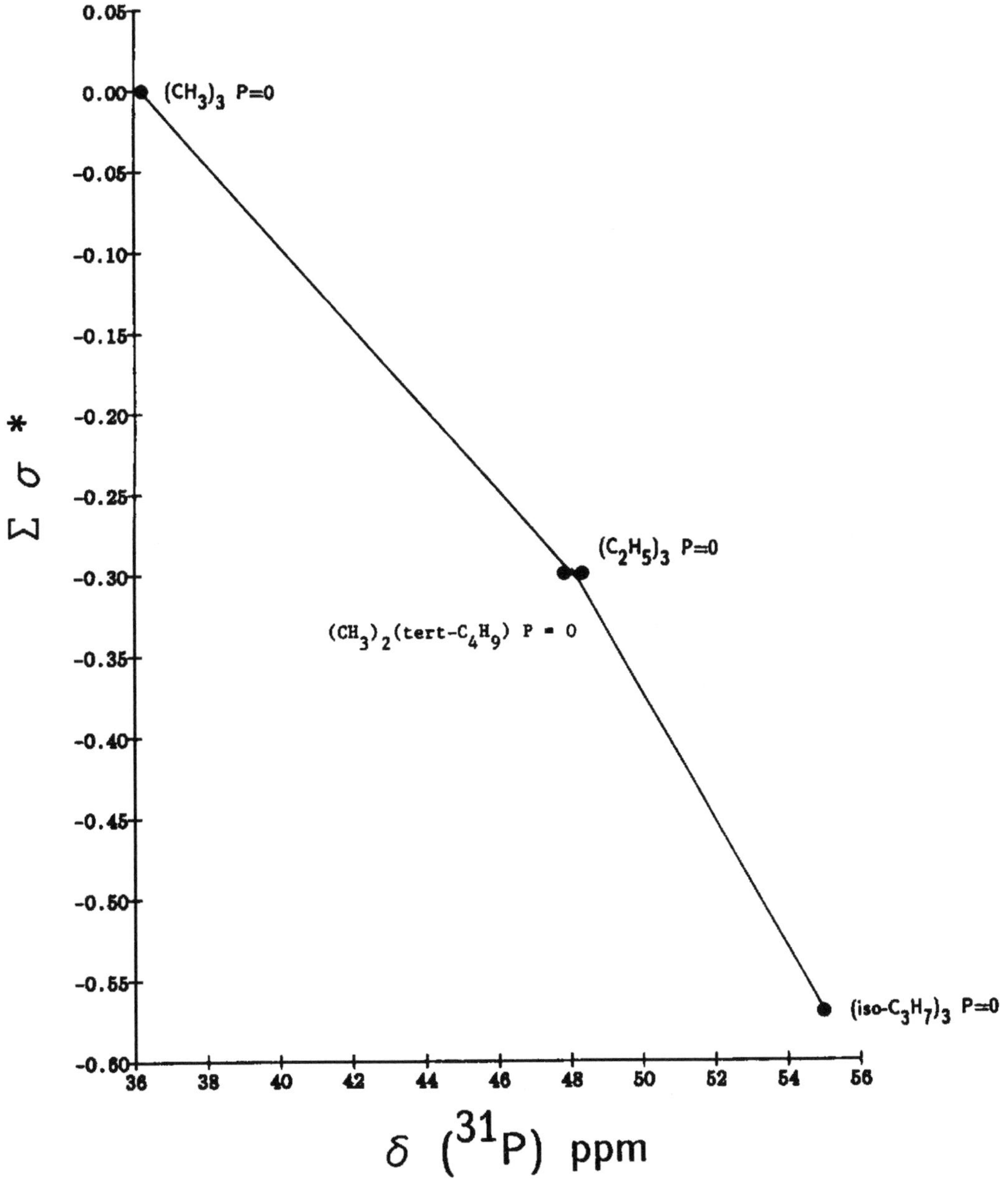

FIGURE 11.62 A plot of the sum of Taft σ^* values δ ^{31}P for trialkylphosphine oxides.

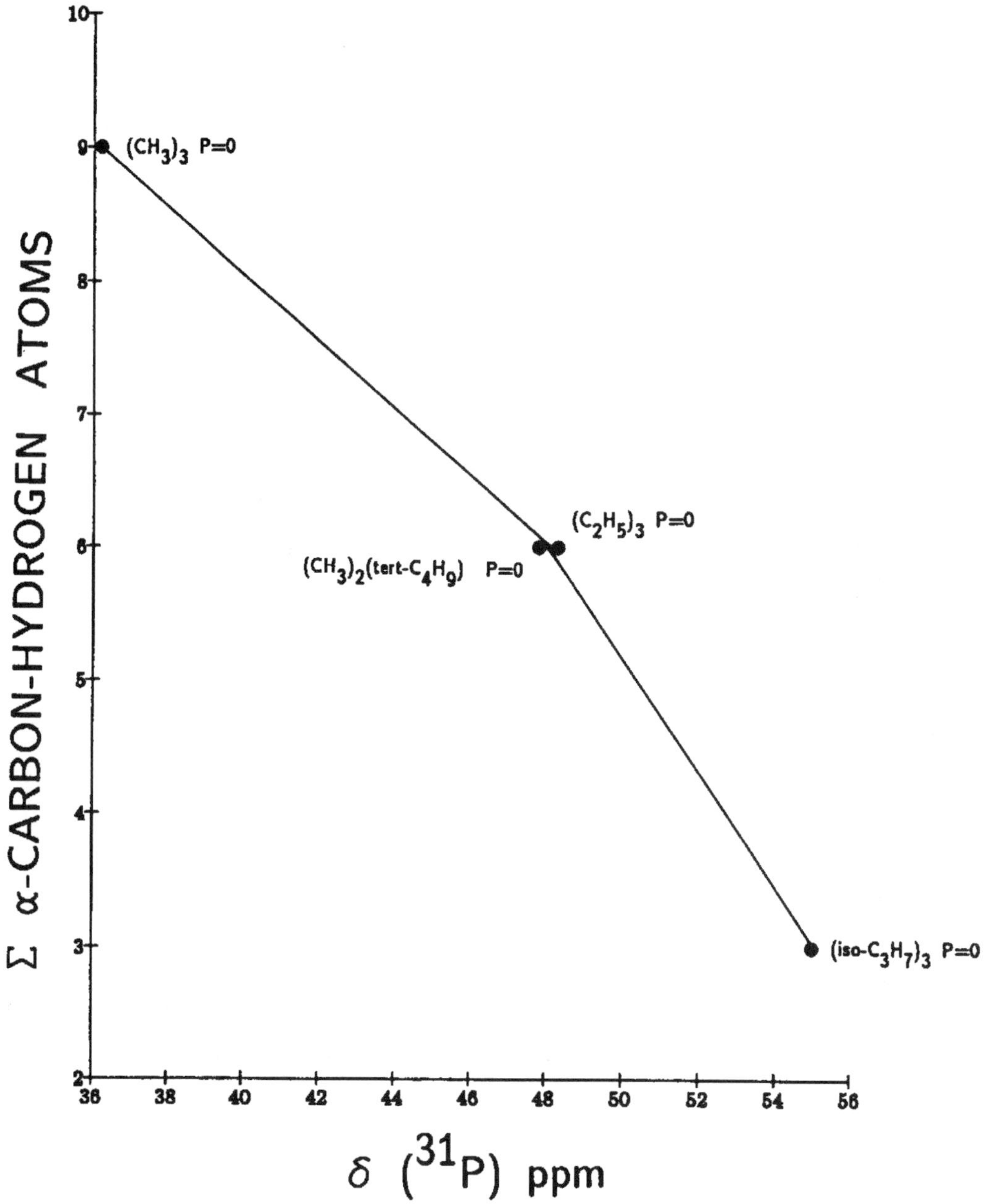

FIGURE 11.63 A plot of the sum of the number of protons on the α-C—P carbon atoms vs δ ^{31}P for trialkylphosphine oxides.

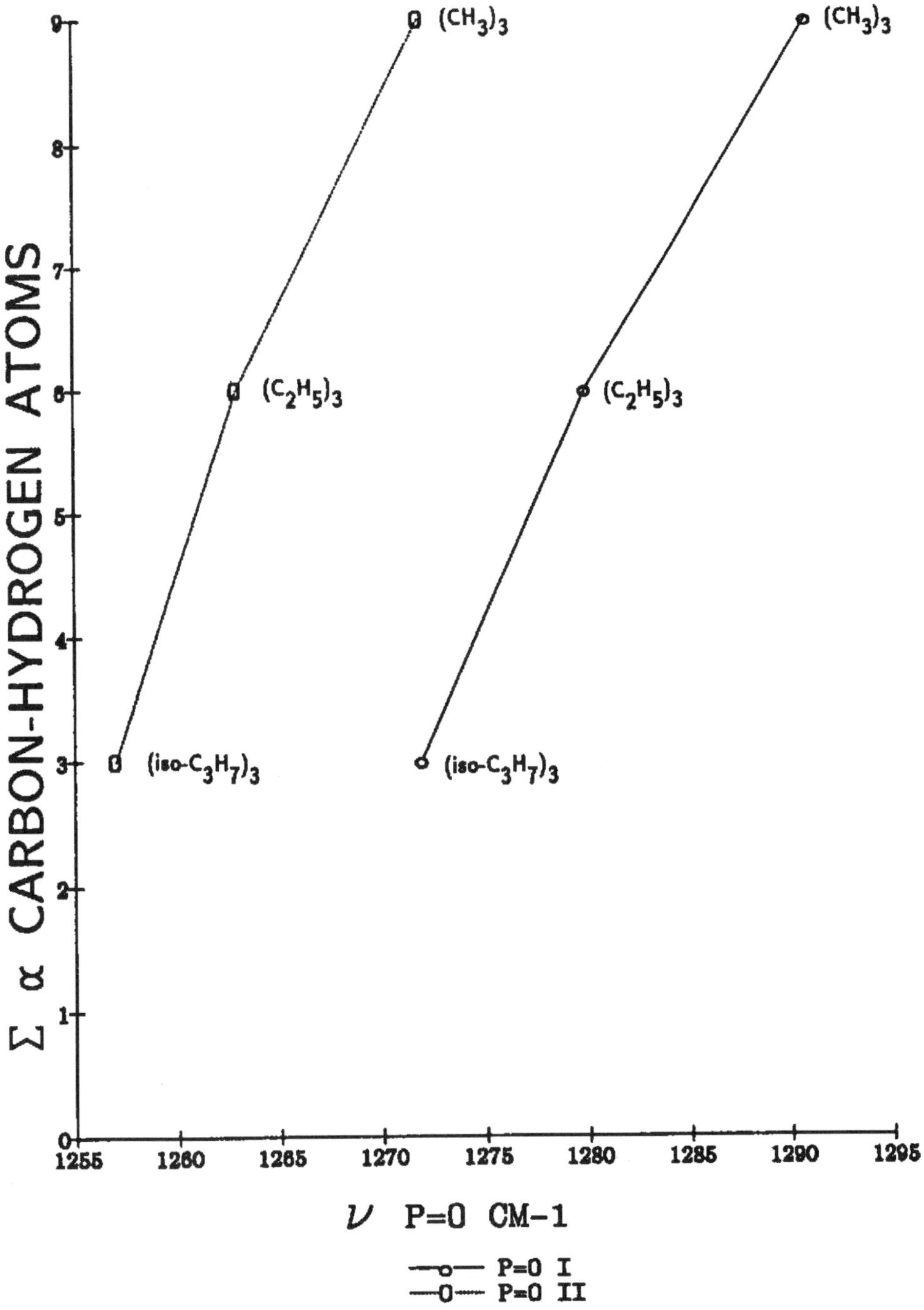

FIGURE 11.64 Plots of ν P=O rotational conformers vs the sum of the number of protons on the α-C—O—P carbon atoms for trialkyl phosphates.

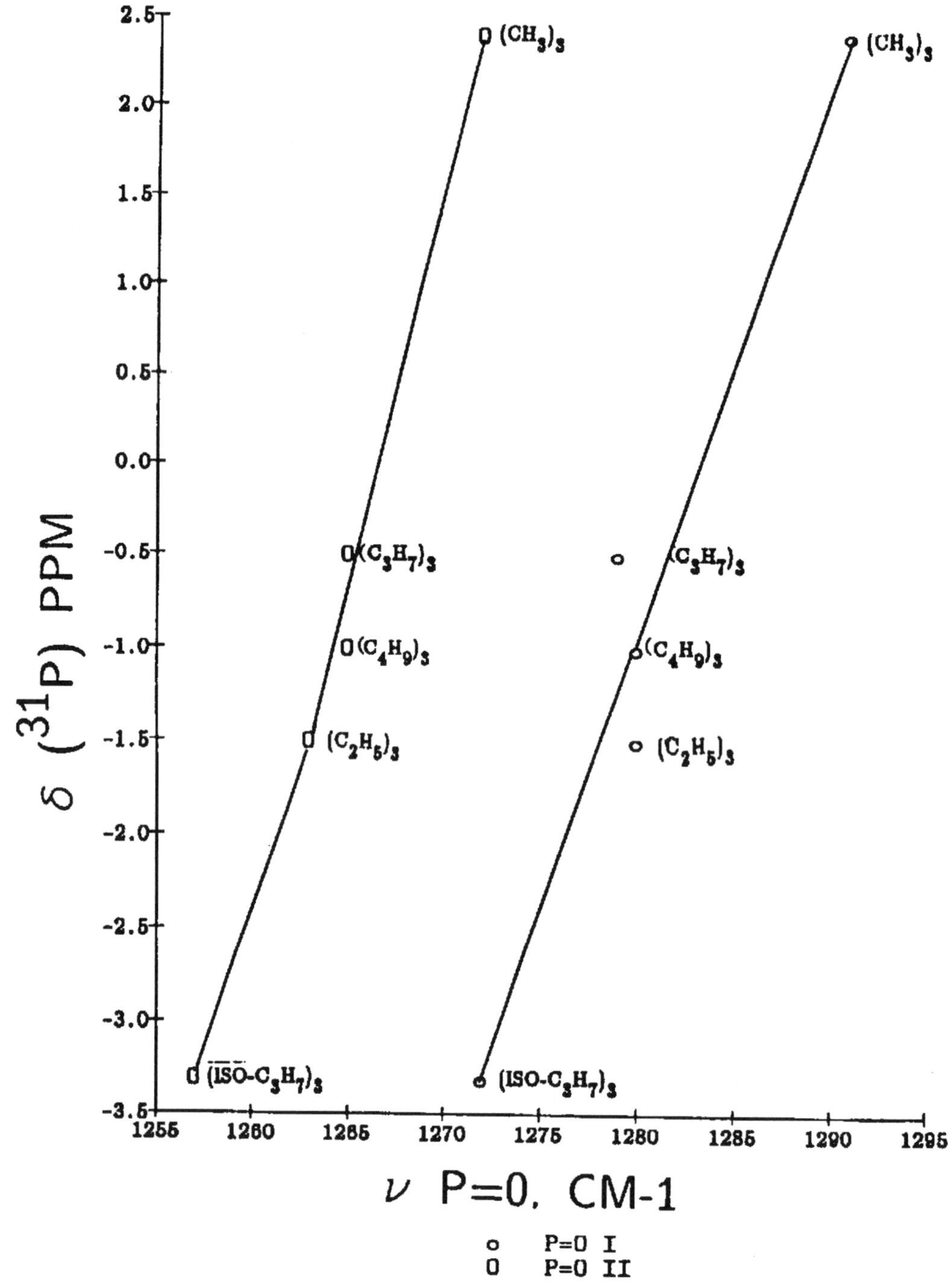

FIGURE 11.65 Plots of ν P=O rotational conformers vs δ ^{31}P for trialkyl phosphates.

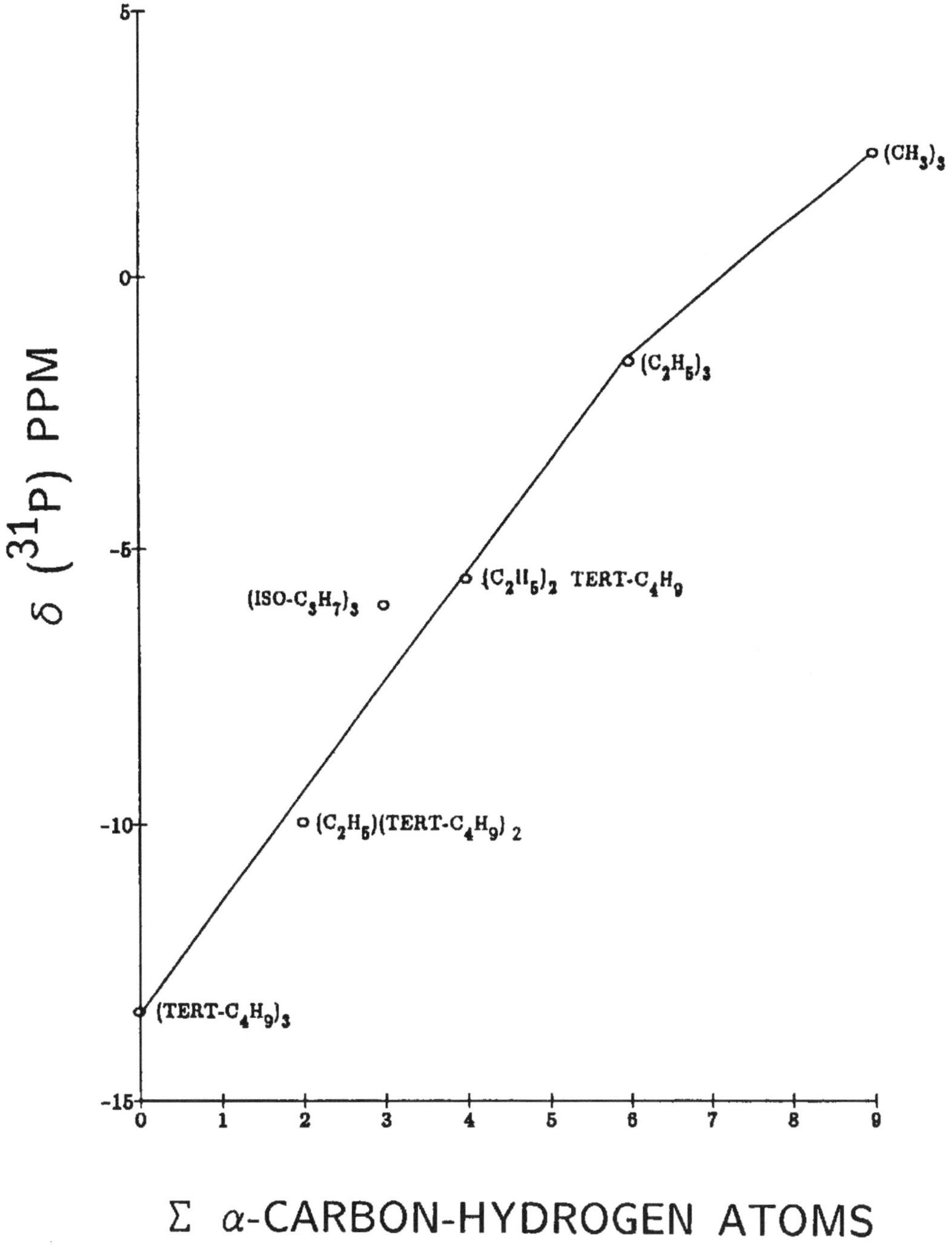

FIGURE 11.66 A plot of δ ^{31}P vs the sum of the number of protons on the α-C—O—P carbon atoms for trialkyl phosphates.

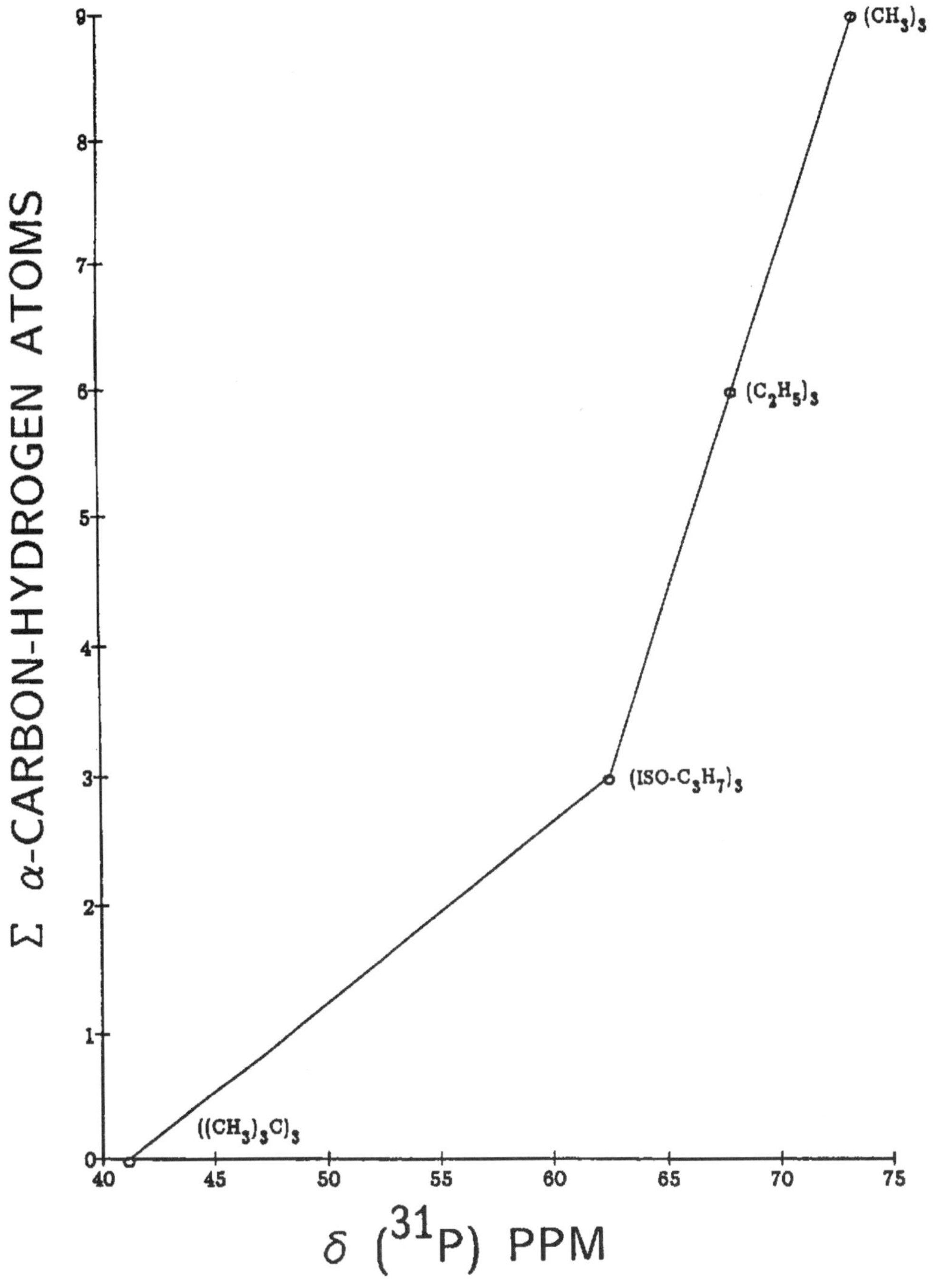

FIGURE 11.67 A plot of the sum of the protons on the α-C—O—P carbon atoms vs δ ^{31}P for O,O,O-trialkyl phosphorothioates.

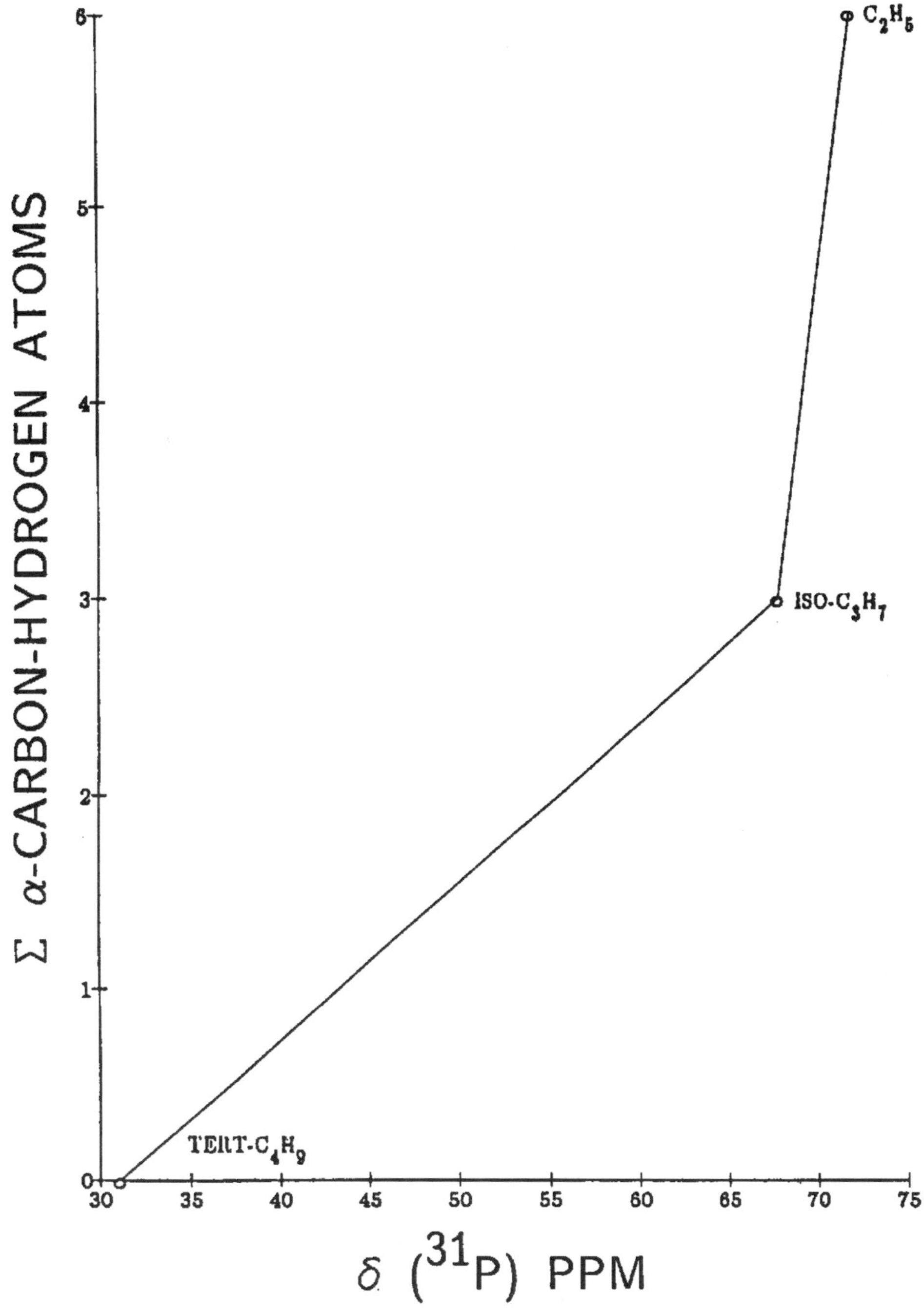

FIGURE 11.68 A plot of the sum of the number of protons on the α-C—O—P carbon atoms vs $\delta\ ^{31}P$ for O,O,O-trialkyl phosphoroselenates.

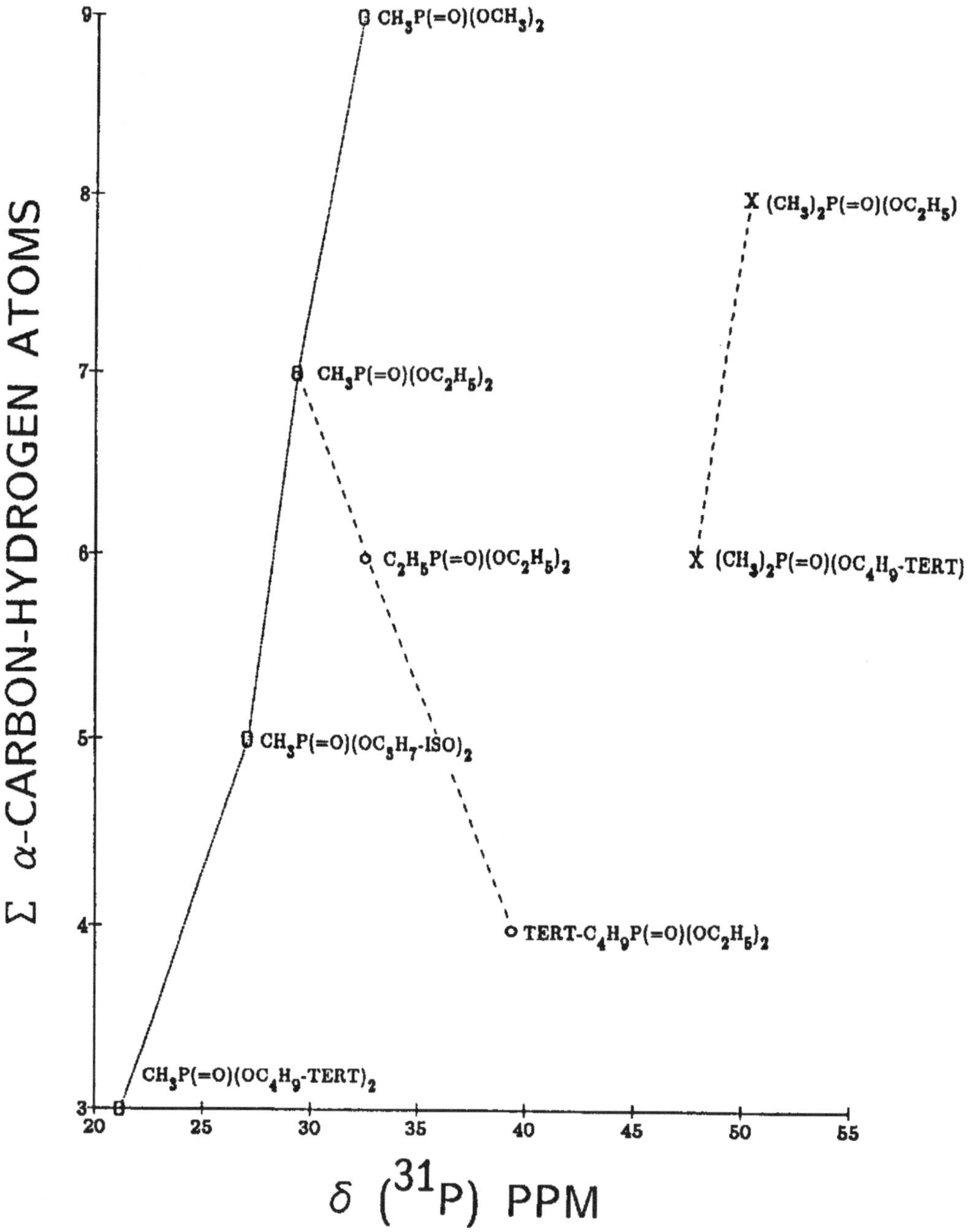

FIGURE 11.69 Plots of δ ^{31}P vs the sum of the number of protons on the α-C–P carbon atoms and the α-C–O–P carbon atoms for $RP(=O)(OC_2H_5)_2$ and $(R)_2P(=O)(OR)$.

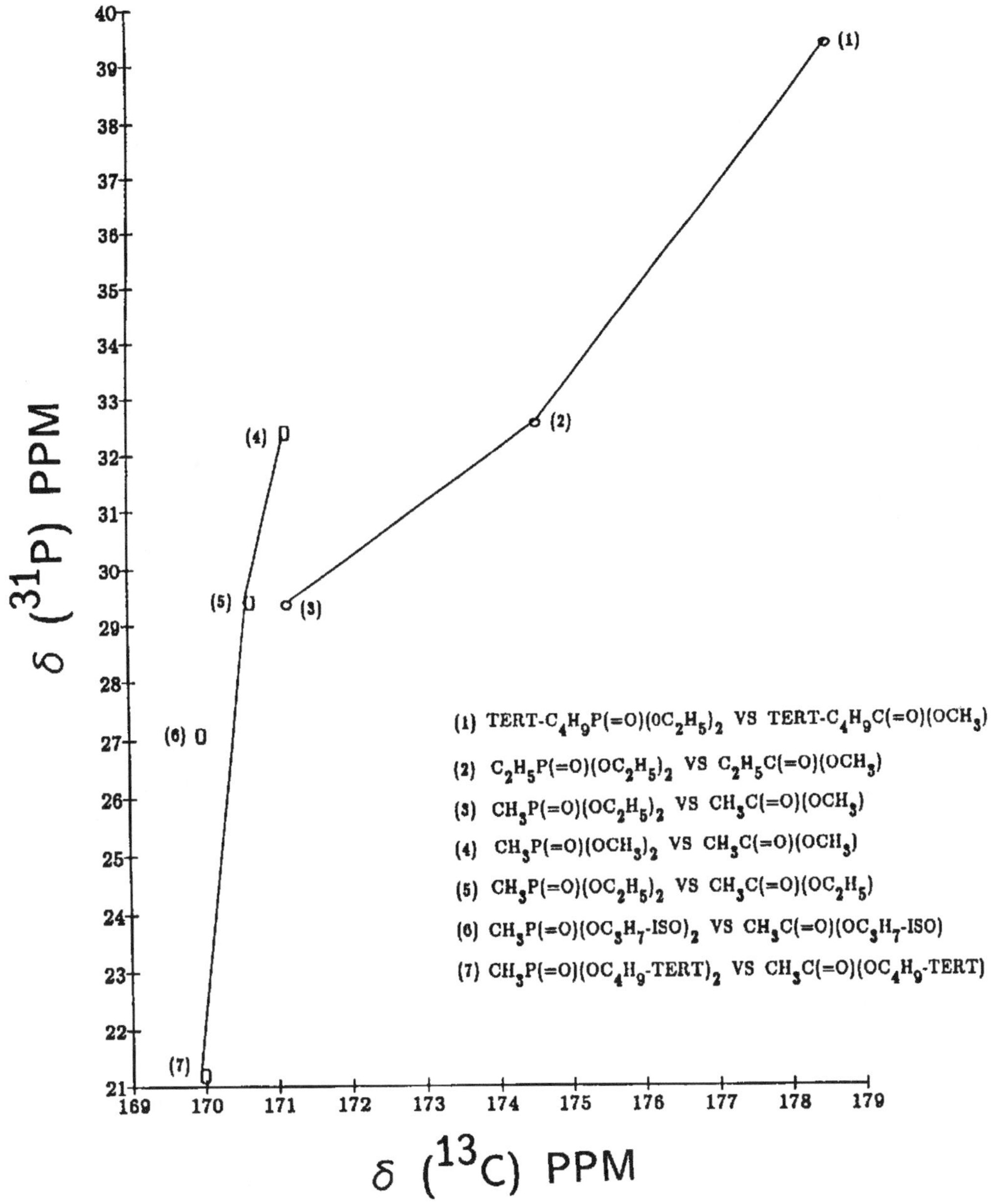

FIGURE 11.70 Plots of $\delta\ ^{31}P$ for $RP(=O)(OC_2H_5)_2$ vs $\delta\ ^{13}C{=}O$ for $RC(=O)(OCH_3)$ and $\delta\ ^{31}P$ for $CH_3P(=O)(OR)_2$ vs $\delta\ ^{13}C{=}O$ for $CH_3C(=O)(OR)$.

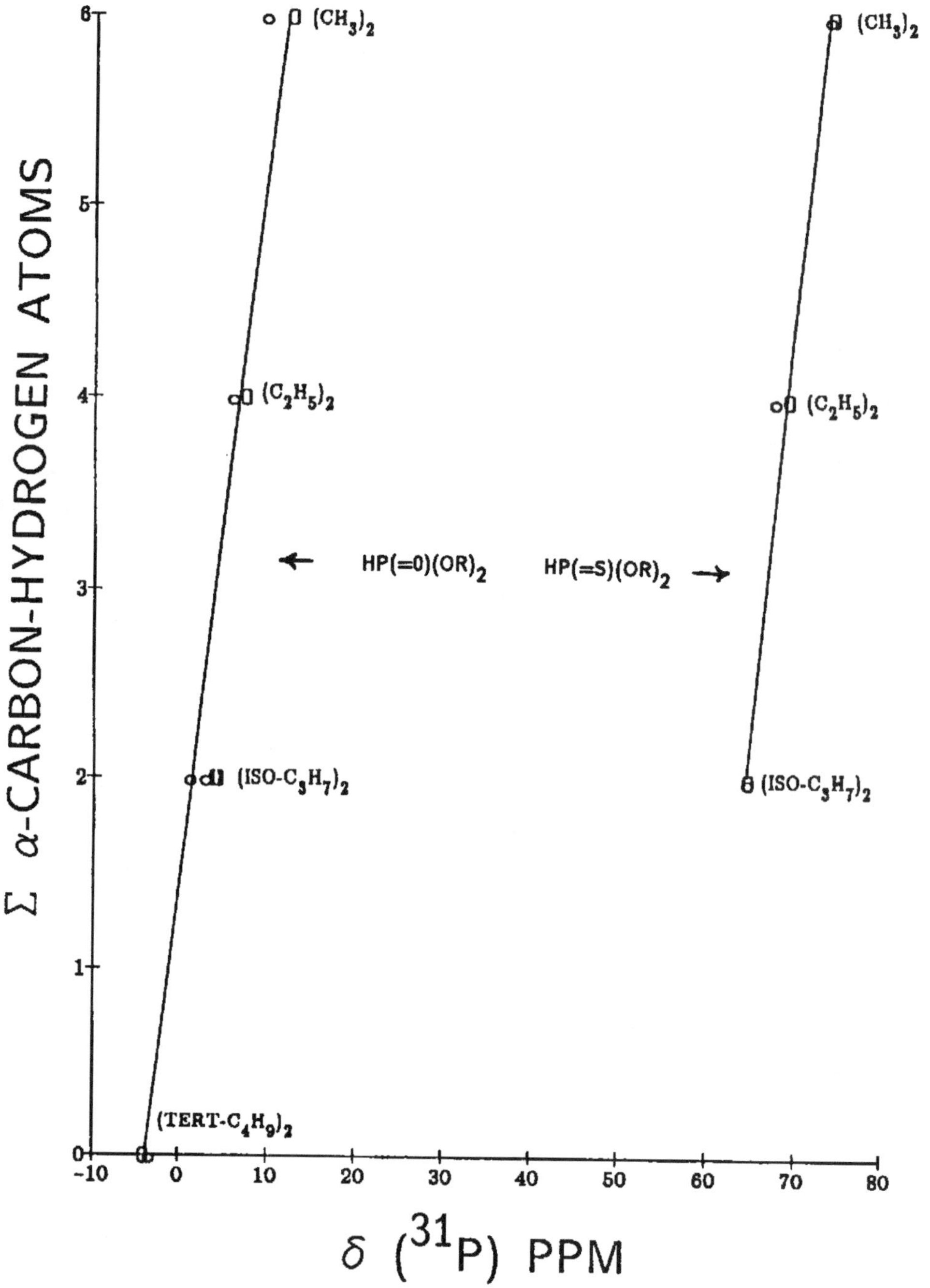

FIGURE 11.71 Plots of δ ^{31}P for $(RO)_2P(=O)H$ and for $(RO)_2P(=S)H$ vs the sum of the protons on the α-C—O—P carbon atoms.

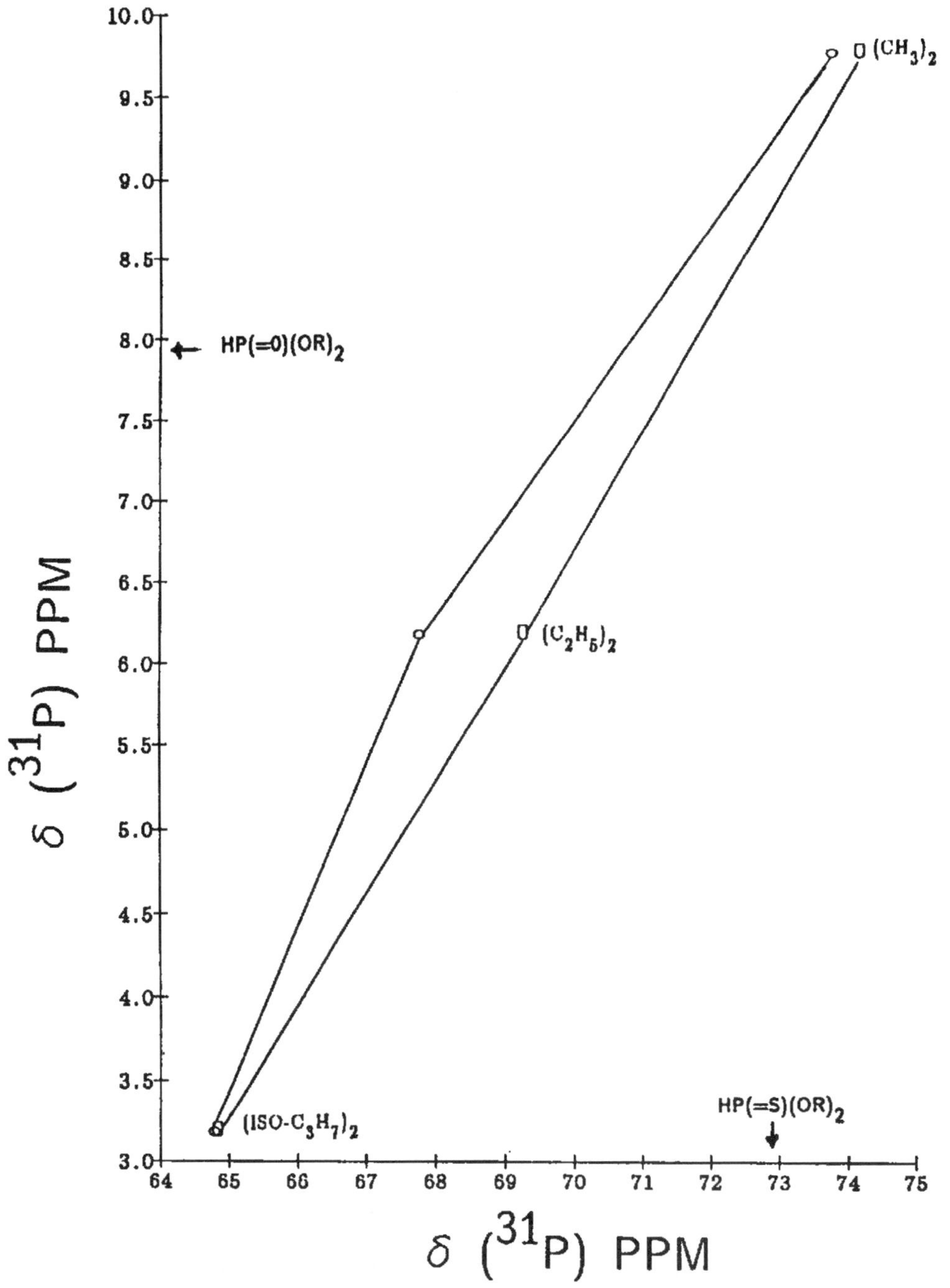

FIGURE 11.72 A plot of δ ^{31}P for $(RO)_2P(=O)H$ analogs vs δ ^{31}P for the corresponding $(RO)_2P(=S)H$ alkyl analogs.

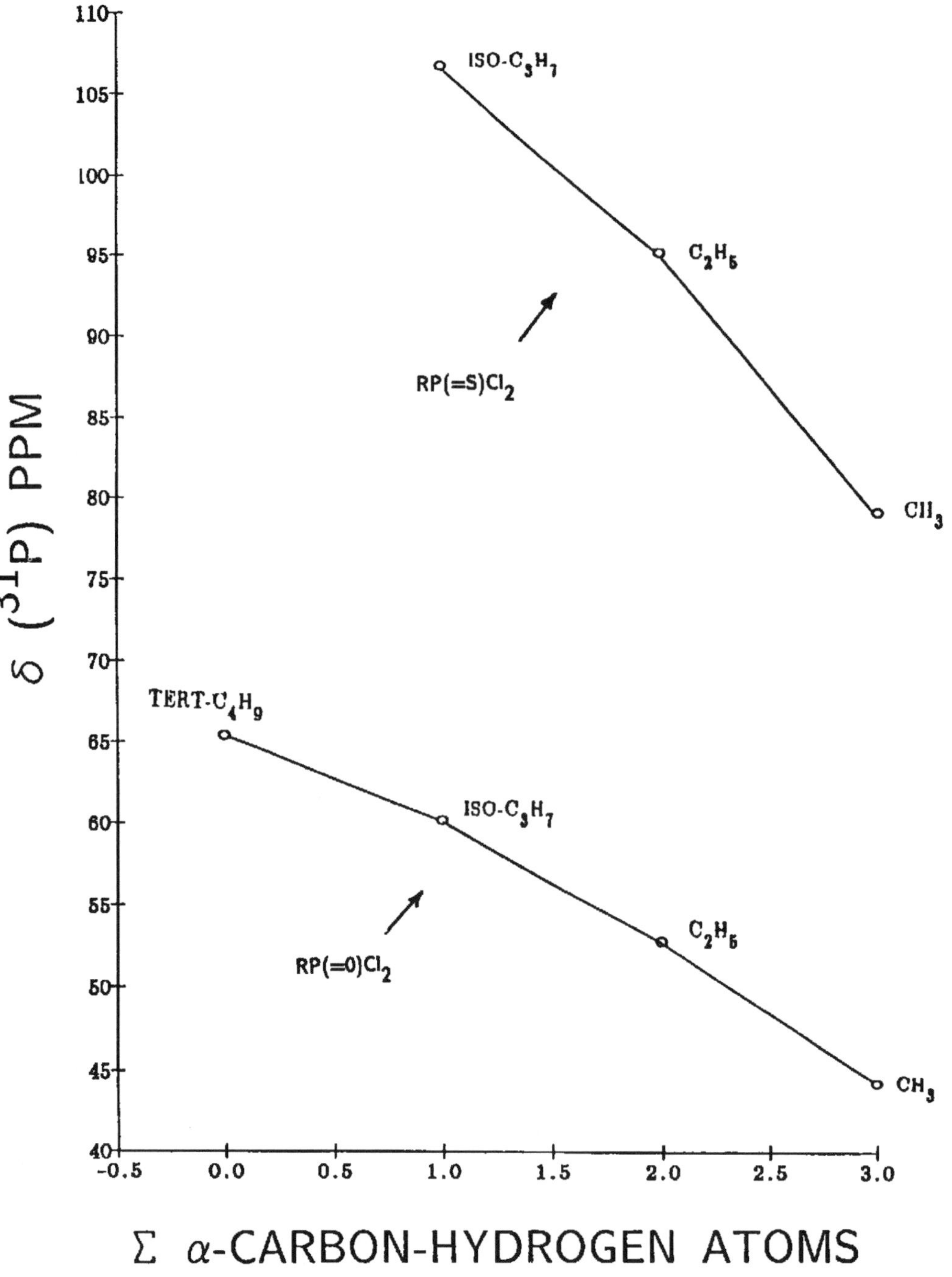

FIGURE 11.73 Plots of the number of protons on the α-CP carbon atoms vs δ ^{31}P for $RP(=O)Cl_2$ and $RP(=S)Cl_2$.

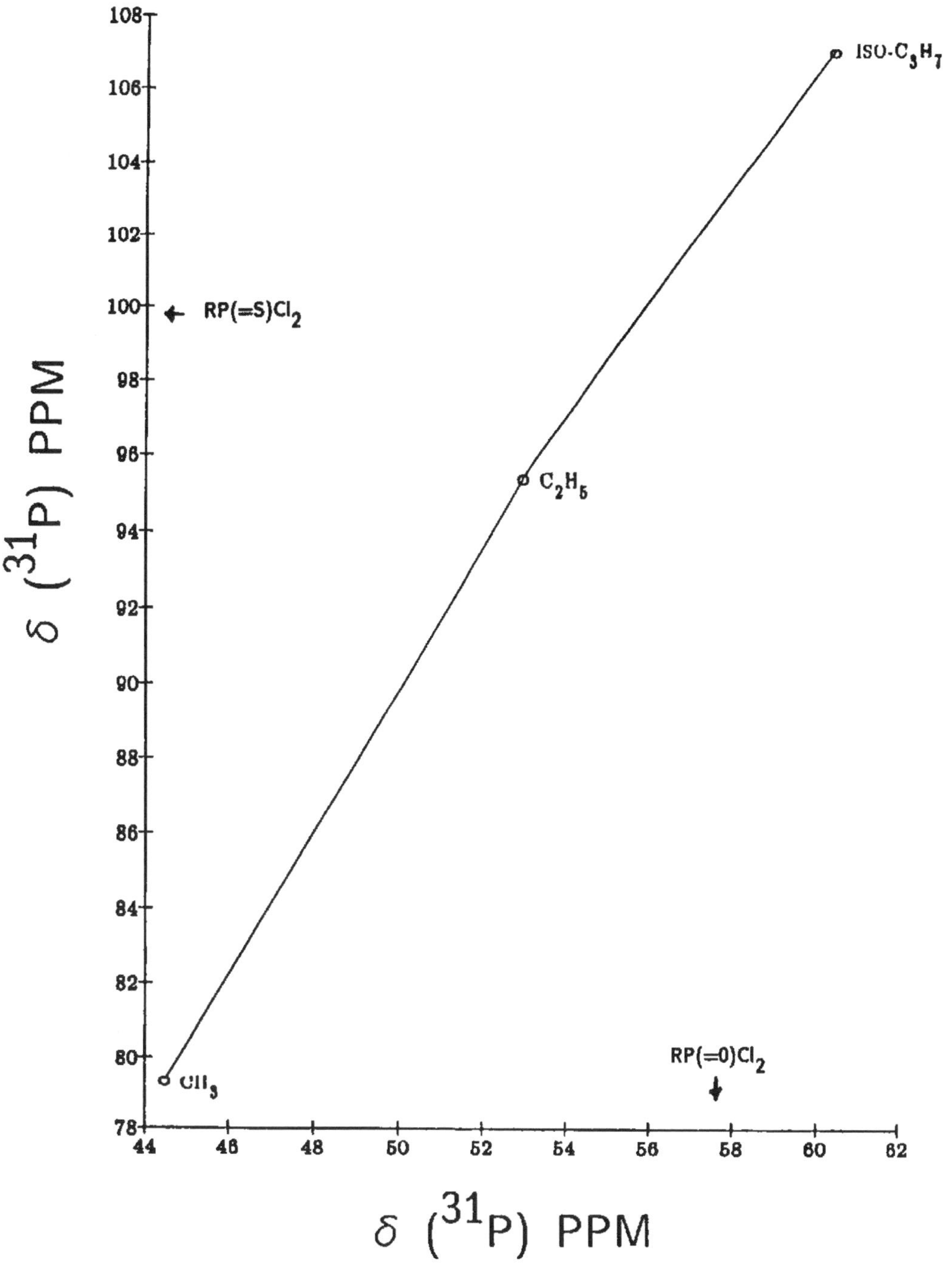

FIGURE 11.74 A plot of δ ^{31}P for $RP(=S)Cl_2$ vs δ ^{31}P for $RP(=O)Cl_2$.

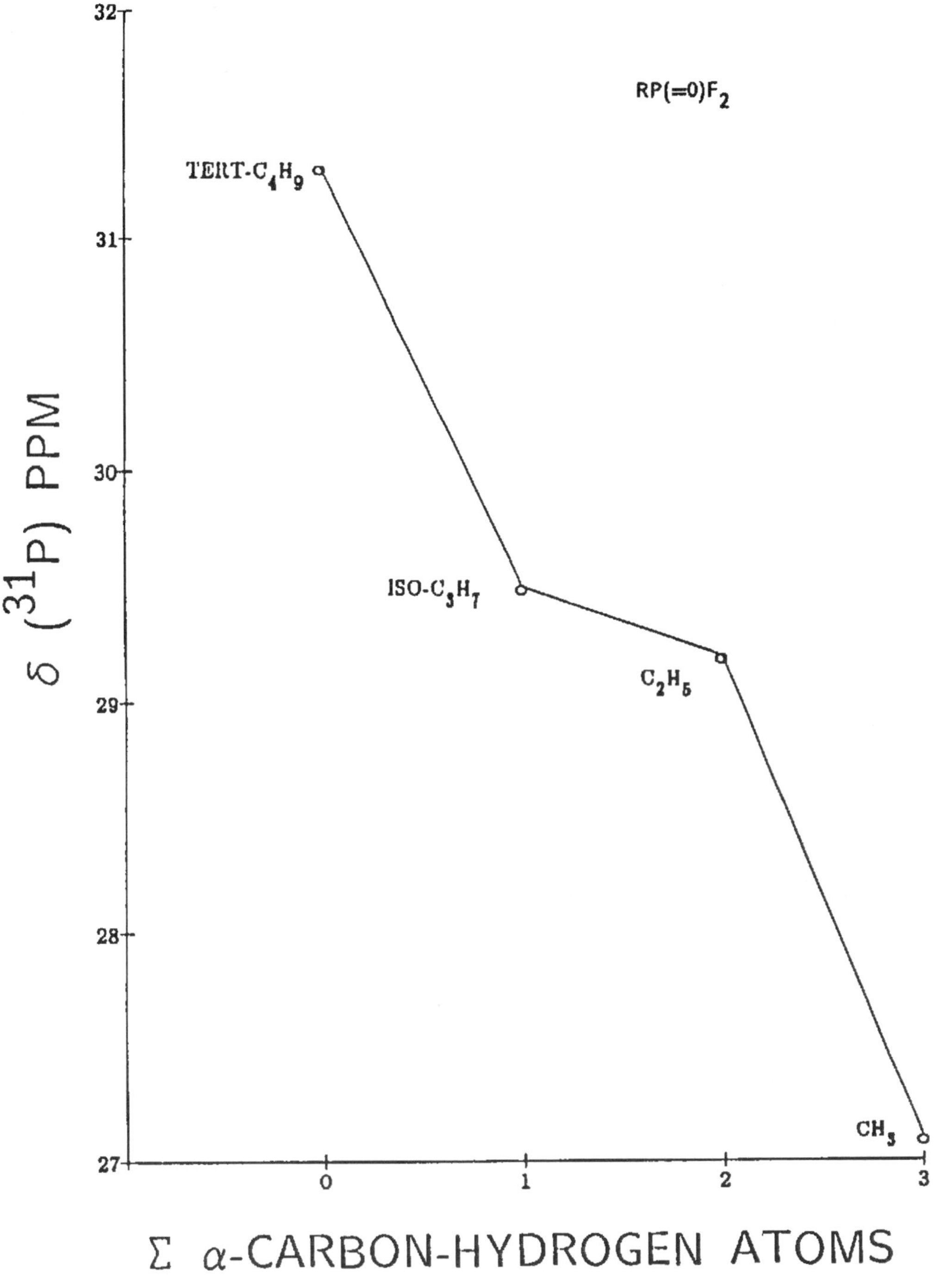

FIGURE 11.75 A plot of δ ^{31}P vs the number of protons on the α-C—P carbon atom for RP(=O)F_2.

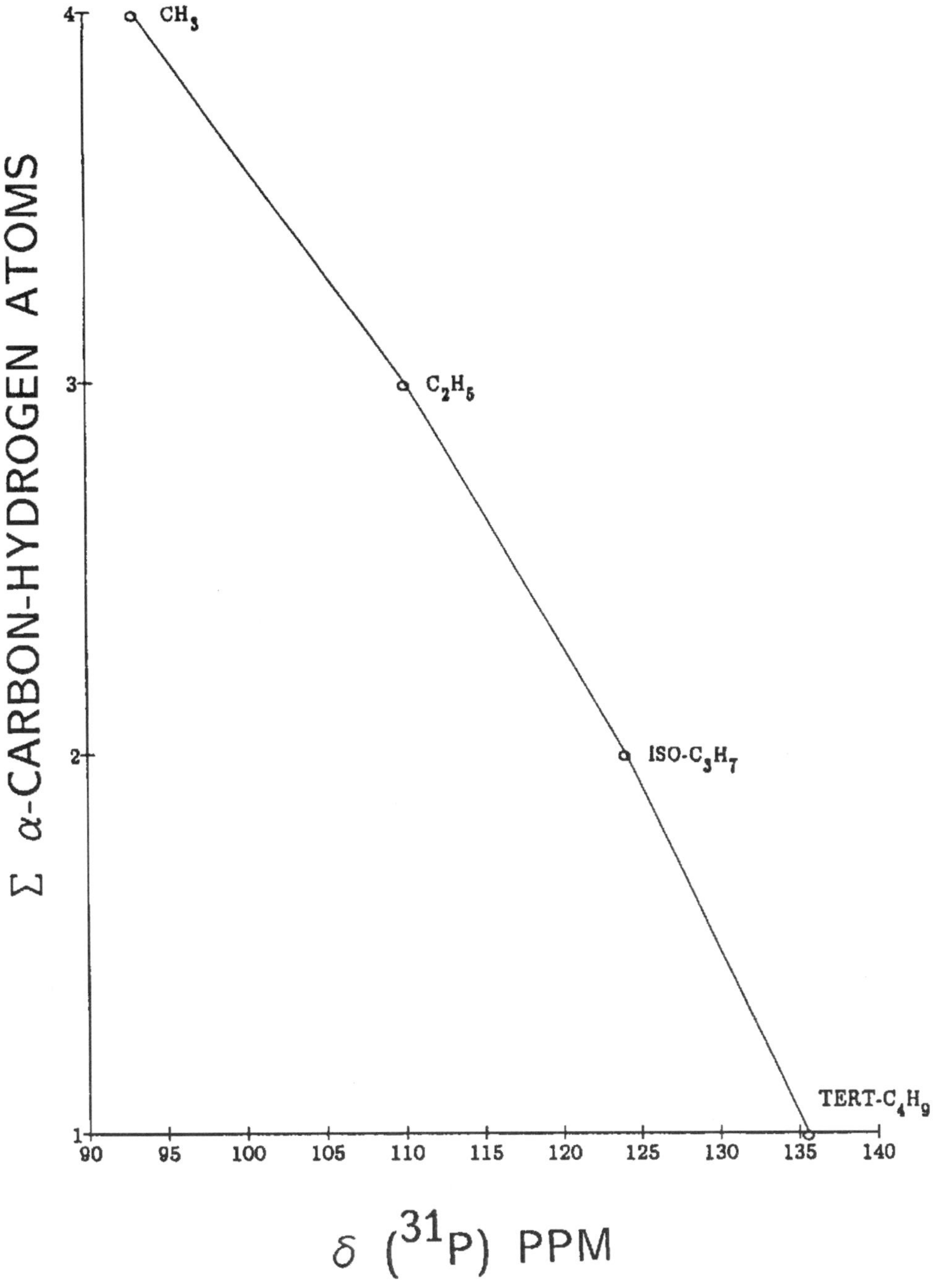

FIGURE 11.76 A plot of δ ^{31}P vs the sum of the number of protons on the α-C—P carbon atom and the number of protons on the α-C—O—P carbon atom for RP(=S)Br($OC_3H_{7\text{-iso}}$).

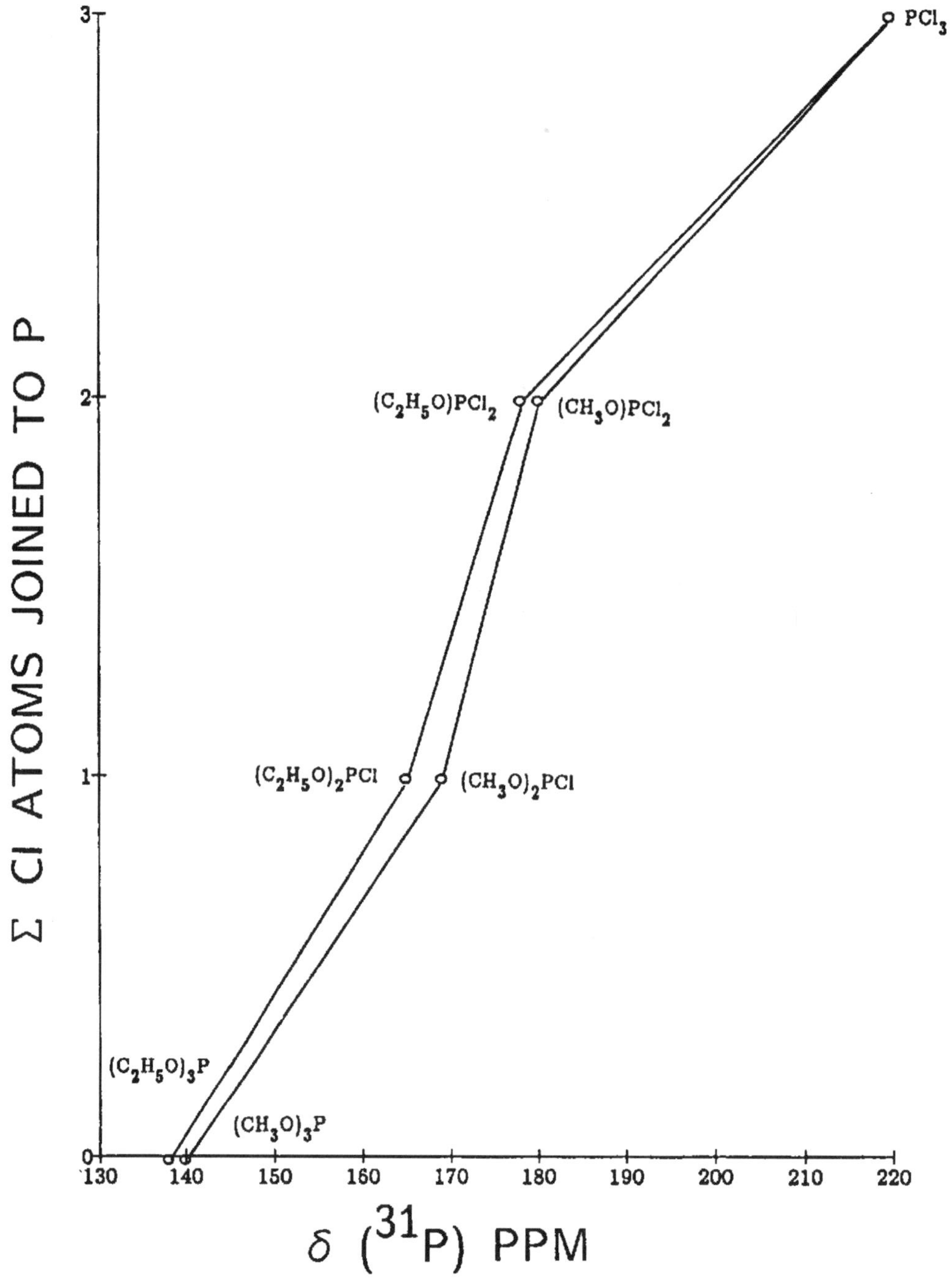

FIGURE 11.77 Plots of δ ^{31}P vs the number of chlorine atoms joined to phosphorus for $PCl_{3-x}(OR)_x$.

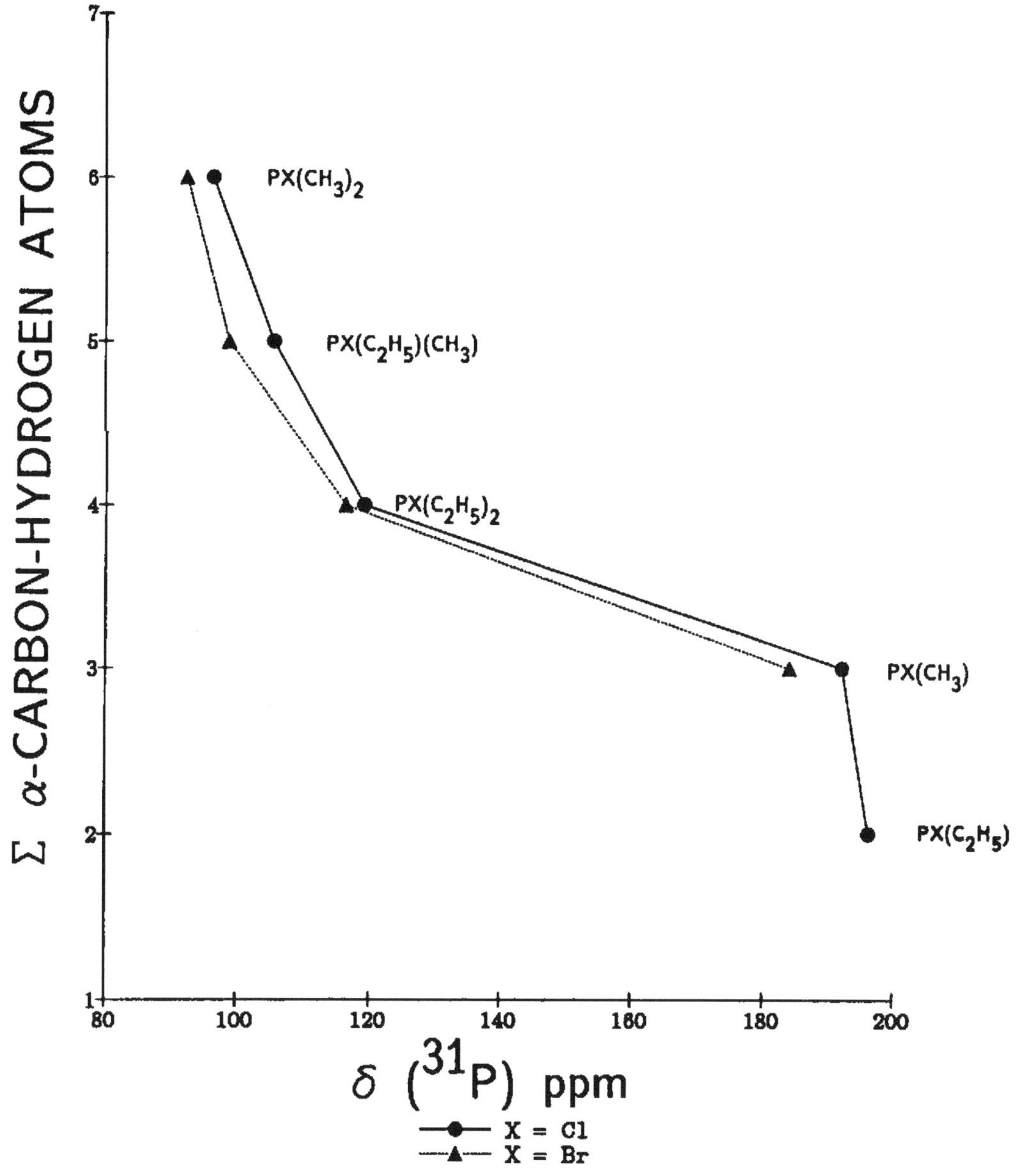

FIGURE 11.78 Plots of the sum of the number of protons on the α-C—P carbon atom(s) vs δ ^{31}P for $PX_{3-n}R_n$.

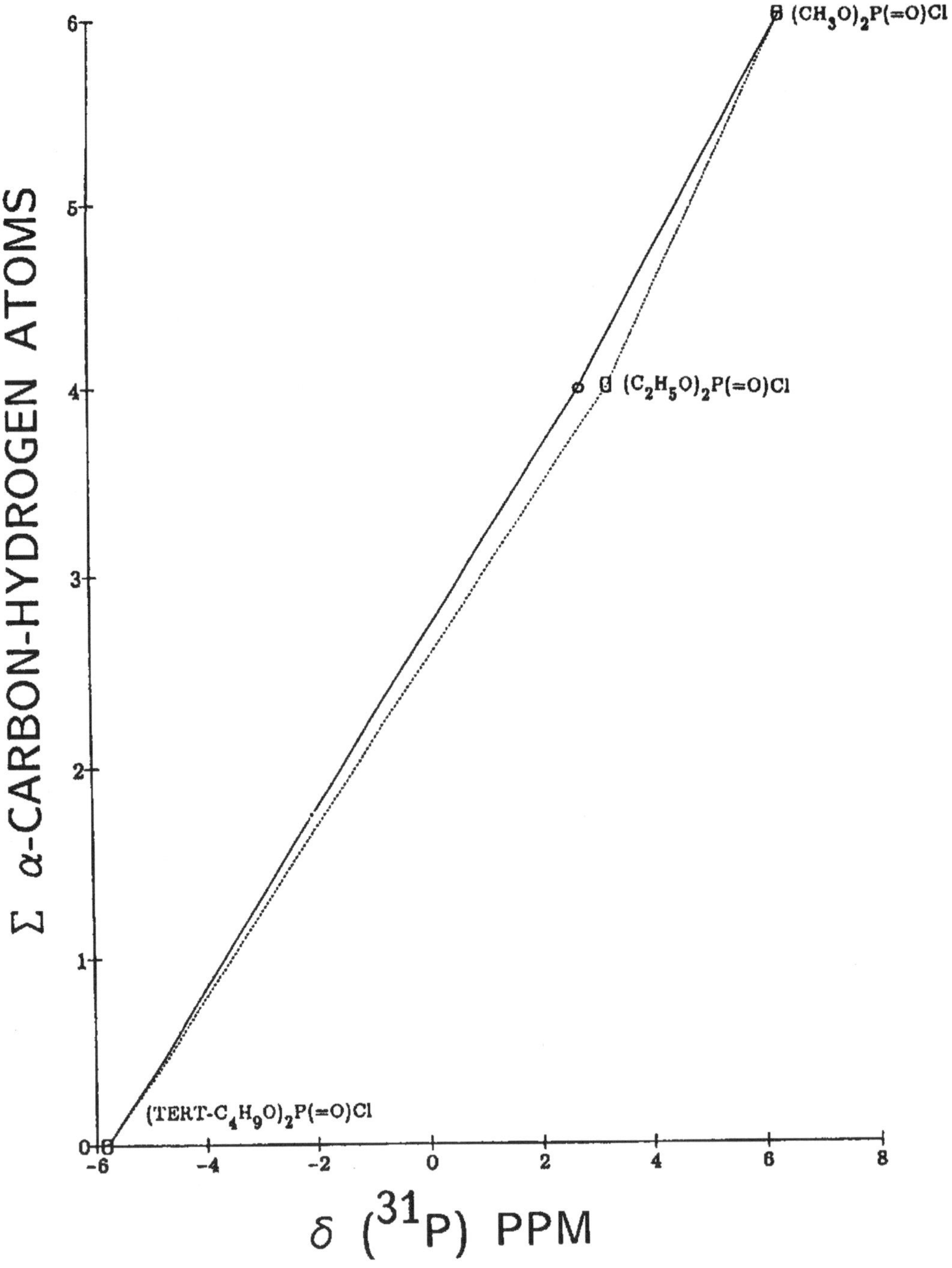

FIGURE 11.79 A plot of δ ^{31}P vs the number of protons on the α-C—O—P carbon atoms for $(R)_2P(=O)Cl_2$.

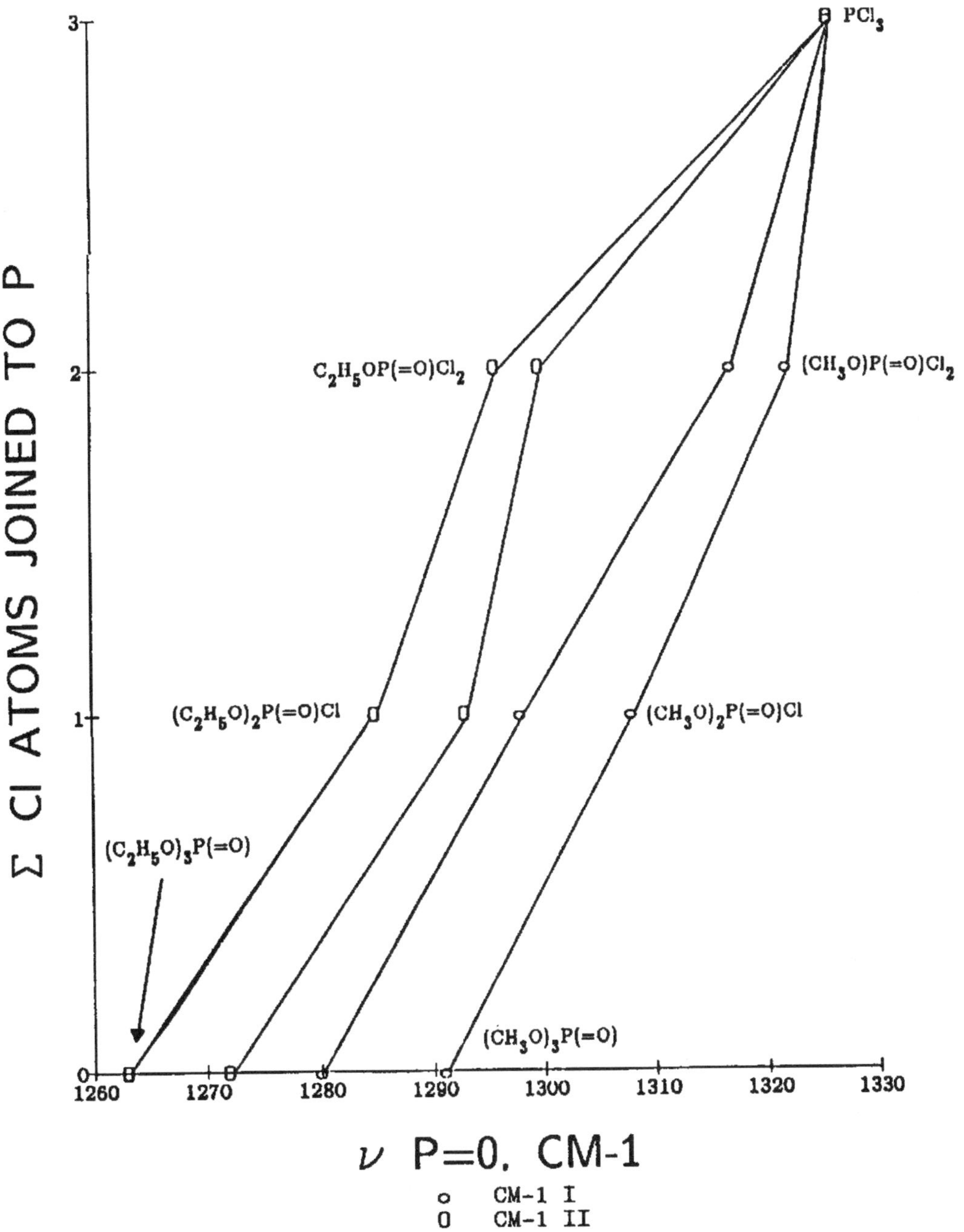

FIGURE 11.80 Plots of the number of Cl atoms joined to phosphorus vs ν P=O and ν P=O rotational conformers for $P(=O)Cl_3$ through $P(=O)Cl_{3-x}(OR)_x$ where R is CH_3 or C_2H_5.

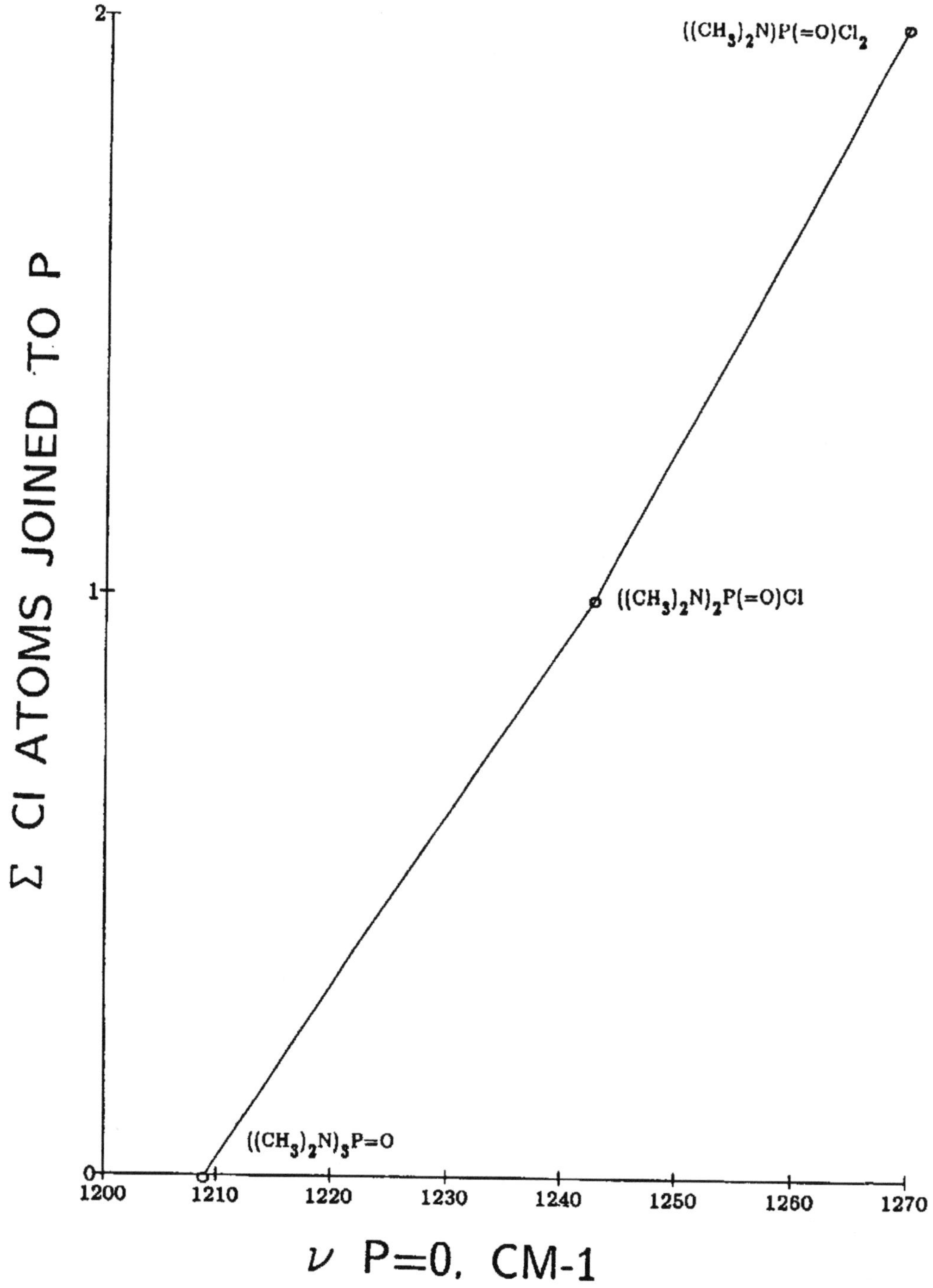

FIGURE 11.81 A plot of the number of Cl atoms joined to phosphorus vs ν P=O for $P(=O)Cl_{3-x}[N(CH_3)_2]_x$.

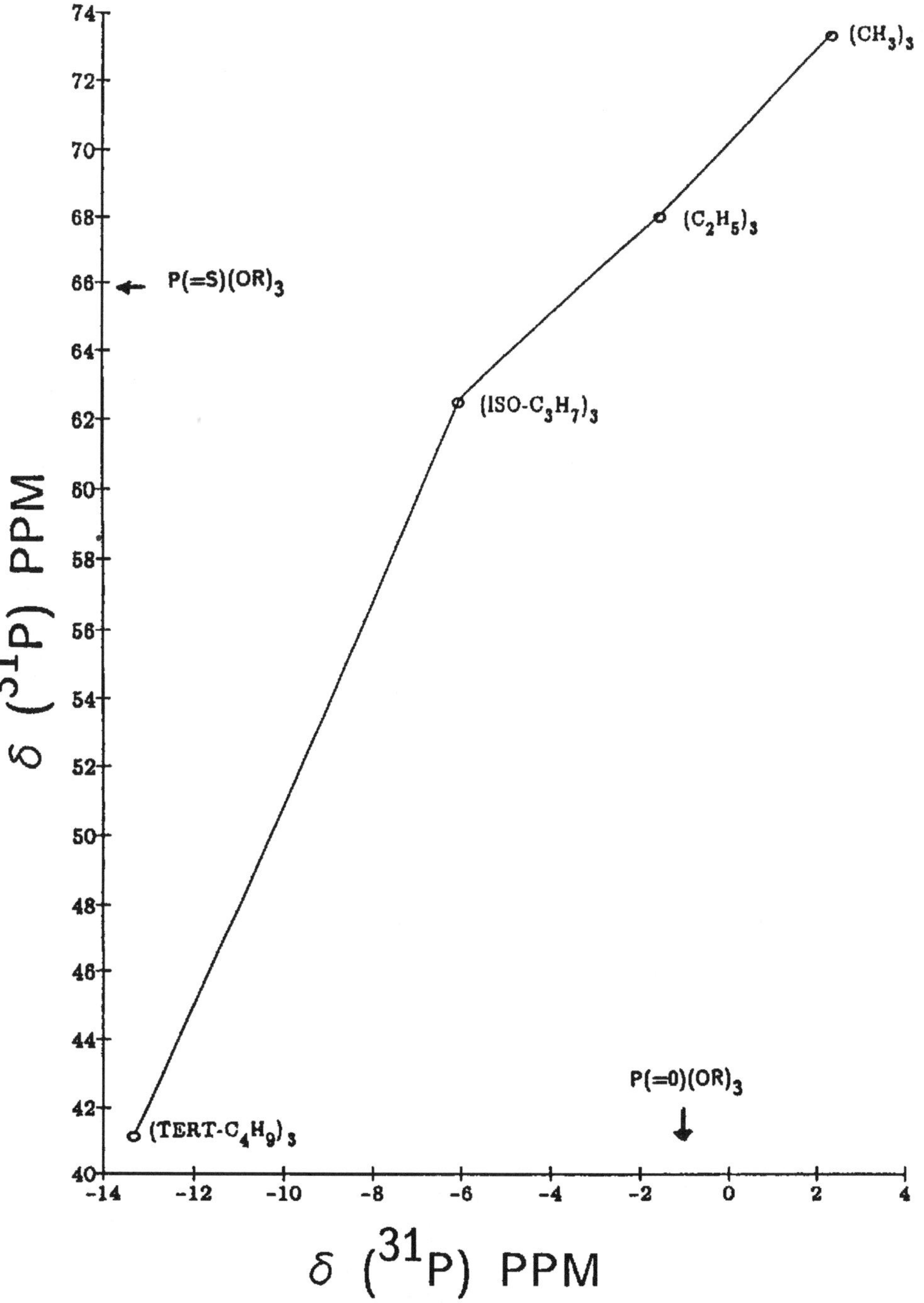

FIGURE 11.82 A plot of δ ^{31}P for $P(=S)(OR)_3$ vs δ ^{31}P for the corresponding alkyl analogs of $P(=O)(OR)_3$.

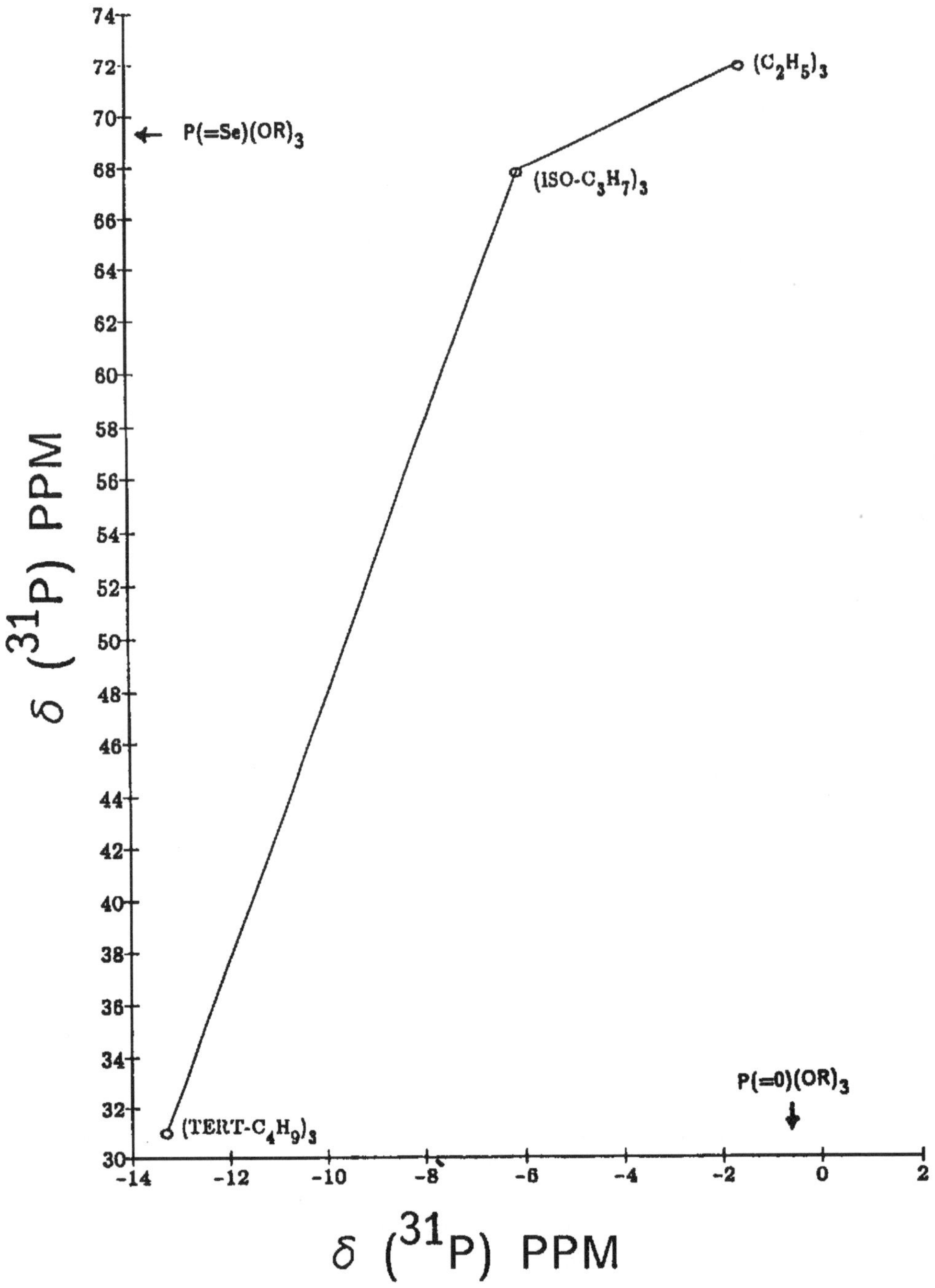

FIGURE 11.83 A plot of δ ^{31}P for $P(=Se)(OR)_3$ vs δ ^{31}P for corresponding alkyl analogs of $P(=O)(OR)_3$.

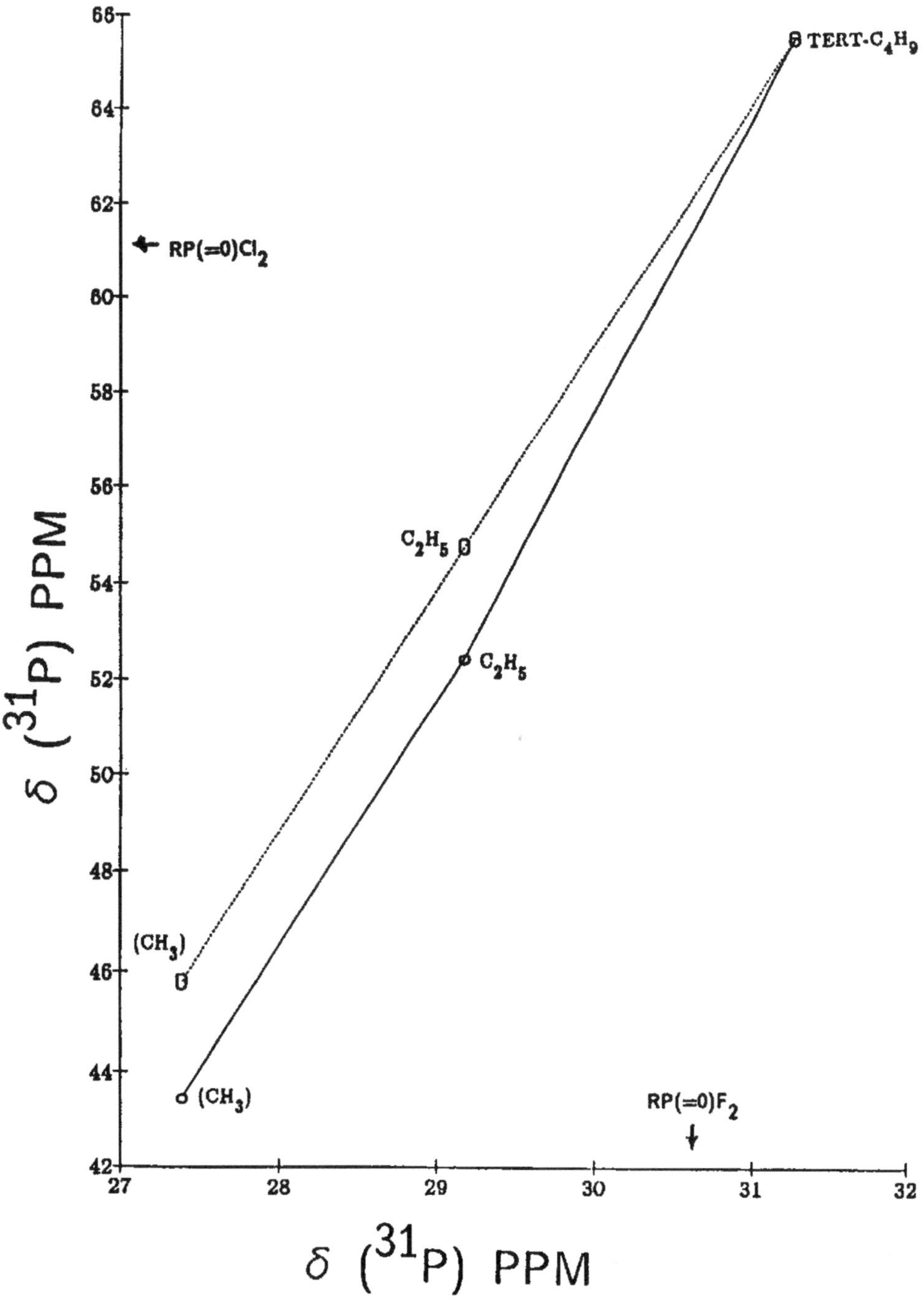

FIGURE 11.84 Plots of δ ^{31}P for $RP(=O)Cl_2$ vs δ ^{31}P for corresponding alkyl analogs of $RP(O)F_2$). The double plot is the δ ^{31}P range given in the literature for the $RP(=O)Cl_2$ analogs.

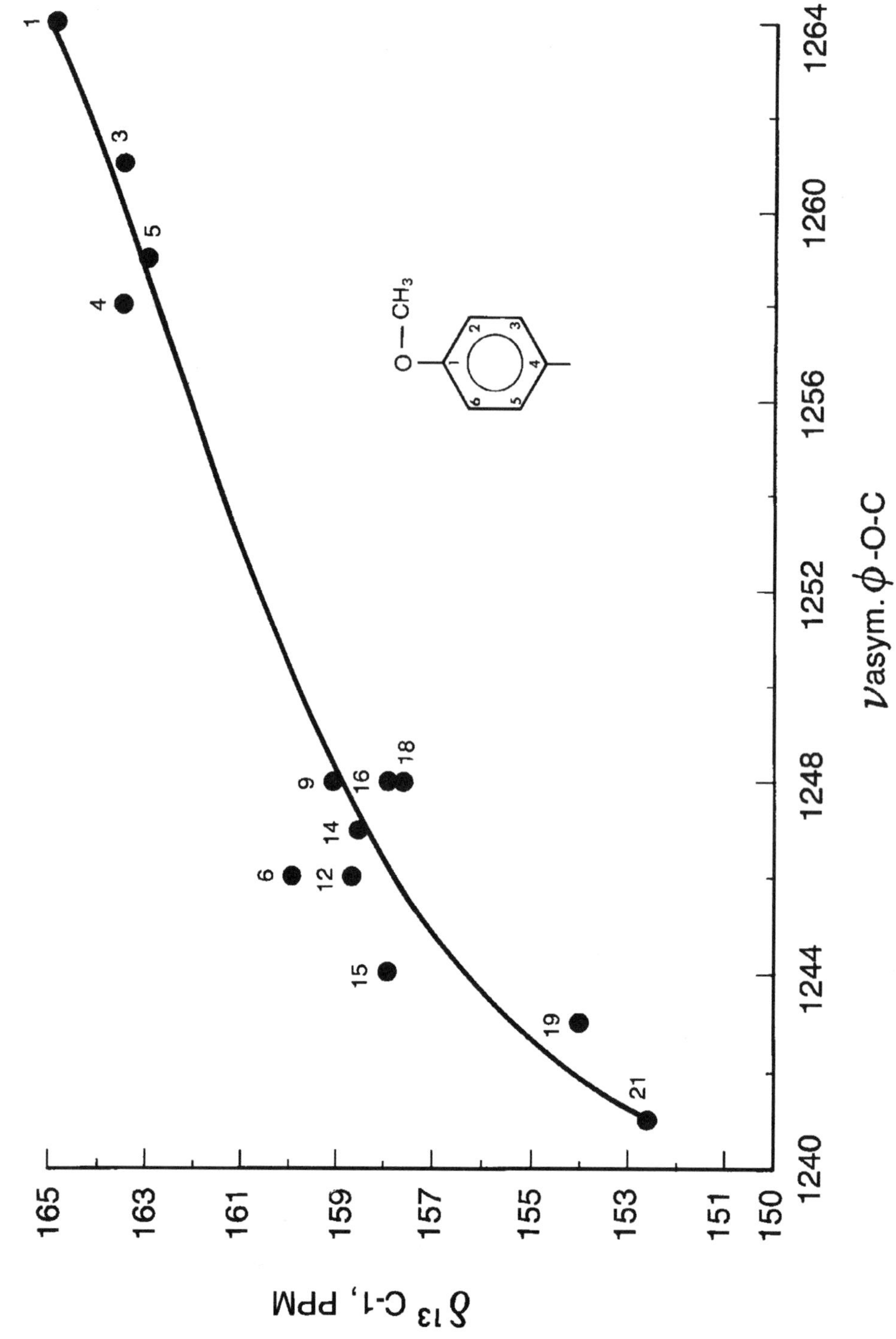

FIGURE 11.85 A plot of ν asym. ϕ-O–C vs δ ^{13}C-1 for 4-x-anisoles.

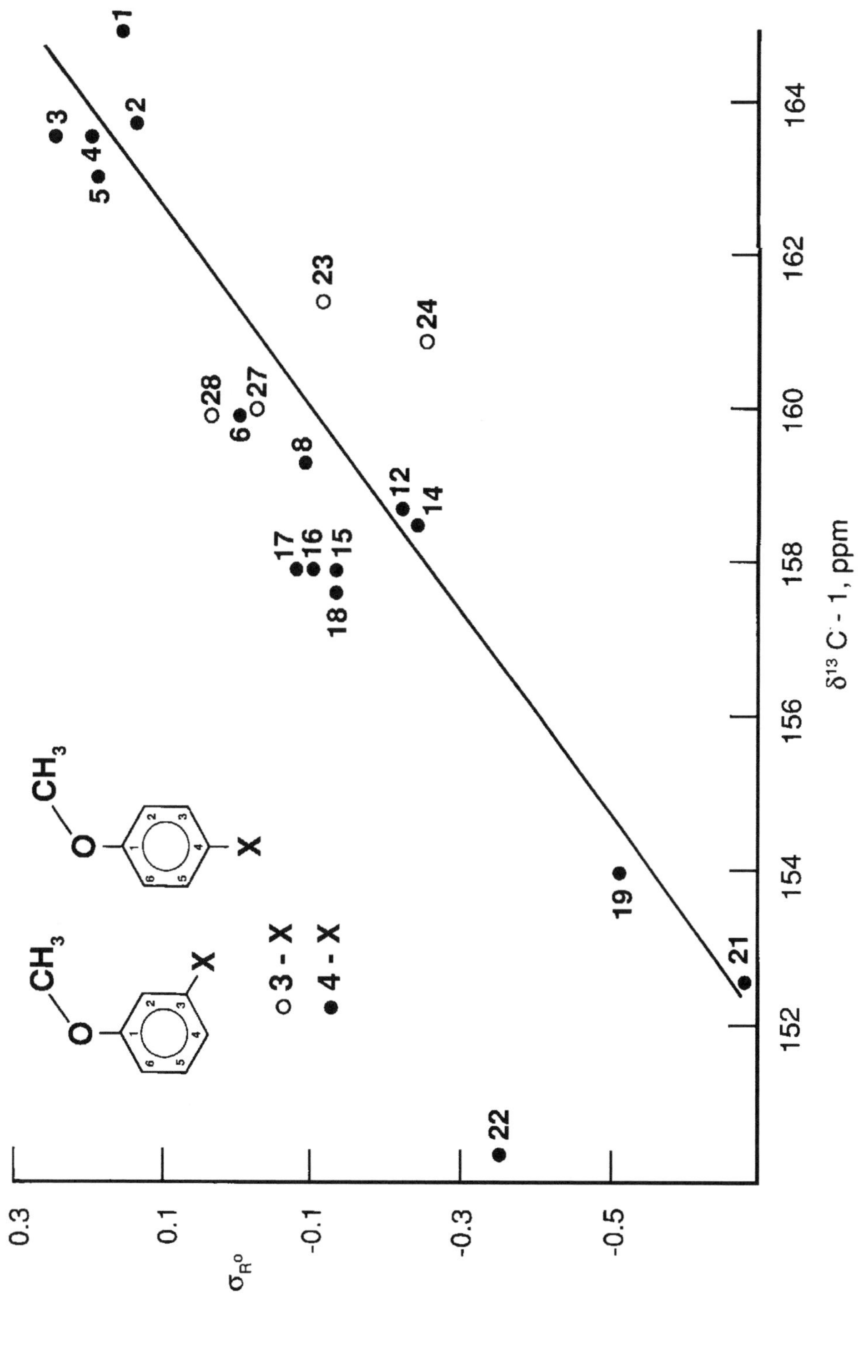

FIGURE 11.86 A plot of δ ^{13}C-1 vs Taft σ_{R^o} for 3-x- and 4-x-anisoles.

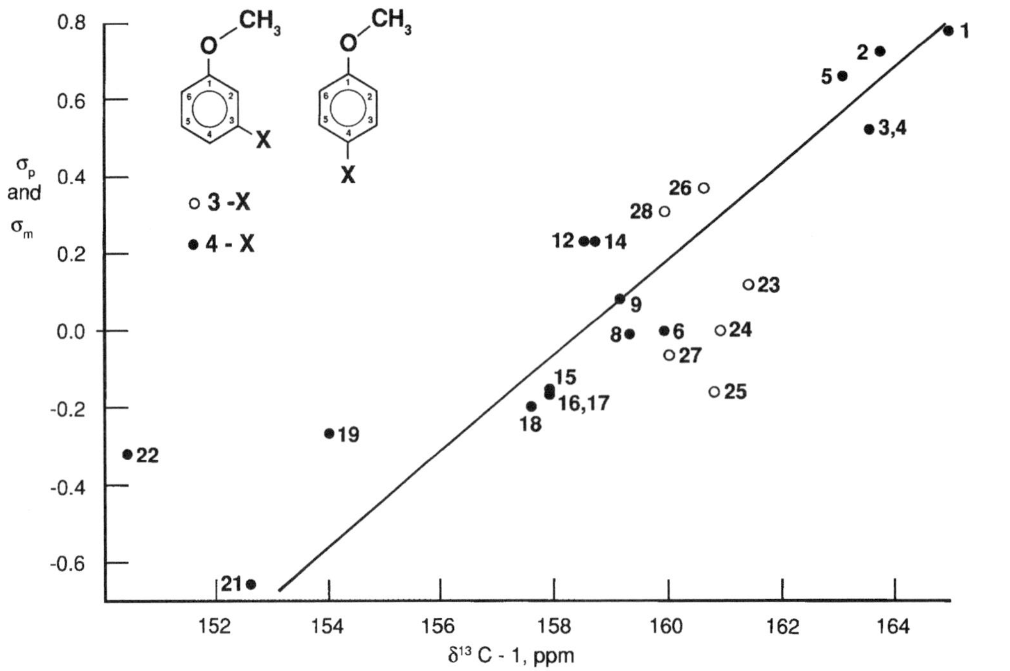

FIGURE 11.87 A plot of δ ^{13}C-1 vs Hammett σ values for 3-x^{v}- and 4-x-anisoles.

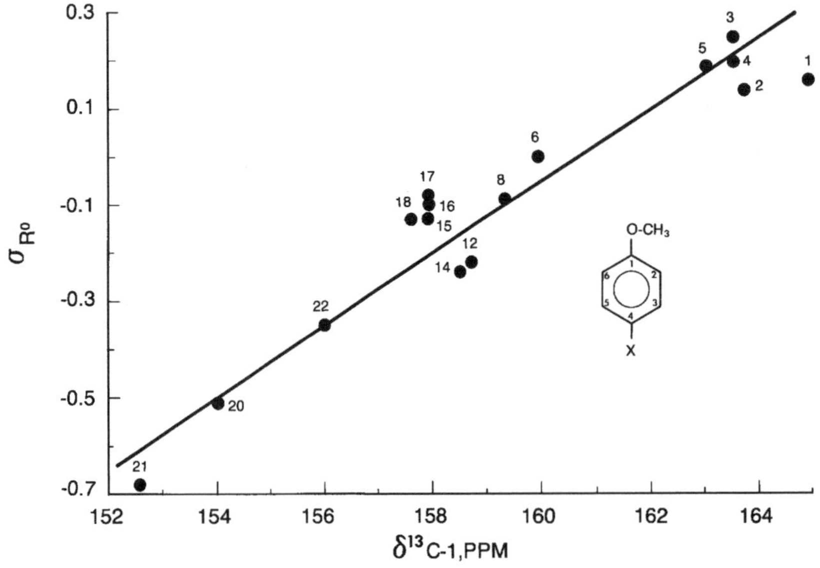

FIGURE 11.88 A plot of Taft $\sigma_{R^{\circ}}$ values vs σ ^{13}C-1 for 4-x-anisoles.

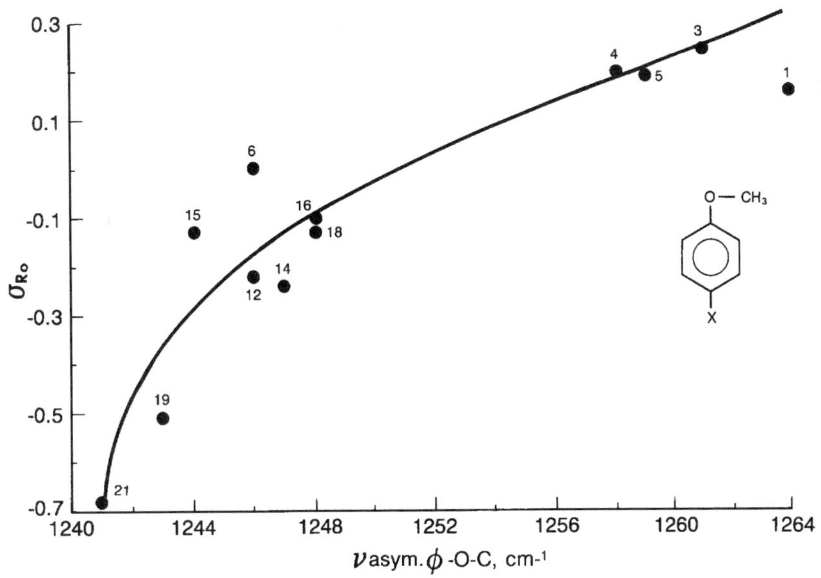

FIGURE 11.89 A plot of Taft $\sigma_{R^{\circ}}$ values vs ν asym. ϕ-O—C for 4-x-anisoles.

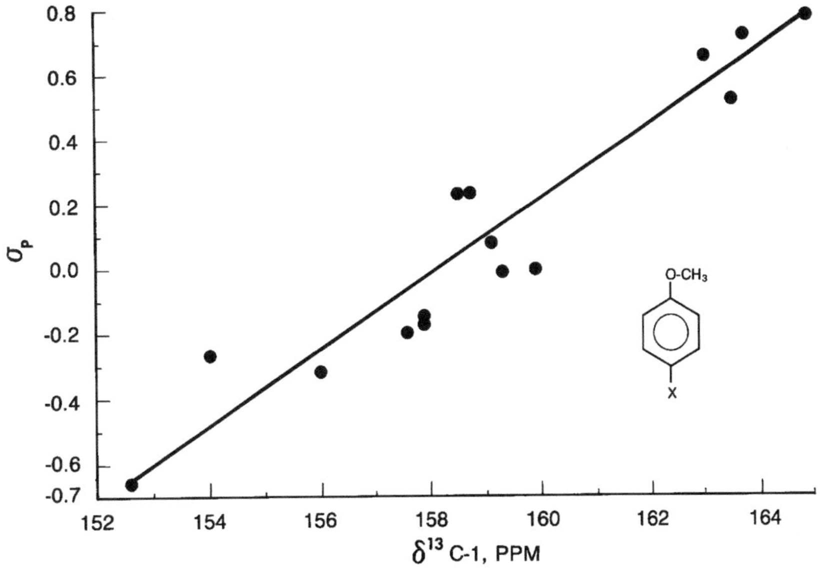

FIGURE 11.90 A plot of δ ^{13}C-1 vs Hammett σ_p values for 4-x-anisoles.

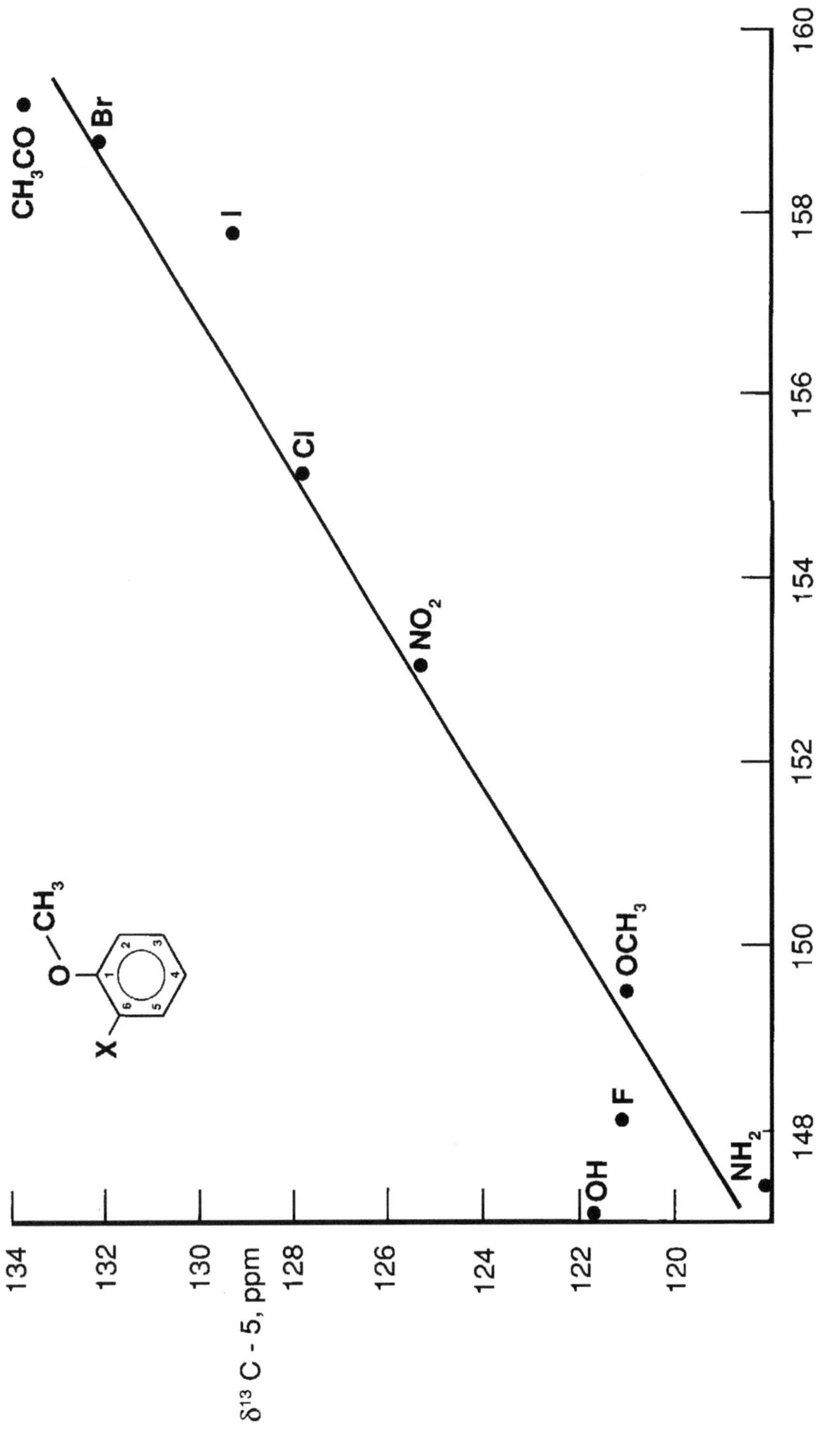

FIGURE 11.91 A plot of δ ^{13}C-1 vs δ ^{13}C-5 for 2-x-anisoles.

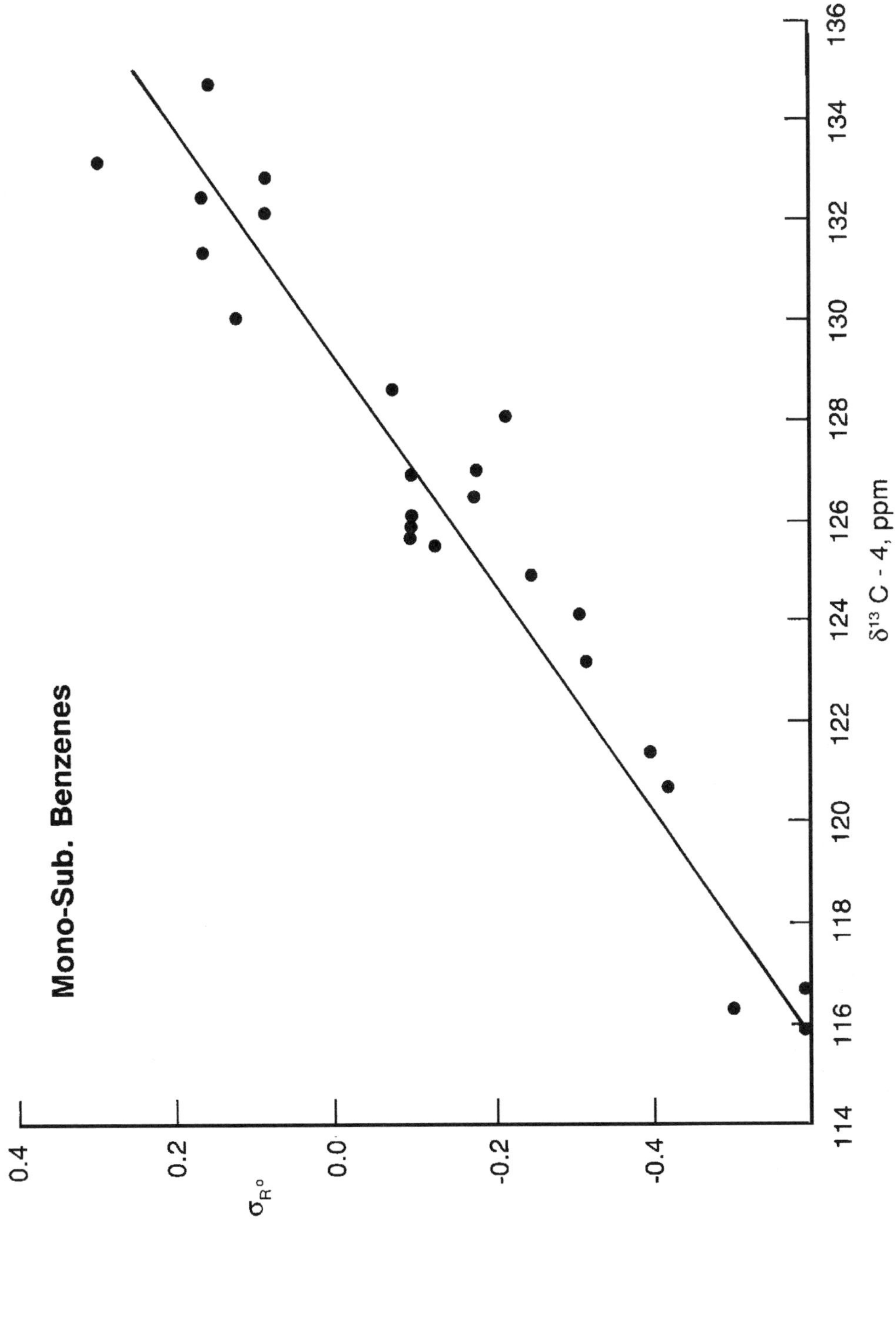

FIGURE 11.92 A plot of δ ^{13}C-4 for mono-substituted benzenes vs Taft σ_{R° values.

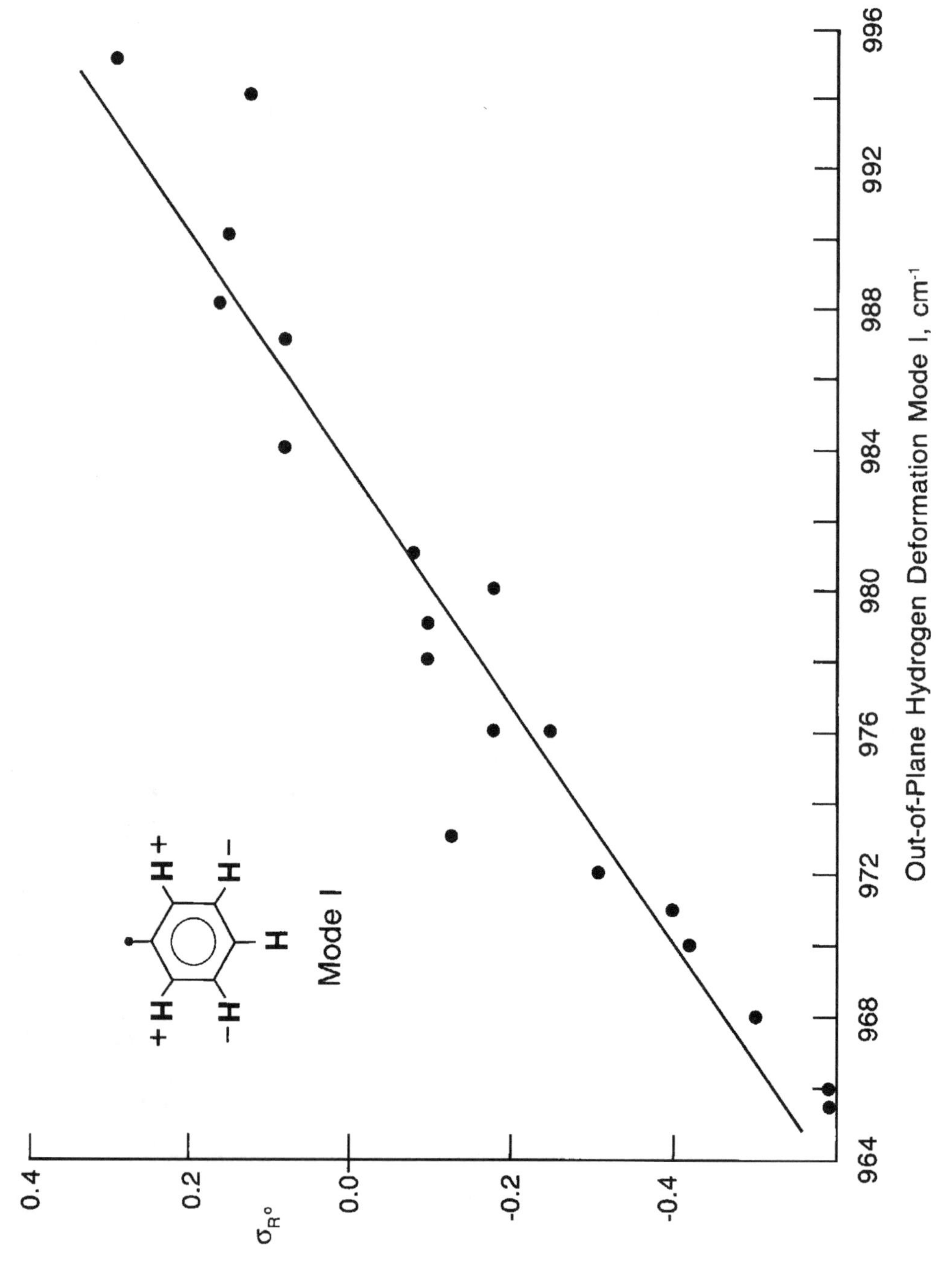

FIGURE 11.93 A plot of the out-of-plane hydrogen deformation mode I for mono-substituted benzenes vs Taft σ_{R° values.

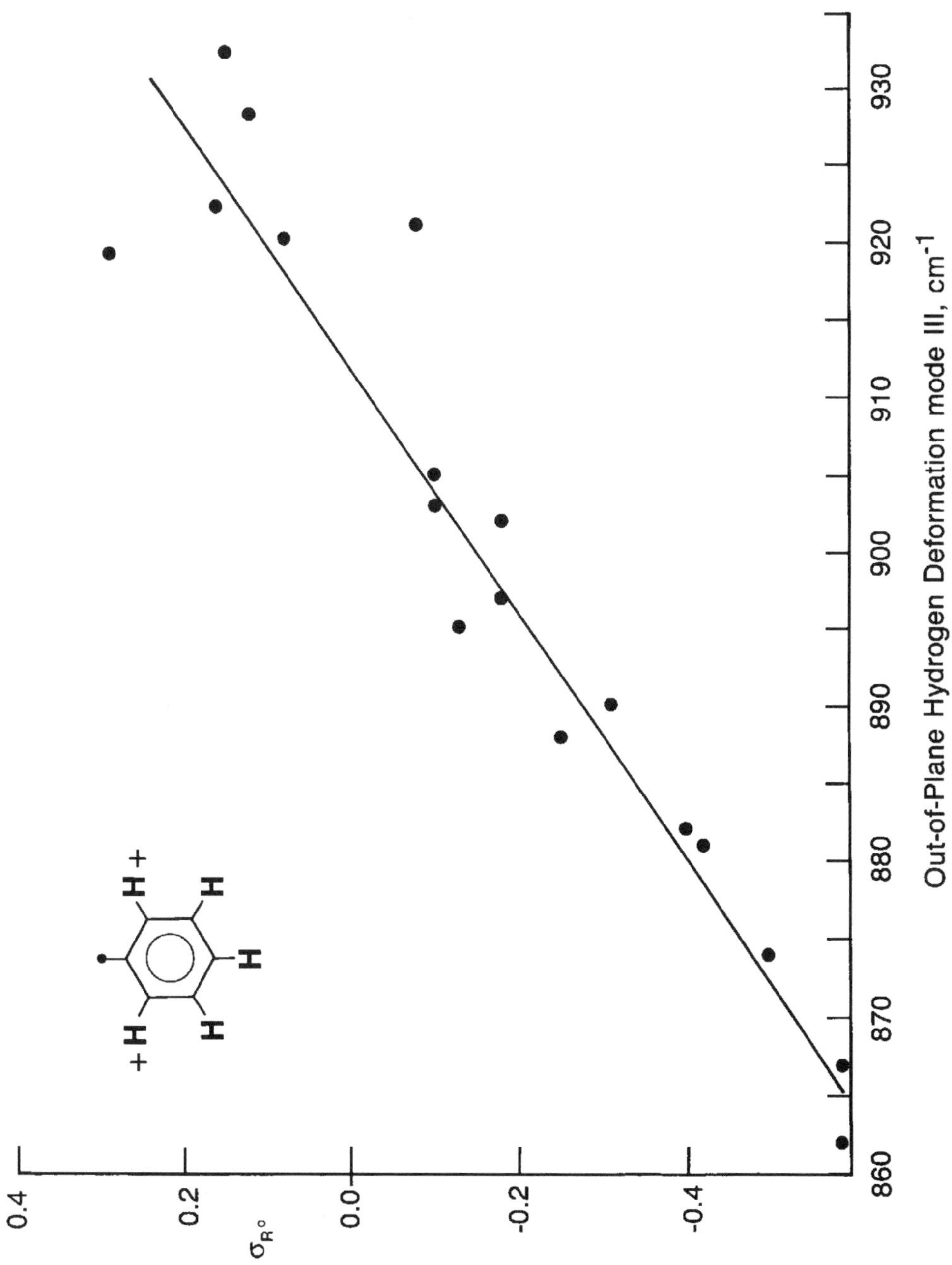

FIGURE 11.94 A plot of the out-of-plane hydrogen deformation mode III for mono-substituted benzenes vs Taft σ_{R° values.

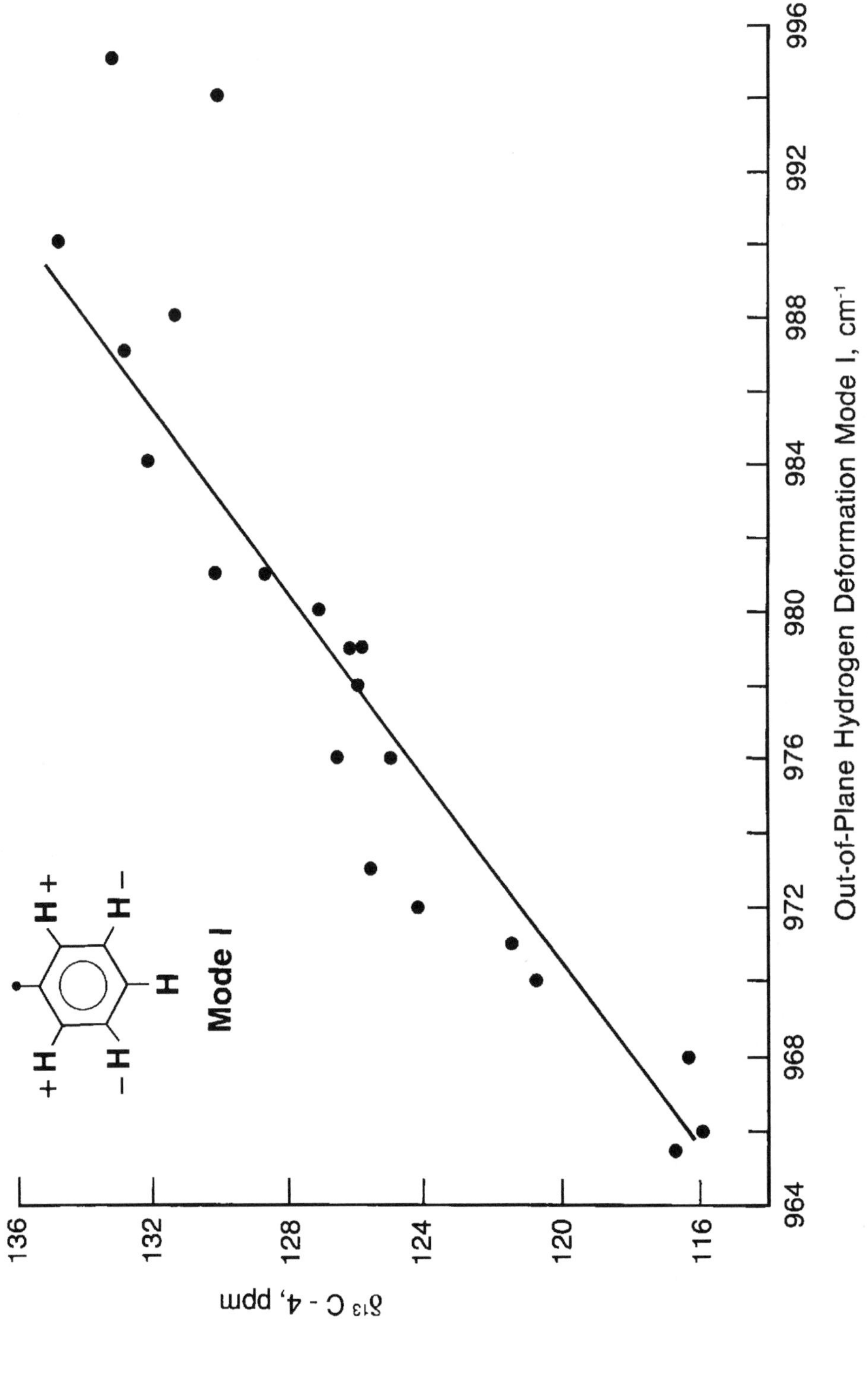

FIGURE 11.95 A plot of the out-of-plane hydrogen deformation mode I for mono-substituted benzenes vs δ ^{13}C-4 for mono-substituted benzenes.

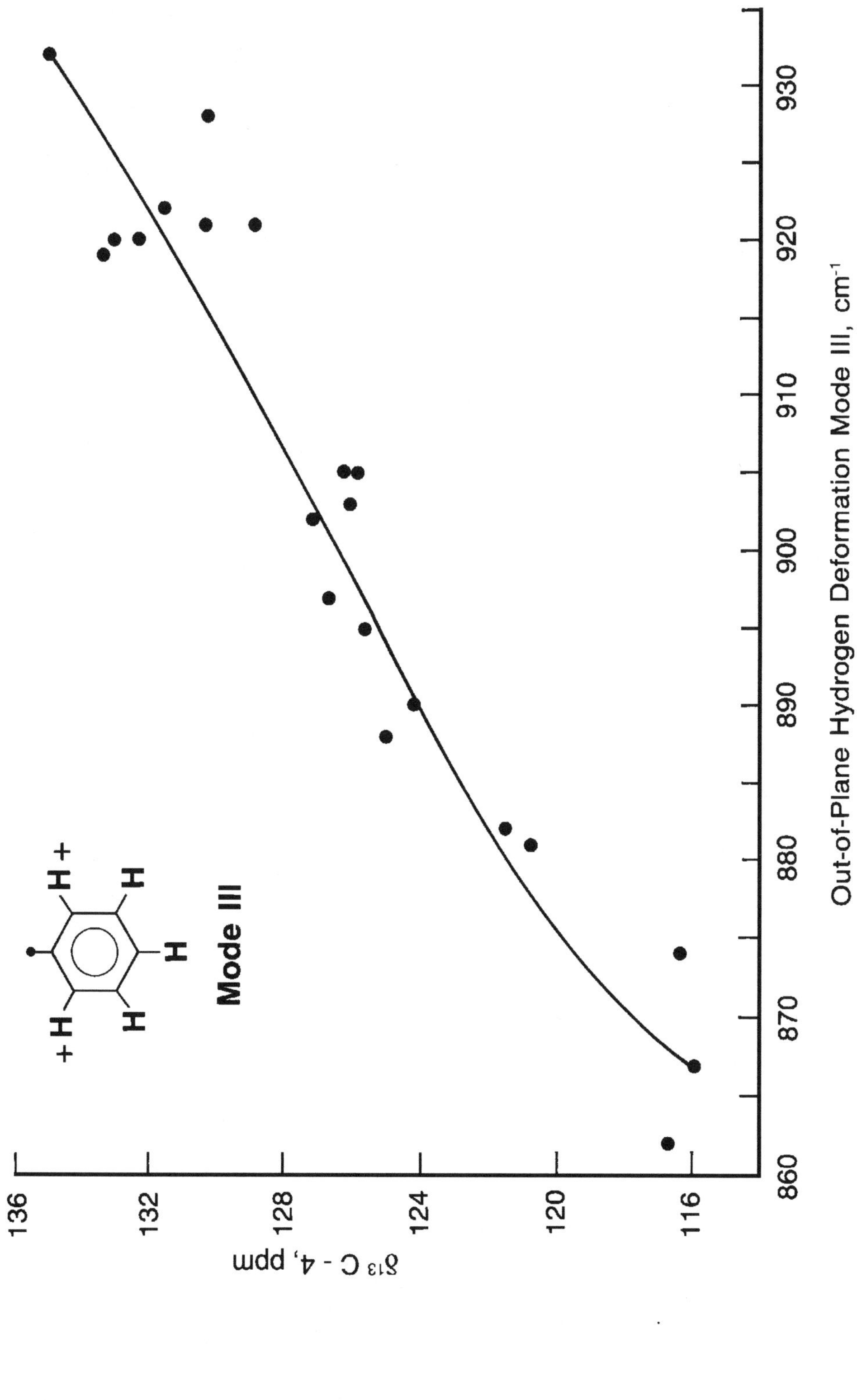

FIGURE 11.96 A plot of the out-of-plane hydrogen deformation mode III vs δ ^{13}C-4 for mono-substituted benzenes.

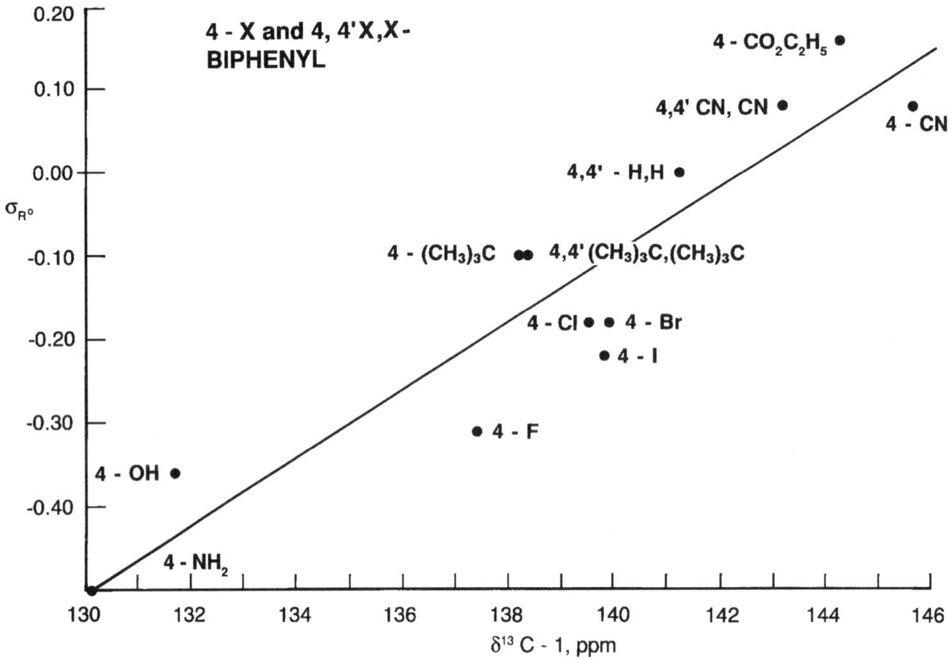

FIGURE 11.97 A plot of δ ^{13}C-1 vs Taft σ_{R^o} values for 4-x and 4,4′-x,x-biphenyls.

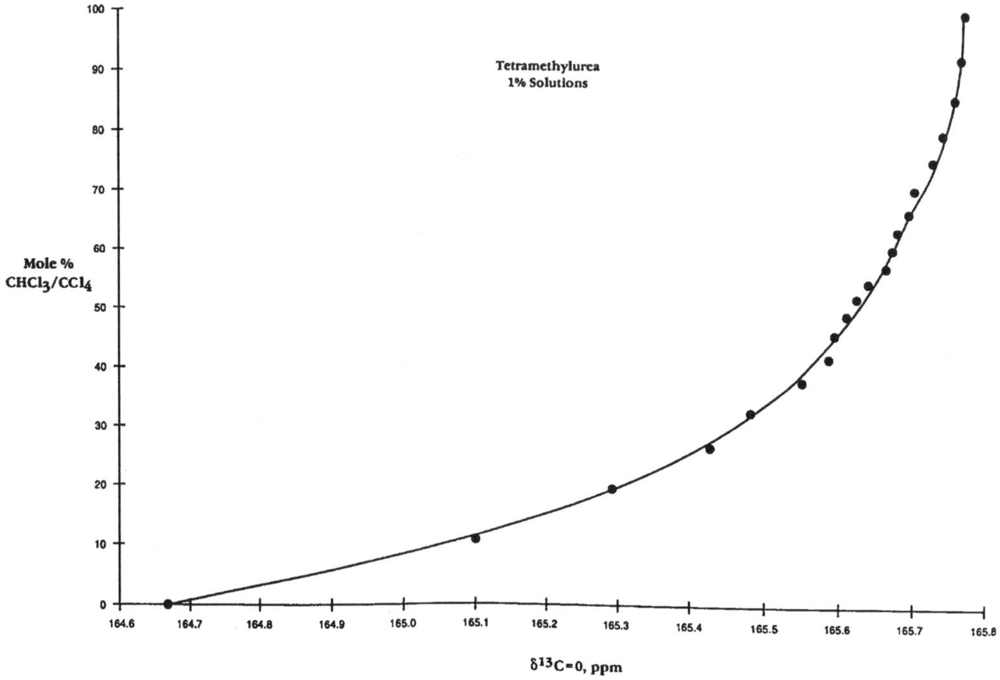

FIGURE 11.98 A plot of δ ^{13}C=O for tetramethylurea (TMU) 1 wt./vol. % solutions vs mole % $CHCl_3/CCl_4$.

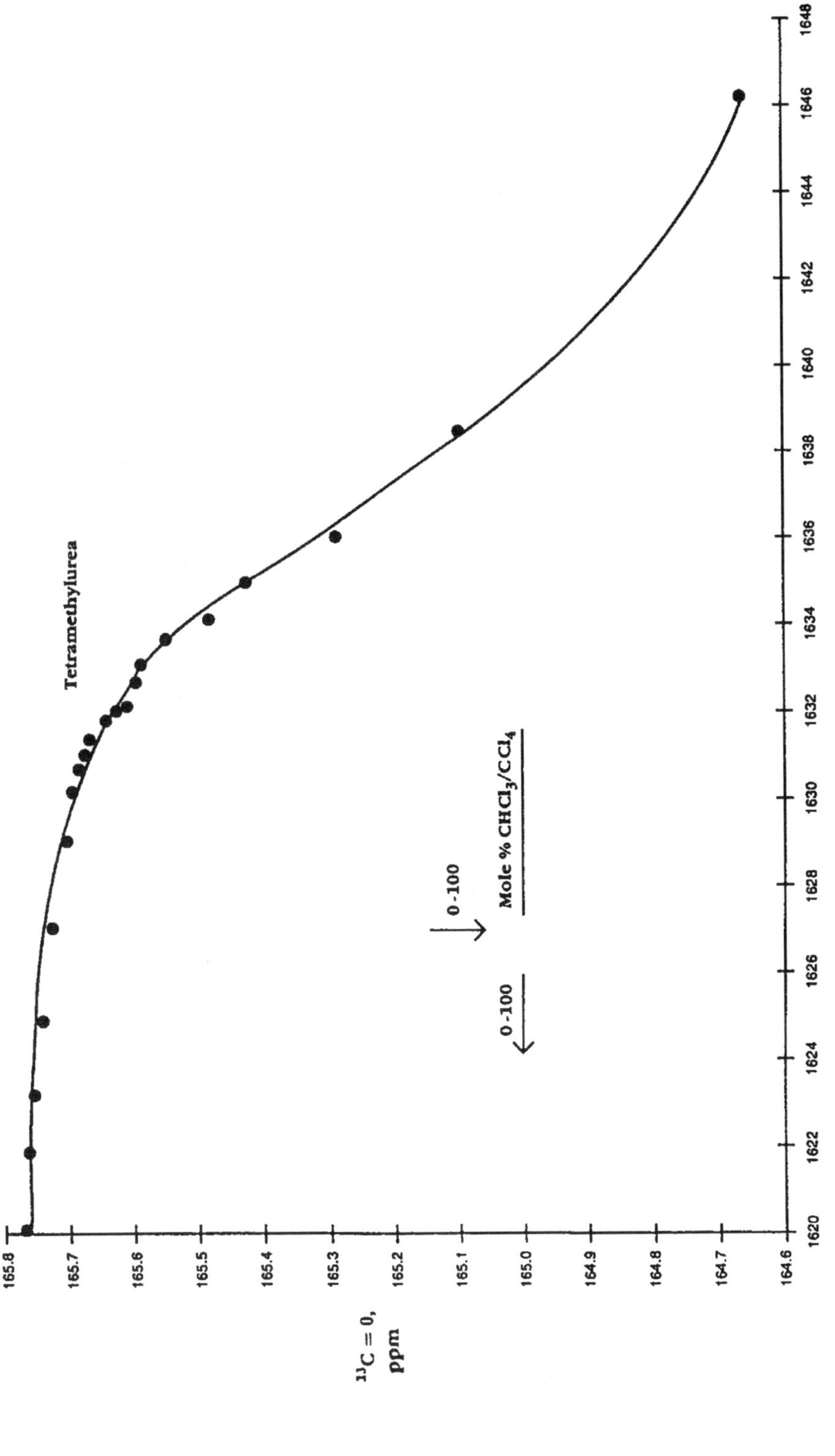

FIGURE 11.99 A plot of ν C=O vs δ ^{13}C=O for TMU 0–100 mol % $CHCl_3/CCl_4$ solutions.

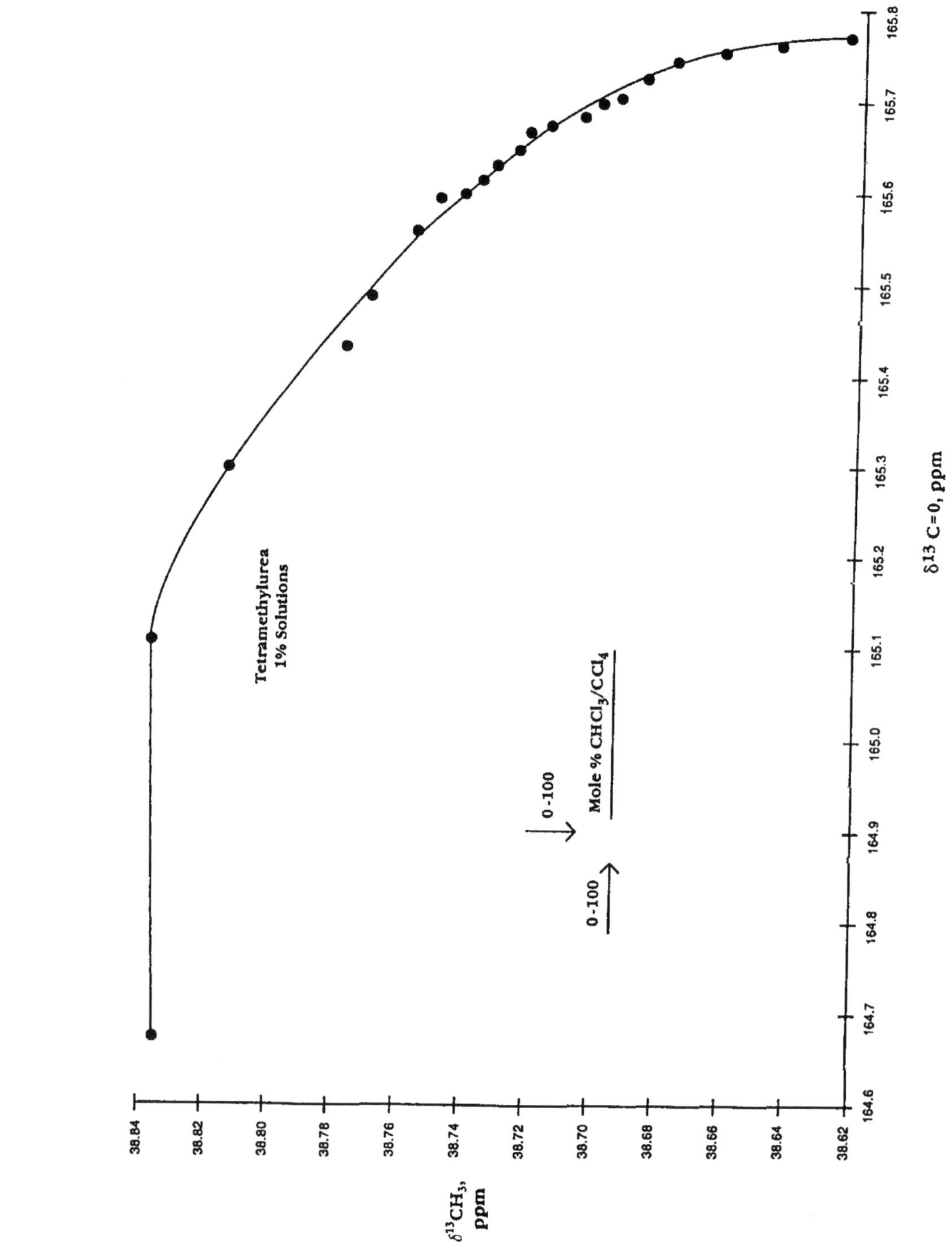

FIGURE 11.100 A plot of δ $^{13}C{=}O$ vs δ $^{13}CH_3$ for TMU in 0–100 mol % $CHCl_3/CCl_4$ solutions.

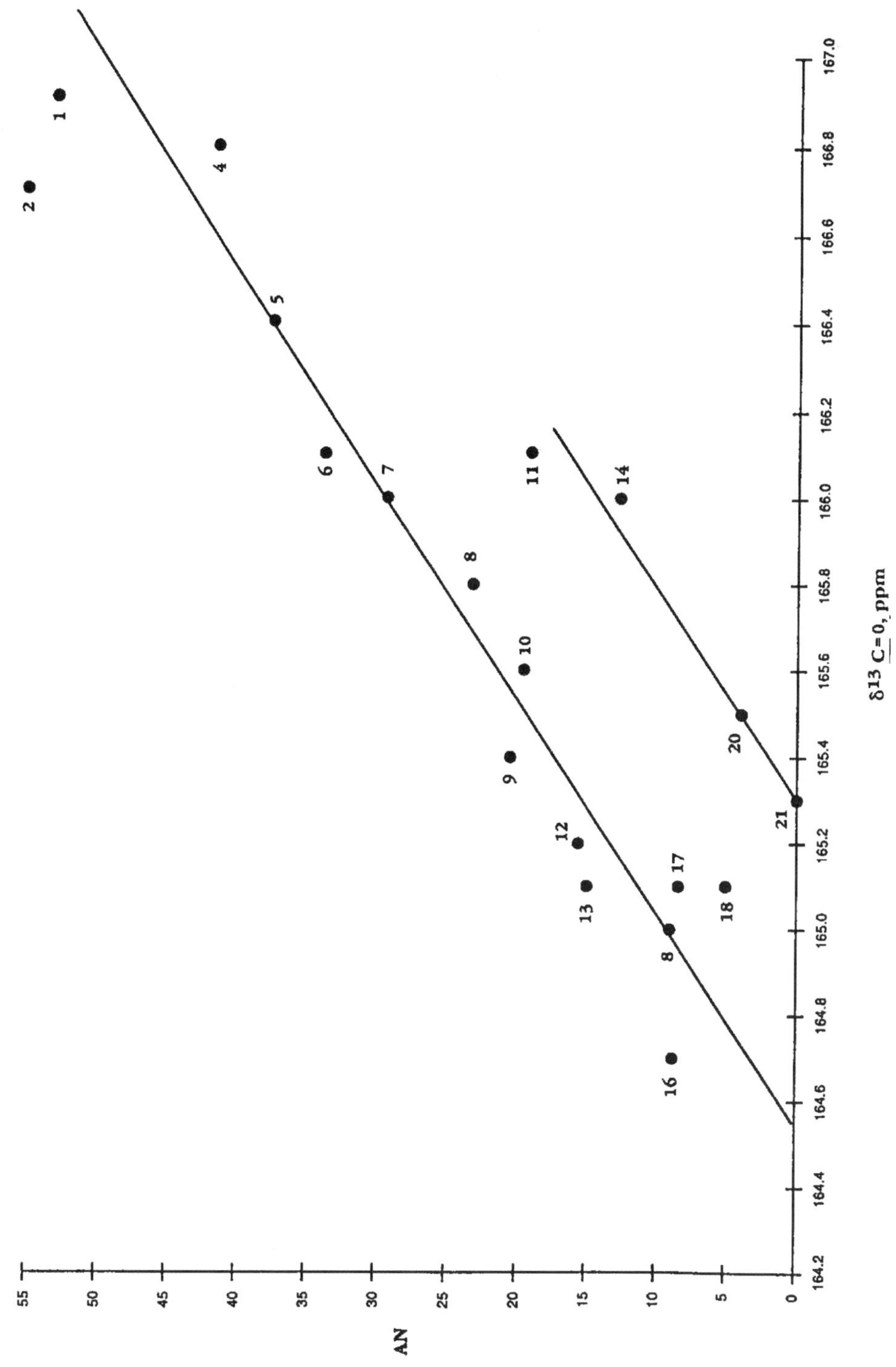

FIGURE 11.101 A plot of δ $^{13}C{=}O$ for TMU vs the solvent acceptor number (AN) for different solvents.

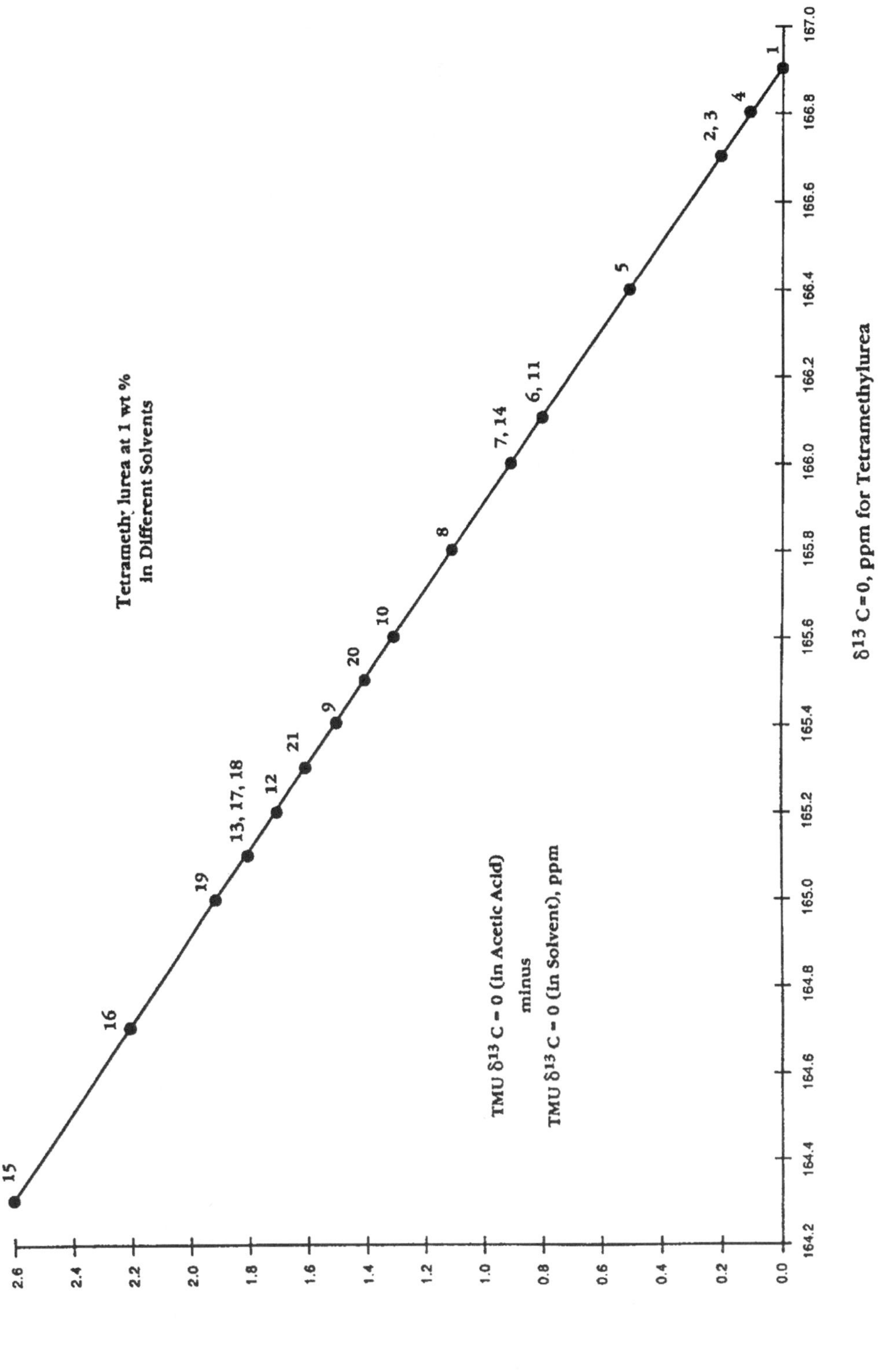

FIGURE 11.102 A plot of the δ $^{13}C{=}O$ chemical shift difference between TMU in solution with acetic acid and in solution with each of the other solvents.

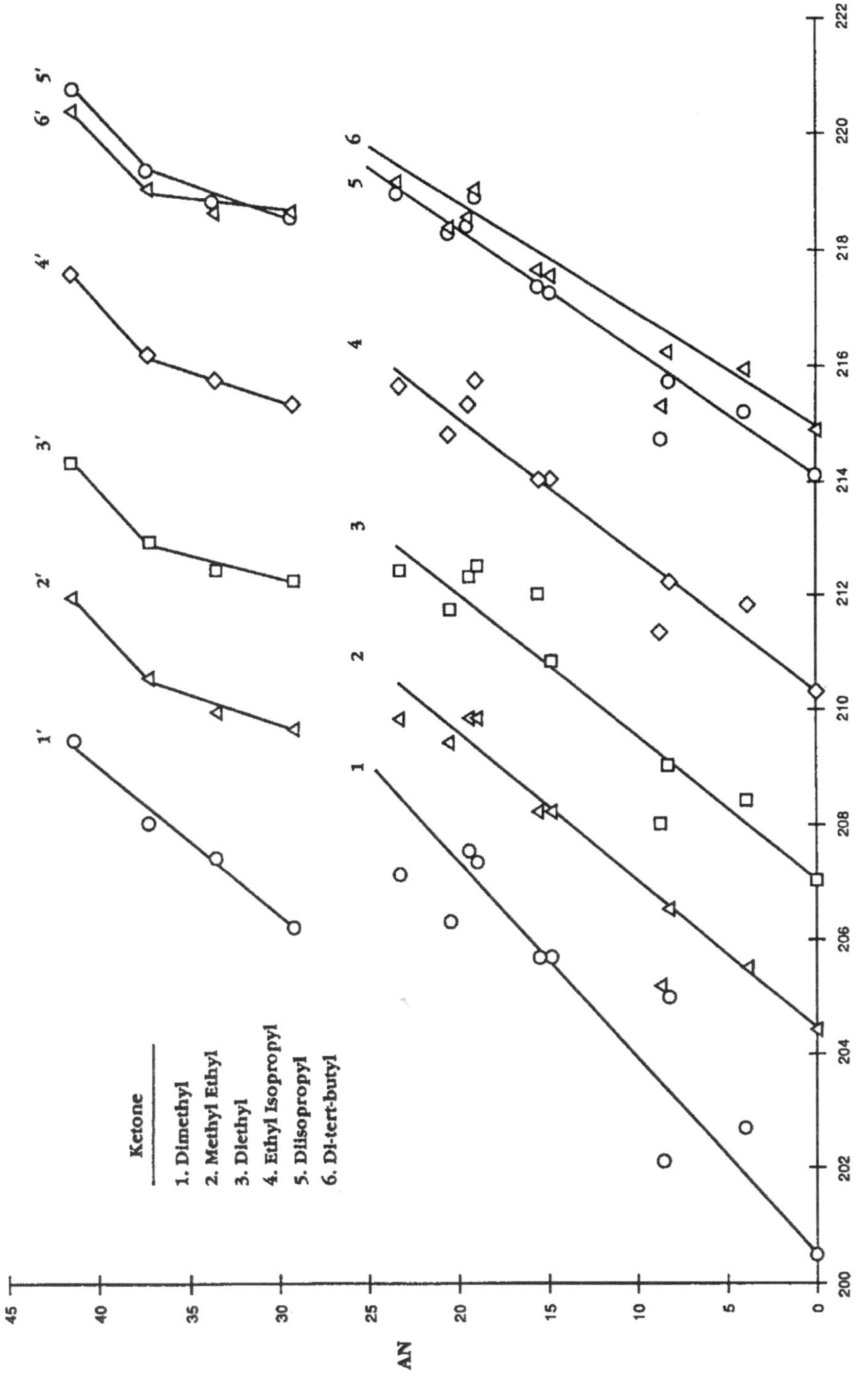

FIGURE 11.103 Plots of δ $^{13}C{=}O$ for dialkylketones at 1 wt./vol. % in different solvents vs the AN for each solvent.

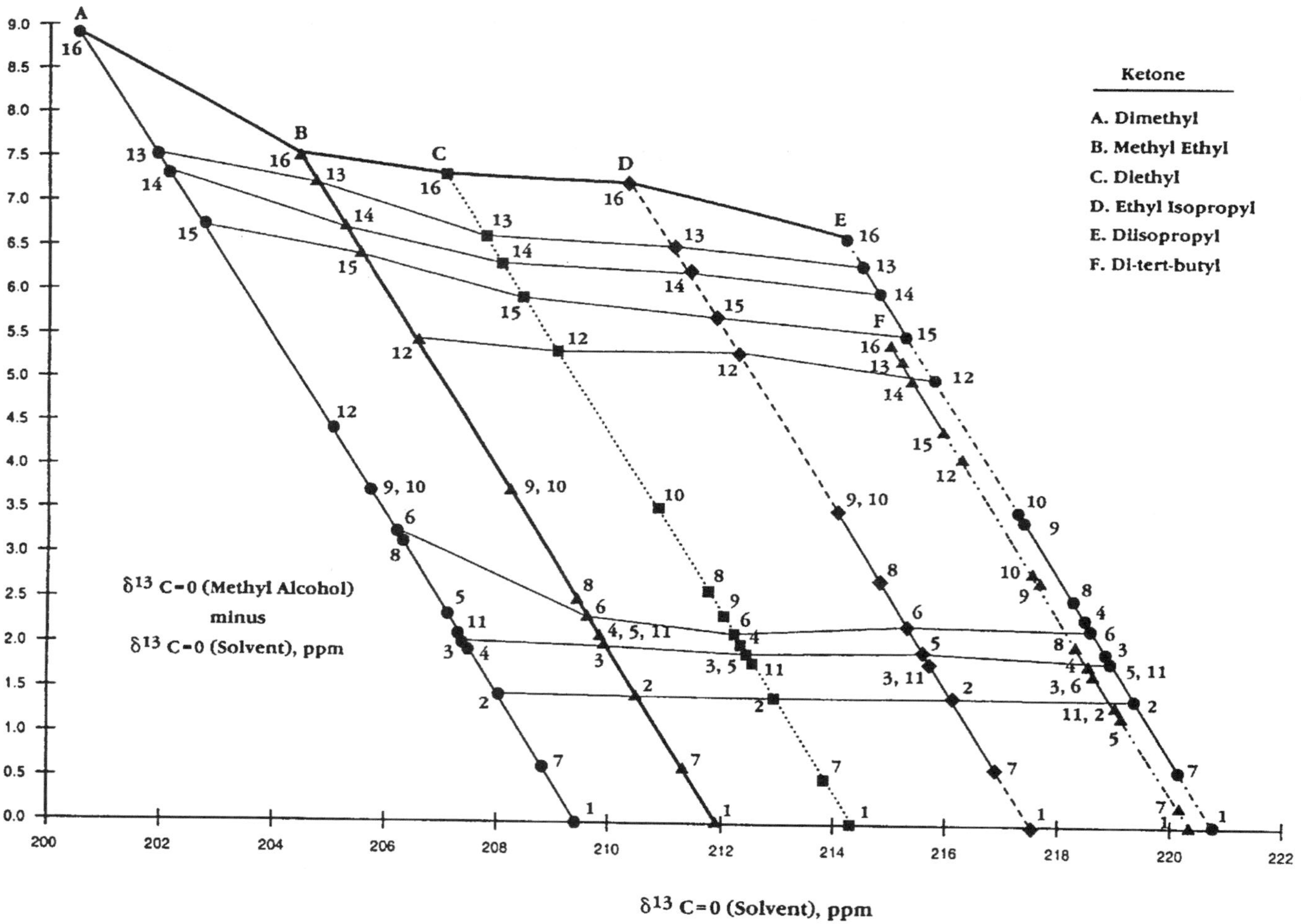

FIGURE 11.104 Plots of the δ ^{13}C=O chemical shift difference between each dialkyl-ketone in methanol and the same dialkylketone in solution with each of the other solvents.

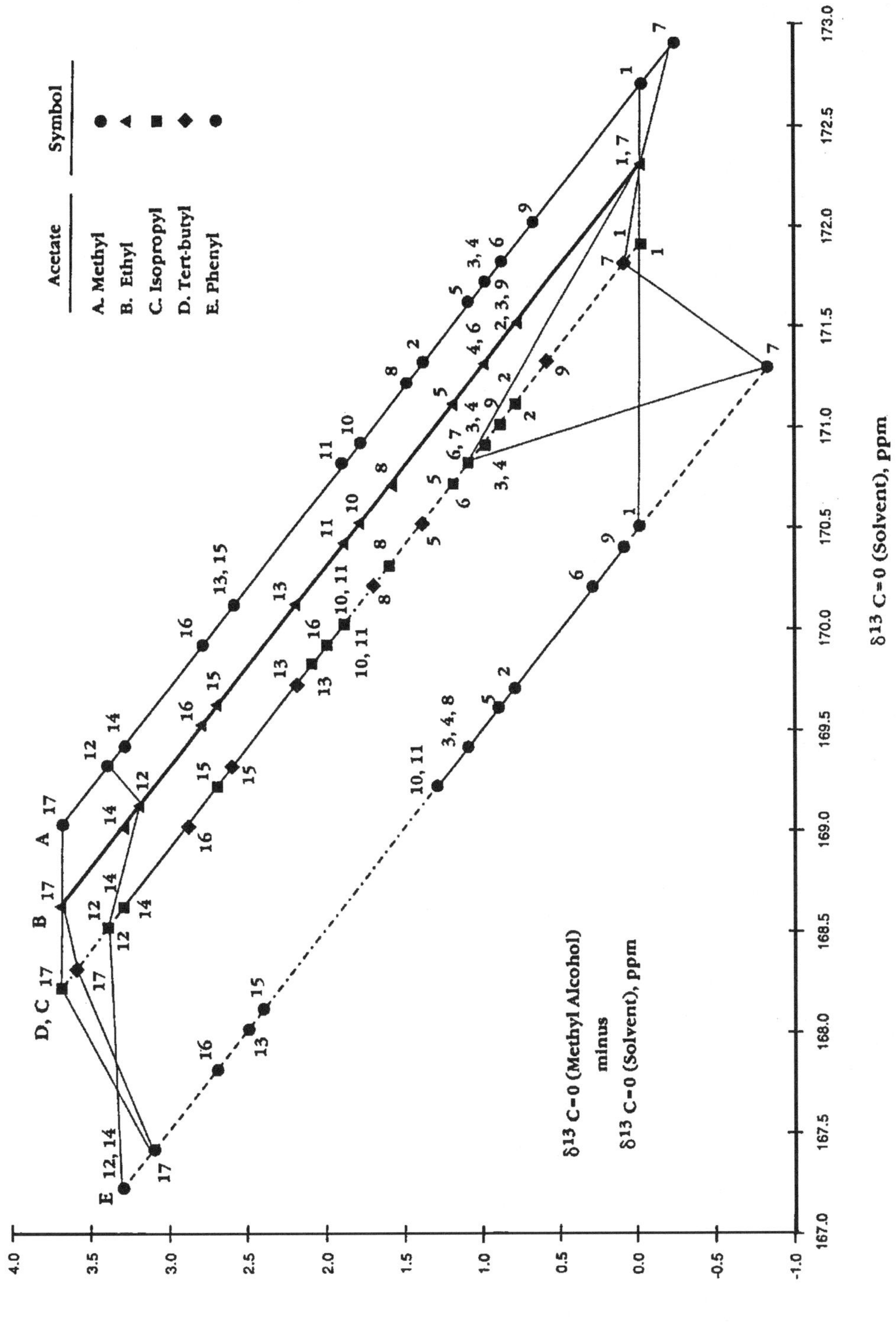

FIGURE 11.105 A plot of δ $^{13}C{=}O$ for alkyl acetates and phenyl acetate vs the difference of δ $^{13}C{=}O$ in methanol and δ $^{13}C{=}O$ in each of the other solvents.

TABLE 11.1 The NMR ^{13}C solution-phase chemical shift data for 4-X-anilines

4-X-Aniline 4-X	Chemical shift δ ^{13}C-1 ppm	Chemical shift δ ^{13}C-2 ppm	Chemical shift δ ^{13}C-3 ppm	Chemical shift δ ^{13}C-4 ppm
NH_2	138.8	116.1	116.1	138.8
OH	139.6	115.7	115.9	149.1
OCH_3	140.5	116.3	114.9	152.6
OC_2H_5	140.4	116.4	115.8	152
OC_3H_7	140.5	116.4	115.8	152.1
OC_4H_9	140.5	116.4	115.7	152.2
OC_6H_{13}	140.3	116.3	115.7	152.2
CH_3	144.3	115.2	129.7	127.2
n-C_3H_7	144.5	115.3	129.2	132.4
n-C_4H_9	144.4	115.3	129.1	132.7
iso-C_3H_7	144.6	115.3	127	138.7
cyclo-C_3H_5	144.3	115.3	126.8	133.4
t-C_4H_9	143.9	114.9	125.8	140.9
F	143.1	116.1	115.7	156.4
Cl	145.2	116.3	129	122.8
Br	145.6	116.7	131.9	109.8
I	147.4	116.7	137.1	76.8
SCH_3	145.4	115.6	130.8	125
CN	151.1	114.5	133.8	120.5
NO_2	155.1	112.8	126.3	136.9
NO_2,HCl	145.4	119.6	125.4	142.4
SO_3H	150.6	116.1	128	133.2
CH_3SO_2	153.2	113.2	128.9	126.2
CF_3	150.2	114.5	127	120.1
CH_3CO	153.8	113.9	131.4	168.8

TABLE 11.2 Infrared and NMR data for 3-x- and 4-x-benzoic acids

3-x or 4-x benzoic acid	σ m	Chemical shift ^{13}C=O ppm	Chemical shift ^{13}C-1 ppm	Chemical shift ^{13}CC-2 ppm	Chemical shift ^{13}C-3 ppm	Chemical shift ^{13}C-4 ppm	Chemical shift ^{13}C-5 ppm	Chemical shift ^{13}C-6 ppm	C=O str. vapor cm^{-1}	C=O str. CCl_4 soln. cm^{-1}	C=O str. [v-CCl_4 soln.] cm^{-1}
3-x											
F	0.34	167.3	133.5	116.5	162.6	119.8	130.3	125.7	1768	1748	20
CH_3	−0.07	172.9	129.5	128.4	138.3	134.5	130.8	127.5	1761		
NH_2		169.2	132.4	115.9	149.3	119.5	130	118.6	1761		
SO_2F		165.5	133.4	130.5	133.1	131.8	129.2	136.7			
CF_3	0.42	166.9	132.7	126.6	130.8	129.2	129.6	133.4	1766		
NO_2	0.71	166.1	133	127.1	148.2	127.1	129.9	135.5	1768	1752	16
CN	0.62	165.9	132.5	133	112.4	135.8	129.7	133.8	1765		
4-x	σ p										
OCH_3	−0.27	167.4	123.3	131.5	113.5	162.9			1760	1737	23
OH	−0.36	169.1	121.3	132.1	115.4	161.8			1755		
C_4H_9NH		168.1	117.1	131.2	110.7	152.7			1751		
OC_2H_5		167.4	123	131.4	113.9	162.4			1755		
t-C_4H_9	−0.2	172.7	126.8	130.2	125.4	157.6			1758		
OC_4H_9		167.5	123.1	131.5	114	162.7					
Cl	0.23	168.8	129.8	131.1	128.5	138.3			1765	1745	20
H	0	172.7	129.5	130.3	128.5	133.8			1762	1744	18
CO_2H		166.9	134.6	129.3							
Range		165.9–172.9	117.1–134.6	116.5–133	110.7–162.6	119.8–162.9	129.2–130.8	118.6–136.7	1751–1768	1737–1752	16–23

TABLE 11.3 Infrared and NMR data for 4-x-acetanilides in solution*

4-x-Acetanilides X	Chemical shift $^{13}C{=}O$ ppm	Chemical shift $^{13}C{=}1$ ppm	Chemical shift $^{13}C{=}2$ ppm	Chemical shift $^{13}C{=}3$ ppm	Chemical shift $^{13}C{=}4$ ppm	C=O str. $[CHCl_4]$ cm^{-1}	C=O str. $[CHCl_3]$ cm^{-1}	C=O str. $[CCl_4\text{-}CHCl_3]$ cm^{-1}
NH_2	167.8	129.2	121.5	114.4	144.1			
OH	168.6	130.6	121.8	115.2	153.6		1686	
H	168.7	139.2	119.7	128.5	123.3	1705		
F	168.5	135.6	121.2	115	158.4			
Br	168.4	138.5	121	114.8	131.2	1709	1694	15
I	168.9	138.9	137.3	137.3	86.2	1709	1696	13
iso-C_3H_7	169.4	136.1	120.8	126.6	144.8			
$CH_3C{=}O$	168.9	143.7	118.4	129.2	131.6	1713	1702	11
NO_2	169.3	145.5	118.6	124.6	142.3	1718	1711	7

*See References 3 and 6.

TABLE 11.4 Infrared and NMR data for 4-X-benzaldehydes in CCl_4 and/or $CHCl_3$ solutions

4-x-Benzaldehyde X	Chemical shift ^{13}CH ppm	Chemical shift $^{13}C{=}O$ ppm	Chemical shift C-1 ppm	C=O str. CCl_4 soln. cm^{-1}	C=O str. $CHCl_3$ soln. cm^{-1}	C=O str. $[CCl_4{-}CHCl_3]$ cm^{-1}	σ p	σ l	σ_{R°
$(CH_3)_2N$	9.66	189.5	125.2	1696.9	1682.5	14.4	−0.6	0.06	−0.59
OH	9.89	190.8	128.7	1701.1	1688.2	12.9	−0.37		−0.4
CH_3O	9.8	190.5	130.2	1697.1	1688.8	8.3	−0.268	0.27	−0.42
F	10	190.2	132.7	1706.7	1700.6	6.1	0.062	0.5	−0.31
C_6H_5O	9.92	190.3	131	1703.3	1696.6	6.7	−0.32	0.38	−0.32
Cl	9.91	190.5	134.9	1710.3	1702.9	7.4	0.226	0.46	−0.18
Br	9.97	190.6	135.2	1710.4	1704.3	6.1	0.232	0.44	−0.18
CH_3	9.97	191.4	134.4	~1711			−0.17	−0.1	−0.13
C_6H_5	10	191.9	135.1	1706.7	1700.3	6.4	−0.01	0.12	−0.1
H	9.94	192	136.6	1708.5	1701.9	6.6	0	0	0
CN	10.13	190.7	138.9	1713.7	1708.1	5.6	0.66	0.56	0.08
CF_3	10.12	191	138.5	1714.6	1707.1	7.5	0.54	0.421	0.08
NO_2	10.18	190.3	139.9	1714.4	1710.4	4.4	0.778	0.65	0.15
CH_3S		190.8	132.6	1700.2	1691.1	9.1	0		−0.25
Range	9.66–10.18	189.5–192	125.2–139.9	1699.9–1714.6	1682.5–1710.4	4.4–14.4			

TABLE 11.5 Infrared and NMR data for 4-x-acetophenones in CCl_4 and/or $CHCl_3$ solutions

4-x-Acetophenone X	Chemical shift $^{13}C{=}O$ ppm	Chemical shift $^{13}CH_3$ ppm	Chemical shift ^{13}C-1 ppm	Chemical shift ^{13}C-4 ppm	C=O str. CCl_4 soln. cm^{-1}	C=O str. $CHCl_3$ soln. cm^{-1}	C=O str. $[CCl_4-CHCl_3]$ cm^{-1}
Cl	196.1	26.3	135.5	139.2	1705	1694	11
CH_3	196.9	26.2	134.9	143.5	1701	1690	11
H	197.4	26.3	137.3	133	1709	1692	17
CH_3O	196	26	130.5	163.5		1684	
NO_2	196.3	26.9	141.7	150.4	1710	1700	10
C_6H_5O	195.8	26.1	132.2	161.7		1686	
Br	196.3	26.3	135.8	128	1705	1695	10
CH_3CO	197.3	26.7	140.3	140.3			
C_6H_5	197.5	26.5	136	145.7		1692	
CN	196.6	26.8	140	116.3		1697.5	
OH	196.6	26	129	162.2	1700	1686	14
CF_3	97	26.8	140.1	134.6		1698	
$N(CH_3)_2$	196.4	26	125.3	153.3		1671	
Range	195.8–197.5	26.0–26.9	125.3–140.3	116.3–163.5		1671–1700	10–17

TABLE 11.6 Vapor-phase IR and NMR $CHCl_3$ solution-phase data for 4-x and 4,4′-x,x-benzophenones

4-x- or 4,4-x,x-Benzophenone	Chemical shift $^{13}C{=}O$ ppm	Chemical shift ^{13}C-1 ppm	Chemical shift ^{13}C-2 ppm	Chemical shift ^{13}C-3 ppm	Chemical shift ^{13}C-4 ppm	Chemical shift $^{13}C'$-1 ppm	Chemical shift $^{13}C'$-2 ppm	Chemical shift $^{13}C'$-3 ppm	Chemical shift $^{13}C'$-4 ppm	C=O str. vapor cm^{-1}
X										
H	196.1	137.6	129.8	128.2	132.2					1681
OCH_3	194.6	130.1	132.4	113.6	163.3	138.4	129.6	128.2	131.8	1675
Cl	194.5	135.7	131.5	128.5	138.1	129.6	129.6	128.5	138.1	1682
Br	195.2	136.3	131.5	131.5	127.4	137.2	129.9	128.3	132.6	1682
CH_3	195.6	135	128.9	130.1	142.9	138	129.7	1282.1	131.9	1680
NO_2	194.3	142.6	130.5	123.4	136.1	136.1	129.8	128.9	133.3	1687
$N(C_2H_5)_2$,$N(C_2H_5)_2$	193.1	125.8	132.4	110	150.3					
$N(CH_3)_2$,$N(CH_3)_2$	192.4	125.6	131.5	110.5	152.5					
OCH_3,OCH_3	194.3	130.9	132.2	113.5	162.9					
CH_3,OCH_3	194.6	130.4	129.8	113.5	163					1672
CH_3,CH_3	195.4	135.4	130.1	128.9	142.6					1675
Cl,Cl										1680
F										1680
OH										1678
Range	193.1–195.6	125.6–142.8	128.8–132.9	110–132.3	127.4–163.3	129.6–138.4	129.6–129.9	128.1–128.9	131.8–138.1	1672–1687

TABLE 11.7 Infrared and NMR data for alkyl 3-x- and 4-x-benzoates

Alkyl 3-x or 4-x Benzoate	Chemical shift $^{13}C{=}O$ ppm	Chemical shift ^{13}C-1 ppm	Chemical shift ^{13}c-2 ppm	Chemical shift ^{13}C-3 ppm	Chemical shift ^{13}C-4 ppm	Chemical shift ^{13}C-5 ppm	Chemical shit ^{13}C-6 ppm	Chemical shift $^{13}CH_3O$ ppm	Chemical shift $^{13}CH_2$ ppm	Chemical shift $^{13}CH_3$ ppm	C=O str. vapor cm^{-1}
Methyl											
3-x											
NH_2	167.4	131	115.7	147.1	119.3	129.3	119.3	51.9			1748
Cl	165.5	132.1	129.7	134.5	132.9	129.7	127.7	52.2			
I	165.2	132.1	138.4	93.8	141.6	129.9	128.7	52.2			1750
NO_2											
4-x											
OH	170.6	113.8	131.3	117.2	167			50.9			1744
I	166	129.5	130.9	137.6	100.6			52			
Ethyl											
3-x											
OH	167.6	131.4	116.5	156.5	120.6	129.6	121.6		61.6	14.1	1741
NH_2	167.2	131.6	115.7	147.5	119.2	129.2	119.2		61	14.3	
4-x											
OH	167.5	122.3	132	115.4	160.7				61.1	14.3	1738
NH_2	166.6	117.5	131.5	113.3	153.3				60	14.4	1735
$N(CH_3)_2$	166.9	117.3	131.2	110.7	153.3				60	14.5	
t-C_4H_9	166.2	128.1	129.6	125.2	156.2				60.5	14.4	1740
Br	165.4	129.5	131.1	131.6	127.8				61.1	14.4	1741
NO_2	165.5	135.1	130.7	123.5	150.6				62	14.3	1749

TABLE 11.8 Summary of IR and NMR data for the C=O group

Compound type	Chemical shift ^{13}C-1 $CHCl_3$ soln. ppm	Chemical shift ^{13}C=O $CHCl_3$ soln. ppm	C=O str. cm^{-1}	Solvent or vapor	Solvent $CHCl_3$
4-x-Benzaldehyde	125.2–139.9	189.9–192.0	1696.9–1714.6	CCl_4	1682.5–1710.4
4-x-Acetophenone	125.3–140.3	195.8–197.5	1671–1700	CCl_4	
4-x-Acetanilide	129.2–145.5	167.8–169.4	1705–1718	CCl_4	1686–1711
4-x- and 4,4′-x,x′-Benzophenone	125.6–142.8	193.1–195.6	1672–1687	vapor	
3-x and 4-x-Benzoic acid	117.1–133.4	165.9–172.9	1751–1768	vapor	1737–1752
3-x- and 4-x-Methyl benzoate	113.8–132.1	165.2–170.6	1744–1750	vapor	
3-x- and 4-x-Ethyl benzoate	117.3–136.1	164.5–167.6	1735–1749	vapor	

TABLE 11.9 IR and NMR data for the C=O group of acetone in $CHCl_3/CCl_4$ solutions

$CHCl_3/CCl_4$	IR C=O str. cm^{-1}	NMR $\delta(^{13}C{=}O)$ (ppm)
0	1717.5	
1.49	1717.1	202.29
3	1716.7	302.56
10.8	1716.1	
15.4	1715.9	203.27
16.9	1715.3	203.72
23.2	1715.1	203.97
26.7	1714.9	204.16
37.7	1714.2	204.74
42.1	1713.9	
45.9	1713.7	
49.2	1713.4	
52.2	1713.2	
55.7	1713.1	205.55
57.4	1712.9	
57.6	1712.8	
60.2	1712.7	
63.4	1712.6	
66.9	1712.3	
70.8	1712.1	205.96
75.2	1711.9	
80.2	1711.6	
85.8	1711.3	
92.4	1710.9	
100	1710.5	

TABLE 11.10 Infrared and NMR data for N,N′-dimethylacetamide in $CHCl_3/CCl_4$ solutions

Dimethylacetamide [1 wt./vol. %]	NMR $^{13}C{=}O$ ppm	NMR $^{13}CH_3$, anti ppm	NMR $^{13}CH_3$, syn ppm	NMR $^{13}CH_3$ ppm	IR C=O str. cm^{-1}
Mole % $CHCl_3/CCl_4$					
0	168.4	37.8	35.1	21.3	1660.5
10.74	169.3	37.9	35.2	21.4	1649
19.4	169.6	38	35.2	21.5	1646.7
26.54	169.9	38	35.2	21.5	1645.1
32.5	170	38	35.3	21.5	1645.2
37.58	170.1	38	35.3	21.5	1642
41.94	170.2	38.1	35.3	21.5	1640.8
45.73	170.2	38.1	35.3	21.5	1640.3
49.06	170.3	38.1	35.3	21.5	1639.6
52	170.3	38.1	35.3	21.6	1639.1
54.62	170.4	38.1	35.3	21.6	1638.7
57.21	170.4	38.1	35.3	21.6	1638.5
60.07	170.4	38.1	35.3	21.6	1638.1
63.23	170.5	38.1	35.3	21.6	1637.8
66.74	170.5	38.1	35.3	21.6	1637.3
70.65	170.6	38.1	35.3	21.6	1336.8
75.05	170.6	38.1	35.3	21.6	1636
80.05	170.6	38.1	35.3	21.6	1635.9
85.75	170.7	38.1	35.3	21.6	1635
92.33	170.8	38.1	35.3	21.6	1634.7
100	170.8	38.1	35.3	21.6	1634.2

TABLE 11.11 The NMR and IR data for maleic anhydride 1 wt./vol. % in CHCl3/CCl4 solutions

Maleic anhydride [1 wt./vol. %]	Chemical shift $^{13}C{=}O$, ppm	Chemical shift $^{13}C{=}C$, ppm	Out-of-phase $(C{=}O)_2$ str. corrected for Fermi Res.(FR) cm^{-1}	In-phase $C{=}O)_2$ str. cm^{-1}	B1 Combination tone corrected for FR cm^{-1}
Mole % $CHCl_3/CCl_4$					
0	163.2	136.1	1787.1	1851.7	1786.8
10.74	163.3	136.2	1786.8	1852	1787.6
19.4	163.4	136.2	1786.6	1852	1787
26.54	163.5	136.3	1786.5	1851.97	1788
32.5	163.6	136.3	1786.4	1851.97	1788.2
37.58	163.7	136.3	1786.35	1851.92	1788.3
41.94	163.7	136.3	1786.29	1851.93	1788.4
45.73	163.7	136.4	1786.26	1851.95	1788.5
49.06	163.8	136.4	1786.2	1851.96	1788.53
52	163.8	136.4	1786.16	1851.93	1788.58
54.62	163.8	136.4	1786.16	1851.9	1788.58
57.21	163.8	136.4	1786.14	1851.9	1788.59
60.07	163.8	136.4	1786.11	1851.87	1788.65
63.23	163.9	136.4	1786.09	1851.91	1788.68
66.74	163.9	136.4	1786.04	1851.88	1788.72
70.65	163.9	136.4	1785.64	1851.87	1788.64
75.05	163.9	136.4	1785.81	1851.85	1788.69
80.05	164	136.4	1785.72	1851.78	1788.77
85.75	164	136.5	1785.64	1851.75	1788.86
92.33	164.1	136.5	1785.4	1851.66	1789.09
100	164.1	136.5	1785.36	1851.68	1789.14
δ	0.9 ppm	0.4 ppm	[$-1.75\ cm^{-1}$]	[$0.06\ cm^{-1}$]	[$2.38\ cm^{-1}$]

TABLE 11.12 Infrared and NMR data for 3-x- and 4-x-benzonitriles

3-x or 4-x Benzonitrile	NMR Chemical shift ^{13}CN ppm	NMR Chemical shift ^{13}C-1 ppm	NMR Chemical shift ^{13}C-2 ppm	NMR Chemical shift ^{13}C-3 ppm	NMR Chemical shift ^{13}C-4 ppm	IR CN str. vapor cm^{-1}	IR CN str. neat cm^{-1}		
3-x								σ m	σ_{R°
NH_2	119.3	112.7			120.6		2212		
F	117.5	114.3				2241		0.34	−0.16
Br	117.2	114.3			136.1		2215	0.39	−0.06
	116.8	113.6			127.1			0.71	0.08
4-x								σ p	
$N(CH_3)_2$	120.6	96.9	132.2	111.4	152.5			−0.6	−0.59
NH_2	120.5	99.5	133.8	114.5	151.1			−0.66	−0.52
OH	119.6	101.9	134	116.5	161.4	2238		−0.37	−0.61
OC_6H_4CN	118.2	108.1	134.5	119.7	159.3			[0.100]*	−0.32
Cl	117.8	111	133.4	129.6	139.3		2220	0.226	−0.18
H	118.8	112.4	132.1	129.2	132.8		2211	0	0
	117.7	117.7	132.9	130	138.9			0.216	0.22
CH_3CO	116.3	117.9	132.6	128.8	140	2236		0.502	0.16
NO_2	116.9	118.4	133.5	124.3	150.2	2240		0.778	0.15
Range	116.3–120.6	99.5–118.4	132.1–134.5	111.4–130	127.1–161.4				

* Estimated.

TABLE 11.13 Infrared, Raman and NMR data for organonitriles

Compound	Empirical structure	Raman data CN str., cm^{-1}	$\delta(^{13}CN)$ ppm	IR data[vapor] CN str., cm^{-1}	CN str. [v-neat] cm^{-1}
Acetonitrile	CH_3CN	2249	117.7	2280	31
Propionitrile	CH_3CH_2CN	2242	120.8		
Isobutyronitrile	$(CH_3)_2CHCN$	2238	123.7	2255	17
Pivalonitrile	$(CH_3)_3CCN$	2236	125.1		
Chloroacetonitrile	$ClCH_2CN$	2258	115.5		
Trichloroacetonitrile	Cl_3CCN	2250	113		
Acrylonitrile	$CH_2{=}CHCN$	2222	117.5		
Methacrylonitrile	$CH_2{=}C(CH_3)CN$	2230	119.3		
Benzonitrile	COH_5CN	2230	118.7		
2-Chloroacrylonitrile	$CH_2{-}CClCN$	2234	114.5		

TABLE 11.14 A comparison of δ(N) chemical shift data for primary, secondary and tertiary amines

Amine	δ(N) ppm	No. protons on the α-carbon atom	C-N str. cm^{-1}	Sum Taft σ^*
Methyl	−377.3	3	1043	0
Ethyl	−355.4	2	1083	−0.1
Isopropyl	−338.1	1	1170	−0.19
Tert-butyl	−324.3	0	1237	−0.3
			wN-H cm^{-1}	
Dimethyl	−369.5	6	715	0
Ethyl methyl	−352	5	[—]	−0.1
Diethyl	−333.7	4	698	−0.2
Diisopropyl	−306.5	2	688	n-0.38
Trimethyl	363.1	9		0
Dimethyl ethyl	−351.3	8		−0.1
Dimethyl isopropyl	−340.5	7		−0.19
Dimethyl tert-butyl	−342.5	6		−0.3

TABLE 11.15 The NMR data for organophosphorus and organonitrogen compounds

Compound	[27,28] Chemical shift ^{31}P ppm	[23] Sum σ^*	[18,25] Chemical shift ^{15}N	Compound
$P(CH_3)_3$	−64	0	−366.9	$N(CH_3)_3$
$P(CH_3)_2(C_2H_5)$	−48	−0.1	−355.2	$N(CH_3)_2(C_2H_5)$
$P(CH_3)(C_2H_5)_2$	−34	−0.2	−343.1	$N(CH_3)(C_2H_5)_2$
$P(C_2H_5)_3$	−19.5	−0.3	−333.6	$N(C_2H_5)_3$
$P(iso\text{-}C_3H_7)_3$	19.4	−0.57		
$P(tert\text{-}C_4H_9)_3$	63	−0.9		
$PH_2(CH_3)$	−163.5		−378.7	$NH_2(CH_3)$
$PH_2(C_2H_5)$	−128		−355.1	$NH_2(C_2H_5)$
$PH(CH_3)_2$	−99		−371	$NH(CH_3)_2$
$PH(CH_3)(C_2H_5)$	−77		−352.8	$NH(CH_3)(C_2H_5)$
$PH(C_2H_5)_2$	−55		−333	$NH(C_2H_5)_2$

TABLE 11.16 Infrared and NMR data for organophosphorus compounds

Compound	[28] Chemical shift ^{31}P ppm	[31] P=O str. cm^{-1}	Compound	[28] Chemical shift ^{31}P ppm	[31] P=O str. cm^{-1} [CCl_4]/[$CHCl_3$]	Compound	[28] Chemical shift ^{31}P ppm	[31] P=O str. cm^{-1}
$(CH_3)_3P{=}O$	36.2		$(CH_3)(Cl)_2P{=}O$	44.5	1278.5/1268.9	$(Cl)_3P$	219.5	
$(C_2H_5)_3P{=}O$	48.3		$(C_2H_5)(Cl)_2P{=}O$	53		$(CH_3O)(Cl)_2P$	180	
$(CH_3)_2$(tert-C_4H_9)P=O	47.8		(iso-C_3H_7)$(Cl)_2P{=}O$	60.4		$(CH_3O)_2(Cl)P$	169	
(iso-$C_3H_7)_3P{=}O$	55		(tert-C_4H_9)$(Cl)_2P{=}O$	65.6	1266.4/1255.7	$(C_2H_5O)(Cl)_2P$	178	
			$(CH_3)(Cl)_2P{=}S$	79.4		$(C_2H_5O)_2(Cl)P$	165	
$(CH_3O)_3P{=}O$	2.4	1291/1272*[1]	$(C_2H_5)(Cl)_2P{=}S$	95.4		$(CH_3O)_3P$	140	
$(C_2H_5O)_3P{=}O$	−1.5	1280/1263*[1]	(iso-C_3H_7)$(Cl)_2P{=}S$	107		$(C_2H_5O)_3P$	138	
$(C_3H_7O)_3P{=}O$	−0.8	1279/1265*[1]						
$(C_4H_9O)_3P{=}O$	−1	1280/1265*[1]	$(CH_3)(F)_2P{=}O$	27.1		$(C_2H_5)(Cl)_2P$	196.3	
(iso-$C_3H_7O)_3P{=}O$	−6	1272/1257*[1]	$(C_2H_5)(F)_2P{=}O$	29.2		$(CH_3)(Cl)_2P$	192	
(tert-$C_4H_9O)_3P{=}O$	−13.3		(iso-C_3H_7)$(F)_2P{=}O$	29.5		$(C_2H_5)_2(Cl)P$	119	
$(C_2H_5O)_2$ (tert-C_4H_9O)P=O	−5.5		(tert-$C_4H_9(F)_2P{=}O$	31.3		$(CH_3)(C_2H_5)$ (Cl)P	105.2	
(C_2H_5O) (tert-$C_4H_9O)_2P{=}O$	−9.9					$(CH_3)_2(Cl)P$	96	
		[P=S str.]						
			(iso-C_3H_7O)(Br) $(CH_3)P{=}S$	93.2		$(CH_3)(Br)_2P$	184	
$(CH_3O)_3P{=}S$	73.4	~605 VP	(iso-C_3H_7O)(Br) $(C_2H_5)P{=}S$	110.2		$(C_2H_5)_2(Br)P$	116.2	
$(C_2H_5O)_3P{=}S$	68	613 vp	(iso-C_3H_7O)(Br) (iso-C_3H_7)P=S	124.1		$(CH_3)(C_2H_5)$ (Br)P	98.5	
(iso-$C_3H_7O)_3P{=}S$	62.5		(iso-C_3H_7O)(Br) (tert-C_4H_9)P=S	135.6		$(CH_3)_2(Br)P$	92 or 87	
(tert-$C_4H_9O)_3P{=}S$	41.2							
						$(CH_3O)_3P$	140	
$(CH_3O)_3P{=}Se$		[P=Se]				$(C_2H_5O)_3P$	[137–139]	
$(C_2H_5O)_3P{=}Se$	72.1					(iso-$C_3H_7O)_3P$	137	
(iso-$C_3H_7O)_3P{=}Se$	67.9					(tert-$C_4H_9O)_3P$	138.2	
(tert-$C_4H_9O)_3P{=}Se$	31.3							

(continued)

Table 11.16 (*continued*)

				Chemical shift $^{13}C=O$ ppm	$(CH_3O)_2$	6.4	1308/1293*[1]
					$(Cl)P=O$	2.8–3.3	1298/1285*[1]
					$(C_2H_5O)_2$ $(Cl)P=O$		
					(tert-$C_4H_9O)_2$ $(Cl)P=O$	−5.8	
$(C_2H_5O)_2(CH_3)P=O$	29.4		(CH_3O) $(CH_3C=O$	171.2			
$(C_2H_5O)_2(C_2H_5)P=O$	32.6	1265	(CH_3O) $(C_2H_5)C=O$	174.6			
$(C_2H_5O)_2$(tert-$C_4H_9)P=O$	39.4		(CH_3O) (tert-$C_4H_9(C=O$	178.6			
(iso-$C_3H_7O)_2(CH_3)P=O$	27.1		(iso-$C_3H_7O)$ $(CH_3C=O$	170			
(tert-$C_4H_9O)_2(CH_3)P=O$	21.2		(tert-$C_4H_9O)$ $(CH_3)C=O$	170			
		P=O str. cm^{-1}					
$(CH_3O)_2(H)P=O$	9.8 to 12.8	[1290 cm^{-1} vp*[2]] 1283sh/1266*[1]					
$(C_2H_5O)_2(H)P=O$	6.2 to 7.6	[1281 cm^{-1} vp*[2]] 1275sh/1262*[1]					
(iso-$C_3H_7O)_2(H)P=O$	3.2 to 4.2						
(C_2H_5O)(tert-$C_4H_9O)(H)P=O$	1.4 to 4.5						
(tert-$C_4H_9O)_2(H)P=O$	[−3.2 to −3.9]						
$(CH_3O)_2(H)P=S$	73.8 to 74.2						
$(C_2H_5O)_2(H)P=S$	67.8 to 69.3						
(iso-$C_3H_7O)_2(H)P=S$	64.8 to 644.85						

*[1] Rotational conformers.
*[2] vp is the abbreviation for vapor phase here.

TABLE 11.17 NMR data for 2-x-, 3-x- and 4-x-anisoles in $CHCl_3$ solutions

Anisole	^{13}C-1 ppm	^{13}C-2 ppm	^{13}C-3 ppm	^{13}C-4 ppm	^{13}C-5 ppm	^{13}C-6 ppm	$^{13}CH_3O$ ppm	ES
4-x								
NO_2	164.9	114.2	125.8	141.7			56.1	
CH_3SO_2	163.7	114.6	129.5	132.4			55.8	
CH_3CO	163.5	113.8	130.5	130.5			55.3	
$CO_2C_2H_5$	163.5	113.7	131.6	123.2			55.3	
CN	163	114.9	133.9	103.9			55.5	
H	159.9	114.1	129.5	120.7			54.8	
$CH{=}CH_2$	159.7	114	127.4	130.5			54.9	
C_6H_5	159.3	114.2	128	133.6			55	
CH_2OH	159.1	114	128.5	133.6			55.2	
iso-C_3H_7	158.9	114	127	131.2			54.9	
2-ClC_2H_4	158.8	114.1	129.8	130.3			55.1	
Br	158.7	115.7	132.1	112.7			55.1	
2-BrC_2H_4	158.6	113.9	129.5	130.9			54.9	
Cl	158.5	115.3	129.3	125.5			55.3	
CH_3	157.9	113.9	129.9	129.9			54.9	
C_2H_5	157.9	113.9	128.8	136.4			55.1	
n-C_3H_7	157.9	113.8	129.3	134.7			55	
t-C_4H_9	157.6	113.5	126.1	143.1			54.8	
CH_3O	154	114.8	114.8	154			55.6	
$C_6H_5CH_2O$	154.1	114.7	115.9	153.1			55.5	
NH_2	152.6	114.9	116.3	140.5			55.6	
C_6H_5O	156	115	120.7	150.4			55.4	
3-x								
CH_3O	161.4	100.8	161.4	106.4	130	106.4	55.1	
OH	160.9	102	157	108.5	130.5	106.9	55.3	
NH_2	160.8	100.9	148.4	107.9	130.1	103.6	54.8	
Cl	160.6	114.6	135	120.9	130.3	112.6	55.3	
CH_3	160	115	139.4	121.6	129.3	111.1	54.8	
CH_3CO	159.9	112.7	138.7	121	129.6	119.3	55.2	
3-BrC_3H_6	159.8	114.3	142.1	120.8	129.4	111.4	55	
2-x								
CH_3CO	159.1	128.3	130.2	120.5	133.7	111.8	55.4	
Br	158.7	115.7	132.1	112.7	132	115.7	55.1	0
I	157.7	86	139.1	122.2	129.3	110.9	56.1	−0.2
Cl	155.1	122.4	130.2	121.2	127.8	112.2	55.8	0.18
NO_2	153	140	134.4	120.4	125.3	114	56.5	−0.75
CH_3O	149.5	149.5	112	121	121	112	55.7	0.99
F	148.1	152.9	116.1	124.5	121.1	113.9	56	0.49
NH_2	147.4	136.7	114.9	121.2	118.1	110.8	55.3	
OH	147.1	146.1	115	120.3	121.7	114.4	55.9	

TABLE 11.17A Infrared data for 3-x- and 4-x-substituted anisoles

Anisole	Phenyl-O−C asym. str. neat cm^{-1}	Phenyl-O−C sym. Str. neat cm^{-1}	Phenyl-O−C asym. str. CS_2 soln. cm^{-1}
4-x			
NO_2	1262	1022	1264
CH_3SO_2	1260	1021	[–]
CH_3CO	1249	1021	1261
$CO_2C_2H_5$	1252	1028	1258
CN	1240	1024	1259
H	1248	1042	1246
$CHCH_2$	1248	1040	[–]
C_6H_5	1249	1032	[–]
CH_2OH	1245	1030	1248
CH_3CHCH_2	1249	1037	[–]
2-ClC_2H_4	1240	1031	[–]
Br	1240	1029	1246
2-C_2H_4	1245	1031	[–]
Cl	1242	1032	1247
CH_3	1242	1028	1244
C_2H_5	1238	1035	1248
tert-C_4H_9	1244	1038	[–]
CH_3O	1246	1029	1243
$C_6H_5CH_2O$	1230	1033	[–]
NH_2	1238	1034	1241
C_6H_5O	1230	1039	[–]
3-x			
CH_3O	1211	1052	[–]
OH	1200	1043	[–]
NH_2	1198	1029	[–]
Cl	1237	1029	[–]
CH_3	1259	1042	[–]
CH_3CO	1270	1040	[–]
3-BrC_3H_6	1250	1034	[–]

TABLE 11.18 Infrared and NMR data for mono-x-benzenes*

x-Benzene x	Chemical shift ^{13}C-1 ppm	Chemical shift ^{13}C-2 ppm	Chemical shift ^{13}C-3 ppm	Chemical shift ^{13}C-4 ppm	σ p	σ_{R°	Mode I cm^{-1}	Mode III cm^{-1}
H	128.5				0	0		
CH_3	137.8	129.2	128.4	125.5	−0.17	−0.13	973	895
C_2H_5	144.3	128.1	128.6	125.9	−0.15	−0.1	978	903
iso-C_3H_7	148.8	126.6	128.6	126.1	−0.15	−0.1	979	905
tert-C_4H_9	150.9	125.4	128.3	125.7	−0.2	−0.1	979	905
n-C_4H_9	143.3	129	128.2	125.7	−0.15	−0.1		
s-C_4H_9	148.4	127.9	129.3	126.8	−0.15			
iso-C_4H_9	148.8	126.6	128.6	126.1	−0.15			
F	163.6	114.2	129.4	124.1	0.06	−0.31	972	890
Cl	134.9	128.7	129.5	126.5	0.23	−0.18	976	897
Br	122.6	131.5	130	127	0.23	−0.18	980	92
I	96.6	138.4	131.1	128.1	0.18	−0.22		
OH	155.1	115.7	130	121.4	−0.27	−0.4	971	882
OCH_3	159.9	114.1	129.5	120.7	−0.66	−0.42	970	881
NH_2	148.7	114.4	129.1	116.3		−0.5	968	874
$NHCH_3$	150.4	112.1	129.1	115.9		−0.59	966	867
$N(CH_3)_2$	150.7	112.7	129	116.7		−0.59		
C_6H_5	140.6	126.7	128.4	126.9	−0.01	−0.1		
C_6H_5O	157.7	119.1	129.9	123.2	0.32	−0.32		
CH_2Cl	137.9	128.9	128.8	128.6	0.18	−0.08	981	921
CN	112.8	132.1	129.2	132.8	0.66	0.08	987	920
CF_3	131.5	125.5	129	132.1	0.54	0.08	984	920
CCl_3	144.1	125.3	128.1	130.1	0.44		981	921
CO_2H	131.4	129.8	128.9	133.1	0.45	0.29	995	919
$CO_2O_2H_5$	131	129.5	128	132.4	0.52	0.16		
$COCH_3$	136.3	128.1	128.1	131.3	0.52	0.16	988	922
NO_2	149.1	124.2	129.8	134.7	0.78	0.15	990	932
CH_3SO_2	145.1	122.6	129.6	130	0.72	0.12	994	928
CH_3S	138.6	126.7	128.7	124.9	0	−0.25	976	888
Range	128.5–163.6	112.1–138.4	128.1–131.1	115.9–134.7			965–995	862–932

* See References 4 and 31.

TABLE 11.19 The NMR data for 4-x- and 4,4′-x,x-biphenyls in $CHCl_3$ solutions

Biphenyl 4-x or 4,4′-x,x	Chemical shift ^{13}C-1 ppm	Chemical shift ^{13}C-2 ppm	Chemical shift ^{13}C-3 ppm	Chemical shift ^{13}C-4 ppm	Chemical shift ^{13}C-1′ ppm	Chemical shift ^{13}C-2′ ppm	Chemical shift ^{13}C-3′ ppm	Chemical shift ^{13}C-4′ ppm
F	137.4	127	128.6	162.5	140.2	127	128.8	127.3
OH	131.7	127.8	115.9	157.2	140.7	126.2	128.6	126.2
t-C_4H_9	138.3	125.6	126.9	150.1	150.1	126.9	138.3	126.9
t-C_4H_9,t-C_4H_9	138.2	126.6	125.5	149.7				
NH_2,NH_2	130.1	126.4	114.9	146				
C_6H_5				135.1				
Cl	139.5	128.8	133.3	128.3	139.9	126.8	128.8	127.6
$CO_2C_2H_5$,$CO_2C_2H_5$	144.2	127.1	130.1	130.1				
H,H	141.2	127.1	128.7	127.1				
Br	139.9	128.5	131.7	121.5	139.9	126.7	128.7	127.5
CN,CN	143.1	127.9	132.7	111.8				
CN	145.6	129.1	132.5	111	139.1	127.7	128.7	127.2
I	139.8	128.8	137.7	93	140.5	126.7	128.8	127.5

TABLE 11.20 The NMR data for tetramethylurea in $CHCl_3/CCl_4$ solutions

Tetramethylurea [1 wt./vol. % solns.	Chemical shift ^{13}C=O, ppm	Chemical shift $^{13}CH_3$, ppm
Mole % $CHCl_3/CCl_4$		
0	164.7	38.8
10.74	165.1	38.8
19.4	165.3	38.8
26.53	165.4	3838
30.5	165.5	3838
37.57	165.5	38.8
41.93	165.6	38.7
45.73	165.6	38.7
49.06	165.6	38.7
52	165.6	38.7
54.62	165.6	38.7
57.22	165.7	38.7
60.07	165.7	38.7
63.28	165.7	38.7
66.74	165.7	38.7
70.65	165.7	38.7
75.06	165.7	38.7
80.05	165.7	38.7
85.85	165.8	38.7
92.33	165.8	38.6
100	165.8	38.6
δ ppm	1.1	−0.2

TABLE 11.21 The NMR data for tetramethylurea 1 wt./vol. % in various solvents

Tetraamethylurea [1 wt./vol. %]	Chemical shift $^{13}C{=}O$ ppm	Chemical shift $^{13}CH_3$ ppm	[$^{13}C{=}O$ in acetic acid]–[$^{13}C{=}O$ in solvent] δ ppm	AN
Solvents				
Acetic acid	166.9	38.5	0	52.9
Water	166.7	38.1	0.2	54.8
Nitromethane	166.7	38.8	0.2	
Methyl alcohol	166.8	38.2	0.1	41.3
Ethyl alcohol	166.4	38.4	0.5	37.1
Isopropyl alcohol	166.1	38.7	0.8	33.5
t-Butyl alcohol	166	38.5	0.9	29.1
Chloroform	165.8	38.6	1.1	23.1
Methylene chloride	165.4	38.3	1.5	20.4
Dimethyl sulfoxide	165.6	38.3	1.3	19.3
Acetonitrile	166.1	38.7	0.8	18.9
Benzonitrile	165.2	38.4	1.7	15.5
Nitrobenzene	165.1	38.2	1.8	14.8
Acetone	166	38.9	0.9	12.5
Carbon disulfide	164.3	39	2.6	
Carbon tetrachloride	164.7	38.8	2.2	8.6
Benzene	165.1	38.1	1.8	8.2
Methyl t-butyl ether	165.1	38.5	1.8	5
Tetrahydrofuran	165	38.1	1.9	8.8
Diethyl ether	165.5	38.5	1.4	3.9
Hexane	165.3	38.5	1.6	0

TABLE 11.22 The NMR data for dialkylketones in various solvents at 1 wt./vol. % concentration

Solvent [1 wt./vol. %]	DMK $^{13}C{=}O$ ppm	MEK $^{13}C{=}O$ ppm	DEK $^{13}C{=}O$ ppm	EIK $^{13}C{=}O$ ppm	DIK $^{13}C{=}O$ ppm	DTBK $^{13}C{=}O$ ppm	AN
Methyl alcohol	209.4	211.9	214.3	217.5	220.7	220.3	41.3
Ethyl alcohol	208	210.5	212.9	216.1	219.3	219	37.1
Isopropyl alcohol	207.4	209.9	212.4	215.7	218.8	218.6	33.5
Dimethyl sulfoxide	207.5	209.8	212.3	215.3	218.4	218.5	19.3
Chloroform	207.1	209.8	212.4	215.6	218.9	219.1	23.1
t-Butyl alcohol	206.2	209.6	212.2	215.3	218.5	218.6	29.1
Nitromethane	208.8	211.3	213.8	216.9	220.1	220.1	
Methylene chloride	206.3	209.4	211.7	214.8	218.2	218.3	20.4
Benzonitrile	205.7	208.2	212	214	217.3	217.6	15.5
Nitrobenzene	205.3	208.2	210.8	214	217.2	217.5	14.8
Acetonitrile	207.1	209.8	212.5	215.7	218.9	219	18.9
benzene	205	206.5	209	212.2	215.7	216.2	8.2
Carbon disulfide	201.9	204.7	207.7	211	214.4	215.1	
Carbon tetrachloride	202.1	205.2	208	211.3	214.7	215.3	8.6
Diethyl ether	202.7	205.5	208.4	211.8	215.2	215.9	3.9
Hexane	200.5	204.4	207	210.3	214.1	214.9	0

TABLE 11.23 The NMR data for alkyl acetatews and phenyl acetate 1 wt./vol. % in various solvents

Solvent [1 wt./vol. %]	Methyl acetate $^{13}C{=}O$ ppm	Ethyl acetate $^{13}C{=}O$ ppm	Isopropyl acetate $^{13}C{=}O$ ppm	t-Butyl acetate $^{13}C{=}O$ ppm	Phenyl acetate $^{13}C{=}O$ ppm
Methyl alcohol	172.7	172.3	171.9	171.9	170.5
Ethyl alcohol	171.3	171.5	171.1	171.1	169.7
Isopropyl alcohol	171.7	171.5	170.9	170.9	169.4
t-Butyl alcohol	171.7	171.3	170.9	170.9	169.4
Chloroform	171.6	171.1	170.7	170.5	169.6
Dimethyl sulfoxide	171.8	171.3	170.8	170.7	170.2
Nitromethane	172.9	172.3	170.8	171.8	171.3
Methylene chloride	171.2	170.7	170.3	170.2	169.4
Acetonitrile	172	171.5	171	171.3	170.4
Nitrobenzene	170.9	170.5	170	170	169.2
Benzonitrile	170.8	170.4	170	170	169.2
Carbon disulfide	169.3	169.1	168.5	168.5	167.2
Benzene	170.1	170.1	169.8	169.7	168
Carbon tetrachloride	169.4	169	168.6	168.6	167.2
Diethyl ether	170.1	169.6	169.2	169.3	168.1
Methyl t-butyl ether	169.9	169.5	169.9	169	167.8
Hexane	169	168.6	168.2	168.3	167.4

TABLE 11.24 Infrared and NMR data for alkyl acrylates and alkyl methacrylates

Chemical shift	Alkyl acrylates CCl_4 soln.	Alkyl acrylates $CHCl_3$ soln.	Alkyl methacrylate CCl_4 soln.	Alkyl methacrylate $CHCl_3$ soln.
δ $^{13}C{=}O$,ppm	164.2–165.2	165.7–166.7	165.6–166.7	167.1–167.8
δ $^{13}CH{=}$,ppm	128.8–130.1	128.1–130.8		
δ $^{13}CCH_2{=}$,ppm	129.6–130.7	130.4–131.4	136.2–136.9	135.9–136.8
δ $^{13}CC{=}$,ppm			123.5–125.5	125.2–126.3
δ $^{13}CCH_3{=}$,ppm			18.6–18.7	18.3–18.3
Group frequency				
C=O str., cm^{-1}	1722.9–1734.1	1713.8–1724.5	1719.5–1726.0	1709–1718.0
C=C str.,s-trans,cm^{-1}	1635.3–1637.0	1635.3–1637.9		
C=C str.,s-cis,cm^{-1}	1619.2–1620.4	1618.5–1619.9		
C=C str., cm^{-1}			1637.3–1638.0	1635.7–1637.3

List of Tables

CHAPTER 1

1.1	Vibrational assignments for ethylene oxide	15
1.2	Characteristic epoxy ring modes	16
1.2a	Vibrational assignments for 1-halo-1,2- epoxypropanes (or epihalohydrins)	17
1.2b	A comparison of IR data for epoxy ring modes and CH_3, $(CH_3)_2$ and $(CH_3)_3$ bending modes	18
1.3	Vibrational assignments for the CH_2X groups for 3-halo-1,2-epoxypropanes, 3-halopropynes, 3-halopropenes, and PCH_2X-containing compounds	19
1.4	Vapor-phase IR data for the oxirane ring vibrations of 1,2-epoxyalkenes	20
1.5	Correlations for the ring stretching vibrations for cyclic ethers	21
1.6	Infrared data for ethers	22
1.7	Infrared data for the a and s Aryl-O-R stretching vibrations for 3-X- and 4-X-anisoles in CS_2 solution, vapor, and in the neat phase	23
1.8	Raman data and assignments for vinyl ethers	24
1.9	Vapor-phase IR data and assignments for the alkyl group of vinyl alkyl ethers	25

CHAPTER 2

2.1	Infrared data for nitriles and cyanogen halides in the vapor, neat or CCl_4 solution phase	39
2.2	Infrared data for acetonitrile in various solvents	40
2.3	A comparison of the IR ν CN stretching frequencies for acetonitrile (corrected for Fermi resonance) with those for benzonitrile	40
2.4	The CN stretching frequency for 4-cyanobenzaldehyde in 0 to 100 mol % $CHCl_3/CCl_4$ solutions (1% wt./vol. solutions)	41
2.5	Raman data for the C=N and C=C groups for some organonitriles in the neat phase	42
2.6	A comparison of infrared data for organonitriles vs organoisonitriles	42

2.7 A comparison of the infrared data for organothiocyanates in the vapor and neat phases 43

CHAPTER 3

3.1 Infrared and Raman data and assignments for the $(C{=}N{-})_2$ antisymmetric and symmetric stretching vibrations for azines 57
3.2 The symmetric and/or asymmetric N=C=O stretching frequencies for alkyl and aryl isocyanates in various physical phases 58
3.3 Infrared data for alkyl isocyanates in 0 to 100 mol % $CHCl_3/CCl_4$ in 0.5% wt./vol. solutions 59
3.4 Infrared data for the antisymmetrical N=C=O stretching frequency and two combination tones for alkyl isocyanates in $CHCl_3$ and CCl_4 solutions 60
3.5 Infrared and Raman data for alkyl isocyanates 60
3.6 Infrared data for 1% wt./vol. alkyl isothiocyanates in 0 to 100 mol % $CHCl_3/CCl_4$ solutions [the *v* asym. N=C=S and the first overtone of *v* C–N frequencies are in Fermi resonance] 61
3.6a Vibrational data for organoisothiocyanates 62
3.7 Infrared vapor- and neat-phase data for dialkyl and diaryl carbodiimides 63

CHAPTER 4

4.1 Infrared and Raman data for alkanethiols and benzenethiols 74
4.1a Raman data for organic thiols 75
4.2 Vapor-phase infrared data for alkanethiols, alkane sulfides, and alkane disulfides 76
4.3 Infrared and Raman data for organic sulfides and disulfides 78
4.4 Vibrational assignments for 4-benzenethiol and 1,4- dichlorobenzene 79
4.5 The P=S, P–S, and S-H stretching frequencies for O,O-dialkyl phosphorodithioate and O,O-bis-(aryl) phosphorodithioate 80
4.6 A comparison of alkyl group joined to sulfur, oxygen, or halogen 81

CHAPTER 5

5.1a A comparison of the S=O stretching frequency for S=O containing compounds 104
5.1b A comparison of asym. SO_2 and sym. SO_2 stretching frequencies in different physical phases, and asym. N=S=O and sym. N=S=O frequencies in CS_2 solution 105
5.1c Vapor-phase infrared data for dimethyl sulfoxide and dialkyl sulfones 107
5.1d Vapor- and solid-phase infrared data for dimethyl sulfoxide and dialkyl sulfones 108
5.2 Vapor- and solid-phase infrared data for diaryl sulfones 109
5.3 Infrared data for phenoxarsine derivatives containing the $AsS(SO_2)R$ and $AsO(SO_2)R$ groups in CS_2 solution 109

5.4 Infrared data for the SO_2 stretching vibrations for compounds in $CHCl_3$ and CCl_4 solutions 110
5.5 Infrared data for dialkyl sulfites 111
5.6 Infrared data for primary sulfonamides in the vapor phase 111
5.6a Infrared data for some NH and NH_2 vibrations for compounds containing SO_2NH and SO_2NH_2 groups in different physical phases 112
5.7 Infrared data for secondary and tertiary sulfonamides 112
5.8 Infrared data for organic sulfonates 113
5.9 Infrared data for organosulfonyl chlorides 114
5.9a Infrared data for organosulfonyl fluorides 115
5.10 Infrared data for organosulfur compounds containing SO_4, SO_3, SO_2N, and SO_2X groups in different physical phases 116
5.11 Infrared data for sulfones and sulfoxides in different physical phases 117
5.12 Infrared data for N-sulfinyl—4-X-anilines and N,N′-disulfinyl-*p*-phenylenediamine 117

CHAPTER 6

6.1 Infrared and Raman data for the methylene halides 164
6.2 Raman and infrared data for trihalomethane and tetrahalomethane 164
6.2a A comparison of CX, CX_2, CX_3 and CX_4 stretching frequencies 165
6.3 Vapor- and liquid-phase infrared data for l- haloalkanes 166
6.4 Vapor- and liquid-phase infrared data for 2- halobutane and tert-butyl halide 167
6.5 Vapor and liquid-phase infrared and Raman liquid- phase data for 1-cyclohexanes 168
6.6 Vapor- and liquid-phase infrared data for primary, primary dialkanes 169
6.7 Raman data for methyl halides and IR and Raman data for tetrabromoalkanes 170
6.8 Carbon halogen stretching frequencies for ethylene propyne, 1,2-epoxypropane, and propadiene derivatives 171

CHAPTER 7

7.1 The ν asym. NO_2 and ν sym. NO_2 frequency shifts of substituted nitro compounds from those for nitromethane in the liquid phase 206
7.1a Vapor-phase infrared data for nitroalkanes 207
7.2 Vapor-phase infrared data for nitroalkanes 208
7.3 Vapor- and liquid-phase infrared data for nitroalkanes 209
7.4 Vapor-phase infrared data for 4-X- nitrobenzenes 209
7.5 Infrared data for 4-X-nitrobenzenes in different phases 210
7.6 A comparison of the frequency differences between ν asym. NO_2 and ν sym. NO_2 in the vapor and $CHCl_3$ solution and in the vapor and neat or solid phases 211
7.7 Infrared data for the ν asym. NO_2 and ν sym. NO_2 frequencies for 3-X and 4-X-nitrobenzenes in CCl_4 and $CHCl_3$ solutions 212
7.8 Infrared data for the ν asym. NO_2 and ν sym. NO_2 frequencies for 4-X-nitrobenzenes in the vapor, CCl_4, and $CHCl_3$ solution phases 213

7.9 Infrared data for ν asym. NO_2 and ν sym. NO_2 frequencies of 4-nitrobenzaldehyde 1% wt./vol. in 0 to 100 mol % $CHCl_3/CCl_4$ solutions 214
7.10 A comparison of infrared data for nitromethane vs nitrobenzene in various solvents 215
7.11 Vapor-phase infrared data for 3-X-nitrobenzenes 216
7.12 Infrared data for 3-X-nitrobenzenes in different phases 216
7.13 A comparison of the frequency difference between ν asym. NO_2 and ν sym. NO_2 for 3-X-nitrobenzenes in the vapor and $CHCl_3$ solution and in the vapor, and solid or neat phases 217
7.14 Vapor-phase infrared data for 2-X-nitrobenzenes 218
7.15 Infrared data for 2-X-nitrobenzenes in different phases 218
7.16 A comparison of the frequency differences between ν asym. NO_2 and ν sym. NO_2 2-X-nitrobenzenes in the vapor and $CHCl_3$ solution and in the vapor, neat or solid phases 219
7.17 Infrared data for nitrobenzenes in the solid and CCl_4 solution phases 220
7.18 Vapor-phase infrared data for 2,5- and 2,6-X,Y-nitrobenzenes 221
7.19 A comparison of the frequency difference between ν asym. NO_2 and ν sym. NO_2 in the vapor, neat or solid phases 221
7.20 Vapor-phase infrared data for tri-X,Y,Z, WXWZ-tetra, and VWXYZ-penta-substituted nitrobenzenes 222
7.21 Infrared data for 4-X-nitrobenzenes in CCl_4 and $CHCl_3$ solutions (1% wt./vol. or less) 223
7.22 Infrared data and assignments for alkyl nitrates 224
7.23 Infrared data for alkyl nitrates in various phases 224
7.24 Infrared data for ethyl nitrate, nitroalkanes, and nitrobenzenes in CCl_4 and $CHCl_3$ solutions 225
7.25 Vapor-phase infrared data for alkyl nitrites 226
7.26 Vapor-phase infrared data of the characteristic vibrations of alkyl nitrites* 226
7.27 Raman data for organonitro compounds 227
7.28 Infrared and Raman data for nitroalkanes in different physical phases 227
7.29 Infrared and Raman data for nitrobenzenes in different physical phases 228
7.30 A comparison of the frequency separation between ν asym. NO_2 and ν sym. NO_2 [vapor-phase data minus $CHCl_3$ solution data] and [vapor-phase data minus neat-phase data] for nitrobenzenes 229
7.31 Infrared data for organonitro compounds in different physical phases 229
7.32 Infrared and Raman data for organonitrates, organonitrites, and organonitrosamines in different physical phases 230
7.33 The ν N=O frequency for nitrosamines in different physical phases 230

CHAPTER 8

8.1 Phosphorus halogen stretching frequencies for inorganic compounds 318
8.2 The PX_3 bending frequencies for compounds of forms PX_3, PXY_2, and XYZ 319
8.3 Vibrational data and assignments for PX_3, $P(=O)X_3$, and $P(=S)X_3$ 320

8.4	The asym. and sym. PCl_2 stretching frequencies for $XPCl_2$, $XP(=O)Cl_2$, and $XP(=S)Cl_2$ groups	321
8.5	Vibrational assignments for $F_2P(=S)Cl$, $(CH_3-O-)_2P(=S)Cl$, and $(CD_3-O-)_2P(=S)Cl$	322
8.5a	The PCl_2 stretching frequencies for methyl phosphorodichloridate and O-methyl phosphorodichloridothioate	323
8.6	P-Cl stretching frequencies	323
8.7	Vibrational assignments for the CH_3, CD_3, C_2H_5, CH_3CD_2 and CD_3CH_2 groups of $R-O-R(=O)Cl_2$ and $R-O-P(=S)Cl_2$ analogs	324
8.8	The P=O stretching frequencies for inorganic and organic phosphorus compounds	325
8.8a	A comparison of P=O stretching frequencies in different physical phases	326
8.9	Infrared data for the rotational conformer PO stretching frequencies of O,O-dimethyl O-(2-chloro-4-X-phenyl) phosphate	326
8.10	IR observed and calculated PO stretching compared*	327
8.11	The PO and PS stretching frequencies for phenoxarsine derivatives	327
8.12	The PO and PS stretching frequencies for $P(O)X_3$- and $P(S)X_3$-type compounds	328
8.13	The PS stretching frequencies for inorganic and organic phosphorus compounds	329
8.14	Vibrational assignments for the skeletal modes of S-methyl phosphorothiodichloridate, and $P(O)Cl_3$	331
8.15	Assignments of the $P(C-O-)_3$ skeletal vibrations for trimethyl phosphite and trimethyl phosphate	331
8.16	Vibrational assignments for the CH_3 groups of trimethyl phosphite	332
8.17	The PS, P—S and S—H stretching frequencies for O,O-dialkyl phosphorodithioate and O,O-diaryl phosphorodithioate	332
8.18	The PO, PH stretching, and PH bending vibrations for O,O-dialkyl hydrogenphosphonates*	333
8.19	The "C—O" and "P—O" stretching frequencies for the C—O—P group	334
8.20	The "C—O" and "P—O" stretching frequencies for compounds containing C—O—PO, C—O—PS, and C—O—PSe groups	336
8.21	The aryl-O stretching frequencies for O-methyl O-(X-phenyl) N-methylphosphoramidate	338
8.22	The C—P stretching frequencies for organophosphorus compounds	338
8.23	The trans and/or cis N—H stretching frequencies for compounds containing the P—NH—R group	339
8.24	The NH_2, NHD, ND_2, NH, and ND frequencies for O,O-dimethyl phosphoramidothioate and O,O-diethyl phosphoramidothioate, and N-methyl O, O-phosphoramidothioate.	340
8.25	The N—H and N—D stretching frequencies for O-methyl O-(2,4,5-trichlorophenyl) N-alkylphosphoramidate and the N-alkyl phosphoramidothioate analogs	341
8.26	The cis and trans NH stretching frequencies for compounds containing OP—NH—R or SP—NH—R groups*	342
8.27	Vibrational assignments for N-alkyl phosphoramidodichloridothioate and the CD_3NH, CD_3ND and CH_3ND analogs	343

8.28 Vibrational assignments for O,O- dimethyl O-(2,4,5-trichlorophenyl) phosphorothioate and its PO and $(CD_3{-}O)_2$ analogs 344
8.29 Vibrational assignments for 1- fluoro-2,4,5-trichlorobenzene and the ring modes for O,O-dimethyl O-(2,4,5-trichlorophenyl) phosphorothioate and its PO and $(CD_3{-}O{-})_2$ analogs 345
8.30 Infrared data for O,O-dialkyl phosphorochloridothioate and O,O,O-trialkyl phosphorothioate in different physical phases 346
8.31 Infrared data for organophosphates and organohydrogenphosphorates in different physical phases 347
8.32 Infrared data for O-alkyl phosphorodichloridothioates and S-alkyl phosphorodichloridothioates in different physical phases 348
8.33 Infrared data for O,O-diethyl N- alkylphosphoramidates in different physical phases 349
8.34 Vibrational assignments for $[CH_3{-}PO_3]^{2-}$, $[CD_3{-}PO_3]^{2-}$, $[HPO_3]^{2-}$, and $[PO_4]^{3-}$ 349
8.35 Infrared data and assignments for the $(CH_3)_2\ PO_2^-$ anion 350

CHAPTER 9

9.1 Vibrational data for chlorobenzene vs chlorinated biphenyls 403
9.2 Vibrational data for 1,2-dichlorobenzene vs chlorinated biphenyls 404
9.3 Vibrational data for 1,3-dichlorobenzene vs 2,3,3′,5,6-pentachlorobiphenyl 405
9.4 Vibrational data for 1,4-dichlorobenzene vs chlorinated biphenyls 406
9.5 Vibrational data for 1,3,5-trichlorobenzene vs chlorinated biphenyls 407
9.6 Vibrational data for 1,2,3-trichlorobenzene vs 2,2′,3-trichlorobiphenyl, 2,2′,3′,4,5-pentachlorobiphenyl, and 2,2′,4,6,′-pentachlorobiphenyl 408
9.7 Vibrational data for 1,2,4-trichlorobenzene vs chlorinated biphenyls 409
9.8 Vibrational data for 1,2,4,5-tetrachlorobenzene vs chlorinated biphenyls 411
9.9 Vibrational data for 1,2,3,5-tetrachlorobenzene vs chlorinated biphenyls 413
9.10 Infrared data for 1,2,3,4-tetrachlorobenzene vs chlorinated biphenyls 414
9.11 Vibrational data for pentachlorobenzene vs 2,2′,3,4,6-pentachloro-biphenyl, and 2,3,4,4′,6-pentachlorobiphenyl 415
9.12 Vibrational data for 1,2,3,4,5,6-hexachlorobenzene vs 1,2,3,4,5-pentachlorobiphenyl 416
9.13 Raman data for monosubstituted benzenes 417
9.14 Infrared and Raman data for an A_1 fundamental for mono-x-benzenes 418
9.15 Vibrational assignments for bromodichlorobenzenes 418
9.16 Raman data and assignments for some in-plane ring modes of 1,2-disubstituted benzenes 419
9.17 Raman data and assignments for some in-plane ring modes for 1,3-disubstituted benzenes 419
9.18 Raman data and assignments for some in-plane ring modes for 1,4-disubstituted benzenes 420

9.19 Raman data and tentative assignments for decabromobiphenyl and bis-(pentabromophenyl) ether 420
9.20 Vibrational assignments for benzene, benzene-d_6, benzyl alcohol, benzyl-2,3,4,5,6-d_5 alcohol, pyridene, and pyridene-d_5 421
9.21 Infrared data for the out-of-plane deformations for mono-x-benzenes 422
9.22 Summary of the out-of-plane hydrogen deformations for substituted benzenes, and the out-of-plane ring deformation for mono-substituted benzenes 422
9.23 Combination and overtones of the out-of-plane hydrogen deformations for substituted benzenes 423
9.24 Hexachlorobenzene vs 2,3,4,5,6-pentachlorobiphenyl 423

CHAPTER 10

10.1 Aliphatic hydrocarbons—the Nyquist Rule 432
10.2 Anilines—the Nyquist Rule 433
10.3 Anhydrides, imides, and 1,4- benzoquinones—the Nyquist Rule 434
10.4 Substituted hydantoins—the Nyquist Rule 435
10.5 1,4-Diphenylbutadiyne and 1-halopropadiene—the Nyquist Rule 435
10.6 3-Nitrobenzenes and 4-nitrobenzenes—the Nyquist Rule 436
10.7 Organic sulfate, sulfonate, sulfonyl chloride, and sulfones—the Nyquist Rule 437

CHAPTER 11

11.1 NMR ^{13}C solution- phase chemical shift data for 4-x-anilines 560
11.2 Infrared and NMR data for 3-x- and 4-x-benzoic acids 561
11.3 Infrared and NMR data for 4-x- acetanilides in solution 562
11.4 Infrared and NMR data for 4-x- benzaldehydes in CCl_4 and/or $CHCl_3$ solutions 563
11.5 Infrared and NMR data for 4-x- acetophenones in CCl_4 and/or $CHCl_3$ solutions 564
11.6 Vapor-phase IR and NMR $CHCl_3$ solution-phase data for 4-x- and 4,4′-x,x-benzophenones 565
11.7 Infrared and NMR data for methyl and ethyl 3-x- and 4-x-benzoates 566
11.8 Summary of IR and NMR data for the CO group 567
11.9 Infrared and NMR data for acetone in $CHCl_3/CCl_4$ solutions 567
11.10 Infrared and NMR data for N,N′-dimethylacetamide in $CHCl_3/CCl_4$ solutions 568
11.11 Infrared and NMR data for 1 wt./vol. % maleic anhydride in $CHCl_3/CCl_4$ solutions 569
11.12 Infrared and NMR data for 3-x- and 4-x-benzonitriles 570
11.13 Infrared, Raman, and NMR data for organonitriles 570
11.14 A comparison of $\delta^{15}N$ chemical shift data for primary, secondary and tertiary amines 571
11.15 The NMR data for organophosphorus and organonitrogen compounds 571

11.16 Infrared and NMR data for organophosphorus compounds 572
11.17 The NMR data for 2-x, 3-x, and 4-x-anisoles in $CHCl_3$ solutions 574
11.17a Infrared data for 3-x- and 4-x-substituted anisoles 575
11.18 Infrared and NMR data for mono-x-benzenes 576
11.19 The NMR data for 4-x- and 4,4′-x,x′-biphenyls in $CHCl_3$ solutions 577
11.20 The NMR data for TMU in $CHCl_3/CCl_4$ solutions 577
11.21 The NMR data for TMU in various solvents 578
11.22 The NMR data for dialkylketones in various solvents at 1 wt./vol. % concentration 578
11.23 The NMR data for alkyl acetates and phenyl acetate in 1 wt./vol. % in various solvents 579
11.24 Infrared and NMR data for alkyl acrylates and alkyl methacrylates in CCl_4 and $CHCl_3$ solutions 579

List of Figures

CHAPTER 1

1.1 Vapor-phase IR spectrum of ethylene oxide. 8
1.2 Vapor-phase IR spectrum of propylene oxide. 9
1.3 A plot of the epoxy ring breathing mode frequency for styrene oxide vs the mole % $CDCl_3/CCl_4$ (4). 10
1.4 A plot of the symmetric ring deformation frequency for styrene oxide vs mole % $CDCl_3/CCl_4$ (4). 10
1.5 A plot of the antisymmetric ring deformation frequency for styrene oxide vs mole % $CDCl_3/CCl_4$ (4). 11
1.6 Vapor-phase IR spectrum of trans-2,3-epoxybutane (or trans-1,2-dimethyl ethylene oxide). 12
1.7 A plot of the absorbance (A) for the oxirane ring breathing mode divided by (A) for the oxirane antisymmetric CH_2 stretching mode vs the number of carbon atoms for the 1,2-epoxyalkanes (ethylene oxide is the exception). 13
1.8 A plot of (A) for oxirane antisymmetric CH_2 stretching divided by (A) for antisymmetric CH_2 stretching for the alkyl group vs the number of carbon atoms in the 1,2-epoxyalkanes. 13
1.9 Vapor-phase IR spectrum of tetrahydrofuran. 14

CHAPTER 2

2.1 A plot of $\nu C{\equiv}N$ for alkanonitriles vs the number of protons on the α-carbon atom. 32
2.2 A plot of unperturbed $\nu C{\equiv}N$, cm^{-1} (1% wt./vol.) vs AN (The solvent acceptor number). 32
2.3 Plots of $\nu C{\equiv}N$, cm^{-1} for acetonitrile (1 wt./vol.%) vs ($\nu C{\equiv}N$ in methyl alcohol) minus ($\nu C{\equiv}N$ in another solvent). The two plots represent perturbed and unperturbed $\nu C{\equiv}N$. 33
2.4 A plot of $\nu C{\equiv}N$ for 1 wt./vol. % vs $\nu C{\equiv}N$ for 1% wt./vol. in 15 different solvents. 34
2.5 A plot of unperturbed $\nu C{\equiv}N$ for acetonitrile vs $\nu C{\equiv}N$ for benzonitrile. Both compounds were recorded at 1% wt./vol. separately in each of the 15 solvents. 35

2.6 A plot of $\nu C{\equiv}N$ for 4-cyanobenzaldehyde in cm^{-1} vs mole % $CHCl_3/CCl_4$. 36
2.7 A plot of the number of protons on the alkyl α-C—N≡ atom vs $\nu N{\equiv}C$ for alkyl isonitriles. 37
2.8 A plot of Taft's σ^* vs $\nu N{\equiv}C$ for alkyl isonitriles. 38

CHAPTER 3

3.1 Plots of ν asym. N=C=O and the combination tone ν (C_α—N) + ν sym. N=C=O and ν (C_α—N) + δ sym. CH_3 for methyl isocyanate all corrected for Fermi resonance. 51
3.2 Plots of the three observed IR bands for methyl isocyanate occurring in the region 2250–2320 vs mole % $CHCl_3/CCl_4$. 51
3.3 Plots of the ν asym. N=C=O frequencies for n-butyl, isopropyl and tert-butyl isocyanate and of the frequencies of the most intense IR band for ν asym. N=C=O in FR (uncorrected for FR) for methyl, ethyl and n-propyl isocyanate vs mole% $CHCl_3/CCl_4$. 52
3.4 Plots of unperturbed ν asym. N=C=O for the alkyl isocyanates vs mole % $CHCl_3/CCl_4$. 52
3.5 Plots of ν asym N=C=O frequencies for alkyl isocyanates in CCl_4 solution and in $CHCl_3$ solution vs σ^* (The inductive release value of the alkyl group.) 53
3.6 Plots of ν asym. N=C=O frequencies for alkyl isocyanates in CCl_4 solution and in $CHCl_3$ solution vs E_s (The stearic parameter of the alkyl group.) 53
3.7 Plots of ν asym. N=C=O frequencies for alkyl isocyanates in CCl_4 solution and in $CHCl_3$ solution vs (E_s) (σ^*). 54
3.8 A plot of perturbed ν asym. N=C=S (not corrected for FR) for five alkyl isothiocyanates vs mole % $CHCl_3/CCl_4$). 54
3.9 A plot of unperturbed ν asym. N=C=S (corrected for FR) for four alkyl isothiocyanates vs mole % $CHCl_3/CCl_4$ and $CDCl_3/CCl_4$. 55
3.10 A plot of perturbed 2ν C—N and perturbed ν asym. N=C=S and unperturbed 2ν C—N and unperturbed ν asym, N=C=S vs mole % $CHCl_3/CCl_4$. 56
3.11 A plot of ν C—N for methyl isothiocyanate vs mole % $CHCl_3/CCl_4$. 56

CHAPTER 4

4.1 Infrared spectra of O,O-bis-(2,4,5-trichlorophenyl) phosphorodithioate in 5 wt./vol. in CS_2 solution (2700–2400 cm^{-1}) at temperatures ranging from 29 to −100 °C. 71
4.2 Top: Liquid-phase IR spectrum of methyl (methylthio) mercury between KBr plates in the region 3800–450 cm^{-1}. Bottom: Liquid-phase IR spectrum of methyl (methylthio) mercury between polyethylene plates in the region 600–45 cm^{-1}. The IR band near 72 cm^{-1} is due to absorbance from poly (ethylene). 72
4.3 Top: Raman liquid-phase spectrum of methyl (methylthio) mercury in a glass capillary tube. The sample was positioned perpendicularly to both the laser beam

and the optical axis of he spectrometer. Bottom: Same as top except that the plane of polarization of the incident beam was rotated 90°. 73

CHAPTER 5

5.1 Plots of ν S=O and the mean average of (ν asym. SO_2 + ν sym. SO_2) vapor-phase frequencies for compounds containing the S=O or SO_2 group vs $\Sigma\sigma'$. 96

5.2 A plot of ν asym; SO_2 vs ν sym. SO_2 vapor-phase frequencies for a variety of compounds containing the SO_2 group. 97

5.3 Plots of ν asym. SO_2 vapor-phase frequencies for a variety of compounds containing a SO_2 group vs $\Sigma\sigma'$. 98

5.4 Plots of ν sym. SO_2 vapor-phase frequencies for a variety of compounds containing the SO_2 group vs $\Sigma\sigma'$. 99

5.5 Plots of ν asym. and ν sym. SO_2 for methyl phenyl sulfone in 10 wt./vol. solvent vs mole % $CHCl_3/CCl_4$. 100

5.6 A plot of ν asym. SO_2 vs ν sym. SO_2 for methyl phenyl sulfone in 1% wt./vol. solvent vs mole % $CHCl_3/CCl_4$. 101

5.7 Plots of ν asym. SO_2 and sym. SO_2 for dimethyl sulfate in 1% wt./vol. solvent vs mole % $CDCl_3/CCl_4$. 102

5.8 A plot of ν asym. SO_2 vs ν sym. SO_2 for dimethyl sulfate 1% wt./vol. solvent vs mole % $CDCl_3/CCl_4$. 103

CHAPTER 6

6.1* Methyl chloride (200-mm Hg sample) (27). 129

6.2* Methyl bromide (100-mm Hg sample) (27). 130

6.3* Methyl iodide (200-mm Hg sample) (27). 131

6.4* Methylene chloride (20- and 100-mm Hg sample) (27). 132

6.5* Methylene bromide (30-mm Hg sample) (27). 133

6.6* Trichloromethane (chloroform) (10- and 50-mm Hg sample) (27). 134

6.7 Tetrafluoromethane (freon 14) (2 and 100 Hg sample) (26) 135

6.8* Tetrachloromethane (carbon tetrachloride) (2- and 100-mm Hg sample) (27). 136

6.9* Chlorotrifluoromethane (10- and 50-mm Hg sample) (27). 137

6.10* Dichlorodifluoro methane (10- and 100-mm Hg sample) (27). 138

6.11* Trichlorofluoromethane (f and 40-mm Hg sample) (27). 139

6.12* Trichlorobromomethane (5- and 30-mm Hg sample) (27). 140

6.13* 1,2-Dichloroethane (ethylene dichloride) (50-mm Hg sample) (27). 141

6.14* Infrared spectra of 3-halopropenes (allyl halides) in CCl_4 solution (3800–1333 cm^{-1}) (133–400 cm^{-1}) (12). 142

6.15 Vapor-phase infrared spectra of 3-halopropenes (allyl halides) (12). 143

6.16a Vapor-phase IR spectrum of 3-fluoropropyne in a 5-cm KBr cell (50-mm Hg sample). 144

6.16b Vapor-phase IR spectrum of 3-chloropropyne in a 5-cm KBr cell (vapor pressure at −10 and 25 °C samples). 144
6.16c Vapor-phase IR spectrum of 3-bromopropyne in a 5-cm KBr cell (vapor pressure at 0 and 25 °C samples). 144
6.16d Vapor-phase IR spectrum of 3-iodopropyne in a 15-cm KBr cell (∼8-mm Hg sample). 145
6.17a Top: Liquid-phase IR spectrum of 3-chloropropyne-1-d in a 0.023-mm KBr cell. Bottom: Liquid-phase IR spectrum of 3-chloropropyne-1-d in a 0.1-mm polyethylene cell. 146
6.17b Top: Vapor-phase IR spectrum of 3-chloropropyne-1-d in a 10-cm KBr cell (33- and 100-mm Hg sample); middle: 3-chloropropyne-1-d in a 10-cm polyethylene cell. Bottom: Solution-phase IR spectrum of 3-chloropropyne-1-d in 10% wt./vol. CCl_4 (3800–1333 cm^{-1}) and 10% wt./vol. in CS_2 (1333–400 cm^{-1}) using 0.1-mm KBr cells. Bands marked with X are due to 3-chloropropyne. 147
6.18 Top: A Raman liquid-phase spectrum of 3-chloropropyne-1-d. Bottom: A Raman polarized liquid-phase spectrum of 3-chloropropyne-1-d. Some 3-chloropropyne is present (15). 148
6.19 Top: Solution-phase IR spectrum of 3-bromopropyne-1-d in 10% wt./vol. in CCl_4 (3800–1333 cm^{-1}) and 10% wt./vol. in CS_2 (1333–450 cm^{-1}) using 0.1-mm KBr cells (16). Bottom: A vapor-phase IR spectrum of 3-bromopropyne-1-d in a 10-cm KBr cell (40-mm Hg sample). Infrared bands marked with X are due to the presence of 3-bromopropyne (16). 149
6.20 Top: Raman spectrum of 3-bromopropyne-1-d. Bottom: Polarized Raman spectrum of 3-bromopropyne-1-d. Infrared bands marked with X are due to the presence of 3-bromopropyne (16). 150
6.21 An IR spectrum of 1,3-dichloropropyne in 10% wt./vol. CCl_4 solution (3800–1333 emsp14;cm^{-1}) and in CS_2 solution (1333–450 cm^{-1}) using 0.1-mm NaCl and KBr cells, respectively. Infrared bands at 1551 and 1580 cm^{-1} are due to CCl_4 and the IR band at 858 cm^{-1} is due to CS_2 (17). 151
6.22 Vapor-phase IR spectrum of 1,3-dichloropropyne (ambient mm Hg sample at 25 °C in a 12.5-cm KBr cell) (17). 152
6.23 Solution-phase IR spectrum of 1,3-dibromopropyne in 10% wt./vol. in CCl_4 (3800–1333 cm^{-1}) and in CS_2 solution (1333–450 cm^{-1}) using 0.1-mm NaCl and KBr cells, respectively. Infrared bands at 1551 and 1580 cm^{-1} are due to CCl_4 and the IR band at 858 cm^{-1} to CS_2 (17). 153
6.24 Approximate normal modes for propyne, 3-halopropynes, and 1,3-dihalopropynes. 154
6.25 Vapor-phase IR spectrum for 1-bromopropyne in a 12.5-cm KBr cell. The weak IR band at 734 cm^{-1} is due to an impurity. The 1-bromopropyne decomposes rapidly in the atmosphere (25). 154
6.26 Infrared vapor spectrum for 1-iodopropyne in a 12.5-cm KBr cell (25). 155
6.27 Top: Vapor-phase IR spectrum of 1-bromo-1-chloroethylene using a 10-cm KBr cell (10-mm Hg sample). Bottom: Same as upper (100-mm Hg sample) (19). 155
6.28 Raman liquid-phase spectrum of 1-homo-1-chloroethylene. Top: Parallel polarization. Bottom: Perpendicular polarization. 156
6.29 Vapor-phase IR spectrum of 1-chloropropadiene in a 12.5-cm KBr cell (50- and 100-mm Hg sample) (23). 156

6.30 Vapor-phase IR spectrum of 1-bromopropadiene in a 12.5-cm KBr cell (50- and 100-mm Hg sample). 157
6.31 Vapor-phase IR spectrum of 1-iodopropadiene in a 5-cm KBr cell (vapor pressure at 25 °C). Bands at 1105 and 1775 cm^{-1} are due to the presence of an impurity. 157
6.32 Infrared spectrum of 1-iodopropadiene in 10% wt./vol. CCl_4 solution (3800–1333 cm^{-1}) and 10% wt./vol. CS_2 solution (1333–450 cm^{-}) using NaCl and KBr cells, respectively. 157
6.33 Vapor-phase IR spectrum of 1-bromopropadiene-1-d in a 12.5-cm KBr cell (50- and 100-mm Hg sample) (16). 158
6.34 Solution-phase IR spectrum of 1-bromopropadiene-1-d 10% wt./vol. in CCl_4 (3800–1333 cm^{-1}) and in CS_2 solution using 0.1-mm KBr cells. Infrared bands marked with X are due to the presence of 1-bromopropadiene (16). 158
6.35 Top: Raman spectrum of 1-bromopropadiene-1-d using a capillary tube. Bottom: Polarized Raman spectrum of 1-bromopropadiene-1-d (16). 159
6.36* Vapor-phase IR spectrum of tetrafluoroethylene (8- and 50-mm Hg samples) (27). 160
6.37* Vapor-phase IR spectrum of tetrachloroethylene (13-mm Hg sample) (27). 161
6.38* Vapor-phase IR spectrum of 1,1-dichloro-2,2-difluoroethylene (10- and 60-mm Hg samples) (26). 162
6.39* Vapor-phase IR spectrum of trichloroethylene (10- and 50-mm Hg samples). 163

*Those vapor-phrase infrared spectra figures for Chapter 6 with an asterisk following the figure number have a total vapor pressure of 600-mm Hg with nitrogen (N_2), in a 5-cm KBr cell. The mm Hg sample is indicated in each figure.

CHAPTER 7

7.1 A plot of the observed ν asym. NO_2 frequencies vs the calculated ν asym. NO_2 frequencies for nitroalkanes using the equation ν asym. $NO_2 = 1582\ cm^{-1} + \Sigma\Delta R$. 186
7.2 A plot of the observed ν sym. NO_2 frequencies vs the calculated ν sym. NO_2 frequencies using the equation ν sym. $NO_2 = 1397\ cm^{-1} + \Sigma\Delta R$. 187
7.3 Vapor-phase IR spectrum for nitromethane in a 5-cm KBr cell (5 and 20 mm Hg sample to 600 mm Hg with N_2). 188
7.4 Vapor-phase IR spectrum for 2-nitropropane in a 5-cm KBr cell (20 mm Hg sample to 600 mm Hg with N_2). 189
7.5 Plots of ν asym. NO_2 and ν C=O for 4-nitrobenzaldehyde vs mole % $CHCl_3/CCl_4$. 190
7.6 Plots of ν sym. NO_2 for 4-nitrobenzaldehyde vs mole % $CHCl_3/CCl_4$. 190
7.7 A plot of ν asym. NO_2 for 4-X-nitrobenzenes in CCl_4 solution vs ν asym. NO_2 in $CHCl_3$ solution. 191
7.8 A plot of ν asym. NO_2 for 4-X-nitrobenzenes in CCl_4 solution vs σ_p. 192
7.9 A plot of ν asym. NO_2 for 4-X-nitrobenzenes in $CHCl_3$ solution vs σ_p. 193
7.10 A plot of ν asym. NO_2 for 4-X-nitrobenzenes in CCl_4 solution vs σ_{R^+}. 194
7.11 A plot of ν asym. NO_2 for 4-X-nitrobenzenes in $CHCl_3$ solution vs σ_{R^+}. 195
7.12 A plot of ν asym. NO_2 for 4-X-nitrobenzenes in CCl_4 solution vs σ_R. 196
7.13 A plot of ν asym. NO_2 for 4-X-nitrobenzenes in $CHCl_3$ solution vs σ_R. 197
7.14 A plot of ν asym. NO_2 for 4-X-nitrobenzenes in CCl_4 solution vs σ_I. 198
7.15 A plot of ν asym. NO_2 for 4-X-nitrobenzenes in $CHCl_3$ solutions vs σ_I. 199

7.16 Vapor-phase IR spectrum for ethyl nitrate in a 5-cm KBr cell (5 and 30 mm Hg sample to 600 mm Hg with N_2). 200
7.17 Vapor-phase IR spectrum for *n*-butyl nitrite in a 5-cm KBr cell (10 and 80 mm Hg sample to 600 mm Hg with N_2). 201
7.18 Plots of cis ν N=O and trans ν N=O for alkyl nitrites vs σ^*. 202
7.19 A plot of trans ν N–O for alkyl nitrites vs σ^*. 203
7.20 A plot of trans ν N=O vs trans ν N–O for alkyl nitrites. 204
7.21 Plots of the absorbance ratio (A) trans ν N=O/(A) cis ν N=O vs σ^*. 205

CHAPTER 8

8.1 Top: Infrared spectrum of thiophosphoryl dichloride fluoride 10% wt./vol. in CCl_4 solution (3800–1333 cm^{-1}) and 10% wt./vol. in CS_2 solution. Bottom: Vapor-phase spectrum of thiophosphoryl dichloride fluoride in the region 3800–450 cm^{-1} (9). 255
8.2 Top: Infrared spectrum of thiophosphoryl dichloride fluoride in hexane solution using polyethylene windows in the region 600–40 cm^{-1}. Bottom: Vapor-phase IR spectrum of thiophosphoryl dichloride fluoride using polyethylene windows in the region 600–150 cm^{-1} (9). 256
8.3 Top: Infrared spectrum of phosphoryl chloride in CCl_4 solution (3800–1333 cm^{-1}) and in CS_2 solution (1333–45 cm^{-1}). Bottom: Infrared spectrum of phosphoryl chloride in hexane solution (9). 257
8.4 A comparison of the approximate normal vibrations of thiophosphoryl dichloride fluoride vs those for phosphoryl chloride (9). 258
8.5 Top: Infrared spectrum of O-methyl phosphorodichloridothioate in 10 wt./vol. % in CCl_4 solution (3800–1333 cm^{-1}) and in 10 wt./vol. % in CS_2 solution (1333–400 cm^{-1}) using 0.1-mm KBr cells. The weak band at ~752 cm^{-1} is due to the presence of a trace amount of $P(-S)Cl_3$. The band at 659 cm^{-1} is due to the presence of ~4% O,O-dimethyl phosphorochloridothioate. Middle: Infrared spectrum of O-methyl-d_3 phosphorodichloride-thioate in 10 wt./vol. % in CCl_4 solution (3800–1333 cm^{-1}) and 10 + 2 wt./vol. % in CS_2 solution (1333–400 cm^{-1}) using 0.1-mm KBr cells. Bottom: Infared spectrum of O-methyl phosphorodichloridate in 10 wt./vol. % in CCl_4 solution (3800–1333 cm^{-1}) and 10 wt./vol. % in CS_2 solution (1333–450 cm^{-1}) using 0.1-mm KBr cells (21). 259
8.6 Top: Infrared spectrum of O-methyl phosphorodichloridothiate in 10 wt./vol. % in hexane solution in a 1-mm polyethylene cell. Bottom: Infrared spectrum of O-methyl phosphorodichloridothioate in 25 wt./vol. % in hexane solution in a 2-mm polyethylene cell (25). 260
8.7 Top: Infrared spectrum of O-methyl-d_3 phosphorodichloridothioate in 10 wt./vol. % hexane solution in a 1-mm polyethylene cell. Bottom: Infrared spectrum of O-methyl-d_3 phosphorodichloridothioate in 25 wt./vol. % hexane solution in a 2-mm polyethylene cell and compensated with polyethylene (25). 261
8.8 Assumed normal vibrations of organophosphorus and inorganophosphorus compounds (25). 262

8.9 Top: Infrared spectrum of O-ethyl phosphorodichlorodiothioate 10 wt./vol. % in CCl_4 solution (3800–1333 cm^{-1}) and 10 and 2 wt./vol. % in CS_2 solution (1333–400 cm^{-1}) in 0.1-mm KBr cells. The solvents were compensated. Bottom: Infrared spectrum of O-ethyl phosphorodichloridothioate in 25 and 2 wt./vol. % in hexane solution (600–45 cm^{-1}) in a 1-mm polyethylene cell (22). 263
8.10 Top: Infrared spectrum of O-ethyl-1,1-d_2 phosphorodichloridothioate in 10 wt./vol. % in CCl_4 solution (3800–1333 cm^{-1}) and 10 and 2.5 wt./vol. % in CS_2 solution (1333–400 cm^{-1}) in 0.1-mm KBr cells. The solvents are compensated. Bottom: Infrared spectrum of O-ethyl-1,1-d_2 phosphorodichloridothioate in 10 wt./vol. hexane solution (600–45 cm^{-1}) in a 1-mm polyethylene cell (22). 264
8.11 Top: Infrared spectrum of O-ethyl-2,2,2-d_3 phosphorodichloridothioate in 10 wt./vol. % CCl_4 solution (3800–1333 cm^{-1}) and 10 wt./vol. % CS_2 solution (1333–400 cm^{-1}) in 0.1-mm KBr cells. The solvents are not compensated. Bottom: Infrared spectrum of O-ethyl-2,2,2-d_3 phosphorodichloridothioate in 10 wt./vol. hexane solution in a 1-mm polyethylene cell (22). 265
8.12 Top: Infrared spectrum of O,O-dimethyl phosphorochloridothioate in 10 wt./vol. % in CCl_4 solution (3800–1333 cm^{-1}) and 10 and 2 wt./vol. % in CS_2 solution (1333–450 cm^{-1}) in 0.1-mm KBr cells. The solvents have not been compensated. Bottom: Liquid-phase IR spectrum of O,O-dimethyl phosphorochloridothioate between KBr plates (27). 266
8.13 Top: Infrared spectrum of O,O-dimethyl-d_6 phosphorochloridothioate in 10 wt./vol. % in CCl_4 solution (3800–1333 cm^{-1}) and 10 and 1 wt./vol. % in CS_2 solution (1333–400 cm^{-1}) in 0.1-mm KBr cells. Traces of toluene and methylene chloride are present. Bottom: Liquid-phase IR spectrum of O,O-dimethyl-d_6 phosphorochloridothioate between KBr plates (27). 267
8.14 Top: Infrared spectrum of O,O-dimethyl phosphorochloridothioate in 2-mm polyethylene cells. The polyethylene has been compensated. Bottom: Infrared spectrum of O,O-dimethyl-d_6 phosphorochloridothioate in 10 wt./vol. % n-hexane in 1- and 2-mm polyethylene cells, respectively (27). 268
8.15 Top: Infrared spectrum of S-methyl phosphorodichloridothioate in 10 wt./vol. % in CCl_4 solution (3800–1333 cm^{-1}) and in 10 and 2 wt./vol. % in CS_2 solution (1333–400 cm^{-1}) in 0.1-mm KBr cells. Bottom: Infrared liquid phase spectrum of S-methyl phosphorodichloridothioate in 10 wt./vol. % hexane solution (600–45 cm^{-1}) in a 1-mm polyethylene cell (23). 269
8.16 Upper: Raman liquid-phase spectrum of S-methyl phosphorodichloridothioate. Lower: Polarized Raman liquid-phase spectrum of S-methyl phosphorodichloridothioate (23). Bottom: upper: Raman liquid-phase spectrum of phosphoryl chloride. Lower: Polarized Raman spectrum of phosphoryl chloride (23). 270
8.17 Top: Infrared spectrum of trimethyl phosphite 10 wt./vol. %13334 cm^{-1}) and 10 wt./vol. % in CS_2 solution (1333–400 cm^{-1}) in 0.1-mm KBr cells. The solvents have not been compensated. Bottom: Infrared spectrum of trimethyl phosphite in 1 wt./vol. % CCl_4 solution (3800–1333 cm^{-1}) and in 1 wt./vol. % CS_2 solution (1333–400 cm^{-1}) using 0.1-mm KBr cells. The solvents have not been compensated (36). 271

8.18 Infrared spectrum of trimethyl phosphite in 10 wt./vol. % hexane solution in a 0.1-mm polyethylene cell. The IR band at $\sim 70\,cm^{-1}$ in a lattice mode for polyethylene (36). 272
8.19 Vapor-phase IR spectrum of trimethyl phosphite in a 10-cm KBr cell (36). 272
8.20 Top: Infrared spectrum of O,O-dimethyl phosphorodithioic acid in 10 wt./vol. % in CCl_4 solution (3800–1333 cm^{-1}) and 10 and 2 wt./vol. % in CS_2 solution (1333–450 cm^{-1}) in 0.1-mm KBr cells. Bottom: Infrared spectrum of O,O-dimethyl phosphorochloridothioate in CCl_4 solution (3800–1333 cm^{-1} and 10 and 2 wt./vol. % in CS_2 solutions (1333–450 cm^{-1}) in 0.1-mm KBr cells (32). 273
8.21 Top: Infrared spectrum of O,O-diethyl phosphorodithioic acid in 10 wt./vol. % in CCl_4 solution (3800–1333 cm^{-1}) and 10 and 2 wt./vol. % in CS_2 solutions (1333–400 cm^{-1}) in 0.1-mm KBr cells. The solvents have been compensated. Bottom: Infrared spectrum of O,O-diethyl phosphorochloridothioate in 10 wt./vol.% in CCl_4 solution and in 10 and 2 wt./vol. % CS_2 solutions (1333–450 cm^{-1}) in 0.1-mm KBr cells (32). 273
8.22 Top: Infrared spectrum of dimethyl hydrogenphosphonate in 10 wt./vol. % in CCl_4 solution (3800–1333 cm^{-1}) and in 10 and 2 wt./vol. % CS_2 solutions in 0.1-mm KBr cells. The solvents have not been compensated. Bottom: Infrared spectrum of dimethyl deuterophosphonate in 10% wt./vol. CCl_4 solution (3800–1333 cm^{-1}) and in 10 and 2 wt./vol. % solutions in CS_2 solution (1333–400 cm^{-1}). The solvents have been compensated (32). 274
8.23 Top: Infrared spectrum of dimethyl hydrogenphosphonate saturated in hexane solution (600–35 $cm^{-1}-1$) in a 2-mm polyethylene cell (32). 274
8.24 Top: Infrared spectrum of diethyl hydrogenphosphonate in 10 wt./vol. % in CCl_4 solution (3800–1333 cm^{-1}) and in 10 and 2 wt./vol. % in CS_2 solutions (1333–400 cm^{-1}) in 0.1-mm KBr cells. The solvents have not been compensated. Bottom: Infrared spectrum of diethyl deuterophosphonate in 10% wt./vol. % in CCl_4 solution (3800–1333 cm^{-1}) and 10 and 2 wt./vol. % CS_2 solutions (1333–400 cm^{-1}) in 0.1-mm KBr cells. The solvents have been compensated (32). 275
8.25 Top: Infrared spectrum for diethyl hydrogenphosphonate in 0.25 and 5 wt./vol. % hexane solutions (600–35 cm^{-1}) in a 2-mm polyethylene cell. Bottom: Infrared spectrum for diethyl deuterophosphonate in 10 wt./vol. % hexane solution (600–35 cm^{-1}) in a 1-mm polyethylene cell. 275
8.26 Top: Raman liquid-phase spectrum of dimethyl hydrogenphosphonate in a 2.5-ml multipass cell, gain 7, spectral slit-width 0.4 cm^{-1}. Bottom: Same as top but with the plane of polarization of the incident beam rotated through 90° (32). 276
8.27 Top: Infrared spectrum of (chloromethyl) phosphonic dichloride in 4 wt./vol. % CS_2 solution (1500–400 cm^{-1}) at 0 °C in a 0.1-mm KBr cell. Bottom: Same as above except at −75 °C. 277
8.28 Top: Infrared spectrum of (chloromethyl) phosphonic dichloride in 10 wt./vol. % in CCl_4 solution (3800–1333 cm^{-1}) and 10 wt./vol. % in CS_2 solution (1333–400 cm^{-1}) in 0.1-mm KBr cells. Bottom: Infrared spectrum of (chloromethyl) phosphonic dichloride in 20 wt./vol. % hexane solution (600–45 cm^{-1}) in a 1-mm polyethylene cell (17). 278

8.29 Top: Infrared spectrum of (chloromethyl) phosphonothioic dichloride in 10 wt./vol. % CCl_4 solution (3800–1333 cm^{-1}) and in 10 wt./vol. % CS_2 solution (1333–400 cm^{-1}) in 0.1-mm KBr cells. Bottom: Infrared spectrum of (chloromethyl) phosphonothioic dichloride in 20 wt./vol. % hexane solution (600–45x%cm^{-1}) in a 1-mm polyethylene cell (17). 279
8.30 Infrared spectrum of (bromomethyl) phosphonic dibromide in 10 wt./vol. % CCl_4 solution (3800–1333 cm^{-1}) and 10 wt./vol. % in CS_2 solution (1333–400 cm^{-1}) in 0.1-mm KBr cells (17). 279
8.31 Top: Infrared spectrum of N-methyl phosphoramidodichloridothioate 10 wt./vol. % in CCl_4 solution (3800–1333 cm^{-1}) and 10 wt./vol. % in CS_2 solution (1333–45 cm^{-1}) in 0.1-mm KBr cells. The solvents have not been compensated. Bottom: Infrared spectrum of N-methyl phosphoramidodichloridothioate in 10 wt./vol. hexane solution (600–45 cm^{-1}) in a 1-mm polyethylene cell (24). 280
8.32 Top: Infrared spectrum of N-methyl-d_3 phosphoramidodichloridothioate in 10 wt./vol. % in CCl_4 (3800–1333 cm^{-1}) and 10 wt./vol. % in CS_2 solution (1333–45 cm^{-1}) in 0.1-mm KBr cells. Bottom:Infrared spectrum of N-methyl-d_3 phosphoramidodichloridothioate in 10 wt./vol. % in hexane solution (600–45 cm^{-1}) in a 1-mm polyethylene cell (24). 281
8.33 Top: Infrared spectrum of N-D, N-methyl phosphoramidodichloridothioate in 10 wt./vol. % CCl_4 solution (3800–1333 cm^{-1}) and in 10 wt./vol. % in CS_2 solution (1333–45 cm^{-1}) in 0.1-mm KBr cells. The solvents have been compensated. Bottom: Infrared spectrum of N-D, N-methyl phosphoramidodichloridothioate in 10 wt./vol. % hexane solution in a 1-mm polyethylene cell. Both samples contain N-methyl phosphoramidodichloridothioate as an impurity (24). 282
8.34 Top: Infrared spectrum for N-D, N-methyl-d_3 phosphoramidodichloridothioate in 10 wt./vol. % CCl_4 solution (3800–1333 cm^{-1}) and 10 and 2 wt./vol. % in CS_2 solutions (1333–400 cm^{-1}) in 0.1-mm KBr cells. The solvent bands are compensated. Bottom: Infrared spectrum of N-D, N-methyl-d_3 phosphoramidodichloridothioate in 10 wt./vol. % hexane solution in a 1-mm polyethylene cell. This sample contains N-methyl-d_3 phosphoramidodichloridothioate as an impurity (24). 283
8.35 Infrared spectrum 1 is for O-alkyl O-aryl N-methylphosphoramidate in 10 wt./vol. % CCl_4 solution (3500–3000 cm^{-1}) in a 0.1-mm NaCl cell. Infrared spectrum 2 is for O-alkyl O-aryl N-methylphosphoramidothioate in 10 wt./vol. % in CCl_4 solution (3500–300 cm^{-1}) in a 0.1-mm NaCl cell (42). 283
8.36 A plot of N–H stretching frequencies of O-alkyl O-aryl N-alkylphosphoramidates in the region 3450–3350 cm^{-1} vs an arbitrary assignment of one for each proton joined to the N-α-carbon atom (42). 284
8.37 Plots of cis and trans N–H stretching frequencies of O-alkyl O-aryl N-alkylphosphoramidothioates in the region 3450–3350 cm^{-1} vs an arbitrary assignment of one for every proton joined to the N-α-carbon atom (42). 284
8.38 Infrared spectra of the cis and trans N–H stretching absorption bands of O-alkyl O-aryl N-alkylphosphoramidothioates in 0.01 molar or less in CCl_4 solutions (3450–3350 cm^{-1}) in a 14-mm NaCl cell. The N-alkyl group for A is methyl, B is

ethyl, C is n-propyl, D is isobutyl, E is n-butyl, F is isopropyl, G is sec-butyl, and H is tert-butyl. Spectra I and J are O,O-dialkyl N-methylphosphoramidothioate and O,O-dialkyl N-propylphosphoramidothioate, respectively (42). 285

8.39 Infrared spectra of the N—H stretching absorption bands of O-alkyl N,N′-dialkyl phosphorodiamidothioates. The N,N′-dialkyl group for A is dimethyl, B is diethyl, C is dipropyl, D is dibutyl, E is diisopropyl, F is di-sec-butyl, G is dibenzyl. The H and I are IR spectra for the N—H stretching bands of O-aryl N,N′-diisopropyl-phosphordiamidothioate and O-aryl N-isopropyl, N-methylphosphorodiamido-thioate, respectively. Spectra A through J were recorded for 0.01 molar or less in CCl_4 solutions using 3-mm KBr cells. Spectra D and J are for O-aryl N,N′-dibutylphosphorodiamidothioate in 0.01 molar CCl_4 solution in a 3-mm cell and 10 wt./vol. % in CCl_4 solution in a 0.1-mm KBr cell (42). 285

8.40 Infrared spectrum R gives the N—H stretching absorption bands for O-alkyl O-aryl phosphoramidothioate in 0.01 molar CCl_4 solution in a 3-mm NaCl cell. Infrared spectrum S gives O,O-dialkyl phosphoramidothioate in 0.01 molar CCl_4 solution in a 3-mm NaCl cell (42). 286

8.41 Top: Infrared spectrum of O,O-diethyl methylphosphoramidothioate in 10 wt./vol. % CCl_4 solution (3800–1333 cm^{-1}) and 10 and 2 wt./vol. % CS_2 solutions (1333–400 cm^{-1}) in 0.1-mm KBr cells. Middle: Infrared spectrum of O,O-diethyl N-methyl, N-D-phosphoramidothioate in 2 wt./vol. % in CCl_4 solution (3800–1333 cm^{-1}) and in 2 wt./vol. % in CS_2 solution (1333–400 cm^{-1}) in 0.1-mm KBr cells. Bottom: Infrared spectrum of O,O-diethyl N-methyl, N-D-phosphoramidothioate 10 wt./vol. % in CCl_4 solution (3800–1333 cm^{-1}) and in 10 wt./vol. % CS_2 solution (1333–400 cm^{-1}) in 0.1-mm KBr cells (37). 287

8.42 Top: Infrared spectrum of O,O-diethyl phosphoramidothioate in 10 wt./vol. % in CCl_4 solution (3800–1333 cm^{-1}) and 2 and 10 wt./vol. % in CS_2 solutions (1333–400 cm^{-1}) in 0.1-mm KBr cells. Middle: Infrared spectrum of O,O-diethyl phosphoramidothioate-N-D_2 in 2 wt./vol. % in CCl_4 solution (3800–1333 cm^{-1}) and in 2 wt./vol. % CS_2 solution (1333–400 cm^{-1}) in 0.1-mm KBr cells. Bottom: Infrared spectrum of O,O-diethyl phosphoramidothioate-N-D_2 10 wt./vol. % in CCl_4 solution (3800–1333 cm^{-1}) and 10 wt./vol. % in CS_2 solution (1333–400 cm^{-1}) in 0.1-mm KBr cells (37). 288

8.43 Top: Infrared spectrum of O,O-diethyl N-methylphosphoramidothioate 10 wt./vol. % in CCl_4 solution (3800–1333 cm^{-1}) and 10 and 2 wt./vol. % in CS_2 solutions (1333–400 cm^{-1}) in 0.1-mm KBr cells. Middle: Infrared spectrum of O,O-diethyl N-methyl, N-D phosphoramidothioate in 2 wt./vol. % in CS_2 solution (1333–400 cm^{-1}) in 0.1-mm KBr cells. Bottom: Infrared spectrum of O,O-diethyl N-methyl, N-D-phosphoramidothioate in 10 wt./vol. % CCl_4 solution (1333–400 cm^{-1}) in 0.1-mm KBr cells. The solvents are not compensated (37). 289

8.44 Infrared spectrum of O,O-dimethyl N-methylphosphoramidothioate in 10 wt./vol. % in CCl_4 solution (3800–1333 cm^{-1}) and 10 and 2 wt./vol. % in CS_2 solutions (1333–400 cm^{-1}) in 0.1-mm KBr cells (37). 290

8.45 Top: Infrared spectrum of O,O-dimethyl O-(2,4,5-trichlorophenyl) phosphor-othioate 10 wt./vol. % in CCl_4 solution (3800–1333 cm^{-1}) and 10 and 2 wt./vol. % in CS_2 solutions (1333–400 cm^{-1}) in 0.1-mm KBr cells. Bottom: Infrared spectrum

of O,O-dimethyl O-(2,4,5-trichlorophenyl) phosphate in 10 wt./vol. % in CCl_4 solution (3800–1333 cm^{-1}) in 0.1-mm KBr cells. The solvents are not compensated (43). 291
8.46 Top: Infrared spectrum for O,O-dimethyl-d_6 O-(2,4,5-trichlorophenyl) phosphorothioate in 10 wt./vol. % CCl_4 solution (3800–1333 cm^{-1}) and in 10 and 2 wt./vol. % CS_2 solution (1333–400 cm^{-1}) in 0.1-mm KBr cells. Bottom: Infrared spectrum for O,O-dimethyl-d_6 O-(2,4,5-trichlorophenyl) phosphate in 10 wt./vol. % CS_2 solutions (1333–450 cm^{-1}) in 0.1-mm KBr cells. The solvents have not been compensated. 292
8.47 Top: Infrared spectrum of disodium methanephosphonate. Bottom: Infrared spectrum of disodium methane-d_3-phosphonate. In fluorolube oil mull (3800–1333 cm^{-1}) and in Nujol oil mull (1333–400 cm^{-1}) (40). These mulls were placed between KBr plates. 293
8.48 Top: Infrared spectrum of disodium methanephosphonate in water solution between AgCl plates. Bottom: Infrared spectrum of disodium methane-d_3-phosphonate in water solution between AgCl plates (40). 294
8.49 Top: Infrared spectrum of disodium methanephosphonate. Middle: Infrared spectrum of disodium methane-d_3-phosphonate. Bottom: Infrared spectrum of dipotassium methanephosphonate. These spectra were recorded from samples prepared as Nujol mulls between polyethylene film (600–45 cm^{-1}) (40). 295
8.50 Infrared spectrum of disodium n-octadecanephosphonate prepared as a fluoroluble mull (3800–1333 cm^{-1}) prepared as a Nujol mull (1333–450 cm^{-1}) (40). 296
8.51 Top: Infrared spectrum of sodium dimethylphosphinate prepared as a fluoroluble mull (3800–1333 cm^{-1}) and prepared as a Nujol mull (1333–400 cm^{-1}) between KBr plates. Bottom: Infrared spectrum of sodium dimethylphosphinate saturated in water solution between AgCl plates (41). 297
8.52 Top: Infrared spectrum of sodium dimethylphosphinate prepared as Nujol mull between polyethylene film (600–45 cm^{-1}). Bottom: Infrared spectrum of potassium dimethylphosphinate prepared as a Nujo mull between polyethylene film (600–45 cm^{-1}) (41). 298
8.53 Top: Raman saturated water solution of sodium dimethylphosphinate using a 0.25-ml multipass cell, gain 13.4, spectral slit 10 cm^{-1}. Bottom: Same as above, except with the plane of polarization of the incident beam rotated 90° (41). 299
8.53a (Top): IR spectrum for sodium diheptylphosphinate. 300
8.53a (Bottom): IR spectrum for sodium dioctylphosphinate. 300
8.54 Plots of ν P=O for $P(=O)Cl_3$ vs mole % solvent system. The open triangles represent CCl_4/C_6H_{14} as the solvent system. The open circles represent $CHCl_3/C_6H_{14}$ as the solvent system; and the open diamonds represent $CHCl_3/CCl_4$ as the solvent system (26). 301
8.55 A plot of ν P=O for $CH_3P(=O)Cl_2$ vs mole % $CDCl_3/CCl_4$ as represented by open circles, and plots of ν P=O for $(CH_3)_3C$ $P(db;O)Cl_2$ vs mole % $CHCl_3/CCl_4$ (open triangles) and mole % ($CDCl_3/CCl_4$ (open diamonds) (26). 301
8.56 Plots of ν P–O rotational conformers 1 and 2 for CH_3O $P(=O)Cl_2$ (conformer 1, open circles; conformer 2, open squares), and for $C_2H_5OP(=O)Cl_2$ (conformer 1, open triangles; conformer 2, open diamonds) vs mole % $CHCl_3/CCl_4$ (26). 302

8.57 Plots of v P=O rotational conformers 1 and 2 for $(CH_3O)_3P{=}O$ vs mole % $CHCl_3/CCl_4$ (conformer 1, solid circles, and conformer 2, solid triangles) (55). 302
8.58 Plots of v P=O rotational conformers 1 and 2 for $(C_2H_5O)_3P{=}O$ vs mole % $CDCl_3/CCl_4$ (conformer 1, solid triangles, conformer 2, closed circles) (55). 303
8.59 Plots of v P=O rotational conformers 1 and 2 for $(C_4H_9O)_3P{=}O$ vs mole % $CDCl_3/CCl_4$ (conformer 1, solid circles, conformer 2, solid triangles) (55). 303
8.60 Plots of v asym. PCl_3 for $P({=}O)Cl_3$ vs mole % solvent system. The open circles represent the $CHCl_3/CCl_4$ solvent system; the open triangles represent the $CHCl_3/C_6H_{14}$ solvent system; and the open diamonds represent the CCl_4/C_6H_{14} solvent system (26). 304
8.61 Plots of v asym. PCl_2 frequencies for $CH_3P({=}O)Cl_2$ (open circles) and for $(CH_3)_3CP({=}O)Cl_2$ (open triangle and open diamonds) vs mole % $CDCl_3/CCl_4$ and $CHCl_3/CCl_4$ (26). 304
8.62 Plots of v asym. PCl_2 rotational conformers 1 and 2 frequencies for $CH_3OP({=}O)Cl_2$ (conformer 1, open circles; conformer 2, open squares) and for $C_2H_5OP({-}O)Cl_2$ (conformer 1, open triangles, conformer 2, open diamonds) frequencies vs mole % $CHCl_3/CCl_4$ (26). 305
8.63 Plots of v sym. PCl_3 frequencies for $P({=}O)Cl_3$ vs mole % solvent system. The open circles represent $CHCl_3/CCl_4$ as the solvent system, the open triangles represent the $CHCl_3/C_6H_{14}$ solvent system, and the open diamonds represent the CCl_4/C_6H_{14} solvent system. 305
8.64 A plot of v sym. PCl_2 frequencies for $CH_3P({=}O)Cl_2$ vs mole % $CDCl_3/CCl_4$ (open circles) and plots of v sym. PCl_2 frequencies for $(CH_3)_3CP({=}O)Cl_2$ vs mole % $CHCl_3/CCl_4$ (open triangles) and mole % $CDCl_3/CCl_4$ (open diamonds) (26). 306
8.65 Plots of v sym. PCl_2 rotational conformers 1 and 2 vs mole % $CHCl_3CCl_4$. The open circles and open squares are for $CH_3OP({=}O)Cl_2)$ conformers 1 and 2, respectively. The open triangles and open diamonds are for $C_2H_5OP({=}O)Cl_2$ conformers 1 and 2, respectively (26). 306
8.66 Plots of v asym. PCl_2 frequencies vs v sym. PCl_2 frequencies for $CH_3OP({=}O)Cl_2$ (conformer 1, solid squares); $CH_3OP({=}O)Cl_2$ (conformer 2, open squares); $C_2H_5OP({=}O)Cl_2$ (conformer 1, solid triangles) $C_2H_5OP({=}O)Cl_2$ (conformer 2, open triangles); $(CH_3)_3CP({=}O)Cl_2$ (solid diamonds); and $CH_3P({=}O)Cl_2$ (solid circles) (26). 307
8.67 Plots of the absorbance (A) ratio of rotational conformer 1 and 2 band pairs for v P=O (solid circles), v COP (solid triangles), v asym. PCl_2 (solid squares), and v sym. PCl_2 (solid diamonds) for $CH_3OP({=}O)Cl_2$ vs mole % $CHCl_3/CCl_4$ (26). 307
8.68 Plots of the absorbance (A) ratio rational conformer 1 and 2 band pairs for v P=O (solid circles), v COP (solid triangles), v asym. PCl_2 (solid squares), and v sym. PCl_2 (solid diamonds) for $C_2H_5OP({=}O)Cl_2$ vs mole % $CHCl_3/CCl_4$ (26). 308
8.69 A plot of δ sym. CH_3 frequencies for $CH_3P({=}O)Cl_2$ vs mole % $CHCl_3/CCl_4$ (26). 308
8.70 A plot of ρCH_3 frequencies for $CH_3P({=}O)Cl_2$ vs mole % $CHCl_3/CCl_4$ (26). 309
8.71 Plots of the in-phase δ sym. $(CH_3)_3$ frequencies (open circles) and the out-of-phase δ sym. $(CH_3)_3$ frequencies (open triangles) vs mole % $CHCl_3/CCl_4$ (26). 309
8.72 Plots of the a′ and a ρCH_3 frequencies for $C_{2_{H5}}O$ $P({=}O)Cl_2$ vs mole % $CHCl_3/CCl_4$ (26). 310

8.73 Plots of the a′ and a ρCH_3 modes of $(C_2H_5O)_3P{=}O$ vs mole % $CDCl_3/CCl_4$ (55). 310
8.74 A plot of ν CC for $C_2H_5OP({=}O)Cl_2)$ vs mole % $CHCl_3/CCl_4$ (26). 311
8.75 A plot of ν CC for $(C_2H_5O)_3$ P=O vs mole % $CDCl_3/CCl_4$ (55). 311
8.76 Plots of ν COP rotational conformers 1 and 2 for CH_3O $P({=}O)Cl_2$ and $C_2H_5OP({=}O)Cl_2$ vs mole % $CDCl_3/CCl_4$ (26). 312
8.77 Plots of ν COP rotational conformers 1 and 2 frequencies for $(CH_3O)_3P{=}O$ vs mole % $CHCl_3/CCl_4$. The solid triangles represent conformer 1 and the solid circles represent conformer 2 (55). 312
8.78 A plot of ν COP frequencies for $(CH_3O)_3P{=}O$ vs mole percnt; $CHCl_3/CCl_4$ (55). 313
8.79 A plot of ν (ϕOP), "$\nu\phi$-O", frequencies for triphenyl phosphate vs mole % $CHCl_3/CCl_4$ (55). 313
8.80 A plot of ν (ϕOP), "νP-O", frequencies for triphenyl phosphate vs mole % $CHCl_3/CCl_4$ (55). 314
8.81 A plot of the in-plane hydrogen deformation frequencies for the phenyl groups of triphenyl phosphate vs mole % $CHCl_3/CCl_4$ (55). 314
8.82 A plot of the out-of-plane ring deformation frequencies for the phenyl groups of triphenyl phosphate vs mole % $CHCl_3/CCl_4$ (55). 315
8.83 A plot of ν CD frequencies (for the solvent system $CDCl_3/CCl_4$) vs mole % $CDCl_3/CCl_4$ (55). 315
8.84 A plot of the ν CD frequencies for $CDCl_3$ vs mole % $CDCl_3/CCl_4$ for the $CDCl_3/CCl_4$ solvent system containing 1 wt./vol. % triphenylphosphate (55). 316
8.85 A plot of ν CD for $CDCl_3$ vs mole % $CDCl_3/CCl_4$. 316
8.86 A plot of ν CD for $CDCl_3$ vs mole % $CDCl_3/CCl_4$ containing 1% triethylphosphate. 317

CHAPTER 9

9.1 Infrared spectrum for 2,3,4,5,6-pentachlorobiphenyl. 368
9.2 Infrared spectrum for 2,2′,3-trichlorobiphenyl. 368
9.3 Infrared spectrum for 2′3,4-trichlorobiphenyl. 369
9.4 Infrared spectrum for 2,2′,3,4,6-pentachlorobiphenyl. 369
9.5 Infrared spectrum for 2,3,4,4′,6-pentachlorobiphenyl. 370
9.6 Infrared spectrum for 2,3′,4,5′,6-pentachlorobiphenyl. 370
9.7 Infrared spectrum for 2,2′,3′,4,5-pentachlorobiphenyl. 371
9.8 Infrared spectrum for 2,2′,4,6,6′-pentachlorobiphenyl. 371
9.9 Infrared spectrum for 2,3′,4,4′,6-pentachlorobiphenyl. 372
9.10 Infrared spectrum for 2,2′,3,4′,6-pentachlorobiphenyl. 372
9.11 Infrared spectrum for 2,2′,3,5′,6-pentachlorobiphenyl. 373
9.12 Infrared spectrum for 2,2′,3,4′,5-pentachlorobiphenyl. 373
9.13 Infrared spectrum for 2,2′,3,5,5′-pentachlorobiphenyl. 374
9.14 Infrared spectrum for 2,2′,4,4′,5-pentachlorobiphenyl. 374
9.15 Infrared spectrum for 2,2′,4,5,5′-pentachlorobiphenyl. 375
9.16 Infrared spectrum for 2,2′,3,4,5′-pentachlorobiphenyl. 375
9.17 A plot of an A_1 fundamental for mono-substituted benzenes vs the number of protons on the ring α-carbon atom. 376

9.18 A plot of an A_1 fundamental for mono-substituted benzenes vs Tafts σ^*. 377
9.19 An IR spectrum for a uniaxially stretched syndiotactic polystyrene film (perpendicular polarization). 378
9.20 An IR spectrum for a uniaxially stretched syndiotactic polystyrene film (parallel polarization). 378
9.21 An IR spectrum of syndiotactic polystyrene film cast from boiling 1,2-dichlorobenzene onto a KBr plate. 379
9.22 An IR spectrum of the same film used to record the IR spectrum shown in Fig. 9.21except that the film was heated to 290 °C then allowed to cool to ambient temperature before recording the IR spectrum of syndiotactic polystyrene. 380
9.23 An IR spectrum syndiotactic styrene (98%)-4-methyl-styrene (2%) copolymer cast from boiling 1,2-dichlorobenzene onto a C_SI plate. 381
9.24 An IR spectrum of a styrene (93%)-4-methylstyrene (7%) copolymer cast from boiling 1,2-dichlorbenzene onto a C_SI plate. 382
9.25 A plot of the IR band intensity ratio A(1511 cm^{-1})/A (900–904 cm^{-1}) vs the weight % 4-methylstyrene in styrene −4-methylstyrene copolymers. 383
9.26 A plot of the IR band intensity ratio A(815–817 cm^{-1})/A(900–904 cm^{-1}) vs the weight % 4-methylstyrene in styrene −4-methylstyrene copolymers. 384
9.27 Top: Infrared spectrum of styrene (92%) −acrylic acid (8 %) copolymer recorded at 35 °C. Bottom: Infrared spectrum of styrene (92%) −acrylic acid (8%) copolymer recorded at 300 °C. 385
9.28 Styrene (92%) – acrylic acid (8%) copolymer absorbance ratios at the indicated frequencies vs copolymer film temperatures in °C; R_1, R_2, and R_4 indicate an increase in CO_2H concentrations at temperatures > 150 °C; R_3 and R_5 indicate a decrease in $(CO_2H)_2$ concentrations at > 150 °C. Base line tangents were drawn from 1630–1780 cm^{-1}, 1560–1630 cm^{-1} and 720–800 cm^{-1} in order to measure the absorbance values at 1746 cm^{-1} and 1700 cm^{-1}, 1600 cm^{-1} and 752 cm^{-1}, respectively. 386
9.29 Top: Infrared spectrum of styrene-acrylamide copolymer recorded at 27 °C. Bottom: Infrared spectrum of styrene-acrylamide copolymer recorded at 275 °C. 387
9.30 An IR spectrum for a styrene-2-isopropenyl-2-oxazoline copolymer (SIPO) cast from methylene chloride onto a KBr plate. 388
9.31 A plot of the weight % IPO in the SIPO copolymer vs the absorbance ratio (A)(1656 cm^{-1})/(A)(1600 cm^{-1}). 388
9.32 Top: Vapor-phase IR spectrum of ethynylbenzene in a 4-m cell (vapor pressure is in an equilibrium with the liquid at 25 °C). Bottom: Vapor-phase IR spectrum of ethynylbenzene-d in a 4-m cell (vapor pressure is in an equilibrium with the liquid at 25 °C). 389
9.33 Top: An IR solution spectrum for ethynylbenzene (3800–1333 cm^{-1} in CCl_4 (0.5 M) solution in a 0.1 mm NaCl cell), (1333–450 cm^{-1} in CS_2 (0.5 M) solution in a 0.1 mm KBr cell), and in hexane (0.5 M) solution using a 2 mm cis I cell. The IR band at 1546 and 853 cm^{-1} is due to the solvents. Bottom: An IR solution spectrum for ethynylbenzene-d recorded under the same conditions used to record the top spectrum. 390

9.34 Top: Liquid-phase IR spectrum ethynylbenzene between KBr plates in the region 3800–450 cm^{-1}, and between C_SI plates in the region 450–300 cm^{-1}. Bottom: Liquid-phase IR spectrum of ethynylbenzene-d recorded under the same conditions as the top spectrum. 391
9.35 Top: Raman spectra for ethynylbenzene in the liquid phase. Bottom: Raman spectra for ethynylbenzene-d in the liquid phase. 392
9.36 A correlation chart for substituted benzenes in the region 5–6 μ (after Young, DuVall, and Wright). 393
9.37 Summary of out-of-plane hydrogen deformations and their combination and overtones for mono-substituted benzenes. 394
9.38 Summary of out-of-plane hydrogen deformations and their combination and overtones for 1,2-disubstituted benzenes. 395
9.39 Summary of out-of-plane hydrogen deformations and their combination and overtones for 1,3-disubstituted benzenes. 396
9.40 Summary of out-of-plane hydrogen deformations and their combination and overtones for 1,4-disubstituted benzenes. 397
9.41 Summary of out-of-plane hydrogen deformations and their combination and overtones for 1,3,5-trisubstituted benzenes. 398
9.42 Summary of out-of-plane hydrogen deformations and their combination and overtones for 1,2,3-trisubstituted benzenes. 399
9.43 Summary of out-of-plane hydrogen deformations and their combination and overtones for 1,2,4-trisubstituted benzenes. 399
9.44 Summary of out-of-plane hydrogen deformations and their combination and overtones for 1,2,3,4-tetrasubstituted benzenes. 400
9.45 Summary for out-of-plane hydrogen deformations and their combination and overtones for 1,2,3,5-tetrasubstituted benzenes. 400
9.46 Summary of out-of-plane hydrogen deformations and their combination and overtones for 1,2,4,5-tetrasubstituted benzenes. 401
9.47 Summary of out-of-plane hydrogen deformation and its first overtone for 1,2,3,4,5-pentasubstituted benzenes. 401
9.48 Infrared correlation chart for out-of-plane hydrogen deformations and their combination and overtones for substituted benzenes. 402

CHAPTER 11

11.1 A plot of the NMR δ ^{13}C-1 chemical shift data for 4-x-anilines in $CHCl_3$ solution vs Hammett σ_p values for the 4-x atom or group. 459
11.2 A plot of the NMR δ ^{13}C-1 chemical shift data for 4-x-anilines in $CHCl_3$ solution vs Taft σ_{R° values for the 4-x atom or group. 460
11.3 A plot of ν asym. NH_2 frequencies for 4-x-anilines in the vapor phase vs δ ^{13}C-1 chemical shift data for 4-x-anilines in $CHCl_3$ solution. 461
11.4 A plot of ν asym. NH_2 frequencies for 3-x- and 4-x-anilines in hexane solution vs δ ^{13}C-1 chemical shift data for 3-x- and 4-x-anilines in $CHCl_3$ solution. 462

11.5 A plot of ν asym. NH_2 frequencies for 3-x- and 4-x-anilines in CCl_4 solution vs δ ^{13} C-1 chemical shift data for 3-x- and 4-x-anlines in $CHCl_3$ solution. 463
11.6 A plot of the ν asym. NH_2 frequencies for 3-x- and 4-x-anilines in $CHCl_3$ solution vs δ ^{13}C-1 chemical shift data for 3-x- and 4-x-anilines in $CHCl_3$ solution. 464
11.7 A plot of the ν sym. NH_2 frequencies for 4-x-anilines in the vapor phase vs δ ^{13}C-1 chemical shift data for 3-x- and 4-x-anilines. 465
11.8 A plot of ν sym. NH_2 frequencies for 3-x- and 4-x-anilines in hexane solution vs δ ^{13}C-1 chemical shift data for 3-x- and 4-x-anilines in $CHCl_3$ solution. 466
11.9 A plot of ν sym. NH_2 frequencies for 3-x- and 4-x-anilines in CCl_4 solution vs δ ^{13}C-1 chemical shift data for 3-x- and 4-x-anilines vs the δ ^{13}C-1 chemical shift data for 3-x- and 4-x-anilines in $CHCl_3$ solution. 467
11.10 A plot of the ν sym. NH_2 frequencies for 3-x- and 4-x-anilines in $CHCl_3$ solution vs
δ ^{13}C-1 chemical shift data for 3-x- and 4-x-anilnies in $CHCl_3$ solution. 468
11.11 A plot of the frequency difference between ν asym. NH_2 and ν sym. NH_2 for 3-x- and 4-x-anilines in hexane solution vs δ ^{13}C-1 chemical shift data for 3-x- and 4-x-anilines in $CHCl_3$ solution. 469
11.12 A plot of the frequency difference between ν asym. NH_2 and ν sym. NH_2 for 3-x- and 4-x-anilines in CCl_4 solution vs δ ^{13}C-1 chemical shift data for 3-x- and 4-x-anilines in $CHCl_3$ solution. 470
11.13 A plot of the frequency difference between ν asym. NH_2 and ν sym. NH_2 for 3-x- and 4-x-anilines in $CHCl_4$ solution vs δ ^{13}C-1 chemical shift data for 3-x- and 4-x-anilines in $CHCl_3$ solution. 471
11.1.4 A plot of the absorbance ratio A(ν asym. NH_2)/A(ν sym. NH_2) for 3-x- and 4-x-anilines in CCl_4 solution vs δ ^{13}C-1 chemical shift data for 3-x- and 4-x-anilines in $CHCl_3$ solution. 472
11.1.5 A plot of δ ^{13}C-1 chemical shift data for 4-x-anilines in $CHCl_3$ solution vs Taft σ_{R° values for the 4-x atom or group. 473
11.16 A plot of δ ^{13}C-1 for 3-x- and 4-x-benzoic acids vs Hammett σ values for the x atom or group. 474
11.17 Plots of ν C=O vs Hammett σ values of 3-x- and 4-x-benzoic acids. The solid circles are for IR data in the vapor phase. The open circles are for IR dilute solution data for unassociated 3-x- and 4-x-benzoic acids. 475
11.18 Plots of ν C=O vs δ ^{13}C-1 for 3-x- and 4-x-benzoic acids. The NMR data are for $CDCl_3$ solutions. The plot with closed circles includes vapor-phase IR data. The plot with open circles includes IR CCl_4-solution data. 476
11.19 A plot of δ ^{13}C-1 for 4-x-acetanilides in $CDCl_3$ solution vs Hammett σ_p values for the x-atom or group. 477
11.20 A plot of ν C=O for 4-x-acetanilides vs Hammett σ_p values. The solid circles represent IR CCl_4 solution data and the solid squares IR $CHCl_3$ solution data. 478
11.21 A plot of δ ^{13}C-1 for 4-x-benzaldehydes vs Taft σ_{R° values for the 4-x atom or group. 479
11.22 A plot of ν C=O for 4-x-benzaldehydes vs Taft σ_{R° values for the 4-x atom or group. The solid circles represent IR data in CCl_4 solutions and the solid triangles represent IR data in $CHCl_3$ solutions. 480

11.23 A plot of ν C=O for 4-x-benzaldehydes vs Hammett σ_p values for the 4-x atom or group. The solid circles represent IR data in CCl_4 solution and the solid triangles IR data in $CHCl_3$ solutions. 481

11.24 Plots of ν C=O vs δ ^{13}C-1 for 4-x-benzaldehydes. The δ ^{13}C-1 data are for $CDCl_3$ solutions. The plot with filled-in circles includes IR CCl_4 solution data and the plot with the filled-in triangles includes IR $CHCl_3$ solution data. 482

11.25 A plot of δ ^{13}C-1 for 4-x-acetophenones in $CDCl_3$ solutions vs Taft $\sigma_{R°}$ values for the 4-x atom or groups. 483

11.26 A plot of ν C=O for 4-x-acetophenones vs Taft $\sigma_{R°}$ values for the 4-x atom or group. The IR data is that for the 4-x-acetophenones in $CHCl_3$ solution. 484

11.27 A plot of ν C=O ($CHCl_3$ solution) vs δ ^{13}C-1 ($CDCl_3$ solution) for 4-x-acetophenones. 485

11.28 A plot of δ ^{13}C-1 vs Taft $\sigma_{R°}$ values for the 4-x and 4,4′-x,x atoms or groups for 4-x and 4,4′-x,x-benophenones. These data are for $CHCl_3$ solutions. 486

11.29 A plot of ν C=O in the vapor phase vs δ ^{13}C-1 in CCl_3 solution for 4-x and 4,4′-x,x-benzophenones. 487

11.30 A plot of ν C=O in the vapor phase vs the sum of Hammett σ_p for the x-atom or group for 4-x- and 4,4′-x,x-benzophenones. 488

11.31 A plot of δ ^{13}C-1 in $CHCl_3$ solution vs Hammett σ values for the x atom or group for methyl and ethyl 3-x- and 4-x-benzoates. 489

11.32 Plots of the range of ν C=O for the compounds studied in different physical phases vs the range of δ ^{13}C=O for the compounds studied in $CHCl_3$ solutions. 490

11.33 A plot of δ ^{13}C=O for acetone vs mole % $CHCl_3/CCl_4$ solutions. 491

11.34 A plot of ν C=O vs δ ^{13}C=O for acetone in mole % $CHCl_3/CCl_4$ solutions. 492

11.35 A plot of ν C=O for dimethylacetamide in CCl_4 and/or $CHCl_3$ solution vs mole % $CHCl_3/CCl_4$. 493

11.35a A plot of δ ^{13}C=O for dimethylacetamide in CCl_4 and/or $CHCl_3$ solution vs mole % $CHCl_3/CCl_4$. 493

11.35b A plot of δ ^{13}C=O for maleic anhydride vs mole % $CHCl_3/CCl_4$. 494

11.35c A plot of δ ^{13}C=O vs δ ^{13}C=C for maleic anhydride in 0 to 100 mole % $CHCl_3/CCl_4$ solutions. 494

11.36 A plot of δ ^{13}C-1 vs Taft $\sigma_{R°}$ values for the 3-x and 4-x atom or group for 3-x- and 4-x-benzonitriles in $CDCl_3$ solution. 495

11.37 A plot of δ ^{13}C-1 vs Hammett σ values for the 3-x or 4-x atom or group for 3-x- and 4-x-benzonitriles in $CHCl_3$ solution. 496

11.38 A plot of δ ^{13}CN vs Hammett σ values for the 3-x or 4-x atom or group for 3-x- and 4-x-benzonitriles in $CHCl_3$ solution. 497

11.39 A plot of ν CN vs δ ^{13}CN for organonitriles. 498

11.40 A plot of δ ^{13}CN vs the number of protons on the -carbon atom of alkyl nitriles. 498

11.41 A plot of Taft σ^* values for the alkyl group vs δ ^{15}N for alkyl isonitriles. 499

11.42 A plot of δ ^{15}N vs ν NC for alkyl isonitriles. 500

11.43 Plots of the number of protons in the alkyl α-carbon atom vs ν sym. NO_2 and ν asym. NO_2 for nitroalkanes in the vapor phase. 501

11.44 Plots of δ ^{13}C ($CHCl_3$ solution) for the alkyl α-carbon atom vs ν sym. NO_2 and ν asym. NO_2 (vapor phase) for nitroalkanes. 502

11.45 Plots of δ ^{15}N (in 0.3 M acetone) vs ν sym. NO_2 and ν asym. NO_2 (vapor phase) for nitroalkanes. 503
11.46 A plot of Taft σ^* values for the alkyl α-carbon atom group vs δ ^{13}C for nitroalkanes. 504
11.47 A plot of Taft σ^* values for the alkyl α-carbon atom vs δ ^{15}N for nitroalkanes. 505
11.48 A plot of δ ^{15}N vs δ ^{13}C for the alkyl α-carbon atom for nitroalkanes. 506
11.49 A plot of δ ^{15}N vs ν asym. N=C=O for alkyl isocyanates. 507
11.50 Plots of Taft σ^* values for the alkyl groups for alkyl isocyanates and alkyl isothiocyanates. 508
11.51 Plots of δ ^{13}C for the N=C=O group vs the number of protons on the alkyl α-carbon atom for alkyl isocyanates and alkyl diisocyanates. 509
11.52 Plots of ν asym. N=C=O vs δ ^{13}C for the N=C=O group of alkyl isocyanates and alkyl diisocyanates. 509
11.53 A plot of δ ^{15}N vs ν C–N for primary alkylamines. 510
11.54 A plot of δ ^{15}N vs ωN–H for dialkylamines. 511
11.55 A plot of ν C–N vs Taft σ^* for the alkyl group of primary alkylamines. 512
11.56 A plot of δ ^{15}N vs Taft σ^* for the alkyl group of primary alkylamines. 513
11.57 Plots of δ ^{15}N vs the number of protons on the alkyl α-carbon atom(s) for mono-, di-, and trialkylamines. 514
11.58 A plot of δ ^{31}P for trialkylphosphines vs the sum of the number of protons on the alkyl α-C–P carbon atoms. 515
11.59 A plot of δ ^{31}P for trialkylphosphines vs Taft σ^* values for the alkyl groups. 516
11.60 Plots of δ ^{31}P for R–PH_2 vs δ ^{15}N for R–NH_2, δ ^{31}P for $(R\text{-})_2PH$ vs δ ^{15}N for $(R\text{-})_2NH$, and δ ^{31}P for $(R\text{-})_3P$ vs δ ^{15}N for $(R\text{-})_3N$. 517
11.61 Plots of the number of protons on the α-C–P atom(s) vs δ ^{31}P for PH_3, R–PH_2, $(R\text{-})_2PH$. 518
11.62 A plot of the sum of Taft σ^* values δ ^{31}P for trialkylphosphine oxides. 519
11.63 A plot of the sum of the number of protons on the α C–P carbon atoms vs δ ^{31}P for trialkylphosphine oxides. 520
11.64 Plots of ν P=O rotational conformers vs the sum of the number of protons on the α-C–O–P carbon atoms for trialkyl phosphates. 521
11.65 Plots of ν P=O rotational conformers vs δ ^{31}P for trialkyl phosphates. 522
11.66 A plot of δ ^{31}P vs the sum of the number of protons on the α-C–O–P carbon atoms for trialkyl phosphates. 523
11.67 A plot of the sum of the protons on the α-C–O–P carbon atoms vs δ ^{31}P for O,O,O-trialkyl phosphorothioates. 524
11.68 A plot of the sum of the number of protons on the α-C–O–P carbon atoms vs δ ^{31}P for O,O,O-trialkyl phosphoroselenates. 525
11.69 Plots of δ ^{31}P vs the sum of the number of protons on the α-C–P carbon atoms and the α-C–O–P carbon atoms for $RP(=O)(OC_2H_5)_2$ and $(R)_2P(=O)(OR)$. 526
11.70 Plots of δ ^{31}P for $RP(=O)(OC_2H_5)_2$ vs δ ^{13}C=O for $RC(=O)(OCH_3)$ and δ ^{31}P for $CH_3P(=O)(OR)_2$ vs δ ^{13}C=O for $CH_3C(=O)(OR)$. 527
11.71 Plots of δ ^{31}P for $(RO)_2P(=O)H$ and for $(RO)_2P(=S)H$ vs the sum of the protons on the α-C–O–P carbon atoms. 528

11.72 A plot of δ ^{31}P for $(RO)_2P(=O)H$ analogs vs δ ^{31}P for the corresponding $(RO)_2P(=S)H$ alkyl analogs. 529
11.73 Plots of the number of protons on the α-CP carbon atoms vs δ ^{31}P for $RP(=O)Cl_2$ and $RP(=S)Cl_2$. 530
11.74 A plot of δ ^{31}P for $RP(=S)Cl_2$ vs δ ^{31}P for $RP(=O)Cl_2$. 531
11.75 A plot of δ ^{31}P vs the number of protons on the α-C—P carbon atom for $RP(=O)F_2$. 532
11.76 A plot of δ ^{31}P vs the sum of the number of protons on the αC—P carbon atom and the number of protons on the α-C—O—P carbon atom for $RP(=S)Br(OC_3H_{7\text{-iso}})$. 533
11.77 Plots of δ ^{31}P vs the number of chlorine atoms joined to phosphorus for $PCl_{3-x}(OR)_x$. 534
11.78 Plots of the sum of the number of protons on the α-C—P carbon atom(s) vs δ ^{31}P for $PX_{3-n}R_n$. 535
11.79 A plot of δ ^{31}P vs the number of protons on the α-C—O—P carbon atoms for $(R)_2P(=O)Cl_2$. 536
11.80 Plots of the number of Cl atoms joined to phosphorus vs ν P=O and ν P=O rotational conformers for $P(=O)Cl_3$ through $P(=O)Cl_{3-x}(OR)_x$ where R is CH_3 or C_2H_5. 537
11.81 A plot of the number of Cl atoms joined to phosphorus vs ν P=O for $P(=O)Cl_{3-x}[N(CH_3)_2]_x$. 538
11.82 A plot of δ ^{31}P for $P(=S)(OR)_3$ vs δ ^{31}P for the corresponding alkyl analogs of $P(=O)(OR)_3$. 539
11.83 A plot of δ ^{31}P for $P(=Se)(OR)_3$ vs δ ^{31}P for corresponding alkyl analogs of $P(=O)(OR)_3$. 540
11.84 Plots of δ ^{31}P for $RP(=O)Cl_2$ vs δ ^{31}P for corresponding alkyl analogs of $RP(O)F_2$). The double plot is the δ ^{31}P range given in the literature for the $RP(=O)Cl_2$ analogs. 541
11.85 A plot of ν asym. ϕ-O—C vs δ ^{13}C-1 for 4-x-anisoles. 542
11.86 A plot of δ ^{13}C-1 vs Taft σ_{R° for 3-x- and 4-x-anisoles. 543
11.87 A plot of δ ^{13}C-1 vs Hammett σ values for 3-x^v and 4-x-anisoles. 544
11.88 A plot of Taft σ_{R° values vs σ ^{13}C-1 for 4-x-anisoles. 544
11.89 A plot of Taft σ_{R° values vs ν asym. ϕ-O—C for 4-x-anisoles. 545
11.90 A plot of δ ^{13}C-1 vs Hammett σ_p values for 4-x-anisoles. 545
11.91 A plot of δ ^{13}C-1 vs δ ^{13}C-5 for 2-x-anisoles. 546
11.92 A plot of δ ^{13}C-4 for mono-substituted benzenes vs Taft σ_{R° values. 547
11.93 A plot of the out-of-plane hydrogen deformation mode I for mono-substituted benzenes vs Taft σ_{R° values. 548
11.94 A plot of the out-of-plane hydrogen deformation mode III for mono-substituted benzenes vs Taft σ_{R° values. 549
11.95 A plot of the out-of-plane hydrogen deformation mode I for mono-substituted benzenes vs δ ^{13}C-4 for mono-substituted benzenes. 550
11.96 A plot of the out-of-plane hydrogen deformation mode III vs δ ^{13}C-4 for mono-substituted benzenes. 551
11.97 A plot of δ ^{13}C-1 vs Taft σ_{R° values for 4-x and 4,4′-x,x-biphenyls. 552

11.98 A plot of δ $^{13}C{=}O$ for tetramethylurea (TMU) 1 wt./vol. % solutions vs mole % $CHCl_3/CCl_4$. 552
11.99 A plot of ν C=O vs δ $^{13}C{=}O$ for TMU 0–100 mol % $CHCl_3/CCl_4$ solutions. 553
11.100 A plot of δ $^{13}C{=}O$ vs δ $^{13}CH_3$ for TMU in 0–100 mol % $CHCl_3/CCl_4$ solutions. 554
11.101 A plot of δ $^{13}C{=}O$ for TMU vs the solvent acceptor number (AN) for different solvents. 555
11.102 A plot of the δ $^{13}C{-}O$ chemical shift difference between TMU in solution with acetic acid and in solution with each of the other solvents. 556
11.103 Plots of δ $^{13}C{=}O$ for dialkylketones at 1 wt./vol. % in different solvents vs the AN for each solvent. 557
11.104 Plots of the δ $^{13}C{=}O$ chemical shift difference between each dialkyl-ketone in methanol and the same dialkylketone in solution with each of the other solvents. 558
11.105 A plot of δ $^{13}C{=}O$ for alkyl acetates and phenyl acetate vs the difference of δ $^{13}C{=}O$ in methanol and δ $^{13}C{=}O$ in each of the other solvents. 559

Name index

B

Bellamy, L. J., 31n16, 94n3, 185nn5, 9, 14, 458n13
Bender, P., 94n12
Bentley, F. K., 50n1, 94n2
Berger, S., 457n5
Berry, R. J., 128n10
Bigorne, M., 254n53
Black, F., 253n18
Blair, E. H., 253nn37, 39
Blake, B. H., 128n26
Bouquet, M., 94n28
Bouquet, S., 254n53
Brame, E. G., Jr., 31n15
Braun, S., 457n5
Breitmaier, E., 458n35
Brown, H. C., 458n33
Brownlee, R.T.C., 185n11

C

Camiade, M., 128n9
Chassaing, G., 94n28
Chittenden, R. A., 253n30
Chrisman, R. W., 367nn17, 21
Clark, J. W., 253nn7, 11, 33
Clark, R.J.H., 95n42
Coburn, W. C., Jr., 50n10
Cole, K. C., 50n5
Colthup, N. B., xvnn3, 10, 31n4, 94n4, 128n1, 254n49
Corbridge, D.E.C., 254n46
Corset, J., 94n28
Craver, C. D., 254n48
Cross, P. C., xvn2
Crutchfield, M. M., 458n28

D

Daasch, L. W., 254n54
Daly, L. H., xvn3, 31n4
Deady, L., 458n15
Decius, J. C., xvn2
de Haseth, J. A., xvn8
Delwaulle, M. L., 253nn2, 8, 12, 13, 15, 34
Detoni, S., 94n16
Di Yorio, J., 253n19
Dollish, F. R., 50n1, 94n2
Dunbar, J. E., 94n36, 253n31
Dungan, C. H., 458n28
Durig, J. R., 128n7, 253nn7, 11, 18, 19, 33
Du Vall, R. B., 367n2

E

Ekejiuba, I.O.C., 128n8
Erley, D. S., 128n26
Eucken, A., 253nn5, 14
Evans, J. C., 7n8, 69nn4, 7, 9, 10, 128nn13, 23, 367nn4, 5, 20, 431n10

F

Fassel, V. A., 50n8, 458n22
Fateley, W. G., xvn10, 50n1, 94nn2, 4, 128n1, 254n49
Favort, J., 94n28
Fawcett, A. H., 94n18
Fee, S., 94n18
Fiedler, S. L., 6n4, 431nn2, 7
Folt, V. L., 128n4
Forel, M., 128n9
Forneris, R., 94n14, 253n6
Fouchea, H. A., 50n17
Francois, F., 253nn2, 8, 12, 13, 15, 34

G

Gaufres, R., 95n42
Geisler, G., 94n15, 185n4
George, W. O., 128n5
Gerding, H., 253n10
Glockler, G., 128n19
Goldman, G. K., 458n30
Goodfield, J. E., 128n5
Grasselli, J. G., xvn10, 31n15, 94n4, 128n1, 254n49
Griffiths, P. R., xvn8
Gutowwsky, H. S., 253n3

H

Hadzi, D., 94n16
Hallam, H. E., 128n8

Ham, N. S., 50n12
Hanai, K., 94nn17, 30, 31, 34
Hanschmann, G., 94n15
Hasha, D. L., 50nn14, 17, 457n4, 458n9
Hellwege, K. H., 253nn5, 14
Herrail, F., 253n38
Herzberg, G., xvn1, 7n14, 50n15, 128n25, 254n45
Hirschmann, R. P., 50n8, 458n22
Hoffman, G. A., 50n17
Hoyer, H., 50n11

I

Igarashi, M., 6n5, 7n6

J

Jeschek, G., 458nn12, 36
Jewett, G. L., 458n24
Johnson, A. L., 128n16
Joshi, M., 94n29
Joshi, U. C., 94n29
Joyner, P., 128n19

K

Kabachnik, M. I., 70n15
Kagel, R. O., xvn12, 7n18, 69n2, 254n51, 367n9
Kalasinsky, V. J., 128n10
Kalinowski, H. O., 457n5
Kallos, G. J., 7nn9, 10, 128nn14, 15, 431n11
Kanesaki, I., 94n35
Karriker, J. M., 128n7
Katritzky, A. R., 458n15
Kawai, K., 94n35
Kesler, G., 185n4
Killingsworth, R. B., 50n3
King, S. T., 50n6, 254n47
Kirrman, A., 50n2
Kitaev, Yu. P., 50n4
Klaeboe, P., 31n6, 94n7
Kloster-Jensen, E., 31n6
Kniseley, R. N., 50n8, 458n22
Koster, D. F., 50n9
Krimm, S., 128n4

L

Lagemann, R. T., 94n13
Landolt-Bornstein, I., 253nn5, 14
Lau, C. L., 367n10
Lehman, H., 458n30
Leugers, M. A., xvn11, 31n5, 94nn9, 19, 95n39, 254n51, 367n13
Levin, J. W., 253n18
Liehr, A. D., 253n3
Limougi, J., 94n28
Lin-Vien, D., xvn10, 94n4, 128n1, 254n49
Little, T. S., 128n10
Lo, Y.-S., 128nn16, 23, 431n10
Loy, B. R., 367nn17, 21
Lunn, W. H., 185n2
Luoma, D. A., 31nn13, 14, 50n13, 185n6, 431nn3, 8, 457n7, 458nn10, 37

M

Machida, K., 94nn17, 30, 31, 33
Maddams, W. F., 128n5
Malanga, M., 367n16
Mann, J. R., 70n16
Mark, V., 458n28
Martz, D. E., 94n13
Mavel, G., 458n27
Mayants, L. S., 70n15
McLachlan, R. D., 7n11, 128n11, 367n13
Mitsch, R. A., 50n6
Muelder, W. W., 69n10, 253nn21, 22, 24, 27, 35, 254n43, 367n5

N

Nakamoto, K., xvnn13, 14
Neely, W. B., 253n39
Newman, M. S., 31nn8, 9, 11
Nielsen, J. R., 185n3
Nivorozhkin, L. E., 50n4
Nyquist, R. A., xvnn5–7, 11, 12, 6nn2–4, 7nn7–13, 15, 16, 18, 19, 31nn1, 2, 5, 7, 12–14, 17, 50nn7, 13, 14, 16–19, 69nn1, 2, 4, 7, 70nn12–14, 16, 94nn1, 5, 6, 9–11, 19–27, 36, 95nn37, 39, 40, 43, 128nn2, 11–18, 22–24, 185nn1, 6–8, 253nn9, 16, 17, 21–27, 31, 32, 35–37, 39–41, 254nn42, 43, 47, 48, 50–52, 55, 367nn6, 7, 9, 11, 13–21, 431nn1–12, 457nn1, 2, 4, 6, 7, 458nn8–12, 17, 18, 24, 26, 29, 31, 34, 36–39

O

Ogden, P. H., 50n6
Ogilivie, J. F., 50n5
Okamoto, Y. J., 458n33
Okuda, T., 94nn30, 34
Osborne, D. W., 253n37
Overend, J., 50n6, 69n10, 128n21, 367n5
Owen, N. L., 7n20

P

Paetzold, R., 94n8
Pan, C. Y., 185n3
Pauling, L., 458n32
Peters, T. L., 50n7
Peterson, D. P., 367n7
Platt, A. E., 367n18
Plegontov, S. A., 50n4
Polo, S. R., 253nn1, 4
Popov, E. M., 70n15
Potts, W. J., Jr., xvn4, 6n1, 254n50, 458n31
Priddy, D. B., 367n18
Puehl, C. W., 50n16, 253n26, 254n55, 431n3
Putzig, C. L., xvn11, 6n2, 7n7, 31n5, 50nn13, 14, 94nn9, 19, 95n39, 128n12, 254n51, 367nn7, 13, 17, 431nn8, 9, 458n9

R

Raevskii, O. A., 50n4
Rao, C.N.R., 458nn14, 30
Reder, T. L., 7nn9, 10, 128n14, 431n11
Reid, C., 50n15
Rey-Lafon, M., 128n9
Ronsch, E., 94n8
Rothschild, W. G., 128n6
Rouffi, C., 128n9

S

Saito, T., 458n16
Saito, Y., 94nn32, 33
Scherer, J. R., 69nn9–11, 128n21, 367nn3–5
Schrader, B., 128
Schuetz, J. E., 367n19
Settineri, S. E., 31nn13, 14, 185nn6, 8, 431n12, 457n7
Shanks, R. A., 458n15
Sheppard, N., 7n20, 69n5
Shipman, J. J., 128n4
Simons, W. W., 69n3
Singh, R. N., 94n29
Skelly, N. E., 7n7, 128n12
Sloane, H. J., 94n36, 253n31
Smith, D. C., 185n3, 254n54
Snyder, R. G., 367n10
Socrates, G., xvn9
Sportouch, S., 95n42
Stammreich, H., 94n14, 253n6
Steaniak, L., 458nn19, 21, 25
Stec, F. F., 7n10, 128n15, 431n11
Stephenson, C. V., 50n10
Stojilykovic, A., 69n8
Streck, R., 458nn12, 36, 39
Strycker, S. J., 94n36, 253n31
Stuckey, M., 94n18

T

Taft, R. W., Jr., 185n10, 253n29, 367nn12, 22, 458n23
Takasuka, M., 458n16
Tanka, Y., 94nn32, 33
Tarte, P., 185n12
Tavares, Y., 94n14, 253n6
Thill, B., 367n13
Thomas, L. C., 253n30
Thompson, J. W., 128n18
Titova, S. Z., 50n4
Topsom, R. D., 185n11, 458n15
Tsubio, M., 254n44

U

Unger, I., 253n39
Uno/Ubo, T., 94nn17, 30, 31

V

Van Wazer, J. R., 458n28
Varsanyi, G., 367n1
Venkataraghavan, R., 458n14
Voelter, W., 458n35

W

Walkden, P., 94n18
Ward, G. R., 7n9, 128n14
Wass, M. N., 253nn22, 24
Webb, G. A., 458nn19, 21, 25
Wertz, D. M., 128n7
West, W., 50n3
Westrick, R., 253n10
Whiffen, D. H., 69n8
Wiberley, S. E., xvn3, 31n4
Wilcox, W. S., 50n10
Willis, J. B., 50n12
Wilson, E. B., Jr., xvn2
Wilson, M. K., 253nn1, 4
Winter, F., 128n20
Witanowski, M., 458nn19, 21, 25
Wood, J. M., Jr., 94n12
Wright, N., 367n2
Wurrey, C. J., 128n10
Wyckoff, R., 31n10

Y

Yamakawa, M., 458n16
Yeh, Y. Y., 128n10
Young, C. W., 367n2

Subject index

A

Acetamide, dimethyl, 445, 493, 568
Acetanilides, 4-x, 443, 477, 478, 562
Acetates, 456, 559, 579
Acetone, 445, 491, 492, 567
Acetonitrile, 28, 29, 32, 33, 35, 40
Acetophenones, 4-x, 444, 483–485, 564
Acrylates
  glycidyl, 4
  spectra structure correlations, 456–457, 579
Acrylonitrile, 27
Aliphatic ethers, 5–6, 22, 23
Aliphatic hydrocarbons, 426–427, 432
Alkanedithiols, 66
Alkanes
  dihaloalkanes, 123–124, 169
  1-haloalkanes, 121–122, 166
  2-haloalkanes, 122, 167
  halocycloalkanes, 121, 123, 168
  nitro, 174–176, 186, 187, 206, 207, 208, 209, 447–448, 501–506
  phosphonic bromide, O-isopropyl, 533
  phosphonic dichloride, 530, 531
  phosphonic difluoide, 532, 541
  phosphonothioic dichloride, 530, 531, 541
  tetrabromoalkanes, 125, 170
Alkanethiols
  C–S stretching, 66
  rotational conformers, 66
  S–S stretching, 65–66, 74, 76–77
Alkanonitriles, 32
Alkylamines, 448–449, 510–513, 571
Alkyl isocyanates, 448, 507–509
Alkyl isonitriles, 447, 499, 500
Alkyl group
  joined to sulfur, oxygen, or halogen, 68–69, 81–83
  vibrations and solvent effects, 250–251
Amides, styrene-acrylamide copolymer, 359, 387
Amines, alkyl, 448–449, 510–513, 571
Anhydrides, 428, 434
  maleic, 446, 494, 569
Anilines
  4-x-, 441–442, 459–473, 560
  Nyquist rule, 427–428, 433
  sulfinyl, 93, 117
Anisoles, 5–6, 23
  spectra structure correlations, 453–454, 542–546, 574, 575
Antisymmetric ring deformation, 2, 3, 11
Aromatic ethers, 5–6, 22, 23
Aryl-O stretching, 241, 338
Azines
  aldehyde, 46
  benzaldehyde, 46, 57
  empirical structures, 45
  Fermi resonance, 46
  inductive effect, 46
  ketone, 46

B

Benzaldehydes
  azine, 46, 57
  4-x, 443, 479–482, 563
Benzene-d6, 361, 421
Benzenes
  See also under type of
  correlation chart, 366
  hexachlorobenzene, 366, 423
  in- and out-of plane vibrations, 353
  infrared data and assignments, 361, 421
  isotatic polystyrene, 355
  mono-substituted, 362, 393, 394, 402, 422, 423
  mono-substituted, spectra structure correlations, 454, 547–551, 576
  out-of-plane deformations, 362–366
  pentabromophenyl ether, 361, 420
  pentachlorobiphenyl, 366, 423
  potential energy distribution and ring modes, 354
  potential energy distribution and toluene, 355
  quantitative analyses, use of, 356–358
  styrene-4-methylstyrene copolymer, 355–358, 381, 382
  toluene, 353, 355
Benzenes, disubstituted (out-of-plane hydrogen deformations)
  1,2-, 363–364, 393, 395, 402, 423
  1,3-, 364, 393, 396, 402, 423
  1,4-, 364, 393, 397, 402, 423

Benzenes, disubstituted (Raman data and vibrations)
1,2-, Raman data and vibrations, 360, 419
1,3-, Raman data and vibrations, 361, 419
1,4-, Raman data and vibrations, 361, 420
Benzenes, pentasubstituted (out-of-plane hydrogen deformations), 366, 393, 401, 402, 423
Benzenes, tetrasubstituted (out-of-plane hydrogen deformations)
1,2,3,4-, 365, 393, 400, 402, 423
1,2,3,5-, 365, 393, 400, 402, 423
1,2,4,5-, 365, 393, 401, 402, 423
Benzenes, trisubstituted (out-of-plane hydrogen deformations)
1,2,3-, 364–365, 393, 399, 402, 423
1,2,4-, 365, 393, 399, 402, 423
1,3,5-, 364, 393, 398, 402, 423
Benzenethiols, 66, 74, 79
chlorobenzenethiol, 67–68, 79
Benzoates, alkyl 3-x and 4-x, 444, 489, 566
Benzoic acids, 3-x and 4-x, 442–443, 474–476, 561
Benzonitriles, 27, 29, 34, 40
3-x and 4-x, 446, 495–497, 570
Benzophenones, 444, 486–488, 565
Benzoquinones, 428, 434
Benzyl alcohol, 361, 421
Benzyl-2,3,4,5,6-d5 alcohol, 361, 421
Biphenyls
chlorinated, spectra structure correlations, 454, 552, 577
decabromo, 361, 420
diffuse reflectance, 354
polychloro, 353–355
Bromodichlorobenzenes, 360, 418
Butyl halides, tertiary, 122–125, 167

C

Carbodiimides, dialkyl and diaryl, 50, 63
Carboxylic acid, styrene-acrylic acid copolymer, 358–359, 385
C—C mode for C_2H_5OP group, 251
C—D mode for $CDCl_3$ and CCl_4 solutions, 252, 316, 317, 359–360
$CD_3{-}PO_3^{2-}$, 247–248, 349
Chlorobenzenethiol, 67–68, 79
C—H mode, 359–360
C=O group, summary of, 444–445, 490, 567
COP group, 251–252
C—O—P stretching, 241, 334–337
C—P stretching, νC—P, 241–242, 338
C—S stretching, alkanethiols, 66
Cyanamides, dialkyl, 31
Cyanobenzaldehyde, 29–30, 36, 41
Cyanogen halides, 28, 39
Cyclic ethers, 5, 21

D

Dialkyl cyanamides, 31
Dialkylketones, 455–456, 557, 558, 578
steric factor, 456
Dielectric constants, 86
Diffuse reflectance infrared Fourier transform. *See* DRIFT
Dimethylacetamide, 445, 493, 568
Dimethyl ether, 5
Dimethyl sulfoxide, 29
1,4-diphenylbutadiyne, 429–430, 435
Dipolar interaction, 3, 28
Disulfides
alkane, 66–67, 76–77, 78
dialkyl, aryl alkyl, and diaryl, 67
DRIFT (diffuse reflectance infrared Fourier transform) technique, 353, 354

E

Epibromohydrin, 3
Epichlorohydrin, 3
Epihalohydrins, 3, 17
Epoxides, 1–4
Epoxyalkanes, 2, 4
Epoxybutane, 3, 12
Epoxypropane, 125–126, 171
Ethers
aliphatic and aromatic, 5–6, 22, 23
cyclic, 5, 21
glycidyl, 4
pentabromophenyl, 361
vinyl, 6, 24–26
Ethylene oxide, 1–2, 8, 15
tetrachloro, 3
trans, 3, 12
Ethylenes, 125–126, 171
chloro- (vinyl chloride), 125, 126
fluoro- (vinyl fluoride), 125
tetrahalo-, 126
trichloro-, 126
Ethyl nitrates, 183, 225
Ethynylbenzene, 359–360, 389, 390, 391
out-of-plane hydrogen deformations, 363
Ethynylbenzene-d, 359–360, 389, 390, 391

F

Fermi resonance
acetonitrile, 29
azines, 46

carbodiimides, 50
isocyanates, 47, 51
isothiocyanates, 49
nitroalkanes, 174
nitrobenzenes, 177, 182
nitropropane, 175
styrene-acrylic acid copolymer, 358
Field effects
correction, 49
N,N-dimethylacetamide, 445
isocyanates, 47
nitrile, 30
out-of-plane ring deformations, 252
solvent system and P=O stretching, 248–249
1-Fluoro-2,4,5-trichlorobenzene, 345

G
Glycidyl acrylate, 4
Glycidyl ethers, 4

H
1-haloalkanes, 121–122, 166
2-haloalkanes, 122, 167
Halogen, alkyl groups joined to, 68–69, 81–83
Halogenated hydrocarbons
ethylene, propyne, epoxypropane, and propadiene, 125–126, 171
1-haloalkanes, 121–122, 166
2-haloalkanes, 122, 167
halopropadienes, 127
methanes, 120–121, 127, 164
tertiary butyl halides, 122–125, 167
Halopropadienes, 127, 429–430, 435
Hammett σ_p values, 182
Hexachlorobenzene, 366, 423
Hydantoins, 428–429, 435
Hydrogen bonding, intermolecular
alkanethiols, 65, 66
anhydrides, 428
anilines, 427–428
benzoquinones, 428
N,N-dimethylacetamide, 445
1,4-diphenylbutadiyne, 429–430
1-halopropadiene, 429–430, 435
hydantoins, 429
hydrogenphosphonate, dialkyl and diphenyl, 240–241
imides, 428
methyl phenyl sulfone, 92–93
nitrobenzaldehyde, 178–179
nitrobenzenes, 178, 181, 430
O-alkyl O-aryl N-methylphosphor amidate and N-methylphosphor amidothioate, 244
phosphoramidates, 247
P=O stretching, 249
styrene-acrylic acid copolymer, 358–359
sulfonamides, 89
trimethyl phosphate, 249–250
Hydrogen bonding, intramolecular
benzenethiols, 66
nitrile, 30
2-X-nitrobenzenes, 180
phosphorodithioates, 68
secondary sulfonamides, 90
sulfonyl fluorides, 91–92, 115
Hydrogen phosphonate, 528–529
dialkyl and diphenyl, 240–241, 333
Hydrogen phosphonthioate, 528, 529

I
Imides, 428, 434
Inductive effects
alkyl nitriles, 28
alkyl nitrites, 184
azines, 46
carbodiimides, 50
cyanogen halides, 28
isocyanates, 47
isothiocyanates, 49
nitroalkanes, 174
P-NHR, 244
sulfonamides, seconary, 90
sulfoxides, 86
Infrared spectra
1-bromo-1-chloroethylene, 155
1-bromopropadiene, 157
1-bromopropadiene-1-d, 158
1-bromopropyne, 154
3-bromopropyne, 144
3-bromopropyne-1-d, 149
bromotrichloromethane, 140
n-butyl nitrite, 201
carbon tetrachloride, 136
1-chloropropadiene, 156
3-chloropropyne, 144
3-chloropropyne-1-d, 146, 147
chlorotrifluoromethane (Freon-13), 137
1,3-dibromopropyne, 153
1,1-dichloro-2,2-difluoroethylene, 162
dichlorodifluoromethane (Freon-12), 138
1,2-dichloroethane (ethylene dichloride), 141
1,3-dichloropropyne, 151, 152
diethyl deuterophosphonate, 275
diethyl hydrogenphosphonate, 275
dimethyl deuterophosphonate, 274
dimethyl hydrogenphosphonate, 274

Infrared spectra (*continued*)
disodium methane-d_3-phosphonate, 293, 294, 295
disodium methanephosphonate, 293, 294, 295
disodium n-octadecanephosphonate, 296
ethylene oxide, 1, 8
ethyl nitrate, 200
ethynylbenzene, 389, 390, 391
ethynylbenzene-d, 389, 390, 391
3-fluoropropyne, 144
halopropenes, 142, 143
1-iodopropadiene, 157
1-iodopropyne, 155
3-iodopropyne, 145
methyl bromide, 130
methyl chloride, 129
methylene bromide, 133
methylene chloride, 132
methyl iodide, 131
methyl (methylthio) mercury, 72
nitromethane, 188
2-nitropropane, 189
N–D, N-methyl phosphoramidodi chloridothioate, 282, 283
N-methyl-d_3 phosphoramidodichlorido thioate, 281
N-methyl phosphoramidodichlorido thioate, 280
O-alkyl N,N′-diaryl phosphordiamido thioate, 285
O-alkyl O-aryl N-alkylphosphoramido thioate, 285
O-alkyl O-aryl phosphoramidothioate, 286
O-alkyl O-aryl N-methylphosphor amidate, 283
O-alkyl O-aryl N-methylphosphoramido thioate, 283
O-ethyl-1,1-d_2 phosphorodichlorido thioate, 236, 264
O-ethyl-2,2,2-d_3 phosphorodichlorido thioate, 236, 265
O-ethyl phosphorodichloridothioate, 236, 263
O-methyl-d_3 phosphorodichlorido-thioate, 259, 261
O-methyl phosphorodichloridothioate, 236, 259, 260
O,O-dialkyl phosphoramidothioate, 286
O,O-diethyl N-methyl phosphoramido thioate, 289
O,O-diethyl N-methyl phosphoramido thioate N-d, 289
O,O-diethyl phosphoramidothioate, 288
O,O-diethyl phosphoramidothioate N-d_2, 288
O,O-diethyl phosphorodithioic acid, 273
O,O-dimethyl-d_6 O-(2,4,5-trichlorophenyl) phosphate, 292
O,O-dimethyl-d_6 O-(2,4,5-trichloro phenyl) phosphorothioate, 292
O,O-dimethyl-d_6 phosphorochlorido thioate, 267, 268
O,O-dimethyl methylphosphoramido thioate, 287
O,O-dimethyl methylphosphoramido thioate N-d_2, 287
O,O-dimethyl N-methyl phosphoramido thioate, 290
O,O-dimethyl O-(2,4,5-trichlorophenyl) phosphate, 291
O,O-dimethyl O-(2,4,5-trichlorophenyl) phosphorothioate, 291
O,O-dimethyl phosphorochloridothioate, 236–237, 266, 268, 273
O,O-dimethyl phosphorodithioic acid, 273
2,2′,3,4′,5-pentachlorobiphenyl, 373
2,2′,3,4,5′-pentachlorobiphenyl, 375
2,2′,3′,4,5-pentachlorobiphenyl, 371
2,2′,3,4,6-pentachlorobiphenyl, 369
2,2′,3,4′,6-pentachlorobiphenyl, 372
2,2′,3,5,5′-pentachlorobiphenyl, 374
2,2′,3,5′,6-pentachlorobiphenyl, 373
2,2′,4,4′,5-pentachlorobiphenyl, 374
2,2′,4,5,5′-pentachlorobiphenyl, 375
2,2′,4,6,6′-pentachlorobiphenyl, 371
2,3,4,4′,6-pentachlorobiphenyl, 370
2,3′,4,4′,6-pentachlorobiphenyl, 372
2,3,4,5,6-pentachlorobiphenyl, 368
2,3′,4,5′,6-pentachlorobiphenyl, 370
phosphonic dibromide (bromomethyl), 279
phosphonic dichloride (chloromethyl), 277, 278, 279
phosphorodithioates, 68, 71
phosphoryl chloride, 257
potassium dimethylphosphinate, 298
propylene oxide, 2, 9
S-methyl phosphorodichloridothioate, 239, 269, 331
sodium diheptylphosphinate, 300
sodium dimethylphosphinate, 297, 298
sodium dioctylphosphinate, 300
styrene-acrylamide copolymer, 387
styrene-acrylic acid copolymer, 385
styrene-2-isopropenyl-2-oxazoline copolymer, 388
styrene-4-methylstyrene copolymer, 381, 382
syndiotactic polystyrene, 378, 379, 380
tetrafluoroethylene, 160
tetrafluoromethane, 135
tetrahydrofuran, 5, 14
thiophosphoryl dichloride fluoride, 255, 256
trans-1,2-dimethyl ethylene oxide, 3, 12
2,2′,3-trichlorobiphenyl, 368
2′,3,4-trichlorobiphenyl, 369
trichloroethylene, 161
trichlorofluoromethane (Freon-11), 139
trichloromethane, 134
1,1,2-trichloro-1,2,2-trifluoroethane, 163
trimethyl phosphite, 271, 272
Infrared spectra structure correlations
acetamide, dimethyl, 445, 493, 568
acetanilides, 4-x, 443, 477, 478, 562
acetone, 445, 491, 492, 567
acetophenones, 4-x, 444, 483–485, 564
acrylates, 456–457, 579

amines, alkyl, 448–449, 510–513, 571
anisoles, 453–454, 542–546, 574, 575
benzaldehydes, 4-x, 443, 479–482, 563
benzenes, mono-substituted, 454, 547–551, 576
benzoates, alkyl 3-x and 4-x, 444, 489, 566
benzoic acids, 3-x and 4-x, 442–443, 474–476, 561
benzonitriles, 3-x and 4-x, 446, 495–497, 570
benzophenones, 444, 486–488, 565
isocyanates, 448, 507–509
maleic anhydride, 446, 494, 569
methacrylates, 456–457, 579
organonitriles, 446–447, 498, 570
phosphates, trialkyl, 521, 522, 523, 539, 572
phosphoramidates, 538
phosphorodichloridate, monoalkyl, 537
tetramethylurea (TMU), 454–455, 553
Isocyanates, alkyl and aryl, 46–49, 51, 52, 53, 54, 58, 59, 60, 448, 507–509
Isonitriles, 37, 38
alkyl, 30, 447, 499, 500
Isothiocyanates, alkyl and aryl, 49–50, 54, 55, 56, 61–62

K

Ketones, dialkyl, 455–456, 557, 558, 578
steric factor, 456

M

Maleic anhydride, 446, 494, 569
Methanes, 120–121, 127, 164
tetra, 120, 164
tri, 120, 164
Methyl halides, 120
Methyl (methylthio) mercury, 69, 72, 73

N

NH_2 stretching, 359
Nitrates
alkyl, 182–183, 224
ethyl, 183, 225
Nitriles, 27–30, 39
acetonitrile, 28, 29, 32, 33, 35, 40
benzonitrile, 27, 29, 34, 40, 446, 495–497, 570
field effect, 30
Hammett σ values, 30
intramolecular hydrogen bonding, 30
isonitriles, 30, 37, 38, 447, 499, 500
organonitriles, 30, 42, 446–447, 498, 570
Nitrites, alkyl, 183–184, 226
Nitroalkanes, 174–176, 186, 187, 206–209
empirical correlations, 174
ethyl nitrates versus, 183, 225
$(NO_2)_4$, 176–177
Raman data for, 184, 227
spectra-structure correlations, 447–448, 501–506
trans and gauche conformers, 175
vapor versus liquid-phase data, 176
Nitrobenzaldehyde, 178–179, 214
Nitrobenzenes, 177–178, 209–213, 215, 221–222
band intensity ratios, 180
ethyl nitrates versus, 183, 225
Hammett σ_p values, 182
intermolecular hydrogen bonding, 178, 430
non-coplanar, 181
Nyquist rule, 430, 436
Raman data for, 184, 228
solvent effects on NO_2, 181
2-X-, 180–181, 218, 219
3-X-, 179–180, 216, 217
4-X-, 181–182, 220, 223
Nitrogen compounds, summary of correlations, 449–450
Nitrosamines, 185, 230
NMR spectra structure correlations
acetamide, dimethyl, 445, 493, 568
acetanilides, 4-x, 443, 477, 478, 562
acetates, 456, 559, 579
acetone, 445, 491, 492, 567
acetophenones, 4-x, 444, 483–485, 564
acrylates, 456–457, 579
alkyl isonitriles, 447, 499, 500
amines, alkyl, 448–449, 510–513, 571
anilines, 3-x and 4-x, 441–442, 459–473, 560
anisoles, 453–454, 542–546, 574, 575
benzaldehydes, 4-x, 443, 479–482, 563
benzenes, mono-substituted, 454, 547–551, 576
benzoates, alkyl 3-x and 4-x, 444, 489, 566
benzoic acids, 3-x and 4-x, 442–443, 474–476, 561
benzonitriles, 3-x and 4-x, 446, 495–497, 570
benzophenones, 444, 486–488, 565
biphenyls, chlorinated, 454, 552, 577
hydrogen phosphonates, 528–529
isocyanates, 448, 507–509
isonitriles, 447, 499, 500
ketones, dialkyl, 455–456, 557, 558, 578
maleic anhydride, 446, 494, 569
methacrylates, 456–457, 579
nitroalkanes, 447–448, 501–506
organonitriles, 446–447, 498, 570
phosphates, trialkyl, 521, 522, 523, 539, 572
phosphine oxides, trialkyl, 519, 520, 572
phosphines, dialkyl, 517, 518
phosphines, monoalkyl, 518
phosphines, trialkyl, 515, 516, 517, 518
phosphonates, 526–529
phosphonic bromide, alkane O-isopropyl, 533
phosphonic dichloride, alkane, 530, 531

NMR spectra structure correlations (*continued*)
phosphonic difluoide, alkane, 532, 541
phosphonothioic dichloride, alkane, 530, 531, 541
phosphoroselenates, trialkyl, 525, 540, 572
phosphorothioates, trialkyl, 524, 539, 540, 572
tetramethylurea (TMU), 454–455, 552–556, 577, 578
thiophosphonothioic dichloride, alkane, 530, 531, 541
N=S=O, 93
Nyquist Frequency, defined, 425
Nyquist Vibrational Group Frequency Rule
aliphatic hydrocarbons, 426–427, 432
anhydrides, 428, 434
anilines, 427–428, 433
benzoquinones, 428, 434
CH_2X_2, 120
defined, 425–426
1,4-diphenylbutadiyne, 429–430, 435
1-halopropadiene, 429–430, 435
hydantoins, 428–429, 435
imides, 428, 434
nitrobenzenes, 430, 436
SO_2 vibrations 87, 88, 89, 92–93, 426
sulfates, 430–431, 437
sulfonates, 430–431, 437
sulfones, 430–431, 437
sulfonyl chloride, 430–431, 437

O

Organoisothiocyanates, 62
Organonitriles, 30, 42, 446–447, 498, 570
Organophosphorus compounds, 450–452, 515, 516, 541, 571, 572–573
Organophosphorus halides, phosphorus halogen stretching for, 235, 321
Organothiocyanates, 31, 43
Oxirane, 4, 13, 20
Oxygen, alkyl groups joined to, 68–69, 81–83

P

PCl_3 and PCl_2 vibrations, 250
P−Cl stretching, 237, 323
Pentabromophenyl ether, 361, 420
Pentachlorobiphenyl, 366, 423
Phenetole, 5
Phenoxarsine derivatives
P=O and P=S stretching for, 238–239, 327
sulfate and thiosulfate, 88–89, 109, 110
Phenylacetylene. *See* Ethynylbenzene
Phosphates
trialkyl, 521, 522, 523, 539, 572
trimethyl, 240, 331
Phosphinate
potassium, 248, 350
sodium dialkyl, 248, 350
sodium dimethyl, 248, 350
Phosphine oxides, trialkyl, 519, 520, 572
Phosphines
dialkyl, 517, 518
monoalkyl, 518
trialkyl, 515, 516, 517, 518
Phosphites
dialkyl chloro, 534, 535
monoalkyl dichloro, 534
trialkyl, 534
trimethyl, 240, 271, 272, 331, 332
Phosphonates, 526–529
hydrogen, 528, 529
Phosphonthioate, hydrogen, 528, 529
Phosphoramidates
O-alkyl O-aryl N-methyl, 243–244, 283
O-methyl O-(2,4,5-trichlorophenyl) N-alkyl, 245, 341
O,O-diethyl N-alkyl, 247, 349
spectra structure correlations, 538
Phosphoramidodichloridothioate, N-alkyl, 245, 343
Phosphoramidothioates
O-alkyl O-aryl N-methyl, 243–244, 283
primary, $P(=S)NH_2$, 244, 245, 340
Phosphorochloridothioate, O,O-dialkyl, 246, 346
Phosphorodichloridate, monoalkyl, 537
Phosphorodichloridothioate
O-alkyl, 247, 348
O-ethyl, 236, 263
O-ethyl-1,1-d_2, 236, 264
O-ethyl-2,2,2-d_3, 236, 265
O-methyl, 236, 259, 260
O,O-dimethyl, 236–237, 266, 268, 273
S-alkyl, 247, 348
S-methyl, 239, 269, 331
Phosphorodithioates, 68, 71, 80
Phosphoroselenates, trialkyl, 525, 540, 572
Phosphorothioates
O,O-dimethyl O-(2,4,5-trichlorophenyl), 245–246, 344, 345
O,O,O-trialkyl, 246, 346, 524, 539, 540, 572
Phosphorus compounds
absorbance ratios, 250
alkyl group vibrations, 250–251
applications, 233
C−C mode for C_2H_5OP group, 251
C−D mode for $CDCl_3$ and CCl_4 solutions, 252
$CD_3-PO_3^{2-1}$, 247–248, 349
$CH_3-PO_3^{2-1}$, 247–248, 349
$(CH_3)_2-PO_2^{1}$, 248, 297, 298, 350
containing P−NH-R groups, 242–243, 339
COP group, 251–252
C−O−P stretching, 241, 334–337

C—P stretching, νC—P, 241–242, 338
halogen, 234–235, 318
halogen stretching for organo phosphorus halides, 235, 321
PCl_3 and PCl_2 vibrations, 250
P—Cl stretching, 237, 323
PNHR and PNH_2, 245, 342
P=O stretching, νP=O, 237–238, 239, 240–241, 325, 326, 327
$P(=S)NH_2$, 244, 245, 340
P—S stretching, 240
P=S stretching, νP=S, 238–239, 240, 327, 329–330
PX coupling, 234–235
PX stretching, 234
rotational conformers, 235–243, 246, 249–250, 251–252
S—H stretching, 240
solvent effects, 248–250, 301, 304–307
summary of correlations, 452–453
Phosphorus oxyhalides, analysis of, 239
P—NH-R groups, 242–243, 339
$P(-O-C)_3$, 240
Polychlorobiphenyls, 353–355
Polystyrene
out-of-plane hydrogen deformations, 362–363
styrene-acrylamide copolymer, 359, 387
styrene-acrylic acid copolymer, 358–359, 385
styrene/2-isopropenyl-2-oxazoline copolymer (SIPO), 359–360, 388
styrene-4-methylstyrene copolymer, 355–358, 381, 382
syndiotatic, 355, 362
P=O stretching, νP=O
in different physical phases, 237, 326
for hydrogenphosphonate, dialkyl and diphenyl, 240–241, 333
for O,O-dimethyl O-(2-chloro 4-X-phenyl) phosphate, 238, 326
for phenoxarsine derivatives, 238–239, 327
for phosphorus compounds, 237–238, 325, 326, 327
predicting frequencies, 238, 327
solvent effects, 248–250, 304–307
versus P=S, 239, 328
Propadiene analogs, 125–126, 171
halo, 127
Propenes, 4, 19
Propylene oxide, 2, 9
Propynes, 4, 19, 125–126, 171
$P(=S)NH_2$, 244, 245, 340
P—S stretching, 240
P=S stretching, νP=S, 240
for phenoxarsine derivatives, 238–239, 327
for phosphorus compounds, 239, 329–330
versus P=O, 239, 328

Q

Quantitative analyses, use of, 356–358

R

Raman spectra
1-bromo-1-chloroethylene, 156
1-bromopropadiene-1-d, 159
3-bromopropyne-1-d, 150
3-chloropropyne-1-d, 148
dimethyl hydrogenphosphonate, 276
ethynylbenzene, 392
ethynylbenzene-d, 392
methyl (methylthio) mercury, 73
S-methyl phosphorodichloridothioate, 270
sodium dimethylphosphinate, 299
Reaction field (RF), 3
benzenes, 362
defined, 86
Reflectance, diffuse. *See* DRIFT (diffuse reflectance infrared Fourier transform)
Refractive indices, 86
Resonance. *See* Fermi resonance
Ring breathing, 2–3, 10
Ring formation, ethylene oxide, 1–2
Rotational conformers
alkanethiols, 66
$ClCH_2P(=O)Cl_2$, 242
dialkyl sulfides, 67
1-haloalkanes, 121–122, 166
2-haloalkanes, 122, 167
phosphorodithioates, 68, 80
phosphorus compounds, 235–243, 246, 249–250, 251–252
P=O conformers for O-methyl phosphorodichloridate, 249, 302
$(RO)_2P(=S)SH$, 240

S

S—H stretching, 240
alkanethiols, 65–66, 74, 76–77
benzenethiols, 66, 74, 79
phosphorodithioates, 68, 80
Solvent effects, phosphorus compounds and, 248–250, 301, 304–307
S=O stretching
frequencies, 86–87, 96, 104
summary of, 92, 116–117
SO_2 stretching
calculated, 93
correlations, 92
frequencies, 87, 105–106
summary of, 92, 116–117

SO_2 stretching (*continued*)
vibrations, 87, 88, 89, 92–93
Spectra-structure correlations
acetamide, dimethyl, 445, 493, 568
acetanilides, 4-x, 443, 477, 478, 562
acetates, 456, 559, 579
acetones, 445, 491, 492, 567
acetophenones, 4-x, 444, 483–485, 564
acrylates, 456–457, 579
alkane phosphonic bromide, O-isopropyl, 533
alkane phosphonic dichloride, 530, 531
alkane phosphonic difluoide, 532, 541
alkane phosphonothioic dichloride, 530, 531, 541
alkylamines, 448–449, 510–513, 571
alkyl isocyanates, 448, 507–509
alkyl isonitriles, 447, 499, 500
amines, 448–449, 510–513, 571
anilines, 3-x and 4-x, 441–442, 459–473, 560
anisoles, 453–454, 542–546, 574, 575
benzaldehydes, 4-x, 443, 479–482, 563
benzenes, mono-substituted, 454, 547–551, 576
benzoates, alkyl 3-x and 4-x, 444, 489, 566
benzoic acids, 3-x and 4-x, 442–443, 474–476, 561
benzonitriles, 3-x and 4-x, 446, 495–497, 570
benzophenones, 444, 486–488, 565
biphenyls, chlorinated, 454, 552, 577
C=O group, summary of, 444–445, 490, 567
dialkylketones, 455–456, 557, 558, 578
dimethylacetamide, 445, 493, 568
hydrogen phosphonates, 528, 529
hydrogen phosphonthioate, 528, 529
isocyanates, 448, 507–509
isonitriles, 447, 499, 500
maleic anhydride, 446, 494, 569
methacrylates, 456–457, 579
nitroalkanes, 447–448, 501–506
nitrogen compounds, summary of correlations, 449–450
organonitriles, 446–447, 498, 570
organophosphorus compounds, 450–452, 515, 516,–541, 571, 572–573
phosphates, trialkyl, 521, 522, 523, 539, 572
phosphine oxides, trialkyl, 519, 520, 572
phosphines, dialkyl, 517, 518
phosphines, monoalkyl, 518
phosphines, trialkyl, 515, 516, 517, 518
phosphites, dialkyl chloro, 534, 535
phosphites, monoalkyl dichloro, 534
phosphites, trialkyl, 534
phosphonates, 526–529
phosphonic bromide, alkane O-isopropyl, 533
phosphonic dichloride, alkane, 530, 531
phosphonic difluoide, alkane, 532, 541
phosphonothioic dichloride, alkane, 530, 531, 541
phosphoramidates, 538
phosphorodichloridate, monoalkyl, 537
phosphoroselenates, trialkyl, 525, 540, 572
phosphorothioates, trialkyl, 524, 539, 540, 572
phosphorus compounds summary of correlations, 452–453, 534
physical phases, 490
role of, 441
tetramethylurea (TMU), 454–455, 552–556, 577, 578
thiophosphonates, 528, 529
thiophosphonothioic dichloride, alkane, 530, 531, 541
Steric factor
alkanethiols, 66
alkyl nitriles, 28
alkyl nitrites, 183
cyanogen halides, 28
dialkyl sulfoxides, 86
isocyanates, 48
ketones, 456
P=O stretching, 249
sulfonamides, 89
Styrene-acrylamide copolymer, 359, 387
Styrene-acrylic acid copolymer, 358–359, 385
Styrene/2-isopropenyl-2-oxazoline copolymer (SIPO), 359–360, 388
Styrene-4-methylstyrene copolymer, 355–358, 381, 382
Styrene oxide, 2–3, 10, 11
dichloro-, 3
Sulfates
Nyquist rule, 430–431, 437
phenoxarsine derivatives, 88–89, 109, 110
Sulfides
alkane, 66–67, 76–77
dialkyl, 67, 78
Sulfinyl anilines, 93, 117
Sulfites, dialkyl, 89, 111
Sulfonamides
primary, 89–90, 111, 112
secondary and tertiary, 90, 112
Sulfonates, organic, 91, 113
Nyquist rule, 430–431, 437
Sulfones
dialkyl, 87–88, 107, 108
diaryl, 88, 109
Nyquist rule, 430–431, 437
Sulfonyl chlorides, 91, 114
Nyquist rule, 430–431, 437
Sulfonyl fluorides, 91–92, 115
Sulfoxides, dimethyl, 87–88, 107, 108
reaction field, 86
Sulfur, alkyl groups joined to, 68–69, 81–83
Symmetric ring deformation, 2, 3, 10

T

Temperature effects
  chloromethyl phosphonic dichloride, 242
  isotatic polystyrene, 355
  O-methyl phosphorodichloridothioate, 236
  phosphorodithioates, 68
  styrene-acrylamide copolymer, 359, 387
  styrene-acrylic acid copolymer, 358–359, 385
Tetrachloroethylene oxide, 3
Tetrahalomethane, 120, 164
Tetrahydrofuran, 5, 14
Tetramethylurea (TMU), 454–455, 552–556, 577, 578
Tetranitromethane, 176–177
Thiocyanates, 31
Thiols
  See also Alkanethiols
  benzenethiols, 66, 74, 79
  chlorobenzenethiol, 67–68, 79
Thiosulfates, phenoxarsine derivatives, 88–89, 109, 110
Toluene, 353, 355
2,4,5-trichlorophenyl, 245–246, 344, 345
Trihalomethane, 120, 164

U

Urea, tetramethyl (TMU), 454–455, 552–556, 577, 578

V

Vinyl ethers, 6, 24–26

ISBN 0-12-523470-8
90065 >
9 780125 234702